Perturbed Semi-Markov Type Processes I

Dmitrii Silvestrov

Perturbed Semi-Markov Type Processes I

Limit Theorems for Rare-Event Times and Processes

Springer

Dmitrii Silvestrov
Department of Mathematics
Stockholm University
Stockholm, Sweden

ISBN 978-3-030-92405-8 ISBN 978-3-030-92403-4 (eBook)
https://doi.org/10.1007/978-3-030-92403-4

Mathematics Subject Classification: 60J10, 60J22, 60J27, 60K05, 60K15, 60K20, 65C40

This Springer imprint is published by the registered company Springer Nature Switzerland AG
The registered company address is: Gewerbestrasse 11, 6330 Cham, Switzerland

Preface

This book is the first volume of a two-volume monograph devoted to the study of limit and ergodic theorems for regularly and singularly perturbed Markov chains, semi-Markov processes, and alternating regenerative processes with semi-Markov modulation.

The first volume presents necessary and sufficient conditions for weak convergence for first-rare-event times and convergence in the topology J for first-rare-event processes defined on regularly perturbed finite Markov chains and semi-Markov processes and also new asymptotic recurrent algorithms of phase space reduction and effective conditions of weak convergence for distributions of hitting times and convergence of expectations of hitting times for regularly and singularly perturbed finite Markov chains and semi-Markov processes.

The second volume presents new super-long, long, and short time ergodic theorems for perturbed multi-alternating regenerative processes modulated by regularly or singularly perturbed finite semi-Markov processes.

Models of perturbed Markov chains and semi-Markov processes, in particular for the most complex cases of the so-called singularly perturbed processes, attracted the attention of researchers in the middle of the twentieth century. Interest to these models has been stimulated by applications to control and queuing systems, information networks, epidemic models, and models of mathematical genetics and population dynamics. As a rule, Markov-type processes with singular perturbations arise as a natural tool for the mathematical analysis of multi-component systems with weakly interacting components.

Semi-Markov processes are a natural generalisation of discrete and continuous time Markov chains. These jump processes possess the Markov property at the moments of jumps and can have arbitrary distributions concentrated on a positive half-line for the times between jumps. In fact, this combination of basic properties makes semi-Markov processes a very flexible and effective tool for constructing new classes of stochastic processes, for example, alternating regenerative processes with semi-Markov modulation, and describing various applied stochastic models such as queuing, reliability and bio-stochastic systems, stochastic networks, financial and insurance processes, etc.

As for Markov chains, hitting times play a very important role in the theory of semi-Markov processes and their applications. These random functionals are also known by names such as first-rare-event times, first passage times, absorption times, in theoretical studies, and also lifetimes, failure times, extinction times, etc., in applications. Hitting times and their expectations play a key role in ergodic theorems because of the dual relations connecting the corresponding stationary distributions and expectations of the return times, the laws of large numbers, central limit theorems, and more general limit theorems for additive, extremal, and other types of functionals for Markov type processes.

The most deeply investigated are cases of Markov chains and semi-Markov processes with finite phase spaces. There is a huge bibliography of works that contain limit theorems for hitting times and related functionals for perturbed finite Markov chains and semi-Markov type processes. However, the theory of such limit and ergodic theorems is still far from completion. The results presented in the book, I hope, confirm this point of view well.

The first volume includes an introduction, 11 chapters grouped in two parts, and two appendices.

In Introduction (Chap. 1), simple examples, models of stochastic processes, conditions, and results are presented in an informal way. It also provides chapter-by-chapter content and additional information for potential readers.

Part I (Chaps. 2–7) presents necessary and sufficient conditions for weak convergence for the first-rare-event times and convergence in the topology J for the first-rare-event processes defined on regularly perturbed finite Markov chains and semi-Markov processes. Similar results are also obtained for counting processes generated by flows of rare events defined on regularly perturbed finite Markov chains and semi-Markov processes. These results are extended to some more general models, such as vector rare-event times and processes, first-rare-event times for Markov renewal process with transition period and extending phase spaces, and others. The above results are illustrated by applications to risk processes and perturbed queuing systems.

Part II (Chaps. 8–12) presents new asymptotic recurrent algorithms of phase space reduction and effective conditions of weak convergence for distributions of hitting times and convergence of expectations for hitting times for regularly and singularly perturbed semi-finite Markov processes. In such models, the phase space is one class of communicative states for embedded Markov chains of pre-limiting perturbed semi-Markov processes, but it can have an arbitrary communicative structure, i.e., it can consist of one (regular perturbations) or several (singular perturbations) closed classes of communicative states and, possibly, the class of transient states for the corresponding limiting embedded Markov chain. Efficient asymptotic forward and backward recurrent algorithms are presented that allow one to find suitable normalisation functions, obtain convergence relations for distributions and expectations of hitting times, and compute limiting Laplace transforms for distributions and limits for expectations of hitting times.

Appendix A contains some limit theorems for randomly stopped stochastic processes, which are essential in the book.

Appendix B contains some methodological and bibliographical notes, as well as comments on the new results presented in the book and some new problems for future research.

I hope that the publication of this new book on asymptotic problems for perturbed stochastic processes will be a useful contribution to the ongoing intensive research in this area. In addition to being used for research and reference purposes, the book can also be used in special courses on this topic and as additional reading to general courses on stochastic processes. In this regard, it can be useful for both specialists as well as doctoral and senior students.

I would also like to thank my colleagues at the Department of Mathematics, Stockholm University and at the Division of Mathematics and Physics, School of Education, Culture and Communication, Mälardalen University for creating an inspiring research environment and a friendly atmosphere, which stimulated my work.

Stockholm, Sweden
October 2021

Dmitrii Silvestrov

Contents

List of Symbols

$\Rightarrow$	Symbol of weak convergence
$\stackrel{\mathsf{d}}{\longrightarrow}$	Symbol of convergence in distribution
$\stackrel{\mathsf{P}}{\longrightarrow}$	Symbol of convergence in probability
$\stackrel{\text{a.s.}}{\longrightarrow}$	Symbol of almost sure convergence
$\stackrel{\mathsf{J}}{\longrightarrow}$	Symbol of convergence in topology J
$\stackrel{\mathsf{U}}{\longrightarrow}$	Symbol of convergence in topology U
$\sim$	Symbol of asymptotic equivalence
ε	Perturbation parameter
$\varepsilon \to 0$	A shorter variant of symbol $0 < \varepsilon \to 0$
$\stackrel{\mathsf{d}}{=}$	Symbol of equality for distributions
$\stackrel{\text{a.s.}}{=}$	Symbol of almost sure equality
$*$	Symbol of convolution
$(*n)$	Symbol of n-fold convolution
$[u]$	Integer part of a real number u
$\mathbb{R}_+$	Interval $[0, \infty)$
$\mathcal{B}_+$	Borel σ-algebra of subsets of $\mathbb{R}_+$
$m(A)$	Lebesgue measure on σ-algebra $\mathcal{B}_+$
$\mathbb{X}$	A finite phase space $\{1, \ldots, m\}$
$\mathcal{B}_{\mathbb{X}}$	The σ-algebra of all subsets of the space $\mathbb{X}$
$\mathbb{Z}$	A phase space of a stochastic process (an arbitrary set)
$\mathcal{B}_{\mathbb{Z}}$	A σ-algebra of measurable subsets of space $\mathbb{Z}$
$Exp_a(\cdot)$	Exponential distribution function with parameter a

$Geo_p(\cdot)$	Geometrical distribution function with parameter p
$Con_a(\cdot)$	Distribution function with the unit jump in point a
A, **B**, **C**, ...	Conditions
$\eta_{\varepsilon,n}$	Embedded Markov chain
${}_k\eta_{\varepsilon,n},\ {}_{\bar{k}_n}\eta_{\varepsilon,n}, \ldots$	Reduced embedded Markov chains
$\kappa_{\varepsilon,n}$	Transition time
$\rho_{\varepsilon,n}$	Indicator of rare event
$\zeta_{\varepsilon,n}$	Moment of jumps
$\eta_\varepsilon(t)$	Semi-Markov process
${}_k\eta_\varepsilon(t),\ {}_{\bar{k}_n}\eta_\varepsilon(t), \ldots$	Reduced semi-Markov processes
$\xi_\varepsilon(t)$	Rare-event-time process
$N_\varepsilon(t),\ \bar{N}_\varepsilon(t), \ldots$	Counting processes generated by flows of rare events
ν_ε	First-rare-event time for embedded Markov chain
ξ_ε	First-rare-event time for semi-Markov process
$\nu_{\varepsilon,\mathbb{D}}$	Hitting time for embedded Markov chain
${}_k\nu_\varepsilon,\ {}_{\bar{k}_n}\nu_{\varepsilon,\mathbb{D}}$	Hitting times for reduced embedded Markov chains
$\tau_{\varepsilon,\mathbb{D}}$	Hitting time for semi-Markov process
${}_k\tau_{\varepsilon,\mathbb{D}},\ {}_{\bar{k}_n}\tau_{\varepsilon,\mathbb{D}}, \ldots$	Hitting times for reduced semi-Markov processes
$\dot{\tau}_{\varepsilon,\mathbb{D}}$	Return time for semi-Markov process
$Q_{\varepsilon,ij}(t)$	Transition probability of semi-Markov process
$e_{ij}(\varepsilon)$	Expectation of transition time for semi-Markov processes
$\psi_{\varepsilon,ij}(s)$	Laplace transform of transition probability of semi-Markov process
${}_kQ_{\varepsilon,ij}(t),\ {}_{\bar{k}_n}Q_{\varepsilon,ij}(t), \ldots$	Transition probabilities for reduced semi-Markov processes
${}_ke_{\varepsilon,ij},\ {}_{\bar{k}_n}e_{\varepsilon,ij}, \ldots$	Expectations of transition time for reduced semi-Markov processes
${}_k\psi_{\varepsilon,ij}(s),\ {}_{\bar{k}_n}\psi_{\varepsilon,ij}(s), \ldots$	Laplace transform of transition probability for reduced semi-Markov processes
$p_{\varepsilon,ij}$	Transition probability for embedded Markov chain
${}_kp_{\varepsilon,ij},\ {}_{\bar{k}_n}p_{\varepsilon,ij}, \ldots$	Transition probabilities for reduced embedded Markov chains
$F_{\varepsilon,ij}(t)$	Conditional distribution function of transition time
$f_{\varepsilon,ij}$	Conditional expectation of transition time

$\phi_{\varepsilon,ij}(s)$	Laplace transform for conditional distribution function of transition time
${}_kF_{\varepsilon,ij}(t)$, ${}_{\bar{k}_n}F_{\varepsilon,ij}(t), \ldots$	Conditional distribution functions of transition times for reduced semi-Markov processes
${}_kf_{\varepsilon,ij}$, ${}_{\bar{k}_n}f_{\varepsilon,ij}, \ldots$	Conditional expectations of transition times for reduced semi-Markov processes
${}_k\phi_{\varepsilon,ij}(s)$, ${}_{\bar{k}_n}\phi_{\varepsilon,ij}(s), \ldots$	Laplace transform for conditional distribution functions of transition times for reduced semi-Markov processes
$P_{\varepsilon,\mathbb{D},ij}$	Hitting probability for semi-Markov process
$G_{\varepsilon,\mathbb{D},ij}(t)$	Distribution function of hitting time
$E_{\varepsilon,\mathbb{D},ij}$	Expectation of hitting time
$\Psi_{\varepsilon,\mathbb{D},ij}(s)$	Laplace transform for distribution function of hitting time
${}_kP_{\varepsilon,\mathbb{D},ij}$, ${}_{\bar{k}_n}P_{\varepsilon,\mathbb{D},ij}$	Hitting probabilities for reduced semi-Markov processes
${}_kG_{\varepsilon,\mathbb{D},ij}(t)$, ${}_{\bar{k}_n}G_{\varepsilon,\mathbb{D},ij}(t), \ldots$	Distribution functions of hitting times for reduced semi-Markov processes
${}_kE_{\varepsilon,\mathbb{D},ij}$, ${}_{\bar{k}_n}E_{\varepsilon,\mathbb{D},ij}$	Expectations of hitting times for reduced semi-Markov processes
${}_k\Psi_{\varepsilon,\mathbb{D},ij}(s)$, ${}_{\bar{k}_n}\Psi_{\varepsilon,\mathbb{D},ij}(s), \ldots$	Laplace transforms for distribution functions of hitting times for reduced semi-Markov processes
MC	Markov chain
MRP	Markov renewal process
SMP	Semi-Markov process

Chapter 1
Introduction

This book is devoted to the study of limit theorems for first-rare-event times and hitting times for regularly and singularly perturbed semi-Markov processes, which is the main objects of study in the book. The basic fact concerned these processes can be found in the books listed in Sect. B.2.7. The introduction aims to informally present the main problems, methods, and algorithms that make up the content of the book. We give simple examples, illustrated by figures, and try to show the logic and ideas underlying the methods of asymptotic analysis of perturbed semi-Markov type processes developed in the book, as well as explain the meaning of the results presented in the book. We also describe the content of the book in parts and chapters and provide additional information for potential readers of the book.

1.1 Part I: First-Rare-Event Times for Regularly Perturbed Semi-Markov Processes

In Sect. 1.1, we outline the circle of problems and illustrate results presented in Part I.

1.1.1 Part I: Contents, Examples, Models, Results

1.1.1.1 Content of Part I. In this part, which includes Chaps. 2–7, we give necessary and sufficient condition of convergence in distribution (equivalent to weak convergence of distributions) for first-rare-event times and J-convergence for first-rare-event time processes defined on regularly perturbed Markov chains and semi-Markov processes and counting processes generated by the corresponding flows of rare events. These results are also illustrated by some applications to be perturbed risk processes and perturbed queuing systems.

D. Silvestrov, *Perturbed Semi-Markov Type Processes I*,
https://doi.org/10.1007/978-3-030-92403-4_1

Limit theorems for random functionals of first-rare-event time type for Markov type processes have been studied by many researchers. The case of Markov chains and semi-Markov processes with finite phase spaces is the most deeply investigated. The main features for the most previous results are that they give sufficient conditions of convergence for such functionals. As a rule, these conditions involve assumptions, which imply weak convergence for sums of i.i.d. random variables distributed as sojourn times for the semi-Markov process (for every state) to some infinitely divisible laws, plus some ergodicity condition for the embedded Markov chain, plus condition of vanishing probabilities of occurring a rare event during one transition step for the semi-Markov process.

The results of Part I relate to the model of regularly perturbed semi-Markov processes with finite phase space. Instead of conditions based on "individual" distributions of sojourn times, we use more general and weaker conditions imposed on distributions sojourn times averaged over the stationary distributions of the corresponding embedded Markov chains. Moreover, we show that these conditions are not only sufficient but also necessary conditions for convergence in distribution of first-rare-event times and convergence in Skorokhod J-topology of first-rare-event time processes. These results give some kind of a "final solution" for limit theorems for first-rare-event times and first-rare-event time processes for regularly perturbed semi-Markov process with a finite phase space.

1.1.1.2 A Model Example. Let $(\eta_{\varepsilon,n}, \kappa_{\varepsilon,n}, \rho_{\varepsilon,n})$, $n = 0, 1, \ldots$ be, for every $\varepsilon \in (0, 1]$,[1] a Markov renewal process, i.e., a discrete time, homogenous Markov chain with a phase space $\mathbb{Z} = \mathbb{X} \times [0, \infty) \times \{0, 1\}$, where $\mathbb{X} = \{1, 2\}$, and transition probabilities defined for $i, j \in \mathbb{X}$, $s, t \geq 0$, $\imath, \jmath = 0, 1$,

$$\begin{aligned} Q_{\varepsilon,ij}(t, \jmath) &= \mathrm{I}(t \geq e_{\varepsilon,i,\jmath}) p_{\varepsilon,ij} p_{\varepsilon,i}(\jmath) \\ &= \mathsf{P}\{\eta_{\varepsilon,1} = j, \kappa_{\varepsilon,1} \leq t, \rho_{\varepsilon,1} = \jmath / \eta_{\varepsilon,0} = i, \kappa_{\varepsilon,0} = s, \rho_{\varepsilon,0} = \imath\}, \end{aligned} \tag{1.1}$$

where: (a) $e_{\varepsilon,i,\jmath} > 0, i \in \mathbb{X}, \jmath = 0, 1$, (b) $p_{\varepsilon,ij} \geq 0, i, j \in \mathbb{X}, p_{\varepsilon,i1} + p_{\varepsilon,i2} = 1, i \in \mathbb{X}$, (c) $p_{\varepsilon,i}(\jmath) \in [0, 1], i \in \mathbb{X}, \jmath = 0, 1, p_{\varepsilon,i}(0) + p_{\varepsilon,i}(1) = 1, i \in \mathbb{X}$.

Let $\mathbf{S}_\varepsilon$ be a queuing system, the functioning of which is described by the Markov renewal process $(\eta_{\varepsilon,n}, \kappa_{\varepsilon,n}, \rho_{\varepsilon,n})$.

The system $\mathbf{S}_\varepsilon$ can operate in two modes say 1 and 2 (which, for example, can be interpreted as "low" and "high" intensity modes) switching modes at moments $\zeta_{\varepsilon,n} = \kappa_{\varepsilon,1} + \cdots + \kappa_{\varepsilon,n}, n = 0, 1, \ldots$. If $\eta_{\varepsilon,n} = i \in \mathbb{X}$ and $\rho_{\varepsilon,n+1} = 0$, then the working period $[\zeta_{\varepsilon,n}, \zeta_{\varepsilon,n+1})$ is "successful" and its duration is $\kappa_{\varepsilon,n+1} = e_{\varepsilon,i,0}$. If $\eta_{\varepsilon,n} = i \in \mathbb{X}$ and $\rho_{\varepsilon,n+1} = 1$, then the working period $[\zeta_{\varepsilon,n}, \zeta_{\varepsilon,n+1})$ is "unsuccessful," its duration is $\kappa_{\varepsilon,n+1} = e_{\varepsilon,i,1}$, and a failure of the system $\mathbf{S}_\varepsilon$ is registered at the end moment of this working period $\zeta_{\varepsilon,n+1}$. According to the above remarks, the random variable $\eta_{\varepsilon,n}$ is the indicator of mode in the $(n + 1)$-th working period $[\zeta_{\varepsilon,n}, \zeta_{\varepsilon,n+1})$. The random variable $\kappa_{\varepsilon,n+1}$ is the duration of this working period. The binary random variable $\rho_{\varepsilon,n+1}$ indicates successful/unsuccessful type of the $(n + 1)$-th working period. It can be referred as a failure indicator.

---

[1] Here and below, ε plays the role of a perturbation parameter.

According to the above description, $e_{\varepsilon,i,j}$ is a duration of working period with the work regime i and successful/unsuccessful type of working period j, $p_{\varepsilon,ij}$ is the conditional probability of switching the type i of working period to the type j for the next working period, and $p_{\varepsilon,i}(1)$ is the failure probability for the working period of the type i.

The object of our interest is the random functional,

$$\xi_\varepsilon = \sum_{n=1}^{\nu_\varepsilon} \kappa_{\varepsilon,n}, \tag{1.2}$$

where

$$\nu_\varepsilon = \min(n \geq 1 : \rho_{\varepsilon,n} = 1). \tag{1.3}$$

The random variable ξ_ε can be referred to as the first failure time for the queuing system $\mathbf{S}_\varepsilon$.

The characteristic property of the Markov renewal process $(\eta_{\varepsilon,n}, \kappa_{\varepsilon,n}, \rho_{\varepsilon,n})$ is that its transition probabilities depend only on the current state of the first component and do not depend on states of other components. In this case, the first component $\eta_{\varepsilon,n}$ is itself a homogeneous Markov chain with the phase space $\mathbb{X}$ and transition probabilities $p_{\varepsilon,ij}, i, j \in \mathbb{X}$. Also, the Markov renewal process $(\eta_{\varepsilon,n}, \kappa_{\varepsilon,n}, \rho_{\varepsilon,n})$ can be used for construction of the semi-Markov process defined as $\eta_\varepsilon(t) = \eta_{\varepsilon,n}$, for $\zeta_{\varepsilon,n} \leq t < \zeta_{\varepsilon,n+1}, n = 0, 1, \ldots$, where $\zeta_{\varepsilon,n} = \kappa_{\varepsilon,1} + \cdots + \kappa_{\varepsilon,n}$ are moments of jumps for the process $\eta_\varepsilon(t)$. As far as the binary random variables $\rho_{\varepsilon,n}$ are concerned, they are used as indicators of "rare" events $\{\rho_{\varepsilon,n} = 1\}$.

The random variable ξ_ε can be referred to as the first-rare-event time for the semi-Markov process $\eta_\varepsilon(t)$. The possibility to interpret events $\{\rho_{\varepsilon,n} = 1\}$ as rare is supported by the "rare-event" condition:

(1) The failure probabilities $0 < p_{\varepsilon,i}(1) \to 0$ as $\varepsilon \to 0$,[2] for $i \in \mathbb{X}$.

We also assume that the Markov chains $\eta_{\varepsilon,n}$ are asymptotically uniformly ergodic. This means that the following condition is satisfied:

(2) The regime switching probabilities $p_{\varepsilon,12}, p_{\varepsilon,21}$ are asymptotically separated from 0, i.e., $\underline{\lim}_{\varepsilon\to 0}\, p_{\varepsilon,12},\ \underline{\lim}_{\varepsilon\to 0}\, p_{\varepsilon,21} > 0$.

The condition **(2)** implies that the Markov chain $\eta_{\varepsilon,n}$ is ergodic for ε small enough and its stationary probabilities take the following forms:

$$\pi_{\varepsilon,1} = \frac{p_{\varepsilon,21}}{p_{\varepsilon,12} + p_{\varepsilon,21}}, \quad \pi_{\varepsilon,2} = \frac{p_{\varepsilon,12}}{p_{\varepsilon,12} + p_{\varepsilon,21}}. \tag{1.4}$$

Moreover, the above stationary probabilities are asymptotically separated from 0 and 1, i.e.,

$$0 < \underline{\lim}_{\varepsilon\to 0}\, \pi_{\varepsilon,i} \leq \overline{\lim}_{\varepsilon\to 0}\, \pi_{\varepsilon,i} < 1, \text{ for } i \in \mathbb{X}. \tag{1.5}$$

---

[2] In what follows, $\varepsilon \to 0$ is a shorten version of the symbol $0 < \varepsilon \to 0$.

Let us define, for $\varepsilon \in (0, 1]$, the averaged (over the above stationary distribution) failure probability,

$$p_\varepsilon = p_{\varepsilon,1}(1)\,\pi_{\varepsilon,1} + p_{\varepsilon,2}(1)\,\pi_{\varepsilon,2}, \tag{1.6}$$

and the averaged duration of successful working period,

$$e_\varepsilon = e_{\varepsilon,1,0}\,\pi_{\varepsilon,1} + e_{\varepsilon,2,0}\,\pi_{\varepsilon,2}. \tag{1.7}$$

Finally, we also assume that the duration of unsuccessful working period makes an asymptotically negligible contribution in the first failure time. This means that the following condition is satisfied:

(3) $p_\varepsilon e_{\varepsilon,i,1} \to 0$ as $\varepsilon \to 0$, for $i \in \mathbb{X}$.

The asymptotic results presented in Part I let us state that, under the assumption that conditions **(1)**–**(3)** are satisfied, the asymptotic relation,

$$e_\varepsilon \to e_0 \in (0, \infty) \text{ as } \varepsilon \to 0, \tag{1.8}$$

is the *necessary* and *sufficient* condition for the fulfilment of the following asymptotic relation,

$$p_\varepsilon \xi_\varepsilon \xrightarrow{\mathsf{d}} \xi_0 \text{ as } \varepsilon \to 0, \tag{1.9}$$

where ξ_0 is a proper non-negative random variable, which has distribution not concentrated in 0. Moreover, in this case, $\mathsf{E}e^{-s\xi_0} = (1 + e_0 s)^{-1}, s \geq 0$, i.e., the random variable ξ_0 is exponentially distributed with parameter e_0^{-1}.

Importantly, neither the convergence of the switching probabilities $p_{\varepsilon,12}$ and $p_{\varepsilon,21}$ nor the convergence of the parameters $e_{\varepsilon,1,0}$, as $\varepsilon \to 0$, is not required. These parameters can fluctuate as functions of $\varepsilon \to 0$. The only convergence of one "leading " parameter e_ε, as $\varepsilon \to 0$, is required.

It is also worth commenting on the role of the condition **(3)**.

If the duration of working period does not depend on the value of the failure indicator, i.e., $e_{\varepsilon,i,1} = e_{\varepsilon,i,0}, \varepsilon \in 0, 1]$, for $i \in \mathbb{X}$, then the first failure time can be represented in the form,

$$\xi_\varepsilon = \sum_{k=1}^{\nu_\varepsilon} \kappa_{\varepsilon,k} = \sum_{k=1}^{\nu_\varepsilon} e_{\varepsilon,\eta_{\varepsilon,k-1},0} = \xi_{\varepsilon,0}. \tag{1.10}$$

Under the assumption that the conditions **(1)**–**(2)** are satisfied, any of the relations (1.8) or (1.9) implies that the condition **(3)** is satisfied, and, thus, this condition can be omitted in the above statement about the equivalence of relations (1.8) and (1.9).

If the duration of working period depends on the value of the failure indicator, then the condition **(3)** cannot be omitted.

For example, let us assume that $e_{\varepsilon,1}(1) = e_{\varepsilon,2}(1) = e_\varepsilon(1), \varepsilon \in (0, 1]$. In this case, the first failure time can be represented in the form,

$$\xi_\varepsilon = \sum_{k=1}^{\nu_\varepsilon} \kappa_{\varepsilon,k} = \sum_{k=1}^{\nu_\varepsilon - 1} e_{\varepsilon,\eta_{\varepsilon,k-1},0} + e_\varepsilon(1)$$
$$= \xi_{\varepsilon,0} - e_{\varepsilon,\eta_{\varepsilon,\nu_\varepsilon-1},0} + e_\varepsilon(1). \tag{1.11}$$

The normalised first failure times $p_\varepsilon \xi_\varepsilon$ may not converge in distribution to a proper random variable.

Indeed, according to the above remarks, the conditions **(1)**–**(2)** and the relation (1.8) imply that (a) $p_\varepsilon \xi_{\varepsilon,0} \stackrel{\mathsf{d}}{\longrightarrow} \xi_0$ as $\varepsilon \to 0$. Also, the relations (1.5) and (1.8) imply that (b) $p_\varepsilon e_{\varepsilon,\eta_{\varepsilon,\nu_\varepsilon-1},0} \le p_\varepsilon \max(e_{\varepsilon,1,0}, e_{\varepsilon,2,0}) \to 0$ as $\varepsilon \to 0$. The asymptotic relations (a) and (b) imply that the random variables $p_\varepsilon \xi_\varepsilon$ do not converge in distribution to a proper random variable if the quantities $p_\varepsilon e_\varepsilon(1)$ do not converge to a finite limit, as $\varepsilon \to 0$. Also, $p_\varepsilon \xi_\varepsilon \stackrel{\mathsf{P}}{\longrightarrow} \infty$ as $\varepsilon \to 0$, if $p_\varepsilon e_\varepsilon(1) \to \infty$ as $\varepsilon \to 0$.

In conclusion, let us mention that it is possible to interpret the first-rare-event time ξ_ε as the first hitting time for some semi-Markov process.

Indeed, the two-component random sequence $(\eta_{\varepsilon,n}, \rho_{\varepsilon,n}), n = 0, 1, \ldots$ is also a homogeneous Markov chain with the phase space $\hat{\mathbb{X}} = \{(1,0), (2,0), (1,1), (2,1)\}$ and transition probabilities $p_{\varepsilon,(i,\imath),(j,\jmath)} = p_{\varepsilon,ij} p_{\varepsilon,i}(\jmath), (i,\imath), (j,\jmath) \in \hat{\mathbb{X}}$. The corresponding two-component semi-Markov process can be defined as $(\eta_\varepsilon(t), \rho_\varepsilon(t)) = (\eta_{\varepsilon,n}, \rho_{\varepsilon,n})$, for $\zeta_{\varepsilon,n} \le t < \zeta_{\varepsilon,n+1}, n = 0, 1, \ldots$, where $\zeta_{\varepsilon,n} = \kappa_{\varepsilon,1} + \cdots + \kappa_{\varepsilon,n}$ are moments of jumps for process $(\eta_\varepsilon(t), \rho_\varepsilon(t))$.

The first-rare-event time ξ_ε is, in fact, the first hitting time in the domain $\mathbb{D} = \{(1,1), (2,1)\}$ for the semi-Markov process $(\eta_\varepsilon(t), \rho_\varepsilon(t))$, i.e.,

$$\xi_\varepsilon = \sum_{k=1}^{\nu_{\varepsilon,\mathbb{D}}} \kappa_{\varepsilon,k}, \tag{1.12}$$

where

$$\nu_{\varepsilon,\mathbb{D}} = \min(n \ge 1 : (\eta_{\varepsilon,n}, \rho_{\varepsilon,n}) \in \mathbb{D}). \tag{1.13}$$

The result presented in the above example is a very particular case of the results given in Part I. In this part, we consider the first-rare-event times (or, equivalently, first hitting times) for perturbed finite semi-Markov processes. Transition times for semi-Markov processes have arbitrary distributions concentrated of positive half-line. The concept of asymptotically uniformly ergodic embedded Markov chains is introduced. In this case, neither the convergence of its transition probabilities nor the convergence of the corresponding stationary probabilities is not required. We give necessary and sufficient conditions for convergence in distribution of first-rare-event times for perturbed semi-Markov processes (based on one-step probabilities of rare events and distributions of sojourn times averaged over the stationary distributions of asymptotically uniformly ergodic embedded Markov chains) and show that the corresponding limiting distribution has the Laplace transform of the form $(1 + A(s))^{-1}$, where $A(s)$ is a cumulant of some infinitely divisible distribution concentrated on the positive half-line. Also, we give necessary and sufficient conditions for convergence in J-topology for first-rare-event processes and related counting processes as well as generalise the above results to more general modifications of first-rare-event times and processes. These results are illustrated by applications to perturbed risk processes and perturbed queuing systems.

1.1.1.3 First-Rare-Event Times and Related Processes for Regularly Perturbed Semi-Markov Processes. Let $(\eta_{\varepsilon,n}, \kappa_{\varepsilon,n}, \rho_{\varepsilon,n})$, $n = 0, 1, \ldots$ be, for every $\varepsilon \in (0, 1]$, a Markov renewal process, i.e., a discrete time, homogenous Markov chain with a phase space $\mathbb{Z} = \mathbb{X} \times [0, \infty) \times \{0, 1\}$, where $\mathbb{X} = \{1, 2, \ldots, m\}$ is a finite set, and transition probabilities, defined for $i, j \in \mathbb{X}, s, t \geq 0, \imath, \jmath = 0, 1$,

$$Q_{\varepsilon,ij}(t, \jmath) = \mathsf{P}\{\eta_{\varepsilon,1} = j, \kappa_{\varepsilon,1} \leq t, \rho_{\varepsilon,1} = \jmath/\eta_{\varepsilon,0} = i, \kappa_{\varepsilon,0} = s, \rho_{\varepsilon,0} = \imath\}. \quad (1.14)$$

The characteristic property of the Markov renewal process is that its transition probabilities depend only on the current state of the first component and do not depend on states of other components. In this case, the first component $\eta_{\varepsilon,n}$ is itself a homogeneous Markov chain with transition probabilities $p_{\varepsilon,ij} = Q_{\varepsilon,ij}(\infty, 0) + Q_{\varepsilon,ij}(\infty, 1), i, j \in \mathbb{X}$. Also, the above Markov renewal process can be used for construction of semi-Markov process defined as $\eta_\varepsilon(t) = \eta_{\varepsilon,n}$, for $\zeta_{\varepsilon,n} \leq t < \zeta_{\varepsilon,n+1}, n = 0, 1, \ldots$, where $\zeta_{\varepsilon,n} = \kappa_{\varepsilon,1} + \cdots + \kappa_{\varepsilon,n}$ are moments of jumps for the semi-Markov process $\eta_\varepsilon(t)$. As far as random variables $\rho_{\varepsilon,n}$ are concerned, they are used as indicators of "rare" events $\{\rho_{\varepsilon,n} = 1\}$.

The object of studies in Part I is the random functional,

$$\xi_\varepsilon = \sum_{n=1}^{\nu_\varepsilon} \kappa_{\varepsilon,n}, \quad (1.15)$$

where

$$\nu_\varepsilon = \min(n \geq 1 : \rho_{\varepsilon,n} = 1). \quad (1.16)$$

The random variable ξ_ε can be referred to as the first-rare-event time for the semi-Markov process $\eta_\varepsilon(t)$. The possibility to interpret events $\{\rho_{\varepsilon,n} = 1\}$ as rare is supported by the "rare-event" condition:

$(\mathcal{A})$: $0 < \max_{i \in \mathbb{X}} p_{\varepsilon,i} \to 0$ as $\varepsilon \to 0$, where $p_{\varepsilon,i} = \sum_{j \in \mathbb{X}} Q_{\varepsilon,ij}(\infty, 1)$, $i \in \mathbb{X}$.

Figure 1.1 illustrates the construction described above. In the example, illustrated by this figure, the first two indicator random variables $\rho_{\varepsilon,1}, \rho_{\varepsilon,2} = 0$, while the indicator random variables $\rho_{\varepsilon,3} = 1$. Thus, $\nu_\varepsilon = 3$ and the first-rare-event time $\xi_\varepsilon = \kappa_{\varepsilon,1} + \kappa_{\varepsilon,2} + \kappa_{\varepsilon,3}$.

Other objects of research in Part I are the first-rare-event processes,

$$\xi_\varepsilon(t) = \sum_{n \leq t\nu_\varepsilon} \kappa_{\varepsilon,k}, \ t \geq 0 \quad (1.17)$$

and the counting processes generated by flows of rare events such as

$$N_\varepsilon(t) = \max(n : \sum_{k=1}^{\nu_\varepsilon(n)} \kappa_{\varepsilon,k} \leq t), \ t \geq 0, \quad (1.18)$$

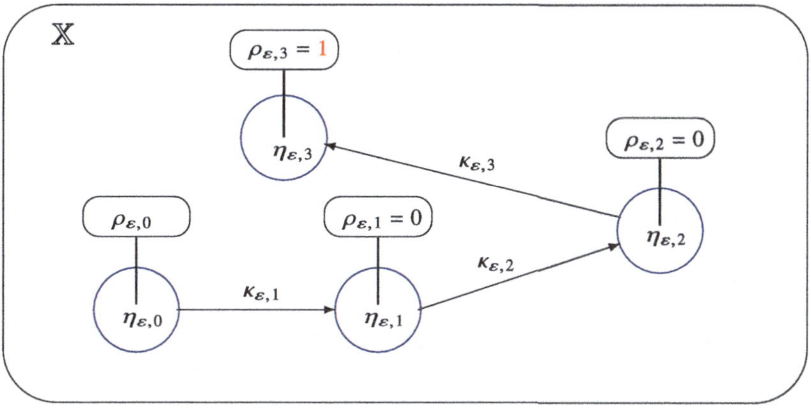

Fig. 1.1 First-rare-event time ξ_ε

where

$$\nu_\varepsilon(n) = \min(k : \sum_{r=1}^{k} \rho_{\varepsilon,r} = n). \tag{1.19}$$

1.1.1.4 Asymptotically Uniformly Ergodic Markov Chains. Let $\eta_{\varepsilon,n}, n = 0, 1, \ldots,$ be, for every $\varepsilon \in (0, 1]$, a finite Markov chain, with phase space $\mathbb{X} = \{1, \ldots, m\}$ and the matrix of transition probabilities $\mathbf{P}_\varepsilon = \|p_{\varepsilon,ij}\|$.

The standard perturbation model is based on the assumption that the transition probabilities $p_{\varepsilon,ij} \to p_{0,ij}$ as $\varepsilon \to 0$, for $i, j \in \mathbb{X}$. In this case, the matrix $\mathbf{P}_0 = \|p_{0,ij}\|$ is also stochastic. Let $\eta_{0,n}, n = 0, 1, \ldots,$ be a Markov chain with the phase space $\mathbb{X}$ and the matrix of transition probabilities $\mathbf{P}_0$. The above convergence relation for the transition probabilities $p_{\varepsilon,ij}$ allows us to consider the Markov chain $\eta_{\varepsilon,n}$ as a perturbed version of the Markov chain $\eta_{0,n}$.

In the regular perturbation model, it is assumed that the phase space $\mathbb{X}$ is one class of communicative states for the limiting Markov chain $\eta_{0,n}$. Thus, the phase space $\mathbb{X}$ also is one class of communicative states for the pre-limiting Markov chain $\eta_{\varepsilon,n}$, for $\varepsilon \in (0, \varepsilon_0]$, for some $\varepsilon_0 \in (0, 1]$. In this case, there exists the unique stationary distribution $\bar{\pi}_\varepsilon = \langle \pi_{\varepsilon,i}, i \in \mathbb{X}\rangle$ for the Markov chain $\eta_{\varepsilon,n}$, for $\varepsilon \in [0, \varepsilon_0]$ and $\pi_{\varepsilon,i} \to \pi_{0,i} > 0$ as $\varepsilon \to 0$, for $i \in \mathbb{X}$.

The concept of asymptotically uniformly ergodic Markov chains let us generalise the regular perturbation model to the model, where it is only assumed that:

($\mathcal{B}$): There exists a chain of states $i_0, i_1, \ldots, i_N = i_0$, which contains all states from the phase space $\mathbb{X}$ and such that $\underline{\lim}_{\varepsilon \to 0} p_{\varepsilon, i_{k-1} i_k} > 0$, for $k = 1, \ldots, N$.

As in the regular perturbation model, the phase space $\mathbb{X}$ is one class of communicative states for the pre-limiting Markov chain $\eta_{\varepsilon,n}$, for $\varepsilon \in (0, \varepsilon_0]$, for some $\varepsilon_0 \in (0, 1]$. In this case, there exists the unique stationary distribution $\bar{\pi}_\varepsilon = \langle \pi_{\varepsilon,i}, i \in \mathbb{X}\rangle$ for Markov chain $\eta_{\varepsilon,n}$, for $\varepsilon \in (0, \varepsilon_0]$. Instead of the convergence relation, the following relation holds: $0 < \underline{\lim}_{\varepsilon\to 0} \pi_{\varepsilon,i} \leq \overline{\lim}_{\varepsilon\to 0} \pi_{\varepsilon,i} < 1$, for $i \in \mathbb{X}$. However, the convergence of transition probabilities $p_{\varepsilon,ij}$ and stationary probabilities $\pi_{\varepsilon,i}$ may not take place.

Figure 1.2 illustrates the concept of asymptotically uniformly ergodic Markov chains. In this figure, transitions along the corresponding chain of states, with transition probabilities separated from 0, as $\varepsilon \to 0$, are indicated by red arrows, while other transitions with positive probabilities are indicated by black arrows.

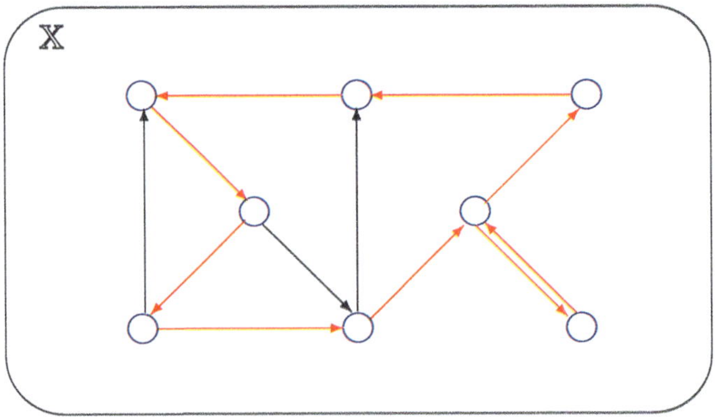

Fig. 1.2 Transition graph for asymptotically uniformly ergodic Markov chains

Obviously, the standard model of regularly perturbed Markov chains is a particular case of the model of asymptotically uniformly ergodic Markov chains.

1.1.1.5 Averaged Characteristics and Necessary and Sufficient Conditions of Convergence. Let $F_{\varepsilon,i}(t) = \sum_{j\in\mathbb{X}, \jmath=0,1} Q_{\varepsilon,ij}(t, \jmath), t \geq 0$ be, for $\varepsilon \in (0, 1]$, the distribution function of sojourn time in state $i \in \mathbb{X}$ for the semi-Markov process $\eta_\varepsilon(t)$.

Let also $\theta_{\varepsilon,i,n}, n = 1, 2, \ldots$ be, for every $\varepsilon \in (0, 1]$ and $i \in \mathbb{X}$, a sequence of i.i.d. random variables with the distribution function $F_{\varepsilon,i}(\cdot)$.

As mentioned above, standard conditions of convergence for random functionals such as first-rare-event times, hitting times, etc., are usually based on the assumption that sums $\sum_{n\leq u_\varepsilon} \theta_{\varepsilon,i,n} \xrightarrow{\mathsf{d}} \theta_{0,i}$ as $\varepsilon \to 0$,[3] for $i \in \mathbb{X}$. Here, $\theta_{0,i}, i \in \mathbb{X}$ are some limiting random variables and $0 < u_\varepsilon \to \infty$ as $\varepsilon \to 0$ is some specially chosen function connected with stopping probabilities $p_{\varepsilon,i}$ defined above in the condition $(\mathcal{A})$.

We, however, use more general and weaker conditions imposed on distributions sojourn times averaged over the stationary distributions of the corresponding embedded Markov chains.

Let $F_\varepsilon(t) = \sum_{i\in\mathbb{X}} F_{\varepsilon,i}(t)\pi_{\varepsilon,i}, t \geq 0$ be, for $\varepsilon \in (0, 1]$, the distribution function of sojourn time $F_{\varepsilon,i}(t)$ averaged over the stationary distribution $\bar{\pi}_\varepsilon$, and $\theta_{\varepsilon,n}, n = 1, 2, \ldots$

[3] $\xrightarrow{\mathsf{d}}$ is the symbol of convergence in distribution for random variables.

be, for $\varepsilon \in (0, 1]$, a sequence of i.i.d. random variables with the distribution function $F_\varepsilon(\cdot)$. Also, let $u_\varepsilon = p_\varepsilon^{-1}$, where $p_\varepsilon = \sum_{i \in \mathbb{X}} p_{\varepsilon,i}\pi_{\varepsilon,i}$.

Necessary and sufficient conditions of convergence in distributions for first-rare-event times ξ_ε presented in Part I are based on three model assumptions (which are analogues of the conditions **(1)**–**(3)** used in the example considered in Sect. 1.1.1.2). The first is the "rare-event" condition $(\mathcal{A})$. The second is the condition $(\mathcal{B})$ of asymptotically uniform ergodicity of Markov chains $\eta_{\varepsilon,n}$. The third is the following condition, which guarantees that the last summand $\kappa_{\varepsilon,\nu_\varepsilon}$ in the random sum ξ_ε is asymptotically negligible:

$(\mathcal{C})$: $\mathsf{P}_i\{\kappa_{\varepsilon,1} > \delta/\rho_{\varepsilon,1} = 1\} \to 0$ as $\varepsilon \to 0$, for $\delta > 0$ and $i \in \mathbb{X}$.

The role of necessary and sufficient condition of convergence in distribution for the first hitting times is played by the following condition:

$(\mathcal{D})$: $\sum_{n \le u_\varepsilon} \theta_{\varepsilon,n} \xrightarrow{\mathsf{d}} \theta_0$ as $\varepsilon \to 0$, where θ_0 is a proper non-negative random variable with distribution not concentrated at zero.

It is well known that, in this case, the random variable θ_0 has an infinitely divisible distribution. Let $A(s)$ be its cumulant, i.e., $\mathsf{E}e^{-s\theta_0} = e^{-A(s)}$, $s \ge 0$. The central criteria of convergence (see, for example, Loève (1977)) also make it possible to reformulate the condition $(\mathcal{D})$ in the equivalent form of convergence relations for characteristics $u_\varepsilon(1 - F_\varepsilon(u))$ and $u_\varepsilon \int_0^u sF_\varepsilon(ds)$.

One of the main results presented in Part I is Theorem 2.1, which states that, under the assumption that the conditions $(\mathcal{A})$, $(\mathcal{B})$, and $(\mathcal{C})$ are satisfied, the condition $(\mathcal{D})$ is the *necessary* and *sufficient* condition for the fulfilment of the asymptotic relation,

$$\xi_\varepsilon(1) \xrightarrow{\mathsf{d}} \xi_0 \text{ as } \varepsilon \to 0, \tag{1.20}$$

where ξ_0 is a proper non-negative random variable with distribution not concentrated at zero. Moreover, the limiting random variable ξ_0 has, in this case, the Laplace transform $\mathsf{E}e^{-s\xi_0} = (1 + A(s))^{-1}$, $s \ge 0$.

Moreover, in this case, the rare-event processes $\xi_\varepsilon(t), t \ge 0 \xrightarrow{\mathsf{J}} \xi_0(t) = \theta_0(t\nu_0), t \ge 0$ as $\varepsilon \to 0$,[4] where (a) ν_0 is a random variable, which has the exponential distribution with parameter 1, (b) $\theta_0(t), t \ge 0$ is a non-negative Lévy process with the Laplace transforms $\mathsf{E}e^{-s\theta_0(t)} = e^{-tA(s)}$, $s, t \ge 0$, and (c) the random variable ν_0 and the process $\theta_0(t), t \ge 0$ are independent.

1.1.2 Part I: Contents by Chapters

Part I includes Chaps. 2–7.

Chapter 2. In this chapter, we introduce first-rare-event times for asymptotically uniformly ergodic perturbed semi-Markov processes and give necessary and suf-

[4] $\xrightarrow{\mathsf{J}}$ is the symbol of convergence in Skorokhod J-topology for càdlàg stochastic processes.

ficient conditions of convergence in distribution for first-rare-event times ξ_ε and J-convergence for first-rare-event time processes $\xi_\varepsilon(t), t \geq 0$.

Chapter 3. In this chapter, we introduce process $N_\varepsilon(t), t \geq 0$, which counts the number of rare events occurring on trajectory of semi-Markov process $\eta_\varepsilon(s)$ in time interval $[0, t]$. Necessary and sufficient conditions of convergence in distribution for counting processes $N_\varepsilon(t)$ are given. Analogous results are also obtained for some more general multivariate counting processes related to rare-event times.

Chapter 4. In this chapter, we generalise results of Chap. 2 to some modifications of first-rare-event times and first-rare-event time processes, in particular, to vector, upper and lower first-rare-event times and first-rare-event time processes, first-rare-event times for Markov renewal process with transition period and extending phase spaces. Also, connection between first-rare-event times and standard and directed hitting times is clarified, and necessary and sufficient conditions of convergence in distribution for first hitting times for asymptotically uniformly ergodic perturbed semi-Markov processes are given.

Chapter 5. In this chapter, we formulate conditions of convergence in distribution for rare-event times represented by geometric type random sums and give necessary and sufficient conditions for weak convergence of non-ruin distribution functions for perturbed risk processes in the models of stable and diffusion approximations for non-ruin distribution functions.

Chapter 6. In this chapter we present necessary and sufficient conditions for convergence in distribution for first-rare-event times for a number of models of perturbed closed M/M-type queuing systems.

Chapter 7. In this chapter we present necessary and sufficient conditions for weak convergence in distribution for first-rare-event times for a number of models of perturbed M/M-type queuing systems with bounded and unbounded buffers.

1.2 Part II: Hitting Times and Phase Space Reduction for Perturbed Semi-Markov Processes

In Sect. 1.2, we outline the circle of problems and illustrate results presented in Part II.

1.2.1 Part II: Contents, Examples, Models, Results

1.2.1.1 Content of Part II. In this part, which includes Chaps. 8–12, we present new asymptotic recurrent algorithms of phase space reduction for regularly and singularly perturbed finite semi-Markov processes. Using these algorithms, we formulate effective conditions of weak convergence for distributions and convergence of expectations for hitting times and get effective recurrent formulas for computing the corresponding normalisation functions, Laplace transforms for limiting distributions, and limits for expectations.

The phase space reduction algorithms presented in Part II are universal. They can be applied to perturbed semi-Markov processes with an arbitrary asymptotic communicative structure and are computationally effective due to recurrent character of computational procedures. What is also important, for some models, for example, for perturbed birth–death-type semi-Markov processes, the above-mentioned algorithms of phase space reduction preserve birth–death structure of the original semi-Markov processes.

1.2.1.2 A Model Example. Let $\mathbf{S}_\varepsilon$ be, for every $\varepsilon \in (0, 1]$, a queuing system, the functioning of which is described by a continuous time, homogeneous Markov chain $\eta_\varepsilon(t), t \geq 0$, with the phase space $\mathbb{X} = \{0, 1, 2\}$ and continuous from the right trajectories.

The system can operate in two "work" modes or in the "failure" mode.

If $\eta_\varepsilon(t) = 1$ or $\eta_\varepsilon(t) = 2$, then the system is, respectively, in the operating mode 1 or 2 at the moment t. If $\eta_\varepsilon(t) = 0$, then the system is in the failure mode at the moment t.

The corresponding generator matrix has the following form, for $\varepsilon \in (0, 1]$,

$$\mathbf{Q}_\varepsilon = \left\| \begin{array}{ccc} -\lambda_{\varepsilon,0} & \lambda_{\varepsilon,01} & \lambda_{\varepsilon,02} \\ \lambda_{\varepsilon,10} & -\lambda_{\varepsilon,1} & \lambda_{\varepsilon,12} \\ \lambda_{\varepsilon,20} & \lambda_{\varepsilon,21} & -\lambda_{\varepsilon,2} \end{array} \right\| \quad \begin{array}{c} \Longleftarrow 0 \\ \Longleftarrow 1 \\ \Longleftarrow 2 \end{array} \tag{1.21}$$

where: (a) $\lambda_{\varepsilon,ij} > 0$, for $i \neq j, i, j \in \mathbb{X}$, (b) $\lambda_{\varepsilon,i} = \sum_{j \neq i} \lambda_{\varepsilon,ij}$, for $i \in \mathbb{X}$.

Figure 1.3 shows the transition graph, the transition intensities, and the parameters of exponentially distributed transition times for the Markov chain $\eta_\varepsilon(t)$.

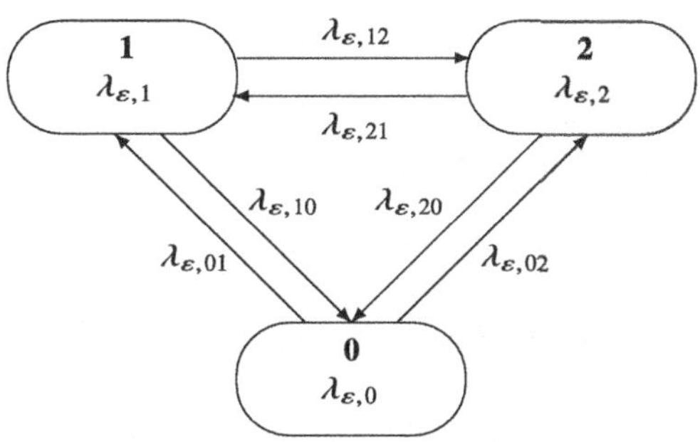

Fig. 1.3 The transition graph for the Markov chain $\eta_\varepsilon(t)$

Suppose that the initial state of the system $\eta_\varepsilon(0) \neq 0$ and describe asymptotic behaviour of the random variables that can be interpreted as the first failure times,

$$\xi_\varepsilon = \inf(t > 0 : \eta_\varepsilon(t) = 0). \tag{1.22}$$

Note that, in the case where the initial state $\eta_\varepsilon(0) \neq 0$, the distribution of random variable ξ_ε does not depend on the intensities $\lambda_{\varepsilon,0i}, i = 1, 2$. Therefore, for simplicity, we can assume that $\lambda_{\varepsilon,0i} = 1, i = 1, 2$, for $\varepsilon \in (0, 1]$.

Let $\kappa_{\varepsilon,n}, n = 1, 2, \ldots$ be sequential transition times, $\zeta_{\varepsilon,n} = \sum_{k=1}^{n} \kappa_{\varepsilon,k}, n = 0, 1, \ldots$ be sequential moments of jumps, and $\eta_{\varepsilon,n} = \eta_\varepsilon(\zeta_{\varepsilon,n}), n = 0, 1, \ldots$ be states of the Markov chain $\eta_\varepsilon(t)$ at moments of jumps.

As is known, the random sequence $\eta_{\varepsilon,n}$ is a discrete time (embedded) Markov chain, with transition probabilities,

$$p_{\varepsilon,ij} = \begin{cases} \frac{\lambda_{\varepsilon,ij}}{\lambda_{\varepsilon,i}} & \text{for } j \neq i, \ i, j \in \mathbb{X}, \\ 0 & \text{for } j = i, \ i \in \mathbb{X}. \end{cases} \tag{1.23}$$

Note that the probabilities $p_{\varepsilon,ij} > 0$, for $j \neq i, \ i, j \in \mathbb{X}$.

In what follows, we use limits for different sums of products of the intensities $\lambda_{\varepsilon,ij}$ and for quotients of such sums and should guarantee the existence of such limits. Using the concept of a complete family of asymptotically comparable functions let us solve this problem.

A family $\mathcal{H}$ of positive functions $h(\varepsilon)$ defined on interval $(0, 1]$ is a complete family of asymptotically comparable functions if: (1) it is closed with respect to operations of summation, multiplication, and division and (2) the functions $h(\cdot) \in \mathcal{H}$ have limits taking values in the interval $[0, \infty]$, as $\varepsilon \to 0$.

The corresponding comments and examples of complete families of asymptotically comparable functions and the corresponding operating rules and formulas are given in Chap. 8.

Let us assume that the following condition is satisfied:

(4) The intensities $\lambda_{\varepsilon,ij}, j \neq i, i = 1, 2$ belong to a complete family of comparable functions $\mathcal{H}$ and $\lim_{\varepsilon \to 0} \lambda_{\varepsilon,ij} = \lambda_{0,ij} \in [0, \infty)$, for $j \neq i, i = 1, 2$.

The condition **(4)** implies that

$$p_{\varepsilon,ij} \to p_{0,ij} \in [0, 1] \text{ as } \varepsilon \to 0, \ \text{ for } j \neq i, \ i, j = 1, 2. \tag{1.24}$$

This creates a variety of possible variants for asymptotics of the first failure times ξ_ε.

The Markov chain $\eta_\varepsilon(t)$ can be considered as a semi-Markov process with transition probabilities $Q_{\varepsilon,ij}(t) = p_{\varepsilon,ij}(1 - e^{-\lambda_{\varepsilon,i}t}), \ t \geq 0, i, j \in \mathbb{X}$.

The distribution functions of transition times take the following form, $F_{\varepsilon,ij}(t) = F_{\varepsilon,i}(t) = p_{\varepsilon,ij}^{-1} Q_{\varepsilon,ij}(t) = 1 - e^{-\lambda_{\varepsilon,i}t}, \ t \geq 0, j \neq i, i, j \in \mathbb{X}$, where intensities $\lambda_{\varepsilon,i}$ are given by the relation (1.21). The corresponding Laplace transforms for the above exponential distribution functions take the form, $\phi_{\varepsilon,ij}(s) = \phi_{\varepsilon,i}(s) = \frac{\lambda_{\varepsilon,i}}{\lambda_{\varepsilon,i}+s}, \ s \geq 0, \ j \neq i, i, j \in \mathbb{X}$.

Let us introduce the normalisation functions $u_{\varepsilon,i} = \lambda_{\varepsilon,i}^{-1}$, for $i \in \mathbb{X}$.

Obviously, the following relation holds, for $\varepsilon \in (0, 1]$ and $j \neq i, i, j \in \mathbb{X}$,

$$\phi_{\varepsilon,ij}(s/u_{\varepsilon,i}) = \phi_{\varepsilon,i}(s/u_{\varepsilon,i}) = \phi_0(s) = \frac{1}{1+s}, \ s \geq 0. \tag{1.25}$$

The states 1 and 2 can be compared by their absorption rates. Suppose, for example, that the state 1 is less absorbing than the state 2. This means that the following limit (which exists according to the condition **(4)**)

$$\lim_{\varepsilon \to 0} \frac{u_{\varepsilon,1}}{u_{\varepsilon,2}} = w_{0,1,2} \in [0, \infty). \tag{1.26}$$

We now exclude state 1 from the phase space $\mathbb{X} = \{0, 1, 2\}$ and construct a new reduced semi-Markov process ${}_1\eta_\varepsilon(t)$ using the following time-space screening procedure. Let $0 = {}_1\zeta_{\varepsilon,0} < {}_1\zeta_{\varepsilon,1} < \cdots$ be successive moments of hitting (resulted by jumps) in the reduced phase space ${}_1\mathbb{X} = \{0, 2\}$ by the process $\eta_\varepsilon(t)$. Then, the reduced semi-Markov process ${}_1\eta_\varepsilon(t)$ is defined as ${}_1\eta_\varepsilon(t) = \eta_\varepsilon({}_1\zeta_{\varepsilon,n})$ for ${}_1\zeta_{\varepsilon,n} \leq t < {}_1\zeta_{\varepsilon,n+1}, n = 0, 1, \ldots$.

Figure 1.4 illustrates the above phase space reduction procedure.

It is important that the first hitting times in state 0 coincides for the semi-Markov processes $\eta_\varepsilon(t)$ and ${}_1\eta_\varepsilon(t)$, i.e.,

$$\xi_\varepsilon = {}_1\xi_\varepsilon = \inf(t > 0 : \ {}_1\eta_\varepsilon(t) = 0). \tag{1.27}$$

The semi-Markov process ${}_1\eta_\varepsilon(t)$ has the phase space ${}_1\mathbb{X} = \{0, 2\}$ and the transition probabilities ${}_1Q_{\varepsilon,ij}(t) = Q_{\varepsilon,ij}(t) + Q_{\varepsilon,i1}(t) * Q_{\varepsilon,1j}(t), \ t \geq 0, \ i, j \in {}_1\mathbb{X}$. The corresponding embedded Markov chain ${}_1\eta_{\varepsilon,n}$ has the transition probabilities, ${}_1p_{\varepsilon,ij} = p_{\varepsilon,ij} + p_{\varepsilon,i1}p_{\varepsilon,1j}, \ i, j \in {}_1\mathbb{X}$. Note that ${}_1p_{\varepsilon,ij} > 0, \ i, j \in {}_1\mathbb{X}$, for $\varepsilon \in (0, 1]$ and

$${}_1p_{\varepsilon,ij} \longrightarrow {}_1p_{0,ij} = p_{0,ij} + p_{0,i1}p_{0,1j} \text{ as } \varepsilon \longrightarrow 0, \ \text{ for } i, j \in {}_1\mathbb{X}. \tag{1.28}$$

The distribution functions of transition times take the following form, ${}_1F_{\varepsilon,ij}(t) = {}_1p^{-1}_{\varepsilon,ij}\,{}_1Q_{\varepsilon,ij}(t), \ t \geq 0, \ i, j \in {}_1\mathbb{X}$, and the Laplace transforms of the above distribution functions take the following form, for $i, j \in {}_1\mathbb{X}$,

$${}_1\phi_{\varepsilon,ij}(s) = {}_1p^{-1}_{\varepsilon,ij}(\phi_{\varepsilon,ij}(s)p_{\varepsilon,ij} + \phi_{\varepsilon,i1}(s)\phi_{\varepsilon,1j}(s)p_{\varepsilon,i1}p_{\varepsilon,1j}), \ s \geq 0. \tag{1.29}$$

The condition **(4)** implies that the probabilities $p_{\varepsilon,ij}, i, j \in \mathbb{X}$ and ${}_1p_{\varepsilon,ij}, i, j \in {}_1\mathbb{X}$ also belong to the family of asymptotically comparable functions $\mathcal{H}$, and thus, there exist limits, for $i, j \in {}_1\mathbb{X}$,

$$\lim_{\varepsilon \to 0} \frac{p_{\varepsilon,ij}}{{}_1p_{\varepsilon,ij}} = {}_1\hat{q}_{0,ij} \in [0, 1]. \tag{1.30}$$

The relations (1.25), (1.26), and (1.28)–(1.30) imply that the following relation holds, for $j \in {}_1\mathbb{X}$ and $s \geq 0$,

$$\begin{aligned} {}_1\phi_{\varepsilon,2j}(s/u_{\varepsilon,2}) &\longrightarrow {}_1\phi_{0,2j}(s) \\ &= {}_1\hat{q}_{0,2j}\phi_0(s) + \phi_0(s)\phi_0(sw_{0,1,2})(1 - {}_1\hat{q}_{0,2j}) \text{ as } \varepsilon \longrightarrow 0. \end{aligned} \tag{1.31}$$

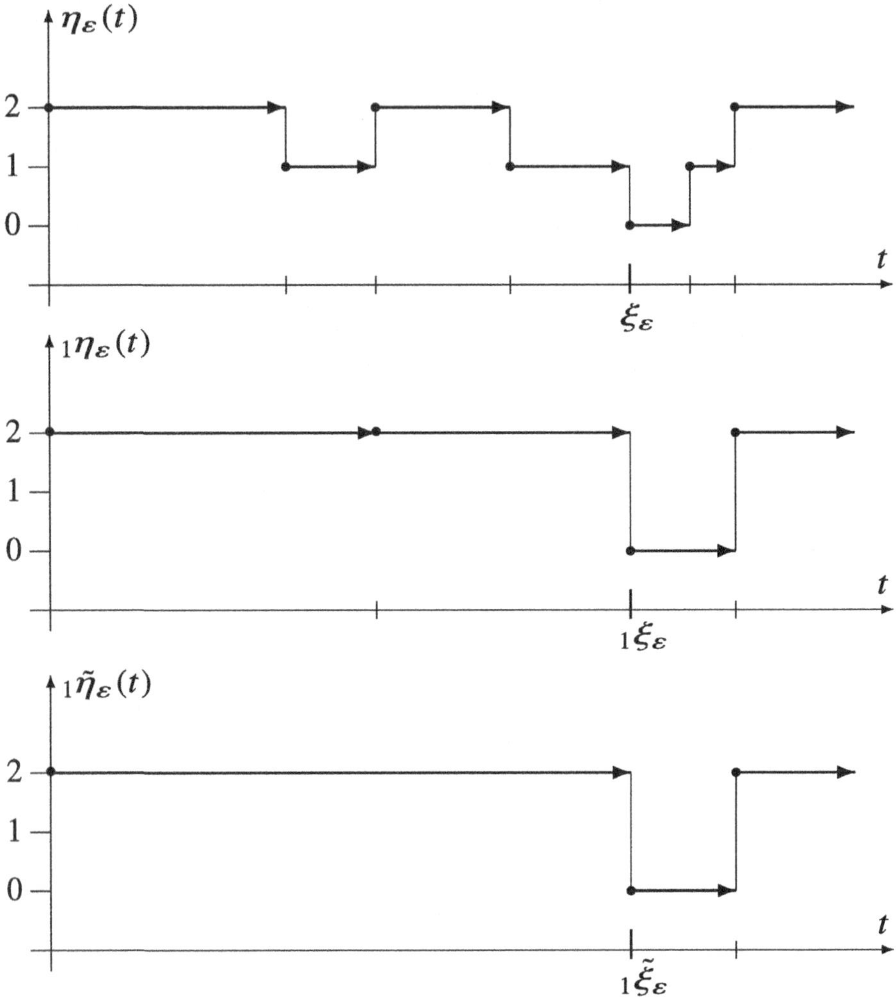

Fig. 1.4 Phase space reduction and removing of virtual transitions for semi-Markov process $\eta(t)$

The next step is to remove virtual transitions from trajectories of the semi-Markov process ${}_1\eta_\varepsilon(t)$ and construct a new reduced semi-Markov process ${}_1\tilde{\eta}_\varepsilon(t)$ using the following time-space screening procedure. Let $0 = {}_1\tilde{\zeta}_{\varepsilon,0} < {}_1\tilde{\zeta}_{\varepsilon,1} < \cdots$ be sequential instants of change of state (resulted by jumps) by the process ${}_1\eta_\varepsilon(t)$. Then the reduced semi-Markov process ${}_1\tilde{\eta}_\varepsilon(t)$ is defined as ${}_1\tilde{\eta}_\varepsilon(t) = {}_1\eta_\varepsilon({}_1\tilde{\zeta}_{\varepsilon,n})$ for ${}_1\tilde{\zeta}_{\varepsilon,n} \leq t < {}_1\tilde{\zeta}_{\varepsilon,n+1}, n = 0, 1, \ldots$.

Figure 1.4 also illustrates the above procedure of removing virtual transitions.

It is important that the first hitting times in the state 0 coincide for the semi-Markov processes $\eta_\varepsilon(t)$, ${}_1\eta_\varepsilon(t)$, and ${}_1\tilde{\eta}_\varepsilon(t)$, i.e.,

$$\xi_\varepsilon = {}_1\xi_\varepsilon = {}_1\tilde{\xi}_\varepsilon = \inf(t > 0 : {}_1\tilde{\eta}_\varepsilon(t) = 0). \tag{1.32}$$

Moreover, the first hitting time in state 0 for the process ${}_1\tilde{\eta}_\varepsilon(t)$ obviously coincides with the moment of its first jump, i.e.,

$$ {}_1\tilde{\xi}_\varepsilon = {}_1\tilde{\zeta}_{\varepsilon,1}. \tag{1.33} $$

The semi-Markov process ${}_1\tilde{\eta}_\varepsilon(t)$ has the phase space ${}_1\mathbb{X} = \{0, 2\}$ and the transition probabilities, ${}_1\tilde{Q}_{\varepsilon,ij}(t) = 0, t \geq 0$, for $t \geq 0$, $j = i, i \in {}_1\mathbb{X}$, and ${}_1\tilde{Q}_{\varepsilon,ij}(t) = \sum_{n=0}^{\infty} {}_1Q^{(*n)}_{\varepsilon,ii}(t) * {}_1Q_{\varepsilon,ij}(t), t \geq 0$, for $j \neq i, i \in {}_1\mathbb{X}$. The corresponding embedded Markov chain ${}_1\tilde{\eta}_{\varepsilon,n}$ has the transition probabilities, ${}_1\tilde{p}_{\varepsilon,ij} = \mathrm{I}(j \neq i), i, j \in {}_1\mathbb{X}$.

The distribution functions of transition times take the following form, ${}_1\tilde{F}_{\varepsilon,ij}(t) = {}_1\tilde{Q}_{\varepsilon,ij}(t)$, $t \geq 0$, for $j \neq i, i \in {}_1\mathbb{X}$, and the Laplace transforms of the above distribution functions take the following form, for $j \neq i, i \in {}_1\mathbb{X}$,

$$ {}_1\tilde{\phi}_{\varepsilon,ij}(s) = \frac{{}_1\phi_{\varepsilon,ij}(s)\, {}_1p_{\varepsilon,ij}}{1 - {}_1p_{\varepsilon,ii}\, {}_1\phi_{\varepsilon,ii}(s)}, \ s \geq 0. \tag{1.34} $$

Let us introduce the Laplace transforms $\Psi_{\varepsilon,i}(s) = \mathsf{E}_i e^{-s\xi_\varepsilon}, s \geq 0$, for $i = 1, 2$.

Two cases should be considered.

(I) ${}_1p_{0,20} > 0$.

In this case, the normalisation function $u_{\varepsilon,2}$ should be used.

The relations (1.28), (1.31), and (1.34) imply that the following relation holds, for $s \geq 0$,

$$ \begin{aligned} \Psi_{\varepsilon,2}(s/u_{\varepsilon,2}) &= {}_1\tilde{\phi}_{\varepsilon,20}(s/u_{\varepsilon,2}) \\ &\to \Psi'_{0,2}(s) = \frac{{}_1\phi_{0,20}(s)\, {}_1p_{0,20}}{1 - {}_1p_{0,22}\, {}_1\phi_{0,22}(s)} \ \text{as } \varepsilon \to 0, \end{aligned} \tag{1.35} $$

where the Laplace transforms ${}_1\phi_{0,2i}(s), i = 0, 2$ are given by the relation (1.31).

The above asymptotic relation implies that the following relation holds, for the case where $\eta_\varepsilon(0) = 2$,

$$ \xi_\varepsilon / u_{\varepsilon,2} \xrightarrow{\mathsf{d}} \xi'_{0,2} \ \text{as } \varepsilon \to 0, \tag{1.36} $$

where the random variable $\xi'_{0,2}$ has the Laplace transform $\Psi'_{0,2}(s)$.

(II) ${}_1p_{0,20} = 0$.

In this case, the normalisation function $\tilde{u}_{\varepsilon,2} = u_{\varepsilon,2}\, {}_1p^{-1}_{\varepsilon,20}$ should be used.

The relations (1.25), (1.31), and (1.34) imply that the following relation holds, for $s \geq 0$,

$$ \begin{aligned} \Psi_{\varepsilon,2}(s/\tilde{u}_{\varepsilon,2}) &= {}_1\tilde{\phi}_{\varepsilon,20}({}_1p_{\varepsilon,20}s/u_{\varepsilon,2}) \\ &= \frac{{}_1\phi_{\varepsilon,20}({}_1p_{\varepsilon,20}\, s/u_{\varepsilon,2})}{1 + {}_1p_{\varepsilon,22} \frac{1 - {}_1\phi_{\varepsilon,22}({}_1p_{\varepsilon,20}\, s/u_{\varepsilon,2})}{{}_1p_{\varepsilon,20}}} \\ &\to \Psi''_{0,2}(s) = (1 + (1 + w_{0,1,2})s)^{-1} \ \text{as } \varepsilon \to 0. \end{aligned} \tag{1.37} $$

Here, it is taken into account that, by the relations (1.23) and (1.28), the quantity ${}_1\hat{q}_{0,22} = 0$.

The above asymptotic relation implies that the following relation holds, for the case where $\eta_\varepsilon(0) = 2$,

$$\xi_\varepsilon / \tilde{u}_{\varepsilon,2} \xrightarrow{\mathsf{d}} \xi''_{0,2} \text{ as } \varepsilon \to 0, \tag{1.38}$$

where $\xi''_{0,2}$ is a random variable with the Laplace transform $\Psi''_{0,2}(s)$, i.e., it is an exponentially distributed random variable, with parameter $(1 + w_{0,1,2})^{-1}$.

The following backward relation connects the Laplace transforms $\Psi_{\varepsilon,i}(s)$ for $i = 1$ and $i = 2$,

$$\Psi_{\varepsilon,1}(s) = \phi_{\varepsilon,10}(s)p_{\varepsilon,10} + \Psi_{\varepsilon,2}(s)\phi_{\varepsilon,12}(s)p_{\varepsilon,12}, \ s \geq 0. \tag{1.39}$$

Three cases should be considered.

(III) ${}_1p_{0,20} > 0$ and $p_{0,12} > 0$.

In this case, the relations (1.25), (1.26), (1.35), and (1.39) imply that the following relation relation holds, for $s \geq 0$,

$$\begin{aligned} \Psi_{\varepsilon,1}(s/u_{\varepsilon,2}) &\to \Psi'_{0,1}(s) \\ &= \phi_0(s w_{0,1,2})p_{0,10} + \Psi'_{0,2}(s)\phi_0(s w_{0,1,2})p_{0,12} \text{ as } \varepsilon \to 0, \end{aligned} \tag{1.40}$$

where the Laplace transform $\Psi'_{0,2}(s)$ is given by the relation (1.35).

The relation (1.40) implies that the following relation holds, for the case where $\eta_\varepsilon(0) = 1$,

$$\xi_\varepsilon / u_{\varepsilon,2} \xrightarrow{\mathsf{d}} \xi'_{0,1} \text{ as } \varepsilon \to 0, \tag{1.41}$$

where $\xi'_{0,1}$ is a random variable with the Laplace transform $\Psi'_{0,1}(s)$ given by the relation (1.40).

(IV) ${}_1p_{0,20} = 0$ and $p_{0,12} > 0$.

In this case, the relations (1.25), (1.26), (1.37), and (1.39) imply that the following relation holds, for $s \geq 0$,

$$\Psi_{\varepsilon,1}(s/\tilde{u}_{\varepsilon,2}) \to \Psi''_{0,1}(s) = p_{0,10} + \Psi''_{0,2}(s)p_{0,12} \text{ as } \varepsilon \to 0, \tag{1.42}$$

where the Laplace transform $\Psi''_{0,2}(s)$ is given by relation (1.37).

Relation (1.42) implies that the following relation holds, for the case where $\eta_\varepsilon(0) = 1$,

$$\xi_\varepsilon / \tilde{u}_{\varepsilon,2} \xrightarrow{\mathsf{d}} \xi''_{0,1} \text{ as } \varepsilon \to 0, \tag{1.43}$$

where $\xi''_{0,1}$ is a random variable with the Laplace transform $\Psi''_{0,1}(s)$ given by the relation (1.42).

(V) $p_{0,12} = 0$.

In this case, the relations (1.35), (1.37), and (1.39) imply that the following relation relation holds, for $s \geq 0$,

$$\Psi_{\varepsilon,1}(s/u_{\varepsilon,1}) \to \Psi'''_{0,1}(s) = \frac{1}{1+s} \text{ as } \varepsilon \to 0. \tag{1.44}$$

The relation (1.44) implies that the following relation holds, for the case where $\eta_\varepsilon(0) = 1$,

$$\xi_\varepsilon / u_{\varepsilon,1} \xrightarrow{\mathsf{d}} \xi'''_{0,1} \text{ as } \varepsilon \to 0, \tag{1.45}$$

where $\xi'''_{0,1}$ is a random variable exponentially distributed with parameter 1.

Having in mind applications to ergodic theorems for semi-Markov type process, we are also interested in conditions, under which expectations of first hitting times normalised by the same functions (as in relations of convergence in distribution) converge to the first moments of the corresponding limiting random variables.

In the example considered above, such "simultaneous" convergence takes place in the cases **(I)**–**(IV)**. While, in the case **(V)**, the answer is determined by the limit, $\hat{w}'_{0,1,2} = \lim_{\varepsilon \to 0} \frac{p_{\varepsilon,12}u_{\varepsilon,2}}{u_{\varepsilon,1}}$, if ${}_1p_{0,20} > 0$, or by the limit, $\hat{w}''_{0,1,2} = \lim_{\varepsilon \to 0} \frac{p_{\varepsilon,12}\tilde{u}_{\varepsilon,2}}{u_{\varepsilon,1}}$, if ${}_1p_{0,20} = 0$ (these limits exist due to the condition **(4)**). The above-mentioned simultaneous convergence takes place, if the corresponding limit is 0, or does not take place, if the corresponding limit is in the range $(0, \infty]$.

The results presented in Part II generalise and essentially expand the above asymptotic results in several directions. First, we consider hitting times for perturbed finite semi-Markov processes. In this case, transition times have arbitrary distributions concentrated on a positive half-line. Second, we consider regularly and singularly perturbed semi-Markov processes with an arbitrary communicative structure of the phase space for the corresponding limiting process. Third, we present effective asymptotic recurrent algorithms (of phase space reduction for perturbed semi-Markov processes) for finding normalisation functions, Laplace transforms for limiting distributions, and limit for expectations of hitting times.

1.2.1.3 Hitting Times for Perturbed Semi-Markov Processes. Let $(\eta_{\varepsilon,n}, \kappa_{\varepsilon,n})$, $n = 0, 1, \ldots$ be, for every $\varepsilon \in (0, 1]$, a Markov renewal process, i.e., a homogenous Markov chain with a phase space $\mathbb{Z} = \mathbb{X} \times [0, \infty)$, where $\mathbb{X} = \{1, \ldots, m\}$ is a finite set, and transition probabilities, defined for $i, j \in \mathbb{X}$, $s, t \geq 0$,

$$Q_{\varepsilon,ij}(t) = F_{\varepsilon,ij}(t)p_{\varepsilon,ij} = \mathsf{P}\{\eta_{\varepsilon,1} = j, \kappa_{\varepsilon,1} \leq t / \eta_{\varepsilon,0} = i, \kappa_{\varepsilon,0} = s\}, \tag{1.46}$$

where: (a) $p_{\varepsilon,ij} = Q_{\varepsilon,ij}(\infty)$, $i, j \in \mathbb{X}$ and (b) $F_{\varepsilon,ij}(\cdot)$, $i, j \in \mathbb{X}$ are proper distribution functions, which are not concentrated in zero.

The characteristic property of the Markov renewal process is that its transition probabilities depend only on the current state of the first component and do not depend on the current state of the second component. In this case, the first component $\eta_{\varepsilon,n}$ is itself a homogeneous Markov chain with transition probabilities $p_{\varepsilon,ij}, i, j \in \mathbb{X}$. Also, the above Markov renewal process can be used for construction of a semi-Markov process defined as $\eta_\varepsilon(t) = \eta_{\varepsilon,n}$, for $\zeta_{\varepsilon,n} \leq t < \zeta_{\varepsilon,n+1}, n = 0, 1, \ldots$, where $\zeta_{\varepsilon,n} = \kappa_{\varepsilon,1} + \cdots + \kappa_{\varepsilon,n}$ are moments of jumps for process $\eta_\varepsilon(t)$. In this case, $F_{\varepsilon,ij}(\cdot)$ are conditional distribution functions of transition (inter-jump) times for the semi-Markov process $\eta_\varepsilon(t)$.

It is worth to mention two important cases.

The semi-Markov process $\eta_\varepsilon(t)$ is a continuous time Markov chain, if its transition probabilities have the following form,

$$Q_{\varepsilon,ij}(t) = (1 - e^{-\lambda_{\varepsilon,i}t})p_{\varepsilon,ij}, \ t \geq 0, \ i, j \in \mathbb{X}, \tag{1.47}$$

where $\lambda_{\varepsilon,i} > 0, i \in \mathbb{X}$.

The semi-Markov process $\eta_\varepsilon(t)$ is a discrete time Markov chain embedded in continuous time, if its transition probabilities have the following form:

$$Q_{\varepsilon,ij}(t) = \mathrm{I}(t \geq 1)p_{\varepsilon,ij}, \ t \geq 0, \ i, j \in \mathbb{X}. \tag{1.48}$$

The main difference between the models of Markov chains and semi-Markov processes can be described informally as follows. The Markov property (the future, $\eta_\varepsilon(s), s \geq t$, depends on the past, $\eta_\varepsilon(s), s \leq t$, only through the present, $\eta_\varepsilon(t)$) holds for Markov chains at any time moment $t \geq 0$ and, moreover, at any moment of jump, $\zeta_{\varepsilon,n}, n = 0, 1, \ldots$, while for general semi-Markov processes, the Markov property is guaranteed only at the moments of jumps $\zeta_{\varepsilon,n}, n = 0, 1, \ldots$.

The basic fact concerned semi-Markov processes can be found in the books listed in Sect. B.2.7.

The object of studies in Part II is the random functional, which is the hitting time of semi-Markov process $\eta_\varepsilon(t)$ in a non-empty domain $\mathbb{D} \subseteq \mathbb{X}$,

$$\tau_{\varepsilon,\mathbb{D}} = \sum_{n=1}^{\nu_{\varepsilon,\mathbb{D}}} \kappa_{\varepsilon,n}, \tag{1.49}$$

where

$$\nu_{\varepsilon,\mathbb{D}} = \min(n \geq 1 : \eta_{\varepsilon,n} \in \mathbb{D}). \tag{1.50}$$

Figure 1.5 illustrates the definition given in relation (1.49).

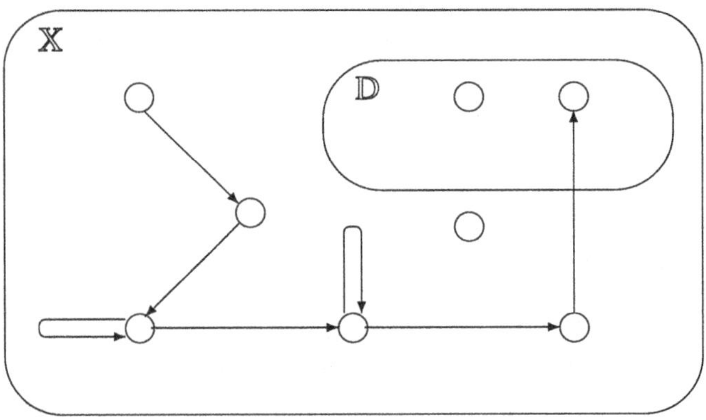

Fig. 1.5 Hitting time $\tau_{\varepsilon,\mathbb{D}}$

We carry out a detailed asymptotic analysis of the distributions $G_{\varepsilon,\mathbb{D},ij}(t) = \mathsf{P}_i\{\tau_{\varepsilon,\mathbb{D}} \leq t, \eta_\varepsilon(\tau_{\varepsilon,\mathbb{D}}) = j\}$ and the expectations $E_{\varepsilon,\mathbb{D},ij} = \mathsf{E}_i\tau_{\varepsilon,\mathbb{D}}\mathrm{I}(\eta_\varepsilon(\tau_{\varepsilon,\mathbb{D}}) = j)$, for regularly and singularly perturbed semi-Markov processes.

1.2.1.4 Regularly and Singularly Perturbed Markov Chains and Semi-Markov Processes. Let $\eta_{\varepsilon,n}, n = 0, 1, \ldots$, be, for every $\varepsilon \in [0, 1]$, a homogeneous Markov chain, with the phase space $\mathbb{X} = \{1, \ldots, m\}$ and the matrix of transition probabilities $\mathbf{P}_\varepsilon = \|p_{\varepsilon,ij}\|$.

In Part II, the standard model of perturbed Markov chains is considered, where the transition probabilities $p_{\varepsilon,ij} \longrightarrow p_{0,ij}$ as $\varepsilon \longrightarrow 0$, for $i, j \in \mathbb{X}$.

It is assumed that the phase space $\mathbb{X}$ is one class of communicative states for the pre-limiting Markov chains $\eta_{\varepsilon,n}$, while $\mathbb{X}$ can possess an arbitrary communicative structure for the limiting Markov chain $\eta_{0,n}$ (with the matrix of transition probabilities $\mathbf{P}_0 = \|p_{0,ij}\|$), i.e., it can consist of one class of communicative states (models with *regular* perturbations), or one class of communicative states plus a class of transient states (models with *semi-regular* perturbations), or several closed classes of communicative states plus possibly a class of transient states (models with *singular* perturbations).

Figure 1.6 illustrates the most complex model with singular perturbations. In this figure, black and white circles indicate, respectively, positive and zero elements in the corresponding matrices of transition probabilities. Here, the phase space $\mathbb{X} = \{1, \ldots, 7\}$ is, for every $\varepsilon \in (0, 1]$, one class of communicative states, for the pre-limiting Markov chains $\eta_{\varepsilon,n}$, while the phase space $\mathbb{X}$ is divided into two closed classes of communicative states $\mathbb{X}_1 = \{1, 2\}$ and $\mathbb{X}_2 = \{3, 4, 5\}$, and the class of transient states $\mathbb{X}_0 = \{6, 7\}$, for the limiting Markov chain $\eta_{0,n}$.

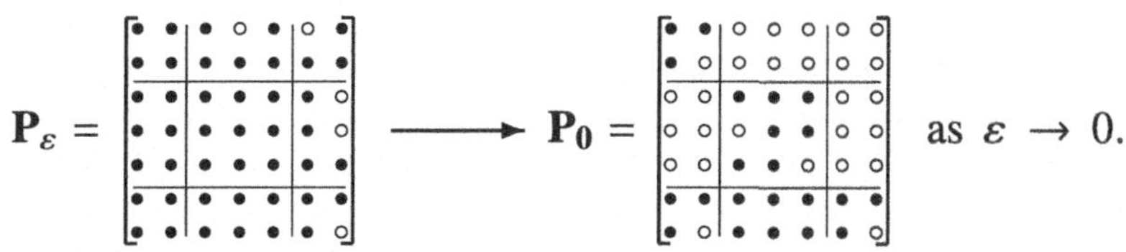

Fig. 1.6 Singularly perturbed Markov chains

Perturbed semi-Markov processes are classified according to the types of perturbations for their embedded Markov chains.

1.2.1.5 Asymptotic Algorithms of Phase Space Reduction for Perturbed Semi-Markov Processes. Our approach to limit theorems for hitting times is based on natural conditions imposed on original perturbed semi-Markov processes.

First, it is the condition of convergence for the transition probabilities of embedded Markov chains:

($\mathcal{E}$) $p_{\varepsilon,ij} \longrightarrow p_{0,ij}$ as $\varepsilon \longrightarrow 0$, for $i, j \in \mathbb{X}$

Second, it is the condition of weak convergence for the distribution functions of transition times:

($\mathcal{F}$) $F_{\varepsilon,ij}(\cdot\, u_{\varepsilon,i}) \Rightarrow F_{0,ij}(\cdot)$ as $\varepsilon \longrightarrow 0$,[5] for $i, j \in \mathbb{X}$, where: (a) $F_{0,ij}(\cdot)$ are distribution functions, which are not concentrated in zero; (b) $u_{\varepsilon,i} \in (0, \infty), i \in \mathbb{X}$ are some normalisation functions such that $u_{\varepsilon,i} \longrightarrow u_{0,i} \in (0, \infty]$ as $\varepsilon \longrightarrow 0$.

Third, this is the condition of convergence for the first moments of transition times:

[5] $\Rightarrow$ is the symbol of weak convergence for distribution functions.

($\mathcal{G}$) $f_{\varepsilon,ij}/u_{\varepsilon,i} = \int_0^\infty u F_{\varepsilon,ij}(du)/u_{\varepsilon,i} \to f_{0,ij} = \int_0^\infty u F_{0,ij}(du) \in (0, \infty)$ as $\varepsilon \to 0$, for $i, j \in \mathbb{X}$.

Note that condition ($\mathcal{G}$) is implied by condition ($\mathcal{F}$) in the example considered in Sect. 1.2.1.2, since distribution functions $F_{\varepsilon,ij}(\cdot)$ are exponential in this example.

The asymptotic algorithm of phase space reduction presented in Part II is based on recurrent application of two procedures. First, all virtual transitions (of the form $i \to i$) are excluded from trajectories of the semi-Markov process $\eta_\varepsilon(t)$, and this semi-Markov process is transformed into the semi-Markov process $\tilde{\eta}_\varepsilon(t)$ without virtual transitions. The inter-jump times for the new semi-Markov process $\tilde{\eta}_\varepsilon(t)$ are times between sequential moments of state change (transitions of the form $i \to j \neq i$), for the semi-Markov process $\eta_\varepsilon(t)$, and states of the semi-Markov process $\tilde{\eta}_\varepsilon(t)$ at instants of jumps coincide with the corresponding states for the semi-Markov process $\eta_\varepsilon(t)$. Second, some state $k \in \bar{\mathbb{D}}$ is chosen, and the semi-Markov process $\tilde{\eta}_\varepsilon(t)$ is transformed into the semi-Markov process ${}_k\eta_\varepsilon(t)$, with the new reduced phase space ${}_k\mathbb{X} = \mathbb{X} \setminus \{k\}$. The initial state of semi-Markov process ${}_k\eta_\varepsilon(t)$ coincides with the state of semi-Markov process $\tilde{\eta}_\varepsilon(t)$ at the time of its first hitting in the space ${}_k\mathbb{X}$, the inter-jump times for the semi-Markov process ${}_k\eta_\varepsilon(t)$ are times between further sequential hitting times of the semi-Markov process $\tilde{\eta}_\varepsilon(t)$ in the space ${}_k\mathbb{X}$, and states of the semi-Markov process ${}_k\eta_\varepsilon(t)$ at moments of jumps coincide with the corresponding states for the semi-Markov process $\tilde{\eta}_\varepsilon(t)$.

We show that under some additional conditions described below, the basic conditions ($\mathcal{E}$)–($\mathcal{G}$) are also satisfied for new semi-Markov processes $\tilde{\eta}_\varepsilon(t)$ and ${}_k\eta_\varepsilon(t)$, and we give explicit formulas for calculating the transition characteristic for these semi-Markov processes, normalisation functions, and the corresponding limits of transition probabilities, distribution functions of transition times, and their first moments.

This allows us to apply the above-described algorithm of phase space reduction to the reduced semi-Markov processes ${}_k\eta_\varepsilon(t)$, etc. Using this technique, we construct the asymptotic recurrent algorithm of phase space reduction for perturbed semi-Markov processes.

To be able to compute all the limits that appear in the above procedures, two additional conditions are also assumed. The first condition is based on the concept of a complete family of asymptotically comparable functions introduced in Sect. 1.2.1.2.

Typical examples of such families, including the family of asymptotically comparable power-type functions (satisfying relations $h(\varepsilon)/a_h\varepsilon^{b_h} \to a[h(\cdot)] \in [0, \infty]$ as $\varepsilon \to 0$, where constants $a_h > 0$ and $b_h \in (-\infty, \infty)$), are given in Chap. 8.

The first additional condition mentioned above:

($\mathcal{H}$) There exists a complete family of asymptotically comparable functions $\mathcal{H}$ such that: **(a)** $p_{\varepsilon,ij}$ either equal to zero for $\varepsilon \in (0, 1]$ or belongs to family $\mathcal{H}$, for $i, j \in \mathbb{X}$, **(b)** $u_{\varepsilon,i}$ belongs to family $\mathcal{H}$, for $i \in \mathbb{X}$.

The main case when studying the asymptotics of the hitting times $\tau_{\varepsilon,\mathbb{D}}$ is, where the initial state $\eta_\varepsilon(0) \in \bar{\mathbb{D}}$.

The second additional condition imposes some natural restriction on the choice of states $k_1, k_2, \dots k_{\bar{m}-1} \in \bar{\mathbb{D}}$ (here $\bar{m}$ is the number of states in the domain $\bar{\mathbb{D}}$), which should be sequentially excluded from the domain $\bar{\mathbb{D}}$. At each step of the recurrent algorithm described above, the reduced semi-Markov process ${}_{\bar{k}_n}\eta_\varepsilon(t)$ (with states from the sequence $\bar{k}_n = \langle k_1, \ldots, k_n \rangle$ excluded from the phase space $\mathbb{X}$) must be constructed in such a way that the state k_n is the least absorbing state in the reduced phase space ${}_{\bar{k}_{n-1}}\mathbb{X} = \mathbb{X} \setminus \{k_1, \ldots, k_{n-1}\}$ of the semi-Markov process ${}_{\bar{k}_{n-1}}\eta_\varepsilon(t)$ (here, ${}_{\bar{k}_0}\eta_\varepsilon(t) = \eta_\varepsilon(t)$ is the original semi-Markov process).

The corresponding normalisation functions used in the analogue of condition ($\mathcal{E}$) for the semi-Markov process ${}_{\bar{k}_{n-1}}\eta_\varepsilon(t)$ are given by formulas,

$$
{}_{\bar{k}_{n-1}}u_{\varepsilon,i} = \prod_{r=0}^{n-1}(1 - {}_{\bar{k}_r}p_{\varepsilon,ii})^{-1}u_{\varepsilon,i}, \ i \in {}_{\bar{k}_{n-1}}\mathbb{X}, \tag{1.51}
$$

where ${}_{\bar{k}_r}p_{\varepsilon,ij}, i, j \in {}_{\bar{k}_r}\mathbb{X}$ are the transition probabilities of the embedded Markov chain for the semi-Markov process ${}_{\bar{k}_r}\eta_\varepsilon(t)$.

The second additional condition mentioned above:

($\mathcal{J}$): $\lim_{\varepsilon\to 0} \frac{{}_{\bar{k}_{n-1}}u_{\varepsilon,k_n}}{{}_{\bar{k}_{n-1}}u_{\varepsilon,i}} = w_{0,k_n,i} \in [0, \infty)$, for $i \in_{\bar{k}_{n-1}} \mathbb{X}, n = 1, \ldots, \bar{m} - 1$.

The condition ($\mathcal{H}$) implies that all normalisation functions ${}_{\bar{k}_{n-1}}u_{\varepsilon,i}$ belong to the family $\mathcal{H}$ and guarantees the existence of limits appearing in the condition ($\mathcal{J}$), as well as the existence of a sequence $k_{\bar{m}-1} = \langle k_1, \ldots, k_{\bar{m}-1} \rangle$ such that the condition ($\mathcal{J}$) is satisfied. Moreover, these limits can be computed using operational rules for asymptotically comparable functions given in Chap. 8.

What is important is that, in the case where the initial state $\eta_\varepsilon(0) \in {}_{\bar{k}_n}\mathbb{X}$, the hitting times in the domain $\mathbb{D}$ coincide for the semi-Markov processes $\eta_\varepsilon(t)$ and ${}_{\bar{k}_n}\eta_\varepsilon(t)$, as well as for the semi-Markov process ${}_{\bar{k}_n}\tilde{\eta}_\varepsilon(t)$ constructed by removing virtual transitions from trajectories of the semi-Markov process ${}_{\bar{k}_n}\eta_\varepsilon(t)$.

However, the domain $\bar{\mathbb{D}} \setminus \{k_1, \ldots, k_{\bar{m}-1}\} = \{k_{\bar{m}}\}$ contains only one state. The semi-Markov process ${}_{\bar{k}_{\bar{m}-1}}\tilde{\eta}_\varepsilon(t)$ hits domain $\mathbb{D}$ immediately after the first jump. Thus, the hitting time in the domain $\mathbb{D}$ for the semi-Markov process ${}_{\bar{k}_{\bar{m}-1}}\tilde{\eta}_\varepsilon(t)$ coincides with the first transition time for this process. The normalisation function and the weak limit for the distribution function of the normalised first transition time for the process ${}_{\bar{k}_{\bar{m}-1}}\tilde{\eta}_\varepsilon(t)$ (calculated using the asymptotic recurrent phase space reduction algorithm described above) coincide with the normalisation function and the weak limit for the distribution of the normalised hitting time $\tau_{\varepsilon,\mathbb{D}}$ for the original semi-Markov processes $\eta_\varepsilon(t)$, for the case where the initial state $\eta_\varepsilon(0) = k_{\bar{m}}$.

We also build the backward recurrent algorithm (based on relations connecting the Laplace transforms $\mathsf{E}_i e^{-s\tau_{\varepsilon,\mathbb{D}}}$ for the states $i = k_{\bar{m}}, k_{\bar{m}-1}, \ldots, k_1$), which allows us to sequentially find normalisation functions and weak limits for the distribution functions of normalised hitting times for the cases, where the initial state $\eta_\varepsilon(0) = k_i$, for $i = \bar{m} - 1, \ldots, 1$.

Figure 1.7 illustrates the transformation of trajectories for perturbed semi-Markov processes $\eta_\varepsilon(t)$ into the above-described above recurrent algorithm of phase space reduction.

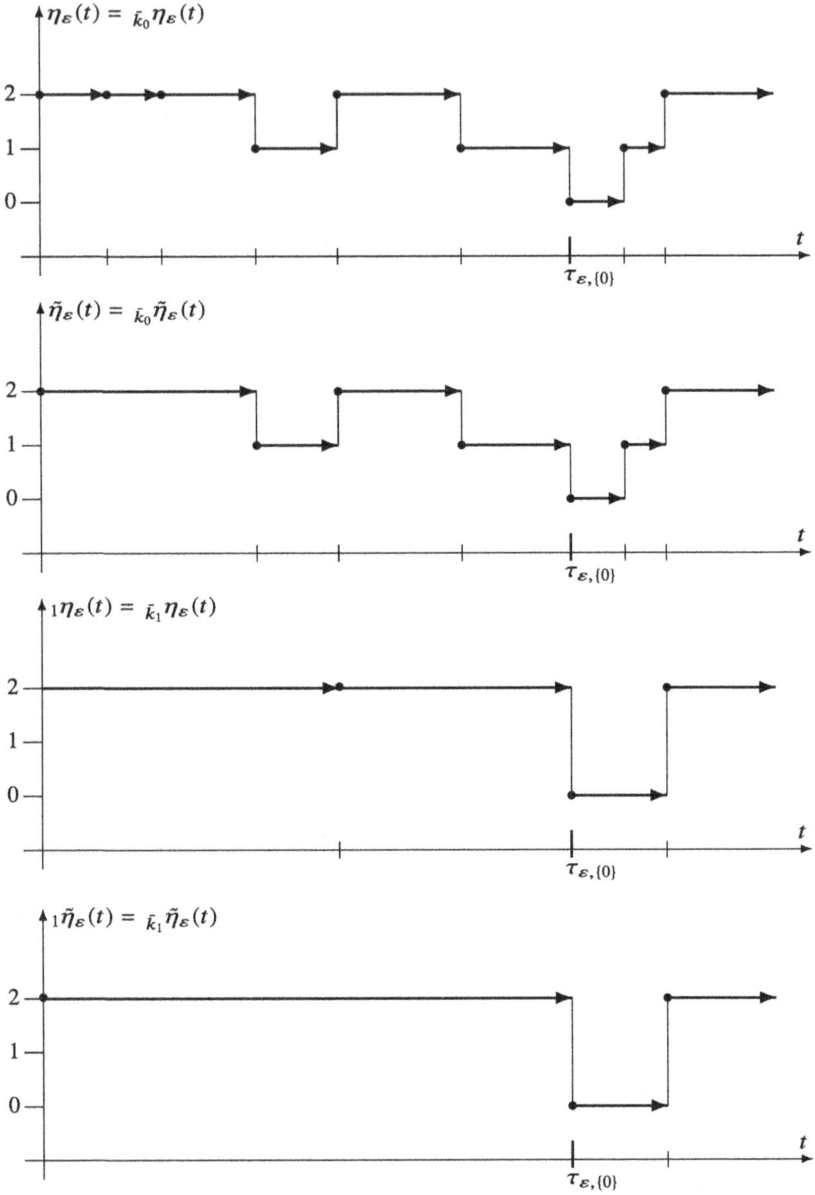

Fig. 1.7 Removing virtual transitions and phase space reduction for the semi-Markov process $\eta(t)$

In this example, the original semi-Markov process $\eta_\varepsilon(t)$ has the phase space $\mathbb{X} = \{0, 1, 2\}$. The domain $\mathbb{D} = \{0\}$. The initial state $\eta_\varepsilon(0) = 2 \in \bar{\mathbb{D}}$.

At the first sub-step, the virtual transitions are excluded from trajectories of the semi-Markov process $\eta_\varepsilon(t) = {}_{\bar{k}_0}\eta_\varepsilon(t)$ and this process is transformed into the semi-Markov process ${}_{\bar{k}_0}\tilde{\eta}_\varepsilon(t)$.

At the second sub-step, state $k_1 = 1$ is excluded from the phase space $\mathbb{X}$, and the semi-Markov process ${}_{\bar{k}_0}\tilde{\eta}_\varepsilon(t)$ is transformed into the semi-Markov process ${}_{\bar{k}_1}\eta_\varepsilon(t)$, with the phase space ${}_{\bar{k}_1}\mathbb{X} = \{0, 2\}$. Finally, the virtual transitions are excluded from trajectories of the semi-Markov process ${}_{\bar{k}_1}\eta_\varepsilon(t)$, and this process is transformed into the semi-Markov process ${}_{\bar{k}_1}\tilde{\eta}_\varepsilon(t)$.

The hitting time $\tau_{\varepsilon,\{0\}}$ is invariant with the respect to the above procedures, that is, the hitting time in the domain $\mathbb{D} = \{0\}$ coincides for the original semi-Markov process $\eta_\varepsilon(t)$ and the resulting semi-Markov process ${}_{\bar{k}_1}\tilde{\eta}_\varepsilon(t)$.

It is useful to comment the difference between examples illustrated by Figs. 1.4 and 1.7. In the former example, the initial semi-Markov process $\eta_\varepsilon(t) = {}_{\bar{k}_0}\eta_\varepsilon(t)$ is, in fact, a continuous time Markov chain, which does not make virtual transitions. In this case, the semi-Markov process ${}_{\bar{k}_0}\eta_\varepsilon(t)$ coincides with the semi-Markov process ${}_{\bar{k}_0}\tilde{\eta}_\varepsilon(t)$. In the latter example, the initial semi-Markov process $\eta_\varepsilon(t) = {}_{\bar{k}_0}\eta_\varepsilon(t)$ admits virtual transitions.

Keeping in mind the applications of limit theorems for hitting times to ergodic theorems for regularly and singularly perturbed semi-Markov processes and alternating regenerating processes modulated by perturbed semi-Markov processes, we are also interested in conditions under which distribution functions of normalised hitting times weakly converge to some limiting distribution functions not concentrated in 0 and, also, their first moments normalised by the same functions converge to the first moments of the corresponding limiting distribution functions.

1.2.2 Part II: Contents by Chapters

Part II includes Chaps. 8–12.

Chapter 8. In this chapter, we introduce the notion of a complete family of asymptotically comparable functions and present operational rules and formulas for such families. We also consider typical examples of such families, namely, families of asymptotically comparable power-type, power-exponential-type, and power-logarithmic-type functions.

Chapter 9. In this chapter, we introduce models of regularly and singularly perturbed finite semi-Markov processes, formulate basic perturbation conditions, and describe procedures of removing virtual transitions and one-state reduction of phase space as well as the asymptotic recurrent algorithm of sequential phase space reduction for perturbed semi-Markov processes.

Chapter 10. In this chapter, we introduce hitting times $\tau_{\varepsilon,\mathbb{D}}$ and some related random functionals, present forward and backward recurrent relations for distributions, Laplace transforms of hitting times, describe the variant of asymptotic recurrent algorithm of sequential phase space reduction, with respect to which the hitting times are invariant in distribution, and get weak limit theorems for hitting times provided by explicit recurrent formulas for computing normalisation functions and limiting Laplace transforms.

Chapter 11. In this chapter, we present forward and backward recurrent relations for expectations of hitting times, get limit theorems for expectation of hitting times, as well as present conditions, which guarantee that pre-limiting expectations normalised by functions used in weak limit theorems for distributions of hitting times converge to the first moments of the corresponding limiting distributions.

Chapter 12. In this chapter, we present some generalisations of limit theorems for hitting times for regularly and singularly perturbed semi-Markov processes to vector and real-valued hitting times and rewards, present the variant of the asymptotic recurrent algorithm of phase space reduction for birth–death-type semi-Markov processes, and consider some numerical examples.

1.3 Appendices and Conclusion

1.3.1 Appendix A: Limit Theorems for Randomly Stopped Stochastic Processes

In this appendix, we present basic functional limit theorems for stochastic processes and some limit theorems for randomly stopped stochastic processes, which play an important role in obtaining necessary and sufficient conditions in limit theorems for first-rare-event times and rare-event time processes.

1.3.2 Appendix B: Methodological and Bibliographical Notes

This appendix includes methodological and historical notes related to the results presented in the book, comments on new results presented in the book and related new problems for future research, as well as bibliographic notes to an extensive bibliography of works in the field of asymptotic problems for perturbed stochastic processes and systems of Markov type.

1.3.3 Conclusion

The asymptotic analysis for perturbed processes of Markov type and related problems are subjects of intensive studies during several decades. However, the development of the theory, in particular, its parts connected with necessary and sufficient conditions of convergence for regularly and singularly perturbed models, and computational algorithms for singularly perturbed models are still far from the completion. The book is concentrated in this area.

Conditions in limit theorems for first-rare-event times and processes for regularly perturbed finite semi-Markov processes presented in the book have the final form, without gaps between its necessary and sufficient parts. New computational asymptotic algorithms of phase space reduction presented in the book have a universal character. They can be applied to perturbed semi-Markov processes with an arbitrary asymptotic communicative structure of phase spaces. The corresponding algorithms are computationally effective due to recurrent character of computational procedures and give effective conditions of weak convergence for distributions and convergence of expectations for hitting times.

The results presented in the book will be interesting to specialists, who work in such areas of the theory of stochastic processes such as ergodic, limit, and large deviation theorems, analytical and computational methods for Markov chains, Markov, semi-Markov, regenerative, and other classes of Markov type stochastic processes and their queuing, network, bio-stochastic and other applications. I hope that the book will also attract attention of those researchers, who are interested in new analytical methods of analysis for perturbed stochastic processes and systems. It can also be useful for doctoral and senior students. This gives me hope that the book will find a sufficient number of readers interested in stochastic processes and their applications.

Part I
First-Rare-Event Times for Regularly Perturbed Semi-Markov Processes

Chapter 2
Asymptotics of First-Rare-Event Times for Regularly Perturbed Semi-Markov Processes

Chapter 2 plays the key role in Part I. In this chapter, we present necessary and sufficient conditions for convergence in distribution for first-rare-event times and convergence in Skorokhod J-topology for first-rare-event processes for regularly perturbed Markov chains and semi-Markov processes with a finite phase space. This chapter includes four sections.

In Sect. 2.1, we define first-rare-event times and first-rare-event processes defined on perturbed semi-Markov processes. The notion of asymptotically uniformly ergodic Markov chains is introduced. The basic model assumptions and the main Theorem 2.1, which gives necessary and sufficient conditions of convergence in distribution for first-rare-event times and convergence in topology J for first-rare-event processes defined on asymptotically uniformly ergodic semi-Markov processes, are formulated.

In Sect. 2.2, we formulate and prove Theorem 2.2, which gives necessary and sufficient conditions of convergence in distribution and convergence in J-topology for step-sum reward processes defined on asymptotically uniformly ergodic Markov chains. This theorem plays the key role in the proof of Theorem 2.1 but also has its own value.

In Sect. 2.3, we formulate and prove Lemmas 2.7 and 2.8, which give necessary and sufficient conditions of convergence in distribution for first-rare-event times and convergence in topology J for first-rare-event reward processes defined on asymptotically uniformly ergodic Markov chains.

In Sect. 2.4, we complete the proof of Theorem 2.1. The integral part of this proof is Lemma 2.9, which establishes the asymptotic independence of the first-rare-event times and the step-sum reward processes defined on asymptotically uniformly ergodic semi-Markov processes, Theorem 2.2, and Lemmas 2.7–2.8.

D. Silvestrov, *Perturbed Semi-Markov Type Processes I*,
https://doi.org/10.1007/978-3-030-92403-4_2

2.1 First-Rare-Event Times for Perturbed Semi-Markov Processes

In this section, we introduce first-rare-event times and processes defined on perturbed semi-Markov processes, formulate the basic model assumptions, and give necessary and sufficient conditions of convergence in distribution for first-rare-event times and convergence in J-topology for rare-event processes defined on asymptotically regularly perturbed semi-Markov processes.

2.1.1 First-Rare-Event Times

Let $(\eta_{\varepsilon,n}, \kappa_{\varepsilon,n}, \rho_{\varepsilon,n})$, $n = 0, 1, \ldots$ be, for every $\varepsilon \in (0, 1]$, a Markov renewal process, i.e., a homogenous Markov chain with a phase space $\mathbb{Z} = \mathbb{X} \times [0, \infty) \times \{0, 1\}$, where $\mathbb{X} = \{1, 2, \ldots, m\}$ is a finite set, an initial distribution $\hat{q}_\varepsilon = \langle q_{\varepsilon,i,s,\imath} = \mathsf{P}\{\eta_{\varepsilon,0} = i, \kappa_{\varepsilon,0} \leq s, \rho_{\varepsilon,0} = \imath\}, (i, s, \imath) \in \mathbb{Z}\rangle$, and transition probabilities defined by the following relation, for $(i, s, \imath), (j, t, \jmath) \in \mathbb{Z}$,

$$
\begin{aligned}
\mathsf{P}\{\eta_{\varepsilon,1} = j, \kappa_{\varepsilon,1} \leq t, \rho_{\varepsilon,1} = \jmath / \eta_{\varepsilon,0} = i, \kappa_{\varepsilon,0} = s, \rho_{\varepsilon,0} = \imath\} \\
= \mathsf{P}\{\eta_{\varepsilon,1} = j, \kappa_{\varepsilon,1} \leq t, \rho_{\varepsilon,1} = \jmath / \eta_{\varepsilon,0} = i\} \\
= Q_{\varepsilon,ij}(t, \jmath).
\end{aligned}
\tag{2.1}
$$

The characteristic property of the Markov renewal process $(\eta_{\varepsilon,n}, \kappa_{\varepsilon,n}, \rho_{\varepsilon,n})$ is that its transition probabilities depend only on the current state i of the first component $\eta_{\varepsilon,n}$ but do not depend on the current states s and $\imath$ of the second and third components $\kappa_{\varepsilon,n}$ and $\rho_{\varepsilon,n}$.

As is known, the first component $\eta_{\varepsilon,n}$ of the above Markov renewal process is also a homogenous Markov chain, with the phase space $\mathbb{X}$, the initial distribution $\bar{q}_\varepsilon = \langle q_{\varepsilon,i} = \mathsf{P}\{\eta_{\varepsilon,0} = i\}, i \in \mathbb{X}\rangle$, and the transition probabilities defined by the following relation, for $i, j \in \mathbb{X}$,

$$
p_{\varepsilon,ij} = Q_{\varepsilon,ij}(+\infty, 0) + Q_{\varepsilon,ij}(+\infty, 1). \tag{2.2}
$$

Also, the random sequence $(\eta_{\varepsilon,n}, \rho_{\varepsilon,n})$, $n = 0, 1, \ldots$ is a Markov renewal process with the phase space $\mathbb{X} \times \{0, 1\}$, the initial distribution $\dot{q}_\varepsilon = \langle q_{\varepsilon,i\imath} = \mathsf{P}\{\eta_{\varepsilon,0} = i, \rho_{\varepsilon,0} = \imath\}, (i, \imath) \in \mathbb{X} \times \{0, 1\}\rangle$, and the transition probabilities defined by the following relation, for $(i, \imath), (j, \jmath) \in \mathbb{X} \times \{0, 1\}$,

$$
p_{\varepsilon,(i,\imath),(j,\jmath)} = Q_{\varepsilon,ij}(+\infty, \jmath). \tag{2.3}
$$

The random variables $\kappa_{\varepsilon,n}, n = 1, 2, \ldots$ can be interpreted as transition times and the random variables $\zeta_{\varepsilon,n} = \kappa_{\varepsilon,1} + \cdots + \kappa_{\varepsilon,n}, n = 1, 2, \ldots, \zeta_{\varepsilon,0} = 0$ as moments of jumps for a semi-Markov process $\eta_\varepsilon(t), t \geq 0$, which is defined by the following relation:

$$\eta_\varepsilon(t) = \eta_{\varepsilon,n} \text{ for } \zeta_{\varepsilon,n} \le t < \zeta_{\varepsilon,n+1},\ n = 0, 1, \dots \tag{2.4}$$

This semi-Markov process has the phase space $\mathbb{X}$ and the transition probabilities, which are defined by the following relation for $i, j \in \mathbb{X},\ s, t \ge 0$,

$$\begin{aligned} \mathsf{P}\{\eta_{\varepsilon,n+1} &= j, \kappa_{\varepsilon,n+1} \le t/\eta_{\varepsilon,n} = i, \kappa_{\varepsilon,n} = s\} \\ &= \mathsf{P}\{\eta_{\varepsilon,n+1} = j, \kappa_{\varepsilon,n+1} \le t/\eta_{\varepsilon,n} = i\} \\ &= Q_{\varepsilon,ij}(t) = Q_{\varepsilon,ij}(t, 0) + Q_{\varepsilon,ij}(t, 1). \end{aligned} \tag{2.5}$$

As far as the random variables $\rho_{\varepsilon,n}, n = 1, 2, \dots$ are concerned, they can be interpreted as the so-called indicator variables and used to record events $\{\rho_{\varepsilon,n} = 1\}$, which can be interpreted as "rare" events.

Let us introduce random variables, for $\varepsilon \in (0, \varepsilon'_0]$,

$$\xi_\varepsilon = \sum_{n=1}^{\nu_\varepsilon} \kappa_{\varepsilon,n}, \tag{2.6}$$

where

$$\nu_\varepsilon = \min(n \ge 1 : \rho_{\varepsilon,n} = 1). \tag{2.7}$$

The random variable ν_ε counts the number of transitions for the embedded Markov chain $\eta_{\varepsilon,n}$ until the first "rare" event occurs. Accordingly, the random variable ξ_ε can be interpreted as the first-rare-event time for the semi-Markov process $\eta_\varepsilon(t)$.

We also consider the first-rare-event processes,

$$\xi_\varepsilon(t) = \sum_{n=1}^{[t\nu_\varepsilon]} \kappa_{\varepsilon,n},\ t \ge 0. \tag{2.8}$$

We use the symbol $\xrightarrow{\mathsf{d}}$ to denote convergence in distribution for random variables (equivalent to weak convergence of their distribution functions, indicated usually by the symbol $\Rightarrow$) or stochastic processes (equivalent to weak convergence of their finitely dimensional distributions), the symbol $\xrightarrow{\mathsf{P}}$ to denote convergence of random variables in probability, and the symbols $\xrightarrow{\mathsf{U}}$ and $\xrightarrow{\mathsf{J}}$ to denote convergence, respectively, in the uniform U-topology and Skorokhod J-topology for càdlàg stochastic processes defined on the time interval $[0, \infty)$ and taking values in $\mathbb{R}_k$.

We refer to Appendix A, where some basic definitions and theorems related to conditions of convergence in U- and J-topologies for càdlàg stochastic processes are formulated, and references to some books, where readers can find the detailed presentation of functional limit theorems for càdlàg stochastic processes.

The main objective of this chapter is to describe the class $\mathcal{G}$ of all possible random variables ξ_0, which can appear in the corresponding limit theorem given in the form of the asymptotic relation, $\xi_\varepsilon \xrightarrow{\mathsf{d}} \xi_0$ as $\varepsilon \to 0$, and give the necessary and sufficient conditions for the fulfilment of the indicated asymptotic relation with a specific (with respect to the distribution function) limiting random variable ξ_0 from the class $\mathcal{G}$.

We also describe the class $\mathcal{G}_\infty$ of all possible càdlàg processes $\xi_0(t), t \geq 0$, which can appear in the corresponding functional limit theorem given in the form of the asymptotic relation, $\xi_\varepsilon(t), t \geq 0 \xrightarrow{\mathsf{J}} \xi_0(t), t \geq 0$ as $\varepsilon \to 0$, and give necessary and sufficient conditions the fulfilment of the above asymptotic relation with a specific (by its finite-dimensional distributions) limiting stochastic process $\xi_0(t), t \geq 0$ from the class $\mathcal{G}_\infty$.

The problems formulated above are solved under three general model assumptions.

We introduce the probabilities of occurrence of a rare event at one transient step of the semi-Markov process $\eta_\varepsilon(t)$,

$$p_{\varepsilon,i} = \mathsf{P}_i\{\rho_{\varepsilon,1} = 1\}, \ i \in \mathbb{X}. \tag{2.9}$$

Hereinafter, P_i and E_i denote, respectively, the conditional probability and expectation under the condition $\eta_{\varepsilon,0} = i$.

The first model assumption, imposed on the probabilities $p_{\varepsilon,i}$, refines the interpretation of the events $\{\rho_{\varepsilon,n} = 1\}$ as "rare" events:

$\mathbf{A}_1$: **(a)** There exist $\varepsilon_0' \in (0, 1]$ such that $\max_{i \in \mathbb{X}} p_{\varepsilon,i} > 0$, for $\varepsilon \in (0, \varepsilon_0']$, **(b)** $p_{\varepsilon,i} \to 0$ as $\varepsilon \to 0$, for $i \in \mathbb{X}$.

2.1.2 Asymptotically Uniformly Ergodic Markov Chains

Let us introduce random variables,

$$\mu_{\varepsilon,i}(n) = \sum_{k=1}^{n} \mathrm{I}(\eta_{\varepsilon,k-1} = i), \ n = 0, 1, \ldots, \ i \in \mathbb{X}. \tag{2.10}$$

If the phase space $\mathbb{X}$ of the Markov chain $\eta_{\varepsilon,n}$ is one class of communicative states for this Markov chain, then this Markov chain is ergodic and its stationary distribution is given by the following ergodic relation,

$$\frac{\mu_{\varepsilon,i}(n)}{n} \xrightarrow{\mathsf{P}} \pi_{\varepsilon,i} \text{ as } n \to \infty, \text{ for } i \in \mathbb{X}. \tag{2.11}$$

The ergodic relation (2.11) holds for any initial distribution $\bar{q}_\varepsilon$, and the stationary distribution $\pi_{\varepsilon,i}, i \in \mathbb{X}$ does not depend on the initial distribution. Also, all stationary probabilities are positive, i.e., $\pi_{\varepsilon,i} > 0, i \in \mathbb{X}$.

As is known, the stationary probabilities $\pi_{\varepsilon,i}, i \in \mathbb{X}$ are the unique solution for the following system of linear equations,

$$\pi_{\varepsilon,i} = \sum_{j \in \mathbb{X}} \pi_{\varepsilon,j} p_{\varepsilon,ji}, \ i \in \mathbb{X}, \ \sum_{i \in \mathbb{X}} \pi_{\varepsilon,i} = 1. \tag{2.12}$$

The second model assumption is a condition of asymptotically uniform ergodicity of the embedded Markov chains $\eta_{\varepsilon,n}$:

$\mathbf{B}_1$: There exists a ring chain of states $i_0, i_1, \ldots, i_N = i_0$, which contains all states from the phase space $\mathbb{X}$ and such that $\underline{\lim}_{\varepsilon\to 0} p_{\varepsilon,i_{k-1}i_k} > 0$, for $k = 1, \ldots, N$.

Let $\tilde{\eta}_{\varepsilon,n}$ be, for every $\varepsilon \in (0, 1]$, a Markov chain with the phase space $\mathbb{X}$ and a matrix of transition probabilities $\|\tilde{p}_{\varepsilon,ij}\|$.

We shall use the following condition:

$\mathbf{B}_2$: $p_{\varepsilon,ij} - \tilde{p}_{\varepsilon,ij} \to 0$ as $\varepsilon \to 0$, for $i, j \in \mathbb{X}$.

If the transition probabilities $\tilde{p}_{\varepsilon,ij} \equiv p_{0,ij}, i, j \in \mathbb{X}$ do not depend on ε, then the condition $\mathbf{B}_2$ reduces to the following condition:

$\mathbf{B}_3$: $p_{\varepsilon,ij} \to p_{0,ij}$ as $\varepsilon \to 0$, for $i, j \in \mathbb{X}$.

Lemma 2.1 *Let the condition $\mathbf{B}_1$ be satisfied for the Markov chains $\eta_{\varepsilon,n}$. Then:*

(i) *There exists $\varepsilon_0'' \in (0, 1]$ such that the Markov chain $\eta_{\varepsilon,n}$ is ergodic, for every $\varepsilon \in (0, \varepsilon_0'']$ and $0 < \underline{\lim}_{\varepsilon\to 0} \pi_{\varepsilon,i} \leq \overline{\lim}_{\varepsilon\to 0} \pi_{\varepsilon,i} < 1$, for $i \in \mathbb{X}$.*

(ii) *If, in addition, the condition $\mathbf{B}_2$ is satisfied, then there exists $\tilde{\varepsilon}_0 \in (0, \varepsilon_0'']$ such that the Markov chain $\tilde{\eta}_{\varepsilon,n}$ is ergodic, for every $\varepsilon \in (0, \tilde{\varepsilon}_0]$, and its stationary distribution $\tilde{\pi}_{\varepsilon,i}, i \in \mathbb{X}$ satisfies the asymptotic relation, $\pi_{\varepsilon,i} - \tilde{\pi}_{\varepsilon,i} \to 0$ as $\varepsilon \to 0$, for $i \in \mathbb{X}$.*

(iii) *If condition $\mathbf{B}_3$ is satisfied, then matrix $\|p_{0,ij}\|$ is stochastic, condition $\mathbf{B}_1$ is equivalent to the assumption that the Markov chain $\eta_{0,n}$, with the matrix of transition probabilities $\|p_{0,ij}\|$, is ergodic, and the following asymptotic relation holds, $\pi_{\varepsilon,i} \to \pi_{0,i}$ as $\varepsilon \to 0$, for $i \in \mathbb{X}$, where $\pi_{0,i}, i \in \mathbb{X}$ is the stationary distribution of the Markov chain $\eta_{0,n}$.*

Proof First, we prove the statement **(iii)**. The condition $\mathbf{B}_3$ obviously implies that the matrix $\|p_{0,ij}\|$ is stochastic. The conditions $\mathbf{B}_1$ and $\mathbf{B}_3$ imply that $\lim_{\varepsilon\to 0} p_{\varepsilon,i_{k-1}i_k} = p_{0,i_{k-1}i_k} > 0, k = 1, \ldots, N$, for the ring chain appearing in the condition $\mathbf{B}_1$. Thus, the Markov chain $\eta_{0,n}$, with the matrix of transition probabilities $\|p_{0,ij}\|$, is ergodic. Vice versa, the assumption that the Markov chain $\eta_{0,n}$, with the matrix of transition probabilities $\|p_{0,ij}\|$, is ergodic implies that there exists a ring chain of states $i_0, \ldots, i_N = i_0$, which contains all states from the phase space $\mathbb{X}$ and such that $p_{0,i_{k-1}i_k} > 0, k = 1, \ldots, N$. In this case, this condition $\mathbf{B}_3$ implies that $\lim_{\varepsilon\to 0} p_{\varepsilon,i_{k-1}i_k} = p_{0,i_{k-1}i_k} > 0, k = 1, \ldots, N$, and, thus, the condition $\mathbf{B}_1$ is satisfied. Let us assume that the convergence relation for stationary distributions penetrating the statement **(iii)** does not hold. In this case, there exist $\delta > 0$ and a sequence $0 < \varepsilon_n \to 0$ as $n \to \infty$ such that $\underline{\lim}_{n\to\infty} |\pi_{\varepsilon_n,i'} - \pi_{0,i'}| \geq \delta$, for some $i' \in \mathbb{X}$. Since the sequences $\pi_{\varepsilon_n,i}, n = 1, 2, \ldots, i \in \mathbb{X}$ are bounded, there exists a subsequence $0 < \varepsilon_{n_k} \to 0$ as $k \to 0$ such that $\pi_{\varepsilon_{n_k},i} \to \pi'_{0,i}$ as $k \to \infty$, for $i \in \mathbb{X}$. This relation, the condition $\mathbf{B}_3$, and the relation (2.12) imply that the limits $\pi'_{0,i}, i \in \mathbb{X}$ satisfy the system of linear equations given in (2.12). This is impossible, since the inequality $|\pi'_{0,i'} - \pi_{0,i'}| \geq \delta$ should hold, while the stationary distribution $\pi_{0,i}, i \in \mathbb{X}$ is the unique solution of the system (2.12).

Let us now prove the statement **(i)**. The condition $\mathbf{B}_1$ obviously implies that there exists $\varepsilon_0'' \in (0, 1]$ such that $p_{\varepsilon, i_{k-1} i_k} > 0, k = 1, \ldots, N$ for the ring chain penetrating condition $\mathbf{B}_1$ and $\varepsilon \in (0, \varepsilon_0'']$. Thus, the Markov chain $\eta_{\varepsilon,n}$ is ergodic, for every $\varepsilon \in (0, \varepsilon_0'']$. Assume that $\underline{\lim}_{\varepsilon \to 0} \pi_{\varepsilon, i'} = 0$, for some $i' \in \mathbb{X}$. In this case, there exists a sequence $0 < \varepsilon_n \to 0$ as $n \to \infty$ such that $\pi_{\varepsilon_n, i'} \to 0$ as $n \to \infty$. Since the sequences $p_{\varepsilon_n, ij}, n = 1, 2, \ldots, i, j \in \mathbb{X}$ are bounded, there exists a subsequence $0 < \varepsilon_{n_k} \to 0$ as $k \to 0$ such that $p_{\varepsilon_{n_k}, ij} \to p_{0,ij}$ as $k \to \infty$, for $i, j \in \mathbb{X}$. By the statement **(iii)**, the matrix $\|p_{0,ij}\|$ is stochastic, the Markov chain $\eta_{0,n}$, with the matrix of transition probabilities $\|p_{0,ij}\|$, is ergodic, and its stationary distribution $\pi_{0,i}, i \in \mathbb{X}$ satisfies the asymptotic relation, $\pi_{\varepsilon_{n_k}, i} \to \pi_{0,i}$ as $k \to \infty$, for $i \in \mathbb{X}$. This is impossible since the equality $\pi_{0,i'} = 0$ should hold, while all stationary probabilities $\pi_{0,i}, i \in \mathbb{X}$ are positive. Thus, $\underline{\lim}_{\varepsilon \to 0} \pi_{\varepsilon,i} > 0$, for $i \in \mathbb{X}$. This implies also that $\overline{\lim}_{\varepsilon \to 0} \pi_{\varepsilon,i} < 1$, for $i \in \mathbb{X}$, since $\sum_{i \in \mathbb{X}} \pi_{\varepsilon,i} = 1$, for $\varepsilon \in (0, \tilde{\varepsilon}_0]$.

Finally, let us prove the statement **(ii)**. The conditions $\mathbf{B}_1$ and $\mathbf{B}_2$ obviously imply that $\underline{\lim}_{\varepsilon \to 0} \tilde{p}_{\varepsilon, i_{k-1} i_k} = \underline{\lim}_{\varepsilon \to 0} p_{\varepsilon, i_{k-1} i_k} > 0, k = 1, \ldots, N$, for the ring chain penetrating condition $\mathbf{B}_1$. Thus, the condition $\mathbf{B}_1$ is also satisfied for the Markov chains $\tilde{\eta}_{\varepsilon,n}$, and there exist $\tilde{\varepsilon}_0 \in (0, \varepsilon_0]$ such that Markov chain $\tilde{\eta}_{\varepsilon,n}$ is ergodic, for every $\varepsilon \in (0, \tilde{\varepsilon}_0]$. Let us assume that the convergence relation for stationary distributions penetrating the statement **(ii)** does not hold. In this case, there exist $\delta > 0$ and a sequence $0 < \varepsilon_n \to 0$ as $n \to \infty$ such that $\underline{\lim}_{n \to \infty} |\pi_{\varepsilon_n, i'} - \tilde{\pi}_{\varepsilon_n, i'}| \geq \delta$, for some $i' \in \mathbb{X}$. Since the sequences $p_{\varepsilon_n, ij}, n = 1, 2, \ldots, i, j \in \mathbb{X}$ are bounded, there exists a subsequence $0 < \varepsilon_{n_k} \to 0$ as $k \to 0$ such that $p_{\varepsilon_{n_k}, ij} \to p_{0,ij}$ as $k \to \infty$, for $i, j \in \mathbb{X}$. These relations and the condition $\mathbf{B}_2$ imply that, also, $\tilde{p}_{\varepsilon_{n_k}, ij} \to p_{0,ij}$ as $k \to \infty$, for $i, j \in \mathbb{X}$. By the statement **(iii)**, the matrix $\|p_{0,ij}\|$ is stochastic, the Markov chain $\eta_{0,n}$, with the matrix of transition probabilities $\|p_{0,ij}\|$, is ergodic, and its stationary distribution $\pi_{0,i}, i \in \mathbb{X}$ satisfies the asymptotic relations, $\pi_{\varepsilon_{n_k}, i} \to \pi_{0,i}$ as $k \to \infty$, for $i \in \mathbb{X}$ and $\tilde{\pi}_{\varepsilon_{n_k}, i} \to \pi_{0,i}$ as $k \to \infty$, for $i \in \mathbb{X}$. This is impossible, since the relation $\underline{\lim}_{k \to \infty} |\pi_{\varepsilon_{n_k}, i'} - \tilde{\pi}_{\varepsilon_{n_k}, i'}| \geq \delta$ should hold. □

Remark 2.1 The statement **(iii)** of Lemma 2.1 implies that, in the case, where the transition probabilities $p_{\varepsilon,ij} = p_{0,ij}, i, j \in \mathbb{X}$ do not depend on parameter ε or $p_{\varepsilon,ij} \to p_{0,ij}$ as $\varepsilon \to 0$, for $i, j \in \mathbb{X}$, the condition $\mathbf{B}_1$ reduces to the standard assumption that the Markov chain $\eta_{0,n}$, with the matrix of transition probabilities $\|p_{0,ij}\|$, is ergodic.

2.1.3 Necessary and Sufficient Conditions for Convergence in Distribution of First-Rare-Event Times

It should be noted that the conditions $\mathbf{A}_1$ and $\mathbf{B}_1$ obviously imply that $\mathsf{P}_i\{\nu_\varepsilon < \infty\} = 1$, for $i \in \mathbb{X}$ and $\varepsilon \in (0, \varepsilon_0]$, where $\varepsilon_0 = \varepsilon_0' \wedge \varepsilon_0''$, and, thus, the first-rare-event time ξ_ε and the first-rare-event process $\xi_\varepsilon(t), t \geq 0$ are well defined, for $\varepsilon \in (0, \varepsilon_0]$.

The third model assumption is the following condition that guarantees that the last summand $\kappa_{\varepsilon,\nu_\varepsilon}$ in the random sum ξ_ε is asymptotically negligible:

$\mathbf{C}_1$: $\mathsf{P}_i\{\kappa_{\varepsilon,1} > \delta/\rho_{\varepsilon,1} = 1\} \to 0$ as $\varepsilon \to 0$, for $\delta > 0$ and $i \in \mathbb{X}$.

Note that we define the probability $\mathsf{P}_i\{\kappa_{\varepsilon,1} > \delta/\rho_{\varepsilon,1} = 1\} = 0$ in the cases where $\mathsf{P}_i\{\rho_{\varepsilon,1} = 1\} = 0$. In fact, both probabilities appear together in calculations only in the form of the product $\mathsf{P}_i\{\kappa_{\varepsilon,1} > \delta/\rho_{\varepsilon,1} = 1\}\mathsf{P}_i\{\rho_{\varepsilon,1} = 1\}$.

Let us introduce the step-sum stochastic process, for $\varepsilon \in (0, \varepsilon_0]$,

$$\kappa_\varepsilon(t) = \sum_{n=1}^{[t\nu_\varepsilon]} \kappa_{\varepsilon,n},\ t \geq 0. \tag{2.13}$$

The random variables $\kappa_\varepsilon(t)$ can be interpreted as rewards accumulated on trajectories of the Markov chain $\eta_{\varepsilon,n}$. Respectively, the random variable ξ_ε can be interpreted as the reward accumulated on the trajectories of the Markov chain $\eta_{\varepsilon,n}$ before the first occurrence of a "rare" event.

Asymptotics of the step-sum reward processes $\kappa_\varepsilon(t), t \geq 0$ have its own value. At the same, the corresponding result formulated below in Theorem 2.2 plays an important role in the proof of Theorem 2.1.

Let us define the probability that is the result of averaging the probabilities of occurrence of a rare event (in one transition step of the semi-Markov process $\eta_\varepsilon(t)$) $p_{\varepsilon,i}$ over the stationary distribution of the embedded Markov chain $\eta_{\varepsilon,n}$,

$$p_\varepsilon = \sum_{i\in\mathbb{X}} \pi_{\varepsilon,i} p_{\varepsilon,i} \ \text{ and } \ u_\varepsilon = p_\varepsilon^{-1}. \tag{2.14}$$

We introduce the distribution function of the sojourn time $\kappa_{\varepsilon,1}$ for the semi-Markov process $\eta_\varepsilon(t)$,

$$F_{\varepsilon,i}(t) = \mathsf{P}_i\{\kappa_{\varepsilon,1} \leq t\},\ t \geq 0,\ i \in \mathbb{X}. \tag{2.15}$$

Let us define the distribution function $F_\varepsilon(t)$, which is a result of averaging the distribution functions of sojourn times $F_{\varepsilon,i}(t)$ over the stationary distribution of the embedded Markov chain $\eta_{\varepsilon,n}$,

$$F_\varepsilon(t) = \sum_{i\in\mathbb{X}} \pi_{\varepsilon,i} F_{\varepsilon,i}(t),\ t \geq 0. \tag{2.16}$$

Let also $\theta_{\varepsilon,n}, n = 1, 2, \ldots$ be i.i.d. random variables with the distribution function $F_\varepsilon(t)$.

We can now formulate a necessary and sufficient condition for convergence in distribution for first-rare-event times:

$\mathbf{D}_1$: $\theta_\varepsilon = \sum_{n=1}^{[u_\varepsilon]} \theta_{\varepsilon,n} \xrightarrow{\mathsf{d}} \theta_0$ as $\varepsilon \to 0$, where θ_0 is a non-negative random variable with distribution function not concentrated at zero.

As well known, $(\mathbf{d}_1)$, the limiting random variable θ_0 penetrating the condition $\mathbf{D}_1$ should be infinitely divisible and, thus, its Laplace transform has the form, $\mathsf{E}e^{-s\theta_0} = e^{-A(s)}$, where $A(s) = \pi s + \int_0^\infty (1 - e^{-sv})\pi(dv), s \geq 0, \pi$ is a non-negative constant and $\pi(dv)$ is a measure on the Borel σ-algebra of the interval $(0, \infty)$ such that $\int_{(0,\infty)} \frac{v}{1+v}\pi(dv) < \infty$; **(d$_2$)** $\pi + \int_{(0,\infty)} \frac{v}{1+v}\pi(dv) > 0$ (this is equivalent to the assumption that $\mathsf{P}\{\theta_0 = 0\} < 1$).

Let us also define a homogeneous step-sum process with independent increments (the summands are i.i.d. random variables),

$$\theta_\varepsilon(t) = \sum_{n=1}^{[tu_\varepsilon]} \theta_{\varepsilon,n}, t \geq 0. \tag{2.17}$$

According to Remark A.3 and Theorem A.3, the condition $\mathbf{D}_1$ is necessary and sufficient for the following asymptotic relation to hold,

$$\theta_\varepsilon(t) = \sum_{n=1}^{[tu_\varepsilon]} \theta_{\varepsilon,n}, t \geq 0 \xrightarrow{\mathsf{J}} \theta_0(t), t \geq 0 \text{ as } \varepsilon \to 0, \tag{2.18}$$

where $\theta_0(t), t \geq 0$ is a non-negative Lévy process (a càdlàg homogeneous process with independent increments) with the Laplace transforms $\mathsf{E}e^{-s\theta_0(t)} = e^{-tA(s)}$, $s, t \geq 0$.

Let us define the Laplace transforms,

$$\phi_{\varepsilon,i}(s) = \mathsf{E}_i e^{-s\kappa_{\varepsilon,1}} = \int_0^\infty e^{-st} F_{\varepsilon,i}(dt),\ s \geq 0,\ i \in \mathbb{X}, \tag{2.19}$$

and

$$\phi_\varepsilon(s) = \mathsf{E}e^{-s\theta_{\varepsilon,1}} = \int_0^\infty e^{-st} F_\varepsilon(dt) = \sum_{\in \mathbb{X}} \pi_{\varepsilon,i}\phi_{\varepsilon,i}(s),\ s \geq 0. \tag{2.20}$$

Condition $\mathbf{D}_1$ can be reformulated (see, for example, Feller (1971)) in the following equivalent form in terms of Laplace transforms:

D$_2$: $u_\varepsilon(1 - \phi_\varepsilon(s)) \to A(s)$ as $\varepsilon \to 0$, for $s > 0$, where the limiting function $A(s) > 0$, for $s > 0$ and $A(s) \to 0$ as $s \to 0$.

In this case, **(d$_3$)** $A(s)$ is the cumulant of a non-negative random variable with a distribution function not concentrated at zero. Moreover, **(d$_4$)** $A(s)$ must be the cumulant of the infinitely divisible distribution function of the form specified in the above conditions **(d$_1$)** and **(d$_2$)**.

The following condition, which is a variant of the so-called central criterion of convergence (see, for example, Loève (1977)), is equivalent to the condition $\mathbf{D}_1$, with the Laplace transform of the limiting random variable θ_0 given in the above conditions **(d$_1$)** and **(d$_2$)**:

D$_3$: **(a)** $u_\varepsilon(1 - F_\varepsilon(u)) \to \pi(u)$ as $\varepsilon \to 0$ for all continuity points $u > 0$ of the limiting function $\pi(u)$, which is non-negative, non-increasing, and right continuous function defined on the interval $(0, \infty)$, having a limiting value $\pi(+\infty) = 0$, and is connected with the measure $\pi(dv)$ by the relation $\pi((u', u'']) = \pi(u') - \pi(u'')$,

$0 < u' \leq u'' < \infty$; **(b)** $u_\varepsilon \int_{(0,u]} v F_\varepsilon(dv) \longrightarrow \pi + \int_{(0,u]} v\pi(dv)$ as $\varepsilon \to 0$ for some $u > 0$ that is a continuity point of the function $\pi(u)$.

It is useful to note that **($\mathbf{d}_5$)** the asymptotic relation appearing in the condition $\mathbf{D}_3$ **(b)** is satisfied, under the condition $\mathbf{D}_2$ **(a)**, for any $u > 0$ that is a continuity point of the function $\pi(u)$.

The main result of this chapter is the following theorem.

Theorem 2.1 *Let the conditions* $\mathbf{A}_1$, $\mathbf{B}_1$, *and* $\mathbf{C}_1$ *are satisfied. Then:*

(i) *The condition* $\mathbf{D}_1$ *is necessary and sufficient for the fulfilment (for some or any initial distributions* $\bar{q}_\varepsilon$, *respectively, in the statements of necessity and sufficiency) of the asymptotic relation,* $\xi_\varepsilon = \xi_\varepsilon(1) \stackrel{\mathsf{d}}{\longrightarrow} \xi_0$ *as* $\varepsilon \to 0$, *where* ξ_0 *is a non-negative random variable with distribution function not concentrated at zero.*

(ii) *The distribution function* $G(u) = \mathsf{P}\{\xi_0 \leq u\}, u \geq 0$ *of the limiting random variable* ξ_0 *has the Laplace transform* $\phi(s) = \int_0^\infty e^{-su} G(du) = \frac{1}{1+A(s)}, s \geq 0$, *where* $A(s)$ *is the cumulant of infinitely divisible distribution appearing in the condition* $\mathbf{D}_1$ *(*$\mathbf{D}_2$*).*

(iii) *The stochastic processes* $\xi_\varepsilon(t), t \geq 0 \stackrel{\mathsf{J}}{\longrightarrow} \xi_0(t) = \theta_0(t\nu_0), t \geq 0$ *as* $\varepsilon \to 0$, *where* **(a)** ν_0 *is a random variable, which has the exponential distribution with parameter* 1, **(b)** $\theta_0(t), t \geq 0$ *is a non-negative Lévy process with the Laplace transforms* $\mathsf{E}e^{-s\theta_0(t)} = e^{-tA(s)}, s, t \geq 0$, *and* **(c)** *the random variable* ν_0 *and the process* $\theta_0(t), t \geq 0$ *are independent.*

Remark 2.2 According to Theorem 2.1, the class $\mathcal{G}$ of all possible non-negative random variables ξ_0 with distribution functions not concentrated in zero, and such that the asymptotic relation, $\xi_\varepsilon \stackrel{\mathsf{d}}{\longrightarrow} \xi_0$ as $\varepsilon \to 0$, holds, coincides with the class of limiting random variables described in the statement **(ii)**. The condition $\mathbf{D}_1$ is the necessary and sufficient condition for holding of the asymptotic relation given in the statements **(i)**–**(ii)**. Also, the class $\mathcal{G}_\infty$ of all possible non-negative, non-decreasing, càdlàg, stochastically continuous processes $\xi_0(t), t \geq 0$ with distribution functions of random variables $\xi_0(t), t > 0$ not concentrated at zero, and such that the asymptotic relation, $\xi_\varepsilon(t), t \geq 0 \stackrel{\mathsf{J}}{\longrightarrow} \xi_0(t), t \geq 0$ as $\varepsilon \to 0$, holds, coincides with the class of limiting processes described in the statement **(iii)**. The condition $\mathbf{D}_1$ is the necessary and sufficient condition for the fulfilment not only of the asymptotic relation appearing in the statements **(i)**–**(ii)** but also of the stronger asymptotic relation appearing in the statement **(iii)**.

Remark 2.3 The note "for some or any initial distributions $\bar{q}_\varepsilon$, respectively, in the statements of necessity and sufficiency" used in the formulation of Theorem 2.1 should be understood in the sense that the asymptotic relation penetrating the statement **(i)** must hold for at least one family of initial distributions $\bar{q}_\varepsilon, \varepsilon \in (0, \varepsilon_0]$, in the statement of necessity, and for any family of initial distributions $\bar{q}_\varepsilon, \varepsilon \in (0, \varepsilon_0]$, in the statement of sufficiency.

Remark 2.4 It is possible to modify the condition $\mathbf{D}_1$ and to assume that the random variable θ_0 appearing in this condition has no atom at zero. In terms of the corresponding cumulant $A(s) = \pi s + \int_0^\infty (1 - e^{-sv})\pi(dv)$ this additional assumption is satisfied if and only if $A(s) \longrightarrow \infty$ as $s \longrightarrow \infty$ or, equivalently, $\pi > 0$ or $\pi = 0$ but $\pi((0, \infty)) = \infty$. In this case, the random variable ξ_0 appearing in the statement **(i)** of Theorem 2.1 must also have a distribution function without an atom at zero.

In what follows, the relation $a(\varepsilon) \sim b(\varepsilon)$ as $\varepsilon \longrightarrow 0$ means that $a(\varepsilon)/b(\varepsilon) \longrightarrow 1$ as $\varepsilon \longrightarrow 0$.

Remark 2.5 Due to Lemma 2.1, the asymptotic relation appearing in the condition $\mathbf{D}_2$ can, under the conditions $\mathbf{A}_1$, $\mathbf{B}_1$ and $\mathbf{B}_2$, be rewritten in an equivalent form, where the stationary probabilities $\pi_{\varepsilon,i}, i \in \mathbb{X}$ are replaced by the stationary probabilities $\tilde{\pi}_{\varepsilon,i}, i \in \mathbb{X}$. Indeed, in this case, $\pi_{\varepsilon,i} - \tilde{\pi}_{\varepsilon,i} \longrightarrow 0$ as $\varepsilon \longrightarrow 0$, for $i \in \mathbb{X}$, and, thus,

$$\begin{aligned} u_\varepsilon(1 - \phi_\varepsilon(s)) &= \sum_{i \in \mathbb{X}} \pi_{\varepsilon,i} u_\varepsilon (1 - \phi_{\varepsilon,i}(s)) \\ &\sim \sum_{i \in \mathbb{X}} \tilde{\pi}_{\varepsilon,i} u_\varepsilon (1 - \phi_{\varepsilon,i}(s)) \\ &\longrightarrow A(s) \text{ as } \varepsilon \longrightarrow 0, \text{ for } s > 0. \end{aligned} \tag{2.21}$$

Remark 2.6 It is worth noting that the condition $\mathbf{C}_1$ cannot be omitted. If this condition is not satisfied, the last summand $\kappa_{\varepsilon,\nu_\varepsilon}$ may be a dominating contribution in the first-rare-event time ξ_ε. As shown in the example considered in Sect. 1.1.1.2, in this case it is possible that the conditions $\mathbf{A}_1$, $\mathbf{B}_1$ are satisfied, but the random variables ξ_ε either converge in distribution to ∞ as $\varepsilon \longrightarrow 0$ or do not converge at all in distribution.

However, the condition $\mathbf{C}_1$ can be omitted in Theorem 2.1, if the sojourn times $\kappa_{\varepsilon,n}$ and the indicator random variables $\rho_{\varepsilon,n}$ are conditionally independent, that is:

$\mathbf{C}_2$: $\mathsf{P}_i\{\kappa_{\varepsilon,1} \leq t, \rho_{\varepsilon,1} = 1\} = \mathsf{P}_i\{\kappa_{\varepsilon,1} \leq t\}\mathsf{P}_i\{\rho_{\varepsilon,1} = 1\}$, for $t \geq 0$, $i \in \mathbb{X}$.

In this case, the condition $\mathbf{C}_1$ is equivalent to the assumption that the following asymptotic relation holds, for $\delta > 0, i \in \mathbb{X}$,

$$\mathsf{P}_i\{\kappa_{\varepsilon,1} > \delta\} = 1 - F_{\varepsilon,i}(\delta) \longrightarrow 0 \text{ as } \varepsilon \longrightarrow 0. \tag{2.22}$$

The conditions $\mathbf{A}_1$, $\mathbf{B}_1$, and $\mathbf{D}_1$ imply the fulfilment of the above asymptotic relation. Indeed, the condition $\mathbf{D}_3$ (equivalent to the condition $\mathbf{D}_1$) implies that $1 - F_\varepsilon(\delta) = \sum_{i \in \mathbb{X}} \pi_{\varepsilon,i}(1 - F_{\varepsilon,i}(\delta)) \longrightarrow 0$ as $\varepsilon \longrightarrow 0$, for $\delta > 0$. Also, the condition $\mathbf{B}_1$ implies that $\underline{\lim}_{\varepsilon \to 0} \pi_{\varepsilon,i} > 0$, for $i \in \mathbb{X}$. Thus, the relation (2.22) holds.

At the same time, the proof of necessity statement of Theorem 2.2 is based on the conditions $\mathbf{A}_1$, $\mathbf{B}_1$ but does not use the condition $\mathbf{C}_1$. Once the condition $\mathbf{D}_1$ is proved, the above arguments again lead to the fulfilment of the relation (2.22).

2.2 Asymptotics of Step-Sum Reward Processes

In this section, we give necessary and sufficient conditions of convergence in distribution and convergence in the J-topology for step-sum reward processes defined on asymptotically uniformly ergodic Markov chains.

2.2.1 Necessary and Sufficient Conditions for Convergence in Distribution for Step-Sum Reward Processes

In what follows, we always assume that asymptotic relations for random variables and processes, defined on trajectories of the Markov renewal processes $(\eta_{\varepsilon,n}, \kappa_{\varepsilon,n}, \rho_{\varepsilon,n})$, hold for any initial distributions $\bar{q}_\varepsilon$, if such distributions are not specified.

It is useful to note that the indicator variables $\rho_{\varepsilon,n}$ are not involved in the definition of the processes $\kappa_\varepsilon(t)$. This let us replace function $u_\varepsilon = p_\varepsilon^{-1}$ by an arbitrary function $0 < u_\varepsilon \to \infty$ as $\varepsilon \to 0$, in the condition $\mathbf{D}_1$, Theorem 2.2 and Lemmas 2.3–2.6 formulated below and their proofs, which are integral parts of the proof of Theorem 2.1.

Theorem 2.2 *Let the condition* $\mathbf{B}_1$ *be satisfied. Then*:

(i) *The condition* $\mathbf{D}_1$ *is the necessary and sufficient condition for the fulfilment (for some or any initial distributions* $\bar{q}_\varepsilon$*, respectively, in the statements of necessity and sufficiency) of the asymptotic relation,* $\kappa_\varepsilon(1) \xrightarrow{\mathsf{d}} \theta_0$ *as* $\varepsilon \to 0$*, where* θ_0 *is a non-negative random variable with distribution function not concentrated at zero.*

(ii) *The limiting random variable* θ_0 *has the infinitely divisible distribution with the Laplace transform* $\mathsf{E}e^{-s\theta_0} = e^{-A(s)}$*,* $s \geq 0$*, where* $A(s)$ *is the cumulant of infinitely divisible distribution appearing in the condition* $\mathbf{D}_1$.

(iii) *The stochastic processes* $\kappa_\varepsilon(t), t \geq 0 \xrightarrow{\mathsf{J}} \theta_0(t), t \geq 0$ *as* $\varepsilon \to 0$*, where* $\theta_0(t), t \geq 0$ *is a non-negative Lévy process with the Laplace transforms* $\mathsf{E}e^{-s\theta_0(t)} = e^{-tA(s)}$*,* $s, t \geq 0$.

Remark 2.7 According to Theorem 2.2, the class of non-negative, non-decreasing, càdlàg, stochastically continuous processes $\theta_0(t), t \geq 0$ with distributions of random variables $\theta_0(t), t > 0$ not concentrated in zero, and such that the asymptotic relation, $\kappa_\varepsilon(t), t \geq 0 \xrightarrow{\mathsf{J}} \theta_0(t), t \geq 0$ as $\varepsilon \to 0$, holds, coincides with the class of non-negative Lévy processes. The condition $\mathbf{D}_1$ is the necessary and sufficient condition for the fulfilment of the asymptotic relation appearing in the statements **(i)**–**(ii)** as well as for the stronger asymptotic relation appearing in the statement **(iii)**.

The proof of Theorem 2.2 is based on three lemmas formulated below.

According to Lemma 2.1, the condition $\mathbf{B}_1$ implies that there exists $\varepsilon_0 \in (0, 1]$ such that the phase space $\mathbb{X}$ is one class of communicative states for the Markov chain $\eta_{\varepsilon,n}$, for every $\varepsilon \in (0, \varepsilon_0]$. In what follows, we assume that $\varepsilon \in (0, \varepsilon_0]$.

Let $\alpha_{\varepsilon,i,0} = 0$ and $\alpha_{\varepsilon,i,n} = \min(k > \alpha_{\varepsilon,i,n-1} : \eta_{\varepsilon,k} = i), n = 1, 2, \ldots$ be the sequential moments of hitting in state $i \in \mathbb{X}$ for the Markov chain $\eta_{\varepsilon,n}$.

Also, let us recall the random variables $\mu_{\varepsilon,i}(n)$ introduced in the relation (2.10).

Lemma 2.2 *Let the condition* $\mathbf{B}_1$ *be satisfied. Then, for any initial distributions* $\bar{q}_\varepsilon$,

$$\alpha^*_{\varepsilon,i}(t) = \frac{\alpha_{\varepsilon,i,[\pi_{\varepsilon,i} t u_\varepsilon]}}{u_\varepsilon}, t \geq 0 \stackrel{\mathsf{U}}{\longrightarrow} t, t \geq 0 \text{ as } \varepsilon \to 0, \tag{2.23}$$

and

$$\mu^*_{\varepsilon,i}(t) = \frac{\mu_{\varepsilon,i}([t u_\varepsilon])}{\pi_{\varepsilon,i} u_\varepsilon}, t \geq 0 \stackrel{\mathsf{U}}{\longrightarrow} t, t \geq 0 \text{ as } \varepsilon \to 0. \tag{2.24}$$

Proof Let $\alpha_{\varepsilon,j} = \min(n > 0 : \eta_{\varepsilon,n} = j)$ be the moment of first hitting into the state $j \in \mathbb{X}$ for the Markov chain $\eta_{\varepsilon,n}$. The condition $\mathbf{B}_1$ implies that there exist $p \in (0, 1)$ and $\varepsilon_p \in (0, 1]$ such that $\prod_{k=1}^{N} p_{\varepsilon,i_{k-1}i_k} > p$, for $\varepsilon \in (0, \varepsilon_0]$. The following inequalities obviously hold, for $\varepsilon \in (0, \varepsilon_p]$ and $k \geq 1, i, j \in \mathbb{X}$,

$$\mathsf{P}_i\{\alpha_{\varepsilon,j} > kN\} \leq (1 - p)^k. \tag{2.25}$$

These inequalities imply that there exists $K_p \in (0, \infty)$ such that, for $\varepsilon \in (0, \varepsilon_p]$ and $i, j \in \mathbb{X}$,

$$\mathsf{E}_i\alpha^2_{\varepsilon,j} \leq K_p < \infty. \tag{2.26}$$

Also, as well known, $\mathsf{E}_i\alpha_{\varepsilon,i} = \pi^{-1}_{\varepsilon,i}, i \in \mathbb{X}$, for $\varepsilon \in (0, \varepsilon_p]$.

Let $\beta_{\varepsilon,i,n} = \alpha_{\varepsilon,i,n} - \alpha_{\varepsilon,i,n-1}, n = 1, 2, \ldots$, where $\alpha_{\varepsilon,i,0} = 0$. The random variables $\beta_{\varepsilon,i,n}, n \geq 1$ are independent and identically distributed for $n \geq 2$.

The relations (2.25) and (2.26) imply that, for $i \in \mathbb{X}$,

$$\alpha_{\varepsilon,i,1}/u_\varepsilon \stackrel{\mathsf{P}}{\longrightarrow} 0 \text{ as } \varepsilon \to 0, \tag{2.27}$$

and, for $\varepsilon \in (0, \varepsilon_p], i \in \mathbb{X}$, and $\delta > 0, t \geq 0$,

$$\mathsf{P}_i\{u^{-1}_\varepsilon|\alpha_{\varepsilon,i,[tu_\varepsilon]} - \pi^{-1}_{\varepsilon,i}[tu_\varepsilon]| > \delta\} \leq tK_p/\delta^2 u_\varepsilon. \tag{2.28}$$

The relations (2.27) and (2.28) imply that, for $i \in \mathbb{X}$ and $t \geq 0$,

$$\alpha_{\varepsilon,i,[tu_\varepsilon]}/\pi^{-1}_{\varepsilon,i}u_\varepsilon \stackrel{\mathsf{P}}{\longrightarrow} t \text{ as } \varepsilon \to 0 \tag{2.29}$$

and

$$\alpha_{\varepsilon,i,[\pi_{\varepsilon,i}tu_\varepsilon]}/u_\varepsilon \stackrel{\mathsf{P}}{\longrightarrow} t \text{ as } \varepsilon \to 0. \tag{2.30}$$

The dual identities $\mathsf{P}\{\mu_{\varepsilon,i}(r) \geq k\} = \mathsf{P}\{\alpha_{\varepsilon,i,k} \leq r\}, r, k = 0, 1, \ldots$ allow converting the asymptotic relation (2.30) into the following equivalent relation, for $i \in \mathbb{X}$ and $t \geq 0$,

$$\mu^*_{\varepsilon,i}(t) = \mu_{\varepsilon,i}([tu_\varepsilon])/\pi_{\varepsilon,i}u_\varepsilon \stackrel{\mathsf{P}}{\longrightarrow} t \text{ as } \varepsilon \to 0. \tag{2.31}$$

Since the processes $\alpha^*_{\varepsilon,i}(t)$ and $\mu^*_{\varepsilon,i}(t), t \geq 0$ are non-decreasing and the corresponding limiting function is continuous, the asymptotic relations (2.29) and (2.31) are, according to Remark A.1, equivalent to the asymptotic relations (2.23) and (2.24) given in Lemma 2.2. □

Let us introduce random variables, which are successive moments of hitting in the state $i \in \mathbb{X}$ by the Markov chain $\eta_{\varepsilon,n}$,

$$\tau_{\varepsilon,i,n} = \begin{cases} \min(k \geq 0, \eta_{\varepsilon,k} = i) & \text{for } n = 1, \\ \min(k > \tau_{\varepsilon,i,n-1}, \eta_{\varepsilon,k} = i) & \text{for } n \geq 2. \end{cases} \tag{2.32}$$

Let us also define random variables,

$$\kappa_{\varepsilon,i,n} = \kappa_{\varepsilon,\tau_{\varepsilon,i,n}+1}, \ n = 1, 2, \ldots, \ i \in \mathbb{X}. \tag{2.33}$$

The following lemma describes useful properties of the above family of random variables.

Lemma 2.3 *Let the condition* $\mathbf{B}_1$ *be satisfied. Then, for every* $\varepsilon \in (0, \varepsilon_0]$:

(i) *The random variables* $\kappa_{\varepsilon,i,n}, n = 1, 2, \ldots, i \in \mathbb{X}$ *are mutually independent.*

(ii) $\mathsf{P}\{\kappa_{\varepsilon,i,n} \leq t\} = F_{\varepsilon,i}(t)$, $t \geq 0$, *for* $n = 1, 2, \ldots, \ i \in \mathbb{X}$.

Proof Let us take an arbitrary sequence of different pairs $(i_1, k_1), (i_2, k_2), \ldots$ taking values in the space $\mathbb{X}\times\{1, 2 \ldots\}$. Please note that the random variables $\tau_{\varepsilon,i_1,k_1}, \tau_{\varepsilon,i_2,k_2}, \ldots$, by definition, take different values.

The hitting times $\tau_{\varepsilon,i,k}$ are Markov moments for the embedded Markov chain $\eta_{\varepsilon,n}$ as, therefore, also for the Markov renewal process $(\eta_{\varepsilon,n}, \kappa_{\varepsilon,n})$. Also, any event $\{\tau_{\varepsilon,i_1,k_1} < \cdots < \tau_{\varepsilon,i_r,k_r} < \min_{r<l\leq n} \tau_{\varepsilon,i_l,k_l}\}$ is determined by the random variables $\eta_{\varepsilon,0}, \ldots, \eta_{\varepsilon,\tau_{\varepsilon,i_r,k_r}}$.

Using the above remarks, we get the following relation, for $u_1, u_2 \geq 0$,

$$\begin{aligned} & \mathsf{P}\{\kappa_{\varepsilon,i_r,k_r} \leq u_r, r = 1, 2, \tau_{\varepsilon,i_1,k_1} < \tau_{\varepsilon,i_2,k_2}\} \\ & \quad = \mathsf{P}\{\kappa_{\varepsilon,\tau_{\varepsilon,i_2,k_2}+1} \leq u_2 / \kappa_{\varepsilon,\tau_{\varepsilon,i_1,k_1}+1} \leq u_1, \tau_{\varepsilon,i_1,k_1} < \tau_{\varepsilon,i_2,k_2}\} \\ & \qquad \times \mathsf{P}\{\kappa_{\varepsilon,\tau_{\varepsilon,i_1,k_1}+1} \leq u_1, \tau_{\varepsilon,i_1,k_1} < \tau_{\varepsilon,i_2,k_2}\} \\ & \quad = F_{\varepsilon,i_2}(u_2)\mathsf{P}\{\kappa_{\varepsilon,\tau_{\varepsilon,i_1,k_1}+1} \leq u_1, \tau_{\varepsilon,i_1,k_1} < \tau_{\varepsilon,i_2,k_2}\} \\ & \qquad \times \mathsf{P}\{\tau_{\varepsilon,i_1,k_1} < \tau_{\varepsilon,i_2,k_2}\} \\ & \quad = F_{\varepsilon,i_1}(u_1)F_{\varepsilon,i_2}(u_2)\mathsf{P}\{\tau_{\varepsilon,i_1,k_1} < \tau_{\varepsilon,i_2,k_2}\}. \end{aligned} \tag{2.34}$$

Also a similar relation, in which the random variables $\tau_{\varepsilon,i_1,k_1}$ and $\tau_{\varepsilon,i_2,k_2}$ swap places with each other, takes place. By adding the above two relations, we get, for $u_1, u_2 \geq 0$, the relation, which proves the pairwise independence of the random variables $\kappa_{\varepsilon,i,n}, n = 1, 2, \ldots, i \in \mathbb{X}$,

$$\mathsf{P}\{\kappa_{\varepsilon,i_r,k_r} \leq u_r, r = 1, 2\} = F_{\varepsilon,i_1}(u_1)F_{\varepsilon,i_2}(u_2). \tag{2.35}$$

Similarly, we can get the following relation,

$$\begin{aligned}
&\mathsf{P}\{\kappa_{\varepsilon,i_r,k_r} \leq u_r, r = 1,2,3, \tau_{\varepsilon,i_1,k_1} < \tau_{\varepsilon,i_2,k_2} < \tau_{\varepsilon,i_3,k_3}\} \\
&\quad = F_{\varepsilon,i_2}(u_2)F_{\varepsilon,i_3}(u_3)\mathsf{P}\{\kappa_{\varepsilon,i_1,k_1} \leq u_1, \tau_{\varepsilon,i_1,k_1} < \tau_{\varepsilon,i_2,k_2} < \tau_{\varepsilon,i_3,k_3}\}
\end{aligned} \tag{2.36}$$

and a similar relation, in which the random variables $\tau_{\varepsilon,i_2,k_2}$ and $\tau_{\varepsilon,i_3,k_3}$ swap places with each other.

By adding the above-mentioned two relations, we get the following relation:

$$\begin{aligned}
&\mathsf{P}\{\kappa_{\varepsilon,i_r,k_r} \leq u_r, r = 1,2,3, \tau_{\varepsilon,i_1,k_1} < \min(\tau_{\varepsilon,i_2,k_2}, \tau_{\varepsilon,i_3,k_3})\} \\
&\quad = F_{\varepsilon,i_2}(u_2)F_{\varepsilon,i_3}(u_3)\mathsf{P}\{\kappa_{\varepsilon,i_1,k_1} \leq u_1, \tau_{\varepsilon,i_1,k_1} < \min(\tau_{\varepsilon,i_2,k_2}, \tau_{\varepsilon,i_3,k_3})\} \\
&\quad = F_{\varepsilon,i_1}(u_1)F_{\varepsilon,i_2}(u_2)F_{\varepsilon,i_3}(u_3)\mathsf{P}\{\tau_{\varepsilon,i_1,k_1} < \min(\tau_{\varepsilon,i_2,k_2}, \tau_{\varepsilon,i_3,k_3})\}.
\end{aligned} \tag{2.37}$$

Also, two similar relations, in which the random variables $\tau_{\varepsilon,i_1,k_1}, \tau_{\varepsilon,i_2,k_2}, \tau_{\varepsilon,i_3,k_3}$ are exchanged, respectively, by the random variables $\tau_{\varepsilon,i_2,k_2}, \tau_{\varepsilon,i_3,k_3}, \tau_{\varepsilon,i_1,k_1}$ or by $\tau_{\varepsilon,i_3,k_3}, \tau_{\varepsilon,i_1,k_1}, \tau_{\varepsilon,i_2,k_2}$, take place. By adding the above-mentioned three relations, we get the relation, which proves the triplet-wise independence of the random variables $\kappa_{\varepsilon,i,n}, n = 1, 2, \ldots, i \in \mathbb{X}$,

$$\mathsf{P}\{\kappa_{\varepsilon,i_r,k_r} \leq u_r, r = 1,2,3\} = F_{\varepsilon,i_1}(u_1)F_{\varepsilon,i_2}(u_2)F_{\varepsilon,i_3}(u_3). \tag{2.38}$$

The proof of independence of the random variables $\tau_{\varepsilon,i_1,k_1}, \tau_{\varepsilon,i_2,k_2}, \ldots, \tau_{\varepsilon,i_n,k_n}$ for any $n > 3$ is similar.

The statement **(ii)** follows from the above remarks. □

The following representation takes place for the process $\kappa_\varepsilon(t)$,

$$\kappa_\varepsilon(t) = \sum_{n=1}^{[tu_\varepsilon]} \kappa_{\varepsilon,n} = \sum_{i\in\mathbb{X}} \sum_{n=1}^{\mu_{\varepsilon,i}([tu_\varepsilon])} \kappa_{\varepsilon,i,n}, t \geq 0. \tag{2.39}$$

The relation (2.39) is obvious because it only represents two alternative ways of grouping summands in the same random sum.

It is useful to note that the families of random variables $\langle \mu_{\varepsilon,i}(n), n = 0, 1, \ldots, i \in \mathbb{X}\rangle$ and $\langle \kappa_{\varepsilon,i,n}, n = 1, 2, \ldots, i \in \mathbb{X}\rangle$ are not independent.

Let now introduce step-sum processes with independent increments,

$$\tilde{\kappa}_\varepsilon(t) = \sum_{i\in\mathbb{X}} \sum_{n=1}^{[t\pi_{\varepsilon,i}u_\varepsilon]} \kappa_{\varepsilon,i,n}, t \geq 0. \tag{2.40}$$

Lemmas 2.2 and 2.3 let us presume that processes $\tilde{\kappa}_\varepsilon(t)$ can be good approximations for processes $\kappa_\varepsilon(t)$.

Lemma 2.4 *Let the condition* $\mathbf{B}_1$ *be satisfied. Then:*

(i) *The condition* $\mathbf{D}_1$ *is satisfied if and only if the following relation holds,* $\tilde{\kappa}_\varepsilon(1) \xrightarrow{d} \theta_0$ *as* $\varepsilon \to 0$, *where* θ_0 *is a non-negative random variable with distribution not concentrated at zero.*

(ii) *The random variable θ_0 has an infinitely divisible distribution with the Laplace transform $\mathsf{E}e^{-s\theta_0} = e^{-A(s)}$, $s \geq 0$ with the cumulant $A(s)$ defined in the condition* $\mathbf{D}_1$.

(iii) *The stochastic processes $\tilde{\kappa}_\varepsilon(t), t \geq 0 \stackrel{\mathsf{J}}{\longrightarrow} \theta_0(t), t \geq 0$ as $\varepsilon \to 0$, where $\theta_0(t), t \geq 0$ is a non-negative Lévy process with the Laplace transforms $\mathsf{E}e^{-s\theta_0(t)} = e^{-tA(s)}$, $s, t \geq 0$.*

Proof (Of Lemma 2.4 and Theorem 2.2) First, let us prove that the condition $\mathbf{D}_1$ implies the fulfilment of the asymptotic relations given in Lemma 2.4 and Theorem 2.2.

Let $\hat{\eta}_{\varepsilon,n}, n = 1, 2, \ldots$ be, for every $\varepsilon \in (0, 1]$, a sequence of random variables such that: (a) it is independent of the Markov chain $(\eta_{\varepsilon,n}, \kappa_{\varepsilon,n}), n = 0, 1, \ldots$ and (b) it is a sequence of i.i.d. random variables taking value i with probability $\pi_{\varepsilon,i}$, for $i \in \mathbb{X}$.

Note that, in this case, the sequence of random variables $\hat{\eta}_{\varepsilon,n}, n = 1, 2, \ldots$ is also independent of the families of random variables $\langle \mu_{\varepsilon,i}(n), n = 0, 1, \ldots, i \in \mathbb{X}\rangle$ and $\langle \kappa_{\varepsilon,i,n}, n = 1, 2, \ldots, i \in \mathbb{X}\rangle$.

Let us define random variables,

$$\hat{\mu}_{\varepsilon,i}(n) = \sum_{k=1}^{n} \mathrm{I}(\hat{\eta}_{\varepsilon,n} = i), \ n = 0, 1, \ldots, \ i \in \mathbb{X}, \tag{2.41}$$

and stochastic processes,

$$\hat{\kappa}_\varepsilon(t) = \sum_{i \in \mathbb{X}} \sum_{n=1}^{\hat{\mu}_{\varepsilon,i}([tu_\varepsilon])} \kappa_{\varepsilon,i,n}, t \geq 0. \tag{2.42}$$

Let us also consider the sequence of random variables $\theta_{\varepsilon,n} = \kappa_{\varepsilon,\hat{\eta}_{\varepsilon,n},n}, n = 1, 2, \ldots$. This is the sequence of i.i.d. random variables, which follows from the above definition of the sequence of random variables $\hat{\eta}_{\varepsilon,n}, n = 1, 2, \ldots$ and Lemma 2.3 describing the family of random variables $\kappa_{\varepsilon,i,n}, n = 1, 2, \ldots, i \in \mathbb{X}$. Also,

$$\mathsf{P}\{\theta_{\varepsilon,1} \leq t\} = \sum_{i \in \mathbb{X}} \pi_{\varepsilon,i} F_{\varepsilon,i}(t) = F_\varepsilon(t), \ t \geq 0. \tag{2.43}$$

We also define for $\varepsilon \in (0, 1]$ a homogeneous step-sum càdlàg processes with independent increments (the relation (2.43) allows us to use the same notation as for the process introduced in the relation (2.17)),

$$\theta_\varepsilon(t) = \sum_{n=1}^{[tu_\varepsilon]} \theta_{\varepsilon,n}, \ t \geq 0. \tag{2.44}$$

As well known, the condition $\mathbf{D}_1$ is equivalent to the following relation,

$$\theta_\varepsilon(t), t \geq 0 \stackrel{\mathsf{d}}{\longrightarrow} \theta_0(t), \ t \geq 0 \text{ as } \varepsilon \to 0. \tag{2.45}$$

From the definitions of the sequence of random variables $\hat{\eta}_{\varepsilon,n}, n = 1, 2, \ldots$ and the family of random variables $\kappa_{\varepsilon,i,n}, n = 1, 2, \ldots, i \in \mathbb{X}$, Lemma 2.3, and the independence of the above sequence and family, it follows that the following relation holds,

$$\hat{\kappa}_\varepsilon(t), t \geq 0 \stackrel{\mathsf{d}}{=} \theta_\varepsilon(t), t \geq 0. \tag{2.46}$$

The relation (2.46) implies that $\hat{\kappa}_\varepsilon(t), t \geq 0$ is a homogeneous step-sum càdlàg process with independent increments and that the condition $\mathbf{D}_1$ is equivalent to the following relation:

$$\hat{\kappa}_\varepsilon(t), t \geq 0 \xrightarrow{\mathsf{d}} \theta_0(t), t \geq 0 \text{ as } \varepsilon \to 0. \tag{2.47}$$

The random variables $\mathrm{I}(\hat{\eta}_{\varepsilon,n} = i), n = 1, 2, \ldots$ are, for every $i \in \mathbb{X}$, i.i.d. random variables taking values 1 and 0 with probabilities, respectively, $\pi_{\varepsilon,i}$ and $1 - \pi_{\varepsilon,i}$. According to the statement **(i)** of Lemma 2.1, $0 < \underline{\lim}_{\varepsilon \to 0} \pi_{\varepsilon,i} \leq \overline{\lim}_{\varepsilon \to 0} \pi_{\varepsilon,i} < 1$, for every $i \in \mathbb{X}$.

Taking into account the above remarks, we get the following relations, $\mathsf{E}|\hat{\mu}^*_{\varepsilon,i}(t) - [tu_\varepsilon]/u_\varepsilon| \leq [tu_\varepsilon]/(\pi_{\varepsilon,i}u_\varepsilon)^2 \to 0$ as $\varepsilon \to 0$ and $[tu_\varepsilon]/u_\varepsilon| - t \to 0$ as $\varepsilon \to 0$. These relations obviously imply that $\hat{\mu}^*_{\varepsilon,i}(t) \xrightarrow{\mathsf{P}} t$ as $\varepsilon \to 0$, for $t \geq 0$. This relation and Theorem A.7 imply that the following relation holds, for $i \in \mathbb{X}$,

$$\hat{\mu}^*_{\varepsilon,i}(t) = \frac{\hat{\mu}_{\varepsilon,i}([tu_\varepsilon])}{\pi_{\varepsilon,i}u_\varepsilon}, t \geq 0 \xrightarrow{\mathsf{d}} \mu_{0,i}(t) = t, t \geq 0 \text{ as } \varepsilon \to 0. \tag{2.48}$$

Since the processes $\hat{\mu}^*_{\varepsilon,i}(t), t \geq 0$ are monotonic, relation (2.48) implies, by Remark A.1, that the following relation holds, for $i \in \mathbb{X}$,

$$\hat{\mu}^*_{\varepsilon,i}(t) = \frac{\hat{\mu}_{\varepsilon,i}([tu_\varepsilon])}{\pi_{\varepsilon,i}u_\varepsilon}, t \geq 0 \xrightarrow{\mathsf{U}} \mu_{0,i}(t) = t, t \geq 0 \text{ as } \varepsilon \to 0. \tag{2.49}$$

Let us choose some $0 < u < 1$.

By the definition, the processes $\tilde{\kappa}_\varepsilon(t)$, $\hat{\kappa}_\varepsilon(t)$, and $\hat{\mu}^*_{\varepsilon,i}(t), i \in \mathbb{X}$ are non-negative and non-decreasing. Taking this into account, we obtain, for $x > 0$,

$$\begin{aligned}\mathsf{P}\{\tilde{\kappa}_\varepsilon(u) > x\} &\leq \mathsf{P}\{\tilde{\kappa}_\varepsilon(u) > x, \hat{\mu}^*_{\varepsilon,i}(1) > u, i \in \mathbb{X}\} \\ &\quad + \sum_{i \in \mathbb{X}} \mathsf{P}\{\tilde{\kappa}_\varepsilon(u) > x, \hat{\mu}^*_{\varepsilon,i}(1) \leq u\} \\ &\leq \mathsf{P}\{\hat{\kappa}_\varepsilon(1) > x\} + \sum_{i \in \mathbb{X}} \mathsf{P}\{\hat{\mu}^*_{\varepsilon,i}(1) \leq u\}.\end{aligned} \tag{2.50}$$

The relations (2.47) and (2.49) imply the distributions of random variables $\tilde{\kappa}_\varepsilon(u)$ are tight, as $\varepsilon \to 0$. Indeed,

$$\lim_{x\to\infty}\overline{\lim_{\varepsilon\to 0}}\,\mathsf{P}\{\tilde{\kappa}_\varepsilon(u) > x\} \le \lim_{x\to\infty}\overline{\lim_{\varepsilon\to 0}}(\mathsf{P}\{\hat{\kappa}_\varepsilon(1) > x\}$$
$$+ \sum_{i\in\mathbb{X}} \mathsf{P}\{\hat{\mu}^*_{\varepsilon,i}(1) \le u\}) \le \lim_{x\to\infty} \mathsf{P}\{\theta_0(1) > x/2\} = 0. \tag{2.51}$$

Let us also introduce homogeneous step-sum càdlàg processes with independent increments, for $i \in \mathbb{X}$,

$$\tilde{\kappa}_{\varepsilon,i}(t) = \sum_{n=1}^{[t\pi_{\varepsilon,i}u_\varepsilon]} \kappa_{\varepsilon,i,n},\ t \ge 0. \tag{2.52}$$

Lemma 2.3 implies that, for every $\varepsilon \in (0, \varepsilon_0]$, the processes $\langle\tilde{\kappa}_{\varepsilon,i}(t), t \ge 0\rangle, i \in \mathbb{X}$ are independent.

Since $\tilde{\kappa}_{\varepsilon,i}(u) \le \tilde{\kappa}_\varepsilon(u)$, for $i \in \mathbb{X}$, the relation (2.51) implies that distributions of random variables $\tilde{\kappa}_{\varepsilon,i}(1)$ are also tight, as $\varepsilon \to 0$, for every $i \in \mathbb{X}$, that is,

$$\lim_{x\to\infty}\overline{\lim_{\varepsilon\to 0}}\,\mathsf{P}\{\tilde{\kappa}_{\varepsilon,i}(u) > x\} \le \lim_{x\to\infty}\overline{\lim_{\varepsilon\to 0}}\,\mathsf{P}\{\tilde{\kappa}_\varepsilon(u) > x\} = 0. \tag{2.53}$$

This implies that the above distributions are relatively compact and, thus, any sequence $0 < \varepsilon_n \to 0$ as $n \to \infty$ contains a subsequence $0 < \varepsilon_{n_k} \to 0$ as $k \to \infty$ such that, for $i \in \mathbb{X}$,

$$\tilde{\kappa}_{\varepsilon_{n_k},i}(u) \xrightarrow{\mathsf{d}} \theta_{0,i,u} \text{ as } k \to \infty, \tag{2.54}$$

where $\theta_{0,i,u}, i \in \mathbb{X}$ are proper non-negative random variables with distributions possibly depending on the choice of subsequence ε_{n_k}.

Moreover, by the central criterion of convergence (see, for example, Loève (1977)), the random variables $\theta_{0,i,u}, i \in \mathbb{X}$ have infinitely divisible distributions. Let $\mathsf{E}e^{-s\theta_{0,i,u}} = e^{-A_{i,u}(s)}, s \ge 0, i \in \mathbb{X}$ be their Laplace transforms.

The relation (2.54) implies, by Remark A.3 and Theorem A.3, that

$$\tilde{\kappa}_{\varepsilon_{n_k},i}(t), t \ge 0 \xrightarrow{\mathsf{J}} \theta_{0,i}(t), t \ge 0 \text{ as } k \to \infty, \text{ for } i \in \mathbb{X}, \tag{2.55}$$

where $\theta_{0,i}(t), t \ge 0, i \in \mathbb{X}$ are non-negative Lévy processes with Laplace transforms $\mathsf{E}e^{-s\theta_{0,i}(t)} = e^{-tA_i(s)}, s, t \ge 0, i \in \mathbb{X}$, possibly depending on the choice of subsequence ε_{n_k}. Moreover, in this case, the cumulants $A_{i,u}(s)$ and $A_i(s)$ are connected by the following relation, $A_{i,u}(s) = uA_i(s), s \ge 0$.

Since the processes $\langle\tilde{\kappa}_{\varepsilon,i}(t), t \ge 0, \rangle, i \in \mathbb{X}$ are independent, the weak convergence of the vector processes $(\tilde{\kappa}_{\varepsilon_{n_k},1}(t), \dots, \tilde{\kappa}_{\varepsilon_{n_k},m}(t)), t \ge 0$ also takes place,

$$(\tilde{\kappa}_{\varepsilon_{n_k},1}(t), \dots, \tilde{\kappa}_{\varepsilon_{n_k},m}(t)), t \ge 0$$
$$\xrightarrow{\mathsf{d}} (\theta_{0,1}(t), \dots, \theta_{0,m}(t)), t \ge 0 \text{ as } k \to \infty, \tag{2.56}$$

where $\theta_{0,i}(t), t \ge 0, i \in \mathbb{X}$ are independent non-negative Lévy processes with Laplace transforms $\mathsf{E}e^{-s\theta_{0,i}(t)} = e^{-tA_i(s)}, s, t \ge 0, i \in \mathbb{X}$, possibly depending on the choice of subsequence ε_{n_k}.

Also (see, for example, Whitt (1980) or Silvestrov (2004)) J-compactness of the vector processes $(\tilde{\kappa}_{\varepsilon_{n_k},1}(t), \ldots, \tilde{\kappa}_{\varepsilon_{n_k},m}(t))$ follows from J-compactness of their components $\tilde{\kappa}_{\varepsilon_{n_k},i}(t), i \in \mathbb{X}$, since the corresponding limiting processes $\langle\theta_{0,i}(t), t \geq 0\rangle$, $i \in \mathbb{X}$ are stochastically continuous and independent and, therefore, their trajectories do not have joint discontinuity points with probability 1.

Therefore, the J-convergence of the vector processes $(\tilde{\kappa}_{\varepsilon_{n_k},1}(t), \ldots, \tilde{\kappa}_{\varepsilon_{n_k},m}(t))$, $t \geq 0$ also takes place,

$$
\begin{aligned}
&(\tilde{\kappa}_{\varepsilon_{n_k},1}(t), \ldots, \tilde{\kappa}_{\varepsilon_{n_k},m}(t)), t \geq 0 \\
&\qquad \stackrel{\mathsf{J}}{\longrightarrow} (\theta_{0,1}(t), \ldots, \theta_{0,m}(t)), t \geq 0 \text{ as } k \to \infty,
\end{aligned} \tag{2.57}
$$

where $\theta_{0,i}(t), t \geq 0, i \in \mathbb{X}$ are independent non-negative Lévy processes with Laplace transforms $\mathsf{E}e^{-s\theta_{0,i}(t)} = e^{-tA_i(s)}$, $s, t \geq 0$, $i \in \mathbb{X}$, possibly depending on the choice of subsequence ε_{n_k}.

The relation (2.57) and the above remarks obviously imply the fulfilment of the following relation:

$$
\tilde{\kappa}_{\varepsilon_{n_k}}(t) = \sum_{i \in \mathbb{X}} \tilde{\kappa}_{\varepsilon_{n_k},i}(t), t \geq 0 \stackrel{\mathsf{J}}{\longrightarrow} \theta_0'(t) = \sum_{i \in \mathbb{X}} \theta_{0,i}(t), t \geq 0 \text{ as } k \to \infty, \tag{2.58}
$$

where $\theta_{0,i}(t), t \geq 0, i \in \mathbb{X}$ are independent non-negative Lévy processes described in the relation (2.56).

Since the limiting processes in (2.24) and (2.49) are nonrandom functions, the relations (2.24), (2.49), and (2.58) imply, by Theorem A.9, that

$$
\begin{aligned}
&(\mu^*_{\varepsilon_{n_k},1}(t), \ldots, \mu^*_{\varepsilon_{n_k},m}(t), \tilde{\kappa}_{\varepsilon_{n_k},1}(t), \ldots, \tilde{\kappa}_{\varepsilon_{n_k},m}(t)), t \geq 0 \\
&\qquad \stackrel{\mathsf{d}}{\longrightarrow} (\mu_{0,1}(t), \ldots, \mu_{0,m}(t), \theta_{0,1}(t), \ldots, \theta_{0,m}(t)), t \geq 0 \text{ as } k \to \infty
\end{aligned} \tag{2.59}
$$

and

$$
\begin{aligned}
&(\hat{\mu}^*_{\varepsilon_{n_k},1}(t), \ldots, \hat{\mu}^*_{\varepsilon_{n_k},m}(t), \tilde{\kappa}_{\varepsilon_{n_k},1}(t), \ldots, \tilde{\kappa}_{\varepsilon_{n_k},m}(t)), t \geq 0 \\
&\qquad \stackrel{\mathsf{d}}{\longrightarrow} (\mu_{0,1}(t), \ldots, \mu_{0,m}(t), \theta_{0,1}(t), \ldots, \theta_{0,m}(t)), t \geq 0 \text{ as } k \to \infty,
\end{aligned} \tag{2.60}
$$

where $\mu_{0,i}(t) = t, t \geq 0, i \in \mathbb{X}$ and $\theta_{0,i}(t), t \geq 0, i \in \mathbb{X}$ are independent non-negative Lévy processes defined in the relation (2.56).

We can now apply Theorem A.6, which gives conditions of J-convergence for vector compositions of càdlàg stochastic processes, and get the following asymptotic relations,

$$
\begin{aligned}
&(\tilde{\kappa}_{\varepsilon_{n_k},1}(\mu^*_{\varepsilon_{n_k},1}(t)), \ldots, \tilde{\kappa}_{\varepsilon_{n_k},m}(\mu^*_{\varepsilon_{n_k},m}(t))), t \geq 0 \\
&\qquad \stackrel{\mathsf{J}}{\longrightarrow} (\theta_{0,1}(\mu_{0,1}(t)), \ldots, \theta_{0,m}(\mu_{0,m}(t))) \\
&\qquad = (\theta_{0,1}(t), \ldots, \theta_{0,m}(t)), t \geq 0 \text{ as } k \to \infty,
\end{aligned} \tag{2.61}
$$

and

$$
\begin{aligned}
&(\tilde{\kappa}_{\varepsilon_{n_k},1}(\hat{\mu}^*_{\varepsilon_{n_k},1}(t)), \ldots, \tilde{\kappa}_{\varepsilon_{n_k},m}(\hat{\mu}^*_{\varepsilon_{n_k},m}(t))), t \geq 0 \\
&\quad \xrightarrow{\mathsf{J}} (\theta_{0,1}(\mu_{0,1}(t)), \ldots, \theta_{0,m}(\mu_{0,m}(t))) \\
&\quad = (\theta_{0,1}(t), \ldots, \theta_{0,m}(t)), t \geq 0 \text{ as } k \to \infty,
\end{aligned} \tag{2.62}
$$

where $\theta_{0,i}(t), t \geq 0, i \in \mathbb{X}$ are independent non-negative Lévy processes defined in relation (2.56).

The relations (2.61) and (2.62) obviously imply J-convergence for the sums of components of these processes, i.e., that the following relations hold,

$$
\begin{aligned}
\kappa_{\varepsilon_{n_k}}(t) &= \sum_{i \in \mathbb{X}} \tilde{\kappa}_{\varepsilon_{n_k},i}(\mu^*_{\varepsilon_{n_k},i}(t)), t \geq 0 \\
&\xrightarrow{\mathsf{J}} \theta'_0(t) = \sum_{i \in \mathbb{X}} \theta_{0,i}(t), t \geq 0 \text{ as } k \to \infty,
\end{aligned} \tag{2.63}
$$

and

$$
\begin{aligned}
\hat{\kappa}_{\varepsilon_{n_k}}(t) &= \sum_{i \in \mathbb{X}} \tilde{\kappa}_{\varepsilon_{n_k},i}(\hat{\mu}^*_{\varepsilon_{n_k},i}(t)), t \geq 0 \\
&\xrightarrow{\mathsf{J}} \theta'_0(t) = \sum_{i \in \mathbb{X}} \theta_{0,i}(t), t \geq 0 \text{ as } k \to \infty,
\end{aligned} \tag{2.64}
$$

where $\theta_{0,i}(t), t \geq 0, i \in \mathbb{X}$ are independent non-negative Lévy processes defined in the relation (2.56).

The relation (2.47) implies that,

$$
\theta'_0(t), t \geq 0 \stackrel{\mathsf{d}}{=} \theta_0(t), t \geq 0. \tag{2.65}
$$

Thus, the limiting process $\theta'_0(t) = \sum_{i \in \mathbb{X}} \theta_{0,i}(t), t \geq 0$ has the same finite-dimensional distributions for all subsequences ε_{n_k} described above. Moreover, the cumulant $A(s)$ of the limiting Lévy process $\theta_0(t)$ is connected with the cumulants $A_i(s), i \in \mathbb{X}$ of the Lévy processes $\theta_{0,i}(t)$ by the relation, $A(s) = \sum_{i \in \mathbb{X}} A_i(s), s \geq 0$.

Therefore, the relations (2.58), (2.63), and (2.64) imply that, respectively, the following relations hold:

$$
\tilde{\kappa}_\varepsilon(t) = \sum_{i \in \mathbb{X}} \tilde{\kappa}_{\varepsilon,i}(t), t \geq 0 \xrightarrow{\mathsf{J}} \theta_0(t), t \geq 0 \text{ as } \varepsilon \to 0, \tag{2.66}
$$

and

$$
\kappa_\varepsilon(t) = \sum_{i \in \mathbb{X}} \tilde{\kappa}_{\varepsilon,i}(\mu^*_{\varepsilon,i}(t)), t \geq 0 \xrightarrow{\mathsf{J}} \theta_0(t), t \geq 0 \text{ as } \varepsilon \to 0, \tag{2.67}
$$

as well as,

$$\hat{\kappa}_\varepsilon(t) = \sum_{i \in \mathbb{X}} \tilde{\kappa}_{\varepsilon,i}(\hat{\mu}^*_{\varepsilon,i}(t)), t \geq 0 \xrightarrow{\mathsf{J}} \theta_0(t), t \geq 0 \text{ as } \varepsilon \to 0. \tag{2.68}$$

It is useful to note that the relation (2.68) for the homogeneous step-sum processes $\hat{\kappa}_\varepsilon(t)$ follows directly from the relation (2.47).

The relation (2.68) was obtained in the way described above only to prove that the limiting process in the relations (2.58), (2.63), and (2.64) is the same and does not depend on the choice of subsequences ε_{n_k} described above. This allows us to write the relations (2.66) and (2.67).

Let us now prove that the asymptotic relation appearing in the statement **(i)** of Theorem 2.2 or in the statement **(i)** of Lemma 2.4 implies that the condition $\mathbf{D}_1$ is satisfied.

In both cases, the first step is to prove that distributions of random variables $\tilde{\kappa}_\varepsilon(u)$ are tight, as $\varepsilon \to 0$, for some $u > 0$.

Let us choose some $0 < u < 1$.

By the definition, the processes $\kappa_\varepsilon(t)$, $\tilde{\kappa}_\varepsilon(t)$, and $\mu^*_{\varepsilon,i}(t), i \in \mathbb{X}$ are non-negative and non-decreasing. Taking this into account, we obtain, for any $x > 0$,

$$\begin{aligned} \mathsf{P}\{\tilde{\kappa}_\varepsilon(u) > x\} &\leq \mathsf{P}\{\tilde{\kappa}_\varepsilon(u) > x, \mu^*_{\varepsilon,i}(1) > u, i \in \mathbb{X}\} \\ &\quad + \sum_{i \in \mathbb{X}} \mathsf{P}\{\tilde{\kappa}_\varepsilon(u) > x, \mu^*_{\varepsilon,i}(1) \leq u\} \\ &\leq \mathsf{P}\{\kappa_\varepsilon(1) > x\} + \sum_{i \in \mathbb{X}} \mathsf{P}\{\mu^*_{\varepsilon,i}(1) \leq u\}. \end{aligned} \tag{2.69}$$

The asymptotic relation appearing in the statement **(i)** of Theorem 2.2, the relation (2.24), and the inequality (2.69) imply that

$$\begin{aligned} \lim_{x \to \infty} \overline{\lim_{\varepsilon \to 0}} \mathsf{P}\{\tilde{\kappa}_\varepsilon(u) > x\} &\leq \lim_{x \to \infty} \overline{\lim_{\varepsilon \to 0}} (\mathsf{P}\{\kappa_\varepsilon(1) > x\} \\ &\quad + \sum_{i \in \mathbb{X}} \mathsf{P}\{\mu^*_{\varepsilon,i}(1) \leq u\}) \leq \lim_{x \to \infty} \mathsf{P}\{\theta_0 > x/2\} = 0. \end{aligned} \tag{2.70}$$

Note that the necessity statement requires the fulfilment of the asymptotic relation appearing in the statement **(i)** of Theorem 2.2 only for at least one family of initial distributions $\bar{q}_\varepsilon, \varepsilon \in (0, \varepsilon_0]$.

The asymptotic relation appearing in the statement **(i)** of Lemma 2.4 implies that

$$\begin{aligned} \lim_{x \to \infty} \overline{\lim_{\varepsilon \to 0}} \mathsf{P}\{\tilde{\kappa}_\varepsilon(u) > x\} &\leq \lim_{x \to \infty} \overline{\lim_{\varepsilon \to 0}} \mathsf{P}\{\tilde{\kappa}_\varepsilon(1) > x\} \\ &\leq \lim_{x \to \infty} \mathsf{P}\{\theta_0 > x/2\} = 0. \end{aligned} \tag{2.71}$$

The relation (2.70), as well as the relation (2.71), implies that the distributions of random variables $\tilde{\kappa}_\varepsilon(u)$ are tight and, thus, relatively compact as $\varepsilon \to 0$.

Now, we can repeat the part of the above proof related to the relations (2.52)–(2.64).

The relation (2.63) and the asymptotic relation appearing in the statement **(i)** of Theorem 2.2, as well as the relation (2.58) and the asymptotic relation appearing in the statement **(i)** of Lemma 2.4, imply that the random variables $\theta'(1)$ and θ_0, which appear in the above asymptotic relations, have the same distribution,

$$\theta'(1) \stackrel{\mathsf{d}}{=} \theta_0. \tag{2.72}$$

Moreover, the cumulant $A(s)$ of the limiting Lévy process $\theta'_0(t)$ coincides with the cumulant of the random variable θ_0, which, therefore, has an infinitely divisible distribution. Moreover, the relation (2.64) implies that the cumulant $A(s)$ is connected with the cumulants $A_i(s)$, $i \in \mathbb{X}$ of Lévy processes $\theta'_{0,i}(t)$ by the relation $A(s) = \sum_{i \in \mathbb{X}} A_i(s)$, $s \geq 0$.

Thus, the limiting process $\theta'_0(t), t \geq 0 = \sum_{i \in \mathbb{X}} \theta_{0,i}(t)$, $t \geq 0$ has the same finite-dimensional distributions for all subsequences ε_{n_k} described above.

This allows us again to write the relations (2.66)–(2.68).

The relation (2.68) proves, in this case, that the condition $\mathbf{D}_1$ is satisfied.

The relation (2.66) proves the statement **(iii)** of Lemma 2.4.

The relation (2.67) proves the statement **(iii)** of Theorem 2.2. □

2.2.2 Examples of Step-Sum Reward Processes

Let us consider a particular case of the model, with random variables $\kappa_{\varepsilon,n} = f_{\varepsilon,\eta_{\varepsilon,n-1}}, n = 1, 2, \ldots, i \in \mathbb{X}$, where $f_{\varepsilon,i} \geq 0$, $i \in \mathbb{X}$ are nonrandom Non-negative numbers. In this case, the stochastic process $\kappa_\varepsilon(t)$ takes the following form:

$$\kappa_\varepsilon(t) = \sum_{n=1}^{[tu_\varepsilon]} f_{\varepsilon,\eta_{\varepsilon,n-1}}, t \geq 0. \tag{2.73}$$

Also, the Laplace transforms $\phi_{\varepsilon,i}(s)$ and $\phi_\varepsilon(s)$ take the following forms:

$$\phi_{\varepsilon,i}(s) = \mathsf{E}_i e^{-sf_{\varepsilon,\eta_{\varepsilon,0}}} = e^{-sf_{\varepsilon,i}}, s \geq 0, \text{ for } i \in \mathbb{X}$$

and

$$\phi_\varepsilon(s) = \sum_{i \in \mathbb{X}} \pi_{\varepsilon,i} e^{-sf_{\varepsilon,i}}, s \geq 0.$$

The condition $\mathbf{D}_2$ takes, in this case, the form of the following relation:

$$\begin{aligned} u_\varepsilon(1 - \phi_\varepsilon(s)) &= \sum_{i \in \mathbb{X}} \pi_{\varepsilon,i} u_\varepsilon(1 - e^{-sf_\varepsilon(i)}) \\ &\longrightarrow A(s) \text{ as } \varepsilon \longrightarrow 0, \text{ for } s > 0, \end{aligned} \tag{2.74}$$

where the limiting function $A(s) > 0$, for $s > 0$ and $A(s) \longrightarrow 0$ as $s \longrightarrow 0$.

This condition obviously implies that $1 - \phi_{\varepsilon,i}(s) \to 0$ as $\varepsilon \to 0$, for $s > 0, i \in \mathbb{X}$ that is equivalent to the relation, $f_{\varepsilon,i} \to 0$ as $\varepsilon \to 0$, for $i \in \mathbb{X}$. In this case, $1 - \phi_{\varepsilon,i}(s) = s f_{\varepsilon,i} + o(s f_{\varepsilon,i})$ as $\varepsilon \to 0$, for every $s > 0, i \in \mathbb{X}$. These relations allow us to reformulate the condition $\mathbf{D}_2$ for the above model in terms of functions,

$$f_\varepsilon = u_\varepsilon \sum_{i \in \mathbb{X}} \pi_{\varepsilon,i} f_{\varepsilon,i}.$$

The condition $\mathbf{D}_2$ is equivalent to the following condition:

$\mathbf{D}_4$: $f_\varepsilon \to f_0 \in (0, \infty)$ as $\varepsilon \to 0$.

Moreover, in this case the cumulant $A(s) = f_0 s, s \geq 0$.

Theorem 2.2 takes the following form.

Lemma 2.5 *Let the condition* $\mathbf{B}_1$ *holds. Then*:

(i) *The condition* $\mathbf{D}_4$ *is the necessary and sufficient condition for the fulfilment (for some or any initial distributions* $\bar{q}_\varepsilon$, *respectively, in the statements of necessity and sufficiency) of the asymptotic relation,* $\kappa_\varepsilon(1) \xrightarrow{\mathsf{d}} \theta_0$ *as* $\varepsilon \to 0$, *where* θ_0 *is a non-negative random variable with distribution function not concentrated at zero.*

(ii) *The limiting random variable* $\theta_0 \stackrel{\mathsf{d}}{=} f_0$, *i.e., it is a constant.*

(iii) *The stochastic processes* $\kappa_\varepsilon(t), t \geq 0 \xrightarrow{\mathsf{J}} f_0 t, t \geq 0$ *as* $\varepsilon \to 0$.

Let us assume that function f_ε satisfies the following natural assumption:

$\mathbf{D}_5$: There exists $\varepsilon'_0 \in (0, \varepsilon_0]$ such that $f_\varepsilon > 0$ for $\varepsilon \in (0, \varepsilon'_0]$.

In this case, we can describe the asymptotic behaviour of the step-sum processes $\kappa_\varepsilon(t)$ under a condition weaker than $\mathbf{D}_4$:

$\mathbf{D}_6$: $f_\varepsilon \to f_0 \in [0, \infty]$ as $\varepsilon \to 0$.

The following lemma generalises and complements Lemma 2.5.

Lemma 2.6 *Let the conditions* $\mathbf{B}_1$ *and* $\mathbf{D}_5$ *hold. Then*:

(i) $f_\varepsilon^{-1}\kappa_\varepsilon(t), t \geq 0 \xrightarrow{\mathsf{J}} g_0(t) = t, t \geq 0$ *as* $\varepsilon \to 0$.

(ii) *The condition* $\mathbf{D}_6$ *is the necessary and sufficient condition for the fulfilment (for some or any initial distributions* $\bar{q}_\varepsilon$, *respectively, in statements of necessity and sufficiency) of the asymptotic relation,* $\kappa_\varepsilon(1) \xrightarrow{\mathsf{d}} \theta_0$ *as* $\varepsilon \to 0$, *where* θ_0 *is a non-negative proper or improper random variable.*

(iii) *The limiting random variable* $\theta_0 \stackrel{\mathsf{d}}{=} f_0$, *i.e., it is a constant, and, moreover*:

(iv) $\kappa_\varepsilon(t) \xrightarrow{\mathsf{P}} f_0 t$ *as* $\varepsilon \to 0$, *for every* $t > 0$.

(v) *If* $f_0 \in [0, \infty)$, *then* $\kappa_\varepsilon(t), t \geq 0 \xrightarrow{\mathsf{J}} f_0 t, t \geq 0$ *as* $\varepsilon \to 0$.

(vi) *If* $f_0 = \infty$, *then* $\min(T, \kappa_\varepsilon(t)), t > 0 \xrightarrow{\mathsf{J}} h_T(t) = T, t > 0$ *as* $\varepsilon \to 0$, *for every* $T > 0$.

Proof We can use the following representation,

$$\kappa_\varepsilon(t) = \sum_{i \in \mathbb{X}} \mu^*_{\varepsilon,i}(t) u_\varepsilon \pi_{\varepsilon,i} f_{\varepsilon,i},\ t \geq 0, \tag{2.75}$$

where processes $\mu^*_{\varepsilon,i}(t), t \geq 0, i \in \mathbb{X}$ are defined in relation (2.24).

For any sequence $0 < \varepsilon_n \longrightarrow 0$ as $n \longrightarrow \infty$, there exists a subsequence $0 < \varepsilon_{n_k} \longrightarrow 0$ as $k \longrightarrow \infty$ such that

$$\frac{u_{\varepsilon_{n_k}} \pi_{\varepsilon_{n_k},i} f_{\varepsilon_{n_k},i}}{f_{\varepsilon_{n_k}}} \longrightarrow g_i \in [0, 1] \text{ as } k \longrightarrow \infty, \text{ for } i \in \mathbb{X}. \tag{2.76}$$

The constants $g_i, i \in \mathbb{X}$ can depend on the choice of subsequence ε_{n_k} but, obviously, satisfy the following relation:

$$\sum_{i \in \mathbb{X}} g_i = 1. \tag{2.77}$$

Since the limiting processes in relations (2.24) are nonrandom functions, the relations (2.24) and (2.76) obviously imply that

$$f^{-1}_{\varepsilon_{n_k}} \kappa_{\varepsilon_{n_k}}(t), t \geq 0 \stackrel{\mathsf{d}}{\longrightarrow} \sum_{i \in \mathbb{X}} t g_i = t, t \geq 0 \text{ as } k \longrightarrow \infty. \tag{2.78}$$

Moreover, since the processes on the left hand side of the above relation are non-decreasing and the limiting function is continuous, the following relation (according to Remarks A.1 and A.2) holds:

$$f^{-1}_{\varepsilon_{n_k}} \kappa_{\varepsilon_{n_k}}(t), t \geq 0 \stackrel{\mathsf{J}}{\longrightarrow} \sum_{i \in \mathbb{X}} t g_i = t, t \geq 0 \text{ as } k \longrightarrow \infty. \tag{2.79}$$

Since the limiting process is the same for all subsequences ε_{n_k} described above, relation (2.79) implies that the following relation holds:

$$f^{-1}_\varepsilon \kappa_\varepsilon(t), t \geq 0 \stackrel{\mathsf{J}}{\longrightarrow} g_0(t) = t, t \geq 0 \text{ as } \varepsilon \longrightarrow 0. \tag{2.80}$$

The relation (2.80) implies that the random variables $f^{-1}_\varepsilon \kappa_\varepsilon(t) \stackrel{\mathsf{d}}{\longrightarrow} t$ as $\varepsilon \longrightarrow 0$, for every $t \geq 0$. This implies that the random variables $\kappa_\varepsilon(1) = f_\varepsilon \cdot (f^{-1}_\varepsilon \kappa_\varepsilon(1))$ can converge in distribution if and only if $f_\varepsilon \longrightarrow f_0 \in [0, \infty]$ as $\varepsilon \longrightarrow 0$. Moreover, in this case, the limiting (possibly improper) random variable is the constant f_0, and $\kappa_\varepsilon(t) \stackrel{\mathsf{P}}{\longrightarrow} f_0 t$ as $\varepsilon \longrightarrow 0$, for every $t \geq 0$.

If $f_0 \in [0, \infty)$, then the asymptotic relation appearing in the statement **(v)** can be obtained by applying Theorem A.6 to processes $\kappa_\varepsilon(t) = g_\varepsilon(f^{-1}_\varepsilon \kappa_\varepsilon(t)), t \geq 0$, which are compositions of processes $f^{-1}_\varepsilon \kappa_\varepsilon(t), t \geq 0$ and functions $g_\varepsilon(t) = f_\varepsilon t, t \geq 0$.

If $f_0 = \infty$, then the asymptotic relation appearing in the statement **(vi)** can be obtained by applying Theorem A.6 to processes $\min(T, \kappa_\varepsilon(t)) = h_{\varepsilon,T}(f^{-1}_\varepsilon \kappa_\varepsilon(t)), t >$

0, which are composition processes $f_\varepsilon^{-1}\kappa_\varepsilon(t), t > 0$ and functions $h_{\varepsilon,T}(t) = \min(T, f_\varepsilon t), t > 0$.

Let us now assume that the asymptotic relation appearing in the statement **(ii)** holds, but the condition $\mathbf{D}_6$ is not satisfied.

The relation $f_\varepsilon \not\to f_0 \in [0, \infty]$ as $\varepsilon \to 0$ holds if and only if there exist at least two subsequences $0 < \varepsilon'_n, \varepsilon''_n \to 0$ as $n \to \infty$ such that (a) $f_{\varepsilon'_n} \to f'_0 \in [0, \infty]$ as $n \to \infty$, (b) $f_{\varepsilon''_n} \to f''_0 \in [0, \infty]$ as $n \to \infty$, and (c) $f'_0 \neq f''_0$. In this case, $\kappa_{\varepsilon'_n}(1) \xrightarrow{\mathsf{P}} f'_0$ as $n \to \infty$ and $\kappa_{\varepsilon''_n}(1) \xrightarrow{\mathsf{P}} f''_0$ as $n \to \infty$. Therefore, the random variables $\kappa_\varepsilon(1)$ do not converge in distribution. □

2.3 Asymptotics of First-Rare-Event Times for Perturbed Markov Chains

In this section, we give necessary and sufficient conditions of convergence in distribution and convergence in J-topology for first-rare-event times and processes defined on asymptotically uniformly ergodic Markov chains.

2.3.1 Convergence in Distribution for First-Rare-Event Times for Perturbed Markov Chains

The following lemma describes the asymptotics for the first-rare-event times ν_ε for the Markov chains $\eta_{\varepsilon,n}$.

Note that in this section, we use the function $u_\varepsilon = p_\varepsilon^{-1}$.

Lemma 2.7 *Let the conditions* $\mathbf{A}_1$ *and* $\mathbf{B}_1$ *be satisfied. Then, the random variables,*

$$\nu_\varepsilon^* = p_\varepsilon \nu_\varepsilon \xrightarrow{\mathsf{d}} \nu_0 \text{ as } \varepsilon \to 0, \tag{2.81}$$

where ν_0 *is a random variable exponentially distributed with parameter* 1.

Proof Let us define probabilities, for $\varepsilon \in (0, \varepsilon_0]$,

$$P_{\varepsilon,ij} = \mathsf{P}_i\{\eta_{\varepsilon,1} = j, \rho_{\varepsilon,1} = 0\}, \ \tilde{p}_{\varepsilon,ij} = \frac{P_{\varepsilon,ij}}{\sum_{r \in \mathbb{X}} P_{\varepsilon,ir}} = \frac{P_{\varepsilon,ij}}{1 - p_{\varepsilon,i}}, \ i, j \in \mathbb{X}.$$

Let also $\tilde{\eta}_{\varepsilon,n}, n = 0, 1, \ldots$ be a homogeneous Markov chain with phase space $\mathbb{X}$, initial distribution $\bar{q}_\varepsilon = \langle q_{\varepsilon,i}, i \in \mathbb{X}\rangle$, and the matrix of transition probabilities $\|\tilde{p}_{\varepsilon,ij}\|$.

The following relation holds, for $t \geq 0$,

$$
\begin{aligned}
\mathsf{P}\{\nu_\varepsilon^* > t\} &= \sum_{i\in\mathbb{X}} q_{\varepsilon,i} \sum_{i=i_0,i_1,\ldots,i_{[tu_\varepsilon]}\in\mathbb{X}} \prod_{k=1}^{[tu_\varepsilon]} P_{\varepsilon,i_{k-1}i_k} \\
&= \sum_{i\in\mathbb{X}} q_{\varepsilon,i} \sum_{i=i_0,i_1,\ldots,i_{[tu_\varepsilon]}\in\mathbb{X}} \prod_{k=1}^{[tu_\varepsilon]} \tilde{p}_{\varepsilon,i_{k-1}i_k}(1-p_{\varepsilon,i_{k-1}}) \\
&= \mathsf{E}\exp\{-\sum_{k=1}^{[tu_\varepsilon]} -\ln(1-p_{\varepsilon,\tilde{\eta}_{\varepsilon,k-1}})\}.
\end{aligned} \tag{2.82}
$$

The conditions $\mathbf{A}_1$ and $\mathbf{B}_1$ imply that the condition $\mathbf{B}_2$ holds for the transition probabilities of the Markov chains $\tilde{\eta}_{\varepsilon,n}$, since the following relation holds, for $i, j \in \mathbb{X}$,

$$
\begin{aligned}
|p_{\varepsilon,ij} - \tilde{p}_{\varepsilon,ij}| &= \frac{|p_{\varepsilon,ij}(1-p_{\varepsilon,i}) - P_{\varepsilon,ij}|}{1-p_{\varepsilon,i}} \\
&= \frac{|\mathsf{P}_i\{\eta_{\varepsilon,1} = j, \rho_{\varepsilon,1} = 0\} - p_{\varepsilon,ij}p_{\varepsilon,i}|}{1-p_{\varepsilon,i}} \\
&\leq \frac{2p_{\varepsilon,i}}{1-p_{\varepsilon,i}} \to 0 \text{ as } \varepsilon \to 0.
\end{aligned} \tag{2.83}
$$

Thus, by Lemma 2.1, there exists $\tilde{\varepsilon}_0 \in (0, \varepsilon_0]$ such that the Markov chain $\tilde{\eta}_{\varepsilon,n}$ is ergodic, for every $\varepsilon_0 \in (0, \tilde{\varepsilon}_0]$, and its stationary probabilities $\tilde{\pi}_{\varepsilon,i}, i \in \mathbb{X}$ satisfy the following relation:

$$
\tilde{\pi}_{\varepsilon,i} - \pi_{\varepsilon,i} \to 0 \text{ as } \varepsilon \to 0, \text{ for } i \in \mathbb{X}. \tag{2.84}
$$

We can apply Lemma 2.5, which is a special case of Theorem 2.2, to non-negative step-sum processes,

$$
\kappa_\varepsilon^*(t) = \sum_{n=1}^{[tu_\varepsilon]} -\ln(1-p_{\varepsilon,\tilde{\eta}_{\varepsilon,n-1}}), t \geq 0. \tag{2.85}
$$

To do this, we check that the condition $\mathbf{D}_4$ is satisfied for the functions $f_\varepsilon(i) = -\ln(1-p_{\varepsilon,i}), i \in \mathbb{X}$. Indeed, using the condition $\mathbf{A}_1$, $\mathbf{B}_1$, Lemma 2.1, and the relation (2.84), we obtain

$$
\begin{aligned}
f_\varepsilon &= -u_\varepsilon \sum_{i\in\mathbb{X}} \tilde{\pi}_{\varepsilon,i} \ln(1-p_{\varepsilon,i}) \sim u_\varepsilon \sum_{i\in\mathbb{X}} \tilde{\pi}_{\varepsilon,i}\, p_{\varepsilon,i} \\
&\sim u_\varepsilon \sum_{i\in\mathbb{X}} \pi_{\varepsilon,i}\, p_{\varepsilon,i} = u_\varepsilon p_\varepsilon = 1 \text{ as } \varepsilon \to 0.
\end{aligned} \tag{2.86}
$$

This relation is a variant of the condition $\mathbf{D}_4$. In this case, the corresponding limiting constant $f_0 = 1$. By applying the sufficiency statement of Lemma 2.5 to the step-sum processes $\kappa_\varepsilon^*(t)$, we obtain the following relation, for $t \geq 0$,

$$\kappa^*_\varepsilon(t) \xrightarrow{\mathsf{d}} t \text{ as } \varepsilon \to 0. \tag{2.87}$$

The expression on the right hand side of the relation (2.82) is the Laplace transform of the non-negative random variable $\kappa^*_\varepsilon(t)$ at point 1. Thus, the relation (2.87) implies, by the continuity theorem for Laplace transforms, that the following relation holds, for $t \geq 0$,

$$\mathsf{P}\{\nu^*_\varepsilon > t\} = \mathsf{E}e^{-\kappa^*_\varepsilon(t)} \to e^{-t} \text{ as } \varepsilon \to 0. \tag{2.88}$$

The proof is complete. □

2.3.2 J-Convergence for First-Rare-Event Time Processes for Perturbed Markov Chains

Let, as in Lemmas 2.5 and 2.6, $f_{\varepsilon,i}, i \in \mathbb{X}$ be nonrandom non-negative numbers and

$$f_\varepsilon = u_\varepsilon \sum_{i \in \mathbb{X}} \pi_{\varepsilon,i} f_{\varepsilon,i} = p_\varepsilon^{-1} \sum_{i \in \mathbb{X}} \pi_{\varepsilon,i} f_{\varepsilon,i}. \tag{2.89}$$

Let us introduce stochastic processes,

$$\nu_\varepsilon(t) = \sum_{n=1}^{[t\nu_\varepsilon]} f_{\varepsilon,\eta_{\varepsilon,n-1}}, t \geq 0. \tag{2.90}$$

The following lemma generalises Lemma 2.7 and is used in what follows.

Lemma 2.8 *Let the conditions* $\mathbf{A}_1$, $\mathbf{B}_1$, *and* $\mathbf{D}_5$ *be satisfied. Then*:

(i) $f_\varepsilon^{-1}\nu_\varepsilon(t), t \geq 0 \xrightarrow{\mathsf{J}} t\nu_0, t \geq 0$ *as* $\varepsilon \to 0$, *where* ν_0 *is a random variable exponentially distributed with parameter* 1.

(ii) *The condition* **I** *is the necessary and sufficient condition for the fulfilment (for some or any initial distributions* $\bar{q}_\varepsilon$, *respectively, in the statements of necessity and sufficiency) of the asymptotic relation,* $\nu_\varepsilon(1) \xrightarrow{\mathsf{d}} \nu$ *as* $\varepsilon \to 0$, *where* ν *is a non-negative random variable with distribution not concentrated at zero.*

(iii) *The limiting random variable* $\nu \stackrel{\mathsf{d}}{=} f_0\nu_0$.

(iv) *If* $f_0 \in [0, \infty)$, *then* $\nu_\varepsilon(t), t \geq 0 \xrightarrow{\mathsf{J}} f_0\nu_0\, t, t \geq 0$ *as* $\varepsilon \to 0$.

(v) *If* $f_0 = \infty$, *then* $\min(T, \nu_\varepsilon(t)), t > 0 \xrightarrow{\mathsf{J}} h_T(t) = T, t > 0$ *as* $\varepsilon \to 0$, *for every* $T > 0$ *and, thus,* $\nu_\varepsilon(t) \xrightarrow{\mathsf{P}} \infty$ *as* $\varepsilon \to 0$, *for* $t > 0$.

Proof The following representation takes place,

$$\nu_\varepsilon(t) = \kappa_\varepsilon(t\nu^*_\varepsilon), t \geq 0, \tag{2.91}$$

where $\kappa_\varepsilon(t)$ are the processes defined in the relation (2.73).

The relations appearing in the statement **(i)** of Lemma 2.6 and in Lemma 2.7 imply, by Theorem A.9, that the following relation holds,

$$(t\nu^*_\varepsilon, f_\varepsilon^{-1}\kappa_\varepsilon(t)),\ t \geq 0 \xrightarrow{\mathsf{d}} (t\nu_0, t),\ t \geq 0 \text{ as } \varepsilon \to 0. \tag{2.92}$$

The components of the processes on the left hand side of the relation (2.92) are non-decreasing processes, and the process on the right hand side of the relation (2.92) is continuous. This let us apply Theorem A.6 to the processes $f_\varepsilon^{-1}\nu_\varepsilon(t) = f_\varepsilon^{-1}\kappa_\varepsilon(t\nu^*_\varepsilon), t \geq 0$ and to obtain the asymptotic relation appearing in the statement **(i)** of Lemma 2.8.

The asymptotic relation appearing in the statement **(i)** of Lemma 2.8 implies that the random variables

$$f_\varepsilon^{-1}\nu_\varepsilon(1) \xrightarrow{\mathsf{d}} \nu_0 \text{ as } \varepsilon \to 0. \tag{2.93}$$

This relation implies that the random variables $\nu_\varepsilon(1) = f_\varepsilon \cdot (f_\varepsilon^{-1}\nu_\varepsilon(1))$ can converge in distribution if and only if $f_\varepsilon \to f_0 \in [0, \infty]$ as $\varepsilon \to 0$. Moreover, in this case, the limiting (possibly improper) random variable $\nu \stackrel{\mathsf{d}}{=} f_0\nu_0$.

If $f_0 \in [0, \infty)$, then relations appearing in the statement **(iv)** of Lemma 2.6 and in Lemma 2.7 imply, by Theorem A.9, that the following relation holds:

$$(t\nu^*_\varepsilon, \kappa_\varepsilon(t)), t \geq 0 \xrightarrow{\mathsf{d}} (t\nu_0, f_0 t), t \geq 0 \text{ as } \varepsilon \to 0. \tag{2.94}$$

The components of the processes on the left hand side of the relation (2.94) are non-decreasing processes and the process on the right hand side of the relation (2.92) is continuous. This let us apply Theorem A.6 to processes $\nu_\varepsilon(t) = \kappa_\varepsilon(t\nu^*_\varepsilon), t \geq 0$ and to obtain the asymptotic relation appearing in the statement **(iv)** of Lemma 2.8.

If $f_0 = \infty$, then relations appearing in the statement **(v)** of Lemma 2.6 and in Lemma 2.7 imply, by Theorem A.9, that the following relation holds:

$$(t\nu^*_\varepsilon, \min(T, \kappa_\varepsilon(t))), t > 0 \xrightarrow{\mathsf{d}} (t\nu_0, T), t > 0 \text{ as } \varepsilon \to 0, \text{ for } T > 0. \tag{2.95}$$

The components of the processes on the left hand side of the relation (2.95) are non-decreasing processes and the process on the right hand side of the relation (2.92) is continuous. Also the limiting random variable $t\nu_0 > 0$ with probability 1, for every $t > 0$. This let us apply Theorem A.6 to processes $\min(T, \nu_\varepsilon(t)) = \min(T, \kappa_\varepsilon(t\nu^*_\varepsilon)), t > 0$ and to obtain the asymptotic relation given in appearing in the statement **(v)** of Lemma 2.8.

The statement **(vi)** of this lemma is a direct corollary of the statement **(v)**.

Let us now assume that the asymptotic relation appearing in the statement **(ii)** holds, but the condition $\mathbf{D}_6$ is not satisfied.

The relation $f_\varepsilon \not\to f_0 \in [0, \infty]$ as $\varepsilon \to 0$ holds if and only if there exist at least two subsequences $0 < \varepsilon'_n, \varepsilon''_n \to 0$ as $n \to \infty$ such that (a) $f_{\varepsilon'_n} \to f'_0 \in [0, \infty]$ as $n \to \infty$, (b) $f_{\varepsilon''_n} \to f''_0 \in [0, \infty]$ as $n \to \infty$ and (c) $f'_0 \neq f''_0$. In this case, $\nu_{\varepsilon'_n}(1) \xrightarrow{\mathsf{d}} f'_0\nu_0$ as

$n \to \infty$ and $\nu_{\varepsilon''_n}(1) \xrightarrow{\mathsf{d}} f''_0 \nu_0$ as $n \to \infty$ and, thus, the random variables $\nu_\varepsilon(1)$ do not converge in distribution. □

2.4 Asymptotics of First-Rare-Event Times for Perturbed Semi-Markov Processes

In this section, we give the necessary and sufficient conditions of convergence in distribution and convergence in the topology J for step-sum reward processes defined on semi-Markov processes with asymptotically uniformly ergodic embedded Markov chains. Also, we complete the proof of Theorem 2.1.

2.4.1 Convergence in Distribution for First-Rare-Event Times for Perturbed Semi-Markov Processes

We are now ready to complete the proof of Theorem 2.1. First of all, we focus on the statements **(i)** and **(ii)** of Theorem 2.1.

Let us introduce Laplace transforms,

$$\phi_{\varepsilon,ij}(\imath, s) = \mathsf{E}_i \mathrm{I}(\eta_{\varepsilon,1} = j, \rho_{\varepsilon,1} = \imath) e^{-s\kappa_{\varepsilon,1}}, \ s \geq 0, \ \text{for } i, j \in \mathbb{X}, \ \imath = 0, 1,$$

$$\phi_{\varepsilon,i}(\imath, s) = \mathsf{E}_i \mathrm{I}(\rho_{\varepsilon,1} = \imath) e^{-s\kappa_{\varepsilon,1}}, \ s \geq 0, \ \text{for } i \in \mathbb{X}, \ \imath = 0, 1,$$

$$\phi_{\varepsilon,ij}(\imath, s) = \mathsf{E}_i \{\mathrm{I}(\eta_{\varepsilon,1} = j) e^{-s\kappa_{\varepsilon,1}} / \rho_{\varepsilon,1} = \imath\}, \ s \geq 0, \ \text{for } i, j \in \mathbb{X}, \ \imath = 0, 1,$$

and

$$\phi_{\varepsilon,i}(\imath, s) = \mathsf{E}_i \{e^{-s\kappa_{\varepsilon,1}} / \rho_{\varepsilon,1} = \imath\}, \ s \geq 0, \ \text{for } i \in \mathbb{X}, \ \imath = 0, 1.$$

Now, let us define probabilities, for $s \geq 0$,

$$p_{\varepsilon,s,ij} = \frac{\phi_{\varepsilon,ij}(0, s)}{\sum_{r \in \mathbb{X}} \phi_{\varepsilon,ij}(0, s)} = \frac{\phi_{\varepsilon,ij}(0, s)}{\phi_{\varepsilon,i}(0, s)}, \ i, j \in \mathbb{X}.$$

Let $(\eta_{\varepsilon,s,n}, \rho_{\varepsilon,s,n}), n = 0, 1, \ldots$ be, for every $s \geq 0$, a Markov renewal process, with the phase space $\mathbb{X} \times \{0, 1\}$, an initial distribution $\bar{q}_\varepsilon = \langle q_{\varepsilon,i} = \mathsf{P}\{\eta_{\varepsilon,s,0} = i, \rho_{\varepsilon,s,0} = 0\} = \mathsf{P}\{\eta_{\varepsilon,s,0} = i\}, i \in \mathbb{X}\rangle$, and transition probabilities,

$$\begin{aligned} \mathsf{P}\{\eta_{\varepsilon,s,n+1} &= j, \rho_{\varepsilon,s,n+1} = \jmath / \eta_{\varepsilon,s,n} = i, \rho_{\varepsilon,s,n} = \imath\} \\ &= \mathsf{P}\{\eta_{\varepsilon,s,n+1} = j, \rho_{\varepsilon,s,n+1} = \jmath / \eta_{\varepsilon,s,n} = i\} \\ &= p_{\varepsilon,s,ij}(p_{\varepsilon,i}\, \jmath + (1 - p_{\varepsilon,i})(1 - \jmath)), \ i, j \in \mathbb{X}, \ \imath, \jmath = 0, 1. \end{aligned} \tag{2.96}$$

Note that the first component of the Markov renewal process, $\eta_{\varepsilon,s,n}, n = 0, 1, \ldots$, is a homogeneous Markov chain with phase space $\mathbb{X}$, initial distribution $\bar{q}_\varepsilon = \langle q_{\varepsilon,i}, i \in \mathbb{X}\rangle$, and the matrix of transition probabilities $\|p_{\varepsilon,s,ij}\|$.

Let us also introduce random variables,

$$\nu_{\varepsilon,s} = \min(n \geq 1 : \rho_{\varepsilon,s,n} = 1). \tag{2.97}$$

Let us prove that the condition $\mathbf{D}_1$ or the conditions $\mathbf{A}_1$, $\mathbf{B}_1$ and the asymptotic relation appearing in the statement **(i)** of Theorem 2.1 imply that the condition $\mathbf{B}_1$ holds for the transition probabilities of the Markov chain $\eta_{\varepsilon,s,n}$, for every $s \geq 0$.

The condition $\mathbf{D}_2$, which is equivalent to the condition $\mathbf{D}_1$, obviously, implies that, for $i \in \mathbb{X}$,

$$\phi_{\varepsilon,i}(s) \longrightarrow 1 \text{ as } \varepsilon \to 0, \text{ for } s \geq 0. \tag{2.98}$$

Conditions $\mathbf{A}_1$, $\mathbf{B}_1$ and the asymptotic relation penetrating proposition **(i)** of Theorem 2.1 also imply that relation (2.98) holds.

Indeed, the following representation takes place,

$$\xi_\varepsilon = \sum_{i\in\mathbb{X}} \sum_{n=1}^{\mu_{\varepsilon,i}(\nu_\varepsilon)} \kappa_{\varepsilon,i,n} = \sum_{i\in\mathbb{X}} \sum_{n=1}^{[\mu^*_{\varepsilon,i}(\nu_\varepsilon)\pi_{\varepsilon,i}u_\varepsilon]} \kappa_{\varepsilon,i,n}. \tag{2.99}$$

Suppose that relation (2.98) does not hold. This means that there exists $i \in \mathbb{X}$ such that for some $\delta, p > 0$ and $\varepsilon_{\delta,p} \in (0, \varepsilon_0]$, the probability $\mathsf{P}\{\kappa_{\varepsilon,i,1} \geq \delta\} \geq p$, for $\varepsilon \in (0, \varepsilon_{\delta,p}]$. This obviously implies that, for $t > 0$,

$$\tilde{\kappa}_{\varepsilon,i}(t) = \sum_{n=1}^{[t\pi_{\varepsilon,i}u_\varepsilon]} \kappa_{\varepsilon,i,n} \stackrel{\mathsf{P}}{\longrightarrow} \infty \text{ as } \varepsilon \to 0. \tag{2.100}$$

The relation (2.100) implies that

$$\min(T, \tilde{\kappa}_{\varepsilon,i}(t)), t > 0 \stackrel{\mathsf{d}}{\longrightarrow} h_T(t) = T, t > 0 \text{ as } \varepsilon \to 0. \tag{2.101}$$

Since the processes $\tilde{\kappa}_{\varepsilon,i}(t), t > 0$ are non-decreasing and the limiting function $h_T(t) = T, t > 0$, is continuous, the relation (2.101) implies (according to Remark A.1) that

$$\min(T, \tilde{\kappa}_{\varepsilon,i}(t)), t > 0 \stackrel{\mathsf{J}}{\longrightarrow} h_T(t) = T, t \geq 0 \text{ as } \varepsilon \to 0. \tag{2.102}$$

From Lemma 2.8 (applied for the model with functions $f_{\varepsilon,j} = I(j = i)(\pi_{\varepsilon,i}u_\varepsilon)^{-1}$, $j \in \mathbb{X}$), it follows that the following relation holds:

$$\mu^*_{\varepsilon,i}(\nu^*_\varepsilon) \stackrel{\mathsf{d}}{\longrightarrow} \nu_0 \text{ as } \varepsilon \to 0, \tag{2.103}$$

where ν_0 is a random variable exponentially distributed with parameter 1.

The relations (2.102) and (2.103) imply, by Theorem A.9, that

$$(\mu^*_{\varepsilon,i}(\nu^*_\varepsilon), \min(T, \tilde{\kappa}_{\varepsilon,i}(t))), t > 0 \xrightarrow{\mathsf{d}} (\nu_0, h_T(t)), t > 0 \text{ as } \varepsilon \to 0. \tag{2.104}$$

Now we can apply Theorem A.4 to the stochastic processes $\min(T, \tilde{\kappa}_{\varepsilon,i}(t))), t > 0$ stopped at the random moments $\mu^*_{\varepsilon,i}(\nu^*_\varepsilon)$. This gives the following relation, for any $T > 0$,

$$\min(T, \tilde{\kappa}_{\varepsilon,i}(\mu^*_{\varepsilon,i}(\nu^*_\varepsilon))) \xrightarrow{\mathsf{d}} T \text{ as } \varepsilon \to 0. \tag{2.105}$$

This is only possible if

$$\tilde{\kappa}_{\varepsilon,i}(\mu^*_{\varepsilon,i}(\nu^*_\varepsilon))) \xrightarrow{\mathsf{P}} \infty \text{ as } \varepsilon \to 0. \tag{2.106}$$

The relation (2.106) implies that

$$\tilde{\kappa}_{\varepsilon,i}(\mu^*_{\varepsilon,i}(\nu^*_\varepsilon))) = \sum_{n=1}^{\mu_{\varepsilon,i}(\nu_\varepsilon)} \kappa_{\varepsilon,i,n} \leq \xi_\varepsilon \xrightarrow{\mathsf{P}} \infty \text{ as } \varepsilon \to 0. \tag{2.107}$$

This relation contradicts to the asymptotic relation appearing in the statement **(i)** of Theorem 2.1.

The relation (2.98) and the condition $\mathbf{A}_1$ imply that, for $s \geq 0, i, j \in \mathbb{X}$,

$$\phi_{\varepsilon,i}(s, 0) = \mathsf{E}_i \mathrm{I}(\rho_{\varepsilon,1} = 0) e^{-s\kappa_{\varepsilon,1}} \to 1 \text{ as } \varepsilon \to 0 \tag{2.108}$$

and, thus, for $s \geq 0, i, j \in \mathbb{X}$,

$$\begin{aligned} |p_{\varepsilon,ij} - p_{\varepsilon,s,ij}| &= \Big|\frac{p_{\varepsilon,ij}\phi_{\varepsilon,i}(0, s) - \phi_{\varepsilon,ij}(0, s)}{\phi_{\varepsilon,i}(0, s)}\Big| \\ &\leq \frac{|p_{\varepsilon,ij}\phi_{\varepsilon,i}(0, s) - p_{\varepsilon,ij}| + |p_{\varepsilon,ij} - \phi_{\varepsilon,ij}(0, s)|}{\phi_{\varepsilon,i}(0, s)} \\ &\leq \frac{p_{\varepsilon,ij}|\phi_{\varepsilon,i}(0, s) - 1| + \mathsf{E}_i \mathrm{I}(\eta_{\varepsilon,1} = j)|1 - \mathrm{I}(\rho_{\varepsilon,1} = 0)e^{-\kappa_{\varepsilon,1}})|}{\phi_{\varepsilon,i}(0, s)} \\ &\leq \frac{2(1 - \phi_{\varepsilon,i}(0, s))}{\phi_{\varepsilon,i}(0, s)} \to 0 \text{ as } \varepsilon \to 0. \end{aligned} \tag{2.109}$$

Relation (2.109) and Lemma 2.1 imply that, for every $s \geq 0$, there exists $\tilde{\varepsilon}_{0,s} \in (0, \varepsilon_0]$ such that the Markov chain $\tilde{\eta}_{\varepsilon,n,s}$ is ergodic, for every $\varepsilon \in (0, \tilde{\varepsilon}_{0,s}]$, and its stationary probabilities $\pi_{\varepsilon,s,i}, i \in \mathbb{X}$ satisfy the following relation,

$$\pi_{\varepsilon,s,i} - \pi_{\varepsilon,i} \to 0 \text{ as } \varepsilon \to 0, \text{ for } i \in \mathbb{X}. \tag{2.110}$$

Let us assume that the Markov chains $\eta_{\varepsilon,n}$ and $\eta_{\varepsilon,n,s}$ have the same initial distribution $\bar{q}_\varepsilon$.

The following representation takes place for the Laplace transform of the random variables ξ_ε, for $s \geq 0$,

$$
\begin{aligned}
\mathsf{E}e^{-s\xi_\varepsilon} &= \sum_{i \in \mathbb{X}} q_{\varepsilon,i} \sum_{n=1}^{\infty} \sum_{i=i_0,i_1,\ldots,i_n \in \mathbb{X}} \prod_{k=1}^{n-1} \phi_{\varepsilon,i_{k-1}i_k}(0,s)\phi_{\varepsilon,i_{n-1}i_n}(1,s) \\
&= \sum_{i \in \mathbb{X}} q_{\varepsilon,i} \sum_{n=1}^{\infty} \sum_{i=i_0,i_1,\ldots,i_{n-1} \in \mathbb{X}} \prod_{k=1}^{n-1} \phi_{\varepsilon,i_{k-1}i_k}(0,s) \sum_{i_n \in \mathbb{X}} \phi_{\varepsilon,i_{n-1}i_n}(1,s) \\
&= \sum_{i \in \mathbb{X}} q_{\varepsilon,i} \sum_{n=1}^{\infty} \sum_{i=i_0,i_1,\ldots,i_{n-1} \in \mathbb{X}} \prod_{k=1}^{n-1} p_{\varepsilon,s,i_{k-1}i_k} \\
&\quad \times (1 - p_{\varepsilon,i_{k-1}})\phi_{\varepsilon,i_{k-1}}(0,s) p_{\varepsilon,i_{n-1}} \sum_{i_n \in \mathbb{X}} \frac{\phi_{\varepsilon,i_{n-1},i_n}(1,s)}{p_{\varepsilon,i_{n-1}}} \\
&= \sum_{i \in \mathbb{X}} q_{\varepsilon,i} \sum_{n=1}^{\infty} \sum_{i=i_0,i_1,\ldots,i_{n-1} \in \mathbb{X}} \prod_{k=1}^{n-1} p_{\varepsilon,s,i_{k-1}i_k} \\
&\quad \times (1 - p_{\varepsilon,i_{k-1}})\phi_{\varepsilon,i_{k-1}}(0,s) p_{\varepsilon,i_{n-1}} \phi_{\varepsilon,i_{n-1}}(1,s) \\
&= \mathsf{E}\exp\{-\sum_{k=1}^{\nu_{\varepsilon,s}} - \ln \phi_{\varepsilon,\eta_{\varepsilon,s,k-1}}(0,s) \\
&\quad - \ln \phi_{\varepsilon,\eta_{\varepsilon,s,\nu_{\varepsilon,s}-1}}(0,s) + \ln \phi_{\varepsilon,\eta_{\varepsilon,s,\nu_{\varepsilon,s}-1}}(1,s)\}.
\end{aligned} \tag{2.111}
$$

The relation (2.108) and the condition $\mathbf{A}_1$ imply that the following relation holds:

$$
\phi_{\varepsilon,i}(0,s) = \frac{\phi_{\varepsilon,i}(0,s)}{1 - p_{\varepsilon,i}} \longrightarrow 1 \text{ as } \varepsilon \longrightarrow 0, \text{ for } s \geq 0, i \in \mathbb{X}. \tag{2.112}
$$

Also, the condition $\mathbf{C}_1$ is equivalent to the following relation:

$$
\begin{aligned}
\phi_{\varepsilon,i}(1,s) &= \mathsf{E}_i\{e^{-s\kappa_{\varepsilon,1}}/\rho_{\varepsilon,1} = 1\} \\
&\longrightarrow 1 \text{ as } \varepsilon \longrightarrow 0, \text{ for } s \geq 0, i \in \mathbb{X}.
\end{aligned} \tag{2.113}
$$

The relations (2.112) and (2.113) imply that, for $s \geq 0$,

$$
|\ln \phi_{\varepsilon,\eta_{\varepsilon,s,\nu_{\varepsilon,s}-1}}(0,s)| + |\ln \phi_{\varepsilon,\eta_{\varepsilon,s,\nu_{\varepsilon,s}-1}}(1,s)| \stackrel{\mathsf{P}}{\longrightarrow} 0 \text{ as } \varepsilon \longrightarrow 0. \tag{2.114}
$$

Let us introduce random variables, for $s > 0$,

$$
\tilde{\nu}_{\varepsilon,s} = \sum_{n=1}^{\nu_{\varepsilon,s}} - \ln \phi_{\varepsilon,\eta_{\varepsilon,s,k-1}}(0,s). \tag{2.115}
$$

The representation (2.111) and the relation (2.114) imply that, for $s > 0$,

$$
\mathsf{E}e^{-s\xi_\varepsilon} \sim \mathsf{E}e^{-\tilde{\nu}_{\varepsilon,s}} \text{ as } \varepsilon \longrightarrow 0. \tag{2.116}
$$

The relations (2.110), (2.112) and the statement **(i)** of Lemma 2.1 imply that, for $s > 0$,

$$\begin{aligned} A_\varepsilon(s) &= -u_\varepsilon \sum_{i\in\mathbb{X}} \pi_{\varepsilon,s,i} \ln \phi_{\varepsilon,i}(0,s) \\ &\sim u_\varepsilon \sum_{i\in\mathbb{X}} \pi_{\varepsilon,s,i}(1-\phi_{\varepsilon,i}(0,s)) \\ &\sim u_\varepsilon \sum_{i\in\mathbb{X}} \pi_{\varepsilon,i}(1-\phi_{\varepsilon,i}(0,s)) \text{ as } \varepsilon \to 0. \end{aligned} \tag{2.117}$$

Suppose that in addition to the conditions $\mathbf{A}_1$–$\mathbf{C}_1$, the condition $\mathbf{D}_1$ is satisfied.

The condition $\mathbf{D}_1$ is equivalent to the condition $\mathbf{D}_2$. Thus, due to the relations (2.112) and (2.113), the condition $\mathbf{A}_1$, and the statement **(i)** of Lemma 2.1, the condition $\mathbf{D}_2$ is equivalent to the following relation:

$$\begin{aligned} u_\varepsilon(1-\phi_\varepsilon(s)) &= u_\varepsilon \sum_{i\in\mathbb{X}} \pi_{\varepsilon,i}(1-\phi_{\varepsilon,i}(s)) \\ &= v_\varepsilon \sum_{i\in\mathbb{X}} \pi_{\varepsilon,i}(1-(1-p_{\varepsilon,i})\phi_{\varepsilon,i}(0,s) - p_{\varepsilon,i}\phi_{\varepsilon,i}(1,s)) \\ &= u_\varepsilon \sum_{i\in\mathbb{X}} \pi_{\varepsilon,i}((1-p_{\varepsilon,i})(1-\phi_{\varepsilon,i}(0,s)) + p_{\varepsilon,i}(1-\phi_{\varepsilon,i}(1,s)) \\ &\sim u_\varepsilon \sum_{i\in\mathbb{X}} \pi_{\varepsilon,i}(1-p_{\varepsilon,i})(1-\phi_{\varepsilon,i}(0,s)) \\ &\sim u_\varepsilon \sum_{i\in\mathbb{X}} \pi_{\varepsilon,i}(1-\phi_{\varepsilon,i}(0,s)) \to A(s) \text{ as } \varepsilon \to 0, \text{ for } s > 0, \end{aligned} \tag{2.118}$$

where $A(s) > 0$, for $s > 0$, and $A(s) \to 0$ as $s \to 0$.

The relations (2.117) and (2.118) imply that, in this case, for $s > 0$,

$$A_\varepsilon(s) = -u_\varepsilon \sum_{i\in\mathbb{X}} \pi_{\varepsilon,s,i} \ln \phi_{\varepsilon,i}(0,s) \to A(s) \text{ as } \varepsilon \to 0. \tag{2.119}$$

Now, we can apply the sufficiency part of the statement **(iv)** of Lemma 2.8 to the random variables $\tilde{\nu}_{\varepsilon,s}$, for every $s > 0$, This gives the following relation:

$$\tilde{\nu}_{\varepsilon,s} \xrightarrow{\mathsf{d}} A(s)\nu_0 \text{ as } \varepsilon \to 0, \text{ for } s > 0, \tag{2.120}$$

where ν_0 is an exponentially distributed random variable with parameter 1.

This relation implies, by the continuity theorem for Laplace transforms, the following relation,

$$\mathsf{E}e^{-s\xi_\varepsilon} \sim \mathsf{E}e^{-\tilde{\nu}_{\varepsilon,s}} \to \mathsf{E}e^{-A(s)\nu_0} = \frac{1}{1+A(s)} \text{ as } \varepsilon \to 0, \text{ for } s > 0. \tag{2.121}$$

The relation (2.121) proves sufficiency parts of the statements **(i)** and **(ii)** of Theorem 2.1.

Suppose now that the conditions $\mathbf{A}_1$–$\mathbf{C}_1$ are satisfied plus the statement **(i)** of Theorem 2.1 takes place.

The asymptotic relation appearing in the statement **(i)** of Theorem 2.1 can be expressed in terms of Laplace transforms. It takes the form of the following relation (which should be assumed to hold for some initial distributions $\bar{q}_\varepsilon$):

$$\mathsf{E}e^{-s\xi_\varepsilon} \to e^{-A_0(s)} \text{ as } \varepsilon \to 0, \text{ for } s > 0, \tag{2.122}$$

where $A_0(s) > 0$ for $s > 0$ and $A_0(s) \to 0$ as $s \to 0$.

Let us assume that the conditions $\mathbf{A}_1$–$\mathbf{C}_1$ are satisfied, but the condition $\mathbf{D}_1$ is not satisfied.

This means, due to the relation (2.122), that either (a) $A_\varepsilon(s) \to A(s) \in (0, \infty)$ as $s \to 0$, for every $s > 0$, but $A(s) \not\to 0$ as $\varepsilon \to 0$, or (b) $A_\varepsilon(s^*) \not\to A(s^*) \in (0, \infty)$ as $\varepsilon \to 0$, for some $s^* > 0$. The latter relation holds if and only if there exist at least two subsequences $0 < \varepsilon'_n, \varepsilon''_n \to 0$ as $n \to \infty$ such that (b$_1$) $A_{\varepsilon'_n}(s^*) \to A'(s^*) \in [0, \infty]$ as $n \to \infty$, (b$_2$) $A_{\varepsilon''_n}(s^*) \to A''(s^*) \in [0, \infty]$ as $n \to \infty$ and (b$_3$) $A''(s^*) < A'(s^*)$.

In the case (a), we can repeat the part of the above proof presented in the relations (2.117)–(2.126) and, taking into account the relation (2.122), obtain the relation, $\mathsf{E}e^{-s\xi_\varepsilon} \sim \mathsf{E}e^{-\tilde{\nu}_{\varepsilon,s}} \to \frac{1}{1+A(s)} = e^{-A_0(s)}$ as $\varepsilon \to 0$, for $s > 0$. This relation implies that $A(s) \to 0$ as $\varepsilon \to 0$, and thus, the case (a) is impossible.

In the case (b), the first sub-case, $A'(s^*) = \infty$, is impossible. Indeed, as was shown in the proof of Lemma 2.8 (applied to the random variables $\tilde{\nu}_{\varepsilon,s^*}$), in this case, $\tilde{\nu}_{\varepsilon'_n,s^*} \xrightarrow{\mathsf{P}} \infty$ as $n \to \infty$. Thus, $\mathsf{E}e^{-s^*\xi_{\varepsilon'_n}} \sim \mathsf{E}e^{-\tilde{\nu}_{\varepsilon'_n,s^*}} \to 0$ as $n \to \infty$. This relation contradicts to the relation (2.122).

The second sub-case, $A''(s^*) = 0$, is also impossible. Indeed, as was shown in the proof of Lemma 2.8 (applied to the random variables $\tilde{\nu}_{\varepsilon,s^*}$), in this case, $\tilde{\nu}_{\varepsilon''_n,s^*} \xrightarrow{\mathsf{P}} 0$ as $n \to \infty$, and, thus, $\mathsf{E}e^{-s^*\xi_{\varepsilon''_n}} \sim \mathsf{E}e^{-\tilde{\nu}_{\varepsilon''_n,s^*}} \to 1$ as $n \to \infty$. This relation also contradicts to the relation (2.122).

Finally, the third sub-case, $0 < A''(s^*) < A'(s^*) < \infty$, is also impossible. Indeed, the sufficiency statement of Lemma 2.7 applied to the random variables $\tilde{\nu}_{\varepsilon,s^*}$ yields, in this case, two relations $\tilde{\nu}_{\varepsilon'_n,s^*} \xrightarrow{\mathsf{d}} A'(s^*)\nu_0$ as $n \to \infty$ and $\tilde{\nu}_{\varepsilon''_n,s^*} \xrightarrow{\mathsf{d}} A''(s^*)\nu_0$ as $n \to \infty$, where ν_0 is exponentially distributed random variable with parameter 1. These relations imply that $\mathsf{E}e^{-s^*\xi_{\varepsilon'_n}} \sim \mathsf{E}e^{-\tilde{\nu}_{\varepsilon'_n,s^*}} \to \frac{1}{1+A'(s^*)}$ as $n \to \infty$ and $\mathsf{E}e^{-s^*\xi_{\varepsilon''_n}} \sim \mathsf{E}e^{-\tilde{\nu}_{\varepsilon''_n,s^*}} \to \frac{1}{1+A''(s^*)}$ as $n \to \infty$. These relations contradict to the relation (2.122), since $\frac{1}{1+A'(s^*)} \neq \frac{1}{1+A''(s^*)}$.

Therefore, the condition $\mathbf{D}_1$ should hold. This completes the proof of the statements **(i)** and **(ii)** of Theorem 2.1.

2.4.2 J-Convergence for First-Rare-Event Processes for Perturbed Semi-Markov Process

The following lemma combines the asymptotic relations given in Theorem 2.1 and Lemma 2.7. This lemma gives the last intermediate result necessary to complete the proof of the statement **(iii)** in Theorem 2.1.

Lemma 2.9 *Let the conditions* $\mathbf{A}_1$, $\mathbf{B}_1$, $\mathbf{C}_1$, *and* $\mathbf{D}_1$ *be satisfied. Then, the following asymptotic relation holds:*

$$(\nu_\varepsilon^*, \kappa_\varepsilon(t)), t \geq 0 \stackrel{\mathsf{d}}{\longrightarrow} (\nu_0, \theta_0(t)), t \geq 0 \text{ as } \varepsilon \to 0, \tag{2.123}$$

where: **(a)** ν_0 *is a random variable that has the exponential distribution with parameter* 1, **(b)** $\theta_0(t), t \geq 0$ *is a non-negative Lévy process with the Laplace transforms* $\mathsf{E}e^{-s\theta_0(t)} = e^{-tA(s)}, s, t \geq 0$ *and cumulant* $A(s)$ *defined in the condition* $\mathbf{D}_1$, *and* **(c)** *the random variable* ν_0 *and the process* $\theta_0(t), t \geq 0$ *are independent.*

Proof The following representation takes place, for $s, t \geq 0$,

$$\begin{aligned}
\mathsf{E}\mathrm{I}(\nu_\varepsilon^* > t)e^{-s\kappa_\varepsilon(t)} &= \sum_{i \in \mathbb{X}} q_{\varepsilon,i} \sum_{i=i_0,i_1,\ldots,i_{[t\nu_\varepsilon]} \in \mathbb{X}} \prod_{k=1}^{[t\nu_\varepsilon]} \phi_{\varepsilon,i_{k-1}i_k}(0,s) \\
&= \sum_{i \in \mathbb{X}} q_{\varepsilon,i} \sum_{i=i_0,i_1,\ldots,i_{[t\nu_\varepsilon]} \in \mathbb{X}} \prod_{k=1}^{[t\nu_\varepsilon]} p_{\varepsilon,s,i_{k-1}i_k} \\
&\quad \times (1 - p_{\varepsilon,i_{k-1}})\phi_{\varepsilon,i_{k-1}}(0,s) \\
&= \mathsf{E}\exp\{-\sum_{k=1}^{[t\nu_\varepsilon]}(-\ln(1 - p_{\varepsilon,\tilde{\eta}_{\varepsilon,k-1}}) \\
&\quad - \ln \phi_{\varepsilon,\tilde{\eta}_{\varepsilon,k-1}}(0,s))\}.
\end{aligned} \tag{2.124}$$

Using the conditions $\mathbf{A}_1$, $\mathbf{B}_1$, Lemma 2.1, and the relation (2.110), we obtain the following analogue of the relation (2.86), for $s \geq 0$,

$$\begin{aligned}
f_{\varepsilon,s} &= -u_\varepsilon \sum_{i \in \mathbb{X}} \tilde{\pi}_{\varepsilon,s,i} \ln(1 - p_{\varepsilon,i}) \\
&\sim u_\varepsilon \sum_{i \in \mathbb{X}} \tilde{\pi}_{\varepsilon,s,i}\, p_{\varepsilon,i} \\
&\sim u_\varepsilon \sum_{i \in \mathbb{X}} \pi_{\varepsilon,i}\, p_{\varepsilon,i} = u_\varepsilon p_\varepsilon = 1 \text{ as } \varepsilon \to 0.
\end{aligned} \tag{2.125}$$

The relations (2.119) and (2.125) imply that Lemma 2.5 can be applied to the processes, for every $s > 0$,

$$\kappa_{\varepsilon,s}(t) = \sum_{k=1}^{[tu_\varepsilon]} (-\ln(1 - p_{\varepsilon,\tilde{\eta}_{\varepsilon,k-1}}) - \ln \phi_{\varepsilon,\tilde{\eta}_{\varepsilon,k-1}}(0, s)),\ t \geq 0. \tag{2.126}$$

This gives the following relation, for every $s > 0$,

$$\kappa_{\varepsilon,s}(t), t \geq 0 \xrightarrow{\mathsf{d}} t + A(s)t, t \geq 0 \text{ as } \varepsilon \to 0. \tag{2.127}$$

Denote, for $i, j \in \mathbb{X}, n = 0, 1, \ldots, s \geq 0$,

$$\Psi_{\varepsilon,ij}(n, s) = \mathsf{E}_i \mathrm{I}(\nu_\varepsilon > n, \eta_{\varepsilon,n} = j) e^{-s \sum_{k=1}^{n} \kappa_{\varepsilon,k}}, \tag{2.128}$$

and

$$\Psi_{\varepsilon,i}(n, s) = \mathsf{E}_i \mathrm{I}(\nu_\varepsilon > n) e^{-s \sum_{k=1}^{n} \kappa_{\varepsilon,k}} = \sum_{j \in \mathbb{X}} \Psi_{\varepsilon,ij}(n, s). \tag{2.129}$$

The relation (2.127) implies, by the continuity theorem for Laplace transforms, the following relation, for $t \geq 0$,

$$\begin{aligned} \mathsf{E}\mathrm{I}(\nu_\varepsilon^* > t) e^{-s\kappa_\varepsilon(t)} &= \Psi_{\varepsilon,i}([t\nu_\varepsilon], s) \\ &= \mathsf{E} e^{-\kappa_{\varepsilon,s}(t)} \to e^{-t - A(s)t} \\ &= e^{-t} e^{-A(s)t} \text{ as } \varepsilon \to 0, \text{ for } s > 0. \end{aligned} \tag{2.130}$$

We also denote, for $i, j \in \mathbb{X}, n = 0, 1, \ldots, s \geq 0$,

$$\psi_{\varepsilon,ij}(n, s) = \mathsf{E}_i \mathrm{I}(\eta_{\varepsilon,n} = j) e^{-s \sum_{k=1}^{n} \kappa_{\varepsilon,k}}, \tag{2.131}$$

and

$$\psi_{\varepsilon,i}(n, s) = \mathsf{E}_i e^{-s \sum_{k=1}^{n} \kappa_{\varepsilon,k}} = \sum_{j \in \mathbb{X}} \psi_{\varepsilon,ij}(n, s). \tag{2.132}$$

The relation (2.130) easily implies that, for $s > 0$ and $0 \leq t'' \leq t' < \infty$,

$$\begin{aligned} \Psi_{\varepsilon,i}([t'\nu_\varepsilon] - [t''\nu_\varepsilon], s) &\sim \Psi_{\varepsilon,i}([(t' - t'')\nu_\varepsilon], s) \\ &\to e^{-(t'-t'')} e^{-A(s)(t'-t'')} \text{ as } \varepsilon \to 0. \end{aligned} \tag{2.133}$$

Also, the statement **(iii)** of Theorem 2.2 implies that, for $s > 0$ and $0 \leq t'' \leq t' < \infty$.

$$\begin{aligned} \psi_{\varepsilon,i}([t'\nu_\varepsilon] - [t''\nu_\varepsilon], s) &\sim \psi_{\varepsilon,i}([(t' - t'')\nu_\varepsilon], s) \\ &\to e^{-A(s)(t'-t'')} \text{ as } \varepsilon \to 0. \end{aligned} \tag{2.134}$$

The relations (2.133) and (2.134) imply that, for $s > 0$ and $0 \leq t'' \leq t' < \infty$,

$$\sum_{j \in \mathbb{X}} \Psi_{\varepsilon,ij}([t'v_\varepsilon] - [t''v_\varepsilon], s)$$
$$= \Psi_{\varepsilon,i}([t'v_\varepsilon] - [t''v_\varepsilon], s) \to e^{-(t'-t'')} e^{-A(s)(t'-t'')} \text{ as } \varepsilon \to 0 \tag{2.135}$$

and

$$\sum_{j \in \mathbb{X}} \psi_{\varepsilon,ij}([t'v_\varepsilon] - [t''v_\varepsilon], s)$$
$$= \psi_{\varepsilon,i}([t'v_\varepsilon] - [t''v_\varepsilon], s) \to e^{-A(s)(t'-t'')} \text{ as } \varepsilon \to 0. \tag{2.136}$$

The following representation for multivariate joint distributions of the random variable v^*_ε and the increments of stochastic process $\kappa_\varepsilon(t)$ for $0 = t_0 \le t_1 < \cdots t_k = t \le t_{k+1} \le \cdots \le t_n < \infty$, $1 \le k < n < \infty$ and $s_1, \ldots, s_n \ge 0$ take place,

$$\begin{aligned}
&\mathsf{E}\mathrm{I}(v^*_\varepsilon > t_k) \exp\{-\sum_{r=1}^{n} s_r(\kappa_\varepsilon(t_r) - \kappa_\varepsilon(t_{r-1})\} \\
&= \sum_{i_0,\ldots,i_n \in \mathbb{X}} q_{\varepsilon,i_0} \prod_{r=1}^{k} \Psi_{\varepsilon,i_{r-1}i_r}([t_r v_\varepsilon] - [t_{r-1}v_\varepsilon], s_r) \\
&\quad \times \prod_{r=k+1}^{n} \psi_{\varepsilon,i_{r-1}i_r}([t_r v_\varepsilon] - [t_{r-1}v_\varepsilon], s_r) \\
&= \sum_{i_0 \in \mathbb{X}} q_{\varepsilon,i_0} \sum_{i_1 \in \mathbb{X}} \Psi_{\varepsilon,i_0 i_1}([t_1 v_\varepsilon] - [t_0 v_\varepsilon], s_1) \\
&\quad \cdots \times \sum_{i_k \in \mathbb{X}} \Psi_{\varepsilon,i_{k-1}i_k}([t_k v_\varepsilon] - [t_{k-1}v_\varepsilon], s_k) \\
&\quad \times \sum_{i_{k+1} \in \mathbb{X}} \psi_{\varepsilon,i_k i_{k+1}}([t_{k+1} v_\varepsilon] - [t_k v_\varepsilon], s_{k+1}) \\
&\quad \cdots \times \sum_{i_n \in \mathbb{X}} \psi_{\varepsilon,i_{n-1}i_n}([t_n v_\varepsilon] - [t_{n-1}v_\varepsilon], s_n).
\end{aligned} \tag{2.137}$$

Using the relations (2.135), (2.136), and the representation (2.137), we obtain recurrently, for $0 = t_0 \le t_1 < \cdots t_k = t \le t_{k+1} \le \cdots \le t_n < \infty$, $1 \le k < n < \infty$ and $s_1, \ldots, s_n > 0$,

$$\begin{aligned}
&\mathsf{E}\mathrm{I}(v^*_\varepsilon > t_k) \exp\{-\sum_{r=1}^{n} s_r(\kappa_\varepsilon(t_r) - \kappa_\varepsilon(t_{r-1}))\} \\
&\sim \mathsf{E}\mathrm{I}(v^*_\varepsilon > t_k) \exp\{-\sum_{r=1}^{n-1} s_r(\kappa_\varepsilon(t_r) - \kappa_\varepsilon(t_{r-1}))\} e^{-A(s_n)(t_n - t_{n-1})} \\
&\cdots \sim \mathsf{E}\mathrm{I}(v^*_\varepsilon > t_k) \exp\{-\sum_{r=1}^{k} s_r(\kappa_\varepsilon(t_r) - \kappa_\varepsilon(t_{r-1}))\}
\end{aligned}$$

$$
\begin{aligned}
&\times \exp\{\sum_{r=k+1}^{n} -A(s_r)(t_r - t_{r-1})\} \\
&\sim \mathsf{E}\mathrm{I}(\nu_\varepsilon^* > t_{k-1}) \exp\{-\sum_{r=1}^{k-1} s_r(\kappa_\varepsilon(t_r) - \kappa_\varepsilon(t_{r-1}))\} \\
&\times \exp\{-(t_k - t_{k-1})\} \exp\{\sum_{r=k}^{n} -A(s_r)(t_r - t_{r-1})\} \\
&\cdots \sim \exp\{-\sum_{r=1}^{k}(t_r - t_{r-1})\} \exp\{\sum_{r=1}^{n} -A(s_r)(t_r - t_{r-1})\} \\
&= \exp\{-t\} \exp\{\sum_{r=1}^{n} -A(s_r)(t_r - t_{r-1})\} \text{ as } \varepsilon \to 0. \qquad (2.138)
\end{aligned}
$$

The relation (2.138) is equivalent to the asymptotic relation (2.123) given in Lemma 2.9. □

Now, we can complete the proof of Theorem 2.1.

The asymptotic relation given in Lemma 2.9 can, obviously, be rewritten in the following equivalent form:

$$
(t\nu_\varepsilon^*, \kappa_\varepsilon(t)), t \geq 0 \xrightarrow{\mathsf{d}} (t\nu_0, \theta_0(t)), t \geq 0 \text{ as } \varepsilon \to 0, \qquad (2.139)
$$

where the random variable ν_0 and the stochastic process $\theta_0(t), t \geq 0$ are described in Lemma 2.9.

The asymptotic relation appearing in statement **(iii)** of Theorem 2.2 and the relation (2.139) let us apply Theorem A.6 to the compositions of stochastic processes $\kappa_\varepsilon(t), t \geq 0$ and $t\nu_\varepsilon^*, t \geq 0$. This gives the following relation:

$$
\kappa_\varepsilon(t\nu_\varepsilon^*), t \geq 0 \xrightarrow{\mathsf{J}} \theta_0(t\nu_0), t \geq 0 \text{ as } \varepsilon \to 0. \qquad (2.140)
$$

The proof of Theorem 2.1 is complete.

Chapter 3
Flows of Rare Events for Regularly Perturbed Semi-Markov Processes

In this chapter, we introduce counting processes generated by flows of rare events defined on asymptotically uniformly ergodic semi-Markov processes. We present necessary and sufficient conditions of convergence in distribution for these counting processes. This chapter includes three sections.

In Sect. 3.1, we introduce process $N_\varepsilon(t)$ counting the number of rare events in interval $[0, t]$, for $t \geq 0$. In Theorem 3.1, we give necessary and sufficient conditions of convergence in distribution for such processes.

In Sect. 3.2, we introduce Markov renewal processes generated by flow of rare events defined on asymptotically uniformly ergodic semi-Markov processes. In Theorem 3.2, we give necessary and sufficient conditions for weak convergence of transition probabilities for such processes.

In Sect. 3.3, we introduce vectors counting processes, in which components count the numbers of rare events in interval $[0, t]$ occurring at moments of hitting different states by the corresponding semi-Markov processes. In Theorem 3.3, we give necessary and sufficient conditions of convergence in distribution for such processes.

3.1 Counting Processes Generated by Flows of Rare Events

In this section, we introduce counting processes generated by flows of rare events defined on asymptotically uniformly perturbed semi-Markov processes and give necessary and sufficient conditions of convergence in distribution and J-convergence for these processes.

D. Silvestrov, *Perturbed Semi-Markov Type Processes I*,
https://doi.org/10.1007/978-3-030-92403-4_3

3.1.1 Counting Processes for Rare Events

Let us define recurrently random variables,

$$\nu_\varepsilon(k) = \min(n \geq \nu_\varepsilon(k-1) : \rho_{\varepsilon,n} = 1), k = 1, 2, \ldots, \nu_\varepsilon(0) = 0. \tag{3.1}$$

The random variable $\nu_\varepsilon(k)$ counts the number of transitions of the embedded Markov chain $\eta_{\varepsilon,n}$ up to the k-th appearance of the "rare" event $\{\rho_{\varepsilon,\cdot} = 1\}$.

Let us also define inter-rare-event times, for $k = 1, 2, \ldots$,

$$\kappa_\varepsilon(k) = \sum_{n=\nu_\varepsilon(k-1)+1}^{\nu_\varepsilon(k)} \kappa_{\varepsilon,n}. \tag{3.2}$$

We also introduce random variables showing the states of the embedded Markov chain $\eta_{\varepsilon,n}$ at moments $\nu_\varepsilon(k), k = 0, 1, \ldots$,

$$\eta_\varepsilon(k) = \eta_{\varepsilon,\nu_\varepsilon(k)}. \tag{3.3}$$

Obviously, the random sequence $(\eta_\varepsilon(k), \kappa_\varepsilon(k)), k = 0, 1, \ldots$ (here $\kappa_\varepsilon(k) = 0$) is a Markov renewal process, that is, a homogeneous Markov chain with phase space $X \times [0, \infty)$ and transition probabilities, for $i, j \in \mathbb{X},\ s, t \geq 0$,

$$\begin{aligned} &\mathsf{P}\{\eta_\varepsilon(k+1) = j, \kappa_\varepsilon(k+1) \leq t/\eta_\varepsilon(k) = i, \kappa_\varepsilon(k) = s\} \\ &\quad = \mathsf{P}\{\eta_\varepsilon(k+1) = j, \kappa_\varepsilon(k+1) \leq t/\eta_\varepsilon(k) = i\} \\ &\quad = \mathsf{P}_i\{\eta_{\varepsilon,\nu_\varepsilon} = j, \xi_\varepsilon \leq t\} = Q_{ij}^{(\varepsilon)}(t). \end{aligned} \tag{3.4}$$

Let us now define random variables,

$$\xi_\varepsilon(k) = \sum_{n=1}^{\nu_\varepsilon(k)} \kappa_{\varepsilon,n} = \sum_{n=1}^{k} \kappa_\varepsilon(n),\ k = 0, 1, \ldots,\ \xi_\varepsilon(0) = 0. \tag{3.5}$$

The random variable $\xi_\varepsilon(k)$ can be interpreted as the time of occurrence of the k-th rare-event time for the semi-Markov process $\eta(t)$. Obviously, $\xi_\varepsilon(1) = \xi_\varepsilon$, where ξ_ε is the first-rare-event time for the semi-Markov process $\eta_\varepsilon(t)$.

Now, we can define a counting stochastic process generated by a flow of rare events,

$$N_\varepsilon(t) = \max(k \geq 0 : \xi_\varepsilon(k) \leq t),\ t \geq 0. \tag{3.6}$$

We denote by $\mathcal{N}$ the class of integer-valued, non-negative, non-decreasing, step-wise with a finite number of jumps in every finite interval, continuous from the right processes defined on the interval $[0, \infty)$.

Obviously, any process from the class $\mathcal{N}$ is a càdlàg process.

The counting process $N_\varepsilon(t)$ belongs to the class $\mathcal{N}$.

3.1.2 Necessary and Sufficient Conditions of Convergence for Counting Processes Generated by Flows of Rare Events

Let $\kappa(k), k = 1, 2, \ldots$, be a sequence of i.i.d. non-negative random variables with distribution function $G(u) = \mathsf{P}\{\kappa(1) \leq t\}, t \geq 0$ not concentrated at zero.

We also define a sequence of random variables, commonly referred to as the renewal process,

$$\xi(k) = \sum_{n=1}^{k} \kappa(n), k = 0, 1, \ldots, \xi(0) = 0.$$

Let us also define the corresponding standard renewal counting process,

$$N(t) = \max(k \geq 0 : \xi(k) \leq t),\ t \geq 0. \tag{3.7}$$

Any standard renewal counting process $N(t), t \geq 0$ belongs to the class $\mathcal{N}$.

We denote by $\mathbb{N}$ the set of points of stochastic continuity $t > 0$ for the process $N(t)$. The set $\overline{\mathbb{N}}$ is at most countable. Moreover, $\overline{\mathbb{N}} \subseteq \mathbb{K} = \cup_{n \geq 1}\mathbb{K}_n$, where $\mathbb{K}_n$ is the set of discontinuity points $u > 0$ for the distribution function $G^{(*n)}(u)$ of the random variable $\xi(n)$, for $n = 1, 2, \ldots$.

The following theorem takes place.

Theorem 3.1 *Let the conditions* $\mathbf{A}_1$, $\mathbf{B}_1$, *and* $\mathbf{C}_1$ *be satisfied. Then*:

(i) *The condition* $\mathbf{D}_1$ *is necessary and sufficient for the fulfilment (for some or any initial distribution* $\bar{q}_\varepsilon$, *respectively, in the statements of necessity and sufficiency) of the asymptotic relation* $N_\varepsilon(t),\ t \in \mathbb{N} \stackrel{\mathsf{d}}{\longrightarrow} N(t),\ t \in \mathbb{N}$ *as* $\varepsilon \to 0$, *where* $N(t), t \geq 0$ *is some process from class* $\mathcal{N}$ *such that* $\mathsf{P}\{N(t) \geq 1\} = G(t), t \geq 0$ *is the distribution function on* $[0, \infty)$ *not concentrated at zero, and* $\mathbb{N}$ *is the set of stochastic continuity points* $t > 0$ *for this process. In this case*:

(ii) *The limiting process* $N(t), t \geq 0$ *is a standard renewal counting process defined by the relation (3.7) through some sequence of non-negative i.i.d. random variables* $\kappa(k), k = 1, 2, \ldots$, *with the distribution function* $G(u) = \mathsf{P}\{\kappa(1) \leq t\}, t \geq 0$ *not concentrated at zero and the Laplace transform* $\int_0^\infty e^{-st} G(dt) = \phi(s) = \frac{1}{1+A(s)}, s \geq 0$, *where* $A(s)$ *is the cumulant of infinitely divisible distribution appearing in the condition* $\mathbf{D}_1$.

Proof Obviously, the random variable $\kappa_\varepsilon(1)$ has the following conditional distribution function, for $i \in \mathbb{X}$,

$$G_i^{(\varepsilon)}(u) = \mathsf{P}_i\{\kappa_\varepsilon(1) \leq u\} = \mathsf{P}_i\{\xi_\varepsilon \leq u\} = \sum_{j \in \mathbb{X}} Q_{ij}^{(\varepsilon)}(u),\ u \geq 0. \tag{3.8}$$

According to Theorem 2.1, the conditions $\mathbf{A}_1$–$\mathbf{D}_1$ imply that, for $i \in \mathbb{X}$,

$$G_i^{(\varepsilon)}(\cdot) = \mathsf{P}\{\kappa_\varepsilon(1) \leq \cdot\} \Rightarrow G(\cdot) \text{ as } \varepsilon \to 0, \tag{3.9}$$

where $G(u)$ is the distribution function on $[0, \infty)$ not concentrated at zero, with the Laplace transform $\int_0^\infty e^{-st} G(dt) = \phi(s) = \frac{1}{1+A(s)}$ appearing in the condition $\mathbf{D}_1$.

Using the Markov property of the Markov renewal process $(\eta_\varepsilon(k), \kappa_\varepsilon(k))$, we obtain the following formula for the joint conditional distribution functions of random variables $\kappa_\varepsilon(k), k = 1, \ldots, n$, for $u_1, \ldots, u_n \geq 0$, and $i_0 \in \mathbb{X}, n = 1, 2, \ldots$,

$$
\begin{aligned}
&\mathsf{P}_{i_0}\{\kappa_\varepsilon(k) \leq u_k, k = 1, \ldots, n\} \\
&\quad = \sum_{i_{n-1} \in \mathbb{X}} \mathsf{P}_{i_0}\{\kappa_\varepsilon(k) \leq u_k, k = 1, \ldots, n-1, \\
&\qquad\qquad \eta_\varepsilon(n-1) = i_{n-1}\} G^{(\varepsilon)}_{i_{n-1}}(u_n).
\end{aligned} \tag{3.10}
$$

The conditions $\mathbf{A}_1$–$\mathbf{D}_1$ and the relations (3.10) and (3.9) imply that the following convergence relation holds for the joint conditional distribution functions of random variables $\kappa_\varepsilon(k), k = 1, 2$, for all points of continuity for the corresponding limiting two-dimensional distribution function, and $i_0 \in \mathbb{X}$,

$$
\begin{aligned}
&|\mathsf{P}_{i_0}\{\kappa_\varepsilon(k) \leq u_k, k = 1, 2\} - G(u_1)G(u_2)| \\
&\quad \leq |\mathsf{P}_{i_0}\{\kappa_\varepsilon(k) \leq u_k, k = 1, 2\} - \mathsf{P}_{i_0}\{\kappa_\varepsilon(1) \leq u_1\}G(u_2)| \\
&\qquad + |\mathsf{P}_{i_0}\{\kappa_\varepsilon(1) \leq u_1\}G(u_2)| - G(u_1)G(u_2)| \\
&\quad \leq \sum_{i_1 \in \mathbb{X}} \mathsf{P}_{i_0}\{\kappa_\varepsilon(k) \leq u_1, \eta_\varepsilon(1) = i_1\}|G^{(\varepsilon)}_{i_1}(u_2) - G(u_2)| \\
&\qquad + |\mathsf{P}_{i_0}\{\kappa_\varepsilon(1) \leq u_1\} - G(u_1)|G(u_2) \to 0 \text{ as } \varepsilon \to 0.
\end{aligned} \tag{3.11}
$$

Similarly, the conditions $\mathbf{A}_1$–$\mathbf{D}_1$ and the relations (3.9) and (3.10) imply that the following convergence relation holds for the joint conditional distribution functions of random variables $\kappa_\varepsilon(k), k = 1, \ldots, n$, for all points of continuity for the corresponding limiting multi-dimensional distribution functions, and $i_0 \in \mathbb{X}, n = 1, 2, \ldots$,

$$
|\mathsf{P}_{i_0}\{\kappa_\varepsilon(k) \leq u_k, k = 1, \ldots, n\} - \prod_{k=1}^n G(u_k)| \to 0 \text{ as } \varepsilon \to 0. \tag{3.12}
$$

The relation (3.11) means that the inter-renewal times $\kappa_\varepsilon(k), k = 1, 2, \ldots$, are asymptotically independent.

Let $\kappa(k), k = 1, 2, \ldots$, be i.i.d. random variables with distribution function $G(u)$, $\xi(n) = \sum_{k=1}^n \kappa(k), n = 1, 2, \ldots$, and $H(u_1, \ldots, u_n)$ be the joint distribution function of random variables $\xi(k), k = 1, \ldots, n$, for $n = 1, 2, \ldots$.

The conditions $\mathbf{A}_1$–$\mathbf{D}_1$ and the relation (3.11) imply that the following convergence relation holds for the joint conditional distribution functions of random variables $\xi_\varepsilon(k) = \sum_{r=1}^k \kappa_\varepsilon(r), k = 1, \ldots, n$, for all points of continuity for the corresponding limiting multi-dimensional distribution functions, and $i \in \mathbb{X}, n = 1, 2, \ldots$

$$
\begin{aligned}
\mathsf{P}_i\{\xi_\varepsilon(k) \leq u_k, k = 1, \ldots, n\} &\to \mathsf{P}\{\xi(k) \leq u_k, k = 1, \ldots, n\} \\
&= H(u_1, \ldots, u_n) \text{ as } \varepsilon \to 0.
\end{aligned} \tag{3.13}
$$

The conditions $\mathbf{A}_1$–$\mathbf{D}_1$ and the relation (3.13) imply that the following convergence relation holds for (a) integers $0 \leq r_1 \leq r_2 \leq \cdots < \infty$, (b) $0 < t_1 \leq t_2 \leq \cdots < \infty$, which are points of continuity for the corresponding limiting distribution functions of random variables $\xi(k)$, $k = 1, 2, \ldots$, and (c) $i \in \mathbb{X}$, $n = 1, 2, \ldots$,

$$\begin{aligned} \mathsf{P}_i\{N_\varepsilon(t_k) \geq r_k, k = 1, \ldots, n\} &= \mathsf{P}_i\{\xi_\varepsilon(r_k) \leq t_k, k = 1, \ldots, n\} \\ &\to \mathsf{P}\{\xi(r_k) \leq t_k, k = 1, \ldots, n\} \\ &= \mathsf{P}\{N(t_k) \geq r_k, k = 1, \ldots, n\} \text{ as } \varepsilon \to 0. \end{aligned} \tag{3.14}$$

The relation (3.14) implies that the following asymptotic relation holds, for any initial distributions $\bar{q}_\varepsilon$,

$$N_\varepsilon(t),\ t \in \overline{\mathbb{K}} \xrightarrow{\mathsf{d}} N(t),\ t \in \overline{\mathbb{K}} \text{ as } \varepsilon \to 0. \tag{3.15}$$

Note that the process $N(t), t \geq 0$ is non-decreasing. Also, the set $\mathbb{K}$ is at most countable and, thus, the set $\overline{\mathbb{K}} \subseteq \mathbb{N}$ is dense in the interval $[0, \infty)$. Therefore, the relation (3.15) can be extended in the standard way to the asymptotic relation appearing in the statement **(i)**, where the limiting process $N(t)$ is the standard counting process described in the statement **(ii)** of Theorem 3.1.

Suppose now that the conditions $\mathbf{A}_1$–$\mathbf{C}_1$ are satisfied and the asymptotic relation, $N_\varepsilon(t), t \in \mathbb{N} \xrightarrow{\mathsf{d}} N(t), t \in \mathbb{N}$ as $\varepsilon \to 0$, takes place, for some initial distributions $\bar{q}_\varepsilon$, where $N(t)$ is some stochastic process described in proposition **(i)** of Theorem 3.1, and $\mathbb{N}$ is the set of points of stochastic continuity $t > 0$ for this process.

It follows from the above assumptions that the following relation holds, for $t \in \mathbb{N}$,

$$\begin{aligned} \mathsf{P}_{\bar{q}_\varepsilon}\{\xi_\varepsilon \leq t\} &= \mathsf{P}_{\bar{q}_\varepsilon}\{N_\varepsilon(t) \geq 1\} \\ &\to \mathsf{P}\{N(t) \geq 1\} = G(t) \text{ as } \varepsilon \to 0. \end{aligned} \tag{3.16}$$

Since $G(t)$ is the distribution function on $[0, \infty)$ not concentrated at zero and the set $\mathbb{N}$ is dense in $[0, \infty)$, the relation (3.16) implies that

$$\mathsf{P}_{\bar{q}_\varepsilon}\{\xi_\varepsilon \leq \cdot\} \Rightarrow G(\cdot) \text{ as } \varepsilon \to 0. \tag{3.17}$$

The relation (3.17) implies, by Theorem 2.1, that the condition $\mathbf{D}_1$ is satisfied (with the distribution function $G(\cdot)$ playing the role of the distribution function of random variable ξ_0). Thus, the statement **(ii)** of Theorem 3.1 takes place and $N(t)$ is the standard renewal counting process described in this statement. □

Remark 3.1 If the condition $\mathbf{D}_1$ is modified, as described in Remark 2.5, that is, it is assumed that the distribution function of the random variable ξ_0 has no atom at zero, and, thus, $N(0) = 0$ with probability 1, the point 0 can be additionally included in the set $\mathbb{N}$ in the sufficiency statement. In addition, if the distribution function $\mathsf{P}\{N(t) \geq 1\} = G(t)$, which appears in the statement **(i)** of Theorem 3.1, has no atom at zero, then the above-mentioned modified condition $\mathbf{D}_1$ is satisfied.

Remark 3.2 In the case described in Remark 3.1, the conditions, $\mathbf{A}_1$–$\mathbf{D}_1$, imply (by Theorem 4.4.1 from Silvestrov (2004)) that a more general asymptotic relation, $N_\varepsilon(t), t \geq 0 \stackrel{\mathsf{J}}{\longrightarrow} N(t), t \geq 0$ as $\varepsilon \to 0$, takes place.

Remark 3.3 The conditions $\mathbf{A}_1$–$\mathbf{D}_1$ of Theorem 2.1 guarantee that the distribution functions of the first-rare-event times $G^{(\varepsilon)}(\cdot) = \mathsf{P}\{\xi_\varepsilon \leq \cdot\} \Rightarrow G(\cdot)$ as $\varepsilon \to 0$, where $G(\cdot)$ is the distribution function of the limiting random variable appearing in the condition $\mathbf{D}_1$. However, the above weak convergence relation does not guarantee that the probabilities $G^{(\varepsilon)}(0)$ converge to probability $G(0)$ if $G(0) > 0$. This is due to the fact that in this case, 0 is a discontinuity point of the distribution function $G(\cdot)$. Since $\mathsf{P}\{N_\varepsilon(0) = r\} = G^{(\varepsilon)}(0)^r(1 - G^{(\varepsilon)}(0)), r = 0, 1, \ldots$, the point 0 cannot be included in the set of weak convergence $\mathbb{N}$ appearing in the asymptotic relation appearing in the statement **(i)** of Theorem 3.2. Also, in the above case, the J-convergence of the processes $N_\varepsilon(t)$ is not guaranteed.

3.2 Markov Renewal Processes Generated by Flows of Rare Events

In this section, we present Markov renewal processes generated by flows of rare events defined on asymptotically uniformly perturbed semi-Markov processes and give necessary and sufficient conditions for weak convergence of transition probabilities for such process.

3.2.1 Return Times and Rare Events

According to Lemma 2.1, the condition $\mathbf{B}_1$ implies that the Markov chain $\eta_{\varepsilon,n}$ is ergodic for $\varepsilon \in (0, \varepsilon_0]$, for some $\varepsilon_0 \in (0, 1]$. In what follows, we assume that $\varepsilon \in (0, \varepsilon_0]$.

It is interesting that the conditions $\mathbf{A}_1$–$\mathbf{D}_1$ do not guarantee weak convergence of the transition probabilities for the Markov renewal processes $(\eta_\varepsilon(k), \kappa_\varepsilon(k))$. Some additional condition must be included in the necessary and sufficient conditions for the weak convergence of the transition probabilities for the above Markov renewal processes.

Let us formulate and prove several useful lemmas.

Let us define the successive return moments of the Markov chain $\eta_{\varepsilon,n}$ in the state $i \in \mathbb{X}$,

$$\alpha_{\varepsilon,i}(k) = \min(n > \alpha_{\varepsilon,i}(k-1),\ \eta_{\varepsilon,n} = i),\ k = 1, 2, \ldots,\ \alpha_{\varepsilon,0} = 0 \tag{3.18}$$

and define the following probabilities, for $j, i, r \in \mathbb{X}$,

$$q_{\varepsilon,ji}(r) = \mathsf{P}_j\{\nu_\varepsilon \leq \alpha_{\varepsilon,i}(1),\ \eta_{\varepsilon,\nu_\varepsilon} = r\}, \tag{3.19}$$

and

$$q_{\varepsilon,ji} = \mathsf{P}_j\{\nu_\varepsilon \leq \alpha_{\varepsilon,i}(1)\} = \sum_{r\in\mathbb{X}} q_{\varepsilon,ji}(r). \tag{3.20}$$

Let us also introduce the following probabilities, for $i, r \in \mathbb{X}$.

$$p_{\varepsilon,i}(r) = \mathsf{P}_i\{\eta_{\varepsilon,1} = r, \rho_{\varepsilon,1} = 1\}, \tag{3.21}$$

and

$$p_\varepsilon(r) = \sum_{i=1}^{m} \pi_{\varepsilon,i} p_{\varepsilon,i}(r). \tag{3.22}$$

Obviously, for $i \in \mathbb{X}$,

$$p_{\varepsilon,i} = \mathsf{P}_i\{\rho_{\varepsilon,1} = 1\} = \sum_{r\in\mathbb{X}} p_{\varepsilon,i}(r), \tag{3.23}$$

and

$$p_\varepsilon = \sum_{i\in\mathbb{X}} \pi_{\varepsilon,i} p_{\varepsilon,i} = \sum_{i\in\mathbb{X}} \pi_{\varepsilon,i} \sum_{r\in\mathbb{X}} p_{\varepsilon,i}(r) = \sum_{r\in\mathbb{X}} p_\varepsilon(r). \tag{3.24}$$

Lemma 3.1 *Let the conditions* $\mathbf{A}_1$–$\mathbf{C}_1$ *be satisfied. Then, for* $i \in X$,

$$\frac{\pi_{\varepsilon,i} q_{\varepsilon,ii}}{p_\varepsilon} \to 1 \text{ as } \varepsilon \to 0. \tag{3.25}$$

Let $\delta_{\varepsilon,ik}$ be the number of visits of the embedded Markov chain $\eta_{\varepsilon,n}$ to the state k before the first visit to the state i.

Proof Let $\delta_{\varepsilon,ik}$ be the number of visits of the embedded Markov chain $\eta_{\varepsilon,n}$ to the state k before the first visit to the state i, for $i, k \in \mathbb{X}$,

$$\delta_{\varepsilon,ik} = \sum_{n=1}^{\alpha_{\varepsilon,i}(1)} \mathrm{I}(\eta_{\varepsilon,n-1} = k). \tag{3.26}$$

We also define matrices, for $i \in \mathbb{X}$,

$${}_i\mathbf{P}_\varepsilon = \begin{bmatrix} p_{\varepsilon,11} & \cdots & p_{\varepsilon,1\,i-1} & 0 & p_{\varepsilon,1\,i+1} & \cdots & p_{\varepsilon,1\,m} \\ \vdots & & \vdots & \vdots & \vdots & & \vdots \\ p_{\varepsilon,m1} & \cdots & p_{\varepsilon,m\,i-1} & 0 & p_{\varepsilon,m\,i+1} & \cdots & p_{\varepsilon,m\,m} \end{bmatrix}. \tag{3.27}$$

As is known, due to the ergodicity of the Markov chain $\eta_{\varepsilon,n}$, for $j, i, k \in \mathbb{X}$,

$$\mathsf{E}_j \delta_{\varepsilon,ik} < \infty. \tag{3.28}$$

Moreover, there exists the inverse matrix, for $i \in \mathbb{X}$,

$$[\mathbf{I} - {}_i\mathbf{P}_\varepsilon]^{-1} = \left\| \mathsf{E}_j \delta_{\varepsilon,ik} \right\|. \tag{3.29}$$

We also define matrices, for $i \in \mathbb{X}$,

$$
{}_i\tilde{\mathbf{P}}_\varepsilon = \begin{bmatrix} \tilde{p}_{\varepsilon,11} & \cdots & \tilde{p}_{\varepsilon,1\,i-1} & 0 & \tilde{p}_{\varepsilon,1\,i+1} & \cdots & \tilde{p}_{\varepsilon,1\,m} \\ \vdots & & \vdots & \vdots & \vdots & & \vdots \\ \tilde{p}_{\varepsilon,m\,1} & \cdots & \tilde{p}_{\varepsilon,m\,i-1} & 0 & \tilde{p}_{\varepsilon,m\,i+1} & \cdots & \tilde{p}_{\varepsilon,mm} \end{bmatrix}, \tag{3.30}
$$

where, for $j, k \in \mathbb{X}$,

$$
\tilde{p}_{\varepsilon,jk} = \mathsf{P}_j\{\eta_{\varepsilon,1} = k, \rho_{\varepsilon,1} = 0\}. \tag{3.31}
$$

Let $\tilde{\delta}_{\varepsilon,ik}$ be the number of visits of the embedded Markov chain $\eta_{\varepsilon,n}$ to the state k before the first visit to the state i or the occurrence of the first-rare-event, for $i, k \in \mathbb{X}$,

$$
\tilde{\delta}_{\varepsilon,ik} = \sum_{n=1}^{\alpha_{\varepsilon,i}(1)\wedge\nu_\varepsilon} \mathrm{I}(\eta_{\varepsilon,n-1} = k). \tag{3.32}
$$

Obviously, $\tilde{\delta}_{\varepsilon,ik} \leq \delta_{\varepsilon,ik}$ and, thus, for $j, i, k \in \mathbb{X}$,

$$
\mathsf{E}_j\tilde{\delta}_{\varepsilon,ik} \leq \mathsf{E}_j\delta_{\varepsilon,ik} < \infty. \tag{3.33}
$$

The matrix ${}_i\tilde{\mathbf{P}}^n_\varepsilon$ can be represented in the following form, for $i \in \mathbb{X}$ and $n \geq 1$,

$$
{}_i\tilde{\mathbf{P}}^n_\varepsilon = \left\| \mathsf{P}_j\{\eta_{\varepsilon,n} = k, \alpha_{\varepsilon,i}(1) \wedge \nu_\varepsilon > n\} \right\|, \tag{3.34}
$$

and, therefore, there exists the inverse matrix,

$$
\left[\mathbf{I} - {}_i\tilde{\mathbf{P}}_\varepsilon\right]^{-1} = \mathbf{I} + {}_i\tilde{\mathbf{P}}_\varepsilon + {}_i\tilde{\mathbf{P}}^2_\varepsilon + \cdots = \left\| \mathsf{E}_j\tilde{\delta}_{\varepsilon,ik} \right\|. \tag{3.35}
$$

The probabilities $p_{\varepsilon,jk} \in [0, 1], j, k \in \mathbb{X}$, for $\varepsilon \in (0, 1]$. Therefore, any sequence $\varepsilon_n \in (0, 1], n = 1, 2, \ldots$ such that $\varepsilon_n \to 0$ as $n \to \infty$ contains a subsequence $\varepsilon_{n_r}, r = 1, 2, \ldots$ such that, for $j, k \in \mathbb{X}$,

$$
p_{\varepsilon_{n_r},jk} \to p_{0,jk} \in [0, 1] \text{ as } r \to \infty, \tag{3.36}
$$

where the limits $p_{0,jk}, j, k \in \mathbb{X}$ can depend on the choice of the subsequence ε_{n_r}.

According to the above remarks, there exists, for every $i \in \mathbb{X}$, the inverse matrix $[\mathbf{I} - {}_i\mathbf{P}_{\varepsilon_{n_r}}]^{-1}$, for all r such that $\varepsilon_{n_r} \in (0, \varepsilon_0]$.

Obviously, the matrix $\mathbf{P}_0 = \|p_{0,jk}\|$ is stochastic. Moreover, the condition $\mathbf{B}_1$ implies that the phase space $\mathbb{X}$ of a Markov chain $\eta_{0,n}$ with the matrix of transition probabilities $\mathbf{P}_0$ is one class of communicative states, i.e., this Markov chain is ergodic.

This means that for reasons similar to those used for the relations (3.28) and (3.29), there exists the inverse matrix $[\mathbf{I} - {}_i\mathbf{P}_0]^{-1}$, for $i \in \mathbb{X}$. Thus, $\det(\mathbf{I} - {}_i\mathbf{P}_0) \neq 0$.

Elements of the matrices $[\mathbf{I} - {}_i\mathbf{P}_{\varepsilon_{n_r}}]^{-1}$ and $[\mathbf{I} - {}_i\mathbf{P}_0]^{-1}$ are continuous rational functions of elements, respectively, of the matrices $[\mathbf{I} - {}_i\mathbf{P}_{\varepsilon_{n_r}}]$ and $[\mathbf{I} - {}_i\mathbf{P}_0]$. That is why, the relation (3.36) implies that, for $i \in \mathbb{X}$,

$$
[\mathbf{I} - {}_i\mathbf{P}_{\varepsilon_{n_r}}]^{-1} \to [\mathbf{I} - {}_i\mathbf{P}_0]^{-1} \text{ as } \varepsilon \to 0. \tag{3.37}
$$

The condition $\mathbf{A}_1$ obviously implies that, for $j, k \in \mathbb{X}$,

$$|\tilde{p}_{\varepsilon,jk} - p_{\varepsilon,jk}| \leq p_{\varepsilon,j} \to 0 \text{ as } \varepsilon \to 0, \tag{3.38}$$

and, therefore, for $j, k \in \mathbb{X}$,

$$\tilde{p}_{\varepsilon_{n_r},jk} \to p_{0,jk} \in [0,1] \text{ as } r \to \infty. \tag{3.39}$$

The relation (3.39) implies that for reasons similar to those used for the relation (3.37), the following relation holds, for $i \in \mathbb{X}$,

$$[\mathbf{I} - {}_i\tilde{\mathbf{P}}_{\varepsilon_{n_r}}]^{-1} \to [\mathbf{I} - {}_i\mathbf{P}_0]^{-1} \text{ as } \varepsilon \to 0. \tag{3.40}$$

The matrices ${}_i\mathbf{P}_0$, $i \in \mathbb{X}$ depend on the choice of subsequence ε_{n_r}. However, the relations (3.37) and (3.40) imply that, for $i \in \mathbb{X}$,

$$[\mathbf{I} - {}_i\tilde{\mathbf{P}}_{\varepsilon_{n_r}}]^{-1} - [\mathbf{I} - {}_i\mathbf{P}_{\varepsilon_{n_r}}]^{-1} \to \mathbf{0} \text{ as } \varepsilon \to 0, \tag{3.41}$$

where $\mathbf{0}$ is a $m \times m$ matrix with all zero elements.

Since an arbitrary choice of sequence $0 < \varepsilon_n \to 0$ and independence of limit $\mathbf{0}$ in relation (3.41) on the choice of subsequence ε_{n_r}, the relation (3.41) implies that

$$[\mathbf{I} - {}_i\tilde{\mathbf{P}}_{\varepsilon}]^{-1} - [\mathbf{I} - {}_i\mathbf{P}_{\varepsilon}]^{-1} \to \mathbf{0} \text{ as } \varepsilon \to 0 \tag{3.42}$$

or, equivalently, that, for $j, i, k \in \mathbb{X}$,

$$\mathsf{E}_j\tilde{\delta}_{\varepsilon,ik} - \mathsf{E}_j\delta_{\varepsilon,ik} \to 0 \text{ as } \varepsilon \to 0. \tag{3.43}$$

The probabilities $q_{\varepsilon,ji}$, $j \in \mathbb{X}$ satisfy, for every $i \in \mathbb{X}$, the following system of linear equations,

$$q_{\varepsilon,ji} = p_{\varepsilon,j} + \sum_{k \neq i} \tilde{p}_{\varepsilon,jk} q_{\varepsilon,ki}, \ j \in \mathbb{X}. \tag{3.44}$$

The system (3.44) can be rewritten, for every $i \in \mathbb{X}$, in the following matrix form,

$$\mathbf{q}_{\varepsilon,i} = \mathbf{p}_{\varepsilon} + {}_i\mathbf{P}_{\varepsilon}\mathbf{q}_{\varepsilon,i}, \tag{3.45}$$

where

$$\mathbf{q}_{\varepsilon,i} = \begin{bmatrix} q_{\varepsilon,1i} \\ \vdots \\ q_{\varepsilon,mi} \end{bmatrix}, \quad \mathbf{p}_{\varepsilon} = \begin{bmatrix} p_{\varepsilon,1} \\ \vdots \\ p_{\varepsilon,m} \end{bmatrix}. \tag{3.46}$$

The matrix $\mathbf{I} - {}_i\tilde{\mathbf{P}}_{\varepsilon}$ has the inverse matrix, for $\varepsilon \in (0, \varepsilon_0]$ and $i \in \mathbb{X}$. Therefore, the solution of system (3.44) has the following form, for every $i \in X$,

$$\mathbf{q}_{\varepsilon,i} = \left[\mathbf{I} - {}_i\tilde{\mathbf{P}}_{\varepsilon}\right]^{-1} \mathbf{p}_{\varepsilon}. \tag{3.47}$$

The relations (3.35) and (3.47) imply that, for $i \in \mathbb{X}$,

$$q_{\varepsilon,ii} = \sum_{k \in \mathbb{X}} \mathsf{E}_i \tilde{\delta}_{\varepsilon,ik} p_{\varepsilon,k}. \tag{3.48}$$

Since the Markov chain $\eta_{\varepsilon,n}$ is ergodic, the following well-known formula takes place, for $i, k \in \mathbb{X}$,

$$\mathsf{E}_i \delta_{\varepsilon,ik} = \frac{\pi_{\varepsilon,k}}{\pi_{\varepsilon,i}}. \tag{3.49}$$

Using the relations (3.43), (3.48), (3.49), and the statement **(i)** of Lemma 2.1, according to which $\underline{\lim}_{\varepsilon \to 0} \pi_{\varepsilon,k} > 0$, we obtain, for $i \in \mathbb{X}$,

$$\begin{aligned} \frac{|q_{\varepsilon,ii} - \frac{p_\varepsilon}{\pi_{\varepsilon,i}}|}{\frac{p_\varepsilon}{\pi_{\varepsilon,i}}} &\leq \sum_{k \in \mathbb{X}} \left| \mathsf{E}_i \tilde{\delta}_{\varepsilon,ik} - \frac{\pi_{\varepsilon,k}}{\pi_{\varepsilon,i}} \right| \frac{\pi_{\varepsilon,i} p_{\varepsilon,k}}{\sum_{j \in \mathbb{X}} \pi_{\varepsilon,j} p_{\varepsilon,j}} \\ &\leq \sum_{k \in \mathbb{X}} \left| \mathsf{E}_i \tilde{\delta}_{\varepsilon,ik} - \mathsf{E}_i \delta_{\varepsilon,ik} \right| \frac{\pi_{\varepsilon,i}}{\pi_{\varepsilon,k}} \to 0 \text{ as } \varepsilon \to 0. \end{aligned} \tag{3.50}$$

The relation (3.50) implies that the relation (3.25) holds. □

Let us define probabilities, for $i, j \in \mathbb{X}$,

$$Q_{ij}^{(\varepsilon)}(\infty) = \mathsf{P}_i\{\eta_{\varepsilon,\nu_\varepsilon} = j\}. \tag{3.51}$$

Lemma 3.2 *Let the conditions* $\mathbf{A}_1$–$\mathbf{C}_1$ *be satisfied. Then, the following relation holds, for* $i, r \in \mathbb{X}$,

$$Q_{ir}^{(\varepsilon)}(\infty) - \frac{p_\varepsilon(r)}{p_\varepsilon} \to 0 \text{ as } \varepsilon \to 0. \tag{3.52}$$

Proof Taking into account that (a) the Markov renewal process $(\eta_{\varepsilon,n}, \kappa_{\varepsilon,n}, \rho_{\varepsilon,n})$ regenerates at the moments of return of the component $\eta_{\varepsilon,n}$ into any state $i \in \mathbb{X}$ and (b) ν_ε is a Markov moment for this process, we can get the following cyclic representation for probabilities $Q_{ir}^{(\varepsilon)}(\infty)$, for $i, r \in \mathbb{X}$,

$$\begin{aligned} Q_{ir}^{(\varepsilon)}(\infty) &= \sum_{n=0}^{\infty} \mathsf{P}_i\{\alpha_{\varepsilon,i}(n) < \nu_\varepsilon \leq \alpha_{\varepsilon,i}(n+1), \eta_{\varepsilon,\nu_\varepsilon} = r\} \\ &= \sum_{n=0}^{\infty} (1 - q_{\varepsilon,ii})^n q_{\varepsilon,ii}(r) = \frac{q_{\varepsilon,ii}(r)}{q_{\varepsilon,ii}}. \end{aligned} \tag{3.53}$$

The probabilities $q_{\varepsilon,ji}(r), j \in \mathbb{X}$ satisfy, for every $i, r \in \mathbb{X}$, the following system of linear equations,

$$q_{\varepsilon,ji}(r) = p_\varepsilon(r) + \sum_{k \neq i} \tilde{p}_{\varepsilon,jk} q_{\varepsilon,ki}(r), \ j \in \mathbb{X}. \tag{3.54}$$

This system has the matrix of coefficients ${}_i\tilde{\mathbf{P}}_\varepsilon$ (the same as the system of linear equations (3.44)) and differs from this system only by the corresponding free terms. Thus, by repeating reasoning used in the proof of Lemma 3.1, we can get the following formula similar to formula (3.35), for $i, r \in \mathbb{X}$,

$$q_{ii\varepsilon}(r) = \sum_{k=1}^{m} \mathsf{E}_i \tilde{\delta}_{\varepsilon,ik}\, p_\varepsilon(r). \tag{3.55}$$

Using relations (3.43), (3.49), (3.55), the statement **(i)** of Lemma 2.1 (according to which $\underline{\lim}_{\varepsilon \to 0} \pi_{\varepsilon,k} > 0$), and relation (3.25), we get the following relation, for $i \in \mathbb{X}$,

$$\begin{aligned}\left| \frac{q_{\varepsilon,ii}(r)}{q_{\varepsilon,i}} - \frac{p_\varepsilon(r)}{\pi_{\varepsilon,i} q_{\varepsilon,i}} \right| &\leq \sum_{k \in \mathbb{X}} \left| \mathsf{E}_i \tilde{\delta}_{\varepsilon,ik} - \frac{\pi_{\varepsilon,k}}{\pi_{\varepsilon,i}} \right| \cdot \frac{\pi_{\varepsilon,i} p_{\varepsilon,k}(r)}{\sum_{j \in \mathbb{X}} \pi_{\varepsilon,j} p_{\varepsilon,j}} \cdot \frac{p_\varepsilon}{\pi_{\varepsilon,i} q_{\varepsilon,i}} \\ &\leq \sum_{k \in \mathbb{X}} \left| \mathsf{E}_i \tilde{\delta}_{\varepsilon,ik} - \mathsf{E}_i \delta_{\varepsilon,ik} \right| \cdot \frac{\pi_{\varepsilon,i}}{\pi_{\varepsilon,k}} \cdot \frac{p_\varepsilon}{\pi_{\varepsilon,i} q_{\varepsilon,i}} \\ &\to 0 \text{ as } \varepsilon \to 0. \end{aligned} \tag{3.56}$$

Also, we get, using relation (3.24) and relation (3.25) given in Lemma 3.1, for $i \in \mathbb{X}$,

$$\begin{aligned}\left| \frac{p_\varepsilon(r)}{\pi_{\varepsilon,i} q_{\varepsilon,i}} - \frac{p_\varepsilon(r)}{p_\varepsilon} \right| &= \frac{p_\varepsilon(r)}{p_\varepsilon} \left| \frac{p_\varepsilon}{\pi_{\varepsilon,i} q_{\varepsilon,i}} - 1 \right| \\ &\leq \left| \frac{p_\varepsilon}{\pi_{\varepsilon,i} q_{\varepsilon,i}} - 1 \right| \to 0 \text{ as } \varepsilon \to 0. \end{aligned} \tag{3.57}$$

Relations (3.53), (3.56), and (3.57) imply that relation (3.52) holds. □

Let us introduce the following balancing condition:

A$_2$: $\frac{p_\varepsilon(r)}{p_\varepsilon} \to Q_r$ as $\varepsilon \to 0$, for $r \in \mathbb{X}$.

Constants Q_r automatically satisfy the following relations,

$$Q_r \geq 0, r \in \mathbb{X}, \ \sum_{r \in \mathbb{X}} Q_r = 1. \tag{3.58}$$

The following lemma is a direct corollary of Lemma 3.2.

Lemma 3.3 *Let the conditions* **A**$_1$–**C**$_1$ *hold. Then, condition* **A**$_2$ *is necessary and sufficient for holding (for some or every $i \in \mathbb{X}$, respectively, in the statements of necessity and sufficiency) the following relation:*

$$Q^{(\varepsilon)}_{ir}(\infty) \to Q_r \text{ as } \varepsilon \to 0, \textit{ for } r \in \mathbb{X}. \tag{3.59}$$

Let us define Laplace transforms, for $j, i, r \in \mathbb{X}$,

$$\psi_{\varepsilon,jir}(s) = \mathsf{E}\{e^{-s\xi_\varepsilon} \mathrm{I}(\eta_{\varepsilon,\nu_\varepsilon} = r)/\eta_{\varepsilon,0} = j, \nu_\varepsilon \leq \alpha_{\varepsilon,i}(1)\}, \ s \geq 0 \tag{3.60}$$

and

$$\tilde{\psi}_{\varepsilon,jir}(s) = \mathsf{E}_j e^{-s\xi_\varepsilon} \mathrm{I}(\eta_{\varepsilon,\nu_\varepsilon} = r, \nu_\varepsilon \leq \alpha_{\varepsilon,i}(1)), \ s \geq 0. \tag{3.61}$$

The following lemma takes place.

Lemma 3.4 *Let the conditions* $\mathbf{A}_1$–$\mathbf{D}_1$ *be satisfied. Then, for* $i, r \in \mathbb{X}$ *and* $s \geq 0$,

$$\frac{p_\varepsilon(r)}{p_\varepsilon} - \psi_{\varepsilon,iir}(s) \to 0 \text{ as } \varepsilon \to 0. \tag{3.62}$$

Proof We define Laplace transforms, for $j, r \in \mathbb{X}$,

$$\tilde{p}_{\varepsilon,jr}(s) = \mathsf{E}_j e^{-s\kappa_{\varepsilon,1}} \mathrm{I}(\rho_{\varepsilon,1} = 0, \eta_{\varepsilon,1} = r) = \hat{\phi}_{\varepsilon,j}(s) p_{\varepsilon,j}(r), \ s \geq 0, \tag{3.63}$$

and

$$\tilde{p}_{\varepsilon,j}(s) = \mathsf{E}_j e^{-s\kappa_{\varepsilon,1}} \mathrm{I}(\rho_{\varepsilon,1} = 1) = \hat{\varphi}_{\varepsilon,j}(s) p_{\varepsilon,j}, \ s \geq 0, \tag{3.64}$$

where

$$\hat{\phi}_{\varepsilon,j}(s) = \mathsf{E}\{e^{-s\kappa_{\varepsilon,1}} / \eta_{\varepsilon,0} = j, \rho_{\varepsilon,1} = 1\}, \ s \geq 0. \tag{3.65}$$

The functions $\psi_{\varepsilon,jir}(s), j \in \mathbb{X}$ satisfy, for every $s \geq 0$ and $i, r \in X$, the following system of linear equations:

$$\psi_{\varepsilon,jir}(s) = p_{\varepsilon,jr}(s) + \sum_{k \neq i} p_{\varepsilon,jk}(s) \psi_{\varepsilon,kir}(s), \ j \in \mathbb{X}. \tag{3.66}$$

The system (3.66) can be rewritten in the following equivalent matrix form:

$$\tilde{\mathbf{\Psi}}_{\varepsilon,ir}(s) = \tilde{\mathbf{p}}_{\varepsilon,r}(s) + {}_i\tilde{\mathbf{P}}_\varepsilon(s) \tilde{\mathbf{\Psi}}_{\varepsilon,ir}(s) \tag{3.67}$$

where

$$\tilde{\mathbf{\Psi}}_{\varepsilon,ir}(s) = \begin{bmatrix} \tilde{\psi}_{\varepsilon,1ir}(s) \\ \vdots \\ \tilde{\psi}_{\varepsilon,mir}(s) \end{bmatrix}, \ \tilde{\mathbf{p}}_{\varepsilon,r}(s) = \begin{bmatrix} \tilde{p}_{\varepsilon,1r}(s) \\ \vdots \\ p_{\varepsilon,mr}(s) \end{bmatrix}, \tag{3.68}$$

and

$$_i\tilde{\mathbf{P}}_\varepsilon(s) = \begin{bmatrix} \tilde{p}_{\varepsilon,11}(s) & \ldots & \tilde{p}_{\varepsilon,1\,i-1}(s) & 0 & \tilde{p}_{\varepsilon,1\,i+1}(s) & \ldots & \tilde{p}_{\varepsilon,1\,m}(s) \\ \vdots & & \vdots & \vdots & \vdots & & \vdots \\ \tilde{p}_{\varepsilon,m\,1}(s) & \ldots & \tilde{p}_{\varepsilon,m\,i-1} & 0 & \tilde{p}_{\varepsilon,m\,i+1}(s) & \ldots & p_{\varepsilon,m\,m}(s) \end{bmatrix}. \tag{3.69}$$

Let us introduce random variables, for $s \geq 0, i, k \in \mathbb{X}$,

$$\tilde{\delta}_{\varepsilon,ik}(s) = \sum_{n=1}^{\alpha_{\varepsilon,i}(1) \wedge \nu_\varepsilon} e^{-s\tau_{\varepsilon,n}} \mathrm{I}(\eta_{\varepsilon,n-1} = k). \tag{3.70}$$

Obviously, $\tilde{\delta}_{\varepsilon,ik}(s) \leq \tilde{\delta}_{\varepsilon,ik}$ and, therefore, for $s \geq 0$ and $j, i, k \in \mathbb{X}$,

$$\mathsf{E}_j\tilde{\delta}_{\varepsilon,ik}(s) \leq \mathsf{E}_j\tilde{\delta}_{\varepsilon,ik} < \infty. \tag{3.71}$$

Moreover, the matrix ${}_i\tilde{\mathbf{P}}^n_\varepsilon(s)$ can be represented in the following form, for $s \geq 0, i \in \mathbb{X}$ and $n \geq 1$,

$$ {}_i\tilde{\mathbf{P}}^n_\varepsilon(s) = \left\|\mathsf{E}_j e^{-s\tau_{\varepsilon,n}}\mathrm{I}(\eta_{\varepsilon,n} = k, \alpha_{\varepsilon,i}(1) \wedge \nu_\varepsilon > n)\}\right\| \tag{3.72}$$

and, thus, there exists the inverse matrix,

$$\left[\mathbf{I} - {}_i\tilde{\mathbf{P}}_\varepsilon(s)\right]^{-1} = \mathbf{I} + {}_i\tilde{\mathbf{P}}_\varepsilon(s) + {}_i\tilde{\mathbf{P}}^2_\varepsilon(s) + \cdots = \left\|\mathsf{E}_j\tilde{\delta}_{\varepsilon,ik}(s)\right\|. \tag{3.73}$$

Therefore, the solution of the system (3.67) has the following form:

$$\tilde{\mathbf{\Psi}}_{\varepsilon,ir}(s) = \left[\mathbf{I} - {}_i\tilde{\mathbf{P}}_{\varepsilon,i}(s)\right]^{-1}\tilde{\mathbf{p}}_{\varepsilon,r}(s). \tag{3.74}$$

The following part of the proof is similar to those given in the relations (3.36)–(3.37).

Choose a sequence $\varepsilon_n \to 0$ as $n \to \infty$ and a subsequence $\varepsilon_{n_r} \to 0$ as $r \to \infty$ in the same way as it was done in the relation (3.36). In this case, $\det(\mathbf{I} - {}_i\mathbf{P}_0) \neq 0$, i.e., there exists the inverse matrix $[\mathbf{I} - {}_i\mathbf{P}_0]^{-1}$ and the relation (3.37) holds.

The condition $\mathbf{D}_1$ implies that, for $\delta > 0$,

$$1 - F_\varepsilon(\delta) = \sum_{j\in\mathbb{X}} \pi_{\varepsilon,i}(1 - F_{\varepsilon,j}(\delta)) \to 0 \text{ as } \varepsilon \to 0. \tag{3.75}$$

Since, according to condition $\mathbf{B}_1$ and Lemma 2.1, $\underline{\lim}_{\varepsilon\to0}\pi_{\varepsilon,i} > 0$, for $i \in \mathbb{X}$, the relation (3.75) implies that, for $j \in \mathbb{X}$ and $\delta > 0$,

$$\mathsf{P}_j\{\kappa_{\varepsilon,1} > \delta\} = (1 - F_{\varepsilon,j}(\delta)) \to 0 \text{ as } \varepsilon \to 0. \tag{3.76}$$

The relation (3.76) and the condition $\mathbf{B}_1$ imply that, for any $s \geq 0, j, k \in \mathbb{X}$ and $\delta > 0$,

$$\begin{aligned} |p_{\varepsilon,jk} - \tilde{p}_{\varepsilon,jk}(s)| &= |p_{\varepsilon,jk} - \tilde{p}_{\varepsilon,jk}| \\ &\quad + |\mathsf{E}_j(1 - e^{-s\kappa_{\varepsilon,j}})\mathrm{I}(\eta_{\varepsilon,1} = k, \rho_{\varepsilon,1} = 0\}| \\ &\leq p_{\varepsilon,j} + (1 - e^{-\delta}) + \mathsf{P}_j\{\kappa_{\varepsilon,j} > \delta\} \to 1 - e^{-\delta} \text{ as } \varepsilon \to 0. \end{aligned} \tag{3.77}$$

Due to the arbitrary choice of $\delta > 0$ in the relation (3.77), this relation implies that, for $s \geq 0$ and $j, k \in \mathbb{X}$,

$$\tilde{p}_{\varepsilon,jk}(s) - p_{\varepsilon,jk} \in [0, 1] \text{ as } \varepsilon \to 0. \tag{3.78}$$

The relations (3.36) and (3.78) imply that, for $s \geq 0$ and $j, k \in \mathbb{X}$,

$$\tilde{p}_{\varepsilon_{n_r},jk}(s) \to p_{0,jk} \in [0, 1] \text{ as } \varepsilon \to 0. \tag{3.79}$$

The relation (3.79) implies that $\det(\mathbf{I} - {}_i\tilde{\mathbf{P}}_{\varepsilon_{n_r}}(s)) \to \det(\mathbf{I} - {}_i\mathbf{P}_0) \neq 0$ as $r \to \infty$, for $s \geq 0, i \in \mathbb{X}$. Thus, for every $s \geq 0, i \in \mathbb{X}$, there exists the inverse matrix $[\mathbf{I} - {}_i\tilde{\mathbf{P}}_{\varepsilon_{n_r}}(s)]^{-1}$ for r large enough.

Elements of the matrices $[\mathbf{I} - {}_i\tilde{\mathbf{P}}_{\varepsilon_{n_r}}(s)]^{-1}$ and $[\mathbf{I} - {}_i\mathbf{P}_0]^{-1}$ are continuous rational functions of elements, respectively, of the matrices $[\mathbf{I} - {}_i\tilde{\mathbf{P}}_{\varepsilon_{n_r}}(s)]$ and $[\mathbf{I} - {}_i\mathbf{P}_0]$. Therefore, the relation (3.79) implies that, for $s \geq 0, i \in \mathbb{X}$,

$$[\mathbf{I} - {}_i\tilde{\mathbf{P}}_{\varepsilon_{n_r}}(s)]^{-1} \to [\mathbf{I} - {}_i\mathbf{P}_0]^{-1} \text{ as } \varepsilon \to 0. \tag{3.80}$$

The matrices ${}_i\mathbf{P}_0$, $i \in \mathbb{X}$ depend on the choice of subsequence ε_{n_r}. However, the relations (3.37) and (3.80) imply that, for $s \geq 0$ and $i \in \mathbb{X}$,

$$[\mathbf{I} - {}_i\mathbf{P}_{\varepsilon_{n_r}}(s)]^{-1} - [\mathbf{I} - {}_i\mathbf{P}_{\varepsilon_{n_r}}]^{-1} \to \mathbf{0} \text{ as } \varepsilon \to 0, \tag{3.81}$$

where $\mathbf{0}$ is a $m \times m$ matrix with all zero elements.

Since an arbitrary choice of sequence $0 < \varepsilon_n \to 0$ and independence of the limit $\mathbf{0}$ in the relation (3.41) on the choice of sequence ε_n and subsequence ε_{n_r}, the relation (3.81) implies that, for $s \geq 0$ and $i \in \mathbb{X}$,

$$[\mathbf{I} - {}_i\tilde{\mathbf{P}}_\varepsilon(s)]^{-1} - [\mathbf{I} - {}_i\mathbf{P}_\varepsilon]^{-1} \to \mathbf{0} \text{ as } \varepsilon \to 0 \tag{3.82}$$

or equivalently that, for $s \geq 0, j, i, k \in \mathbb{X}$,

$$\mathsf{E}_j\tilde{\delta}_{\varepsilon,ik}(s) - \mathsf{E}_j\delta_{\varepsilon,ik} \to 0 \text{ as } \varepsilon \to 0. \tag{3.83}$$

Taking into account the relations (3.63), (3.64), (3.73), and (3.74), we obtain the following relation, for $s \geq 0$ and $i, r \in \mathbb{X}$,

$$\tilde{\psi}_{\varepsilon,iir}(s) = \sum_{k \in \mathbb{X}} \mathsf{E}_i\tilde{\delta}_{\varepsilon,ik}(s)\hat{\varphi}_{\varepsilon,k}(s)p_{\varepsilon,k}(r) \tag{3.84}$$

The condition $\mathbf{A}_2$ implies that, for every $s \geq 0$ and $k \in \mathbb{X}$,

$$\begin{aligned} 0 \leq \overline{\lim_{\varepsilon \to 0}}(1 - \hat{\phi}_{\varepsilon,k}(s)) &\leq 1 - e^{-s\delta} + \overline{\lim_{\varepsilon \to 0}} \mathsf{P}_k\{\kappa_{\varepsilon,1} > \delta/\rho_{\varepsilon,1} = 1\} \\ &\to 1 - \exp\{-s\delta\} \text{ as } \varepsilon \to 0. \end{aligned} \tag{3.85}$$

Due to the arbitrary choice of $\delta > 0$ in the relation (3.85), the relation (3.85) implies that, for $s \geq 0$ and $j, k \in \mathbb{X}$,

$$\hat{\phi}_{\varepsilon,k}(s) \to 1 \text{ as } \varepsilon \to 0. \tag{3.86}$$

Obviously, for $s \geq 0$ and $i, r \in \mathbb{X}$,

$$\psi_{\varepsilon,iir}(s) = \frac{\tilde{\psi}_{\varepsilon,iir}(s)}{q_{\varepsilon,ii}}. \tag{3.87}$$

Lemma 3.1, the statement **(i)** of Lemma 2.1 (according to which $\underline{\lim}_{\varepsilon\to 0}\pi_{\varepsilon,k} > 0$), and the relations (3.49), (3.83), (3.86) imply that, for every $s \geq 0$ and $i, k \in \mathbb{X}$,

$$\begin{aligned}
|\psi_{\varepsilon,iir}(s) - \frac{p_\varepsilon(r)}{p_\varepsilon}| &\leq |\frac{p_\varepsilon(r)}{\pi_{\varepsilon,i}q_{\varepsilon,ii}} - \frac{p_\varepsilon(r)}{p_\varepsilon}| \\
&\quad + \sum_{k\in\mathbb{X}} |\mathsf{E}\delta_{\varepsilon,iik}(s)\hat{\phi}_{\varepsilon,k}(s)\frac{p_{\varepsilon,k}(r)}{q_{\varepsilon,ii}} - \frac{\pi_{\varepsilon,k}p_{\varepsilon,k}(r)}{\pi_{\varepsilon,i}q_{\varepsilon,ii}}| \\
&\leq \sum_{k\in\mathbb{X}} |\mathsf{E}\delta_{\varepsilon,iik}(s)\hat{\phi}_{\varepsilon,k}(s) - \frac{\pi_{\varepsilon,k}}{\pi_{\varepsilon,i}}|\frac{p_\varepsilon}{\pi_{\varepsilon,i}q_{\varepsilon,ii}}\frac{p_{\varepsilon,k}(r)\pi_{\varepsilon,i}}{p_\varepsilon} \\
&\leq \sum_{k\in\mathbb{X}} |\mathsf{E}\delta_{\varepsilon,iik}(s)\hat{\phi}_{\varepsilon,k}(s) - \frac{\pi_{\varepsilon,k}}{\pi_{\varepsilon,i}}|\frac{p_\varepsilon}{\pi_{\varepsilon,i}q_{\varepsilon,ii}}\frac{\pi_{\varepsilon,i}}{\pi_{\varepsilon,k}} \\
&\to 0 \text{ as } \varepsilon \to 0. \qquad (3.88)
\end{aligned}$$

The proof is complete. □

The following lemma is a direct corollary of Lemma 3.4.

Lemma 3.5 *Let the conditions* $\mathbf{A}_1$–$\mathbf{D}_1$ *hold. Then, the condition* $\mathbf{A}_2$ *is necessary and sufficient for the fulfilment (for some or every* $i \in \mathbb{X}$*, respectively, in the statements of necessity and sufficiency) of the following relation:*

$$\psi_{\varepsilon,iir}(s) \to Q_r \text{ as } \varepsilon \to 0, \text{ for } s \geq 0, r \in \mathbb{X}. \qquad (3.89)$$

3.2.2 Necessary and Sufficient Conditions of Convergence for Markov Renewal Processes Generated by Flows of Rare Events

The following theorem shows that the first-rare-event time ξ_ε and the random functional $\eta_{\varepsilon,\nu_\varepsilon}$ are asymptotically independent and completes the description of the asymptotic behaviour of the transition probabilities $Q_{ij}^{(\varepsilon)}(t)$ for the Markov renewal process $(\eta_\varepsilon(k), \kappa_\varepsilon(k))$.

Theorem 3.2 *Let the conditions* $\mathbf{A}_1$, $\mathbf{B}_1$, *and* $\mathbf{C}_1$ *be satisfied. Then:*

(i) *The conditions* $\mathbf{D}_1$ *and* $\mathbf{A}_2$ *are necessary and sufficient for the fulfilment (for some or every initial distributions* $\bar{q}_\varepsilon$*, respectively, in the statements of necessity and sufficiency) of the asymptotic relation* $(\eta_\varepsilon(1), \xi_\varepsilon(1)) \xrightarrow{\mathsf{d}} (\eta_0(1), \xi_0(1))$ *as* $\varepsilon \to 0$*, where* $(\eta_0(1), \xi_0(1))$ *is a random vector that takes values in the space* $\mathbb{X} \times [0, \infty)$ *and has the joint distribution* $\mathsf{P}\{\eta_0(1) = r, \xi_0(1) \leq u,\} = G(r, u), r \in \mathbb{X}, u \geq 0$ *such that the distribution function* $G(u) = \sum_{r\in\mathbb{X}} G(r, u)$ *is not concentrated at zero.*

(ii) *The distribution function* $G(r, u) = Q_r G(u), r \in \mathbb{X}, u \geq 0$*, where* $G(u), u \geq 0$ *is the distribution function with Laplace transform* $\int_0^\infty e^{-st}G(dt) = \frac{1}{1+A(s)}, s \geq 0$ *and* $A(s)$ *is the cumulant appearing in the condition* $\mathbf{D}_1$*, and* $Q_r, r \in \mathbb{X}$ *are the limits appearing in the condition* $\mathbf{A}_2$.

(iii) *The following relation holds, for any $r_k \in \mathbb{X}$, $k = 1, \dots, n$, points $u_k \geq 0$, $k = 1, \dots, n$, which are points of continuity for the distribution function $G(u)$, and any $i \in \mathbb{X}$, $n = 1, 2, \dots$,*

$$\begin{aligned} \mathsf{P}_i\{\eta_\varepsilon(k)) = r_k, \xi_\varepsilon(k) \leq u_k, k = 1, \dots, n\} \\ \to Q(r_k, u_k, k = 1, \dots, n) = \prod_{k=1}^{n} Q_{r_k} G(u_k) \text{ as } \varepsilon \to 0. \end{aligned} \tag{3.90}$$

Proof The necessity part in the statement **(i)** of Theorem 3.2 follows from Theorem 2.1 and Lemma 3.3, since the asymptotic relation appearing in this statement implies that $\xi_\varepsilon(1) \xrightarrow{\mathsf{d}} \xi_0(1)$ as $\varepsilon \to 0$, where $\xi_0(1)$ is a non-negative random variable with the distribution function $\mathsf{P}\{\xi_0(1) \leq u\} = G(u) = \sum_{r \in \mathbb{X}} G(u, r), u \geq 0$, and $\eta_\varepsilon(1) \xrightarrow{\mathsf{d}} \eta_0(1)$ as $\varepsilon \to 0$, where $\eta_0(1)$ is a random variable taking values in space $\mathbb{X}$ such that $\mathsf{P}\{\eta_0(1) = r\} = Q_r = G(\infty, r), r \in \mathbb{X}$.

Let us prove that the conditions $\mathbf{A}_1$–$\mathbf{C}_1$ and $\mathbf{D}_1$, $\mathbf{A}_2$ imply the fulfilment of the asymptotic relation appearing in the statement **(i)** and the relation appearing in the statement **(ii)** of Theorem 3.2.

We define random variables, for $i \in \mathbb{X}$,

$$\beta_{\varepsilon,i}(k) = \sum_{n=\alpha_{\varepsilon,i}(k-1)+1}^{\alpha_{\varepsilon,i}(k)} \kappa_{\varepsilon,n}, \ k = 1, 2, \dots, \tag{3.91}$$

and Laplace transforms, for $j, i \in \mathbb{X}$,

$$\psi_{\varepsilon,ji}(s) = \mathsf{E}\{e^{-s\beta_{\varepsilon,i}(1)}/\eta_{\varepsilon,0} = j, \nu_\varepsilon > \alpha_{\varepsilon,i}(1)\}, \ s \geq 0. \tag{3.92}$$

The following representation, similar to the representation (3.53), takes place, for $s \geq 0, i \in \mathbb{X}$,

$$\begin{aligned} &\mathsf{E}_i \exp\{-s \sum_{k=1}^{\nu_{\varepsilon,i}} \beta_{\varepsilon,i}(k)\} \\ &= \sum_{n=0}^{\infty} \mathsf{E}_i \exp\{-s \sum_{k=1}^{n} \beta_{\varepsilon,i}(k)\} \mathrm{I}(\alpha_{\varepsilon,i}(n) < \nu_\varepsilon \leq \alpha_{\varepsilon,i}(n+1)) \\ &= \sum_{n=0}^{\infty} (1 - q_{\varepsilon,ii})^n \, \psi_{\varepsilon,ii}(s)^n q_{\varepsilon,ii} \\ &= \frac{q_{\varepsilon,ii}}{1 - (1 - q_{\varepsilon,ii})\psi_{\varepsilon,ii}(s)}. \end{aligned} \tag{3.93}$$

Let us introduce random variables, for $i \in \mathbb{X}$,

$$\beta_{\varepsilon,i} = \sum_{n=\alpha_{\varepsilon,i}(\nu_{\varepsilon,i})+1}^{\nu_\varepsilon} \kappa_{\varepsilon,n}, \tag{3.94}$$

where

$$\nu_{\varepsilon,i} = \max\{n \geq 0 : \alpha_{\varepsilon,i}(n) \leq \nu_\varepsilon\} = \mu_{\varepsilon,i}(\nu_\varepsilon). \tag{3.95}$$

Also, the following representation, similar to the representation (3.53), take place, for $s \geq 0, i, r \in \mathbb{X}$,

$$\begin{aligned}
&\mathsf{E}_i \exp\{-s\beta_{\varepsilon,i}\}\mathrm{I}(\eta_{\varepsilon,\nu_\varepsilon} = r) \\
&= \sum_{n=0}^{\infty} \mathsf{E}_i\{\exp\{-s \sum_{k=\alpha_{\varepsilon,i}(n)+1}^{\nu_e} \kappa_{\varepsilon,k}\}\mathrm{I}(\eta_{\varepsilon,\nu_\varepsilon} = r)/\alpha_{\varepsilon,i}(n) < \nu_\varepsilon \leq \alpha_{\varepsilon,i}(n+1)\} \\
&\quad \times \mathsf{P}\{\alpha_{\varepsilon,i}(n) < \nu_\varepsilon \leq \alpha_{\varepsilon,i}(n+1)\} \\
&= \sum_{n=0}^{\infty} \psi_{\varepsilon,iir}(s)(1 - q_{\varepsilon,ii})^n q_{\varepsilon,ii} = \psi_{\varepsilon,iir}(s).
\end{aligned} \tag{3.96}$$

Let us introduce Laplace transforms, for $i, r \in \mathbb{X}$,

$$\Phi_{\varepsilon,ir}(s) = \mathsf{E}_i e^{-s\xi_\varepsilon}\mathrm{I}(\eta_{\varepsilon,\nu_\varepsilon} = r), \ s \geq 0. \tag{3.97}$$

Finally, the following representation, similar to the representation (3.53), takes place, for $s \geq 0, \ i, r \in \mathbb{X}$,

$$\begin{aligned}
\Phi_{\varepsilon,ir}(s) &= \mathsf{E}_i \exp\{-s\sum_{k=1}^{\nu_{\varepsilon,i}} \beta_{\varepsilon,i}(k) + \beta_{\varepsilon,i}\}\mathrm{I}(\eta_{\varepsilon,\nu_\varepsilon} = r) \\
&= \sum_{n=0}^{\infty} \mathsf{E}_i \exp\{-s\sum_{k=1}^{n} \beta_{\varepsilon,i}(k)\} \\
&\quad \times \exp\{-s\beta_{\varepsilon,i}\}\mathrm{I}(\eta_{\varepsilon,\nu_\varepsilon} = r, \alpha_{\varepsilon,i}(n) < \nu_\varepsilon \leq \alpha_{\varepsilon,i}(n+1)) \\
&= \sum_{n=0}^{\infty}(1 - q_{\varepsilon,ii})^n \, \psi_{\varepsilon,ii}(s)^n q_{\varepsilon,ii}\psi_{\varepsilon,iir}(s) \\
&= \frac{q_{\varepsilon,ii}}{1 - (1 - q_{\varepsilon,ii})\psi_{\varepsilon,ii}(s)} \cdot \psi_{\varepsilon,iir}(s).
\end{aligned} \tag{3.98}$$

The following relation holds, for $s \geq 0, i \in \mathbb{X}$,

$$\mathsf{E}_i \exp\{-s\sum_{n=1}^{\nu_{\varepsilon,i}} \beta_{\varepsilon,i}(n)\} = \mathsf{E}_i \exp\{-s \sum_{n=1}^{\alpha_{\varepsilon,i}(\nu_{\varepsilon,i})} \kappa_{\varepsilon,n}\}. \tag{3.99}$$

The random variable $\alpha_{\varepsilon,i}(\nu_{\varepsilon,i})$ can be represented, for $i \in \mathbb{X}$, in the following form,

$$\alpha_{\varepsilon,i}(\nu_{\varepsilon,i}) = \alpha_{\varepsilon,i}(\mu_{\varepsilon,i}(\nu_\varepsilon)). \tag{3.100}$$

Lemma 2.1 allows us to apply Theorem A.6 to the random variables $\alpha_{\varepsilon,i}(\mu_{\varepsilon,i}([tu_\varepsilon]))$ and get the following asymptotic relation, for $i \in \mathbb{X}$ and the function $u_\varepsilon = p_\varepsilon^{-1}$,

$$\frac{\alpha_{\varepsilon,i}(\mu_{\varepsilon,i}([tp_\varepsilon^{-1}]))}{p_\varepsilon^{-1}} = \alpha^*_{\varepsilon,i}(\mu^*_{\varepsilon,i}(t), t \geq 0 \stackrel{\mathsf{U}}{\longrightarrow} t, t \geq 0 \text{ as } \varepsilon \to 0, \tag{3.101}$$

or, equivalently,

$$\frac{\alpha_{\varepsilon,i}(\mu_{\varepsilon,i}([tp_\varepsilon^{-1}]))}{p_\varepsilon^{-1}} - t = \alpha^*_{\varepsilon,i}(\mu^*_{\varepsilon,i}(t) - t, t \geq 0 \stackrel{\mathsf{U}}{\longrightarrow} 0(t), t \geq 0 \text{ as } \varepsilon \to 0 \tag{3.102}$$

where $0(t) \equiv 0, t \geq 0$.

The relation (3.102), Lemma 2.9, and Theorem A.9 imply that, for $i \in \mathbb{X}$,

$$(\nu^*_\varepsilon, \alpha^*_{\varepsilon,i}(\mu^*_{\varepsilon,i}(t) - t), t \geq 0 \stackrel{\mathsf{d}}{\longrightarrow} (\nu_0, 0(t)), t \geq 0 \text{ as } \varepsilon \to 0. \tag{3.103}$$

The relations (3.102), (3.103) and Theorem A.4 imply that, for $i \in \mathbb{X}$,

$$\alpha^*_{\varepsilon,i}(\mu^*_{\varepsilon,i}(\nu^*_\varepsilon)) - \nu^*_\varepsilon \stackrel{\mathsf{d}}{\longrightarrow} 0 \text{ as } \varepsilon \to 0. \tag{3.104}$$

The relation (3.104), Lemma 2.9, Theorem A.9 imply that, for $i \in \mathbb{X}$,

$$\begin{aligned}(\alpha^*_{\varepsilon,i}(\mu^*_{\varepsilon,i}(\nu^*_\varepsilon)), \kappa_\varepsilon(t)) &= (\alpha^*_{\varepsilon,i}(\mu^*_{\varepsilon,i}(\nu^*_\varepsilon)) - \nu^*_\varepsilon + \nu^*_\varepsilon, \kappa_\varepsilon(t)), t \geq 0 \\ &\stackrel{\mathsf{d}}{\longrightarrow} (\nu_0, \theta_0(t)), t \geq 0 \text{ as } \varepsilon \to 0. \end{aligned} \tag{3.105}$$

The asymptotic relation appearing in the statement **(iii)** of Theorem 2.2 and the relation (3.105) allows us to apply Theorem A.4 to the càdlàgprocesses $\kappa_\varepsilon(t), t \geq 0$ randomly stopped at the moment $\nu^*_\varepsilon = p_\varepsilon^{-1}\nu_\varepsilon$. This gives the following relation:

$$\kappa_\varepsilon(\alpha^*_{\varepsilon,i}(\mu^*_{\varepsilon,i}(\nu^*_\varepsilon))) \stackrel{\mathsf{d}}{\longrightarrow} \xi_0 = \theta_0(\nu_0) \text{ as } \varepsilon \to 0, \tag{3.106}$$

where $\theta_0(\nu_0)$ is the random variable appearing in the condition $\mathbf{D}_1$.

Lemma 3.5 and the relations (3.98), (3.99), and (3.106) imply that the following relation holds, $i, r \in X$ and $s \geq 0$,

$$\Phi_{\varepsilon,ir}(s) = \frac{1}{1 + (1 - q_{\varepsilon,ii})\frac{(1-\psi_{\varepsilon,ii}(s))}{q_{\varepsilon,i}}} \cdot \psi_{\varepsilon,iir}(s) \tag{3.107}$$

$$\to \frac{1}{1 + A(s)} \cdot Q_r \text{ as } \varepsilon \to 0. \tag{3.108}$$

The relation (3.107) is equivalent to the sufficiency part of the statement **(i)** and the statement **(ii)** of Theorem 3.2.

Suppose now that the conditions $\mathbf{A}_1$–$\mathbf{D}_1$ and $\mathbf{A}_2$ are satisfied. Obviously, the random vector $(\eta_\varepsilon(1), \kappa_\varepsilon(1))$ has the following conditional distribution, for $i, r \in \mathbb{X}, u \geq 0$,

$$Q_{ir}^{(\varepsilon)}(u) = \mathsf{P}_i\{\eta_\varepsilon(1) = r, \kappa_\varepsilon(1) \le u\} = \mathsf{P}_i\{\eta_{\varepsilon,\nu_\varepsilon} = r, \xi_\varepsilon \le u\}. \tag{3.109}$$

Using the Markov property of the Markov renewal process $(\eta_\varepsilon(k), \kappa_\varepsilon(k))$, we get the following relation for the joint conditional distribution function of the random variables $\eta_\varepsilon(k), \kappa_\varepsilon(k), k = 1, \ldots, n$, for $i_k \in \mathbb{X}, u_k \ge 0, k = 1, \ldots, n$, and $i_0 \in \mathbb{X}, n = 1, 2, \ldots$,

$$\begin{aligned}
&\mathsf{P}_{i_0}\{\eta_\varepsilon(k) = i_k, \kappa_\varepsilon(k) \le u_k, k = 1, \ldots, n\} \\
&\quad = \mathsf{P}_{i_0}\{\eta_\varepsilon(k) = i_k, \kappa_\varepsilon(k) \le u_k, k = 1, \ldots, n-1\} Q_{i_{n-1},i_n}^{(\varepsilon)}(u_n) \\
&\quad = \cdots = \\
&\quad = \prod_{k=1}^{n} Q_{i_{k-1},i_k}^{(\varepsilon)}(u_k).
\end{aligned} \tag{3.110}$$

According to the statements **(i)** and **(ii)** of Theorem 3.2, the conditions $\mathbf{A}_1$–$\mathbf{D}_1$ and $\mathbf{A}_2$ imply that, for $i, r \in \mathbb{X}$,

$$Q_{ir}^{(\varepsilon)}(\cdot) \Rightarrow G(\cdot)Q_r \text{ as } \varepsilon \to 0, \tag{3.111}$$

where $G(u), u \ge 0$ is the distribution function on $[0, \infty)$ not concentrated at zero, which appears in the condition $\mathbf{D}_1$ and $Q_r, r \in \mathbb{X}$ are limits appearing in the condition $\mathbf{A}_2$.

The conditions $\mathbf{A}_1$–$\mathbf{D}_1$ and $\mathbf{A}_2$ and the relations (3.110) and (3.111) imply that the following convergence relation holds for the joint conditional distribution function of random variables $\eta_\varepsilon(k), \kappa_\varepsilon(k), k = 1, \ldots, n$, for all $r_k \in \mathbb{X}, k = 1, \ldots, n$ and $u_k \ge 0, k = 1, \ldots, n$, which are points of continuity for the distribution function $F(u)$, and $i \in \mathbb{X}, n = 1, 2, \ldots$,

$$\begin{aligned}
&\mathsf{P}_i\{\eta_\varepsilon(k) = r_k, \kappa_\varepsilon(k) \le u_k, k = 1, \ldots, K\} \\
&\quad \to \mathsf{P}\{\eta(k) = r_k, \kappa(k) \le u_k, k = 1, \ldots, n\} \\
&\quad = Q(r_k, u_k, k = 1, \ldots, n) \text{ as } \varepsilon \to 0.
\end{aligned} \tag{3.112}$$

The relation (3.112) completes the proof of the statement **(iii)** of Theorem 3.2.□

3.3 Vector Counting Processes Generated by Flows of Rare Events

In this section, we introduce vectors counting processes, in which components count the numbers of rare events in interval $[0, t]$ occurring at moments of hitting different states by the corresponding semi-Markov processes, and give necessary and sufficient condition of convergence in distribution for such processes.

3.3.1 Vector Counting Processes for Rare Events

Let us consider the following vector counting processes:

$$\bar{M}_\varepsilon(t) = (M_{\varepsilon,r}(t), r \in \mathbb{X}), t \geq 0, \tag{3.113}$$

where, for $r \in \mathbb{X}$,

$$M_{\varepsilon,r}(t) = \sum_{k=1}^{[t]} \mathrm{I}(\eta_\varepsilon(k) = r),\ t \geq 0, \tag{3.114}$$

and

$$\bar{N}_\varepsilon(t) = (N_{\varepsilon,r}(t), r \in \mathbb{X}), t \geq 0, \tag{3.115}$$

where, for $r \in \mathbb{X}$,

$$N_{\varepsilon,r}(t) = M_{\varepsilon,r}(N_\varepsilon(t)),\ t \geq 0. \tag{3.116}$$

Obviously, the following relation takes place:

$$N_\varepsilon(t) = \sum_{r \in \mathbb{X}} N_{\varepsilon,r}(t), t \geq 0. \tag{3.117}$$

Let $\eta(k), \kappa(k), k = 1, 2, \ldots$, be mutually independent random variables such that: (a) the random variable $\eta(k), k = 1, 2, \ldots$, takes values r with probabilities Q_r, for $r \in \mathbb{X}$, (b) $\kappa(k), k = 1, 2, \ldots$, are non-negative random variables with distribution function $G(u)$ not concentrated at zero. We also define random variables $\xi(n) = \sum_{k=1}^n \kappa(k), n = 1, 2, \ldots, \xi(0) = 0$.

Let us now define a vector frequency process, $\bar{M}(t) = (M_r(t), r \in \mathbb{X}), t \geq 0$, where, for $r \in \mathbb{X}$,

$$M_r(t) = \sum_{k=1}^{[t]} \mathrm{I}(\eta(k) = r),\ t \geq 0 \tag{3.118}$$

and a counting process $\bar{N}(t) = (N_r(t), r \in \mathbb{X}), t \geq 0$, where

$$N_r(t) = M_r(N(t)),\ t \geq 0. \tag{3.119}$$

Obviously, the following relation takes place:

$$N(t) = \sum_{r \in \mathbb{X}} N_r(t), t \geq 0. \tag{3.120}$$

Since $N(t), t \geq 0$ and $M_r(t), t \geq 0, r \in \mathbb{X}$ are integer-valued, non-negative, step-wise, non-decreasing, càdlàg processes, $M_r(N(t)), t \geq 0, r \in \mathbb{X}$ also are integer-valued, non-negative, step-wise, non-decreasing, càdlàg processes.

Note also that processes $N(t)$ and $\bar{M}(N(t))$ have the same set $\mathbb{N}$ of points of stochastic continuity $t > 0$.

The role of finite-dimensional distributions for the vector process $N(t), t \geq 0$ is played by the following probabilities (which can be expressed via the joint distributions of random sums $\xi(n) = \sum_{k=1}^{n} \kappa(k)$ of i.i.d. random variables $\kappa(k), k = 1, 2, \ldots$, with the distribution function $G(\cdot)$), for (a) integers $0 \leq m[1] \leq m[2] \leq \cdots < \infty$, and (b) $0 \leq t_1 \leq t_2 \leq \cdots < \infty$,

$$\begin{aligned}
&\mathsf{P}\{N(t_1) = m[1]\} \\
&\quad = \mathsf{P}\{N(t_1) \geq m[1]\} - \mathsf{P}\{N(t_1) \geq m[1] + 1\} \\
&\quad = \mathsf{P}\{\xi(m[1]) \leq t_1\} - \mathsf{P}\{\xi(m[1] + 1) \leq t_1\}, \\
&\mathsf{P}\{N(t_l) = m[l], l = 1, 2\} \\
&\quad = \mathsf{P}\{N(t_1) = m[1], N(t_2) \geq m[2]\} \\
&\qquad - \mathsf{P}\{N(t_1) = m[1], N(t_2) \geq m[2] + 1\} \\
&\quad = \mathsf{P}\{N(t_1) \geq m[1], N(t_2) \geq m[2]\} - \mathsf{P}\{N(t_1) \geq m[1] + 1, N(t_2) \geq m[2]\} \\
&\qquad + \mathsf{P}\{N(1) \geq m[1], N(t_2) \geq m[2] + 1\} \\
&\qquad - \mathsf{P}\{N(t_1) \geq m[1] + 1, N(t_2) \geq m[2] + 1\} \\
&\quad = \mathsf{P}\{\xi(m[1]) \leq t_1, \xi(m[2]) \leq t_2\} - \mathsf{P}\{\xi(m[1] + 1) \leq t_1, \xi(m[2]) \leq t_2\} \\
&\qquad + \mathsf{P}\{\xi(m[1]) \leq t_1, \xi(m[2] + 1) \leq t_2\} \\
&\qquad - \mathsf{P}\{\xi(m[1] + 1) \leq t_1, \xi(m[2] + 1) < t_2\}, \\
&\text{etc.}
\end{aligned} \tag{3.121}$$

In what follows, $\bar{l}_k = (l_{k,r}, r \in \mathbb{X}), k = 1, 2, \ldots$, be vectors with integer components $l_{k,r}, r \in \mathbb{X}$ satisfying the following relation:

$$l_{k,r} \geq 0, r \in \mathbb{X}, \sum_{r \in \mathbb{X}} l_{k,r} = k, \ k = 1, 2, \ldots \tag{3.122}$$

Also, $\bar{p} = (p_r, r \in \mathbb{X})$ be discrete distributions with probabilities $p_r \geq 0, r \in \mathbb{X}$ such that $\sum_{r \in \mathbb{X}} p_r = 1$.

Let $G(k, \bar{l}_k, \bar{p})$ be, for $k = 1, 2, \ldots$, the multinomial probabilities, defined by the following formula, for vectors $\bar{l}_k = (l_{k,r}, r \in \mathbb{X})$ and discrete distributions $\bar{p} = (p_r, r \in \mathbb{X})$ satisfying the above assumptions,

$$G(k, \bar{l}_k, \bar{p}) = \frac{k!}{\prod_{r \in \mathbb{X}} l_{k,r}!} \prod_{r \in \mathbb{X}} p_r^{l_{k,r}}. \tag{3.123}$$

The process $\bar{M}(t), t \geq 0$ is a vector process with independent increments, whose increments $\bar{M}(t) - \bar{M}(s)$ have multinomial joint distributions of the components, i.e., for vectors $\bar{Q} = (Q_r, r \in \mathbb{X})$ and $\bar{l}_{m[s,t]}$, where $m[s, t] = [t] - [s]$, and $0 \leq s \leq t < \infty$,

$$\mathsf{P}\{\bar{M}(t) - \bar{M}(s) = \bar{l}_{m[s,t]}\} = G(m[s, t], \bar{l}_{m[s,t]}, \bar{Q}). \tag{3.124}$$

Note also that the counting process $N(t), t \geq 0$ and the process $\bar{M}(t), t \geq 0$ are independent.

Therefore, the finite-dimensional distributions for the vector process $\bar{N}(t), t \geq 0$ take the following form, for (a) any vectors $\bar{l}_k, k = 1, 2, \ldots$, satisfying the relation (3.122), (b) integer $0 \leq m[1] \leq m[2] \leq \cdots < \infty$, (c) $0 \leq t_1 \leq t_2 \leq \cdots < \infty$, and (d) $n = 1, 2, \ldots$,

$$\begin{aligned}\mathsf{P}\{\bar{N}(t_k) = \bar{l}_{m[k]}, k = 1, \ldots, n\} \\ = \mathsf{P}\{N(t_k) = m[k], k = 1, \ldots, n\} \\ \times \mathsf{P}\{\bar{M}(m[k]) = \bar{l}_{m[k]}, k = 1, \ldots, n\}.\end{aligned} \tag{3.125}$$

Here, the relation (3.120) is taken into account.

3.3.2 Necessary and Sufficient Conditions of Convergence for Vector Counting Process Generated by Flows of Rare Events

The following theorem generalises Theorem 3.2 and describes the asymptotics in distribution for the vector rare-event counting processes $\bar{N}_\varepsilon(t)$.

Theorem 3.3 *Let the conditions* $\mathbf{A}_1$, $\mathbf{B}_1$, *and* $\mathbf{C}_1$ *hold. Then*:

(i) *The conditions* $\mathbf{D}_1$ *and* $\mathbf{A}_2$ *are necessary and sufficient for the fulfilment (for some or any initial distributions* $\bar{q}_\varepsilon$, *respectively, in the statements of necessity and sufficiency) of the asymptotic relation* $\bar{N}_\varepsilon(t), t \in \mathbb{N} \xrightarrow{\mathsf{d}} \bar{N}(t), t \in \mathbb{N}$ *as* $\varepsilon \to 0$, *where* $\bar{N}(t), t \geq 0$ *is a stochastic process with integer-valued, non-negative, non-decreasing, step-wise, càdlàg components such that* $\mathsf{P}\{N(t) = \sum_{r \in \mathbb{X}} N_r(t) \geq 1\} = G(t)$ *is the distribution function on* $[0, \infty)$ *not concentrated at zero, and* $\mathbb{N}$ *is the set of stochastic continuity points* $t > 0$ *for this process.*

(ii) *The limiting process* $\bar{N}(t), t \geq 0$ *is defined by the relations (3.7), (3.118), and (3.116) via some mutually independent random variables* $\eta(k), \kappa(k), k = 1, 2, \ldots$, *where: (a)* $\kappa(k), k = 1, 2, \ldots$ *are non-negative random variables with the distribution function* $G(u)$ *not concentrated at zero and with the Laplace transform* $\int_0^\infty e^{-st} G(dt) = \phi(s) = \frac{1}{1+A(s)}$ *defined via the cumulant* $A(s)$ *of infinitely divisible distribution appearing in the condition* $\mathbf{D}_1$, *(b)* $\eta(k), k = 1, 2, \ldots$ *are random variables taking values* $r \in \mathbb{X}$ *with probabilities* $Q_r, r \in \mathbb{X}$ *appearing in the condition* $\mathbf{A}_2$, *and (c)* $\mathbb{N}$ *is the set of stochastic continuity points* $t > 0$ *for the process* $\bar{N}(t)$.

Proof The conditions $\mathbf{A}_1$–$\mathbf{D}_1$ and $\mathbf{A}_2$, and the relation (3.90) appearing in the statement **(iii)** of Theorem 3.2 imply that the following convergence relation holds for: (a) any vectors $\bar{l}_k, k = 1, 2, \ldots$, satisfying the relation (3.122), (b) $0 < u_1 \leq u_2 \leq \cdots < \infty$, which are points of continuity for the distribution functions of random variables $\xi(k), k = 1, 2, \ldots$, and (c) $i \in \mathbb{X}, n = 1, 2, \ldots$,

$$\begin{aligned}\mathsf{P}_i\{\bar{M}_\varepsilon(k) = \bar{l}_k, \xi_\varepsilon(k) \leq u_k, k = 1, \ldots, n\} \\ \to \mathsf{P}\{\bar{M}(k) = \bar{l}_k, k = 1, \ldots, n\} \mathsf{P}\{\xi(k) \leq u_k, k = 1, \ldots, n\} \text{ as } \varepsilon \to 0.\end{aligned} \tag{3.126}$$

The conditions $\mathbf{A}_1$–$\mathbf{D}_1$ and the relation (3.126) imply that the following convergence relation holds for (a) any vectors $\bar{l}_k, k = 1, 2, \ldots$, satisfying the relation (3.122), (b) $0 < u_1 \leq u_2 \leq \cdots < \infty$, which are points of continuity for the distribution functions of random variables $\xi(k), k = 1, 2, \ldots$, and (c) $i \in \mathbb{X}, n = 1, 2, \ldots$,

$$\begin{aligned}
& \mathsf{P}_i\{\bar{M}_\varepsilon(m_k) = \bar{l}_{m_k}, N_\varepsilon(t_k) \geq r_k, k = 1, \ldots, n\} \\
& \quad = \mathsf{P}_i\{\bar{M}_\varepsilon(m_k) = \bar{l}_{m_k}, \xi_\varepsilon(r_k) \leq t_k, k = 1, \ldots, n\} \\
& \quad \to \mathsf{P}\{\bar{M}(m_k) = \bar{l}_{m_k}, k = 1, \ldots, n\}\mathsf{P}\{\xi(r_k) \leq t_k, k = 1, \ldots, n\} \\
& \quad = \mathsf{P}\{\bar{M}(m_k) = \bar{l}_{m_k}, k = 1, \ldots, n\}\mathsf{P}\{N(t_k) \geq r_k, k = 1, \ldots, n\} \\
& \quad \text{as } \varepsilon \to 0. \qquad (3.127)
\end{aligned}$$

The following relations, similar to (3.121), take place, for (a) any vectors $\bar{l}_k, k = 1, 2, \ldots$, satisfying the relation (3.122), (b) integers $0 \leq m_1 \leq m_2 \leq \cdots < \infty$ and $0 \leq m[1] \leq m[2] \leq \cdots$, (c) $0 < t_1 \leq t_2 \leq \cdots < \infty$, and (d) $i \in \mathbb{X}$,

$$\begin{aligned}
& \mathsf{P}_i\{\bar{M}_\varepsilon(m_1) = \bar{l}_{m_1}, N_\varepsilon(t_1) = m[1]\} \\
& \quad = \mathsf{P}_i\{\bar{M}_\varepsilon(m_1) = \bar{l}_{m_1}, N_\varepsilon(t_1) \geq m[1]\} \\
& \qquad - \mathsf{P}_i\{\bar{M}_\varepsilon(m_1) = \bar{l}_{m_1}, N_\varepsilon(t_1) \geq m[1] + 1\} \\
& \quad = \mathsf{P}_i\{\bar{M}_\varepsilon(m_1) = \bar{l}_{m_1}, \xi_\varepsilon(m[1]) \leq t_1\} \\
& \qquad - \mathsf{P}_i\{\bar{M}_\varepsilon(m_1) = \bar{l}_{m_1}, \xi_\varepsilon(m[1] + 1) \leq t_1\}, \\
& \mathsf{P}_i\{\bar{M}_\varepsilon(m_l) = \bar{l}_{m_l}, N_\varepsilon(t_l) = m[l], l = 1, 2\} \\
& \quad = \mathsf{P}_i\{\bar{M}_\varepsilon(m_l) = \bar{l}_{m_l}, l = 1, 2, N_\varepsilon(t_1) = m[1], N_\varepsilon(t_2) \geq m[2]\} \\
& \qquad - \mathsf{P}_i\{\bar{M}_\varepsilon(m_l) = \bar{l}_{m_l}, l = 1, 2, N_\varepsilon(t_1) = m[1], N_\varepsilon(t_2) \geq m[2] + 1\} \\
& \quad = \mathsf{P}_i\{\bar{M}_\varepsilon(m_l) = \bar{l}_{m_l}, l = 1, 2, N_\varepsilon(t_1) \geq m[1], N_\varepsilon(t_2) \geq m[2]\} \\
& \qquad - \mathsf{P}_i\{\bar{M}_\varepsilon(m_l) = \bar{l}_{m_l}, l = 1, 2, N_\varepsilon(t_1) \geq m[1] + 1, N_\varepsilon(t_2) \geq m[2]\} \\
& \qquad + \mathsf{P}_i\{\bar{M}_\varepsilon(m_l) = \bar{l}_{m_l}, l = 1, 2, N_\varepsilon(t_1) \geq m[1], N_\varepsilon(t_2) \geq m[2] + 1\} \\
& \qquad - \mathsf{P}_i\{\bar{M}_\varepsilon(m_l) = \bar{l}_{m_l}, l = 1, 2, N_\varepsilon(t_1) \geq m[1] + 1, N_\varepsilon(t_2) \geq m[2] + 1\} \\
& \quad = \mathsf{P}_i\{\bar{M}_\varepsilon(m_l) = \bar{l}_{m_l}, l = 1, 2, \xi_\varepsilon(m[1]) \leq t_1, \xi_\varepsilon(m[2]) \leq t_2\} \\
& \qquad - \mathsf{P}_i\{\bar{M}_\varepsilon(m_l) = \bar{l}_{m_l}, l = 1, 2, \xi_\varepsilon(m[1] + 1) \leq t_1, \xi_\varepsilon(m[2]) \leq t_2\} \\
& \qquad + \mathsf{P}_i\{\bar{M}_\varepsilon(m_l) = \bar{l}_{m_l}, l = 1, 2, \xi_\varepsilon(m[1]) \leq t_1, \xi_\varepsilon(m[2] + 1) \leq t_2\} \\
& \qquad - \mathsf{P}_i\{\bar{M}_\varepsilon(m_l) = \bar{l}_{m_l}, l = 1, 2, \xi_\varepsilon(m[1] + 1) \leq t_1, \xi_\varepsilon(m[2] + 1) < t_2\}, \\
& \text{etc.} \qquad (3.128)
\end{aligned}$$

The relations (3.127 and (3.128) imply that, for (a) any vectors $\bar{l}_k, k = 1, 2, \ldots$, satisfying the relation (3.122), (b) integers $0 \leq m_1 \leq m_2 \leq \cdots < \infty$ and $0 \leq m[1] \leq m[2] \leq \cdots$, (c) $0 < t_1 \leq \cdots \leq t_n < \infty$, which are the points of continuity for the corresponding limiting distribution functions of random variables $\xi(k), k = 1, \ldots$, and (d) $i \in \mathbb{X}, n = 1, 2, \ldots$,

$$
\begin{aligned}
&\mathsf{P}_i\{\bar{M}_\varepsilon(m_l) = \bar{l}_{m_l}, N_\varepsilon(t_l) = m[l], l = 1, \ldots, n\} \\
&\quad \longrightarrow \mathsf{P}\{\bar{M}(m_l) = \bar{l}_{m_l}, l = 1, \ldots, n\} \\
&\qquad \times \mathsf{P}\{N(t_l) = m[l], l = 1, \ldots, n\} \text{ as } \varepsilon \to 0.
\end{aligned} \tag{3.129}
$$

Finally, the conditions $\mathbf{A}_1$–$\mathbf{D}_1$ and $\mathbf{A}_2$, and the relations (3.117) and (3.129), imply that, for (a) any vectors $\bar{l}_k, k = 1, 2, \ldots$, satisfying the relation (3.122), (b) integers $0 \leq m[1] \leq m[2] \leq \cdots$, (c) $0 < t_1 \leq \cdots \leq t_n < \infty$, which are the points of continuity for the corresponding limiting distribution functions of random variables $\xi(k), k = 1, \ldots$, and (d) $i \in \mathbb{X}, n = 1, 2, \ldots$,

$$
\begin{aligned}
&\mathsf{P}_i\{\bar{N}_\varepsilon(t_k) = \bar{l}_{m[k]}, k = 1, \ldots, n\} \\
&\quad = \mathsf{P}_i\{N_\varepsilon(t_k) = m[k], \bar{M}_\varepsilon(m[k]) = \bar{l}_{m[k]}, k = 1, \ldots, n\} \\
&\quad \longrightarrow \mathsf{P}\{N(t_k) = m[k], k = 1, \ldots, n\} \\
&\qquad \times \mathsf{P}\{\bar{M}(m[k]) = \bar{l}_{m[k]}, k = 1, \ldots, n\} \\
&\quad = \mathsf{P}\{\bar{N}(t_k) = \bar{l}_{m[k]}, k = 1, \ldots, n\} \text{ as } \varepsilon \to 0.
\end{aligned} \tag{3.130}
$$

The relation (3.14) implies that the following asymptotic relation holds, for any initial distributions $\bar{q}_\varepsilon$,

$$
\bar{N}_\varepsilon(t), t \in \overline{\mathbb{K}} \xrightarrow{\mathsf{d}} \bar{N}(t), t \in \overline{\mathbb{K}} \text{ as } \varepsilon \to 0. \tag{3.131}
$$

The process $\bar{N}(t), t \geq 0$ is a càdlàgprocess with non-decreasing components. The set $\mathbb{K}$ is at most countable and, thus, the set $\overline{\mathbb{K}} \subseteq \mathbb{N}$ is dense in the interval $[0, \infty)$. Therefore, the relation (3.15) can be extended in a standard way to the asymptotic relation appearing in the statement **(i)**, with the limiting counting process $\bar{N}(t)$ described in the statement **(ii)** of Theorem 3.3.

Suppose now that the conditions $\mathbf{A}_1$–$\mathbf{C}_1$ are satisfied and the following asymptotic relation takes place for some initial distributions $\bar{q}_\varepsilon$,

$$
\bar{N}_\varepsilon(t), t \in \mathbb{N} \xrightarrow{\mathsf{d}} \bar{N}(t), t \in \mathbb{N} \text{ as } \varepsilon \to 0, \tag{3.132}
$$

where $\bar{N}(t)$ is some stochastic process described in the statement **(i)** of Theorem 3.3, and $\mathbb{N}$ is the set of points of stochastic continuity $t > 0$ for this process.

Let us prove that, in this case, the conditions $\mathbf{D}_1$ and $\mathbf{A}_2$ are satisfied.

The relation (3.132) implies that the following relation holds, for the above initial distributions $\bar{q}_\varepsilon$ and $t \in \mathbb{N}$,

$$
N_\varepsilon(t) = \sum_{r \in \mathbb{X}} N_{\varepsilon,r}(t), t \in \mathbb{N} \xrightarrow{\mathsf{d}} N(t) = \sum_{r \in \mathbb{X}} N_r(t), t \in \mathbb{N} \text{ as } \varepsilon \to 0. \tag{3.133}
$$

Obviously, $N(t)$ is an integer-valued, non-negative, non-decreasing, step-wise, càdlàgstochastic process such that $\mathsf{P}\{N(t) \geq 1\} = G(t)$ is the distribution function on $[0, \infty)$ not concentrated at zero, and $\mathbb{N}$ is the set of stochastic continuity points $t > 0$ for this process.

Therefore, according to the statement **(i)** of Theorem 3.1, the condition $\mathbf{D}_1$ is satisfied, and, according to the statement **(ii)** of Theorem 3.1, $G(u)$ is the distribution function with the Laplace transform $\frac{1}{1+A(s)}$, where $A(s)$ is the cumulant appearing in the condition $\mathbf{D}_1$.

Since the functions $p_\varepsilon(r)/p_\varepsilon \in [0, 1]$ for $\varepsilon \in (0, 1]$, $r \in \mathbb{X}$, any sequence $0 < \varepsilon_n \to 0$ as $n \to \infty$ contains a subsequence $0 < \varepsilon_{n_l} \to 0$ as $l \to \infty$ such that the following relation holds:

$$\frac{p_{\varepsilon_{n_l}}(r)}{p_{\varepsilon_{n_l}}} \to Q_r \text{ as } l \to \infty, \text{ for } r \in \mathbb{X}. \tag{3.134}$$

Obviously, the conditions $\mathbf{A}_1$–$\mathbf{C}_1$ are satisfied for the Markov renewal processes $(\xi_{\varepsilon_{n_l}}(k), \eta_{\varepsilon_{n_l}}(k))$. According to the above remarks, the condition $\mathbf{D}_1$ also holds for the Markov renewal processes $(\xi_{\varepsilon_{n_l}}(k), \eta_{\varepsilon_{n_l}}(k))$. The relation (3.134) plays the role of the condition $\mathbf{A}_2$ for the Markov renewal processes $(\xi_{\varepsilon_{n_l}}(k), \eta_{\varepsilon_{n_l}}(k))$.

Therefore, the statement **(ii)** of Theorem 3.3 takes place for the Markov renewal processes $(\xi_{\varepsilon_{n_l}}(k), \eta_{\varepsilon_{n_l}}(k))$. According to this statement, the following asymptotic relation holds, for (a) any vectors $\bar{l}_k$, $k = 1, 2, \ldots$, satisfying the relation (3.122), (b) $t \in \mathbb{N}$, and (c) $i \in \mathbb{X}$,

$$\begin{aligned} \mathsf{P}_i\{\bar{N}_{\varepsilon_{n_l}}(t) = \bar{l}_k\} &= \mathsf{P}_i\{N_{\varepsilon_{n_l}}(t) = k, \bar{M}_{\varepsilon_{n_l}}(k) = \bar{l}_k\} \\ &\to \mathsf{P}\{\bar{N}(t) = \bar{l}_k\} \\ &= \mathsf{P}\{N(t) = k\}\mathsf{P}\{\bar{M}(1) = \bar{l}_k\} \\ &= \mathsf{P}\{N(t) = k\}G(k, \bar{l}_k, \bar{Q}) \text{ as } l \to \infty. \end{aligned} \tag{3.135}$$

Note that, according to the relations (3.132) and (3.133), the probabilities $\mathsf{P}\{\bar{N}(t) = \bar{l}_k\}$ and $\mathsf{P}\{N(t) = k\}$ do not depend on the choice of subsequence ε_{n_l}, while vectors $\bar{Q} = (Q_r, r \in \mathbb{X})$ and, thus, probabilities $G(k, \bar{l}_k, \bar{Q})$ may depend on the choice of subsequence ε_{n_l}.

Obviously, $N(t) \xrightarrow{\mathsf{P}} \infty$ as $t \to \infty$. Therefore, $\mathsf{P}\{N(t) \geq 1\} = \sum_{k \geq 1} \mathsf{P}\{N(t) = k\} > 0$ for t large enough. Thus, there exists $t \in \mathbb{N}$ such that $\mathsf{P}\{N(t) = k\} > 0$ for some $k \geq 1$.

Let us choose an arbitrary $r \in \mathbb{X}$ and, then, a vector $\bar{l}_k = (k\mathrm{I}(l_i = l_r), i \in \mathbb{X})$. In this case, the equality in relation (3.135) takes the following form:

$$\mathsf{P}\{\bar{N}(t) = \bar{l}_k\} = \mathsf{P}\{N(t) = k\}Q_r^k. \tag{3.136}$$

This equality implies that probability Q_r is the same for any sequence ε_n and subsequence ε_{n_l} chosen such that the asymptotic relation (3.134) holds. It is so for any $r \in \mathbb{X}$. Thus, the following asymptotic relation takes place:

$$\frac{p_\varepsilon(r)}{p_\varepsilon} \to Q_r \text{ as } \varepsilon \to 0, \text{ for } r \in \mathbb{X}, \tag{3.137}$$

where $Q_r \geq 0, r \in \mathbb{X}, \sum_{r \in \mathbb{X}} Q_r = 1$.

Therefore, not only the condition $\mathbf{D}_1$, but also the condition $\mathbf{A}_2$ holds. □

Remark 3.4 In the case, where condition $\mathbf{D}_1$ is modified in the way described in Remark 3.3, $N_r(0) = 0$ with probability 1, for $r \in \mathbb{X}$, and, thus, the point 0 can be additionally included in the set $\mathbb{N}$.

Remark 3.5 In the case described in Remark 3.4, the relation (3.11) implies (by Theorem 4.4.1 from Silvestrov (2004)) that, under the conditions, $\mathbf{A}_1$–$\mathbf{D}_1$ and $\mathbf{A}_2$, the more general asymptotic relation, $\bar{N}_\varepsilon(t), t \geq 0 \stackrel{\mathsf{J}}{\longrightarrow} \bar{N}(t), t \geq 0$ as $\varepsilon \to 0$, takes place.

Chapter 4
Generalisations of Limit Theorems for First-Rare-Event Times

In this chapter, we generalise the results of Chap. 2 to some modifications of first-rare-event times and processes. The relationship between first-rare-event times and hitting times is also clarified. This chapter includes two sections.

In Sect. 4.1, we present results generalising Theorem 2.1. Theorem 4.1 presents necessary and sufficient conditions of convergence in distribution for vector first-rare-event times and convergence in the topology J for vector first-rare-event processes defined on asymptotically uniformly ergodic semi-Markov processes. Lemmas 4.2 and 4.3 present necessary and sufficient conditions of convergence in distribution for low and upper first-rare-event times and processes. Theorems 4.2 and 4.3 present necessary and sufficient conditions of convergence in distribution for first-rare-event times and convergence in the topology J for first-rare-event processes for models with transition periods. Finally, Theorem 4.4 presents necessary and sufficient conditions of convergence in distribution for first-rare-event times for models with extending phase spaces of the corresponding semi-Markov processes.

In Sect. 4.2, we clarify connection between first-rare-event times and standard and directed hitting times for semi-Markov processes. In Theorems 4.5 and 4.6 we present necessary and sufficient conditions of convergence in distribution, respectively, for standard and directed hitting times defined on asymptotically uniformly ergodic semi-Markov processes.

4.1 Modifications of First-Rare-Event Times and Rare-Event Time Processes

In this section, we present several generalisations of asymptotic results given in Chap. 2 on vector first-rare-event times and processes, low and upper first-rare-event times and processes for models with transition periods, and first-rare-event times for models with extending phase spaces of the corresponding semi-Markov processes.

D. Silvestrov, *Perturbed Semi-Markov Type Processes I*,
https://doi.org/10.1007/978-3-030-92403-4_4

4.1.1 Vector First-Rare-Event Times and Rewards

4.1.1.1 Necessary and Sufficient Conditions of Convergence for Vector First-Rare-Event Times. Let $(\eta_{\varepsilon,n}, \bar{\kappa}_{\varepsilon,n} = (\kappa_{\varepsilon,1,n}, \ldots, \kappa_{\varepsilon,L,n}), \rho_{\varepsilon,n})$, $n = 0, 1, \ldots$ be, for every $\varepsilon \in (0, 1]$, a Markov renewal process, i.e., a homogenous Markov chain with a phase space $\mathbb{Z} = \mathbb{X} \times \mathbb{R}^+_L \times \{0, 1\}$, where $\mathbb{X} = \{1, 2, \ldots, m\}$ is a finite set and $\mathbb{R}^+_L = \{\bar{x} = (x_1, \ldots, x_L) : x_1, \ldots, x_l \in [0, \infty)\}$, an initial distribution $\hat{q}_\varepsilon = \langle q_{\varepsilon,i,s,\imath} = \mathsf{P}\{\eta_{\varepsilon,0} = i, \bar{\kappa}_{\varepsilon,0} \leq \bar{s}, \rho_{\varepsilon,0} = \imath\}, (i, s, \imath) \in \mathbb{Z}\rangle$, and transition probabilities defined by the following relation, for $(i, \bar{s}, \imath), (j, \bar{t}, \jmath) \in \mathbb{Z}$,

$$
\begin{aligned}
\mathsf{P}\{\eta_{\varepsilon,1} = j, \bar{\kappa}_{\varepsilon,1} \leq \bar{t}, \rho_{\varepsilon,1} = \jmath/\eta_{\varepsilon,0} = i, \bar{\kappa}_{\varepsilon,0} = \bar{s}, \rho_{\varepsilon,0} = \imath\} \\
= \mathsf{P}\{\eta_{\varepsilon,1} = j, \bar{\kappa}_{\varepsilon,1} \leq \bar{t}, \rho_{\varepsilon,1} = \jmath/\eta_{\varepsilon,0} = i\} \\
= Q_{\varepsilon,ij}(\bar{t}, \jmath). \qquad (4.1)
\end{aligned}
$$

In this case, we are interested in vector first-rare-event times,

$$
\begin{aligned}
\bar{\xi}_\varepsilon &= (\xi_{\varepsilon,1}, \ldots, \xi_{\varepsilon,L}) \\
&= (\sum_{n=1}^{\nu_\varepsilon} \kappa_{\varepsilon,1,n}, \ldots, \sum_{n=1}^{\nu_\varepsilon} \kappa_{\varepsilon,L,n}) = \sum_{n=1}^{\nu_\varepsilon} \bar{\kappa}_{\varepsilon,n} \qquad (4.2)
\end{aligned}
$$

and vector first-rare-event time processes,

$$
\begin{aligned}
\bar{\xi}_\varepsilon(t) &= (\xi_{\varepsilon,1}(t), \ldots, \xi_{\varepsilon,L}(t)) \\
&= (\sum_{n=1}^{[t\nu_\varepsilon]} \kappa_{\varepsilon,1,n}, \ldots, \sum_{n=1}^{[t\nu_\varepsilon]} \kappa_{\varepsilon,L,n}) = \sum_{n=1}^{[t\nu_\varepsilon]} \bar{\kappa}_{\varepsilon,n}, \ t \geq 0, \qquad (4.3)
\end{aligned}
$$

where

$$
\nu_\varepsilon = \min(n \geq 1 : \rho_{\varepsilon,n} = 1). \qquad (4.4)
$$

In this model, the components of the random vector $\bar{\xi}_\varepsilon$ can be interpreted in different ways. For example, the first component $\xi_{\varepsilon,1}$ can be interpreted as the first-rare-event time, while the other components $\xi_{\varepsilon,l}, l > 1$ as rewards accumulated on trajectories of Markov chain $\eta_{\varepsilon,n}$ until the first-rare-event time.

Let us recall the stationary distribution $\bar{\pi}_\varepsilon = \langle \pi_{\varepsilon,i}, i \in \mathbb{X}\rangle$ for the imbedded Markov chain $\eta_{\varepsilon,n}$.

Let us also recall the one-step probabilities of occurrence of the first-rare-event,

$$
p_{\varepsilon,i} = \mathsf{P}_i\{\rho_{\varepsilon,1} = 1\}, \ i \in \mathbb{X}, \qquad (4.5)
$$

and

$$
p_\varepsilon = \sum_{i \in \mathbb{X}} \pi_{\varepsilon,i} p_{\varepsilon,i} \text{ and } u_\varepsilon = p_\varepsilon^{-1}. \qquad (4.6)
$$

The conditions $\mathbf{A}_1$ and $\mathbf{B}_1$ do not change.

The condition $\mathbf{C}_1$ takes, in this case, the following form:

$\mathbf{C}_3$: $\mathsf{P}_i\{\kappa_{\varepsilon,l,1} > \delta/\rho_{\varepsilon,1} = 1\} = \mathsf{P}\{\kappa_{\varepsilon,l,1} > \delta/\rho_{\varepsilon,1} = 1, \eta_{\varepsilon,0} = i\} \to 0$ as $\varepsilon \to 0$, for $\delta > 0, l = 1, \ldots, L, i \in \mathbb{X}$.

Note that we define the probability $\mathsf{P}_i\{\kappa_{\varepsilon,l,1} > \delta/\rho_{\varepsilon,1} = 1\} = 0$ in the cases where $\mathsf{P}_i\{\rho_{\varepsilon,1} = 1\} = 0$.

Let us introduce multi-dimensional distribution functions, for $i \in \mathbb{X}$ and $\varepsilon \in (0, 1]$,

$$F_{\varepsilon,i}(\bar{t}) = \mathsf{P}_i\{\bar{\kappa}_{\varepsilon,1} \leq \bar{t}\}, \ \bar{t} \in \mathbb{R}_L^+. \tag{4.7}$$

Let $\bar{\theta}_{\varepsilon,n}, n = 1, 2, \ldots$ be i.i.d. random vectors with the distribution function $F_\varepsilon(\bar{t})$, which is the result of averaging the distribution functions $F_{\varepsilon,i}(\bar{t})$ over the stationary distribution of the imbedded Markov chain $\eta_{\varepsilon,n}$,

$$F_\varepsilon(\bar{t}) = \sum_{i \in \mathbb{X}} \pi_{\varepsilon,i} F_{\varepsilon,i}(\bar{t}), \ t \in \mathbb{R}_L^+. \tag{4.8}$$

Now we can formulate a necessary and sufficient condition for the convergence in the distribution of the vectors of the first-rare-event times:

$\mathbf{D}_7$: $\bar{\theta}_\varepsilon = (\theta_{\varepsilon,1}, \ldots, \theta_{\varepsilon,L}) = \sum_{n=1}^{[u_\varepsilon]} \bar{\theta}_{\varepsilon,n} \stackrel{\mathsf{d}}{\longrightarrow} \bar{\theta}_0 = (\theta_{0,1}, \ldots, \theta_{0,L})$ as $\varepsilon \to 0$, where $\bar{\theta}_0$ is a random vector, with non-negative components not concentrated at zero.

As well known, the limiting random vector $\bar{\theta}_0$ penetrating condition $\mathbf{D}_7$ should be infinitely divisible, i.e., its Laplace transform has the form, $\mathsf{E}e^{-(\bar{s},\bar{\theta}_0)} = \mathsf{E}e^{-\sum_{l=1}^L s_l \theta_{0,l}} = e^{-A(\bar{s})}, \bar{s} = (s_1, \ldots, s_L) \in \mathbb{R}_L^+, t \geq 0$, where $A(\bar{s})$ is the cumulant of the corresponding multi-dimensional infinitely divisible distribution.

Let us also consider a homogeneous step-sum process with independent increments (summands are i.i.d. random vectors),

$$\bar{\theta}_\varepsilon(t) = \sum_{n=1}^{[tu_\varepsilon]} \bar{\theta}_{\varepsilon,n}, t \geq 0. \tag{4.9}$$

According to Theorem A.3, the condition $\mathbf{D}_7$ is necessary and sufficient for holding of the asymptotic relation,

$$\bar{\theta}_\varepsilon(t) = \sum_{n=1}^{[tu_\varepsilon]} \bar{\theta}_{\varepsilon,n}, t \geq 0 \stackrel{\mathsf{J}}{\longrightarrow} \bar{\theta}_0(t), t \geq 0 \text{ as } \varepsilon \to 0, \tag{4.10}$$

where $\theta_0(t) = (\theta_{0,1}(t), \ldots, \theta_{0,L}(t)), t \geq 0$ is the vector Lévy process (a càdlàg homogeneous process with independent increments) with non-negative components not concentrated at zero and with Laplace transforms $\mathsf{E}e^{-(\bar{s},\bar{\theta}_0(t))} = \mathsf{E}e^{-\sum_{l=1}^L s_l \theta_{0,l}(t))} = e^{-tA(\bar{s})}, \bar{s} = (s_1, \ldots, s_L) \in \mathbb{R}_L^+, t \geq 0$, where $A(\bar{s})$ is the cumulant of infinitely divisible random vector with non-negative components not concentrated at zero appearing in condition $\mathbf{D}_7$.

Let us recall the initial distribution $\bar{q}_\varepsilon = \langle q_{\varepsilon,i} = \mathsf{P}\{\eta_{\varepsilon,0} = 0\}, i \in \mathbb{X}\rangle$ for the imbedded Markov chain $\eta_{\varepsilon,n}$.

The following analogue of Theorem 2.1 takes place.

Theorem 4.1 *Let the conditions* $\mathbf{A}_1$, $\mathbf{B}_1$, *and* $\mathbf{C}_3$ *be satisfied. Then:*

(i) *The condition* $\mathbf{D}_7$ *is necessary and sufficient for the fulfilment (for some or any initial distributions* $\bar{q}_\varepsilon$, *respectively, in the statements of necessity and sufficiency) of the asymptotic relation* $\bar{\xi}_\varepsilon = \bar{\xi}_\varepsilon(1) \xrightarrow{\mathsf{d}} \bar{\xi}_0$ *as* $\varepsilon \to 0$, *where* $\bar{\xi}_0$ *is a random vector with non-negative components not concentrated at zero.*

(ii) *The distribution function* $G(\bar{u}) = \mathsf{P}\{\bar{\xi}_0 \leq \bar{u}\}, \bar{u} \in \mathbb{R}_L^+$ *of the limiting random vector* $\bar{\xi}_0$ *has the Laplace transform* $\phi(\bar{s}) = \int_0^\infty e^{-(\bar{s},\bar{u})} G(d\bar{u}) = \frac{1}{1+A(\bar{s})}, s \in \mathbb{R}_L^+$, *where* $A(\bar{s})$ *is the cumulant of multivariate infinitely divisible random vector with non-negative components not concentrated at zero appearing in the condition* $\mathbf{D}_7$.

(iii) *The stochastic processes* $\bar{\xi}_\varepsilon(t), t \geq 0 \xrightarrow{\mathsf{J}} \bar{\xi}_0(t) = \bar{\theta}_0(t\nu_0), t \geq 0$ *as* $\varepsilon \to 0$, *where:* **(a)** ν_0 *is a random variable, which has the exponential distribution with parameter* 1, **(b)** $\bar{\theta}_0(t), t \geq 0$ *is a non-negative Lévy process with the Laplace transforms* $\mathsf{E}e^{-(\bar{s},\bar{\theta}_0(t))} = e^{-tA(\bar{s})}, \bar{s} \in \mathbb{R}_L^+, t \geq 0$, *and* **(c)** *the random variable* ν_0 *and the process* $\bar{\theta}_0(t), t \geq 0$ *are independent.*

The proof can be reduced to the scalar case using the method known as the Wold–Cramér device (see, for example, Silvestrov (2004) or Kallenberg (2021)).

The following "scalar" random variables can be defined, for $\bar{s} = (s_1, \ldots, s_L) \in \mathbb{R}_L^+$,

$$\theta_\varepsilon[\bar{s}] = \sum_{l=1}^{L} s_l \theta_{\varepsilon,l}. \tag{4.11}$$

The condition $\mathbf{D}_7$ can, according to the Wold–Cramér device, be reformulated in the following equivalent form:

$\mathbf{D}_8$: $\theta_\varepsilon[\bar{s}] \xrightarrow{\mathsf{d}} \bar{\theta}_0[\bar{s}] = \sum_{l=1}^L s_l \theta_{0,l}$ as $\varepsilon \to 0$, for $\bar{s} = (s_1, \ldots, s_L) \in \mathbb{R}_L^+$, where $\bar{\theta}_0 = (\theta_{0,1}, \ldots, \theta_{0,L})$ is a random vector, with non-negative components not concentrated at zero.

Then, Theorem 2.1 can be applied to the following random variables, for $\bar{s} = (s_1, \ldots, s_L) \in \mathbb{R}_L^+$,

$$\xi_\varepsilon[\bar{s}] = \sum_{l=1}^{L} s_l \xi_{\varepsilon,l} = \sum_{n=1}^{\nu_\varepsilon} \kappa_{\varepsilon,n}[\bar{s}], \tag{4.12}$$

where

$$\kappa_{\varepsilon,n}[\bar{s}] = \sum_{l=1}^{L} s_l \kappa_{\varepsilon,l,n}, \ n = 1, 2, \ldots. \tag{4.13}$$

In this way, we shall prove that, under the conditions $\mathbf{A}_1$, $\mathbf{B}_1$, and $\mathbf{C}_5$, the condition $\mathbf{D}_7$ is necessary and sufficient for the fulfilment (for some or any initial distributions $\bar{q}_\varepsilon$, respectively, in statements of necessity and sufficiency) of the following relation, for $\bar{s} = (s_1, \ldots, s_L) \in \mathbb{R}_L^+$,

$$\xi_\varepsilon[\bar{s}] \xrightarrow{\mathsf{d}} \xi_0[\bar{s}] = \sum_{l=1}^{L} s_l \xi_{0,l} \text{ as } \varepsilon \to 0, \tag{4.14}$$

where $\bar{\xi}_0 = (\xi_{0,1}, \cdots, \xi_{0,L})$ is a random vector with non-negative components not concentrated at zero.

The relation (4.14) is, according to the Wold–Cramér device, equivalent to the following asymptotic relation:

$$\bar{\xi}_\varepsilon \xrightarrow{\mathsf{d}} \bar{\xi}_0 \text{ as } \varepsilon \to 0. \tag{4.15}$$

The proof of the statement **(iii)** can be accomplished in the following way.

First, the Wold–Cramér device can be used in the manner described above to obtain the following relation of convergence in distribution for stochastic processes $(t\nu^*_\varepsilon, \bar{\kappa}_\varepsilon(t)) = (tp_\varepsilon \nu_\varepsilon, \sum_{k \leq tu_\varepsilon} \bar{\kappa}_{\varepsilon,k}), t \geq 0$, similar to the relation (2.123) given in Lemma 2.9,

$$(t\nu^*_\varepsilon, \bar{\kappa}_\varepsilon(t)), t \geq 0 \xrightarrow{\mathsf{d}} (t\nu_0, \bar{\theta}_0(t)), t \geq 0 \text{ as } \varepsilon \to 0, \tag{4.16}$$

where: (a) ν_0 is a random variable, which has the exponential distribution with parameter 1, (b) $\bar{\theta}_0(t), t \geq 0$ is a non-negative Lévy process with the Laplace transforms $\mathsf{E}e^{-s\theta_0(t)} = e^{-tA(s)}, \bar{s} \in \mathbb{R}^+_L, t \geq 0$, and (c) the random variable ν_0 and the process $\theta_0(t), t \geq 0$ are independent.

Also, the relation of J-convergence,

$$(\bar{\kappa}_\varepsilon(t)), t \geq 0 \xrightarrow{\mathsf{J}} \bar{\theta}_0(t), t \geq 0 \text{ as } \varepsilon \to 0, \tag{4.17}$$

can be proved for the vector case ($L > 1$) in the way analogous to those used in the proof of Theorem 2.2 for the scalar case ($L = 1$).

Finally, the asymptotic relations (4.16) and (4.17) let us apply Theorem A.6 to the compositions of stochastic processes $\kappa_\varepsilon(t), t \geq 0$ and $t\nu^*_\varepsilon$. This gives the following relation:

$$\bar{\kappa}_\varepsilon(t\nu^*_\varepsilon), t \geq 0 \xrightarrow{\mathsf{J}} \bar{\theta}_0(t\nu_0), t \geq 0 \text{ as } \varepsilon \to 0. \tag{4.18}$$

4.1.1.2 Sufficient Conditions of Convergence for Real-Valued First-Rare-Event Rewards. Another natural generalisation concerns the model with real-valued rewards. In this case, we consider a Markov renewal process $(\eta_{\varepsilon,n}, \kappa_{\varepsilon,n}, \rho_{\varepsilon,n}), n = 0, 1, \ldots$ with a phase space $\mathbb{X} \times \mathbb{R}_1 \times \{0, 1\}$ ($\mathbb{X} = \{1, \ldots, m\}$ is a finite set and $\mathbb{R}_1 = (-\infty, \infty)$ is a real line).

The corresponding real-valued first-rare-event process can be defined in the standard way and represented as the difference of two non-negative first-rare-event processes,

$$\xi_\varepsilon(t) = \sum_{n=1}^{\nu_\varepsilon t} \kappa_{\varepsilon,n} = \xi^+_\varepsilon(t) - \xi^-_\varepsilon(t), t \geq 0, \tag{4.19}$$

where

$$\xi^\pm_\varepsilon(t) = \sum_{n=1}^{\nu_\varepsilon t} \kappa^\pm_{\varepsilon,n} \text{ and } \kappa^\pm_{\varepsilon,n} = \pm\kappa_{\varepsilon,n}\mathrm{I}(\pm\kappa_{\varepsilon,n} \geq 0),\ n \geq 1. \tag{4.20}$$

One can consider the vector first-rare-event process $\bar{\xi}_\varepsilon(t) = (\xi^+_\varepsilon(t), \xi^-_\varepsilon(t)), t \geq 0$ and apply to these processes the sufficiency statements of Theorem 4.1. The relations of convergence in distribution and J-convergence for the processes $\xi_\varepsilon(t), t \geq 0$ obviously imply the corresponding convergence relations for the processes $\xi_\varepsilon(t) = \xi^+_\varepsilon(t) - \xi^-_\varepsilon(t)), t \geq 0$, but not vice versa.

The question about necessary and sufficient convergence conditions for processes $\xi_\varepsilon(t), t \geq 0$ (based on the characteristics of the Markov renewal processes $(\eta_{\varepsilon,n}, \kappa_{\varepsilon,n})$ averaged over the stationary distributions of the embedded Markov chains $\eta_{\varepsilon,n}$) is open for the case where the corresponding "external" limiting process is a general Lévy process with a Gaussian component. Our conjecture is that this can be done by using methods similar to those used in the proofs of Theorems 2.1 and 2.2 but combined with a more sophisticated approach to proving the tightness for distributions of the corresponding non-monotone, in this case, step-sum reward processes and the asymptotic independence of these processes and the first-rare-event times. However, this is outside the scope of this book.

4.1.2 Upper and Lower First-Rare-Event Times

Let us define the random variables $\xi_{\varepsilon,+}$ and $\xi_{\varepsilon,-}$, which can be interpreted, respectively, as the upper first-rare-event time and the lower first-rare-event time,

$$\xi_{\varepsilon,+} = \xi_\varepsilon = \sum_{n=1}^{\nu_\varepsilon} \kappa_{\varepsilon,n}, \text{ and } \xi_{\varepsilon,-} = \sum_{n=1}^{\nu_\varepsilon - 1} \kappa_{\varepsilon,n}. \tag{4.21}$$

Lemma 4.1 *Let the condition* $\mathbf{C}_1$ *be satisfied. Then, the following relation holds, for any initial distribution* $\bar{q}_\varepsilon$,

$$\kappa_{\nu_\varepsilon} = \xi_{\varepsilon,+} - \xi_{\varepsilon,-} \xrightarrow{\mathsf{P}} 0 \text{ as } \varepsilon \to 0. \tag{4.22}$$

Proof The Markov property of the Markov renewal process $(\eta_{\varepsilon,n}, \kappa_{\varepsilon,n}, \rho_{\varepsilon,n})$ and the condition $\mathbf{C}_1$ imply that, for $\delta > 0$ and $i \in \mathbb{X}$,

$$\begin{aligned}
\mathsf{P}_i\{\kappa_{\varepsilon,\nu_\varepsilon} > \delta\} &= \sum_{n=1}^{\infty} \mathsf{P}_i\{\kappa_{\varepsilon,n} > \delta, \nu_\varepsilon = n\} \\
&= \sum_{n=1}^{\infty} \sum_{j \in \mathbb{X}} \mathsf{P}_i\{\kappa_{\varepsilon,n} > \delta, \eta_{\varepsilon,n-1} = j, \nu_\varepsilon > n-1, \rho_{\varepsilon,n} = 1\} \\
&= \sum_{n=1}^{\infty} \sum_{j \in \mathbb{X}} \mathsf{P}\{\kappa_{\varepsilon,n} > \delta, \rho_{\varepsilon,n} = 1 / \eta_{\varepsilon,n-1} = j, \nu_\varepsilon > n-1\} \\
&\quad \times \mathsf{P}_i\{\eta_{\varepsilon,n-1} = j, \nu_\varepsilon > n-1\}
\end{aligned}$$

$$
\begin{aligned}
&= \sum_{n=1}^{\infty}\sum_{j\in\mathbb{X}} \mathsf{P}\{\kappa_{\varepsilon,n} > \delta, \rho_{\varepsilon,n} = 1/\eta_{\varepsilon,n-1} = j\} \\
&\qquad \times \mathsf{P}_i\{\eta_{\varepsilon,n-1} = j, \nu_\varepsilon > n-1\} \\
&= \sum_{n=1}^{\infty}\sum_{j\in\mathbb{X}} \mathsf{P}\{\kappa_{\varepsilon,n} > \delta/\rho_{\varepsilon,n} = 1, \eta_{\varepsilon,n-1} = j\} \\
&\qquad \times \mathsf{P}\{\rho_{\varepsilon,n} = 1/\eta_{\varepsilon,n-1} = j\}\mathsf{P}_i\{\eta_{\varepsilon,n-1} = j, \nu_\varepsilon > n-1\} \\
&= \sum_{n=1}^{\infty}\sum_{j\in\mathbb{X}} \mathsf{P}\{\kappa_{\varepsilon,1} > \delta/\rho_{\varepsilon,1} = 1, \eta_{\varepsilon,0} = j\} \\
&\qquad \times \mathsf{P}_i\{\eta_{\varepsilon,n-1} = j, \nu_\varepsilon = n\} \\
&\leq \max_{j\in\mathbb{X}} \mathsf{P}_j\{\kappa_{\varepsilon,1} > \delta/\rho_{\varepsilon,1} = 1\} \\
&\qquad \times \sum_{n=1}^{\infty}\sum_{j\in\mathbb{X}} \mathsf{P}_i\{\eta_{\varepsilon,n-1} = j, \nu_\varepsilon = n\} \\
&= \max_{j\in\mathbb{X}} \mathsf{P}_j\{\kappa_{\varepsilon,1} > \delta/\rho_{\varepsilon,1} = 1\} \to 0 \text{ as } \varepsilon \to 0. \qquad (4.23)
\end{aligned}
$$

The proof is complete. □

The condition $\mathbf{C}_1$ can be reformulated in the following equivalent form:

$\mathbf{C}_1'$: $\mathsf{P}_i\{\kappa_{\varepsilon,1} > \delta, \rho_{\varepsilon,1} = 1\} \leq o_{i,\delta}(\varepsilon)\mathsf{P}_i\{\rho_{\varepsilon,1} = 1\}$, where $0 \leq o_{i,\delta}(\varepsilon) \to 0$ as $\varepsilon \to 0$, for $\delta > 0, i \in \mathbb{X}$.

The Markov inequality $\mathsf{P}_i\{\kappa_{\varepsilon,1} > \delta/\rho_{\varepsilon,1} = 1\} \leq \delta^{-1}\mathsf{E}_i\{\kappa_{\varepsilon,1}/\rho_{\varepsilon,1} = 1\}$, $\delta > 0$ implies that the condition $\mathbf{C}_1$ can also be replaced by the following sufficient condition:

$\mathbf{C}_4$: $\mathsf{E}_i\kappa_{\varepsilon,1}\mathrm{I}(\rho_{\varepsilon,1} = 1) \leq o_i(\varepsilon)\mathsf{P}_i\{\rho_{\varepsilon,1} = 1\}$, where $0 \leq o_i(\varepsilon) \to 0$ as $\varepsilon \to 0$, for $i \in \mathbb{X}$.

Also, the Hölder inequality $\mathsf{E}_i\kappa_{\varepsilon,1}\mathrm{I}(\rho_{\varepsilon,1} = 1) \leq (\mathsf{E}_i\kappa_{\varepsilon,1}^{\alpha})^{\frac{1}{\alpha}}\mathsf{P}_i\{\rho_{\varepsilon,1} = 1\})^{\frac{1}{\beta}}$ (where $\alpha, \beta > 1, \frac{1}{\alpha}+\frac{1}{\beta} = 1$) implies that the following condition is sufficient for the fulfilment of the condition $\mathbf{C}_1$:

$\mathbf{C}_5$: $\mathsf{E}_i\kappa_{\varepsilon,1}^{\alpha}\mathsf{P}_i\{\rho_{\varepsilon,1} = 1\}^{\frac{\alpha}{\beta}} \leq o_{i,\alpha,\beta}(\varepsilon)\mathsf{P}_i\{\rho_{\varepsilon,1} = 1\}^{\alpha}$, where $0 \leq o_{i,\alpha,\beta}(\varepsilon) \to 0$ as $\varepsilon \to 0$, for $i \in \mathbb{X}$.

It is useful to note that the inequality in the condition $\mathbf{C}_5$ automatically holds, if $\mathsf{E}_i\kappa_{\varepsilon,1}^{\alpha} < \infty$ and $\mathsf{P}_i\{\rho_{\varepsilon,1} = 1\} = 0$. In the case, where $\mathsf{P}_i\{\rho_{\varepsilon,1} = 1\} > 0$, this inequality reduces to the inequality,

$$
\mathsf{E}_i\kappa_{\varepsilon,1}^{\alpha} \leq o_{i,\alpha,\beta}(\varepsilon)\mathsf{P}_i\{\rho_{\varepsilon,1} = 1\}^{\frac{\alpha(\beta-1)}{\beta}}. \qquad (4.24)
$$

The relation (4.23) implies that, under the condition $\mathbf{C}_1$, the following relation holds, for any initial distributions $\bar{q}_\varepsilon$,

$$\kappa_{\nu_\varepsilon} = \xi_{\varepsilon,+} - \xi_{\varepsilon,-} \to 0 \xrightarrow{\mathsf{P}} 0 \text{ as } \varepsilon \to 0. \tag{4.25}$$

Relation $(4.25)_1$ implies that the upper first-rare-event times $\xi_{\varepsilon,+} = \xi_\varepsilon$ can be replaced in Theorem 2.1 by the low first-rare-event times $\xi_{\varepsilon,-}$.

Moreover, let $\xi_{\varepsilon,\circ}$ be random variables for which there exists $\varepsilon_\circ \in (0, 1]$ such that the following relation holds, for $\varepsilon \in (0, \varepsilon_\circ]$,

$$\xi_{\varepsilon,-} \le \xi_{\varepsilon,\circ} \le \xi_{\varepsilon,+}. \tag{4.26}$$

The relations (4.25) and (4.26) imply that the following lemma takes place.

Lemma 4.2 *Let $\xi_{\varepsilon,\circ}$ be random variables, for which the relation (4.26) holds. Then, under the conditions* $\mathbf{A}_1$, $\mathbf{B}_1$, *and* $\mathbf{C}_1$ *of Theorem 2.1, the condition* $\mathbf{D}_1$ *is necessary and sufficient (for some or any initial distributions $\bar{q}_\varepsilon$, respectively, in the statements of necessity and sufficiency) for holding of the following relation, $\xi_{\varepsilon,\circ} \xrightarrow{\mathsf{d}} \xi_0$ as $\varepsilon \to 0$, where the limiting random variable ξ_0 is described in Theorem 2.1.*

A similar result can be obtained for the first-rare-event processes.

We define the stochastic processes, which can be interpreted, respectively, as the upper rare-event time process and the lower rare-event time process,

$$\xi_{\varepsilon,+}(t) = \xi_\varepsilon(t) = \sum_{n=1}^{[t\nu_\varepsilon]} \kappa_{\varepsilon,n}, t \ge 0 \text{ and } \xi_{\varepsilon,-}(t) = \sum_{n=1}^{[t(\nu_\varepsilon - 1)]} \kappa_{\varepsilon,n}, t \ge 0. \tag{4.27}$$

Let us also prove that, under the conditions of $\mathbf{A}_1$–$\mathbf{D}_1$ of Theorem 2.1, the following relation holds:

$$\xi_{\varepsilon,+}(t) - \xi_{\varepsilon,-}(t), t \ge 0 \xrightarrow{\mathsf{d}} 0(t) \equiv 0, t \ge 0 \text{ as } \varepsilon \to 0. \tag{4.28}$$

Indeed, the weak convergence relation (2.139) given in the proof of Lemma 2.9 can be rewritten in the following equivalent doubled form:

$$(t\nu^*_\varepsilon, t\nu^*_\varepsilon, \kappa_\varepsilon(t), \kappa_\varepsilon(t)) = (t\frac{\nu_\varepsilon}{u_\varepsilon}, t\frac{\nu_\varepsilon}{u_\varepsilon}, \sum_{k=1}^{[tu_\varepsilon]} \kappa_{\varepsilon,k}, \sum_{k=1}^{[tu_\varepsilon]} \kappa_{\varepsilon,k}), \ t \ge 0$$
$$\xrightarrow{\mathsf{d}} (t\nu_0, t\nu_0, \theta_0(t), \theta_0(t)), \ t \ge 0 \text{ as } \varepsilon \to 0, \tag{4.29}$$

where the limiting random variable ν_0 and the stochastic process $\theta_0(t), t \ge 0$ are described in Lemma 2.9.

Obviously, for $t \ge 0$,

$$t\nu^*_\varepsilon - t(\nu^*_\varepsilon - u_\varepsilon^{-1}) = tu_\varepsilon^{-1} \to 0 \text{ as } \varepsilon \to 0. \tag{4.30}$$

The relations (4.29), (4.30), and Theorem A.10 imply that the following relation holds:

$$(t(\nu_\varepsilon^* - u_\varepsilon^{-1}),\ t\nu_\varepsilon^*,\ \sum_{n=1}^{[tu_\varepsilon]} \kappa_{\varepsilon,n},\ \sum_{n=1}^{[tu_\varepsilon]} \kappa_{\varepsilon,n}),\ t \geq 0$$

$$\stackrel{\mathsf{d}}{\longrightarrow} (t\nu_0,\ t\nu_0,\ \theta_0(t),\ \theta_0(t)),\ t \geq 0 \text{ as } \varepsilon \to 0. \tag{4.31}$$

The relation of J-convergence for processes $\kappa_\varepsilon(t), t \geq 0$ appearing in the statement **(iii)** of Theorem 2.2 can obviously be rewritten in the following doubled form:

$$(\kappa_\varepsilon(t), \kappa_\varepsilon(t)) = (\sum_{n=1}^{[tu_\varepsilon]} \kappa_{\varepsilon,n},\ \sum_{n=1}^{[tu_\varepsilon]} \kappa_{\varepsilon,n})),\ t \geq 0$$

$$\stackrel{\mathsf{J}}{\longrightarrow} (\theta_0(t), \theta_0(t)) \text{ as } \varepsilon \to 0. \tag{4.32}$$

The relations (4.31) and (4.32) imply, by Theorem A.6, that the following relation holds:

$$(\xi_{\varepsilon,-}(t), \xi_{\varepsilon,+}(t)) = (\kappa_\varepsilon(t(\nu_\varepsilon^* - u_\varepsilon^{-1})), \kappa_\varepsilon(t\nu_\varepsilon^*))$$

$$= (\sum_{n=1}^{[t(\nu_\varepsilon - 1)]} \kappa_{\varepsilon,n},\ \sum_{n=1}^{[t\nu_\varepsilon]} \kappa_{\varepsilon,n})),\ t \geq 0$$

$$\stackrel{\mathsf{d}}{\longrightarrow} (\theta_0(t\nu_0), \theta_0(t\nu_0)) \text{ as } \varepsilon \to 0. \tag{4.33}$$

The relation (4.33) obviously implies the fulfilment of the relation (4.28).

Let $\xi_{\varepsilon,\circ}(t), t \geq$ be some càdlàg stochastic processes, for which the following relation takes place, for $0 < \varepsilon \leq \varepsilon_\circ \leq 1$,

$$\xi_{\varepsilon,-}(t) \leq \xi_{\varepsilon,\circ}(t) \leq \xi_{\varepsilon,+}(t), \text{ for } t \geq 0. \tag{4.34}$$

The relations (4.33) and (4.34) imply that the following lemma takes place.

Lemma 4.3 *Let $\xi_{\varepsilon,\circ}(t), t \geq 0$ be càdlàg stochastic processes, for which the relation (4.34) holds. Then, under the conditions* $\mathbf{A}_1$, $\mathbf{B}_1$, *and* $\mathbf{C}_1$ *of Theorem 2.1, the condition* $\mathbf{D}_1$ *is necessary and sufficient (for some or any initial distributions $\bar{q}_\varepsilon$, respectively, in the statements of necessity and sufficiency) for the fulfilment of the relation, $\xi_{\varepsilon,\circ}(t), t \geq 0 \stackrel{\mathsf{d}}{\longrightarrow} \xi_0(t) = \theta_0(t\nu_0)), t \geq 0$ as $\varepsilon \to 0$, where the limiting process $\xi_0(t), t \geq 0$ is described in Theorem 2.1.*

It is useful to note that the relations (4.31) and (4.32) imply, by Theorem A.6, holding of the following relation,

$$\xi_{\varepsilon,\pm}(t), t \geq 0 \stackrel{\mathsf{J}}{\longrightarrow} \theta_0(t\nu_0), t \geq 0 \text{ as } \varepsilon \to 0. \tag{4.35}$$

However, the vector processes $(\xi_{\varepsilon,-}(t), \xi_{\varepsilon,+}(t)), t \geq 0$ may not converge in the topology J. Indeed, the processes $\xi_{\varepsilon,+}(t)$ and $\xi_{\varepsilon,-}(t)$ are connected by the relation $\xi_{\varepsilon,+}(t) = \xi_{\varepsilon.-}(t\frac{\nu_\varepsilon}{\nu_\varepsilon - 1}), t \geq 0$, i.e., the process $\xi_{\varepsilon,-}(t)$ is a shifted in time version of the process $\xi_{\varepsilon,+}(t)$. This can lead to the fact that the vector processes $(\xi_{\varepsilon,-}(t), \xi_{\varepsilon,+}(t))$

can have large increments at close moments $\tau - c \leq \tau' \leq \tau \leq \tau'' \leq \tau + c$ and, therefore, the J-compactness may not take place for these vector processes.

As a consequence, the J-convergence of the processes $\xi_{\varepsilon,\circ}(t), t \geq 0$ also cannot be guaranteed.

The vector processes $(\xi_{\varepsilon,-}(t), \xi_{\varepsilon,+}(t)), t \geq 0$ converge in the topology J if the limiting process $\theta_0(t\nu_0), t \geq 0$ is continuous (in this case, J-convergence is equivalent to U-convergence). This is so if the limiting process $\theta_0(t), t \geq 0$ is a nonrandom linear function, i.e., $\theta_0(t) = a_0 t, t \geq 0$, where $a_0 > 0$ is a positive constant. In this case, the following relation takes place,

$$\xi_{\varepsilon,\circ}(t), t \geq 0 \stackrel{\mathsf{U}}{\longrightarrow} a_0 \nu_0\, t,\ t \geq 0 \text{ as } \varepsilon \to 0. \tag{4.36}$$

4.1.3 First-Rare-Event Times for Markov Renewal Processes with Transition Periods

4.1.3.1 First-Rare-Event Times and Processes for Markov Renewal Process with Transition Period. Let $\mathbb{X} = \{1, \ldots, m\}$ and $\tilde{\mathbb{X}} = \{1, \ldots, \tilde{m}\}$ are two finite sets such that $\mathbb{X} \subseteq \tilde{\mathbb{X}}$.

We define a Markov renewal process $(\tilde{\eta}_{\varepsilon,n}, \tilde{\kappa}_{\varepsilon,n}, \tilde{\rho}_{\varepsilon,n}), n = 0, 1, \ldots,$ with transition period as a Markov chain, with the phase space $\tilde{\mathbb{Z}} = \tilde{\mathbb{X}} \times [0, \infty) \times \{0, 1\}$ for the initial random vector $(\tilde{\eta}_{\varepsilon,0}, \tilde{\kappa}_{\varepsilon,0}, \tilde{\rho}_{\varepsilon,0})$ and the phase space $\mathbb{Z} = \mathbb{X} \times [0, \infty) \times \{0, 1\}$ for random vectors $(\tilde{\eta}_{\varepsilon,n}, \tilde{\kappa}_{\varepsilon,n}, \tilde{\rho}_{\varepsilon,n}), n \geq 1$, initial distribution $\tilde{q}_\varepsilon = \langle q_{\varepsilon,i,s,\imath} = \mathsf{P}\{\tilde{\eta}_{\varepsilon,0} = i, \tilde{\kappa}_{\varepsilon,0} \leq s, \tilde{\rho}_{\varepsilon,0} = \imath\}, (i, s, \imath) \in \tilde{\mathbb{Z}}\rangle$, and transition probabilities of the following form:

$$\begin{aligned}
&\mathsf{P}\{\tilde{\eta}_{\varepsilon,n} = j, \tilde{\kappa}_{\varepsilon,n} \leq t, \tilde{\rho}_{\varepsilon,n} = \jmath / \tilde{\eta}_{\varepsilon,n-1} = i, \tilde{\kappa}_{\varepsilon,n-1} = s, \tilde{\rho}_{\varepsilon,n-1} = \imath\} \\
&\quad = \mathsf{P}\{\tilde{\eta}_{\varepsilon,n} = j, \tilde{\kappa}_{\varepsilon,n} \leq t, \tilde{\rho}_{\varepsilon,n} = \jmath / \tilde{\eta}_{\varepsilon,n-1} = i\} \\
&\quad = \begin{cases} \tilde{Q}_{\varepsilon,ij}(t, \jmath) & \text{for } i \in \tilde{\mathbb{X}}, j \in \mathbb{X},\ s, t \geq 0,\ \imath, \jmath = 0, 1,\ n = 1, \\ Q_{\varepsilon,ij}(t, \jmath) & \text{for } i, j \in \mathbb{X},\ s, t \geq 0,\ \imath, \jmath = 0, 1,\ n > 1. \end{cases}
\end{aligned} \tag{4.37}$$

The differences from the standard definition of the Markov renewal process are, first, that the phase space of the initial state of the first component $\tilde{\eta}_{\varepsilon,0}$ differs from the phase space for the random sequence $\eta_{\varepsilon,n}, n \geq 1$ and, secondly, that the transition probabilities of this Markov renewal process at the first transition step differ from its transition probabilities at subsequent steps.

As in the standard case, the first component $\eta_{\varepsilon,n}$ of the above Markov renewal process is also a Markov chain, with the phase space $\tilde{\mathbb{X}}$ for the initial state $\tilde{\eta}_{\varepsilon,0}$ and the phase space $\mathbb{X}$ for random variables $\tilde{\eta}_{\varepsilon,n}, n \geq 1$, initial distribution $\tilde{q}_\varepsilon = \langle \tilde{q}_{\varepsilon,i} = \mathsf{P}\{\tilde{\eta}_{\varepsilon,0} = i\}, i \in \tilde{\mathbb{X}}\rangle$, and transition probabilities of the following form,

$$\begin{aligned}
&\mathsf{P}\{\tilde{\eta}_{\varepsilon,n} = j / \tilde{\eta}_{\varepsilon,n-1} = i\} \\
&\quad = \begin{cases} \tilde{p}_{\varepsilon,ij} = \tilde{Q}_{\varepsilon,ij}(\infty, 0) + \tilde{Q}_{\varepsilon,ij}(\infty, 1) & \text{for } i \in \tilde{\mathbb{X}}, j \in \mathbb{X}, n = 1, \\ p_{\varepsilon,ij} = Q_{\varepsilon,ij}(\infty, 0) + Q_{\varepsilon,ij}(\infty, 1) & \text{for } i, j \in \mathbb{X},\ n > 1. \end{cases}
\end{aligned} \tag{4.38}$$

The corresponding first-rare-event time and process are defined for this Markov renewal process as in Sect. 2.1.1, i.e.,

$$\tilde{\xi}_\varepsilon = \sum_{n=1}^{\tilde{\nu}_\varepsilon} \tilde{\kappa}_{\varepsilon,n}, \tag{4.39}$$

and

$$\tilde{\xi}_\varepsilon(t) = \sum_{n=1}^{[t\tilde{\nu}_\varepsilon]} \tilde{\kappa}_{\varepsilon,n}, t \geq 0, \tag{4.40}$$

where

$$\tilde{\nu}_\varepsilon = \min(n \geq 1 : \tilde{\rho}_{\varepsilon,n} = 1). \tag{4.41}$$

Let us also consider the shifted Markov renewal process,

$$(\eta_{\varepsilon,n}, \kappa_{\varepsilon,n}, \rho_{\varepsilon,n}) = (\tilde{\eta}_{\varepsilon,n+1}, \tilde{\kappa}_{\varepsilon,n+1}, \tilde{\rho}_{\varepsilon,n+1}), \ n = 0, 1, \dots \tag{4.42}$$

This is a standard Markov renewal process (no transition period) with phase space $\mathbb{Z}$ and transition probabilities $Q_{\varepsilon,ij}(t, J)$.

The first-rare-event time and process are defined for this Markov renewal process as in Sect. 2.1.1, i.e.,

$$\xi_\varepsilon = \sum_{n=1}^{\nu_\varepsilon} \kappa_{\varepsilon,n}, \tag{4.43}$$

and

$$\xi_\varepsilon(t) = \sum_{n=1}^{[t\nu_\varepsilon]} \kappa_{\varepsilon,n}, t \geq 0, \tag{4.44}$$

where

$$\nu_\varepsilon = \min(n \geq 1 : \rho_{\varepsilon,n} = 1). \tag{4.45}$$

The following two relations play a key role in the subsequent analysis,

$$\begin{aligned} \tilde{\xi}_\varepsilon &= \tilde{\kappa}_{\varepsilon,1} + (1 - \tilde{\rho}_{\varepsilon,1}) \cdot \sum_{n=2}^{\nu_\varepsilon+1} \tilde{\kappa}_{\varepsilon,n} \\ &= \tilde{\kappa}_{\varepsilon,1} + (1 - \tilde{\rho}_{\varepsilon,1}) \cdot \sum_{n=1}^{\nu_\varepsilon} \kappa_{\varepsilon,n} \\ &= \tilde{\kappa}_{\varepsilon,1} + (1 - \tilde{\rho}_{\varepsilon,1})\xi_\varepsilon \end{aligned} \tag{4.46}$$

and

$$\tilde{\xi}_\varepsilon(t) = \tilde{\kappa}_{\varepsilon,1} + \tilde{\rho}_{\varepsilon,1} \cdot \sum_{n=2}^{[t]} \tilde{\kappa}_{\varepsilon,n} + (1 - \tilde{\rho}_{\varepsilon,1}) \cdot \sum_{n=2}^{[t(\nu_\varepsilon+1)]} \tilde{\kappa}_{\varepsilon,n}$$

$$= \tilde{\kappa}_{\varepsilon,1} + \tilde{\rho}_{\varepsilon,1} \cdot \sum_{n=1}^{[t]-1} \kappa_{\varepsilon,n} + (1 - \tilde{\rho}_{\varepsilon,1}) \cdot \sum_{n=1}^{[t(\nu_\varepsilon+1)]-1} \kappa_{\varepsilon,n}$$

$$= \tilde{\kappa}_{\varepsilon,1} + \tilde{\rho}_{\varepsilon,1} \cdot \sum_{n=1}^{[t]-1} \kappa_{\varepsilon,n} - \tilde{\rho}_{\varepsilon,1} \cdot \hat{\xi}_\varepsilon(t) + \hat{\xi}_\varepsilon(t), \ t \geq 0, \tag{4.47}$$

where

$$\hat{\xi}_\varepsilon(t) = \sum_{n=1}^{[t(\nu_\varepsilon+1)]-1} \kappa_{\varepsilon,n}, \ t \geq 0. \tag{4.48}$$

4.1.3.2 First-Rare-Event Times and Processes for Markov Renewal Process with Degenerating Transition Period. According to Theorem 2.1, if the conditions $\mathbf{A}_1, \mathbf{B}_1$, and $\mathbf{C}_1$ are satisfied for the shifted Markov renewal processes $(\eta_{\varepsilon,n}, \kappa_{\varepsilon,n}, \rho_{\varepsilon,n})$, then the condition $\mathbf{D}_1$ (for these Markov renewal processes) is necessary and sufficient for the fulfilment of the following relation (for some or any initial distributions $\bar{q}_\varepsilon$ of the random variable $\eta_{\varepsilon,0} = \tilde{\eta}_{\varepsilon,1}$, respectively, in the statements of necessity and sufficiency),

$$\xi_\varepsilon \xrightarrow{\mathsf{d}} \xi_0 \text{ as } \varepsilon \to 0, \tag{4.49}$$

where the limiting random variable ξ_0 is described in Theorem 2.1.

Also, the following relation holds (for any initial distributions $\bar{q}_\varepsilon$ of the random variable $\eta_{\varepsilon,0} = \tilde{\eta}_{\varepsilon,1}$),

$$\xi_\varepsilon(t), t \geq 0 \xrightarrow{\mathsf{d}} \xi_0(t) = \kappa_0(t\theta_0), t \geq 0 \text{ as } \varepsilon \to 0, \tag{4.50}$$

where the limiting stochastic process $\xi_0(t), t \geq 0$ is also described in Theorem 2.1.

In addition, let us assume that the following conditions are satisfied:

$\mathbf{A}_3$: $\mathsf{P}_i\{\tilde{\rho}_{\varepsilon,1} = 1\} \to 0$ as $\varepsilon \to 0$, for $i \in \tilde{\mathbb{X}}$

and

$\mathbf{C}_6$: $\mathsf{P}_i\{\tilde{\kappa}_{\varepsilon,1} > \delta\} \to 0$ as $\varepsilon \to 0$, for $\delta > 0$ and $i \in \tilde{\mathbb{X}}$.

The relation (4.46) and the conditions $\mathbf{A}_3$ and $\mathbf{C}_6$ imply that (for any initial distributions $\tilde{q}_\varepsilon$ of the random variable $\tilde{\eta}_{\varepsilon,0}$)

$$\tilde{\kappa}_{\varepsilon,1} - \tilde{\rho}_{\varepsilon,1}\xi_\varepsilon \xrightarrow{\mathsf{P}} 0 \text{ as } \varepsilon \to 0. \tag{4.51}$$

Note that the initial distribution $\bar{q} = \langle q_{\varepsilon,j} = \mathsf{P}\{\eta_{\varepsilon,0} = j\}, \ j \in \mathbb{X}\rangle$, for the random variable $\eta_{\varepsilon,0} = \tilde{\eta}_{\varepsilon,1}$, is determined by the initial distribution $\tilde{q} = \langle \tilde{q}_{\varepsilon,i} = \mathsf{P}\{\tilde{\eta}_{\varepsilon,0} = i\}, \ i \in \tilde{\mathbb{X}}\rangle$, for the random variable $\tilde{\eta}_{\varepsilon,0}$. Namely, $q_{\varepsilon,j} = \sum_{i \in \tilde{\mathbb{X}}} \tilde{q}_{\varepsilon,i} \mathsf{P}_j\{\tilde{\eta}_{\varepsilon,1} = j\}, j \in \mathbb{X}$.

The relations (4.49) and (4.51) imply, by Theorem A.9, that, under the conditions $\mathbf{A}_3$ and $\mathbf{C}_6$, the relation (4.49) holds if and only if the following relation holds

$$\tilde{\xi}_\varepsilon \xrightarrow{\mathsf{d}} \xi_0 \text{ as } \varepsilon \to 0. \tag{4.52}$$

An analogue of the relation (4.52) also takes place for the first-rare-event processes $\tilde{\xi}_\varepsilon(t), t \geq 0$.

The processes $\tilde{\kappa}_{\varepsilon,1} + \mathrm{I}(\tilde{\rho}_{\varepsilon,1} = 1) \cdot \sum_{n=1}^{[t]-1} \kappa_{\varepsilon,n}, t \geq 0$ and $\tilde{\rho}_{\varepsilon,1}\hat{\xi}_\varepsilon(t), t \geq 0$ are monotonic. Therefore, the conditions $\mathbf{A}_3$ and $\mathbf{C}_6$ imply that (for any initial distributions $\tilde{q}_\varepsilon$ of the random variable $\tilde{\eta}_{\varepsilon,0}$)

$$\tilde{\kappa}_{\varepsilon,1} + \tilde{\rho}_{\varepsilon,1} \cdot \sum_{n=1}^{[t]-1} \kappa_{\varepsilon,n}, t \geq 0 \xrightarrow{\mathsf{U}} 0(t), t \geq 0 \text{ as } \varepsilon \to 0 \tag{4.53}$$

and

$$\tilde{\rho}_{\varepsilon,1}\hat{\xi}_\varepsilon(t), t \geq 0 \xrightarrow{\mathsf{U}} 0(t), t \geq 0 \text{ as } \varepsilon \to 0 \tag{4.54}$$

where $0(t) = 0, t \geq 0$.

The process $\xi_\varepsilon(t) = \sum_{n=1}^{[t\nu_\varepsilon]} \kappa_{\varepsilon,n}, t \geq 0$ is a superposition of two processes, $t\frac{\nu_\varepsilon}{u_\varepsilon}, t \geq 0$, and $\sum_{n=1}^{[tu_\varepsilon]} \kappa_{\varepsilon,n}, t \geq 0$.

The conditions $\mathbf{A}_1$–$\mathbf{D}_1$ (imposed on the Markov renewal processes $(\eta_{\varepsilon,n}, \kappa_{\varepsilon,n}, \rho_{\varepsilon,n})$) and Lemma 2.9 imply that the following relation of J-convergence takes place,

$$\xi_\varepsilon(t) = \sum_{n=1}^{[t\nu_\varepsilon]} \kappa_{\varepsilon,n}, t \geq 0 \xrightarrow{\mathsf{J}} \theta_0(t\nu_0), t \geq 0 \text{ as } \varepsilon \to 0. \tag{4.55}$$

This follows from the statement **(iii)** of Theorem 2.2 and the relation (2.139) taking, in this case, the following form,

$$(t\frac{\nu_\varepsilon}{u_\varepsilon}, \sum_{n=1}^{[t\nu_\varepsilon]} \kappa_{\varepsilon,n}), t \geq 0 \xrightarrow{\mathsf{d}} (t\nu_0, \theta_0(t)), t \geq 0 \text{ as } \varepsilon \to 0, \tag{4.56}$$

where the random variable ν_0 and the stochastic process $\theta_0(t), t \geq 0$ are described in Lemma 2.9.

Obviously, for $t \geq 0$,

$$|\frac{[t(\nu_\varepsilon + 1)] - 1}{u_\varepsilon} - \frac{t\nu_\varepsilon}{u_\varepsilon}| \leq \frac{1+t}{u_\varepsilon} \to 0 \text{ as } \varepsilon \to 0. \tag{4.57}$$

The relations (4.56) and (4.57) imply that

$$(\frac{[t(\nu_\varepsilon + 1)] - 1}{u_\varepsilon}, \sum_{n=1}^{[tu_\varepsilon]} \kappa_{\varepsilon,n+1}), t \geq 0 \xrightarrow{\mathsf{d}} (t\nu_0, \theta_0(t)), t \geq 0 \text{ as } \varepsilon \to 0. \tag{4.58}$$

The relations (4.55) and (4.58) imply, by Theorem A.6, that, under the conditions $\mathbf{A}_1$–$\mathbf{D}_1$ (imposed on the Markov renewal processes $(\eta_{\varepsilon,n}, \kappa_{\varepsilon,n}, \rho_{\varepsilon,n})$),

$$\hat{\xi}_\varepsilon(t) = \sum_{n=1}^{[t(\nu_\varepsilon+1)]-1} \kappa_{\varepsilon,n}, t \geq 0 \xrightarrow{\mathsf{J}} \theta_0(t\nu_0), t \geq 0 \text{ as } \varepsilon \to 0. \tag{4.59}$$

Finally, the relations (4.47), (4.53), (4.54), and (4.59) imply by Theorem A.11, that, under the conditions $\mathbf{A}_1$–$\mathbf{D}_1$ (imposed on the Markov renewal processes $(\eta_{\varepsilon,n}, \kappa_{\varepsilon,n}, \rho_{\varepsilon,n})$) and $\mathbf{A}_3$, $\mathbf{C}_6$, the following relation holds,

$$\tilde{\xi}_\varepsilon(t) = \tilde{\kappa}_{\varepsilon,1} + \tilde{\rho}_{\varepsilon,1} \cdot \sum_{n=1}^{[t]-1} \kappa_{\varepsilon,n} - \tilde{\rho}_{\varepsilon,1}\hat{\xi}_\varepsilon(t) + \hat{\xi}_\varepsilon(t),\ t \geq 0$$

$$\xrightarrow{\mathsf{J}} \theta_0(t\nu_0),\ t \geq 0 \text{ as } \varepsilon \to 0. \tag{4.60}$$

This relation is an analog of the asymptotic relation appearing in the statement **(iii)** of Theorem 2.1.

It follows from the above remarks that the following modified version of Theorem 2.1 takes place.

The remarks made above allow us to formulate a variant of Theorem 2.1 for the first-rare-event times and processes $\tilde{\xi}_\varepsilon$ and $\tilde{\xi}_\varepsilon(t)$ based on the Markov renewal processes with transition period $(\tilde{\eta}_{\varepsilon,n}, \tilde{\kappa}_{\varepsilon,n}, \tilde{\rho}_{\varepsilon,n})$ and the shifted Markov renewal processes $(\eta_{\varepsilon,n}, \kappa_{\varepsilon,n}, \rho_{\varepsilon,n})$ defined by the relation (4.42).

Note that in the theorem formulated below, the conditions $\mathbf{A}_1$, $\mathbf{B}_1$, $\mathbf{C}_1$, and $\mathbf{D}_1$ refer to processes $(\eta_{\varepsilon,n}, \kappa_{\varepsilon,n}, \rho_{\varepsilon,n})$, while the conditions $\mathbf{A}_3$ and $\mathbf{C}_6$ refer to processes $(\tilde{\eta}_{\varepsilon,n}, \tilde{\kappa}_{\varepsilon,n}, \tilde{\rho}_{\varepsilon,n})$.

Theorem 4.2 *Let the conditions* $\mathbf{A}_1$, $\mathbf{B}_1$, $\mathbf{C}_1$ *and* $\mathbf{A}_3$, $\mathbf{C}_6$ *be satisfied; then*:

(i) *The condition* $\mathbf{D}_1$ *is necessary and sufficient for the fulfilment (for some or any initial distributions* $\tilde{q}_\varepsilon$, *respectively, in the statements of necessity and sufficiency) of the asymptotic relation,* $\tilde{\xi}_\varepsilon = \tilde{\xi}_\varepsilon(1) \xrightarrow{\mathsf{d}} \xi_0$ *as* $\varepsilon \to 0$, *where* ξ_0 *is a non-negative random variable with distribution not concentrated in zero.*

(ii) *The distribution function* $G(u) = \mathsf{P}\{\xi_0 \leq u\}, u \geq 0$ *of the limiting random variable* ξ_0 *has the Laplace transform* $\phi(s) = \int_0^\infty e^{-su}G(du) = \frac{1}{1+A(s)}, s \geq 0$, *where* $A(s)$ *is the cumulant of infinitely divisible distribution appearing in the condition* $\mathbf{D}_1$.

(iii) *The stochastic processes* $\tilde{\xi}_\varepsilon(t), t \geq 0 \xrightarrow{\mathsf{J}} \xi_0(t) = \theta_0(t\nu_0), t \geq 0$ *as* $\varepsilon \to 0$, *where* **(a)** ν_0 *is a random variable, which has the exponential distribution with parameter* 1, **(b)** $\theta_0(t), t \geq 0$ *is a non-negative Lévy process with the Laplace transforms* $\mathsf{E}e^{-s\theta_0(t)} = e^{-tA(s)}, s, t \geq 0$, *and* **(c)** *the random variable* ν_0 *and the process* $\theta_0(t), t \geq 0$ *are independent.*

It is useful to note that the statement similar to Lemma 4.2 takes place for random variables $\xi_{\varepsilon,\circ}$ such that, for some $0 < \varepsilon \leq \varepsilon_\circ \leq 1$,

$$\tilde{\xi}_{\varepsilon,-} \leq \tilde{\xi}_{\varepsilon,\circ} \leq \tilde{\xi}_{\varepsilon,+}, \tag{4.61}$$

where

$$\tilde{\xi}_{\varepsilon,+} = \tilde{\xi}_\varepsilon = \sum_{n=1}^{[t\tilde{\nu}_\varepsilon]} \tilde{\kappa}_{\varepsilon,n},\ \tilde{\xi}_{\varepsilon,-} = \sum_{n=1}^{[t(\tilde{\nu}_\varepsilon - 1)]} \tilde{\kappa}_{\varepsilon,n} \tag{4.62}$$

Lemma 4.4 *Let $\tilde{\xi}_{\varepsilon,\circ}$ be random variables, for which relation (4.61) holds. Then, under conditions* $\mathbf{A}_1$, $\mathbf{B}_1$, $\mathbf{C}_1$, *and* $\mathbf{A}_3$, $\mathbf{C}_6$ *of Theorem 4.2, condition* $\mathbf{D}_1$ *(for the Markov renewal processes $(\eta_{\varepsilon,n}, \kappa_{\varepsilon,n}, \rho_{\varepsilon,n})$) is necessary and sufficient (for some or any initial distributions $\bar{q}_\varepsilon$ of the random variables $\tilde{\eta}_{\varepsilon,0}$, respectively, in statements of necessity and sufficiency) for holding the asymptotic relation, $\tilde{\xi}_{\varepsilon,\circ} \xrightarrow{\mathsf{d}} \xi_0$ as $\varepsilon \to 0$, where the limiting random variable ξ_0 is described in Theorem 4.2.*

Also, it is useful to note that an analogue of Lemma 4.3 takes place for càdlàg stochastic processes $\xi_{\varepsilon,\circ}(t), t \geq 0$ such that, for some $0 < \varepsilon \leq \varepsilon_\circ \leq 1$,

$$\tilde{\xi}_{\varepsilon,-}(t) \leq \xi_{\varepsilon,\circ}(t) \leq \tilde{\xi}_{\varepsilon,+}(t), t \geq 0, \tag{4.63}$$

where

$$\tilde{\xi}_{\varepsilon,+}(t) = \tilde{\xi}_\varepsilon(t) = \sum_{n=1}^{[t\tilde{\nu}_\varepsilon]} \tilde{\kappa}_{\varepsilon,n}, \ \tilde{\xi}_{\varepsilon,-}(t) = \sum_{n=1}^{[t(\tilde{\nu}_\varepsilon - 1)]} \tilde{\kappa}_{\varepsilon,n}, \ t \geq 0. \tag{4.64}$$

Lemma 4.5 *Let $\tilde{\xi}_{\varepsilon,\circ}(t), t \geq 0$ be càdlàg stochastic processes, for which relation (4.63) holds. Then, under the conditions* $\mathbf{A}_1$, $\mathbf{B}_1$, $\mathbf{C}_1$, *and* $\mathbf{A}_3$ $\mathbf{C}_6$ *of Theorem 4.2, the condition* $\mathbf{D}_1$ *(for the Markov renewal processes $(\eta_{\varepsilon,n}, \kappa_{\varepsilon,n}, \rho_{\varepsilon,n})$) is necessary and sufficient (for some or any initial distributions $\bar{q}_\varepsilon$ of the random variable $\tilde{\eta}_{\varepsilon,0}$, respectively, in the statements of necessity and sufficiency) for holding the following relation, $\tilde{\xi}_{\varepsilon,\circ}(t), t \geq 0 \xrightarrow{\mathsf{d}} \xi_0(t) = \theta_0(t\nu_0)), t \geq 0$ as $\varepsilon \to 0$, where the limiting process $\xi_0(t)$ is described in Theorem 4.2.*

4.1.3.3 First-Rare-Event Times and Processes for Markov Renewal Process with Non-degenerating Transition Periods. Suppose now that instead of the conditions $\mathbf{A}_3$ and $\mathbf{C}_4$, the following conditions are satisfied:

$\mathbf{A}_4$: **(a)** $\tilde{P}_{\varepsilon,i,\jmath} = \mathsf{P}_i\{\tilde{\rho}_{\varepsilon,1} = \jmath\} \to \tilde{P}_{0,i,\jmath}$ as $\varepsilon \to 0$, for $\jmath = 0, 1$, $i \in \tilde{\mathbb{X}}$, where $P_{0,i,\jmath} \geq 0, l = 0, 1, P_{0,i,0} + P_{0,i,1} = 1$,

and

$\mathbf{C}_7$: $\tilde{F}_{\varepsilon,i,\jmath}(\cdot) = \mathsf{P}_i\{\tilde{\kappa}_{\varepsilon,1} \leq \cdot / \tilde{\rho}_{\varepsilon,1} = \jmath\} \Rightarrow \tilde{F}_{0,i,\jmath}(\cdot)$ as $\varepsilon \to 0$, for $\jmath = 0, 1$, $i \in \tilde{\mathbb{X}}$, where $\tilde{F}_{0,i,\jmath}(\cdot), \jmath = 0, 1, i \in \tilde{\mathbb{X}}$ are some proper distribution functions concentrated on $[0, \infty)$.

We also use the following convergence condition imposed on the initial distributions $\tilde{q}_\varepsilon = \langle \tilde{q}_{\varepsilon,i} = \mathsf{P}\{\tilde{\eta}_{\varepsilon,0} = i\}, i \in \tilde{\mathbb{X}} \rangle$:

$\mathbf{B}_4$: **(a)** $\tilde{q}_{\varepsilon,i} \to \tilde{q}_{0,i}$ as $\varepsilon \to 0$, for $i \in \tilde{\mathbb{X}}$, where $\tilde{q}_{0,i} \geq 0, i \in \tilde{\mathbb{X}}$, $\sum_{i \in \tilde{\mathbb{X}}} \tilde{q}_{0,i} = 1$.

The Markov property of the Markov renewal processes $(\tilde{\eta}_{\varepsilon,n}, \tilde{\kappa}_{\varepsilon,n}, \tilde{\rho}_{\varepsilon,n})$ implies that the following relation takes place, for $u, t \geq 0, i \in \mathbb{X}$,

$$\begin{aligned} & \mathsf{P}_i\{\tilde{\kappa}_{\varepsilon,1} \leq u, \tilde{\rho}_{\varepsilon,1} = \jmath, \xi_\varepsilon \leq t\} \\ & \quad = \sum_{j \in \mathbb{X}} \mathsf{P}_i\{\tilde{\kappa}_{\varepsilon,1} \leq u, \tilde{\rho}_{\varepsilon,1} = \jmath, \tilde{\eta}_{\varepsilon,1} = j\} \mathsf{P}_j\{\xi_\varepsilon \leq t\}. \end{aligned} \tag{4.65}$$

The conditions $\mathbf{A}_4$ and $\mathbf{C}_7$ imply that the following relation holds, for any $u \geq 0$, which is a point of continuity of the distribution function $F_{0,i',J'}(\cdot), i' \in \tilde{\mathbb{X}}, J' = 0, 1$ and $i \in \tilde{\mathbb{X}}, J = 0, 1$,

$$\mathsf{P}_i\{\tilde{\kappa}_{\varepsilon,1} \leq u, \tilde{\rho}_{\varepsilon,1} = J\} \Rightarrow F_{0,i,J}(u)P_{0,i,J} \text{ as } \varepsilon \to 0. \tag{4.66}$$

The relations (4.49) and (4.66) imply that, under the conditions $\mathbf{A}_1$–$\mathbf{C}_1$ assumed to be satisfied for the shifted Markov renewal processes $(\eta_{\varepsilon,n}, \kappa_{\varepsilon,n}, \rho_{\varepsilon,n})$, and the conditions $\mathbf{A}_4$, $\mathbf{B}_4$, and $\mathbf{C}_7$, the following relation takes place, for any $u \geq 0$ and $v \geq 0$, which are points of continuity for distribution functions, respectively, $F_{0,i',J'}(u), i' \in \tilde{\mathbb{X}}, J' = 0, 1$ and $G(v) = \mathsf{P}\{\xi_0 \leq v\}$, and $i \in \tilde{\mathbb{X}}, J = 0, 1$,

$$\begin{aligned}
&\mathsf{P}_{\tilde{q}_\varepsilon}\{\tilde{\kappa}_{\varepsilon,1} \leq u, \tilde{\rho}_{\varepsilon,1} = J, \xi_\varepsilon \leq v\} \\
&\quad = \sum_{i \in \tilde{\mathbb{X}}} \tilde{q}_{\varepsilon,i} \sum_{j \in \mathbb{X}} \mathsf{P}_i\{\tilde{\kappa}_{\varepsilon,1} \leq u, \tilde{\rho}_{\varepsilon,1} = J, \tilde{\eta}_{\varepsilon,1} = j\}\mathsf{P}_j\{\xi_\varepsilon \leq v\} \\
&\quad = \sum_{i \in \tilde{\mathbb{X}}} \tilde{q}_{\varepsilon,i}\Big(\sum_{j \in \mathbb{X}} \mathsf{P}_i\{\tilde{\kappa}_{\varepsilon,1} \leq u, \tilde{\rho}_{\varepsilon,1} = J, \tilde{\eta}_{\varepsilon,1} = j\} \\
&\qquad \times (\mathsf{P}_j\{\xi_\varepsilon \leq t\} - \mathsf{P}\{\xi_0 \leq v\}) + \mathsf{P}_i\{\tilde{\kappa}_{\varepsilon,1} \leq u, \tilde{\rho}_{\varepsilon,1} = J\}\mathsf{P}\{\xi_0 \leq v\}\Big) \\
&\quad \to \sum_{i \in \tilde{\mathbb{X}}} \tilde{q}_{0,i}\tilde{F}_{0,i,J}(u)\tilde{P}_{0,i,J}G(v) \text{ as } \varepsilon \to 0.
\end{aligned} \tag{4.67}$$

The relations (4.46) and (4.67) imply that, under the conditions $\mathbf{A}_1$–$\mathbf{C}_1$ assumed to be satisfied for the shifted Markov renewal processes $(\eta_{\varepsilon,n}, \kappa_{\varepsilon,n}, \rho_{\varepsilon,n})$, and the conditions $\mathbf{A}_4$, $\mathbf{B}_4$, and $\mathbf{C}_7$, the condition $\mathbf{D}_1$ is satisfied (for the Markov renewal processes $(\eta_{\varepsilon,n}, \kappa_{\varepsilon,n}, \rho_{\varepsilon,n})$) and the following relation takes place:

$$\begin{aligned}
\mathsf{P}_{\tilde{q}_\varepsilon}\{\tilde{\xi}_\varepsilon \leq \cdot\} &= \mathsf{P}_{\tilde{q}_\varepsilon}\{\tilde{\kappa}_{\varepsilon,1} + (1 - \tilde{\rho}_{\varepsilon,1})\xi_\varepsilon \leq \cdot\} \\
&\Rightarrow \mathsf{P}_{\tilde{q}_0}\{\tilde{\kappa}_{0,1} + (1 - \tilde{\rho}_{0,1})\xi_0 \leq \cdot\} \\
&= \mathsf{P}_{\tilde{q}_0}\{\tilde{\xi}_0 \leq \cdot\} \text{ as } \varepsilon \to 0,
\end{aligned} \tag{4.68}$$

where: (a) the limiting random vector $(\tilde{\kappa}_{0,1}, \tilde{\rho}_{0,1})$ and the limiting random variable ξ_0 are independent, (b) $\mathsf{P}_{\tilde{q}_0}\{\tilde{\kappa}_{0,1} \leq u, \tilde{\rho}_{0,1} = J\} = \sum_{i \in \tilde{\mathbb{X}}} \tilde{q}_{0,i}\tilde{F}_{0,i,J}(u)\tilde{P}_{0,i,J}, u \geq 0, J = 0, 1$, (c) $\mathsf{P}\{\xi_0 \leq v\} = G(v), v \geq 0$.

Let us consider processes,

$$\hat{\xi}_\varepsilon(t) = \sum_{n=1}^{[t_k(v_\varepsilon+1)]-1} \kappa_{\varepsilon,n}, t \geq 0. \tag{4.69}$$

The Markov property of the Markov renewal processes $(\tilde{\eta}_{\varepsilon,n}, \tilde{\kappa}_{\varepsilon,n}, \tilde{\rho}_{\varepsilon,n})$ implies that the following relation takes place, for $u, u_k, t_k \geq 0, k = 1, 2, \ldots, n, n \geq 1, i \in \mathbb{X}$,

$$
\begin{aligned}
& \mathsf{P}_i\{\tilde{\kappa}_{\varepsilon,1} \leq u, \tilde{\rho}_{\varepsilon,1} = \jmath, \sum_{n=1}^{[t_k(\nu_\varepsilon+1)]-1} \kappa_{\varepsilon,n} \leq u_k, k = 1,\ldots,n\} \\
& \quad = \sum_{j\in\mathbb{X}} \mathsf{P}_i\{\tilde{\kappa}_{\varepsilon,1} \leq u, \tilde{\rho}_{\varepsilon,1} = \jmath, \tilde{\eta}_{\varepsilon,1} = j\} \\
& \quad \times \mathsf{P}_j\{\hat{\xi}_\varepsilon(t_k) \leq u_k, k = 1,\ldots,n\}. \qquad (4.70)
\end{aligned}
$$

The relations (4.50), (4.65), and (4.70) imply that, under the conditions $\mathbf{A}_1$–$\mathbf{C}_1$ assumed to be satisfied for the shifted Markov renewal processes $(\eta_{\varepsilon,n}, \kappa_{\varepsilon,n}, \rho_{\varepsilon,n})$, and the conditions $\mathbf{A}_4$, $\mathbf{B}_4$, and $\mathbf{C}_7$, the following relation takes place, for any $u \geq 0$ and $u_k \geq 0, k \geq 1$, which are points of continuity for distribution functions, respectively, $F_{0,i',\jmath'}(u), i' \in \tilde{\mathbb{X}}, \jmath' = 0, 1$ and $\mathsf{P}\{\xi_0(t_k) \leq u_k, k = 1,\ldots,n\}, n \geq 1$, and $i \in \tilde{\mathbb{X}}, \jmath = 0, 1$,

$$
\begin{aligned}
& \mathsf{P}_{\tilde{q}_\varepsilon}\{\tilde{\kappa}_{\varepsilon,1} \leq u, \tilde{\rho}_{\varepsilon,1} = \jmath, \xi_\varepsilon(t_k) \leq u_k, k = 1,\ldots,n\} \\
& = \sum_{i\in\tilde{\mathbb{X}}} \tilde{q}_{\varepsilon,i} \sum_{j\in\mathbb{X}} \mathsf{P}_i\{\tilde{\kappa}_{\varepsilon,1} \leq u, \tilde{\rho}_{\varepsilon,1} = \jmath, \tilde{\eta}_{\varepsilon,1} = j\} \\
& \quad \times \mathsf{P}_j\{\hat{\xi}_\varepsilon(t_k) \leq u_k, k = 1,\ldots,n\} \\
& = \sum_{i\in\tilde{\mathbb{X}}} \tilde{q}_{\varepsilon,i}\Big(\sum_{j\in\mathbb{X}} \mathsf{P}_i\{\tilde{\kappa}_{\varepsilon,1} \leq u, \tilde{\rho}_{\varepsilon,1} = \jmath, \tilde{\eta}_{\varepsilon,1} = j\} \\
& \quad (\mathsf{P}_j\{\hat{\xi}_\varepsilon(t_k) \leq u_k, k = 1,\ldots,n\} - \mathsf{P}\{\xi_0(t_k) \leq u_k, k = 1,\ldots,n\}) \\
& \quad + \mathsf{P}_i\{\tilde{\kappa}_{\varepsilon,1} \leq u, \tilde{\rho}_{\varepsilon,1} = \jmath\}\mathsf{P}\{\xi_0(t_k) \leq u_k, k = 1,\ldots,n\})\Big) \\
& \to \sum_{i\in\tilde{\mathbb{X}}} \tilde{q}_{0,i}\tilde{F}_{0,i,\jmath}(u)\tilde{P}_{0,i,\jmath}\mathsf{P}\{\xi_0(t_k) \leq u_k, k = 1,\ldots,n\} \text{ as } \varepsilon \to 0. \qquad (4.71)
\end{aligned}
$$

The relations (4.50) and (4.71) imply that, under the conditions $\mathbf{A}_1$–$\mathbf{C}_1$ assumed to be satisfied for the shifted Markov renewal processes $(\eta_{\varepsilon,n}, \kappa_{\varepsilon,n}, \rho_{\varepsilon,n})$, and the conditions $\mathbf{A}_4$, $\mathbf{B}_4$, and $\mathbf{C}_7$, the condition $\mathbf{D}_1$ is satisfied (for the Markov renewal processes $(\eta_{\varepsilon,n}, \kappa_{\varepsilon,n}, \rho_{\varepsilon,n})$) and the following relation takes place,

$$
\begin{aligned}
& \tilde{\kappa}_{\varepsilon,1} + (1 - \tilde{\rho}_{\varepsilon,1}) \sum_{n=1}^{[t(\nu_\varepsilon+1)]-1} \kappa_{\varepsilon,n}, t \geq 0 \\
& \quad \xrightarrow{\mathsf{d}} \tilde{\kappa}_{0,1} + (1 - \tilde{\rho}_{0,1})\xi_0(t), t \geq 0 \text{ as } \varepsilon \to 0, \qquad (4.72)
\end{aligned}
$$

where: (a) the limiting random vector $(\tilde{\kappa}_0, \tilde{\rho}_0)$ and the limiting stochastic process $\xi_0(t), t \geq 0$ are independent, (b) $\mathsf{P}_{\tilde{q}_0}\{\tilde{\kappa}_{0,1} \leq u, \tilde{\rho}_{0,1} = \jmath\} = \sum_{i\in\tilde{\mathbb{X}}} \tilde{q}_{0,i}\tilde{F}_{0,i,\jmath}(u)\tilde{P}_{0,i,\jmath}, u \geq 0, \jmath = 0, 1$, and (c) the process $\xi_0(t) = \theta(t\nu_0), t \geq 0$ is described in Theorem 2.1.

The relation (4.59) implies that the processes $\hat{\xi}_\varepsilon(t), t \geq 0$ are J-compact, as $\varepsilon \to 0$, i.e., $\lim_{c\to 0} \overline{\lim}_{\varepsilon\to 0} \mathsf{P}\{\Delta_{\mathsf{J}}(\hat{\xi}_\varepsilon(\cdot), c, T) > \delta\} = 0, \delta, T > 0$ (see, relation A.3 for the formula defining the modulus of J-compactness). Since $\Delta_{\mathsf{J}}(\tilde{\kappa}_{\varepsilon,1} + (1 - \tilde{\rho}_{\varepsilon,1})\hat{\xi}_\varepsilon(\cdot), c, T) \leq$

$\Delta_{\mathsf{J}}(\hat{\xi}_\varepsilon(\cdot), c, T), T \geq 0$, the processes $\tilde{\kappa}_{\varepsilon,1} + (1 - \tilde{\rho}_{\varepsilon,1})\hat{\xi}_\varepsilon(t), t \geq 0$ are also J-compact, as $\varepsilon \to 0$.

This fact and the relation (4.72) imply that

$$\begin{aligned} \tilde{\kappa}_{\varepsilon,1} + (1 - \tilde{\rho}_{\varepsilon,1}) \sum_{n=1}^{[t(\nu_\varepsilon+1)]-1} \kappa_{\varepsilon,n}, t \geq 0 \\ \xrightarrow{\mathsf{J}} \tilde{\kappa}_{0,1} + (1 - \tilde{\rho}_{0,1})\xi_0(t), t \geq 0 \text{ as } \varepsilon \to 0. \end{aligned} \tag{4.73}$$

Theorem 2.2 implies that, for any $s, t > 0$, and $\delta > 0$

$$\begin{aligned} \mathsf{P}_{\tilde{q}_\varepsilon}\{\tilde{\rho}_{\varepsilon,1} \cdot \sum_{n=1}^{[t]-1} \kappa_{\varepsilon,n} > \delta\} \leq \overline{\lim_{\varepsilon\to 0}} \mathsf{P}_{\tilde{q}_\varepsilon}\{\sum_{n=1}^{[su_\varepsilon]} \kappa_{\varepsilon,n} > \delta\} \\ \leq \mathsf{P}\{\theta_0(s) > \delta/2\} \to 0 \text{ as } s \to 0. \end{aligned} \tag{4.74}$$

Since the processes $\tilde{\rho}_{\varepsilon,1} \cdot \sum_{n=1}^{[t]-1} \kappa_{\varepsilon,n}, t \geq 0$ are monotonic, the relation (4.74) implies, by Theorem A.3 and Remark A.1, that, under the conditions $\mathbf{A}_1$–$\mathbf{C}_1$ assumed to be satisfied for the shifted Markov renewal processes $(\eta_{\varepsilon,n}, \kappa_{\varepsilon,n}, \rho_{\varepsilon,n})$, and the conditions $\mathbf{A}_4$, $\mathbf{B}_4$, and $\mathbf{C}_7$,

$$\tilde{\rho}_{\varepsilon,1} \cdot \sum_{n=1}^{[t]-1} \kappa_{\varepsilon,n}, t \geq 0 \xrightarrow{\mathsf{U}} 0(t), t \geq 0 \text{ as } \varepsilon \to 0. \tag{4.75}$$

Finally, the relations (4.47), (4.73), (4.75) and Theorem A.11 imply that

$$\begin{aligned} \tilde{\xi}_\varepsilon(t), t \geq 0 = \tilde{\rho}_{\varepsilon,1} \cdot \sum_{n=1}^{[t]-1} \kappa_{\varepsilon,n} + \tilde{\kappa}_{\varepsilon,1} + (1 - \tilde{\rho}_{\varepsilon,1})\hat{\xi}_\varepsilon(t), t \geq 0 \\ \xrightarrow{\mathsf{J}} \tilde{\kappa}_{0,1} + (1 - \tilde{\rho}_{0,1})\xi_0(t), t \geq 0 \text{ as } \varepsilon \to 0. \end{aligned} \tag{4.76}$$

The following theorem complements Theorem 4.2.

Theorem 4.3 *Let $(\tilde{\eta}_{\varepsilon,n}, \tilde{\kappa}_{\varepsilon,n}, \tilde{\rho}_{\varepsilon,n})$ be, for every $\varepsilon \in (0, 1]$, a Markov renewal process with transition period and the corresponding phase spaces $\tilde{\mathbb{X}}$ and $\mathbb{X}$, and $(\eta_{\varepsilon,n}, \kappa_{\varepsilon,n}, \rho_{\varepsilon,n})$ be the corresponding shifted Markov renewal process. If the conditions $\mathbf{A}_1$–$\mathbf{C}_1$ (for the Markov renewal processes $(\eta_{\varepsilon,n}, \kappa_{\varepsilon,n}, \rho_{\varepsilon,n})$) and the conditions $\mathbf{A}_4$, $\mathbf{B}_4$, and $\mathbf{C}_7$ are satisfied, then:*

(i) *The first-rare-event times $\tilde{\xi}_\varepsilon \xrightarrow{\mathsf{d}} \tilde{\xi}_0 = \tilde{\kappa}_{0,1} + (1 - \tilde{\rho}_{0,1})\xi_0$ as $\varepsilon \to 0$, where the limiting random vector $(\tilde{\kappa}_{0,1}, \tilde{\rho}_{0,1}, \xi_0)$ has the distribution given by the relation* (4.68).

(ii) *The first-rare-event processes $\tilde{\xi}_\varepsilon(t), t \geq 0 \xrightarrow{\mathsf{J}} \tilde{\xi}_0(t) = \tilde{\kappa}_{0,1} + (1 - \tilde{\rho}_{0,1})\xi_0(t), t \geq 0$ as $\varepsilon \to 0$, where the limiting process is described in the relation* (4.72).

4.1.4 First-Rare-Event Times for Markov Renewal Processes with Extending Phase Spaces

4.1.4.1 First-Rare-Event Times and the Phase Space Reduction for Markov Renewal Processes with Extending Phase Spaces. Let $(\check{\eta}_{\varepsilon,n}, \check{\kappa}_{\varepsilon,n}, \check{\rho}_{\varepsilon,n}), n = 0, 1, \ldots$ be, for every $\varepsilon \in (0, 1]$, a Markov renewal process, with a phase space $\mathbb{Z}_\varepsilon = \mathbb{X}_\varepsilon \times [0, \infty) \times \{0, 1\}$, where $\mathbb{X}_\varepsilon = \{1, 2, \ldots\}$ is a non-empty finite or countable set, and transition probabilities defined for $i, j \in \mathbb{X}_\varepsilon,\ s, t \geq 0,\ \imath, \jmath = 0, 1$,

$$\begin{aligned}
&\mathsf{P}\{\check{\eta}_{\varepsilon,1} = j, \check{\kappa}_{\varepsilon,1} \leq t, \check{\rho}_{\varepsilon,1} = \jmath / \check{\eta}_{\varepsilon,0} = i, \check{\xi}_{\varepsilon,0} = s, \check{\rho}_{\varepsilon,0} = \imath\} \\
&\quad = \mathsf{P}\{\check{\eta}_{\varepsilon,1} = j, \check{\kappa}_{\varepsilon,1} \leq t, \check{\rho}_{\varepsilon,1} = \jmath / \check{\eta}_{\varepsilon,0} = i\} \\
&\quad = \check{Q}_{\varepsilon,ij}(t, \jmath). \qquad (4.77)
\end{aligned}$$

Let us assume that the phase spaces $\mathbb{X}_\varepsilon, \varepsilon \in (0, 1]$ have the following property,

$$\mathbb{X}_{\varepsilon'} \subseteq \mathbb{X}_{\varepsilon''}, \text{ for } 0 < \varepsilon'' \leq \varepsilon' \leq 1. \qquad (4.78)$$

From the relation (4.78) it follows that there exists a set limit,

$$\lim_{\varepsilon \to 0} \mathbb{X}_\varepsilon = \mathbb{X}_0 = \cup_{\varepsilon \in (0,1]} \mathbb{X}_\varepsilon \text{ as } \varepsilon \to 0. \qquad (4.79)$$

It is useful to note that the case, where sets $\mathbb{X}_\varepsilon \equiv \mathbb{X}_0$ for $\varepsilon \in (0, 1]$, is also admitted.

Let $\mathbb{X} = \{1, \ldots, m\} \subseteq \tilde{\mathbb{X}} = \{1, \ldots, \tilde{m}\}$ be two finite sets such that $\tilde{\mathbb{X}} \subseteq \mathbb{X}_0$.

The relation (4.79) implies that there exists $\tilde{\varepsilon} \in (0, 1]$ such that $\tilde{\mathbb{X}} \subseteq \mathbb{X}_\varepsilon$, for $\varepsilon \in (0, \tilde{\varepsilon}]$.

Let $\check{\nu}_{\varepsilon,\mathbb{X}}$ be the first hitting time in the set $\mathbb{X}$ for the Markov chain $\check{\eta}_{\varepsilon,n}$, i.e.,

$$\check{\nu}_{\varepsilon,\mathbb{X}} = \min(n \geq 1 : \check{\eta}_{\varepsilon,n} \in \mathbb{X}). \qquad (4.80)$$

In what follows, we assume that the following condition is satisfied:

$\mathbf{B}_5$: There exists $\check{\varepsilon} \in (0, \tilde{\varepsilon}]$ such that $\mathsf{P}_i\{\check{\nu}_{\varepsilon,\mathbb{X}} < \infty\} = 1, i \in \mathbb{X}_\varepsilon$, for $\varepsilon \in (0, \check{\varepsilon}]$.

We are interested in asymptotic relations for distributions of various functionals defined on the trajectories of Markov chains $\check{\eta}_{\varepsilon,n}$, as $\varepsilon \to 0$. Therefore, we can assume that $\varepsilon \in (0, \check{\varepsilon}]$.

Let us also assume that the initial state $\check{\eta}_{\varepsilon,0} \in \tilde{\mathbb{X}}$, for $\varepsilon \in (0, \check{\varepsilon}]$.

Condition $\mathbf{B}_5$ allows us to recursively define, for each $\varepsilon \in (0, \check{\varepsilon}]$, successive hitting times in the set $\mathbb{X}$ by the Markov chain $\check{\eta}_{\varepsilon,n}$,

$$\check{\nu}_{\varepsilon,\mathbb{X},n} = \min(k > \check{\nu}_{\varepsilon,\mathbb{X},n-1}, \check{\eta}_{\varepsilon,k} \in \mathbb{X}), n = 1, 2, \ldots, \text{ where } \check{\nu}_{\varepsilon,\mathbb{X},0} = 0. \qquad (4.81)$$

Now, we can construct the new Markov renewal process $(\tilde{\eta}_{\varepsilon,n}, \tilde{\kappa}_{\varepsilon,n}, \tilde{\rho}_{\varepsilon,n}), n = 0, 1, \ldots$, with transition period, the reduced phase spaces $\tilde{\mathbb{X}}$ for the initial state $\tilde{\eta}_{\varepsilon,0}$, and the phase space $\mathbb{X}$ for the random variables $\tilde{\eta}_{\varepsilon,n}, n \geq 1$, using the following recurrent relations,

$$\begin{cases} \tilde{\eta}_{\varepsilon,n} &= \check{\eta}_{\varepsilon,\check{\nu}_{\varepsilon,\mathbb{X},n}},\ n = 1, 2, \ldots,\ \tilde{\eta}_{\varepsilon,0} = \check{\eta}_{\varepsilon,0}, \\ \tilde{\kappa}_{\varepsilon,n} &= \sum_{k=\check{\nu}_{\varepsilon,\mathbb{X},n-1}+1}^{\check{\nu}_{\varepsilon,\mathbb{X},n}} \check{\kappa}_{\varepsilon,k},\ n = 1, 2, \ldots,\ \tilde{\kappa}_{\varepsilon,0} = \check{\kappa}_{\varepsilon,0}, \\ \tilde{\rho}_{\varepsilon,n} &= 1 - \prod_{k=\check{\nu}_{\varepsilon,\mathbb{X},n-1}+1}^{\check{\nu}_{\varepsilon,\mathbb{X},n}} (1 - \check{\rho}_{\varepsilon,k}),\ n = 1, 2, \ldots,\ \tilde{\rho}_{\varepsilon,0} = \check{\rho}_{\varepsilon,0}. \end{cases} \tag{4.82}$$

The Markov renewal process $(\tilde{\eta}_{\varepsilon,n}, \tilde{\kappa}_{\varepsilon,n}, \tilde{\rho}_{\varepsilon,n}), n = 0, 1, \ldots$ has the phase space $\tilde{\mathbb{Z}} = \tilde{\mathbb{X}} \times [0, \infty) \times \{0, 1\}$ for the initial random vector $(\tilde{\eta}_{\varepsilon,0}, \tilde{\kappa}_{\varepsilon,0}, \tilde{\rho}_{\varepsilon,0})$ and the phase space $\mathbb{Z} = \mathbb{X} \times [0, \infty) \times \{0, 1\}$ for random vectors $(\tilde{\eta}_{\varepsilon,n}, \tilde{\kappa}_{\varepsilon,n}, \tilde{\rho}_{\varepsilon,n}), n \geq 1$, the initial distribution $\tilde{q}_\varepsilon = \langle q_{\varepsilon,i,s,\imath} = \mathsf{P}\{\tilde{\eta}_{\varepsilon,0} = i, \tilde{\kappa}_{\varepsilon,0} \leq s, \tilde{\rho}_{\varepsilon,0} = \imath\}, (i, s, \imath) \in \tilde{\mathbb{Z}}\rangle$, and transition probabilities of the following form,

$$\begin{aligned} &\mathsf{P}\{\tilde{\eta}_{\varepsilon,n} = j, \tilde{\kappa}_{\varepsilon,n} \leq t, \tilde{\rho}_{\varepsilon,n} = \jmath/\tilde{\eta}_{\varepsilon,n-1} = i, \tilde{\kappa}_{\varepsilon,n-1} = s, \tilde{\rho}_{\varepsilon,n-1} = \imath\} \\ &\quad = \mathsf{P}\{\tilde{\eta}_{\varepsilon,n} = j, \tilde{\kappa}_{\varepsilon,n} \leq t, \tilde{\rho}_{\varepsilon,n} = \jmath/\tilde{\eta}_{\varepsilon,n-1} = i\} \\ &\quad = \begin{cases} \tilde{Q}_{\varepsilon,ij}(t, \jmath) & \text{for } i \in \tilde{\mathbb{X}}, j \in \mathbb{X},\ s, t \geq 0,\ \imath, \jmath = 0, 1,\ n = 1, \\ Q_{\varepsilon,ij}(t, \jmath) & \text{for } i, j \in \mathbb{X},\ s, t \geq 0,\ \imath, \jmath = 0, 1,\ n > 1. \end{cases} \end{aligned} \tag{4.83}$$

where, for $i \in \tilde{\mathbb{X}}, j \in \mathbb{X},\ t \geq 0, \jmath = 0, 1$,

$$\begin{aligned} \tilde{Q}_{\varepsilon,ij}(t, \jmath) &= \mathsf{P}\{\tilde{\eta}_{\varepsilon,1} = j, \tilde{\kappa}_{\varepsilon,1} \leq t, \tilde{\rho}_{\varepsilon,1} = \jmath/\tilde{\eta}_{\varepsilon,0} = i\} \\ &= \mathsf{P}_i\{\check{\eta}_{\varepsilon,\check{\nu}_{\varepsilon,\mathbb{X},1}} = j, \sum_{k=1}^{\check{\nu}_{\varepsilon,\mathbb{X},1}} \check{\kappa}_{\varepsilon,k} \leq t, 1 - \prod_{k=1}^{\check{\nu}_{\varepsilon,\mathbb{X},1}} (1 - \check{\rho}_{\varepsilon,k}) = \jmath\}, \end{aligned} \tag{4.84}$$

and, for $i, j \in \mathbb{X},\ t \geq 0, \jmath = 0, 1$,

$$Q_{\varepsilon,ij}(t, \jmath) = \tilde{Q}_{\varepsilon,ij}(t, \jmath). \tag{4.85}$$

It is useful to note that the Markov renewal process $(\tilde{\eta}_{\varepsilon,n}, \tilde{\kappa}_{\varepsilon,n}, \tilde{\rho}_{\varepsilon,n})$ has no transition period, if the phase spaces $\tilde{\mathbb{X}} = \mathbb{X}$. Indeed, the transition probabilities $\tilde{Q}_{\varepsilon,ij}(t, \jmath)$ in this case coincide with the transition probabilities $Q_{\varepsilon,ij}(t, \jmath)$.

The above embedding procedure allows us to formulate a version of Theorem 4.2 for the first-rare-event times $\tilde{\xi}_\varepsilon$ based on the Markov renewal processes $(\tilde{\eta}_{\varepsilon,n}, \tilde{\kappa}_{\varepsilon,n}, \tilde{\rho}_{\varepsilon,n})$ and the corresponding shifted Markov renewal processes,

$$(\eta_{\varepsilon,n}, \kappa_{\varepsilon,n}, \rho_{\varepsilon,n}) = (\tilde{\eta}_{\varepsilon,n+1}, \tilde{\kappa}_{\varepsilon,n+1}, \tilde{\rho}_{\varepsilon,n+1}), n = 0, 1, \ldots. \tag{4.86}$$

It is also worth noting that Theorem 4.2 allows us to obtain the corresponding asymptotic relations for the first-rare-event times based on the original Markov renewal processes $(\check{\eta}_{\varepsilon,n}, \check{\kappa}_{\varepsilon,n}, \check{\rho}_{\varepsilon,n})$,

$$\check{\xi}_\varepsilon = \sum_{n=1}^{\check{\nu}_\varepsilon} \check{\kappa}_{\varepsilon,n}, \tag{4.87}$$

where

$$\check{\nu}_\varepsilon = \min(n \geq 1 : \check{\rho}_{\varepsilon,n} = 1). \tag{4.88}$$

The point is that the following inequality holds for $\varepsilon \in (0, \check{\varepsilon}]$,

$$\tilde{\xi}_{\varepsilon,-} \leq \check{\xi}_\varepsilon \leq \tilde{\xi}_{\varepsilon,+}, \tag{4.89}$$

where

$$\tilde{\xi}_{\varepsilon,-} = \sum_{n=1}^{\tilde{\nu}_\varepsilon - 1} \tilde{\kappa}_{\varepsilon,n}, \ \tilde{\xi}_{\varepsilon,+} = \tilde{\xi}_\varepsilon = \sum_{n=1}^{\tilde{\nu}_\varepsilon} \tilde{\kappa}_{\varepsilon,n}. \tag{4.90}$$

The remarks made above and in Sect. 4.1.3.2 allow us to formulate a variant of Theorem 4.2 for the first-rare-event times $\check{\xi}_\varepsilon$ based on the Markov renewal processes $(\check{\eta}_{\varepsilon,n}, \check{\kappa}_{\varepsilon,n}, \check{\rho}_{\varepsilon,n})$, the embedded Markov renewal process with transition period $(\tilde{\eta}_{\varepsilon,n}, \tilde{\kappa}_{\varepsilon,n}, \tilde{\rho}_{\varepsilon,n})$ constructed using the recurrent relations (4.82), the shifted Markov renewal processes $(\eta_{\varepsilon,n}, \kappa_{\varepsilon,n}, \rho_{\varepsilon,n})$ defined by the relation (4.87).

Note that in the theorem formulated below, the conditions $\mathbf{A}_1$, $\mathbf{B}_1$, $\mathbf{C}_1$, and $\mathbf{D}_1$ refer to the Markov renewal processes $(\eta_{\varepsilon,n}, \kappa_{\varepsilon,n}, \rho_{\varepsilon,n})$, the conditions $\mathbf{A}_3$ and $\mathbf{C}_6$ refer to the Markov renewal processes $(\tilde{\eta}_{\varepsilon,n}, \tilde{\kappa}_{\varepsilon,n}, \tilde{\rho}_{\varepsilon,n})$, and the condition $\mathbf{B}_5$ refers to the Markov renewal processes $(\check{\eta}_{\varepsilon,n}, \check{\kappa}_{\varepsilon,n}, \check{\rho}_{\varepsilon,n})$.

Theorem 4.4 *Let the conditions* $\mathbf{A}_1$, $\mathbf{B}_1$, $\mathbf{C}_1$ *and* $\mathbf{A}_3$, $\mathbf{B}_5$, $\mathbf{C}_6$ *be satisfied; then*:

(i) *The condition* $\mathbf{D}_1$ *is necessary and sufficient for the fulfilment (for some or any initial distributions* $\tilde{q}_\varepsilon$ *concentrated of the phase space* $\tilde{\mathbb{X}}$, *respectively, in the statements of necessity and sufficiency) of the asymptotic relation,* $\check{\xi}_\varepsilon \xrightarrow{\mathsf{d}} \xi_0$ *as* $\varepsilon \to 0$, *where* ξ_0 *is a non-negative random variable with distribution not concentrated at zero.*

(ii) *The distribution function* $G(u) = \mathsf{P}\{\xi_0 \leq u\}, u \geq 0$ *of the limiting random variable* ξ_0 *has the Laplace transform* $\phi(s) = \int_0^\infty e^{-su} G(du) = \frac{1}{1+A(s)}, s \geq 0$, *where* $A(s)$ *is the cumulant of infinitely divisible distribution appearing in the condition* $\mathbf{D}_1$.

It is also useful to mention the case, where the conditions $\mathbf{A}_3$ and $\mathbf{C}_6$ are satisfied for any finite set $\tilde{\mathbb{X}} = \{1, \ldots, \tilde{m}\} \subseteq \mathbb{X}_0$. In this case, Theorem 4.4 is obviously true for the above-mentioned Markov renewal processes with initial distributions $\tilde{q}_\varepsilon$ concentrated on any finite subset $\tilde{\mathbb{X}}$ of $\mathbb{X}_0$.

4.1.4.2 First-Rare-Event Times and the Phase Space Reduction for Markov Renewal Processes with a Finite Phase Space. Additional useful analysis of the conditions $\mathbf{A}_1$–$\mathbf{C}_1$ and $\mathbf{A}_3$, $\mathbf{B}_5$, and $\mathbf{C}_6$ can be done for a model, in which $\mathbb{X}_0 = \{1, \ldots, m_0\}$ is a finite set and $\mathbb{X}_\varepsilon = \mathbb{X}_0$, for $\varepsilon \in (0, 1]$.

In this case, it is natural to choose the set $\tilde{\mathbb{X}} = \mathbb{X}_0$ and, thus, $\tilde{\varepsilon} = 1$.

The case, where the set $\mathbb{X} = \tilde{\mathbb{X}}$, is covered by Theorem 2.1. Thus, we assume that the set $\mathbb{X} \subset \tilde{\mathbb{X}}$.

Suppose now that the following ergodicity condition is satisfied:

$\mathbf{B}_6$: There exists $\varepsilon' \in (0, 1]$ such that the phase space $\tilde{\mathbb{X}}$ of the Markov chain $\check{\eta}_{\varepsilon,n}$ is one class of communicative states, for $\varepsilon \in (0, \varepsilon']$.

The condition $\mathbf{B}_6$ obviously implies that the condition $\mathbf{B}_5$ is satisfied.

However, the condition $\mathbf{B}_6$ does not guarantee the fulfilment of the condition $\mathbf{B}_1$ (for the Markov renewal processes $(\eta_{\varepsilon,n}, \kappa_{\varepsilon,n}, \rho_{\varepsilon,n})$), i.e., that the Markov chains $\eta_{\varepsilon,n}$ are asymptotically uniformly ergodic.

Let $\check{p}_{\varepsilon,ij}$ be the transition probabilities for the Markov chain $\check{\eta}_{\varepsilon,n}$, i.e.,

$$\check{p}_{\varepsilon,ij} = \mathsf{P}_i\{\check{\eta}_{\varepsilon,1} = j\}, \ i, j \in \tilde{\mathbb{X}}. \tag{4.91}$$

Suppose that the following condition of asymptotically uniform attainability of the set $\mathbb{X}$ is satisfied:

$\mathbf{B}_7$: There exists, for every $i \in \tilde{\mathbb{X}}, j \in \mathbb{X}$, a chain of states $r_{ij,0}, r_{ij,1}, \dots, r_{ij,n_{ij}} \in \tilde{\mathbb{X}}$ such that $r_{ij,0} = i, r_{ij,n_{ij}} = j$ and $\underline{\lim}_{\varepsilon\to 0} \check{p}_{\varepsilon,r_{ij,k-1}r_{ij,k}} > 0$, for $k = 1, \dots, n_{ij}$.

Let us show that the condition $\mathbf{B}_7$ implies that the condition $\mathbf{B}_1$ (for the Markov renewal processes $(\eta_{\varepsilon,n}, \kappa_{\varepsilon,n}, \rho_{\varepsilon,n})$) is satisfied.

Let $p_{\varepsilon,ij}$ be the transition probabilities for the Markov chain $\eta_{\varepsilon,n} = \tilde{\eta}_{\varepsilon,n+1}$, i.e.,

$$p_{\varepsilon,ij} = \mathsf{P}_i\{\eta_{\varepsilon,1} = j\}, \ i, j \in \mathbb{X}. \tag{4.92}$$

Let us $r_{ii+1,0}, r_{ii+1,1}, \dots, r_{ii+1,n_{ii+1}}$ be the sequences of states appearing in the condition $\mathbf{B}_7$, for $i = 1, \dots, m$ (here, we assume that $m + 1$ denotes the state 1 in the set $\mathbb{X} = \{1, \dots, m\}$).

Let also $0 = l_{i,0} < l_{i,1} < \cdots < l_{i,n_i} = n_{ii+1}$ be all indices such that the states $r_{ii+1,l_{i,k}} \in \mathbb{X}$, for $k = 0, \dots, n_i, i = 1, \dots, m$.

Then, for $k = 0, \dots, n_i - 1, i = 1, \dots, m$,

$$p_{\varepsilon,r_{ii+1,l_{i,k}}r_{ii+1,l_{i,k+1}}} \geq \prod_{l=l_{i,k}}^{l_{i,k+1}-1} \check{p}_{\varepsilon,r_{ii+1,l}r_{ii+1,l+1}}. \tag{4.93}$$

The condition $\mathbf{B}_7$ implies that, for $k = 0, \dots, n_i - 1, i = 1, \dots, m$,

$$\underline{\lim}_{\varepsilon\to 0} p_{\varepsilon,r_{ii+1,l_{i,k}}r_{ii+1,l_{i,k+1}}} \geq \underline{\lim}_{\varepsilon\to 0} \prod_{l=l_{i,k}}^{l_{i,k+1}-1} \check{p}_{\varepsilon,r_{ii+1,l}r_{ii+1,l+1}} > 0. \tag{4.94}$$

Due to the relation (4.93), the role of ring of states appearing in the condition $\mathbf{B}_1$ (for the Markov renewal processes $(\eta_{\varepsilon,n}, \kappa_{\varepsilon,n}, \rho_{\varepsilon,n})$) can be played by the sequence of states,

$$\begin{aligned} 1 = r_{12,l_{1,0}}, \dots, r_{12,l_{1,n_1}} = 2 = r_{23,l_{2,0}}, \dots, r_{23,l_{2,n_2}} = 3 \\ = \dots = m = r_{mm+1,l_{m,0}}, \dots, r'_{mm+1,l_{mn_m}} = 1. \end{aligned} \tag{4.95}$$

Denote

$$p(\varepsilon) = \min_{i\in\tilde{\mathbb{X}}, j\in\mathbb{X}} \prod_{k=1}^{n_{ij}} \check{p}_{\varepsilon,r_{ij,k-1}r_{ij,k}}, \ N = \max_{i\in\tilde{\mathbb{X}}, j\in\mathbb{X}} n_{ij}. \tag{4.96}$$

The condition $\mathbf{B}_7$ implies that there exist $\varepsilon'' \in (0, \varepsilon']$ and $p \in (0, 1)$ such that $p(\varepsilon) \geq p$, for $\varepsilon \in (0, \varepsilon'']$.

It is easy to see that, for $\varepsilon \in (0, \varepsilon'']$ and $i \in \tilde{\mathbb{X}}$, $n \geq 1$,

$$\mathsf{P}_i\{\check{\nu}_{\varepsilon,\mathbb{X}} > nN\} \leq (1-p)^n \to 0 \text{ as } n \to \infty. \tag{4.97}$$

Let us also assume that the following condition is satisfied:

$\mathbf{A}_5$: **(a)** There exists $\check{\varepsilon}' \in (0, 1]$ such that $\max_{i \in \tilde{\mathbb{X}}} \mathsf{P}_i\{\check{\rho}_{\varepsilon,1} = 1\} > 0$, for $\varepsilon \in (0, \check{\varepsilon}']$, **(b)** $\mathsf{P}_i\{\check{\rho}_{\varepsilon,1} = 1\} \to 0$ as $\varepsilon \to 0$, for $i \in \tilde{\mathbb{X}}$.

Let us prove that the conditions $\mathbf{A}_5$ and $\mathbf{B}_7$ imply that the condition $\mathbf{A}_3$ is satisfied.

In this case, the condition $\mathbf{A}_3$ takes the form of the following relation, which should hold for $i \in \tilde{\mathbb{X}}$,

$$\mathsf{P}_i\{\tilde{\rho}_{\varepsilon,1} = 1\} \to 0 \text{ as } \varepsilon \to 0. \tag{4.98}$$

Using the relations (4.82) and (4.97), we obtain the following relation, for $\varepsilon \in (0, \varepsilon'']$ and $i \in \tilde{\mathbb{X}}$, $n \geq 1$,

$$\begin{aligned}
\mathsf{P}_i\{\tilde{\rho}_{\varepsilon,1} = 1\} &= \mathsf{P}_i\{\prod_{k=1}^{\check{\nu}_{\varepsilon,\mathbb{X},1}} (1 - \check{\rho}_{\varepsilon,k}) = 0\} \\
&\leq \mathsf{P}_i\{\nu_{\varepsilon,\mathbb{X},1} > nN\} + \sum_{r=1}^{nN} \mathsf{P}_i\{\prod_{k=1}^{r}(1 - \check{\rho}_{\varepsilon,k}) = 0, \nu_{\varepsilon,\mathbb{X},1} = r\} \\
&\leq (1-p)^n + \sum_{r=1}^{nN}\sum_{k=1}^{r} \mathsf{P}_i\{\check{\rho}_{\varepsilon,k} = 1, \nu_{\varepsilon,\mathbb{X},1} = r\} \\
&\leq (1-p)^n + \sum_{r=1}^{nN}\sum_{k=1}^{r}\sum_{j \in \tilde{\mathbb{X}}} \mathsf{P}_i\{\check{\rho}_{\varepsilon,k} = 1, \check{\eta}_{\varepsilon,k-1} = j\} \\
&\leq (1-p)^n + \frac{nN(nN+1)}{2}\tilde{m} \max_{j \in \tilde{\mathbb{X}}} \mathsf{P}_j\{\check{\rho}_{\varepsilon,1} = 1\}.
\end{aligned} \tag{4.99}$$

The relation (4.99) and the condition $\mathbf{A}_5$ **(b)** imply that, for $i \in \tilde{\mathbb{X}}$,

$$\overline{\lim_{\varepsilon \to 0}} \mathsf{P}_i\{\tilde{\rho}_{\varepsilon,1} = 1\} \leq (1-p)^n \to 0 \text{ as } n \to \infty. \tag{4.100}$$

The condition $\mathbf{A}_1$ (for the Markov renewal processes $(\eta_{\varepsilon,n}, \kappa_{\varepsilon,n}, \rho_{\varepsilon,n})$) consists of two sub-conditions.

The first sub-condition, $\mathbf{A}_1$ **(a)**, requires there exists some $\tilde{\varepsilon}' \in (0, 1]$ such that for $\varepsilon \in (0, \tilde{\varepsilon}']$ and $i \in \mathbb{X}$,

$$\max_{i \in \mathbb{X}} \mathsf{P}_i\{\rho_{\varepsilon,1} = 1\} = \mathsf{P}_i\{\tilde{\rho}_{\varepsilon,1} = 1\} > 0. \tag{4.101}$$

The second sub-condition, $\mathbf{A}_1$ **(b)**, requires the fulfilment of the asymptotic relation (4.98), for $i \in \mathbb{X}$.

The latter sub-condition is a special case of the relation (4.98), since $\mathbb{X} \subset \tilde{\mathbb{X}}$.

Let us prove that the conditions $\mathbf{B}_6$ and $\mathbf{A}_5$ imply that the relation (4.101) holds.

Let j be some state such that $\mathsf{P}_j\{\check{\rho}_{\varepsilon,1} = 1\} > 0$, for $\varepsilon \in (0, \check{\varepsilon}']$.

First, suppose $j \in \mathbb{X}$.

In this case, the relation (4.102) implies that, for $\varepsilon \in (0, \check{\varepsilon}']$,

$$\begin{aligned} \mathsf{P}_j\{\rho_{\varepsilon,1} = 1\} &= \mathsf{P}_j\{1 - \prod_{k=1}^{\check{\nu}_{\varepsilon,\mathbb{X}}}(1 - \check{\rho}_{\varepsilon,k}) = 1\} \\ &= \mathsf{P}_j\{\prod_{k=1}^{\check{\nu}_{\varepsilon,\mathbb{X}}}(1 - \check{\rho}_{\varepsilon,k}) = 0\} \\ &= \mathsf{P}_j\{\sum_{k=1}^{\check{\nu}_{\varepsilon,\mathbb{X}}} \check{\rho}_{\varepsilon,k} \geq 1\} \geq \mathsf{P}_j\{\check{\rho}_{\varepsilon,k} = 1\} > 0. \end{aligned} \tag{4.102}$$

Second, suppose $j \in \bar{\mathbb{X}} = \tilde{\mathbb{X}} \setminus \mathbb{X}$.

In this case, condition $\mathbf{B}_6$ implies that there exists, for $\varepsilon \in (0, \varepsilon']$, a chain of states $\langle i = r_{ij,0}, r_{ij,1} \in \bar{\mathbb{X}}, \ldots, r_{ij,n_{ij}-1} \in \bar{\mathbb{X}}, r_{ij,n_{ij}} = j\rangle$ such that

$$\prod_{k=1}^{n_{ij}} p_{\varepsilon, r_{ij,k-1}, r_{ij,k}} > 0. \tag{4.103}$$

Indeed, let us choose an arbitrary state $l \in \mathbb{X}$. If $j \in \tilde{\mathbb{X}} \setminus \mathbb{X}$, then, by the condition $\mathbf{B}_6$, there exists, for $\varepsilon \in (0, \varepsilon']$, a chain of states $\langle l = r'_{lj,0}, r'_{lj,1}, \ldots, r'_{lj,n'_{lj}} = j\rangle$ such that $\prod_{1 \leq k = \leq n'_{lj}} p_{\varepsilon, r'_{lj,k-1}, r'_{lj,k}} > 0$. Define, $k_{lj} = \max(1 \leq k \leq n'_{lj} - 1 : r'_{lj,k} \in \mathbb{X})$. Obviously, the state $i = k_{lj} \in \mathbb{X}$ and the chain of states $\langle r_{ij,0} = r'_{lj,k_{lj}}, r_{ij,1} = r'_{lj,k_{lj}+1}, \ldots, r_{ij,n_{ij}} = r'_{ij,n_{lj}}\rangle$, where $n_{ij} = n'_{lj} - k_{lj}$, satisfies the relation (4.103).

Denote by $\mathcal{A}_{r,\mathbb{X},n}$ the set of all chains $\bar{l}_n = \langle l_0, l_l, \ldots, l_n\rangle$ such that: (a) $l_0 = r$, $l_n \in \mathbb{X}$, (b) $\cup_{\bar{l}_n \in \mathcal{A}_{r,\mathbb{X}n}} \{\eta_{\varepsilon,k} = l_k, k = 0, \ldots, n\} = \{\eta_{\varepsilon,0} = r, \nu_{\varepsilon,\mathbb{X}} = n\}$, for $n \geq 1$ and $\varepsilon \in (0, \varepsilon']$. The chain of states $\langle i = r_{ij,0}, \ldots, r_{ij,n_{ij}}, j = l_0, l_1, \ldots, l_n\rangle \in \mathcal{A}_{i,\mathbb{X},n_{ij}+n}$, for any chain of states $\bar{l} = \langle l_1, \ldots, l_n\rangle \in \mathcal{A}_{j,\mathbb{X},n}$.

The following relation takes place, for $\varepsilon \in (0, \varepsilon'']$ and $i \in \mathbb{X}$,

$$\begin{aligned} \mathsf{P}_i\{\rho_{\varepsilon,1} = 1\} &= \mathsf{P}_i\{\sum_{k=1}^{\check{\nu}_{\varepsilon,\mathbb{X}}} \check{\rho}_{\varepsilon,k} \geq 1\} \\ &= \sum_{n\geq 1} \sum_{\bar{l}_n \in \mathcal{A}_{i,\mathbb{X},n}} \mathsf{P}_i\{\eta_{\varepsilon,k} = l_k, k = 0, \ldots, n, \sum_{k=1}^{n} \check{\rho}_{\varepsilon,k} \geq 1\} \\ &\geq \sum_{n\geq 1} \sum_{\bar{l}_n \in \mathcal{A}_{j,\mathbb{X},n}} \mathsf{P}_i\{\eta_{\varepsilon,k} = r_{ij,k}, k = \ldots, n_{ij}, \end{aligned}$$

$$\eta_{\varepsilon,n_{ij}+r} = l_r, r = 0, \ldots, n, \check{\rho}_{\varepsilon,n_{ij}+1} = 1\}$$
$$\times \sum_{n\geq 1} \sum_{\bar{l}_n \in \mathcal{A}_{j,\mathbb{X},n}} \mathsf{P}_j\{\eta_{\varepsilon,n_{ij}+r} = l_r, r = 0, \ldots, n, \check{\rho}_{\varepsilon,n_{ij}+1} = 1\}$$
$$= \prod_{k=1}^{n_{ij}} p_{\varepsilon,r_{ij,k-1},r_{ij,k}} \mathsf{P}_j\{\eta_{\varepsilon,n_{ij}} = j, \check{\rho}_{\varepsilon,n_{ij}+1} = 1\}$$
$$= \prod_{k=1}^{n_{ij}} p_{\varepsilon,r_{ij,k-1},r_{ij,k}} \mathsf{P}_j\{\check{\rho}_{\varepsilon,1} = 1\} > 0. \tag{4.104}$$

The relations (4.102) and (4.104) imply that the relation (4.101) holds.

Thus, the conditions $\mathbf{A}_5$ and $\mathbf{B}_7$ imply that the condition $\mathbf{A}_1$ (for the Markov renewal processes $(\eta_{\varepsilon,n}, \kappa_{\varepsilon,n}, \rho_{\varepsilon,n})$) is satisfied.

Let us assume that the following condition is satisfied:

$\mathbf{C}_8$: $\mathsf{P}_i\{\check{\kappa}_{\varepsilon,1} > \delta\} \to 0$ as $\varepsilon \to 0$, for $\delta > 0$ and $i \in \tilde{\mathbb{X}}$.

The conditions $\mathbf{B}_7$ and $\mathbf{C}_8$ imply that the condition $\mathbf{C}_6$ is satisfied.

Indeed, using the relations (4.82) and (4.97), we obtain the following relation, for $\varepsilon \in (0, \varepsilon'']$ and $i \in \tilde{\mathbb{X}}$, $n \geq 1$,

$$\begin{aligned}
\mathsf{P}_i\{\tilde{\kappa}_{\varepsilon,1} > \delta\} &= \mathsf{P}_i\{\sum_{k=1}^{\check{\nu}_{\varepsilon,\mathbb{X},1}} \check{\kappa}_{\varepsilon,k} > \delta\} \\
&\leq \mathsf{P}_i\{\nu_{\varepsilon,\mathbb{X},1} > nN\} + \sum_{r=1}^{nN} \mathsf{P}_i\{\sum_{k=1}^{r} \check{\kappa}_{\varepsilon,k} > \delta, \check{\nu}_{\varepsilon,\mathbb{X},1} = r\} \\
&\leq (1-p)^n + \sum_{r=1}^{nN} \sum_{k=1}^{r} \sum_{j\in\tilde{\mathbb{X}}} \mathsf{P}_i\{\check{\kappa}_{\varepsilon,k} > \frac{\delta}{r}, \check{\eta}_{\varepsilon,k-1} = j\} \\
&\leq (1-p)^n + \frac{nN(nN+1)}{2} \tilde{m} \max_{j\in\tilde{\mathbb{X}}} \mathsf{P}_j\{\check{\kappa}_{\varepsilon,1} > \frac{\delta}{nN}\}.
\end{aligned} \tag{4.105}$$

The relation (4.105) and the condition $\mathbf{C}_8$ imply that, for $i \in \tilde{\mathbb{X}}$,

$$\overline{\lim_{\varepsilon\to 0}} \mathsf{P}_i\{\tilde{\kappa}_{\varepsilon,1} > \delta\} \leq (1-p)^n \to 0 \text{ as } n \to \infty. \tag{4.106}$$

We summarise the results of the above analysis in the following lemma.

Lemma 4.6 *Let $(\check{\eta}_{\varepsilon,n}, \check{\kappa}_{\varepsilon,n}, \check{\rho}_{\varepsilon,n})$ be a Markov renewal process defined above, with a finite phase space of the first component $\mathbb{X}_0 = \{1, \ldots, m_0\}$, and $(\tilde{\eta}_{\varepsilon,n}, \tilde{\kappa}_{\varepsilon,n}, \tilde{\rho}_{\varepsilon,n})$ be the Markov renewal process with the reduced phase spaces $\tilde{\mathbb{X}} = \mathbb{X}_0$ and $\mathbb{X} = \{1, \ldots, m\} \subseteq \tilde{\mathbb{X}}$ defined by the recurrent relations (4.82) and $(\eta_{\varepsilon,n}, \kappa_{\varepsilon,n}, \rho_{\varepsilon,n})$ be the shifted Markov renewal process defined by the relation (4.86). Then:*

(i) *The conditions $\mathbf{A}_5$, $\mathbf{B}_6$, $\mathbf{B}_7$, $\mathbf{C}_8$ assumed to be satisfied for the processes $(\check{\eta}_{\varepsilon,n}, \check{\kappa}_{\varepsilon,n}, \check{\rho}_{\varepsilon,n})$ imply that: (a) the condition $\mathbf{B}_5$ is satisfied for the processes*

$(\check{\eta}_{\varepsilon,n}, \check{\kappa}_{\varepsilon,n}, \check{\rho}_{\varepsilon,n})$; *(b) the conditions* $\mathbf{A}_5$, $\mathbf{C}_6$ *are satisfied for the processes* $(\tilde{\eta}_{\varepsilon,n}, \tilde{\kappa}_{\varepsilon,n}, \tilde{\rho}_{\varepsilon,n})$; *(c) the conditions* $\mathbf{A}_1$, $\mathbf{B}_1$ *are satisfied for the processes* $(\eta_{\varepsilon,n}, \kappa_{\varepsilon,n}, \rho_{\varepsilon,n})$.

(ii) *The version of Theorem 4.4 takes place, where the condition of this theorem is replaced by the conditions* $\mathbf{A}_5$, $\mathbf{B}_6$, $\mathbf{B}_7$, $\mathbf{C}_8$ *assumed to be satisfied for the processes* $(\check{\eta}_{\varepsilon,n}, \check{\kappa}_{\varepsilon,n}, \check{\rho}_{\varepsilon,n})$ *and the condition* $\mathbf{C}_1$ *assumed to be satisfied for the processes* $(\eta_{\varepsilon,n}, \kappa_{\varepsilon,n}, \rho_{\varepsilon,n})$.

Section 4.1.2 contains the condition $\mathbf{C}_5$, which entails the fulfilment of the condition $\mathbf{C}_1$. The condition $\mathbf{C}_5$ is formulated in terms of moments $\mathsf{E}_i\kappa^{\alpha}_{\varepsilon,1}, i \in \mathbb{X}$, for some $\alpha > 1$.

Below, we give upper bounds for moments of random variables $\mathsf{E}_i\tilde{\kappa}^{\alpha}_{\varepsilon,1}, i \in \tilde{\mathbb{X}}$ in terms of moments $\mathsf{E}_i\check{\kappa}^{\alpha\beta}_{\varepsilon,1}, i \in \tilde{\mathbb{X}}$ for some $\alpha, \beta > 1$. Since distribution functions $\mathsf{P}_i\{\tilde{\kappa}_{\varepsilon,1} \leq \cdot\} = \mathsf{P}_i\{\kappa_{\varepsilon,1} \leq \cdot\}$, for $i \in \mathbb{X}$, such upper bounds, in combination with the above condition $\mathbf{C}_5$, give useful conditions (formulated in terms of processes $(\check{\eta}_{\varepsilon,n}, \check{\kappa}_{\varepsilon,n}, \check{\rho}_{\varepsilon,n})$) entailing the fulfilment of the condition $\mathbf{C}_1$.

Let us assume that the following condition is satisfied:

$\mathbf{C}_9$: There exist $\varepsilon' \in (0, 1]$ and $\alpha, \beta > 1$ such that, for $\varepsilon \in (0, \varepsilon']$ and $i \in \tilde{\mathbb{X}}$,

$$\check{e}_{\varepsilon,i,\alpha,\beta} = (\mathsf{E}_i\check{\kappa}^{\alpha\beta}_{\varepsilon,1} \vee \max_{j \in \tilde{\mathbb{X}} \setminus \mathbb{X}} \mathsf{E}_j\check{\kappa}^{\alpha\beta}_{\varepsilon,1})^{\frac{1}{\beta}} < \infty.$$

Lemma 4.7 *Let the conditions* $\mathbf{B}_7$ *and* $\mathbf{C}_9$ *be satisfied. Then, the following inequality takes place for* $\varepsilon \in (0, \varepsilon''']$ *(where* $\varepsilon''' = \varepsilon' \wedge \varepsilon''$ *and* ε'' *is defined in relation* (4.97)), $\gamma = \frac{\beta}{1-\beta}$, *and* $i \in \tilde{\mathbb{X}}$,

$$\mathsf{E}_i\tilde{\kappa}^{\alpha}_{\varepsilon,1} \leq \check{e}_{\varepsilon,i,\alpha,\beta} \cdot a_{p,N,\alpha,\gamma}, \tag{4.107}$$

where

$$a_{p,N,\alpha,\gamma} = \sum_{n \geq 1} n^{\alpha-1}(1 + (n-1)(\tilde{m} - m))(1-p)^{\frac{1}{\gamma}[\frac{n-1}{N}]} < \infty. \tag{4.108}$$

Proof The condition $\mathbf{C}_9$ and the relation (4.97) imply that the following relation holds, for $\varepsilon \in (0, \varepsilon' \wedge \varepsilon'']$ and $i \in \tilde{\mathbb{X}}$,

$$\begin{aligned}
\mathsf{E}_i\tilde{\kappa}^{\alpha}_{\varepsilon,1} &= \mathsf{E}_i\Big(\sum_{k=1}^{\check{\nu}_{\varepsilon,\mathbb{X}}} \check{\kappa}_{\varepsilon,k}\Big)^{\alpha} \\
&= \mathsf{E}_i\sum_{n \geq 1}\Big(\sum_{k=1}^{n} \check{\kappa}_{\varepsilon,k}\Big)^{\alpha}\mathrm{I}(\check{\nu}_{\varepsilon,\mathbb{X}} = n) \\
&= \sum_{n \geq 1}\mathsf{E}_i\Big(\sum_{k=1}^{n} \check{\kappa}_{\varepsilon,k}\Big)^{\alpha}\mathrm{I}(\check{\nu}_{\varepsilon,\mathbb{X}} = n) \\
&\leq \sum_{n \geq 1}\mathsf{E}_i\Big(n^{\alpha-1}\sum_{k=1}^{n} \check{\kappa}^{\alpha}_{\varepsilon,k}\mathrm{I}(\check{\nu}_{\varepsilon,\mathbb{X}} = n)\Big)
\end{aligned}$$

$$
\begin{aligned}
&= \sum_{n \geq 1} n^{\alpha-1} \big(\mathsf{E}_i \check{\kappa}_{\varepsilon,1}^{\alpha} \cdot \mathrm{I}(\check{\nu}_{\varepsilon,\mathbb{X}} = n) \\
&\quad + \sum_{k=2}^{n} \sum_{j \in \tilde{\mathbb{X}} \setminus \mathbb{X}} \mathsf{E}_i \check{\kappa}_{\varepsilon,k}^{\alpha} \mathrm{I}(\eta_{\varepsilon,k-1} = j) \cdot \mathrm{I}(\check{\nu}_{\varepsilon,\mathbb{X}} = n) \big) \\
&\leq \sum_{n \geq 1} n^{\alpha-1} \big((\mathsf{E}_i \check{\kappa}_{\varepsilon,1}^{\alpha\beta})^{\frac{1}{\beta}} \mathsf{P}_i \{ \check{\nu}_{\varepsilon,\mathbb{X}} = n \}^{\frac{1}{\gamma}} \\
&\quad + \sum_{k=2}^{n} \sum_{j \in \tilde{\mathbb{X}} \setminus \mathbb{X}} (\mathsf{E}_i \check{\kappa}_{\varepsilon,k}^{\alpha\beta} \mathrm{I}(\eta_{\varepsilon,k-1} = j))^{\frac{1}{\beta}} \mathsf{P}_i \{ \check{\nu}_{\varepsilon,\mathbb{X}} = n \}^{\frac{1}{\gamma}} \big) \\
&= \sum_{n \geq 1} n^{\alpha-1} \big((\mathsf{E}_i \check{\kappa}_{\varepsilon,1}^{\alpha\beta})^{\frac{1}{\beta}} \\
&\quad + \sum_{k=2}^{n} \sum_{j \in \tilde{\mathbb{X}} \setminus \mathbb{X}} (\mathsf{E}_i \{ \check{\kappa}_{\varepsilon,k}^{\alpha\beta} / \eta_{\varepsilon,k-1} = j \} \mathsf{P}_i \{ \eta_{\varepsilon,k-1} = j \})^{\frac{1}{\beta}} \big) \mathsf{P}_i \{ \check{\nu}_{\varepsilon,\mathbb{X}} = n \}^{\frac{1}{\gamma}} \\
&\leq \check{e}_{\varepsilon,i,\alpha,\beta} \sum_{n \geq 1} n^{\alpha-1} \big(1 + (n-1)(\tilde{m} - m) \big) \mathsf{P}_i \{ \check{\nu}_{\varepsilon,\mathbb{X}} = n \}^{\frac{1}{\gamma}} \\
&\leq \check{e}_{\varepsilon,i,\alpha,\beta} \sum_{n \geq 1} n^{\alpha-1} \big(1 + (n-1)(\tilde{m} - m) \big) \mathsf{P}_i \{ \check{\nu}_{\varepsilon,\mathbb{X}} > N [\frac{n-1}{N}] \}^{\frac{1}{\gamma}} \\
&\leq \check{e}_{\varepsilon,i,\alpha,\beta} \cdot a_{p,N,\alpha,\gamma}.
\end{aligned} \tag{4.109}
$$

The proof is complete. □

4.2 First-Rare-Event Times and Hitting Times

In this section, we define standard and directed hitting times for semi-Markov processes, clarify their relationship with the first-rare-event times, and present necessary and sufficient conditions of convergence in distribution for standard and directed hitting times defined on asymptotically uniformly ergodic semi-Markov processes.

4.2.1 Necessary and Sufficient Conditions of Convergence in Distribution for Standard Hitting Times

4.2.1.1 Limit Theorems for Standard Hitting Times – I. Let $(\eta_{\varepsilon,n}, \kappa_{\varepsilon,n})$, $n = 0, 1, \ldots$ be, for every $\varepsilon \in (0, 1]$, a Markov renewal process, with a phase space $\mathbb{X} \times [0, \infty)$, where $\mathbb{X} = \{1, 2, \ldots, m\}$ is a finite set, and transition probabilities defined, for $(i, s), (j, t) \in \mathbb{X} \times [0, \infty)$,

$$
\begin{aligned}
Q_{\varepsilon,ij}(t) &= \mathsf{P}\{\eta_{\varepsilon,1} = j, \kappa_{\varepsilon,1} \leq t / \eta_{\varepsilon,0} = i, \kappa_{\varepsilon,0} = s\} \\
&= \mathsf{P}\{\eta_{\varepsilon,1} = j, \kappa_{\varepsilon,1} \leq t / \eta_{\varepsilon,0} = i\}.
\end{aligned} \tag{4.110}
$$

Let us now introduce the Markov renewal process $(\eta_{\varepsilon,n}, \kappa_{\varepsilon,n}, \rho_{\varepsilon,n}), n = 0, 1, \ldots,$ with the additional third component $\rho_{\varepsilon,n}$ that has the following form,

$$
\rho_{\varepsilon,n} = \mathrm{I}(\eta_{\varepsilon,n} \in \mathbb{D}), \text{ for } n = 1, 2, \ldots, \ \rho_{\varepsilon,0} = 0, \tag{4.111}
$$

where $\mathbb{D}$ is a subset of the phase space $\mathbb{X}$, such that $\emptyset \subset \mathbb{D} \subset \mathbb{X}$.

In this case, the transition probabilities $Q_{\varepsilon,ij}(u, \jmath)$ for the above Markov renewal process take the following form, for $u \geq 0, i, j \in \mathbb{X}, \jmath = 0, 1$,

$$
\begin{aligned}
Q_{\varepsilon,ij}(t, \jmath) &= \mathsf{P}\{\eta_{\varepsilon,1} = j, \kappa_{\varepsilon,1} \leq t, \rho_{\varepsilon,1} = \jmath / \eta_{\varepsilon,0} = i\} \\
&= \begin{cases} Q_{\varepsilon,ij}(u) & \text{for } \jmath = 0, j \in \bar{\mathbb{D}} \text{ or } \jmath = 1, j \in \mathbb{D}, \\ 0 & \text{for } \jmath = 0, j \in \mathbb{D} \text{ or } \jmath = 1, j \in \bar{\mathbb{D}}. \end{cases}
\end{aligned} \tag{4.112}
$$

In this model, the first-rare-event time,

$$
\xi_\varepsilon = \tau_{\varepsilon,\mathbb{D}} = \sum_{n=1}^{\nu_{\varepsilon,\mathbb{D}}} \kappa_{\varepsilon,n}, \tag{4.113}
$$

where

$$
\nu_{\varepsilon,\mathbb{D}} = \min(n \geq 1 : \eta_{\varepsilon,n} \in \mathbb{D}). \tag{4.114}
$$

Let $\eta_\varepsilon(t), t \geq 0$ be the semi-Markov process, defined by the following relation:

$$
\eta_\varepsilon(t) = \eta_{\varepsilon,n} \text{ for } \zeta_{\varepsilon,n-1} \leq t < \zeta_{\varepsilon,n}, \ n = 1, 2, \ldots, \tag{4.115}
$$

where

$$
\zeta_{\varepsilon,n} = \sum_{k=1}^{n} \kappa_{\varepsilon,k}, n = 0, 1, \ldots. \tag{4.116}
$$

The relation (4.113) means that, in this case, the first-rare-event time $\xi_\varepsilon = \tau_{\varepsilon,\mathbb{D}}$ is the first hitting time in the domain $\mathbb{D}$ for the semi-Markov process $\eta_\varepsilon(t)$.

In this case, the one-step probabilities of occurrence of rare event take the following form:

$$
p_{\varepsilon,i} = \mathsf{P}_i\{\eta_{\varepsilon,1} \in \mathbb{D}\} = \sum_{j \in \mathbb{D}} p_{\varepsilon,ij}, \ i \in \mathbb{X}, \tag{4.117}
$$

where

$$
p_{\varepsilon,ij} = \mathsf{P}_i\{\eta_{\varepsilon,1} = j\}, \ i, j \in \mathbb{X}. \tag{4.118}
$$

The condition $\mathbf{A}_1$ requires that $0 < \max_{i \in \mathbb{X}} p_{\varepsilon,i} \to 0$ as $\varepsilon \to 0$.

It is useful to note that, in this case, the distribution functions $G_{\varepsilon,\mathbb{D},i}(u) = \mathsf{P}_i\{\tau_{\varepsilon,\mathbb{D}} \leq u\}, u \geq 0, i \in \bar{\mathbb{D}}$ are determined by the transition probabilities $Q_{\varepsilon,kr}(v), v \geq 0, k \in \bar{\mathbb{D}}, r \in \mathbb{X}$. Thus, the distribution functions $G_{\varepsilon,\mathbb{D},i}(\cdot), i \in \bar{\mathbb{D}}$ do not change if one changes probabilities $Q_{\varepsilon,kr}(v), v \geq 0, k \in \mathbb{D}, r \in \mathbb{X}$.

This allows us to replace the condition $\mathbf{A}_1$ with a weaker condition:

$\mathbf{A}_6$: **(a)** There exists $\hat{\varepsilon}' \in (0, 1]$ such that $\max_{i \in \bar{\mathbb{D}}} p_{\varepsilon,i} > 0$, for $\varepsilon \in (0, \hat{\varepsilon}']$, **(b)** $p_{\varepsilon,i} \to 0$ as $\varepsilon \to 0$, for $i \in \bar{\mathbb{D}}$.

First, we analyse the case, where the initial state $\eta_{\varepsilon,0} \in \bar{\mathbb{D}}$, for $\varepsilon \in (0, 1]$.

Let us introduce the following conditional distribution functions, for $u \geq 0,\ i, j \in \mathbb{X}$,

$$F_{\varepsilon,ij}(u) = \mathsf{P}\{\kappa_{\varepsilon,1} \leq u/\eta_{\varepsilon,0} = i, \eta_{\varepsilon,1} = j\}. \tag{4.119}$$

Obviously, for $u \geq 0, i, j \in \mathbb{X}$,

$$F_{\varepsilon,ij}(u) = \begin{cases} \frac{Q_{\varepsilon,ij}(u)}{p_{\varepsilon,ij}} & \text{if } p_{\varepsilon,ij} > 0, \\ \mathrm{I}(u \geq 0) & \text{if } p_{\varepsilon,ij} = 0. \end{cases} \tag{4.120}$$

Here, we take into account that arbitrary distribution functions $F_{\varepsilon,ij}(u)$ can play the role of the above conditional distribution functions for $i, j \in \mathbb{X}$ such that $p_{\varepsilon,ij} = 0$, and choose, in these cases, the simplest distribution function $\mathrm{I}(u \geq 0), u \geq 0$.

Let us also introduce the following distribution functions, for $u \geq 0,\ i \in \mathbb{X}$,

$$F_{\varepsilon,i}(u) = \mathsf{P}\{\kappa_{\varepsilon,1} \leq u/\eta_{\varepsilon,0} = i\} = \sum_{j \in \mathbb{X}} F_{\varepsilon,ij}(u) p_{\varepsilon,ij}. \tag{4.121}$$

In what follows we denote by $F_1(u) * \cdots * F_n(u)$ the convolution of n proper or improper distribution functions $F_1(u), \ldots, F_n(u)$.

The following representation takes place for the distribution of the first hitting time $\tau_{\varepsilon,\mathbb{D}}$, for $u \geq 0, i_0 \in \bar{\mathbb{D}}$,

$$\begin{aligned}
\mathsf{P}_{i_0}\{\tau_{\varepsilon,\mathbb{D}} \leq u\} &= \sum_{n=1}^{\infty} \mathsf{P}_i\{\tau_{\varepsilon,\mathbb{D}} \leq u, \nu_{\varepsilon,\mathbb{D}} = n\} \\
&= \sum_{n=1}^{\infty} \sum_{i_0,\ldots,i_{n-1} \in \bar{\mathbb{D}}, i_n \in \mathbb{D}} F_{\varepsilon,i_0 i_1}(u) * \cdots * F_{\varepsilon,i_{n-1} i_n}(u) * \prod_{k=1}^{n} p_{\varepsilon,i_{k-1} i_k} \\
&= \sum_{n=1}^{\infty} \sum_{i_0,\ldots,i_{n-1} \in \bar{\mathbb{D}}} F_{\varepsilon,i_0 i_1}(u) * \cdots * F_{\varepsilon,i_{n-2} i_{n-1}}(u) \\
&\quad * \tilde{F}_{\varepsilon,i_{n-1}}(u) \prod_{k=1}^{n-1} (1 - p_{\varepsilon,i_{k-1}}) p_{\varepsilon,i_{n-1}} \prod_{k=1}^{n} \tilde{p}_{\varepsilon,i_{k-1} i_k} \\
&= \sum_{n=1}^{\infty} \sum_{i_0,\ldots,i_{n-1} \in \bar{\mathbb{D}}} F_{\varepsilon,i_0 i_1}(u)(1 - p_{\varepsilon,i_0}) * \cdots * F_{\varepsilon,i_{n-2} i_{n-1}}(u)(1 - p_{\varepsilon,i_{n-2}}) \\
&\quad * \tilde{F}_{\varepsilon,i_{n-1}}(u) p_{\varepsilon,i_{n-1}} \prod_{k=1}^{n} \tilde{p}_{\varepsilon,i_{k-1} i_k},
\end{aligned} \tag{4.122}$$

where

$$\tilde{p}_{\varepsilon,ij} = \begin{cases} \frac{p_{\varepsilon,ij}}{1-p_{\varepsilon,i}} & \text{for } i \in \bar{\mathbb{D}} \text{ such that } p_{\varepsilon,i} < 1, \\ \mathrm{I}(j = i) & \text{for } i \in \bar{\mathbb{D}} \text{ such that } p_{\varepsilon,i} = 1, \end{cases} \tag{4.123}$$

and

$$\tilde{F}_{\varepsilon,i}(u) = \begin{cases} \sum_{j \in \mathbb{D}} \frac{F_{\varepsilon,ij}(u) p_{\varepsilon,ij}}{p_{\varepsilon,i}} & \text{for } u \geq 0, i \in \bar{\mathbb{D}}, \text{ if } p_{\varepsilon,i} > 0, \\ \mathrm{I}(u \geq 0) & \text{for } u \geq 0, i \in \bar{\mathbb{D}}, \text{ if } p_{\varepsilon,i} = 0. \end{cases} \tag{4.124}$$

Let $(\hat{\eta}_{\varepsilon,n}, \hat{\kappa}_{\varepsilon,n}, \hat{\rho}_{\varepsilon,n})$ be a Markov renewal process, with phase space $\hat{\mathbb{Z}} = \hat{\mathbb{X}} \times [0, \infty) \times \{0, 1\}$ and transition probabilities $\hat{Q}_{\varepsilon,ij}(u, \jmath), u \geq 0, \jmath = 0, 1, i, j \in \hat{\mathbb{X}}$.

Let also $\hat{\xi}_\varepsilon$ be the corresponding first-rare-event time, i.e.,

$$\hat{\xi}_\varepsilon = \sum_{n=1}^{\hat{\nu}_\varepsilon} \hat{\kappa}_{\varepsilon,n}, \tag{4.125}$$

where

$$\hat{\nu}_\varepsilon = \max(n \geq 1 : \hat{\rho}_{\varepsilon,n}). \tag{4.126}$$

We define, for $u \geq 0, \ \jmath = 0, 1, \ i, j \in \hat{\mathbb{X}}$,

$$\hat{F}_{\varepsilon,ij,\jmath}(u) = \begin{cases} \mathsf{P}_i\{\hat{\kappa}_{\varepsilon,1} \leq u, \hat{\rho}_{\varepsilon,1} = \jmath / \hat{\eta}_{\varepsilon,1} = j\} & \text{if } \hat{p}_{\varepsilon,ij} > 0, \\ \mathrm{I}(u \geq 0) & \text{if } \hat{p}_{\varepsilon,ij} = 0, \end{cases} \tag{4.127}$$

where $\hat{p}_{\varepsilon,ij}, i, j \in \hat{\mathbb{X}}$ are the transition probabilities of the Markov chain $\hat{\eta}_{\varepsilon,n}$,

$$\hat{p}_{\varepsilon,ij} = \mathsf{P}_i\{\hat{\eta}_{\varepsilon,1} = j\} = \hat{Q}_{\varepsilon,ij}(\infty, 0) + \hat{Q}_{\varepsilon,ij}(\infty, 1), \tag{4.128}$$

and

$$\hat{F}_{\varepsilon,i,1}(u) = \mathsf{P}_i\{\hat{\kappa}_{\varepsilon,1} \leq u, \hat{\rho}_{\varepsilon,1} = 1\} = \sum_{j \in \hat{\mathbb{X}}} \hat{F}_{\varepsilon,ij,1}(u) \hat{p}_{\varepsilon,ij}. \tag{4.129}$$

Using the Markov property of the Markov renewal processes $(\hat{\eta}_{\varepsilon,n}, \hat{\kappa}_{\varepsilon,n}, \hat{\rho}_{\varepsilon,n})$, we obtain the following useful representation for the distribution of the first-rare-event time $\hat{\xi}_\varepsilon$, for $u \geq 0, i_0 \in \hat{\mathbb{X}}$,

$$\begin{aligned} \mathsf{P}_{i_0}\{\hat{\xi}_\varepsilon \leq u\} &= \mathsf{P}_{i_0}\{\sum_{n=1}^{\hat{\nu}_\varepsilon} \hat{\kappa}_{\varepsilon,n} \leq u\} \\ &= \sum_{n=1}^{\infty} \mathsf{P}_{i_0}\{\sum_{k=1}^{n} \hat{\kappa}_{\varepsilon,k} \leq u, \hat{\nu}_\varepsilon = n\} \\ &= \sum_{n=1}^{\infty} \sum_{i_0, \ldots, i_n \in \hat{\mathbb{X}}} \mathsf{P}_{i_0}\{\sum_{k=1}^{n} \hat{\kappa}_{\varepsilon,k} \leq u, \hat{\rho}_{\varepsilon,k} = 0, \\ &\quad \hat{\eta}_{\varepsilon k} = i_k, k = 1, \ldots, n-1, \hat{\rho}_{\varepsilon,n} = 1, \hat{\eta}_{\varepsilon,n} = i_n\} \end{aligned}$$

$$
\begin{aligned}
&= \sum_{n=1}^{\infty} \sum_{i_1,\ldots,i_n \in \hat{\mathbb{X}}} \hat{F}_{\varepsilon,i_0 i_1,0}(u) * \cdots * \hat{F}_{\varepsilon,i_{n-2}i_{n-1},0}(u) \\
&\quad * \hat{F}_{\varepsilon,i_{n-1}i_n,1}(u) \cdot \prod_{k=1}^{n} \hat{p}_{\varepsilon,i_{k-1}i_k} \\
&= \sum_{n=1}^{\infty} \sum_{i_1,\ldots,i_{n-1} \in \hat{\mathbb{X}}} \hat{F}_{\varepsilon,i_0 i_1,0}(u) * \cdots * \hat{F}_{\varepsilon,i_{n-2}i_{n-1},0}(u) \\
&\quad * \hat{F}_{\varepsilon,i_{n-1},1}(u) \cdot \prod_{k=1}^{n-1} \hat{p}_{\varepsilon,i_{k-1}i_k}. \qquad (4.130)
\end{aligned}
$$

Comparison of the relations (4.122) and (4.130) allows us to conclude that, for $i \in \bar{\mathbb{D}}$,

$$
\mathsf{P}_i\{\tau_{\varepsilon,\mathbb{D}} \leq \cdot\} = \mathsf{P}_i\{\hat{\xi}_\varepsilon \leq \cdot\}, \qquad (4.131)
$$

if the Markov renewal process $(\hat{\eta}_{\varepsilon,n}, \hat{\kappa}_{\varepsilon,n}, \hat{\rho}_{\varepsilon,n})$ is constructed in such a way that the following relations hold:

$$
\begin{cases}
\hat{\mathbb{X}} = \bar{\mathbb{D}}, \\
\hat{p}_{\varepsilon,ij} = \tilde{p}_{\varepsilon,ij}, \ i, j \in \bar{\mathbb{D}}, \\
\hat{F}_{\varepsilon,ij,0}(\cdot) = F_{\varepsilon,ij}(\cdot)(1 - p_{\varepsilon,i}), \ i, i \in \bar{\mathbb{D}}, \\
\hat{F}_{\varepsilon,i,1}(\cdot) = \tilde{F}_{\varepsilon,i}(\cdot) p_{\varepsilon,i}, \ i \in \bar{\mathbb{D}}.
\end{cases} \qquad (4.132)
$$

The above relations hold for the Markov renewal process $(\hat{\eta}_{\varepsilon,n}, \hat{\kappa}_{\varepsilon,n}, \hat{\rho}_{\varepsilon,n})$, with phase space $\bar{\mathbb{D}} \times [0, \infty) \times \{0, 1\}$ and the transition probabilities given by the following relation:

$$
\hat{Q}_{\varepsilon,ij}(\cdot, \jmath) = \begin{cases}
F_{\varepsilon,ij}(u)(1 - p_{\varepsilon,i})\tilde{p}_{\varepsilon,ij} & \text{for } u \geq 0, \ \jmath = 0, \ i, j \in \bar{\mathbb{D}}, \\
\tilde{F}_{\varepsilon,i}(u) p_{\varepsilon,i} \tilde{p}_{\varepsilon,ij} & \text{for } u \geq 0, \ \jmath = 1, \ i, j \in \bar{\mathbb{D}}.
\end{cases} \qquad (4.133)
$$

We are going to analyse the asymptotic behaviour of the hitting times $\tau_{\varepsilon,\mathbb{D}} \overset{\mathsf{d}}{=} \hat{\xi}_\varepsilon$, applying Theorem 2.1 to the first-rare-event times $\hat{\xi}_\varepsilon$.

The condition $\mathbf{A}_1$ takes, in this case, the form of the condition $\mathbf{A}_6$.

The condition $\mathbf{B}_1$ takes in this case, the following form:

$\mathbf{B}_8$: There exists a ring chain of states $i_0, i_1, \ldots, i_N = i_0$, which contains all states from set $\bar{\mathbb{D}}$ and such that $\underline{\lim}_{\varepsilon \to 0} \tilde{p}_{\varepsilon,i_{k-1}i_k} > 0$, for $k = 1, \ldots, N$.

It is useful to note that, due to the condition $\mathbf{A}_6$, the following relation holds, for $i, j \in \bar{\mathbb{D}}$,

$$
\tilde{p}_{\varepsilon,ij} - p_{\varepsilon,ij} \to 0 \text{ as } \varepsilon \to 0. \qquad (4.134)
$$

Therefore, the condition $\mathbf{B}_8$ can also be reformulated in the following equivalent form:

$\mathbf{B}_8^\circ$: There exists a ring chain of states $i_0, i_1, \ldots, i_N = i_0$, which contains all states from set $\bar{\mathbb{D}}$ and such that $\underline{\lim}_{\varepsilon \to 0} p_{\varepsilon,i_{k-1}i_k} > 0$, for $k = 1, \ldots, N$.

The condition $\mathbf{C}_1$ takes, in this case, the following form:

$\mathbf{C}_{10}$: $1 - \tilde{F}_{\varepsilon,i}(\delta) \longrightarrow 0$ as $\varepsilon \longrightarrow 0$, for $\delta > 0$ and $i \in \bar{\mathbb{D}}$.

The condition $\mathbf{B}_7$ implies that there exists $\hat{\varepsilon} \in (0, 1]$ such that the Markov chain $\hat{\eta}_{\varepsilon,n}$ is ergodic, for every $\varepsilon \in (0, \hat{\varepsilon}]$.
Let $\hat{\pi}_\varepsilon = \langle \hat{\pi}_{\varepsilon,i}, i \in \bar{\mathbb{D}} \rangle$ be its stationary distribution.
In this case, for $i \in \bar{\mathbb{D}}$,

$$\hat{F}_{\varepsilon,i}(\cdot) = \mathsf{P}_i\{\hat{\kappa}_{\varepsilon,i} \leq \cdot\} = F_{\varepsilon,i}(\cdot)(1 - p_{\varepsilon,i}) + \tilde{F}_{\varepsilon,i}(\cdot) p_{\varepsilon,i}. \tag{4.135}$$

Let $\hat{\theta}_{\varepsilon,n}, n = 1, 2, \ldots$ be a sequence of i.i.d. random variables with the distribution function $\hat{F}_\varepsilon(\cdot)$, where

$$\hat{F}_\varepsilon(\cdot) = \sum_{i \in \bar{\mathbb{D}}} \hat{\pi}_{\varepsilon,i} \hat{F}_{\varepsilon,i}(\cdot). \tag{4.136}$$

The function u_ε takes in this case the following form,

$$u_\varepsilon = p_\varepsilon^{-1}, \text{ where } p_\varepsilon = \sum_{i \in \bar{\mathbb{D}}} \hat{\pi}_{\varepsilon,i} p_{\varepsilon,i}. \tag{4.137}$$

The condition $\mathbf{D}_1$ takes, in this case, the following form:

$\mathbf{D}_9$: $\sum_{n=1}^{[u_\varepsilon]} \hat{\theta}_{\varepsilon,n} \xrightarrow{\mathsf{d}} \hat{\theta}_0$ as $\varepsilon \longrightarrow 0$, where $\hat{\theta}_0$ is a non-negative random variable with distribution not concentrated at zero.

In this case, $\hat{\theta}_0$ is infinitely divisible random variable and its Laplace transform has the form $\mathsf{E}e^{-s\hat{\theta}_0} = e^{-A(s)}, s \geq 0$, where $A(s)$ is the cumulant of corresponding infinitely divisible distribution.

The following theorem, which presents analogues of the statements **(i)** and **(ii)** of Theorem 2.1, takes place.

Theorem 4.5 *Let the conditions* $\mathbf{A}_6$, $\mathbf{B}_8$, *and* $\mathbf{C}_{10}$ *be satisfied. Then:*

(i) *The condition* $\mathbf{D}_9$ *is necessary and sufficient for the fulfilment (for some or any initial distributions* $\bar{q}_\varepsilon$ *of random variable* $\eta_{\varepsilon,0}$ *concentrated on domain* $\bar{\mathbb{D}}$, *respectively, in the statements of necessity and sufficiency) of the asymptotic relation* $\tau_{\varepsilon,\mathbb{D}} \xrightarrow{\mathsf{d}} \xi_0$ *as* $\varepsilon \longrightarrow 0$, *where* ξ_0 *is a non-negative random variable with distribution not concentrated at zero.*

(ii) *The distribution function* $G(u) = \mathsf{P}\{\xi_0 \leq u\}, u \geq 0$ *of the limiting random variable* ξ_0 *has the Laplace transform* $\phi(s) = \int_0^\infty e^{-su} G(du) = \frac{1}{1+A(s)}, s \geq 0$, *where* $A(s)$ *is a cumulant of infinitely divisible distribution appearing in the condition* $\mathbf{D}_9$.

It is worth to note that Theorem 4.5 can also be reformulated in terms of first-rare-event times $\hat{\xi}_\varepsilon$ based on the defined above Markov renewal processes $(\hat{\eta}_{\varepsilon,n}, \hat{\kappa}_{\varepsilon,n}, \hat{\rho}_{\varepsilon,n})$, with the phase space $\bar{\mathbb{D}}$ and transition probabilities given by the relation (4.133). Moreover, an analogue of the statement **(iii)** of Theorem 2.1 can also be formulated for the first-rare-event processes $\hat{\xi}_\varepsilon(t) = \sum_{n=1}^{[t\hat{v}_\varepsilon]} \hat{\kappa}_{\varepsilon,n}, t \geq 0$.

However, it should be borne in mind that the Markov renewal process $(\hat{\eta}_{\varepsilon,n}, \hat{\kappa}_{\varepsilon,n}, \hat{\rho}_{\varepsilon,n})$, differs from the original Markov renewal processes $(\eta_{\varepsilon,n}, \kappa_{\varepsilon,n}, \rho_{\varepsilon,n})$ used in the relation (4.113) to define the hitting time $\tau_{\varepsilon,\mathbb{D}}$.

4.2.1.2 Limit Theorems for Standard Hitting Times – II. Theorem 4.5 describes the asymptotics in distribution for the hitting times $\tau_{\varepsilon,\mathbb{D}}$ for the case, where their initial distributions are concentrated in the domain $\bar{\mathbb{D}}$, i.e., $\mathsf{P}\{\eta_{\varepsilon,0} \in \bar{\mathbb{D}}\} = 1$, for $\varepsilon \in (0,1]$.

Suppose now that the initial state $\eta_{\varepsilon,0} \in \mathbb{D}$, for $\varepsilon \in (0,1]$.

In this case, we can use the following representation, which takes place for $\varepsilon \in (0,1]$,

$$\begin{aligned}\tau_{\varepsilon,i,\mathbb{D}} &\overset{\mathsf{d}}{=} \sum_{j\in\mathbb{D}} \kappa_{\varepsilon,i,1}\mathrm{I}(\eta_{\varepsilon,i,1} = j) \\ &\quad + \sum_{j\in\bar{\mathbb{D}}} (\kappa_{\varepsilon,i,1} + \tau_{\varepsilon,j,\mathbb{D}})\mathrm{I}(\eta_{\varepsilon,i,1} = j) \\ &= \kappa_{\varepsilon,i,1} + \sum_{j\in\bar{\mathbb{D}}} \mathrm{I}(\eta_{\varepsilon,i,1} = j)\tau_{\varepsilon,j,\mathbb{D}}, \ \text{for } i \in \mathbb{D},\end{aligned} \tag{4.138}$$

where: (a) $\tau_{\varepsilon,i,\mathbb{D}}$ is a non-negative random variable, which has the distribution function $\mathsf{P}\{\tau_{\varepsilon,j,\mathbb{D}} \leq u\} = \mathsf{P}_j\{\tau_{\varepsilon,\mathbb{D}} \leq u\}, u \geq 0$, for $i \in \mathbb{X}$; (b) $(\kappa_{\varepsilon,i,1}, \eta_{\varepsilon,i,1})$ is a random vector, which takes values in the set $[0,\infty) \times \mathbb{X}$ and has the distribution $\mathsf{P}\{\kappa_{\varepsilon,i,1} \leq u, \eta_{\varepsilon,i,1} = j\} = Q_{\varepsilon,ij}(u), u \geq 0, j \in \mathbb{X}$, for $i \in \mathbb{X}$; (c) the random vector $(\kappa_{\varepsilon,i,1}, \eta_{\varepsilon,i,1})$ and the random variable $\tau_{\varepsilon,j,\mathbb{D}}$ are independent for every $i, j \in \mathbb{X}$.

Let us assume that the following condition is satisfied:

$\mathbf{C}_{11}$: **(a)** $p_{\varepsilon,ij} \to p_{0,ij}$ as $\varepsilon \to 0$, for $i \in \mathbb{D}, j \in \mathbb{X}$, where $p_{0,ij} \geq 0, \sum_{j\in\mathbb{X}_0} p_{0,ij} = 1, i \in \mathbb{D}, j \in \mathbb{X}$, **(b)** $F_{\varepsilon,ij}(\cdot) \Rightarrow F_{0,ij}(\cdot)$ as $\varepsilon \to 0$, for $i \in \mathbb{D}, j \in \mathbb{X}$, where $F_{0,ij}(\cdot), i \in \mathbb{D}, j \in \mathbb{X}$ are some distribution functions concentrated on the interval $[0,\infty)$.

Let us introduce Laplace transforms, for $\varepsilon \in [0,1]$ and $i \in \mathbb{X}$,

$$\phi_{\varepsilon,ij}(s) = \int_0^\infty e^{-su} F_{\varepsilon,ij}(du), \ s \geq 0 \tag{4.139}$$

and, for $\varepsilon \in (0,1]$ and $i \in \mathbb{X}$,

$$\Phi_{\varepsilon,i,\mathbb{D}}(s) = \int_0^\infty e^{-su}\mathsf{P}_i\{\tau_{\varepsilon,\mathbb{D}} \in du\}, \ s \geq 0. \tag{4.140}$$

The relation (4.138) can be rewritten in the following equivalent form, for $\varepsilon \in (0,1]$,

$$\Phi_{\varepsilon,i,\mathbb{D}}(s) = \sum_{j\in\mathbb{D}} \phi_{\varepsilon,ij}(s)p_{\varepsilon,ij}$$

$$+ \sum_{j \in \bar{\mathbb{D}}} \phi_{\varepsilon,ij}(s)\Phi_{\varepsilon,j,\mathbb{D}}(s)p_{\varepsilon,ij} \text{ for } s \geq 0,\ i \in \mathbb{D}. \tag{4.141}$$

Let us assume that the conditions of Theorem 4.5 and the condition $\mathbf{C}_{11}$ are satisfied.

In this case, the following relation takes place, for $i \in \mathbb{D}$,

$$\begin{aligned}\Phi_{\varepsilon,i,\mathbb{D}}(s) &= \sum_{j \in \mathbb{D}} \phi_{\varepsilon,ij}(s)p_{\varepsilon,ij} + \sum_{j \in \bar{\mathbb{D}}} \phi_{\varepsilon,ij}(s)\Phi_{\varepsilon,j,\mathbb{D}}(s)p_{\varepsilon,ij} \\ &\to \sum_{j \in \mathbb{D}} \phi_{0,ij}(s)p_{0,ij} + \phi(s)\sum_{j \in \bar{\mathbb{D}}} \phi_{0,ij}(s)p_{0,ij} \\ &= \Phi_{0,i,\mathbb{D}}(s) \text{ for } s \geq 0.\end{aligned} \tag{4.142}$$

The relation (4.142) is equivalent to the following relation, for $i \in \mathbb{D}$,

$$G_{\varepsilon,i,\mathbb{D}}(\cdot) = \mathsf{P}_i\{\tau_{\varepsilon,\mathbb{D}} < \cdot\} \Rightarrow G_{0,i,\mathbb{D}}(\cdot) \text{ as } \varepsilon \to 0, \tag{4.143}$$

where $G_{0,i,\mathbb{D}}(\cdot)$ is, for $i \in \mathbb{D}$, the distribution function with the Laplace transform $\Phi_{0,i,\mathbb{D}}(s)$, $s \geq 0$ given by the relation (4.142).

4.2.2 Necessary and Sufficient Conditions of Convergence in Distribution for Directed Hitting Times

4.2.2.1 Directed Hitting Times. Let $(\eta_{\varepsilon,n}, \kappa_{\varepsilon,n}, \rho_{\varepsilon,n}), n = 0, 1, \ldots$ be, for every $\varepsilon \in (0, 1]$, the Markov renewal process, introduced in Sect. 4.2.1.1.

Let us consider the model, where the third component $\rho_{\varepsilon,n}$ of this process has the following form,

$$\rho_{\varepsilon,n} = \mathrm{I}(\eta_{\varepsilon,n-1} \in \bar{\mathbb{D}}, \eta_{\varepsilon,n} \in \mathbb{D}), \text{ for } n = 1, 2, \ldots,\ \rho_{\varepsilon,0} = 0, \tag{4.144}$$

where $\mathbb{D}$ is some non-empty subset of the phase space $\mathbb{X}$.

This form differs from the form of the indicators $\rho_{\varepsilon,n}$ specified in the relation (4.111). The random variables $\rho_{\varepsilon,n}$, defined in the relation (4.144), indicate directed hits of the Markov chain $\eta_{\varepsilon,n}$ in the domain $\mathbb{D}$ from the domain $\bar{\mathbb{D}}$.

Accordingly, the directed hitting time is defined as follows:

$$\tau_{\varepsilon,\to\mathbb{D}} = \sum_{n=1}^{\nu_{\varepsilon,\to\mathbb{D}}} \kappa_{\varepsilon,n}, \tag{4.145}$$

where

$$\nu_{\varepsilon,\to\mathbb{D}} = \max(n \geq 1 : \eta_{\varepsilon,n-1} \in \bar{\mathbb{D}}, \eta_{\varepsilon,n} \in \mathbb{D}). \tag{4.146}$$

If the initial state $\eta_{\varepsilon,0} \in \bar{\mathbb{D}}$, the standard and directed hitting times coincide. Thus, the following relation takes place,

$$\tau_{\varepsilon,j,\to\mathbb{D}} \overset{\mathsf{d}}{=} \tau_{\varepsilon,j,\mathbb{D}}, \ j \in \bar{\mathbb{D}}, \tag{4.147}$$

where the random variables $\tau_{\varepsilon,j,\to\mathbb{D}}$ and $\tau_{\varepsilon,j,\mathbb{D}}$ have the distribution function $\mathsf{P}\{\tau_{\varepsilon,j,\to\mathbb{D}} \leq u\} = \mathsf{P}\{\tau_{\varepsilon,j,\mathbb{D}} \leq u\} = \mathsf{P}_j\{\tau_{\varepsilon,\mathbb{D}} \leq u\}, u \geq 0$, for $j \in \bar{\mathbb{D}}$.

If the initial state $\eta_{\varepsilon,0} \in \mathbb{D}$, the standard and directed hitting times do not coincide. In this case, the following relation takes place,

$$\tau_{\varepsilon,i,\to\mathbb{D}} \overset{\mathsf{d}}{=} \sum_{j\in\bar{\mathbb{D}}} \mathrm{I}(\eta_{\varepsilon,i,\bar{\mathbb{D}}} = j)(\tau_{\varepsilon,i,\bar{\mathbb{D}}} + \tau_{\varepsilon,j,\mathbb{D}}), \ i \in \mathbb{D}, \tag{4.148}$$

where: (a) the random vector $(\tau_{\varepsilon,i,\bar{\mathbb{D}}}, \eta_{\varepsilon,i,\bar{\mathbb{D}}})$ takes values in the set $[0,\infty)\times\bar{\mathbb{D}}$ and has the distribution $P\{\tau_{\varepsilon,\bar{\mathbb{D}}} \leq u, \eta_{\varepsilon,i,\bar{\mathbb{D}}} = j\} = \mathsf{P}_i\{\tau_{\varepsilon,\bar{\mathbb{D}}} \leq u, \eta_{\varepsilon,\nu_{\varepsilon,\bar{\mathbb{D}}}} = j\}, u \geq 0, j \in \bar{\mathbb{D}}$, for $i \in \mathbb{D}$; (b) the random vector $(\tau_{\varepsilon,i,\bar{\mathbb{D}}}, \eta_{\varepsilon,i,\bar{\mathbb{D}}})$ and the random variable $\tau_{\varepsilon,j,\mathbb{D}}$ are independent, for $i \in \mathbb{D}, j \in \bar{\mathbb{D}}$.

The only case, where $\eta_{\varepsilon,0} \in \mathbb{D}$, should be analysed.

Two sub-cases should be considered.

In the first sub-case, the probabilities of one-step transitions $\mathbb{D} \to \bar{\mathbb{D}}$ and $\bar{\mathbb{D}} \to \mathbb{D}$ tend to zero as $\varepsilon \to 0$, i.e., the following condition is satisfied:

$\mathbf{A}_7$: $p_{\varepsilon,ij} \to 0$ as $\varepsilon \to 0$, for $i \in \mathbb{D}, j \in \bar{\mathbb{D}}$ and $i \in \bar{\mathbb{D}}, j \in \mathbb{D}$.

In the second sub-case, the probabilities of one-step transitions $\bar{\mathbb{D}} \to \mathbb{D}$ tend to zero as $\varepsilon \to 0$, i.e., the following condition is satisfied:

$\mathbf{A}_8$: $p_{\varepsilon,ij} \to 0$ as $\varepsilon \to 0$, for $i \in \bar{\mathbb{D}}, j \in \mathbb{D}$.

4.2.2.2 Hitting Times for the Case of Asymptotically Small Probabilities of Transitions Between Domains $\mathbb{D}$ and $\bar{\mathbb{D}}$. Let us consider the case, where the condition $\mathbf{A}_7$ is satisfied.

In this case, Theorem 4.5 can be applied, both to the hitting times $\tau_{\varepsilon,\mathbb{D}}$, for the case where the initial state $\eta_{\varepsilon,0} \in \bar{\mathbb{D}}$, and to the hitting times $\tau_{\varepsilon,\bar{\mathbb{D}}}$, for the case where the initial state $\eta_{\varepsilon,0} \in \mathbb{D}$.

Let us denote by $\mathbf{A}'_6$, $\mathbf{B}'_8$, $\mathbf{C}'_{10}$, and $\mathbf{D}'_9$, respectively, the conditions $\mathbf{A}_6$, $\mathbf{B}_8$, $\mathbf{C}_{10}$, and $\mathbf{D}_9$, for the case, where the domain $\bar{\mathbb{D}}$ and the transition probabilities $Q_{\varepsilon,ij}(u), u \geq 0, i \in \bar{\mathbb{D}}, j \in \mathbb{X}$ are replaced by the domain $\mathbb{D}$ and the transition probabilities $Q_{\varepsilon,ij}(u), u \geq 0, i \in \mathbb{D}, j \in \mathbb{X}$.

Analogously, we denote by ξ'_0 and $\phi'(s) = (1+A'(s))^{-1}, s \geq 0$ the limiting random variable and its Laplace transform, in the asymptotic relation $\tau_{\varepsilon,\bar{\mathbb{D}}} \xrightarrow{\mathsf{d}} \xi'_0$ as $\varepsilon \to 0$, which holds, under the assumption that conditions $\mathbf{A}'_8$, $\mathbf{B}'_8$, $\mathbf{C}'_{10}$, and $\mathbf{D}'_9$ are satisfied (for some or any initial distributions $\bar{q}'_\varepsilon$ of the random variables $\eta_{\varepsilon,0}$ concentrated on the domain $\mathbb{D}$, respectively, in the statements of necessity and sufficiency), according to the corresponding variant of Theorem 4.5.

The relation (4.148), rewritten in terms of Laplace transforms, takes the following form, for any initial distribution $\bar{q}'_\varepsilon$ of the random variable $\eta_{\varepsilon,0}$ concentrated on the domain $\mathbb{D}$,

$$\mathsf{E}_{\bar{q}'_\varepsilon} e^{-s\tau_{\varepsilon,\to\mathbb{D}}} = \sum_{j\in\bar{\mathbb{D}}} \mathsf{E}_{\bar{q}'_\varepsilon} e^{-s\tau_{\varepsilon,\bar{\mathbb{D}}}} \mathrm{I}(\eta_{\varepsilon,\nu_{\varepsilon,\bar{\mathbb{D}}}} = j)\mathsf{E}_j e^{-s\tau_{\varepsilon,\bar{\mathbb{D}}}}, \ s \geq 0. \tag{4.149}$$

If the conditions $\mathbf{A}_6$, $\mathbf{B}_8$, $\mathbf{C}_{10}$, $\mathbf{D}_9$ and $\mathbf{A}'_6$, $\mathbf{B}'_8$, $\mathbf{C}'_{10}$, $\mathbf{D}'_9$ are satisfied, then (a) for any initial distributions $\bar{q}_\varepsilon$ of the random variables $\eta_{\varepsilon,0}$ concentrated on domain $\bar{\mathbb{D}}$, the hitting times $\tau_{\varepsilon,\mathbb{D}} \xrightarrow{\mathsf{d}} \xi_0$ as $\varepsilon \to 0$ and (b) for any initial distributions $\bar{q}'_\varepsilon$ of the random variables $\eta_{\varepsilon,0}$ concentrated on domain $\mathbb{D}$, the hitting times $\tau_{\varepsilon,\bar{\mathbb{D}}} \xrightarrow{\mathsf{d}} \xi'_0$ as $\varepsilon \to 0$, where (c) ξ_0 and ξ'_0 are non-negative random variables with the Laplace transforms, respectively, $\phi(s) = (1 + A(s))^{-1}, s \geq 0$ and $\phi'(s) = (1 + A'(s))^{-1}, s \geq 0$.

Therefore, the following relation holds

$$
\begin{aligned}
\mathsf{E}_{\bar{q}'_\varepsilon} e^{-s\tau_{\varepsilon,\to\mathbb{D}}} &= \sum_{j\in\bar{\mathbb{D}}} \mathsf{E}_{\bar{q}'_\varepsilon} e^{-s\tau_{\varepsilon,\bar{\mathbb{D}}}} \mathrm{I}(\eta_{\varepsilon,\nu_{\varepsilon,\bar{\mathbb{D}}}} = j) \mathsf{E}_j e^{-s\tau_{\varepsilon,\bar{\mathbb{D}}}} \\
&= \sum_{j\in\bar{\mathbb{D}}} \mathsf{E}_{\bar{q}'_\varepsilon} e^{-s\tau_{\varepsilon,\bar{\mathbb{D}}}} \mathrm{I}(\eta_{\varepsilon,\nu_{\varepsilon,\bar{\mathbb{D}}}} = j)(\mathsf{E}_j e^{-s\tau_{\varepsilon,\bar{\mathbb{D}}}} - \phi(s)) \\
&\quad + \mathsf{E}_{\bar{q}'_\varepsilon} e^{-s\tau_{\varepsilon,\bar{\mathbb{D}}}} \phi(s) \\
&\to \phi'(s)\phi(s) \text{ as } \varepsilon \to 0, \text{ for } s \geq 0.
\end{aligned}
\tag{4.150}
$$

The relation (4.150) is equivalent to the following relation (which holds for some or any initial distributions $\bar{q}_\varepsilon$ of the random variables $\eta_{\varepsilon,0}$ concentrated on domain $\mathbb{D}$, respectively, in the statements of necessity and sufficiency),

$$
\mathsf{P}_{\bar{q}_\varepsilon}\{\tau_{\varepsilon,\to\mathbb{D}} < \cdot\} \Rightarrow G_{0,\to\mathbb{D}}(\cdot) \text{ as } \varepsilon \to 0,
\tag{4.151}
$$

where $G_{0,\to\mathbb{D}}(\cdot)$ is, for $i \in \mathbb{D}$, the distribution function with the Laplace transform $\phi_{\to}(s) = \phi(s)\phi'(s) = (1 + A(s))^{-1}(1 + A'(s))^{-1}, s \geq 0$ given by the relation (4.150).

If the conditions $\mathbf{A}_6$, $\mathbf{B}_8$, $\mathbf{C}_{10}$ and $\mathbf{A}'_6$, $\mathbf{B}'_8$, $\mathbf{C}'_{10}$ are satisfied, and, (a) for some initial distributions $\bar{q}_\varepsilon$ of the random variables $\eta_{\varepsilon,0}$ concentrated on the domain $\bar{\mathbb{D}}$, the hitting times $\tau_{\varepsilon,\mathbb{D}} \xrightarrow{\mathsf{d}} \xi_0$ as $\varepsilon \to 0$ and, (b) for some initial distributions $\bar{q}'_\varepsilon$ of the random variables $\eta_{\varepsilon,0}$ concentrated on the domain $\mathbb{D}$, the hitting times $\tau_{\varepsilon,\bar{\mathbb{D}}} \xrightarrow{\mathsf{d}} \xi'_0$ as $\varepsilon \to 0$, where (c) ξ_0 and ξ'_0 are some non-negative random variables with distribution functions not concentrated at zero, then the conditions $\mathbf{D}_9$ and $\mathbf{D}'_9$ are satisfied, and the random variables ξ_0 and ξ'_0 have the Laplace transforms, respectively, $\phi(s) = (1 + A(s))^{-1}, s \geq 0$ and $\phi'(s) = (1 + A'(s))^{-1}, s \geq 0$ appearing in the conditions $\mathbf{D}_9$ and $\mathbf{D}'_9$, respectively.

Therefore, the relations (4.150) and (4.151) take place.

4.2.2.3 Hitting Times for the Case of Asymptotically Small Probabilities of Transitions from Domain $\bar{\mathbb{D}}$ to Domain $\mathbb{D}$. Let us consider the case, where the condition $\mathbf{A}_8$ is satisfied.

Let us first assume that the following condition is satisfied:

$\mathbf{B}_9$: For every $i \in \mathbb{D}$, there exists a chain of states $i = j_0, j_1, \ldots, j_{N_{i,\mathbb{D}}-1} \in \mathbb{D}$, $j_{N_{i,\mathbb{D}}} \in \bar{\mathbb{D}}$ such that $\underline{\lim}_{\varepsilon\to 0}\, p_{\varepsilon,j_{k-1}j_k} > 0$, for $k = 1, \ldots, N_{i,\mathbb{D}}$.

The condition $\mathbf{B}_9$ implies that there exists $\varepsilon_{\mathbb{D}} \in (0, 1]$ such that $p_{i,\mathbb{D}} = \prod_{k=1}^{N_{i,\mathbb{D}}} p_{\varepsilon,j_{k-1}j_k} > 0$, for $\varepsilon \in (0, \varepsilon_{\mathbb{D}}]$.

Denote $N_{\mathbb{D}} = \max_{i\in\mathbb{D}} N_{i,\mathbb{D}} < \infty$ and $p_{\mathbb{D}} = \min_{i\in\mathbb{D}} p_{i,\mathbb{D}} > 0$.

It is easy to see that, for $\varepsilon \in (0, \varepsilon_{\mathbb{D}}]$ and $i \in \mathbb{D}$, $n \geq 1$,

$$\mathsf{P}_i\{\nu_{\varepsilon,\mathbb{D}} > nN_{\mathbb{D}}\} \leq (1 - p_{\mathbb{D}})^n. \tag{4.152}$$

It should be noted that the conditions of Theorem 4.5 (applied to the hitting times $\tau_{\varepsilon,\mathbb{D}}$) imply that $\mathsf{P}_i\{\kappa_{\varepsilon,1} > \delta\} \longrightarrow 0$, as $\varepsilon \longrightarrow 0$, for $\delta > 0$ and $i \in \bar{\mathbb{D}}$.

Therefore, for the states $i \in \mathbb{D}$, it is natural to make a similar assumption, i.e., assume that the following condition is satisfied:

C$_{12}$: $\mathsf{P}_i\{\kappa_{\varepsilon,1} > \delta\} \longrightarrow 0$, as $\varepsilon \longrightarrow 0$, for $\delta > 0$ and $i \in \mathbb{D}$.

The condition **B**$_{10}$ implies that, for $\varepsilon \in (0, \varepsilon_{\mathbb{D}}]$ and $\delta > 0$,

$$\begin{aligned}
\mathsf{P}_i\{\tau_{\varepsilon,\bar{\mathbb{D}}} > \delta\} &= \mathsf{P}_i\{\sum_{k=1}^{\nu_{\varepsilon,\bar{\mathbb{D}}}} \kappa_{\varepsilon,k} > \delta\} \\
&\leq \mathsf{P}_i\{\nu_{\varepsilon,\bar{\mathbb{D}}}\} > nN_{\mathbb{D}}\} + \sum_{r=1}^{nN_{\mathbb{D}}} \mathsf{P}_i\{\sum_{k=1}^{r} \kappa_{\varepsilon,k} > \delta, \nu_{\varepsilon,\bar{\mathbb{D}}} = r\} \\
&\leq (1 - p_{\mathbb{D}})^n + \sum_{r=1}^{nN_{\mathbb{D}}} \sum_{k=1}^{r} \sum_{j \in \mathbb{D}} \mathsf{P}_i\{\kappa_{\varepsilon,k} > \frac{\delta}{r}, \eta_{\varepsilon,k-1} = j\} \\
&\leq (1 - p_{\mathbb{D}})^n + \frac{nN_{\mathbb{D}}(nN_{\mathbb{D}} + 1)}{2} m_{\mathbb{D}} \max_{j \in \mathbb{D}} \mathsf{P}_j\{\kappa_{\varepsilon,1} > \frac{\delta}{nN_{\mathbb{D}}}\},
\end{aligned} \tag{4.153}$$

where $m_{\mathbb{D}}$ is the number of states in the domain $\mathbb{D}$.

The conditions **B**$_9$ and **C**$_{12}$ imply that, for any initial distributions $\bar{q}'_\varepsilon$ of the random variables $\eta_{\varepsilon,0}$ concentrated on the domain $\mathbb{D}$, and $\delta > 0$,

$$\overline{\lim_{\varepsilon \to 0}} \mathsf{P}_{\bar{q}'_\varepsilon}\{\tau_{\varepsilon,\bar{\mathbb{D}}} > \delta\} \leq (1 - p_{\mathbb{D}})^n \longrightarrow 0 \text{ as } n \longrightarrow \infty. \tag{4.154}$$

Therefore, the conditions **B**$_9$ and **C**$_{12}$ imply that, for any initial distributions $\bar{q}'_\varepsilon$ of the random variables $\eta_{\varepsilon,0}$ concentrated on the domain $\mathbb{D}$,

$$\tau_{\varepsilon,\bar{\mathbb{D}}} \stackrel{\mathsf{P}}{\longrightarrow} 0 \text{ as } \varepsilon \longrightarrow 0. \tag{4.155}$$

The above relation implies that the following relation holds, for $j \in \bar{\mathbb{D}}$,

$$\begin{aligned}
&\mathsf{E}_{\bar{q}'_\varepsilon}(1 - e^{-s\tau_{\varepsilon,\bar{\mathbb{D}}}})\mathrm{I}(\eta_{\varepsilon,\nu_{\varepsilon,\bar{\mathbb{D}}}} = j)\mathsf{E}_j e^{-s\tau_{\varepsilon,\bar{\mathbb{D}}}} \\
&\quad \leq \mathsf{E}_{\bar{q}'_\varepsilon}(1 - e^{-s\tau_{\varepsilon,\bar{\mathbb{D}}}}) \\
&\quad = 1 - \mathsf{E}_{\bar{q}'_\varepsilon} e^{-s\tau_{\varepsilon,\bar{\mathbb{D}}}} \longrightarrow 0 \text{ as } \varepsilon \longrightarrow 0, \text{ for } s \geq 0.
\end{aligned} \tag{4.156}$$

If the conditions **A**$_6$, **B**$_8$, **C**$_{10}$, **D**$_9$ and **B**$_9$, **C**$_{12}$ are satisfied, then: (a) for any initial distributions $\bar{q}_\varepsilon$ of the random variables $\eta_{\varepsilon,0}$ concentrated on the domain $\bar{\mathbb{D}}$, the Laplace transforms $\mathsf{E}_{\bar{q}_\varepsilon} e^{-s\tau_{\varepsilon,\mathbb{D}}} \longrightarrow \phi(s) = \frac{1}{1+A(s)}$ as $\varepsilon \longrightarrow 0$, for $s \geq 0$, and (b) the asymptotic relation (4.156) holds.

Let us denote

$$\bar{q}_\varepsilon = \langle q_{\varepsilon,j} = \mathsf{P}_{\bar{q}'_\varepsilon}\{\eta_{\varepsilon,\nu_{\varepsilon,\bar{\mathbb{D}}}} = j\}, j \in \bar{\mathbb{D}}\rangle. \tag{4.157}$$

The above remarks and the relations (4.148) and (4.156) imply that, for any initial distributions $\bar{q}'_\varepsilon$ of the random variables $\eta_{\varepsilon,0}$ concentrated on the domain $\mathbb{D}$,

$$\begin{aligned}
\mathsf{E}_{\bar{q}'_\varepsilon} e^{-s\tau_{\varepsilon,\to\mathbb{D}}} &= \sum_{j\in\bar{\mathbb{D}}} \mathsf{E}_{\bar{q}'_\varepsilon} e^{-s\tau_{\varepsilon,\bar{\mathbb{D}}}} \mathrm{I}(\eta_{\varepsilon,\nu_{\varepsilon,\bar{\mathbb{D}}}} = j)\mathsf{E}_j e^{-s\tau_{\varepsilon,\bar{\mathbb{D}}}} \\
&= -\sum_{j\in\bar{\mathbb{D}}} \mathsf{E}_{\bar{q}'_\varepsilon}(1 - e^{-s\tau_{\varepsilon,\bar{\mathbb{D}}}})\mathrm{I}(\eta_{\varepsilon,\nu_{\varepsilon,\bar{\mathbb{D}}}} = j)\mathsf{E}_j e^{-s\tau_{\varepsilon,\bar{\mathbb{D}}}} \\
&\quad + \sum_{j\in\bar{\mathbb{D}}} \mathsf{E}_{\bar{q}'_\varepsilon} \mathrm{I}(\eta_{\varepsilon,\nu_{\varepsilon,\bar{\mathbb{D}}}} = j)\mathsf{E}_j e^{-s\tau_{\varepsilon,\bar{\mathbb{D}}}} \\
&= -\sum_{j\in\bar{\mathbb{D}}} \mathsf{E}_{\bar{q}'_\varepsilon}(1 - e^{-s\tau_{\varepsilon,\bar{\mathbb{D}}}})\mathrm{I}(\eta_{\varepsilon,\nu_{\varepsilon,\bar{\mathbb{D}}}} = j)\mathsf{E}_j e^{-s\tau_{\varepsilon,\bar{\mathbb{D}}}} \\
&\quad + \mathsf{E}_{\bar{q}_\varepsilon} e^{-s\tau_{\varepsilon,\bar{\mathbb{D}}}} \longrightarrow \phi(s) \text{ as } \varepsilon \to 0, \text{ for } s \geq 0,
\end{aligned} \tag{4.158}$$

or, equivalently,

$$\tau_{\varepsilon,\to\mathbb{D}} \stackrel{\mathsf{d}}{\longrightarrow} \xi_0 \text{ as } \varepsilon \to 0, \tag{4.159}$$

where ξ_0 is a non-negative random variable such that $\mathsf{E}e^{-s\xi_0} = \phi(s) = (1 + A(s))^{-1}, s \geq 0$.

Let us now assume that: (a) the conditions $\mathbf{A}_6$, $\mathbf{B}_8$, $\mathbf{C}_{10}$, and $\mathbf{B}_9$, $\mathbf{C}_{12}$ are satisfied, (b) for some initial distributions $\bar{q}'_\varepsilon$ of the random variable $\eta_{\varepsilon,0}$ concentrated on the domain $\mathbb{D}$, the directed hitting times $\tau_{\varepsilon,\to\mathbb{D}} \stackrel{\mathsf{d}}{\longrightarrow} \xi_{\to,0}$ as $\varepsilon \to 0$, where $\xi_{\to,0}$ is a non-negative random variable with distribution function not concentrated at zero.

The relations (4.148) and (4.156) imply that, for distributions $\bar{q}_\varepsilon$ given by the relation (4.157), there exists a limit

$$\begin{aligned}
\lim_{\varepsilon\to0} \mathsf{E}_{\bar{q}_\varepsilon} e^{-s\tau_{\varepsilon,\bar{\mathbb{D}}}} &= \lim_{\varepsilon\to0} \Big(-\sum_{j\in\bar{\mathbb{D}}} \mathsf{E}_{\bar{q}'_\varepsilon}(1 - e^{-s\tau_{\varepsilon,\bar{\mathbb{D}}}})\mathrm{I}(\eta_{\varepsilon,\nu_{\varepsilon,\bar{\mathbb{D}}}} = j)\mathsf{E}_j e^{-s\tau_{\varepsilon,\bar{\mathbb{D}}}} \\
&\quad + \mathsf{E}_{\bar{q}_\varepsilon} e^{-s\tau_{\varepsilon,\bar{\mathbb{D}}}} \Big) \\
&= \lim_{\varepsilon\to0} \Big(-\sum_{j\in\bar{\mathbb{D}}} \mathsf{E}_{\bar{q}'_\varepsilon}(1 - e^{-s\tau_{\varepsilon,\bar{\mathbb{D}}}})\mathrm{I}(\eta_{\varepsilon,\nu_{\varepsilon,\bar{\mathbb{D}}}} = j)\mathsf{E}_j e^{-s\tau_{\varepsilon,\bar{\mathbb{D}}}} \\
&\quad + \sum_{j\in\bar{\mathbb{D}}} \mathsf{E}_{\bar{q}'_\varepsilon} \mathrm{I}(\eta_{\varepsilon,\nu_{\varepsilon,\bar{\mathbb{D}}}} = j)\mathsf{E}_j e^{-s\tau_{\varepsilon,\bar{\mathbb{D}}}} \Big) \\
&= \lim_{\varepsilon\to0} \sum_{j\in\bar{\mathbb{D}}} \mathsf{E}_{\bar{q}'_\varepsilon} e^{-s\tau_{\varepsilon,\bar{\mathbb{D}}}} \mathrm{I}(\eta_{\varepsilon,\nu_{\varepsilon,\bar{\mathbb{D}}}} = j)\mathsf{E}_j e^{-s\tau_{\varepsilon,\bar{\mathbb{D}}}} \\
&= \lim_{\varepsilon\to0} \mathsf{E}_{\bar{q}'_\varepsilon} e^{-s\tau_{\varepsilon,\to\mathbb{D}}} = \phi_\to(s) \text{ as } \varepsilon \to 0, \text{ for } s \geq 0,
\end{aligned} \tag{4.160}$$

or, equivalently,

$$\tau_{\varepsilon,\mathbb{D}} \stackrel{\mathsf{d}}{\longrightarrow} \xi_{\to,0} \text{ as } \varepsilon \to 0, \tag{4.161}$$

where $\xi_{\to,0}$ is a random variable with the Laplace transform $\mathsf{E}e^{-s\xi_{\to,0}} = \phi_{\to}(s)$, $s \geq 0$.

Therefore, by Theorem 4.5, the condition $\mathbf{D}_9$ is satisfied, and thus, the limiting random variable $\xi_{\to,0}$ (in the relation (4.159)) has the Laplace transform $\mathsf{E}e^{-s\xi_0} = \phi_{\to}(s) = (1 + A(s))^{-1}$, $s \geq 0$, where $A(s)$ is the cumulant of the infinitely divisible distribution appearing in the condition $\mathbf{D}_9$.

The following theorem, which is an analogue of Theorem 4.5 for directed hitting times, summarises the above remarks.

Theorem 4.6 *Let the conditions* $\mathbf{A}_6$, $\mathbf{B}_8$, $\mathbf{C}_{10}$, *and* $\mathbf{B}_9$, $\mathbf{C}_{12}$ *be satisfied. Then*:

(i) *The condition* $\mathbf{D}_9$ *is necessary and sufficient for the fulfilment (for some or any initial distributions* $\bar{q}'_\varepsilon$ *of the random variables* $\eta_{\varepsilon,0}$ *concentrated on the domain* $\mathbb{D}$, *respectively, in the statements of necessity and sufficiency) of the asymptotic relation* $\tau_{\varepsilon,\to\mathbb{D}} \xrightarrow{\mathsf{d}} \xi_{\to,0}$ *as* $\varepsilon \to 0$, *where* $\xi_{\to,0}$ *is a non-negative random variable with distribution not concentrated at zero.*

(ii) *The distribution function* $G_{\to}(u) = \mathsf{P}\{\xi_{\to,0} \leq u\}$, $u \geq 0$ *of the limiting random variable* $\xi_{\to,0}$ *has the Laplace transform* $\phi_{\to}(s) = \int_0^\infty e^{-su} G_{\to}(du) = \frac{1}{1+A(s)}$, $s \geq 0$, *where* $A(s)$ *is a cumulant of infinitely divisible distribution appearing in the condition* $\mathbf{D}_9$.

Chapter 5
First-Rare-Event Times for Perturbed Risk Processes

In this chapter, the results obtained in previous chapters are illustrated by applications to geometric type random sums. Necessary and sufficient conditions of convergence in distribution for first-rare-event times represented by geometric type random sums are formulated. Also, necessary and sufficient conditions of weak convergence for non-ruin distribution functions in the models of stable and diffusion approximations for perturbed risk processes are given. This chapter includes two sections.

In Sect. 5.1, we give alternative definitions for first-rare-event times and first-rare-event processes, which are more natural for the model, in which the basic Markov renewal process does not depend on perturbation parameter. Theorem 5.1 gives necessary and sufficient conditions of convergence in distribution for first-rare-event times and convergence in topology J for first-rare-event processes.

In Sect. 5.2, we consider perturbed risk processes. Theorem 5.2 gives necessary and sufficient conditions of weak convergence for non-ruin distribution functions in the models of stable and diffusion approximations for perturbed risk processes.

5.1 Stable Asymptotics of First-Rare-Event Times

In this section, we give necessary and sufficient condition of stable asymptotics for first-rare-event times and processes defined on a finite ergodic Markov chain.

Let $(\eta_n, \kappa_n, \rho_n)$, $n = 0, 1, \ldots$, be a Markov renewal process, i.e., a homogenous Markov chain with phase space $\mathbb{Z} = \mathbb{X} \times [0, +\infty) \times \mathbb{Y}$ (here, $\mathbb{X} = \{1, 2, \ldots, m\}$ and $\mathbb{Y}$ is a measurable space with σ-algebra of measurable subsets $\mathcal{B}_{\mathbb{Y}}$) and transition probabilities,

$$\begin{aligned} \mathsf{P}\{\eta_{n+1} = j, \kappa_{n+1} \leq t, \rho_{n+1} \in A/\eta_n = i, \kappa_n = s, \rho_n = y\} \\ = \mathsf{P}\{\eta_{n+1} = j, \kappa_{n+1} \leq t, \rho_{n+1} \in A/\eta_n = i\} \\ = Q_{ij}(t, A),\ i, j \in \mathbb{X},\ s, t \geq 0,\ y \in \mathbb{Y}, A \in B_{\mathbb{Y}}. \end{aligned} \tag{5.1}$$

D. Silvestrov, *Perturbed Semi-Markov Type Processes I*,
https://doi.org/10.1007/978-3-030-92403-4_5

The characteristic property, which specifies Markov renewal processes in the class of general multivariate Markov chains $(\eta_n, \kappa_n, \rho_n)$, is that transition probabilities do depend only on the current position of the first component η_n.

As is known, the first component η_n of the Markov renewal process is also a homogenous Markov chain with the phase space $\mathbb{X}$ and transition probabilities $p_{ij} = Q_{ij}(+\infty, \mathbb{Y}), i, j \in \mathbb{X}$.

Also, the first two components of the above Markov renewal process η_n and κ_n can be associated with the semi-Markov process $\eta(t)$, $t \geq 0$ defined as

$$\eta(t) = \eta_n \text{ for } \zeta_n \leq t < \zeta_{n+1},\ n = 0, 1, \ldots, \tag{5.2}$$

where $\zeta_0 = 0$ and $\zeta_n = \kappa_1 + \cdots + \kappa_n$, $n \geq 1$.

The random variables κ_n are times between jumps for the semi-Markov $\eta(t)$. As for the random variables ρ_n, they are used to register "rare" events and can be referred as "rare-event indicators."

Let $\mathbb{D}_\varepsilon$, $\varepsilon \in (0, 1]$ be a family of measurable "small" in some sense subsets of $\mathbb{Y}$. Then events $\{\rho_n \in \mathbb{D}_\varepsilon\}$ can be interpreted as "rare" events.

Let us introduce random variables,

$$\xi_\varepsilon = \sum_{n=1}^{\nu_\varepsilon} \kappa_{\varepsilon,n}, \tag{5.3}$$

where: (a) $\nu_\varepsilon = \min(n \geq 1 : \rho_n \in \mathbb{D}_\varepsilon)$, (b) $\kappa_{\varepsilon,n} = \kappa_n / w_\varepsilon, n = 1, 2, \ldots$, (c) $w_\varepsilon, \varepsilon \in (0, 1]$ is a normalisation function such that $0 < w_\varepsilon \to \infty$ as $\varepsilon \to 0$.

The random variable ν_ε counts the number of transitions of the embedded Markov chain η_n before the first-rare-event occurs, and the random variable ξ_ε can be interpreted as the first-rare-event time for the semi-Markov process $\eta(t)$.

We also consider rare-event time processes,

$$\xi_\varepsilon(t) = \sum_{n=1}^{[t\nu_\varepsilon]} \kappa_{\varepsilon,n},\ t \geq 0. \tag{5.4}$$

The rare-event time model presented above is a special case of the model considered in Sect. 2.1.

The corresponding Markov renewal processes $(\eta_{\varepsilon,n}, \kappa_{\varepsilon,n}, \rho_{\varepsilon,n})$, $n = 0, 1, \ldots$, and $(\eta_n, \kappa_n, \rho_n)$, $n = 0, 1, \ldots$, are connected by the following relation, for $\varepsilon \in (0, 1]$,

$$\eta_{\varepsilon,n} = \eta_n,\ \kappa_{\varepsilon,n} = \kappa_n / w_\varepsilon,\ \rho_{\varepsilon,n} = \mathrm{I}(\rho_n \in \mathbb{D}_\varepsilon),\ n = 0, 1, \ldots. \tag{5.5}$$

The probabilities $p_{\varepsilon,i}$ of occurrence of a rare event at one transient step of the semi-Markov process $\eta_\varepsilon(t)$ take, in this case, the following form:

$$p_{\varepsilon,i} = \mathsf{P}_i\{\rho_n \in \mathbb{D}_\varepsilon\},\ i \in \mathbb{X}. \tag{5.6}$$

The condition $\mathbf{A}_1$ takes in this case the following form:

A$_9$: **(a)** There exists $\dot{\varepsilon} \in (0, 1]$ such that $\max_{i \in \mathbb{X}} p_{\varepsilon,i} > 0$, for $\varepsilon \in (0, \dot{\varepsilon}]$, **(b)** $p_{\varepsilon,i} \to 0$ as $\varepsilon \to 0$, for $i \in \mathbb{X}$.

The condition **B**$_1$ reduces to the following standard condition of ergodicity for the Markov chain η_n:

B$_{10}$: $\eta_n, n = 0, 1, \ldots$, is an ergodic Markov chain (i.e., its phase space $\mathbb{X}$ is one class of communicative states).

Let π_i, $i \in \mathbb{X}$ be the stationary distribution of the Markov chain η_n. This stationary distribution does not depend on ε.

The condition **C**$_1$, which guaranties that the last summand in the random sum ξ_ε is negligible, takes the following form:

C$_{13}$: $\mathsf{P}_i\{\frac{\kappa_1}{w_\varepsilon} > \delta/\rho_1 \in \mathbb{D}_\varepsilon\} \to 0$ as $\varepsilon \to 0$, for $\delta > 0, i \in \mathbb{X}$.

The relation (2.14) defining probability p_ε (which is the result of averaging of the probabilities of occurrence of rare event in one transition step over the stationary distribution of the embedded Markov chain η_n) and its reciprocal u_ε takes in this case the following forms:

$$p_\varepsilon = \sum_{i \in \mathbb{X}} \pi_i p_{\varepsilon,i} \ \text{ and } \ u_\varepsilon = p_\varepsilon^{-1}. \tag{5.7}$$

Let us introduce the distribution functions of the sojourn times for semi-Markov processes $\eta(t)$,

$$F_i(t) = \mathsf{P}\{\kappa_1 \le t\}, \ t \ge 0, \ i \in \mathbb{X}, \tag{5.8}$$

and

$$F(t) = \sum_{i \in \mathbb{X}} \pi_i F_i(t), \ t \ge 0, \ i \in \mathbb{X}. \tag{5.9}$$

Then, the distribution functions $F_{\varepsilon,i}(t)$ and $F_\varepsilon(t)$, introduced in the relations (2.15) and (2.16), take the following forms,

$$F_{\varepsilon,i}(t) = \mathsf{P}_i\{\kappa_1/w_\varepsilon \le t\} = F_i(tw_\varepsilon), \ t \ge 0, \ i \in \mathbb{X} \tag{5.10}$$

and

$$F_\varepsilon(t) = \sum_{i \in \mathbb{X}} \pi_i F_{\varepsilon,i}(t) = F(tw_\varepsilon), \ t \ge 0. \tag{5.11}$$

We will say that a positive function e_ε, $\varepsilon \in (0, 1]$ belongs to the class $\mathcal{E}$ if: **(a)** $e_\varepsilon \to \infty$ as $\varepsilon \to 0$, **(b)** there exists a sequence $0 < \varepsilon_n \to 0$ such that $e_{\varepsilon_{n+1}}/e_{\varepsilon_n} \to 1$ as $n \to \infty$.

The fourth condition is a kind of assumption about the regularity of the corresponding normalisation functions:

E$_1$: $u_\varepsilon = p_\varepsilon^{-1}$, $w_\varepsilon \in \mathcal{E}$.

The condition **E**$_1$ is not restrictive. For example, it is satisfied if u_ε and w_ε are continuous functions of ε.

Now we are now to formulate necessary and sufficient conditions for the weak convergence of distribution functions for the first-rare-event times ξ_ε.

Let $0 < \gamma \leq 1$ and $a > 0$. Let also $\Gamma(\alpha) = \int_0^\infty t^{\alpha-1} e^{-t} dt, t \geq 0$ be a Gamma function.

The above-mentioned necessary and sufficient conditions take the following forms:

$\mathbf{D}_{10}$: $\frac{t[1-F(t)]}{\int_0^t sF(ds)} \longrightarrow \frac{1-\gamma}{\gamma}$ as $t \to \infty$.

$\mathbf{D}_{11}$: $\frac{u_\varepsilon \int_0^{w_\varepsilon} sF(ds)}{w_\varepsilon} \longrightarrow a\frac{\gamma}{\Gamma(2-\gamma)}$ as $\varepsilon \to 0$.

The conditions $\mathbf{D}_{10}$ and $\mathbf{D}_{11}$ replace and are, in this case, equivalent to the condition $\mathbf{D}_3$ (equivalent to the condition $\mathbf{D}_1$).

Moreover, in this case, the cumulant $A(s) = as^\gamma, s \geq 0$. It is the cumulant of a stable law.

Theorem 2.1 takes the following form.

Theorem 5.1 *Let the conditions* $\mathbf{A}_9$, $\mathbf{B}_{10}$, $\mathbf{C}_{13}$, *and* $\mathbf{E}_1$ *be satisfied. Then:*

(i) *The conditions* $\mathbf{D}_{10}$ *and* $\mathbf{D}_{11}$ *are necessary and sufficient for the fulfilment (for some or any initial distributions* $\bar{q}$*, respectively, in the statements of necessity and sufficiency) of the asymptotic relation,* $\xi_\varepsilon = \xi_\varepsilon(1) \xrightarrow{\mathsf{d}} \xi_0$ *as* $\varepsilon \to 0$*, where* ξ_0 *is a non-negative random variable with distribution function not concentrated at zero.*

(ii) *The distribution function* $G(u)$ *of the limiting random variable* ξ_0 *has the Laplace transform* $\phi(s) = \int_0^\infty e^{-su} G(du) = \frac{1}{1+as^\gamma}, s \geq 0$*, where* a *and* γ *are parameters appearing in the conditions* $\mathbf{D}_{10}$ *and* $\mathbf{D}_{11}$.

(iii) *The stochastic processes* $\xi_\varepsilon(t), t \geq 0 \xrightarrow{\mathsf{J}} \xi_0(t) = \theta_0(t\nu_0), t \geq 0$ *as* $\varepsilon \to 0$*, where* **(a)** ν_0 *is a random variable that has the exponential distribution with parameter* 1*,* **(b)** $\theta_0(t), t \geq 0$ *is a non-negative stable Lévy process with the Laplace transforms* $\mathsf{E}e^{-s\theta_0(t)} = e^{-as^\gamma t}, s, t \geq 0$*,* **(c)** *the random variable* ν_0 *and the process* $\theta_0(t), t \geq 0$ *are independent.*

Proof Let us introduce Laplace transforms,

$$\phi_i(s) = \mathsf{E}e^{-s\kappa_1} = \int_0^\infty e^{-st} F_i(dt), \ s \geq 0, \ i \in \mathbb{X} \tag{5.12}$$

and

$$\phi(s) = \int_0^\infty e^{-st} F(dt) = \sum_{i \in \mathbb{X}} \pi_i \phi_i(s), \ s \geq 0, \ i \in \mathbb{X}. \tag{5.13}$$

In this case, the Laplace transforms introduced in relations (2.19) and (2.20) take the following forms:

$$\phi_{\varepsilon,i}(t) = \phi_i(s/w_\varepsilon), \ s \geq 0, \ i \in \mathbb{X}, \tag{5.14}$$

and

$$\phi_\varepsilon(s) = \int_0^\infty e^{-st} F_\varepsilon(dt) = \phi(s/w_\varepsilon),\ s \geq 0,\ i \in \mathbb{X}. \tag{5.15}$$

As mentioned in Sect. 2.1.3, the condition $\mathbf{D}_3$ is equivalent to the condition $\mathbf{D}_2$, which, in this case, takes the form of the following relation,

$$u_\varepsilon(1 - \varphi(s/w_\varepsilon) \longrightarrow A(s) \text{ as } \varepsilon \to 0, \text{ for } s \geq 0, \tag{5.16}$$

where the limiting function $A(s) > 0$, for $s > 0$ and $A(s) \to 0$ as $s \to 0$.

The relation (5.16) holds if and only if the following two asymptotic relations hold,

$$1 - \varphi(s) \sim s^\gamma L(1/s) \text{ as } 0 < s \to 0, \tag{5.17}$$

and

$$u_\varepsilon w_\varepsilon^{-\gamma} L(w_\varepsilon) \longrightarrow a \text{ as } \varepsilon \to 0, \tag{5.18}$$

where: $0 < \gamma \leq 1$, $a > 0$, and $L(t)$ is a slowly varying function.

This statement is well known, and we present its proof here only for the text to be self-sufficient.

If the relations (5.17) and (5.18) hold, then, for $s > 0$,

$$u_\varepsilon(1 - \varphi(s/w_\varepsilon)) \sim u_\varepsilon w_\varepsilon^{-\gamma} L(w_\varepsilon) \cdot s^\gamma \cdot \frac{L(w_\varepsilon/s)}{L(w_\varepsilon)} \longrightarrow as^\gamma \text{ as } \varepsilon \to 0. \tag{5.19}$$

On the other hand, let us assume that the relation (5.16) holds.

Let us define an auxiliary function $\tilde{\phi}(s) = 1 - \phi(1/s)$. The function $\tilde{\phi}(s)$ is monotonically decreasing. Due to the condition $\mathbf{E}_1$, there exists $\varepsilon_n \to 0$ as $n \to \infty$ such that $u_{\varepsilon_{n+1}}/u_{\varepsilon_n} \to 1$ as $n \to \infty$. Relation (5.16) implies that $u_{\varepsilon_n}\tilde{\phi}(w_{\varepsilon_n} s) \to A(1/s) > 0$ as $n \to \infty$, for $s > 0$. Thus, according to the well-known criterion (see, for example, Feller (1966)), the function $\tilde{\phi}(s)$ regularly varies, i.e., $\tilde{\phi}(s) = s^\rho L(s)$, where $L(s)$ is a slowly varying function, and $A(1/s) = as^\rho$, where $-\infty < \rho < +\infty$ and $a > 0$. These representations can be rewritten in the following equivalent forms:

$$1 - \phi(s) = s^\gamma L(1/s), \text{ and } A(s) = as^\gamma,\ s > 0, \tag{5.20}$$

where $-\infty < \gamma = \rho^{-1} < \infty$ and $a > 0$.

Since function $e^{-A(s)} = e^{-as^\gamma}$ should be the Laplace transform of some non-negative and non-zero random variable, the only values $0 < \gamma \leq 1$ are admissible. In this case, e^{-as^γ} is the Laplace transform of a non-negative stable random variable with parameter γ. The case $\gamma \leq 0$ must obviously be excluded. The case $\gamma > 1$ must also be excluded, since, for any non-negative and non-zero random variable ξ, the corresponding Laplace transform $\mathsf{E}e^{-s\xi} \geq e^{-s\delta}\mathsf{P}\{\xi \leq \delta\}$. Therefore, $\mathsf{E}e^{-s\xi}$ cannot decline in s with the super-exponential rate e^{-as^γ}.

The relation (5.17) follows from the relation (5.20) and the above remarks.

To verify the relation (5.18), we simply have to repeat the calculations given in the relation (5.19), assuming that the relations (5.16) and representation (5.20) hold. Indeed, the relations (5.16) and (5.20) imply that, for $s > 0$,

$$
\begin{aligned}
u_\varepsilon(1-\varphi(s/w_\varepsilon)) &= u_\varepsilon w_\varepsilon^{-\gamma}\cdot s^\gamma L(w_\varepsilon/s) \\
&= u_\varepsilon w_\varepsilon^{-\gamma} L(w_\varepsilon) s^\gamma \frac{L(w_\varepsilon/s)}{L(w_\varepsilon)} \\
&\to A(s) = a s^\gamma \text{ as } \varepsilon \to 0. \qquad (5.21)
\end{aligned}
$$

The relation (5.21), where one should take $s = 1$, implies that the relation (5.18) holds.

Let us now prove that the conditions $\mathbf{D}_{10}$ and $\mathbf{D}_{11}$ are equivalent to the assumption that the asymptotic relations (5.17) and (5.18) hold.

First, let us consider the case $\gamma = 1$, which corresponds to the situation where the limiting process $\theta_0(t) = at, t \geq 0$ is a nonrandom linear function.

According to the central criterium of convergence for the sums of i.i.d. random variables (see, for example, Loève (1977)), the necessary and sufficient conditions for the weak convergence of such sums given in the condition $\mathbf{D}_3$ (which is equivalent to the relation (5.16) or, according to the above proof, to the relations (5.17) and (5.18)) take the form of the following two relations:

$$
u_\varepsilon(1 - F_\varepsilon(u)) = u_\varepsilon(1 - F(uw_\varepsilon)) \to 0 \text{ as } \varepsilon \to 0, \text{ for } u > 0, \qquad (5.22)
$$

and

$$
u_\varepsilon \int_0^v sF_\varepsilon(ds) \to a \text{ as } \varepsilon \to 0, \text{ for some } v > 0. \qquad (5.23)
$$

Note that the relation (5.22) implies that the relation (5.23) either holds or does not hold simultaneously for all $v > 0$. Indeed, the relation (5.22) implies that, for any $0 < v' < v'' < \infty$,

$$
u_\varepsilon \int_{v'}^{v''} sF_\varepsilon(ds) \leq u_\varepsilon v''(1 - F_\varepsilon(v') \to 0 \text{ as } \varepsilon \to 0. \qquad (5.24)
$$

Taking into account the relation (5.24), we can transform the relations (5.22) and (5.23) into the following equivalent forms:

$$
\frac{(1 - F_\varepsilon(u))}{\int_0^u sF_\varepsilon(ds)} \to 0 \text{ as } \varepsilon \to 0, \text{ for } u > 0, \qquad (5.25)
$$

and

$$
u_\varepsilon \int_0^1 sF_\varepsilon(ds) \to a \text{ as } \varepsilon \to 0. \qquad (5.26)
$$

Since $1 - F_\varepsilon(u) = 1 - F(uw_\varepsilon)$ and $\int_0^u sF_\varepsilon(ds) = \int_0^{uw_\varepsilon} sF(ds)/w_\varepsilon$, the relations (5.25) and (5.26) can be rewritten as

$$
\frac{w_\varepsilon(1 - F(uw_\varepsilon))}{\int_0^{uw_\varepsilon} tF(dt)} \to 0 \text{ as } \varepsilon \to 0, \text{ for } u > 0, \qquad (5.27)
$$

and

$$\frac{u_\varepsilon \int_0^{w_\varepsilon} sF(ds)}{w_\varepsilon} \to a \text{ as } \varepsilon \to 0. \tag{5.28}$$

The relation (5.28) coincides with the condition $\mathbf{D}_{11}$.

The condition $\mathbf{D}_{10}$ (with parameter $\gamma = 1$) implies that the relation (5.27) holds. This is easy to see by setting $t = uu_\varepsilon$ in the asymptotic relation given in the condition $\mathbf{D}_{10}$ (with the parameter $\gamma = 1$).

It remains to show that the relation (5.27) implies the fulfilment of the condition $\mathbf{D}_{10}$ (with the parameter $\gamma = 1$).

According to the condition $\mathbf{E}_1$, the function $w_\varepsilon \in \mathcal{E}$. Thus, there exists sequence $0 < \varepsilon_n \to 0$ as $n \to \infty$ such that $w_{\varepsilon_{n+1}}/w_{\varepsilon_n} \to 1$ as $n \to \infty$. For any $t, u > 0$, we define $n_u(t) = \max(n : uw_{\varepsilon_n} \le t)$. The definition of $n_u(t)$ implies that $uw_{\varepsilon_{n(t)}} \le t < uw_{\varepsilon_{n(t)+1}}$, for $u, t > 0$, and $n_u(t) \to \infty$ as $t \to \infty$, for $u > 0$. Thus, $w_{\varepsilon_{n_u(t)+1}}/w_{\varepsilon_{n_u(t)}} \to 1$ as $t \to \infty$, for $u > 0$. Using the above asymptotic relation and the relation (5.27), we obtain, for $u > 0$,

$$\begin{aligned} \frac{t(1 - F(t))}{\int_0^t sF(ds)} &\le \frac{uw_{\varepsilon_{n_u(t)+1}}(1 - F(uw_{\varepsilon_{n_u(t)}}))}{\int_0^{uw_{\varepsilon_{n_u(t)}}} sF(ds)} \\ &= \frac{w_{\varepsilon_{n_u(t)+1}}}{w_{\varepsilon_{n_u(t)}}} \cdot \frac{uw_{\varepsilon_{n_u(t)}}(1 - F(uw_{\varepsilon_{n_u(t)}}))}{\int_0^{uw_{\varepsilon_{n_u(t)}}} sF(ds)} \\ &\to 0 \text{ as } t \to \infty. \end{aligned} \tag{5.29}$$

Now consider the case $0 < \gamma < 1$.

Due to the corresponding Tauberian theorem (see, for example, Feller (1971)), the relation (5.17) is equivalent to the following relation:

$$1 - F(t) \sim \frac{t^{-\gamma} L(t)}{\Gamma(1 - \gamma)} \text{ as } t \to \infty. \tag{5.30}$$

Due to the corresponding theorem about regularly varying functions (see, for example, Feller (1971)), the relation (5.30) is equivalent to the following relation:

$$\frac{t(1 - F(t))}{\int_0^t (1 - F(s))ds} \to 1 - \gamma \text{ as } t \to \infty. \tag{5.31}$$

Since $\int\limits_0^t [1 - F(s)]ds = t[1 - F(t)] + \int\limits_0^t sF(ds), t \ge 0$, the relation (5.31) can be rewritten in the following equivalent form:

$$\frac{t(1 - F(t))}{\int_0^t sF(ds)} \to \frac{1 - \gamma}{\gamma} \text{ as } t \to \infty. \tag{5.32}$$

The relation (5.32) coincides with the condition $\mathbf{D}_{10}$ (with the parameter $\gamma \in (0, 1)$) and, as mentioned above, is equivalent to the relation (5.17).

Let us show that, under the condition $\mathbf{D}_{10}$ (with parameter $\gamma \in (0, 1)$), the condition $\mathbf{D}_{11}$ (with parameters $\gamma \in (0, 1)$ and $a > 0$) and the relation (5.18) are equivalent.

Indeed, the relations (5.30) and (5.32), which are equivalent to the condition $\mathbf{D}_{10}$, imply that the following asymptotic relation holds:

$$\begin{aligned}\frac{v_\varepsilon \int_0^{u_\varepsilon} sF(ds)}{u_\varepsilon} &\sim \frac{v_\varepsilon u_\varepsilon (1 - F(u_\varepsilon))}{u_\varepsilon} \cdot \frac{\gamma}{1-\gamma} \\ &\sim \frac{L(u_\varepsilon)}{p_\varepsilon u_\varepsilon^\gamma \Gamma(1-\gamma)} \cdot \frac{\gamma}{1-\gamma} \\ &= \frac{L(u_\varepsilon)}{p_\varepsilon u_\varepsilon^\gamma} \cdot \frac{\gamma}{\Gamma(2-\gamma)} \text{ as } \varepsilon \to 0. \end{aligned} \tag{5.33}$$

This relation completes the proof of Theorem 5.1. □

Remark 5.1 As follows from the proof presented above, the assumption $w_\varepsilon \in \mathcal{E}$ can be omitted in the statements of necessity of Theorem 5.1; the assumption $u_\varepsilon = p_\varepsilon^{-1} \in \mathcal{E}$ in the statement of sufficiency of Theorem 5.1, for the case $\gamma = 1$; and the assumption $w_\varepsilon, u_\varepsilon \in \mathcal{E}$, i.e., the condition $\mathbf{E}_1$, in the statement of sufficiency of Theorem 5.1, for the case $0 < \gamma < 1$.

Remark 5.2 The simplest variant for normalisation functions is where $w_\varepsilon = \varepsilon^{-1}$. In this case, due to the relation (5.18), the function $u_\varepsilon = p_\varepsilon^{-1} = \frac{a\varepsilon^{-\gamma}}{L(\varepsilon^{-1})}$. In this case, both functions, w_ε and u_ε, belong to the class $\mathcal{E}$. In this case, the condition $\mathbf{E}_1$ can be omitted in Theorem 5.1.

Remark 5.3 As follows from the proof of Theorem 5.1, the conditions $\mathbf{D}_{10}$ and $\mathbf{D}_{11}$ can be replaced by the equivalent conditions given in the relations (5.17) and (5.18).

Remark 5.4 In the case $0 < \gamma < 1$, the condition $\mathbf{D}_{10}$ is equivalent to the relation (5.30), which means that the distribution function $F(t)$ belongs to the domain of attraction of a stable law with the parameter γ. In the case $\gamma = 1$, the condition $\mathbf{D}_{10}$ is the necessary and sufficient condition for the distribution $G(t)$ to belong to the domain of attraction of the degenerate law (see, for example, Feller (1971)). In Theorem 5.1, these conditions are presented in a convenient unified form. As for the condition $\mathbf{D}_{11}$, this is the condition that balances the normalisation function w_ε and the function $u_\varepsilon = p_\varepsilon^{-1}$ that determines the number of terms in the sum, representing the first-rare-event time $\xi_\varepsilon(t)$.

Remark 5.5 It follows from the remarks made in Sect. 4.1.1.2 that, under the condition $\mathbf{C}_{12}$, the random variables $\kappa_{\nu_\varepsilon}/w_\varepsilon \xrightarrow{\mathsf{P}} 0$ as $\varepsilon \to 0$. This relation implies that the first-rare-event times $\xi_\varepsilon = \sum_{n=1}^{\nu_\varepsilon} \varkappa_n$ can be replaced in Theorem 5.1 by the modified first-rare-event times $\xi'_\varepsilon = \sum_{n=1}^{\nu_\varepsilon - 1} \varkappa_n$ and, moreover, by any random variable ξ''_ε such that $\xi'_\varepsilon \leq \xi''_\varepsilon \leq \xi_\varepsilon$.

5.2 Necessary and Sufficient Conditions for Stable Approximation of Non-ruin Distribution Functions

Let us apply the results given in Theorem 5.1 to the so-called geometric random sums. This is the reduction of our model for the case when the embedded Markov chain η_n has the degenerate set of states $\mathbb{X} = \{1\}$.

In this case, the first-rare-event time $\xi_\varepsilon = \sum_{n=1}^{\nu_\varepsilon} \kappa_n$ is a geometric random sum.

Indeed, $(\kappa_n, \rho_n), n = 1, 2, \ldots$, is a sequence of i.i.d. random vectors. Therefore, the random variable $\nu_\varepsilon = \min(n \geq 1 : \rho_n \in \mathbb{D}_\varepsilon)$ has a geometric distribution with the success probability $p_\varepsilon = \mathsf{P}\{\rho_n \in \mathbb{D}_\varepsilon\}$.

It is worth noting that ξ_ε is not a standard geometric sum, since in the above model the random summands $\kappa_n, n = 1, 2, \ldots$, and the geometric random index $\nu_\varepsilon = \min(n \geq 1 : \rho_{\varepsilon,n} = 1)$ are dependent random variables. They are rare-event indicators $\rho_{\varepsilon,n} = \mathrm{I}(\rho_n \in \mathbb{D}_\varepsilon), n = 1, 2, \ldots$. The random vectors $(\kappa_n, \rho_n), n = 1, 2, \ldots$, are independent and identically distributed. This implies that $(\kappa_n, \rho_{\varepsilon,n}), n = 1, 2, \ldots$, is, for every $\varepsilon \in (0, 1]$, a sequence of i.i.d. random vectors. However, the components κ_n and ρ_n of the random vector (κ_n, ρ_n) and, thus, the components κ_n and $\rho_{\varepsilon,n}$ of the random vector $(\kappa_n, \rho_{\varepsilon,n})$ may be dependent random variables.

In this case, the condition $\mathbf{B}_{10}$ obviously holds.

The conditions $\mathbf{A}_9$ and $\mathbf{C}_{13}$ take, in this case, the following forms:

$\mathbf{A}_{10}$: There exists $\dot{\varepsilon} \in (0, 1]$ such that $p_\varepsilon > 0$, for $\varepsilon \in (0, \dot{\varepsilon}]$, **(b)** $p_\varepsilon = \mathsf{P}\{\rho_1 \in \mathbb{D}_\varepsilon\} \to 0$
as $\varepsilon \to 0$.

$\mathbf{C}_{14}$: $\mathsf{P}\{\kappa_1 > \delta w_\varepsilon / \rho_1 \in \mathbb{D}_\varepsilon\} \to 0$ as $\varepsilon \to 0$.

The condition $\mathbf{E}_1$ does not change. It should be imposed on the function $u_\varepsilon = p_\varepsilon^{-1}$ appearing in the condition $\mathbf{A}_{10}$ and the normalisation function w_ε.

The conditions $\mathbf{D}_{10}$ and $\mathbf{D}_{11}$ also do not change. They should be imposed on the distribution function $G(t) = \mathsf{P}\{\kappa_1 \leq t\}$ (no averaging is involved) and the functions $u_\varepsilon = p_\varepsilon^{-1}$ and w_ε.

The standard geometric sum is a special case of the model described above, in which $\kappa_n, n = 1, 2, \ldots$, and $\rho_{\varepsilon,n} = \mathrm{I}(\rho_n \in \mathbb{D}_\varepsilon\}, n = 1, 2, \ldots$, are independent sequences of i.i.d. random variables, for each $\varepsilon \in (0, 1]$.

Note that the standard geometric random sum, with any distribution functions $G(t)$ of the summands κ_n and parameter $p_\varepsilon \in (0, 1]$ of the geometric random index ν_ε, can be modelled as follows. Consider the geometric sum $\xi_\varepsilon = \sum_{n=1}^{\nu_\varepsilon} \kappa_n$, where: **(a)** $\kappa_n, n = 1, 2, \ldots$, is a sequence of i.i.d. random variables with the distribution function $G(t)$; **(b)** $\nu_\varepsilon = \max(n \geq 1 : \rho_n \in \mathbb{D}_\varepsilon)$, where $\rho_n, n = 1, 2, \ldots$, is a sequence of i.i.d. random variables uniformly distributed in the interval $[0, 1]$ and the domains $\mathbb{D}_\varepsilon = [0, p_\varepsilon]$; **(b)** the sequences of random variables $\kappa_n, n = 1, 2, \ldots$, and $\rho_n, n = 1, 2, \ldots$, are independent.

In the case of standard geometric sums, the propositions **(i)** and **(ii)** of Theorem 5.1 reduce to the result equivalent to those obtained by Kovalenko (1965). The difference is in the form of necessary and sufficient conditions. In the above paper, conditions based on the Laplace transforms $\phi(s)$ and expressed in the form of the relations (5.17) and (5.18) were used. We use the conditions $\mathbf{D}_{10}$ and $\mathbf{D}_{11}$, which are based on the distribution function $G(t)$ and, as we think, have a more transparent form.

Let us illustrate applications of Theorem 5.1, indicating the necessary and sufficient conditions for stable approximation of non-ruin probabilities.

Let us consider a classical risk process,

$$\Sigma_\varepsilon(t) = c_\varepsilon t - \sum_{n=1}^{N_\lambda(t)} \rho_n, \ t \geq 0. \tag{5.34}$$

Here, c_ε is a positive constant (depending on parameter $\varepsilon \in (0, 1])$ known as gross premium rate, $N_\lambda(t)$, $t \geq 0$ is a Poisson process with parameter λ counting the number of claims against an insurance company for a period of time $[0, t]$, and $\rho_n, n = 1, 2, \ldots$, is a sequence of non-negative i.i.d. random variables independent on the process $N_\lambda(t), t \geq 0$. The random variable ρ_k is the amount of the k-th claim.

An important object for the study in this model is the non-ruin probabilities on infinite time interval for a company with an initial capital $u \geq 0$,

$$G_\varepsilon(u) = \mathsf{P}\{u + \inf_{t\geq 0} \Sigma_\varepsilon(t) \geq 0\}, \ u \geq 0.$$

Let $H(u) = \mathsf{P}\{\rho_1 \leq u\}$ be the claim distribution function.

We assume that the following standard condition is satisfied:

$\mathbf{F}_1$: $\mu = \int_0^\infty sH(ds) < \infty$.

The crucial role is played by the so-called safety loading coefficient $\alpha_\varepsilon = \lambda\mu/c_\varepsilon$. If $\alpha_\varepsilon \geq 1$, then $G(u) = 0, \ u \geq 0$. The only non-trivial case is where $\alpha_\varepsilon < 1$.

We assume that the following condition is satisfied:

$\mathbf{F}_2$: $\alpha_\varepsilon < 1$ for $\varepsilon \in (0, 1]$ and $\alpha_\varepsilon \longrightarrow 1$ as $\varepsilon \longrightarrow 0$.

Let us introduce the normalised non-ruin distribution function $F(uw_\varepsilon), u \geq 0$, where $0 < w_\varepsilon \longrightarrow \infty$ as $\varepsilon \longrightarrow 0$ is some normalisation function.

According to the Pollaczek–Khinchine formula (see, for example, Asmussen (2000) or Asmussen and Albrecher (2010)), the non-ruin distribution function coincides with the distribution function of a geometric random sum, which is slightly different from the standard geometric sums discussed above. Namely,

$$G_\varepsilon(uw_\varepsilon) = \mathsf{P}\{\xi_{\varepsilon,-} \leq u\}, \ u \geq 0, \tag{5.35}$$

where

$$\xi_{\varepsilon,-} = \sum_{n=1}^{\nu_\varepsilon - 1} \kappa_n / w_\varepsilon. \tag{5.36}$$

Here: (**a**) $\kappa_n, n = 1, 2, \ldots$, is a sequence of non-negative i.i.d. random variables with the distribution function $\tilde{H}(u) = \frac{1}{\mu}\int_0^u (1 - H(s))ds, u \geq 0$ (the so-called steady claim distribution); (**b**) $\nu_\varepsilon = \min(n \geq 1, \rho_{\varepsilon,n} = 1)$; (**c**) $\rho_{\varepsilon,n}, n = 1, 2, \ldots$, is a sequence of i.i.d. random variables taking values 1 and 0 with probabilities $p_\varepsilon = 1 - \alpha_\varepsilon$ and $1 - p_\varepsilon$; (**d**) the random sequences $\kappa_n, n = 1, 2, \ldots$, and $\rho_{n\varepsilon}, n = 1, 2, \ldots$, are independent.

The condition $\mathbf{A}_{10}$ implies the fulfilment of the condition $\mathbf{A}_9$.

The condition $\mathbf{B}_{10}$ can be omitted.

The condition $\mathbf{C}_{14}$ is obviously satisfied, since in this case, $\mathsf{P}\{\kappa_1 > \delta w_\varepsilon / \rho_{\varepsilon,1} = 1\} = \mathsf{P}\{\kappa_1 > \delta w_\varepsilon\} \to$ as $\varepsilon \to 0$ for $\delta > 0$.

The condition $\mathbf{E}_1$ takes the following form:

$\mathbf{E}_2$: $w_\varepsilon, u_\varepsilon = p_\varepsilon^{-1} = (1 - \alpha_\varepsilon)^{-1} \in \mathcal{E}$.

According to Remark 5.5, Theorem 5.1 (which is specified for geometric sums as described above) can be applied to the geometric random sums ξ'_ε.

The conditions $\mathbf{D}_{10}$ and $\mathbf{D}_{11}$ ($a > 0$ and $0 < \gamma \leq 1$) take, in this case, the following forms:

$\mathbf{D}_{12}$: $\dfrac{t \int_t^\infty (1-H(s))ds}{\int_0^t s(1-H(s))ds} \to \dfrac{1-\gamma}{\gamma}$ as $t \to \infty$,

and

$\mathbf{D}_{13}$: $\dfrac{\int_0^{w_\varepsilon} s(1-H(s))ds}{(1-\alpha_\varepsilon)\mu w_\varepsilon} \to a \dfrac{\gamma}{\Gamma(2-\gamma)}$ as $\varepsilon \to 0$.

We will summarise the above discussion in the form of the following theorem that gives necessary and sufficient conditions for a stable approximation of non-ruin distribution functions.

Theorem 5.2 *Let the conditions* $\mathbf{E}_2$, $\mathbf{F}_1$, *and* $\mathbf{F}_2$ *be satisfied. Then*:

(i) *The conditions* $\mathbf{D}_{12}$ *and* $\mathbf{D}_{13}$ *are necessary and sufficient for the fulfilment of the asymptotic relation,* $G_\varepsilon(\cdot w_\varepsilon) \Rightarrow G_0(\cdot)$ *as* $\varepsilon \to 0$, *where* $G_0(\cdot)$ *is a distribution function on* $[0, \infty)$ *not concentrated at zero.*

(ii) *The distribution function* $G_0(u)$ *has the Laplace transform* $\int_0^\infty e^{-su} G_0(du) = \frac{1}{1+as^\gamma}$, $s \geq 0$.

The case $\gamma = 1$ corresponds to the so-called diffusion approximations for risk processes. The traditional method for obtaining the diffusion type asymptotics is based on the approximation of risk processes by Wiener processes with a shift. Typical conditions assume finiteness of the second moment of the claim distribution $H(t)$ (this is equivalent to the assumption about finiteness of the expectation for the steady claim distribution $\bar{H}(t) = \frac{1}{\mu} \int_0^t (1 - H(s))ds$):

$\mathbf{F}_3$: $\mu_2 = \int_0^\infty s^2 H(ds) < \infty$.

Obviously, the condition $\mathbf{F}_3$ is sufficient for the fulfilment of the condition $\mathbf{D}_{12}$ (with the parameter $\gamma = 1$). Note that this condition does not require the finiteness of the second moment of the claim distribution.

In this case, the condition $\mathbf{D}_{13}$ takes the following simple equivalent form:

$\mathbf{D}_{14}$: $(1 - \alpha_\varepsilon) w_\varepsilon \to b = \mu_2 / 2\mu a$ as $\varepsilon \to 0$.

The corresponding limiting distribution $G_0(\cdot)$ is in this case the exponential with the parameter a. This is consistent with the classical form of diffusion approximation for ruin probabilities.

The case $\gamma \in (0, 1)$ corresponds to the so-called stable approximation for risk processes.

Recall that, according to the relation (5.30), the condition $\mathbf{D}_{12}$ (with the parameter $\gamma \in (0, 1)$) is equivalent to the following condition, which requires a regular variation for the tail probabilities of the steady claim distribution $\bar{H}(t) = \frac{1}{\mu}\int_0^t (1 - H(s))ds$:

$\mathbf{D}_{15}$: $\frac{1}{\mu}\int_t^\infty (1 - H(s))ds \sim \frac{t^{-\gamma}L(t)}{\Gamma(1-\gamma)}$ as $t \to \infty$.

As follows from theorems on regularly varying functions (see, for example, Feller (1971)), the following condition imposing requirement of regular variation for the tail probabilities of the claim distribution $H(t)$ is sufficient for the fulfilment of the condition $\mathbf{D}_{15}$:

$\mathbf{D}_{16}$: $1 - H(t) \sim \frac{t^{-(\gamma+1)}L(t)\gamma\mu}{\Gamma(1-\gamma)}$ as $t \to \infty$.

As for the condition $\mathbf{D}_{13}$, in this case it can be formulated in the following form, which is equivalent to the relation (5.18):

$\mathbf{D}_{17}$: $\frac{L(w_\varepsilon)}{(1-\alpha_\varepsilon)w_\varepsilon^\gamma} \to a$ as $\varepsilon \to 0$.

Chapter 6
First-Rare-Event Times for Perturbed Closed Queuing Systems

In this chapter, the results obtained in Chaps. 2–4 are illustrated by applications to perturbed closed M/M-type queuing systems. We present necessary and sufficient conditions for convergence in distribution for first-rare-event times for a number of models of perturbed closed M/M-type queuing systems. This chapter includes two sections.

In Sect. 6.1, we present, in Theorem 6.1, necessary and sufficient conditions of convergence in distribution for the first-rare-event times for perturbed M/M-type queuing systems with a finite number of servers and rare failure events caused by too long service times.

In Sect. 6.2, we present, in Theorem 6.2, necessary and sufficient conditions of convergence in distribution for the first-rare-event times for perturbed M/M-type queuing systems with a finite number of servers and rare failure events caused by the absence of working servers in the system with small failure intensities for all servers.

6.1 Queuing Systems with Rare Events Caused by Too Long Service Times

In this section, we present necessary and sufficient conditions of convergence in distribution for the first failure event times for perturbed M/M-type queuing system with a finite number of servers and rare failure events caused by too long service times in some states with a small number of working servers.

6.1.1 Asymptotic Uniform Ergodicity for Birth–Death Markov Chains

Let $\eta_{\varepsilon,n}$, $n = 0, 1, \ldots$, be, for every $\varepsilon \in (0, 1]$, a birth–death finite Markov chain with the phase space $\mathbb{X} = \{0, 1, \ldots, m\}$ and transition probabilities,

D. Silvestrov, *Perturbed Semi-Markov Type Processes I*,
https://doi.org/10.1007/978-3-030-92403-4_6

$$p_{\varepsilon,ij} = \begin{cases} p_{\varepsilon,i,\pm} & \text{if } j = \frac{1\pm1}{2}, \quad \text{for } i = 0, \\ p_{\varepsilon,i,\pm} & \text{if } j = i \pm 1, \quad \text{for } 1 < i < m, \\ p_{\varepsilon,m,\pm} & \text{if } j = m + \frac{-1\pm1}{2}, \text{ for } i = m, \\ 0 & \text{otherwise}, \end{cases} \tag{6.1}$$

where $p_{\varepsilon,i,\pm} \geq 0$, $p_{\varepsilon,i,-} + p_{\varepsilon,i,+} = 1$, for $i \in \mathbb{X}$ and $\varepsilon \in (0, 1]$.

In this case, the condition of asymptotically uniform ergodicity $\mathbf{B}_1$ takes the following form:

$\mathbf{B}_{11}$: **(a)** $\underline{\lim}_{\varepsilon\to0}\, p_{\varepsilon,i,+} > 0$, for $i = 0, \dots, m-1$, **(b)** $\underline{\lim}_{\varepsilon\to0}\, p_{\varepsilon,i,-} > 0$, for $i = 1, \dots, m$.

The condition $\mathbf{B}_{11}$ implies that there exists $\varepsilon_0 \in (0, 1]$ such that the Markov chain $\eta_{\varepsilon,n}$ is ergodic, or every $\varepsilon \in (0, \varepsilon_0]$.

The stationary probabilities $\pi_{\varepsilon,i}, i \in \mathbb{X}$ are the only positive solution for the following system of linear equations,

$$\begin{cases} \pi_{\varepsilon,0} = \pi_{\varepsilon,0}p_{\varepsilon,0,-} + \pi_{\varepsilon,1}p_{\varepsilon,1,-}, \\ \pi_{\varepsilon,i} = \pi_{\varepsilon,i-1}p_{\varepsilon,i-1,+} + \pi_{\varepsilon,i+1}p_{\varepsilon,i+1,-}, \ 1 < i < m, \\ \pi_{\varepsilon,m} = \pi_{\varepsilon,m-1}p_{\varepsilon,m-1,+} + \pi_{\varepsilon,m}p_{\varepsilon,m,+}, \\ \sum_{j\in\mathbb{X}} \pi_{\varepsilon,j} = 1. \end{cases} \tag{6.2}$$

The solution of this system is well known and is given by the following relation:

$$\pi_{\varepsilon,i} = \prod_{r=0}^{i-1} \frac{p_{\varepsilon,r,+}}{p_{\varepsilon,r+1,-}} \Big(\sum_{j=0}^{m} \prod_{k=0}^{j-1} \frac{p_{\varepsilon,k,+}}{p_{\varepsilon,k+1,-}} \Big)^{-1}, \ i \in \mathbb{X}. \tag{6.3}$$

6.1.2 M/M-Type Queuing System with Rare Events Caused by Too Long Service Times

Let $\mathbf{S}_\varepsilon$ be, for every $\varepsilon \in (0, 1]$, a closed M/M-type queuing system, which consists of N servers functioning independently, with exponentially distributed life and repairing times. The failure and repairing intensities, respectively, $\mu_{\varepsilon,-} > 0$ and $\mu_{\varepsilon,+} > 0$, are the same for all servers.

The state of system at the moment t is represented by the random variable $\tilde{\eta}_\varepsilon(t)$, which is the number of working servers at this moment. The process $\tilde{\eta}_\varepsilon(t)$ is a birth–death continuous time Markov chain, with phase space $\mathbb{X} = \{0, \dots, N\}$, continuous from the right step-wise trajectories and generator matrix,

$$\mathbf{G}_\varepsilon = \begin{bmatrix} -\lambda_{\varepsilon,0} & \lambda_{\varepsilon,0} & 0 & 0 & \cdots & 0 \\ \lambda_{\varepsilon,1}p_{\varepsilon,1,-} & -\lambda_{\varepsilon,1} & \lambda_{\varepsilon,1}p_{\varepsilon,1,+} & 0 & \cdots & 0 \\ \cdots & \cdots & \cdots & \cdots & \cdots & \cdots \\ 0 & \cdots & 0\cdots & 0 & \lambda_{\varepsilon,N} & -\lambda_{\varepsilon,N} \end{bmatrix}, \tag{6.4}$$

where, for $k = 0, \dots, N$,

$$p_{\varepsilon,k,+} = \frac{\mu_{\varepsilon,+} \cdot (N-k)}{\lambda_{\varepsilon,k}}, \; p_{\varepsilon,k,-} = \frac{\mu_{\varepsilon,-} \cdot k}{\lambda_{\varepsilon,k}}, \; \lambda_{\varepsilon,k} = \mu_{\varepsilon,+} \cdot (N-k) + \mu_{\varepsilon,-} \cdot k. \quad (6.5)$$

We also assume that the initial state of the system $\eta_\varepsilon(0)$ is a random variable independent of service processes and $\bar{q}_\varepsilon = \langle q_{\varepsilon,0}, \dots q_{\varepsilon,N} \rangle$ is its distribution.

Let $0 = \tilde{\tau}_{\varepsilon,0} < \tilde{\tau}_{\varepsilon,1} < \cdots$ be successive moments of jumps for the Markov chain $\tilde{\eta}_\varepsilon(t)$, $\tilde{\nu}_\varepsilon(t) = \max(n \geq 1 : \tilde{\tau}_{\varepsilon,n} \leq t)$ be the number of jumps in the interval $[0, t]$, and $\tilde{\kappa}_\varepsilon(t) = t - \tilde{\tau}_{\varepsilon,\tilde{\nu}_\varepsilon(t)}$ be the time of continuous stay of the system $\mathbf{S}_\varepsilon$ in the state $\tilde{\eta}_\varepsilon(t)$. Let also $\tilde{\eta}_{\varepsilon,n} = \tilde{\eta}(\tilde{\tau}_{\varepsilon,n}), n = 0, 1, \dots,$ be the corresponding discrete time embedded Markov chain and $\tilde{\kappa}_{\varepsilon,n} = \tilde{\tau}_{\varepsilon,n} - \tilde{\tau}_{\varepsilon,n-1}, n = 1, \dots$ be the inter-jump times for the Markov chain $\tilde{\eta}_\varepsilon(t)$.

The Markov chain $\tilde{\eta}_{\varepsilon,n}$ has the transition probabilities determined in an obvious way by the matrix $\mathbf{G}_\varepsilon$.

The following condition is, in this case, equivalent to the condition $\mathbf{B}_{11}$:

$\mathbf{B}_{12}$: $0 < \underline{\lim}_{\varepsilon \to 0} \mu_{\varepsilon,\pm} \leq \overline{\lim}_{\varepsilon \to 0} \mu_{\varepsilon,\pm} < \infty$.

The corresponding stationary probabilities $\pi_{\varepsilon,i}, i \in \mathbb{X}$ are given by the explicit formulas (6.3), which take the following form:

$$\pi_{\varepsilon,i} = \prod_{r=0}^{i-1} \frac{p_{\varepsilon,r,+}}{p_{\varepsilon,r+1,-}} \Big(\sum_{j=0}^{m} \prod_{k=0}^{j-1} \frac{p_{\varepsilon,k,+}}{p_{\varepsilon,k+1,-}} \Big)^{-1}, \; i \in \mathbb{X}, \quad (6.6)$$

where the transition probabilities $p_{\varepsilon,i,\pm}$, $i \in \mathbb{X}$ are given by the relation (6.5).

We assume that there exist $0 \leq L \leq N$ such that a rare failure event occurs in the system at instant t, if $\tilde{\eta}_\varepsilon(t) = i \leq L$ and $\tilde{\kappa}_\varepsilon(t) = T_{\varepsilon,i}$.

Here, $0 < T_{\varepsilon,i} \to \infty$ as $\varepsilon \to 0$, for $i = 0, \dots, L$.

For example, if $L = 0$, then the first-rare-event occurs at the moment when the time of continuous absence of working servers reaches the level $T_{\varepsilon,0}$ for the first time.

We assume that $T_{\varepsilon,i} \to \infty$ as $\varepsilon \to 0$ with comparable rates, i.e., that the following condition is satisfied:

$\mathbf{A}_{11}$: $T_{\varepsilon,i} = a_{\varepsilon,i} T_\varepsilon$, $i = 0, \dots, L$, where: **(a)** $0 < \underline{\lim}_{\varepsilon \to 0} a_{\varepsilon,i} \leq \overline{\lim}_{\varepsilon \to 0} a_{\varepsilon,i} < \infty$, for $i = 0, \dots, L$, **(b)** $0 < T_\varepsilon \to \infty$ as $\varepsilon \to 0$.

In this case, the corresponding Markov renewal process $(\eta_{\varepsilon,n}, \kappa_{\varepsilon,n}, \rho_{\varepsilon,n}), n = 0, 1, \dots,$ defined in Sect. 2.1.1 has components,

$$\begin{cases} \eta_{\varepsilon,n} = \tilde{\eta}_{\varepsilon,n}, \\ \kappa_{\varepsilon,n} = \tilde{\kappa}_{\varepsilon,n}/w_\varepsilon, \\ \rho_{\varepsilon,n} = \mathrm{I}(\tilde{\eta}_{\varepsilon,n-1} \leq L)\mathrm{I}(\tilde{\kappa}_{\varepsilon,n} \geq T_{\varepsilon,\eta_{\varepsilon,n-1}}) \\ \quad\;\; = \mathrm{I}(\eta_{\varepsilon,n-1} \leq L)\mathrm{I}(\kappa_{\varepsilon,n} \geq T_{\varepsilon,\eta_{\varepsilon,n-1}}/w_\varepsilon), \end{cases} \quad (6.7)$$

where $0 < w_\varepsilon \to \infty$ as $\varepsilon \to 0$ is some normalisation function, which we will define below.

It is natural to define the normalised first failure time for the system $\mathbf{S}_\varepsilon$ as follows:

$$\xi_{\mathbf{S}_\varepsilon} = \frac{1}{w_\varepsilon} \inf(t \geq 0 : \tilde{\eta}_\varepsilon(t) \leq L, \tilde{\kappa}_\varepsilon(t) = T_{\eta_\varepsilon(t)}). \tag{6.8}$$

The first-rare-event time for the above Markov renewal process is defined as

$$\xi_\varepsilon = \sum_{k=1}^{\nu_\varepsilon} \kappa_{\varepsilon,k}, \tag{6.9}$$

where

$$\begin{aligned} \nu_\varepsilon &= \max(n \geq 1 : \rho_{\varepsilon,n} = 1) \\ &= \max(n \geq 1 : \eta_{\varepsilon,n-1} \leq L, \kappa_{\varepsilon,n} \geq T_{\eta_{\varepsilon,n-1}}/w_\varepsilon). \end{aligned} \tag{6.10}$$

Obviously, the following inequality holds, for $\varepsilon \in (0, 1]$,

$$\xi_{\varepsilon,-} \leq \xi_{\mathbf{S}_\varepsilon} \leq \xi_{\varepsilon,+}, \tag{6.11}$$

where

$$\xi_{\varepsilon,-} = \sum_{k=1}^{\nu_\varepsilon - 1} \kappa_{\varepsilon,k}, \ \xi_{\varepsilon,+} = \sum_{k=1}^{\nu_\varepsilon} \kappa_{\varepsilon,k}. \tag{6.12}$$

Taking into account remarks made in Sect. 4.1.2, we can apply Lemma 4.1 to the first-rare-event times $\xi_{\mathbf{S}_\varepsilon}$.

The distribution functions $F_{\varepsilon,i}(t)$ are, in this case, exponential, i.e., for $i = 0, \ldots, N$,

$$\begin{aligned} F_{\varepsilon,i}(t) &= \mathsf{P}_i\{\kappa_{\varepsilon,1} \leq t\} = \mathsf{P}_i\{\tilde{\kappa}_{\varepsilon,1}/w_\varepsilon \leq t\} \\ &= 1 - \exp\{-\lambda_{\varepsilon,i} w_\varepsilon t\}, \ t \geq 0, \end{aligned} \tag{6.13}$$

and, thus, the averaged distribution function $F_\varepsilon(t)$ has the following form:

$$F_\varepsilon(t) = \sum_{i=0}^{N} \pi_{\varepsilon,i}(1 - \exp\{-\lambda_{\varepsilon,i} w_\varepsilon t\}), t \geq 0. \tag{6.14}$$

The probabilities $p_{\varepsilon,i}$ take, in this case, the following form, for $i = 0, \ldots, N$,

$$\begin{aligned} p_{\varepsilon,i} &= \mathrm{I}(i \leq L)\mathsf{P}_i\{\kappa_{\varepsilon,1} \geq T_{\varepsilon,i}/w_\varepsilon\} \\ &= \mathrm{I}(i \leq L)\mathsf{P}_i\{\tilde{\kappa}_{\varepsilon,1} \geq T_{\varepsilon,i}\} \\ &= \mathrm{I}(i \leq L)\exp\{-\lambda_{\varepsilon,i} T_{\varepsilon,i}\} = \mathrm{I}(i \leq L)\exp\{-\lambda_{\varepsilon,i} a_{\varepsilon,i} T_\varepsilon\}. \end{aligned} \tag{6.15}$$

The conditions $\mathbf{A}_{11}$ and $\mathbf{B}_{12}$ imply that there exists $\varepsilon' \in (0, 1]$ such that $p_{\varepsilon,i} > 0$, for $\varepsilon \in (0, \varepsilon']$ and $p_{\varepsilon,i} \to 0$ as $\varepsilon \to 0$, for $i \leq L$. Thus, the condition $\mathbf{A}_1$ holds.

The probability $p_\varepsilon = u_\varepsilon^{-1}$ takes, in this case, the following form:

$$p_\varepsilon = \sum_{i \leq L} \pi_{\varepsilon,i} \exp\{-\lambda_{\varepsilon,i} T_{\varepsilon,i}\} = \sum_{i \leq L} \pi_{\varepsilon,i} \exp\{-\lambda_{\varepsilon,i} a_{\varepsilon,i} T_\varepsilon\}. \tag{6.16}$$

In this case, the Laplace transform $\phi_\varepsilon(s)$ takes the following form:

$$\phi_\varepsilon(s) = \int_0^\infty e^{-st} F_\varepsilon(dt) = \sum_{i=0}^N \pi_{\varepsilon,i} \frac{1}{1 + s/\lambda_{\varepsilon,i} w_\varepsilon}, \ s \geq 0. \tag{6.17}$$

Let us denote, for $\varepsilon \in (0, \varepsilon_0]$,

$$B_\varepsilon = \sum_{i=0}^N \frac{\pi_{\varepsilon,i}}{\lambda_{\varepsilon,i}}. \tag{6.18}$$

The condition $\mathbf{B}_{12}$ implies that

$$0 < \underline{\lim}_{\varepsilon \to 0} B_\varepsilon \leq \overline{\lim}_{\varepsilon \to 0} B_\varepsilon < \infty. \tag{6.19}$$

Let $\frac{1}{1+s} = 1 - s + o(s)$ be the asymptotic Taylor expansion of the first order for the function $\frac{1}{1+s}$ at the point 0. The relation (6.17) and the condition $\mathbf{B}_{12}$ imply that the following asymptotic relation takes place, for $s > 0$,

$$\begin{aligned} u_\varepsilon(1 - \phi_\varepsilon(s)) &= u_\varepsilon \sum_{i=0}^N \pi_{\varepsilon,i} \big(1 - \frac{1}{1 + s/\lambda_{\varepsilon,i} w_\varepsilon}\big) \\ &= u_\varepsilon \sum_{i=0}^N \pi_{\varepsilon,i} \big(s/\lambda_{\varepsilon,i} w_\varepsilon + o(s/\lambda_{\varepsilon,i} w_\varepsilon)\big) \\ &= u_\varepsilon s \sum_{i=0}^N \pi_{\varepsilon,i}/\lambda_{\varepsilon,i} w_\varepsilon + \sum_{i=0}^N \pi_{\varepsilon,i} u_\varepsilon \, o(s/\lambda_{\varepsilon,i} w_\varepsilon) \\ &\sim s B_\varepsilon \frac{u_\varepsilon}{w_\varepsilon} \ \text{as } \varepsilon \to 0. \end{aligned} \tag{6.20}$$

The relation (6.20) implies that the condition $\mathbf{D}_2$ is, in this case, equivalent to the following condition:

$\mathbf{D}_{18}$: $A_\varepsilon = B_\varepsilon \frac{u_\varepsilon}{w_\varepsilon} \to A \in (0, \infty)$ as $\varepsilon \to 0$.

Moreover, in this case, the corresponding limiting cumulant $A(s) = As$, and thus the corresponding limiting distribution $F(u)$, is exponential with parameter A^{-1}.

It is useful also noting that condition $\mathbf{D}_{18}$ automatically holds if to choose the normalisation function $w_\varepsilon = B_\varepsilon u_\varepsilon$. In this case, $A = 1$.

The condition $\mathbf{D}_{18}$ and the relation (6.19) imply that

$$\overline{\lim}_{\varepsilon \to 0} \frac{u_\varepsilon}{w_\varepsilon} < \infty \tag{6.21}$$

and, thus, for $i \leq L$,

$$\begin{aligned}
\frac{T_{\varepsilon,i}}{w_\varepsilon} &= \frac{u_\varepsilon}{w_\varepsilon}\frac{T_{\varepsilon,i}}{u_\varepsilon} \\
&= \frac{u_\varepsilon}{w_\varepsilon} a_{\varepsilon,i} T_\varepsilon \sum_{i\le L} \pi_{\varepsilon,i} \exp\{-\lambda_{\varepsilon,i} a_{\varepsilon,i} T_\varepsilon\} \\
&\le \frac{u_\varepsilon}{w_\varepsilon} \alpha_{\varepsilon,i} T_\varepsilon \exp\{-\min_{i\le L} \lambda_{\varepsilon,i} a_{\varepsilon,i} \cdot T_\varepsilon\} \to 0 \text{ as } \varepsilon \to 0.
\end{aligned} \tag{6.22}$$

The relation (6.21) implies that for $i \le L$ and $\delta > 0$,

$$\begin{aligned}
\mathsf{P}_i\{\kappa_{\varepsilon,1} > \delta / \rho_{\varepsilon,1} = 1\} &= \mathsf{P}_i\{\kappa_{\varepsilon,1} > \delta / \kappa_{\varepsilon,1} \ge T_{\varepsilon,i}/w_\varepsilon\} \\
&= \frac{\mathsf{P}_i\{\kappa_{\varepsilon,1} > \delta,\, \kappa_{\varepsilon,1} \ge T_{\varepsilon,i}/u_\varepsilon\}}{\mathsf{P}\{\kappa_{\varepsilon,1} \ge T_{\varepsilon,i}/w_\varepsilon\}\}} \\
&= \frac{\mathsf{P}_i\{\tilde{\kappa}_{\varepsilon,1} > \delta w_\varepsilon,\, \tilde{\kappa}_{\varepsilon,1} \ge T_{\varepsilon,i}\}}{\mathsf{P}\{\tilde{\kappa}_{\varepsilon,1} \ge T_{\varepsilon,i}\}\}} \\
&= \frac{\exp\{-\lambda_{\varepsilon,i}(\delta w_\varepsilon \vee T_{\varepsilon,i})\}}{\exp\{-\lambda_{\varepsilon,i} T_{\varepsilon,i}\}} \to 0 \text{ as } \varepsilon \to 0.
\end{aligned} \tag{6.23}$$

Therefore, the conditions $\mathbf{A}_{11}$, $\mathbf{B}_{12}$, and $\mathbf{D}_{18}$ imply that the condition $\mathbf{C}_1$ is satisfied.

Therefore, the condition $\mathbf{C}_1$ is not required in the sufficiency statement of the corresponding variant of Theorem 2.1.

At the same time, the proof of the necessity statement of Theorem 2.2 is based on the conditions $\mathbf{A}_1$, $\mathbf{B}_1$ but does not use the condition $\mathbf{C}_1$. Once the condition $\mathbf{D}_1$ is proved, the above arguments again lead to the fact that the condition $\mathbf{C}_1$ is satisfied.

The following theorem is a variant of Lemma 4.2 for the above model.

Theorem 6.1 *Let the conditions* $\mathbf{A}_{11}$ *and* $\mathbf{B}_{12}$ *be satisfied. Then*:

(i) *The condition* $\mathbf{D}_{18}$ *is necessary and sufficient for the fulfilment (for some or any initial distributions* $\bar{q}_\varepsilon$*, respectively, in the statements of necessity and sufficiency) of the asymptotic relation* $\xi_{\mathbf{S}_\varepsilon} \xrightarrow{\mathsf{d}} \xi_0$ *as* $\varepsilon \to 0$*, where* ξ_0 *is a non-negative random variable with distribution not concentrated in zero.*

(ii) *The limiting random variable* ξ_0 *has an exponential distribution with the parameter* A^{-1}.

In conclusion, note that analogues of the statement **(iii)** of Theorem 2.1 and Theorems 3.1–3.3 can also be formulated.

6.2 Queuing Systems with Highly Reliable Servers

In this section, we present necessary and sufficient conditions of convergence in distribution for the first-rare-event times for perturbed closed M/M-type queuing systems with a finite number of servers and rare failure events caused by the absence of working servers.

Let us $\mathbf{S}_\varepsilon$ be, for every $\varepsilon \in (0, 1]$, a closed M/M-type queuing system, which consists of N servers functioning independently of each other, with independent life and repairing times. The failure and repairing times are exponentially distributed, with intensities, respectively, $q_{\varepsilon,i,-}$ and $q_{\varepsilon,i,+}$, for $i = 1, \ldots, N$.

The state of a server i at instant t is given by the random indicator variable $\hat{\eta}_{\varepsilon,i}(t)$ that takes the value 1 if the server i is functioning at this instant, and the value 0 if the server i is being repaired at this moment.

The state of the system at the moment t is given by the random vector $\hat{\eta}_\varepsilon(t) = (\hat{\eta}_{\varepsilon,1}(t), \ldots, \hat{\eta}_{\varepsilon,N}(t)), t \geq 0$. The phase space of the stochastic process $\hat{\eta}_\varepsilon(t)$ is the m-dimensional binary hypercube $\hat{\mathbb{X}} = \{\bar{x} = (x_1, \ldots, x_N) : x_i = 0, 1, i = 1, \ldots, N\}$.

It should be noted that the space $\hat{\mathbb{X}}$ contains $m = 2^N$ states, which is a large number even for relatively small values of N.

It is also useful to note that there is a natural way to order states in the space $\hat{\mathbb{X}}$. In fact, there is a one-to-one correspondence between the states $\bar{x} = (x_1, \ldots, x_m)$ and binary fractions $0.x_1x_2 \cdots x_N$, which can be naturally ordered to be their values,

$$\begin{array}{lcl} (1, \ldots, 1, 1, 1) & \leftrightarrow & 0.1 \cdots 111 & \leftrightarrow & 1, \\ (1, \ldots, 1, 1, 0) & \leftrightarrow & 0.1 \cdots 110 & \leftrightarrow & 2, \\ (1, \ldots, 1, 0, 1) & \leftrightarrow & 0.1 \cdots 101 & \leftrightarrow & 3, \\ (1, \ldots, 1, 0, 0) & \leftrightarrow & 0.1 \cdots 100 & \leftrightarrow & 4, \\ \cdots & & \cdots & & \cdots \\ (0, \ldots, 0, 0, 0) & \leftrightarrow & 0.0 \cdots 000 & \leftrightarrow & m. \end{array} \tag{6.24}$$

The process $\hat{\eta}_\varepsilon(t)$ is a continuous time Markov chain with phase space $\hat{\mathbb{X}}$, continuous from the right step-wise trajectories and transition intensities,

$$q_{\varepsilon,\bar{x},\bar{y}} = \begin{cases} q_{\varepsilon,i,-} & \text{if } y_i = 0, x_i = 1, y_j = x_j, j \neq i, \text{ for some } 1 \leq i \leq N, \\ q_{\varepsilon,i,+} & \text{if } y_i = 1, x_i = 0, y_j = x_j, j \neq i, \text{ for some } 1 \leq i \leq N, \\ 0 & \text{otherwise.} \end{cases} \tag{6.25}$$

We also assume that the initial state of the system $\hat{\eta}_\varepsilon(0)$ is a random variable independent of service processes and $\bar{q}_\varepsilon = \langle q_\varepsilon(\bar{x}), \bar{x} \in \hat{\mathbb{X}} \rangle$ is its distribution.

Let $\hat{\tau}_\varepsilon(n), n = 0, 1, \ldots$, be successive moments of jumps of the Markov chain $\hat{\eta}_\varepsilon(t)$, $t \geq 0$, $\hat{\kappa}_{\varepsilon,n} = \hat{\tau}_{\varepsilon,n} - \hat{\tau}_{\varepsilon,n-1}, n = 1, \ldots$ be inter-jump times for the Markov chain $\hat{\eta}_\varepsilon(t)$, and $\hat{\eta}_{\varepsilon,n} = \hat{\eta}_\varepsilon(\hat{\tau}_\varepsilon(n)), n = 0, 1, \ldots$, be the states of the Markov chain $\hat{\eta}_\varepsilon(t)$ at successive moments of jumps. The random sequence $\hat{\eta}_{\varepsilon,n}, n = 0, 1, \ldots$, is an embedded discrete time Markov chain for the continuous time Markov chain $\hat{\eta}_\varepsilon(t)$.

Consider also the Markov renewal process $(\hat{\eta}_{\varepsilon,n}, \hat{\kappa}_{\varepsilon,n}, \hat{\rho}_{\varepsilon,n}), n = 0, 1, \ldots$, which has the phase space $\mathbb{X} \times [0, \infty) \times \{0, 1\}$ and the third component $\hat{\rho}_{\varepsilon,n}$ of the following form:

$$\hat{\rho}_{\varepsilon,n} = \mathrm{I}(\hat{\eta}_{\varepsilon,n} = \bar{0}), \text{ for } n = 0, 1, \ldots, \; \hat{\rho}_{\varepsilon,0} = 0, \tag{6.26}$$

where: $\bar{0} = (0, \ldots, 0)$.

Let us assume that the following condition is satisfied:

$\mathbf{B}_{13}$: **(a)** $0 < \underline{\lim}_{\varepsilon \to 0} q_{\varepsilon,i,+} \leq \overline{\lim}_{\varepsilon \to 0} q_{\varepsilon,i,+} < \infty$, for $i = 1, \ldots, N$, **(b)** there exists $\varepsilon' \in (0, 1]$ such that $q_{\varepsilon,i,-} > 0$, for $\varepsilon \in (0, \varepsilon']$ and $q_{\varepsilon,i,-} \to 0$ as $\varepsilon \to 0$, for $i = 1, \ldots, N$.

The condition $\mathbf{B}_{13}$ implies that there exist $\varepsilon'' \in (0, \varepsilon']$ such that all intensities $\mu_{\varepsilon,i,\pm} > 0, i = 1, \ldots, N$, for $\varepsilon \in (0, \varepsilon'']$. In what follows, we assume that $\varepsilon \in (0, \varepsilon'']$.

It is easy to see that, for $\bar{x} = (x_1, \ldots, x_N) \in \check{\mathbb{X}}$,

$$\mathsf{P}_{\bar{x}}\{\hat{\kappa}_{\varepsilon,1} \leq u\} = 1 - e^{-q_{\varepsilon,\bar{x}} u}, \ u \geq 0, \tag{6.27}$$

where

$$q_{\varepsilon,\bar{x}} = \sum_{\bar{y} \neq \bar{x}} q_{\varepsilon,\bar{x},\bar{y}} = \sum_{i: x_i = 1} q_{\varepsilon,i,-} + \sum_{i: x_i = 0} q_{\varepsilon,i,+}. \tag{6.28}$$

Also, for $\bar{x} = (x_1, \ldots, x_N), \bar{y} = (y_1, \ldots, y_N) \in \check{\mathbb{X}}$,

$$\mathsf{P}_{\bar{x}}\{\hat{\eta}_{\varepsilon,1} = \bar{y}/\hat{\eta}_{\varepsilon,0} = \bar{x}\} = p_{\varepsilon,\bar{x},\bar{y}}, \tag{6.29}$$

where

$$p_{\varepsilon,\bar{x},\bar{y}} = \frac{q_{\varepsilon,\bar{x},\bar{y}}}{q_{\varepsilon,\bar{x}}}. \tag{6.30}$$

The condition $\mathbf{B}_{13}$ **(b)** means that all servers are asymptotically highly reliable.

The first hitting of the process $\hat{\eta}_\varepsilon(t)$ in the state $\bar{0} = (0, \ldots, 0)$ occurs in the above system at the first moment when all servers are down (are in repair mode).

Such event can be interpreted as a rare, due to the condition $\mathbf{B}_{13}$.

Let us consider the Markov renewal process $(\check{\eta}_{\varepsilon,n}, \check{\kappa}_{\varepsilon,n}, \check{\rho}_{\varepsilon,n}), n = 0, 1, \ldots,$ which has the phase space $\check{\mathbb{X}} \times [0, \infty) \times \{0, 1\}$, where $\check{\mathbb{X}} = \hat{\mathbb{X}}$, and components,

$$\begin{cases} \check{\eta}_{\varepsilon,n} &= \hat{\eta}_{\varepsilon,n}, \\ \check{\kappa}_{\varepsilon,n} &= \hat{\kappa}_{\varepsilon,n}/w_\varepsilon, \\ \check{\rho}_{\varepsilon,n} &= \mathrm{I}(\hat{\eta}_{\varepsilon,n-1} = \bar{0}), \end{cases} \tag{6.31}$$

where $0 < w_\varepsilon \to \infty$ as $\varepsilon \to 0$ is some normalisation function, which we will define below.

It is natural to define the normalised first-rare-event time for the system $\mathbf{S}_\varepsilon$, which can also be called the first failure time, as follows:

$$\xi_{\mathbf{S}_\varepsilon} = \sum_{n=1}^{\check{\nu}_\varepsilon} \check{\kappa}_{\varepsilon,n} = \sum_{n=1}^{\hat{\nu}_\varepsilon} \hat{\kappa}_{\varepsilon,n}/w_\varepsilon, \tag{6.32}$$

where

$$\check{\nu}_\varepsilon = \min(n \geq 1 : \check{\eta}_{\varepsilon,n} = \bar{0}). \tag{6.33}$$

In this case, the condition $\mathbf{B}_1$ is not satisfied for the Markov chains $\check{\eta}_{\varepsilon,n}$, since the condition $\mathbf{B}_{13}$ **(b)** implies that, for $\bar{y} \neq \bar{1} = (1, \ldots, 1)$,

$$p_{\varepsilon,\bar{1},\bar{y}} \to 0 \text{ as } \varepsilon \to 0. \tag{6.34}$$

However, we can use the results presented in Sect. 4.1.4.2.

The condition $\mathbf{B}_{13}$ implies that the phase space $\check{\mathbb{X}}$ is one class of communicative states for the Markov chain $\check{\eta}_{\varepsilon,n}$, for $\varepsilon \in (0, \varepsilon'']$.

Therefore, the condition $\mathbf{B}_6$ is satisfied and, thus, for any subset $\emptyset \neq \mathbb{D} \subseteq \check{\mathbb{X}}$ and $i \in \check{\mathbb{X}}$,

$$\mathsf{P}_i\{\check{\nu}_{\varepsilon,\mathbb{D}} < \infty\} = 1, \tag{6.35}$$

where

$$\check{\nu}_{\varepsilon,\mathbb{D}} = \min(n \geq 1 : \check{\eta}_{\varepsilon,n} \in \mathbb{D}). \tag{6.36}$$

Let us use the simplest variant, where $\mathbb{X}$ is the following one-state set,

$$\mathbb{X} = \{\bar{1} = (1, \ldots, 1)\}. \tag{6.37}$$

The relation (6.35) makes it possible to define the successive times of hitting the Markov chain $\check{\eta}_{\varepsilon,n}$ in the set $\mathbb{X}$,

$$\check{\nu}_{\varepsilon,\mathbb{X},n} = \min(k > \check{\nu}_{\varepsilon,\mathbb{X},n-1}, \check{\eta}_{\varepsilon,k} = \bar{1}), n = 1, 2, \ldots, \text{ where } \check{\nu}_{\varepsilon,\mathbb{X},0} = 0. \tag{6.38}$$

The corresponding Markov renewal process with transition period $(\tilde{\eta}_{\varepsilon,n}, \tilde{\kappa}_{\varepsilon,n}, \tilde{\rho}_{\varepsilon,n})$, $n = 0, 1, \ldots$ (with the phase space $\tilde{\mathbb{X}} = \check{\mathbb{X}}$, for the initial state $\tilde{\eta}_{\varepsilon,0}$, and the phase space $\mathbb{X}$, for the random variables $\tilde{\eta}_{\varepsilon,n}, n \geq 1$) can be constructed using the following recurrent relations:

$$\begin{cases} \tilde{\eta}_{\varepsilon,n} = \check{\eta}_{\varepsilon,\check{\nu}_{\varepsilon,\mathbb{X},n}}, \ n = 1, 2, \ldots, \ \tilde{\eta}_{\varepsilon,0} = \check{\eta}_{\varepsilon,0}, \\ \tilde{\kappa}_{\varepsilon,n} = \sum_{k=\check{\nu}_{\varepsilon,\mathbb{X},n-1}+1}^{\check{\nu}_{\varepsilon,\mathbb{X},n}} \check{\kappa}_{\varepsilon,k}, \ n = 1, 2, \ldots, \ \tilde{\kappa}_{\varepsilon,0} = \check{\kappa}_{\varepsilon,0}, \\ \tilde{\rho}_{\varepsilon,n} = 1 - \prod_{k=\check{\nu}_{\varepsilon,\mathbb{X},n-1}+1}^{\check{\nu}_{\varepsilon,\mathbb{X},n}} (1 - \check{\rho}_{\varepsilon,k}), \ n = 1, 2, \ldots, \ \tilde{\rho}_{\varepsilon,0} = \check{\rho}_{\varepsilon,0}. \end{cases} \tag{6.39}$$

Also, as in Sect. 4.1.4.2, the corresponding shifted Markov renewal process can be defined by the following relation:

$$(\eta_{\varepsilon,n}, \kappa_{\varepsilon,n}, \rho_{\varepsilon,n}) = (\tilde{\eta}_{\varepsilon,n+1}, \tilde{\kappa}_{\varepsilon,n+1}, \tilde{\rho}_{\varepsilon,n+1}), n = 0, 1, \ldots. \tag{6.40}$$

It is obvious that the first component $\eta_{\varepsilon,n} = \bar{1}, n = 0, 1, \ldots$, i.e., it is a trivial nonrandom sequence and, therefore, $(\kappa_{\varepsilon,n}, \rho_{\varepsilon,n}), n = 0, 1, \ldots$, is the sequence i.i.d. random vectors.

Obviously, in this case, the stationary probabilities $\pi_{\varepsilon,\bar{1}} \equiv 1$.

Obviously, the following inequality holds, for $\varepsilon \in (0, \varepsilon'']$,

$$\tilde{\xi}_{\varepsilon,-} \leq \xi_{\mathbf{S}_\varepsilon} \leq \tilde{\xi}_{\varepsilon,+}, \tag{6.41}$$

where

$$\tilde{\xi}_{\varepsilon,-} = \sum_{k=1}^{\tilde{\nu}_\varepsilon - 1} \tilde{\kappa}_{\varepsilon,k}, \ \tilde{\xi}_{\varepsilon,+} = \sum_{k=1}^{\tilde{\nu}_\varepsilon} \tilde{\kappa}_{\varepsilon,k}, \tag{6.42}$$

and

$$\tilde{\nu}_\varepsilon = \min(n \geq 1 : \tilde{\rho}_{\varepsilon,n} = 1). \tag{6.43}$$

Let us show that the conditions $\mathbf{C}_1$ (for the Markov renewal processes $(\eta_{\varepsilon,n}, \kappa_{\varepsilon,n}, \rho_{\varepsilon,n})$) and $\mathbf{A}_5$, $\mathbf{B}_6$, $\mathbf{B}_7$, $\mathbf{C}_8$ are satisfied and, thus, Lemma 4.6 can be applied.

In this case, the condition $\mathbf{B}_1$ is obviously satisfied for Markov chains $\eta_{\varepsilon,n}$, which are asymptotically uniformly ergodic.

The condition $\mathbf{B}_{13}$ obviously implies that the conditions $\mathbf{B}_6$ and $\mathbf{B}_7$ are satisfied.

Let us introduce sets, for $n = 0, \dots, N$,

$$\mathbb{Y}_n = \{\bar{x} = (x_1, \dots, x_N) : \sum_{k=1}^{N} x_k = N - n\}. \tag{6.44}$$

Obviously, the number of states in the set $\mathbb{Y}_n$ is $m_n = \binom{N}{n}$, for $n = 0, 1, \dots, N$.

In this case, the probabilities,

$$\check{p}_{\varepsilon,\bar{x}} = \mathsf{P}_{\bar{x}}\{\check{\rho}_{\varepsilon,1} = 1\} = \begin{cases} p_{\varepsilon,\bar{x},\bar{0}} & \text{for } \bar{x} \in \mathbb{Y}_{N-1}, \\ 0 & \text{for } \bar{x} \notin \mathbb{Y}_{N-1}. \end{cases} \tag{6.45}$$

If $\bar{x} = (x_1, \dots, x_N)$, where $x_i = \mathrm{I}(i = j), i = 1, \dots, N$, for some $1 \leq j \leq N$, then

$$p_{\varepsilon,\bar{x},\bar{0}} = \frac{q_{\varepsilon,j,-}}{q_{\varepsilon,\bar{x}}}, \ q_{\varepsilon,\bar{x}} = q_{\varepsilon,j,-} + \sum_{i \neq j} q_{\varepsilon,i,+}. \tag{6.46}$$

The condition $\mathbf{B}_{13}$ and the relations (6.45)–(6.46) imply that the condition $\mathbf{A}_5$ is satisfied.

Thus, the condition $\mathbf{A}_1$ is also satisfied, i.e., there exists $\varepsilon''' \in (0, \varepsilon'']$ such that, for $\varepsilon \in (0, \varepsilon''']$,

$$p_{\varepsilon,\bar{1}} = \mathsf{P}\{\tilde{\rho}_{\varepsilon,1} = 1\} = \mathsf{P}\{\prod_{k=1}^{\check{\nu}_{\varepsilon,\mathbb{X},1}} \check{\rho}_{\varepsilon,k} \geq 1\} > 0, \tag{6.47}$$

and

$$p_{\varepsilon,\bar{1}} \to 0 \text{ as } \varepsilon \to 0. \tag{6.48}$$

Thus, the function

$$u_\varepsilon = p_{\varepsilon,\bar{1}}^{-1} \to \infty \text{ as } \varepsilon \to 0. \tag{6.49}$$

Finding more explicit asymptotic approximations for the functions $p_{\varepsilon,\bar{1}}$ and u_ε is a rather difficult task.

First of all, note that the states $\bar{x} \in \hat{\mathbb{X}}$ must be renumbered according to the relation (6.24) in order to get the standard representations for the matrices defined below.

Let us introduce the $1 \times m_1$ matrix,

$$\mathbf{P}_{\varepsilon,0,-} = \|p_{\varepsilon,\bar{1},\bar{y}}\|_{\bar{1} \in \mathbb{Y}_0, \bar{y} \in \mathbb{Y}_1}. \tag{6.50}$$

If $\bar{y} = (y_1, \dots, y_N)$, where $y_i = \mathrm{I}(i \neq j), i = 1, \dots, N$, for some $1 \leq j \leq N$, then

$$p_{\varepsilon,\bar{1},\bar{y}} = \frac{q_{\varepsilon,j,-}}{q_{\varepsilon,\bar{1}}}, \ q_{\varepsilon,\bar{1}} = \sum_{i=1}^{N} q_{\varepsilon,i,-}. \tag{6.51}$$

Let us also introduce the $m_n \times m_{n\pm1}$ matrices, for $n = 1, \ldots, N-1$,

$$\mathbf{P}_{\varepsilon,n,\pm} = \| p_{\varepsilon,\bar{x},\bar{y}} \|_{\bar{x} \in \mathbb{Y}_n, \bar{y} \in \mathbb{Y}_{n\mp1}}. \tag{6.52}$$

If $\bar{x} = (x_1, \ldots, x_N) \in \mathbb{Y}_n, \bar{y} = (y_1, \ldots, y_N) \in \mathbb{Y}_{n\mp1}$, and $y_j = \frac{1\pm1}{2} \neq x_j$, $y_i = x_i$, $i \neq j$ for some $1 \leq j \leq N$, then

$$p_{\varepsilon,\bar{x},\bar{y}} = \frac{q_{\varepsilon,j,\pm}}{q_{\varepsilon,\bar{x}}}, \ q_{\varepsilon,\bar{x}} = \sum_{x_i=1} q_{\varepsilon,i,-} + \sum_{x_i=0} q_{\varepsilon,i,+}. \tag{6.53}$$

The relations (6.52), (6.53) and the condition $\mathbf{B}_{13}$ imply that, for $n = 1, \ldots, N-1$,

$$|\mathbf{P}_{\varepsilon,n,\pm}| = \max_{\bar{x} \in \mathbb{Y}_n} \sum_{\bar{y} \in \mathbb{Y}_{n\mp1}} |p_{\varepsilon,\bar{x},\bar{y}}| = \max_{\bar{x} \in \mathbb{Y}_n} \sum_{x_i = \frac{1\mp1}{2}} \frac{q_{\varepsilon,i,\pm}}{q_{\varepsilon,\bar{x}}} < 1. \tag{6.54}$$

Moreover, the relation (6.53) and the condition $\mathbf{B}_{13}$ imply that, for $n = 1, \ldots, N-1$,

$$|\mathbf{P}_{\varepsilon,n,-}| \leq q_\varepsilon = \frac{\sum_{1 \leq i \leq N} q_{\varepsilon,i,-}}{\min_{1 \leq i \leq N} q_{\varepsilon,i,+}} \longrightarrow 0 \text{ as } \varepsilon \longrightarrow 0, \tag{6.55}$$

while

$$|\mathbf{P}_{\varepsilon,n,+}| \longrightarrow 1 \text{ as } \varepsilon \longrightarrow 0. \tag{6.56}$$

Let us also define, for $n = 1, \ldots, N-1$, the matrices,

$$\mathbf{Q}_{\varepsilon,n} = \| Q_{\varepsilon,n,\bar{x},\bar{y}} \|_{\bar{x} \in \mathbb{Y}_n, \bar{y} \in \mathbb{Y}_{n+1}}, \tag{6.57}$$

where, for $\bar{x} \in \mathbb{Y}_n, \bar{y} \in \mathbb{Y}_{n+1}$,

$$Q_{\varepsilon,n,\bar{x},\bar{y}} = \mathsf{P}_{\bar{x}}\{\check{\tau}_{\varepsilon,\mathbb{Y}_{n+1}} < \check{\tau}_{\varepsilon,\mathbb{Y}_0}, \check{\eta}_{\varepsilon,\check{\tau}_{\varepsilon,\mathbb{Y}_{n+1}}} = \bar{y}\}. \tag{6.58}$$

Obviously,

$$\mathbf{Q}_{\varepsilon,1} = \mathbf{P}_{\varepsilon,1,-}. \tag{6.59}$$

The condition $\mathbf{B}_6$ implies that, for $n = 1, \ldots, N-1$,

$$\begin{aligned} |\mathbf{Q}_{\varepsilon,n}| &= \max_{\bar{x} \in \mathbb{Y}_n} \sum_{\bar{y} \in \mathbb{Y}_{n+1}} Q_{\varepsilon,n,\bar{x},\bar{y}} \\ &= \max_{\bar{x} \in \mathbb{Y}_n} \mathsf{P}_{\bar{x}}\{\check{\tau}_{\varepsilon,\mathbb{Y}_{n+1}} < \check{\tau}_{\varepsilon,\mathbb{Y}_0}\} < 1. \end{aligned} \tag{6.60}$$

Let us also define, for $n = 1, \ldots, N-1$, the matrices,

$$\mathbf{R}_{\varepsilon,n} = \| R_{\varepsilon,\bar{x},\bar{y}} \|_{\bar{x} \in \mathbb{Y}_n, \bar{y} \in \mathbb{Y}_n}, \tag{6.61}$$

where, for $\bar{x} \in \mathbb{Y}_n, \bar{y} \in \mathbb{Y}_n$,

$$R_{\varepsilon,n,\bar{x},\bar{y}} = \mathsf{P}_{\bar{x}}\{\check{\tau}_{\varepsilon,\mathbb{Y}_n} < \check{\tau}_{\varepsilon,\mathbb{Y}_0} \wedge \check{\tau}_{\varepsilon,\mathbb{Y}_{n+1}}, \check{\eta}_{\varepsilon,\check{\tau}_{\varepsilon,\mathbb{Y}_n}} = \bar{y}\}. \tag{6.62}$$

Obviously, for $\bar{x} \in \mathbb{Y}_1, \bar{y} \in \mathbb{Y}_1$,

$$R_{\varepsilon,1,\bar{x},\bar{y}} = 0. \tag{6.63}$$

The following relation takes place, for $n = 2, \dots, N-1$,

$$\mathbf{R}_{\varepsilon,n} = \mathbf{P}_{\varepsilon,n,+}\mathbf{Q}_{\varepsilon,n-1}, \tag{6.64}$$

and, thus, by the relations (6.54) and (6.60),

$$|\mathbf{R}_{\varepsilon,n}| \leq |\mathbf{P}_{\varepsilon,n,+}||\mathbf{Q}_{\varepsilon,n-1}| < 1. \tag{6.65}$$

The above relation implies that, for $n = 2, \dots, N-1$, there exists the inverse matrix, $[\mathbf{I}_n - \mathbf{R}_{\varepsilon,n}]^{-1}$, where $\mathbf{I}_n = \|\mathrm{I}(\bar{x} = \bar{y}\|_{\bar{x}\in\mathbb{Y}_n,\bar{y}\in\mathbb{Y}_n}$ is the $m_n \times m_n$ unit matrix, for $n = 1, \dots, N-1$.

The following relation takes place, for $n = 1, \dots, N-1$,

$$\begin{aligned}
\mathbf{Q}_{\varepsilon,n} &= \mathbf{P}_{\varepsilon,n,-} + \mathbf{R}_{\varepsilon,n}\mathbf{P}_{\varepsilon,n,-} + \mathbf{R}^2_{\varepsilon,n}\mathbf{P}_{\varepsilon,n,-} + \cdots \\
&= \left(\mathbf{I}_n + \mathbf{R}_{\varepsilon,n} + \mathbf{R}^2_{\varepsilon,n} + \cdots\right)\mathbf{P}_{\varepsilon,n,-} \\
&= [\mathbf{I}_n - \mathbf{R}_{\varepsilon,n}]^{-1}\mathbf{P}_{\varepsilon,n,-} = [\mathbf{I}_n + \mathbf{R}_{\varepsilon,n}[\mathbf{I}_n - \mathbf{R}_{\varepsilon,n}]^{-1}]\mathbf{P}_{\varepsilon,n,-}.
\end{aligned} \tag{6.66}$$

Let us choose an arbitrary $0 < q < 1$.

The relations (6.54), (6.55), and (6.59) imply that there exists $\varepsilon'_q \in (0, \varepsilon''']$ such that for $\varepsilon \in (0, \varepsilon'_q]$,

$$|\mathbf{Q}_{\varepsilon,1}| = q_\varepsilon \leq q, \tag{6.67}$$

and

$$|\mathbf{Q}_{\varepsilon,1}| \to 0 \text{ as } \varepsilon \to 0. \tag{6.68}$$

The relations (6.60), (6.64), (6.67), and (6.68) imply that, for $\varepsilon \in (0, \varepsilon'_q]$,

$$|\mathbf{R}_{\varepsilon,2}| \leq |\mathbf{Q}_{\varepsilon,1}| = q_\varepsilon \leq q, \tag{6.69}$$

and

$$|\mathbf{R}_{\varepsilon,2}| \to 0 \text{ as } \varepsilon \to 0. \tag{6.70}$$

The relations (6.54), (6.55), and (6.69) imply that there exist $\varepsilon''_q \in (0, \varepsilon'_q]$ such that, for $\varepsilon \in (0, \varepsilon''_q]$,

$$\begin{aligned}
|\mathbf{Q}_{\varepsilon,2}| &\leq |\mathbf{P}_{\varepsilon,2,-}| + |\mathbf{R}_{\varepsilon,2}||\mathbf{P}_{\varepsilon,2,-}| + |\mathbf{R}_{\varepsilon,2}|^2|\mathbf{P}_{\varepsilon,2,-}| + \cdots \\
&\leq \left(1 + q + q^2 + \cdots\right)q_\varepsilon = (1-q)^{-1}q_\varepsilon \leq q
\end{aligned} \tag{6.71}$$

and

$$|\mathbf{Q}_{\varepsilon,2}| \to 0 \text{ as } \varepsilon \to 0. \tag{6.72}$$

The relations (6.60), (6.64), (6.71), and (6.72) imply that, for $\varepsilon \in (0, \varepsilon''_q]$,

$$|\mathbf{R}_{\varepsilon,3}| \leq |\mathbf{Q}_{\varepsilon,2}| \leq q, \tag{6.73}$$

and

$$|\mathbf{R}_{\varepsilon,3}| \to 0 \text{ as } \varepsilon \to 0. \tag{6.74}$$

The relations (6.54), (6.55), and (6.73) imply that, for $\varepsilon \in (0, \varepsilon''_q]$,

$$\begin{aligned} |\mathbf{Q}_{\varepsilon,3}| &\leq |\mathbf{P}_{\varepsilon,3,-}| + |\mathbf{R}_{\varepsilon,2}||\mathbf{P}_{\varepsilon,3,-}| + |\mathbf{R}_{\varepsilon,3}|^2|\mathbf{P}_{\varepsilon,3,-}| + \cdots \\ &\leq \left(1 + q + q^2 + \cdots\right)q_\varepsilon = (1-q)^{-1}q_\varepsilon \leq q \end{aligned} \tag{6.75}$$

and

$$|\mathbf{Q}_{\varepsilon,3}| \to 0 \text{ as } \varepsilon \to 0. \tag{6.76}$$

The above recurrent calculations can be continued and give two series of relations. First, the following relations hold, for $n = 1, \ldots, N-1$ and $\varepsilon \in (0, \varepsilon''_q]$,

$$|\mathbf{Q}_{\varepsilon,n}| \leq q, \tag{6.77}$$

and, for $n = 1, \ldots, N-1$,

$$|\mathbf{Q}_{\varepsilon,n}| \to 0 \text{ as } \varepsilon \to 0. \tag{6.78}$$

Second, the following relation holds, for $n = 2, \ldots, N-1$ and $\varepsilon \in (0, \varepsilon''_q]$,

$$|\mathbf{R}_{\varepsilon,n}| \leq q, \tag{6.79}$$

and, for $n = 1, \ldots, N-1$,

$$|\mathbf{R}_{\varepsilon,n}| \to 0 \text{ as } \varepsilon \to 0. \tag{6.80}$$

The following representation takes place:

$$\begin{aligned} p_{\varepsilon,\bar{1}} &= \mathbf{P}_{\varepsilon,0,-}\,\mathbf{Q}_{\varepsilon,1}\cdots\mathbf{Q}_{\varepsilon,N-1} \\ &= \mathbf{P}_{\varepsilon,0,-}\,\mathbf{P}_{\varepsilon,1,-}\,[\mathbf{I}_2 + \mathbf{R}_{\varepsilon,2}[\mathbf{I}_2 - \mathbf{R}_{\varepsilon,2}]^{-1}]\mathbf{P}_{\varepsilon,2,-} \\ &\quad \cdots [\mathbf{I}_{N-1} + \mathbf{R}_{\varepsilon,N-1}[\mathbf{I}_{N-1} - \mathbf{R}_{\varepsilon,N-1}]^{-1}]\mathbf{P}_{\varepsilon,N-1,-}\,. \end{aligned} \tag{6.81}$$

The relations (6.79) and (6.80) imply that, for $n = 2, \ldots, N-1$ and $\varepsilon \in (0, \varepsilon''_q]$,

$$|\mathbf{R}_{\varepsilon,n}[\mathbf{I}_{N-1} - \mathbf{R}_{\varepsilon,N-1}]^{-1}]| \leq \frac{q}{1-q}, \tag{6.82}$$

and

$$|\mathbf{R}_{\varepsilon,n}[\mathbf{I}_{N-1} - \mathbf{R}_{\varepsilon,N-1}]^{-1}]| \to 0 \text{ as } \varepsilon \to 0. \tag{6.83}$$

Finally, the relations (6.81), (6.82), and (6.83) imply that

$$p_{\varepsilon,\bar{1}} \sim \dot{p}_{\varepsilon,\bar{1}} = \mathbf{P}_{\varepsilon,0,-}\,\mathbf{P}_{\varepsilon,1,-}\,\cdots\,\mathbf{P}_{\varepsilon,N-1,-} \text{ as } \varepsilon \to 0. \tag{6.84}$$

We can use the explicit formulas for elements of the matrices $\mathbf{P}_{\varepsilon,n,-}, n = 0, \ldots, N-1$ given by the relation (6.53) and extract from the product of these matrices all quantities $q_{\varepsilon,j,-}$ and $q_{\varepsilon,\bar{1}}$, which tends to 0 as $\varepsilon \to 0$.

The resulting matrix takes the following form:

$$\begin{aligned} p_{\varepsilon,\bar{1}} &= \mathbf{P}_{\varepsilon,0,-}\,\mathbf{P}_{\varepsilon,1,-} \cdots \mathbf{P}_{\varepsilon,N-1,-} \\ &= \frac{\prod_{j=1}^{N} q_{\varepsilon,j,-}}{\sum_{j=1}^{N} q_{\varepsilon,j,-}} \dot{\mathbf{P}}_{\varepsilon,0,-}\,\dot{\mathbf{P}}_{\varepsilon,1,-} \cdots \dot{\mathbf{P}}_{\varepsilon,N-1,-}, \end{aligned} \tag{6.85}$$

where, for $n = 0, \ldots, N-1$,

$$\dot{\mathbf{P}}_{\varepsilon,n,-} = \begin{cases} \|\mathrm{I}(q_{\varepsilon,\bar{1},\bar{y}} > 0)\|_{\bar{1}\in\mathbb{Y}_0,\bar{y}\in\mathbb{Y}_1} & \text{for } n = 0, \\ \|\frac{\mathrm{I}(q_{\varepsilon,\bar{x},\bar{y}}>0)}{q_{\varepsilon,\bar{x}}}\|_{\bar{x}\in\mathbb{Y}_n,\bar{y}\in\mathbb{Y}_{n+1}} & \text{for } n = 1, \ldots, N-1. \end{cases} \tag{6.86}$$

The matrix $\dot{\mathbf{P}}_{\varepsilon,0,-}\,\dot{\mathbf{P}}_{\varepsilon,1,-} \cdots \dot{\mathbf{P}}_{\varepsilon,N-1,-} = \mathbf{C}_\varepsilon = \|C_\varepsilon\|$ is a 1×1 matrix. The condition $\mathbf{B}_{13}$ implies that

$$0 < \varliminf_{\varepsilon\to 0} C_\varepsilon \leq \varlimsup_{\varepsilon\to 0} C_\varepsilon < \infty. \tag{6.87}$$

Let us now define the new normalisation function,

$$\dot{u}_\varepsilon = \frac{\sum_{j=1}^{N} q_{\varepsilon,j,-}}{\prod_{j=1}^{N} q_{\varepsilon,j,-}} C_\varepsilon^{-1}. \tag{6.88}$$

The relations (6.84), (6.85), and (6.87) imply that

$$u_\varepsilon \sim \dot{u}_\varepsilon \text{ as } \varepsilon \to 0. \tag{6.89}$$

In this case, for $\bar{x} \in \check{\mathbb{X}}$ and $\delta \geq 0$,

$$\mathsf{P}_{\bar{x}}\{\check{\kappa}_{\varepsilon,1} > \delta\} = e^{-\delta w_\varepsilon q_{\varepsilon,\bar{x}}}. \tag{6.90}$$

The condition $\mathbf{B}_{13}$ and the relation (6.28) imply that, for $\bar{x} \neq \bar{1}$,

$$0 < \varliminf_{\varepsilon\to 0} q_{\varepsilon,\bar{x}} \leq \varlimsup_{\varepsilon\to 0} q_{\varepsilon,\bar{x}} < \infty, \tag{6.91}$$

while

$$q_{\varepsilon,\bar{1}} = \sum_{i=1}^{N} q_{\varepsilon,i,-} \to 0 \text{ as } \varepsilon \to 0. \tag{6.92}$$

Let us, also, introduce, for $n = 1, \ldots, N-1$, the random variables,

$$\check{\nu}_{\varepsilon,n} = \sum_{k=1}^{\check{\tau}_{\varepsilon,\mathbb{Y}_0\cup\mathbb{Y}_{n+1}}} \mathrm{I}(\check{\eta}_{\varepsilon,k} \in \mathbb{Y}_n). \tag{6.93}$$

Obviously, for $n = 1, \dots, N-1$ and $k = 0, 1, \dots,$

$$\mathbf{S}_{\varepsilon,n,k} = \|\mathsf{P}_{\bar{x}}\{\check{\nu}_{\varepsilon,n} = k, \check{\eta}_{\varepsilon,\check{\tau}_{\varepsilon,\mathbb{Y}_0\cup\mathbb{Y}_{n+1}}} = \bar{y}\}\|_{\bar{x}\in\mathbb{Y}_n,\bar{y}\in\mathbb{Y}_{n+1}} = \mathbf{R}^k_{\varepsilon,n}\,\mathbf{P}_{\varepsilon,n,-} \tag{6.94}$$

The condition $\mathbf{B}_{13}$ implies that, for $\bar{x} \in \mathbb{Y}_n, n = 1, \dots, N-1$,

$$\mathsf{P}_{\bar{x}}\{\check{\nu}_{\varepsilon,n} < \infty\} = 1. \tag{6.95}$$

Also, the condition $\mathbf{B}_{13}$ obviously implies that the condition $\mathbf{C}_8$ holds, if the function $0 < w_\varepsilon \to \infty$ as $\varepsilon \to 0$ is chosen in such a way that, for $\bar{x} \in \check{\mathbb{X}}$,

$$w_\varepsilon q_{\varepsilon,\bar{1}} \to \infty \text{ as } \varepsilon \to 0. \tag{6.96}$$

The relation (6.31) implies that, for $u \geq 0$ and $\bar{x} \in \check{\mathbb{X}}$,

$$\tilde{F}_{\varepsilon,\bar{x}}(u) = \mathsf{P}_{\bar{x}}\{\tilde{\kappa}_{\varepsilon,1} \leq u\} = \mathsf{P}_{\bar{1}}\{\sum_{k=1}^{\check{\nu}_{\varepsilon,\mathbb{X},1}} \check{\kappa}_{\varepsilon,k} \leq u\} = \hat{F}_\varepsilon(uw_\varepsilon), \tag{6.97}$$

where

$$\hat{F}_\varepsilon(u) = \mathsf{P}_{\bar{1}}\{\sum_{k=1}^{\hat{\nu}_{\varepsilon,\mathbb{X},1}} \hat{\kappa}_{\varepsilon,k} \leq u\}. \tag{6.98}$$

The following stochastic representation takes place and plays a key role in subsequent asymptotic analysis,

$$\tau_{\varepsilon,\bar{1}} \overset{\mathsf{d}}{=} \zeta_\varepsilon + \sum_{\bar{y}\in\mathbb{Y}_1} \mathrm{I}(\iota_\varepsilon = \bar{y})\tau_{\varepsilon,\bar{y}}, \tag{6.99}$$

where: (a) the random variable $\tau_{\varepsilon,\bar{x}}$ has the distribution function $\mathsf{P}_{\bar{x}}\{\tilde{\kappa}_{\varepsilon,1} \leq \cdot\}$, for $\bar{x} \in \tilde{\mathbb{X}}$, (b) the random variable ζ_ε is exponentially distributed with parameter $q_{\varepsilon,\bar{1}}w_\varepsilon$, (c) the random variable ι_ε takes values $\bar{y}$ with probabilities, respectively, $p_{\varepsilon,\bar{1},\bar{y}}$, for $\bar{y} \in \mathbb{Y}_1$, and (d) the random variables $\zeta_\varepsilon, \iota_\varepsilon, \tau_{\varepsilon,\bar{x}}, \bar{x} \in \tilde{\mathbb{X}}$ are mutually independent.

In this case, $\mathsf{P}_{\bar{x}}\{\check{\kappa}_{\varepsilon,1} \leq u\} = 1 - e^{-w_\varepsilon q_{\varepsilon,\bar{x}} u},\ u \geq 0$, for $\bar{x} \in \check{\mathbb{X}}$, and, thus, for $\bar{x} \in \check{\mathbb{X}}$ and $n \geq 1$,

$$\mathsf{E}_{\bar{x}}\check{\kappa}^n_{\varepsilon,1} = \frac{n!}{(w_\varepsilon q_{\varepsilon,\bar{x}})^n}. \tag{6.100}$$

Taking into account the relations (6.99) and (6.100) and calculating expectations and second moments for the random variable $\tau_{\varepsilon,\bar{1}}$, we get the following relations:

$$\mathsf{E}_{\bar{1}}\tilde{\kappa}_{\varepsilon,1} = \frac{1}{w_\varepsilon q_{\varepsilon,\bar{1}}} + \sum_{\bar{y}\in\mathbb{Y}_1} p_{\varepsilon,\bar{1},-}\mathsf{E}_{\bar{y}}\tilde{\kappa}_{\varepsilon,1}, \tag{6.101}$$

and

$$\mathsf{E}_{\bar{1}}\tilde{\kappa}^2_{\varepsilon,1} = \frac{2}{(w_\varepsilon q_{\varepsilon,\bar{1}})^2} + 2\frac{1}{w_\varepsilon q_{\varepsilon,\bar{1}}}\sum_{\bar{y}\in\mathbb{Y}_1} p_{\varepsilon,\bar{1},-}\mathsf{E}_{\bar{y}}\tilde{\kappa}_{\varepsilon,1} + \sum_{\bar{y}\in\mathbb{Y}_1} p_{\varepsilon,\bar{1},-}\mathsf{E}_{\bar{y}}\tilde{\kappa}^2_{\varepsilon,1}. \tag{6.102}$$

The relation (6.100) implies that the condition $\mathbf{C}_9$ is satisfied, for $\bar{x} \in \tilde{\mathbb{X}} \setminus \mathbb{X}$, $\alpha = \beta = 2$ and, thus, by Lemma 4.7, for $\bar{x} \neq \bar{1}$,

$$\mathsf{E}_{\bar{x}}\tilde{\kappa}^2_{\varepsilon,1} \leq \frac{B_\varepsilon}{w_\varepsilon^2}, \tag{6.103}$$

where

$$B_\varepsilon = \check{e}_{\varepsilon,\bar{x},2,2}a_{p,N,2,2} = \max_{\bar{y}\neq\bar{1}} \frac{2}{q^2_{\varepsilon,\bar{y}}} a_{p,N,2,2} \tag{6.104}$$

and the constant $a_{p,N,2,2}$ is given by the relation (4.108).

The condition $\mathbf{B}_{13}$ implies that

$$0 < \varliminf_{\varepsilon\to 0} B_\varepsilon \leq \varlimsup_{\varepsilon\to 0} B_\varepsilon < \infty. \tag{6.105}$$

Let $\hat{\theta}_{\varepsilon,n}, n = 1, 2, \ldots$, be, for $\varepsilon \in (0, \varepsilon''_q]$, i.i.d. random variables with the distribution function $\hat{F}_{\varepsilon,\mathbb{X}}(\cdot) = \hat{F}_\varepsilon(\cdot)$. Let us also consider the sequence of i.i.d. random variables $\theta^{(w)}_{\varepsilon,n} = \hat{\theta}_{\varepsilon,n}/w_\varepsilon, n = 1, 2, \ldots$. Obviously, these random variables have the distribution function $F_\varepsilon(\cdot) = \hat{F}_\varepsilon(uw_\varepsilon)$.

The relation (6.103) also implies that, for $\bar{x} \neq \bar{1}$,

$$\mathsf{E}_{\bar{x}}\tilde{\kappa}_{\varepsilon,1} \leq \sqrt{\mathsf{E}_{\bar{x}}\tilde{\kappa}^2_{\varepsilon,1}} \leq \frac{\sqrt{B_\varepsilon}}{w_\varepsilon}. \tag{6.106}$$

The relations (6.92), (6.103), and (6.106) imply that

$$\mathsf{E}\theta^{(w)}_{\varepsilon,1} = \mathsf{E}_{\bar{1}}\tilde{\kappa}_{\varepsilon,1} \sim \frac{1}{w_\varepsilon q_{\varepsilon,\bar{1}}} \text{ as } \varepsilon \to 0 \tag{6.107}$$

and

$$\mathsf{E}(\theta^{(w)}_{\varepsilon,1})^2 = \mathsf{E}_{\bar{1}}\tilde{\kappa}^2_{\varepsilon,1} \leq \frac{1}{(w_\varepsilon q_{\varepsilon,\bar{1}})^2}(2 + 2\sqrt{B_\varepsilon}q_{\varepsilon,\bar{1}} + B_\varepsilon q^2_{\varepsilon,\bar{1}}). \tag{6.108}$$

Let us consider the random variables,

$$\theta^{(w)}_\varepsilon = \sum_{n=1}^{[u_\varepsilon]} \theta^{(w)}_{\varepsilon,n} = \sum_{n=1}^{[u_\varepsilon]} \hat{\theta}_{\varepsilon,n}/w_\varepsilon. \tag{6.109}$$

Let us, first, choose the normalisation function,

$$w_{\circ,\varepsilon} = \frac{\dot{u}_\varepsilon}{q_{\varepsilon,\bar{1}}}. \tag{6.110}$$

The relations (6.49), (6.89), (6.92), and (6.107) imply that

$$\begin{aligned}
\mathsf{E}\theta_\varepsilon^{(w_\circ)} &= [u_\varepsilon]\mathsf{E}\theta_{\varepsilon,1}^{(w_\circ)} = [u_\varepsilon]\frac{1}{w_{\circ,\varepsilon} q_{\varepsilon,\bar{1}}} \\
&= \frac{[u_\varepsilon]}{u_\varepsilon}\frac{u_\varepsilon}{\dot{u}_\varepsilon}\frac{q_{\varepsilon,\bar{1}}}{q_{\varepsilon,\bar{1}}} \longrightarrow 1 \text{ as } \varepsilon \to 0.
\end{aligned} \tag{6.111}$$

Also, the relations (6.49), (6.92), and (6.108) imply that

$$\begin{aligned}
\mathrm{Var}\theta_\varepsilon^{(w_\circ)} &= [u_\varepsilon]\mathrm{Var}\theta_{\varepsilon,1}^{(w_\circ)} \le [u_\varepsilon]\mathsf{E}(\theta_{\varepsilon,1}^{(w_\circ)})^2 \\
&\le [u_\varepsilon]\frac{1}{(w_{\circ,\varepsilon} q_{\varepsilon,\bar{1}})^2}(2 + 2\sqrt{B_\varepsilon} q_{\varepsilon,\bar{1}} + B_\varepsilon q_{\varepsilon,\bar{1}}^2) \\
&\le \frac{[u_\varepsilon]}{\dot{u}_\varepsilon^2}(2 + 2\sqrt{B_\varepsilon} q_{\varepsilon,\bar{1}} + B_\varepsilon q_{\varepsilon,\bar{1}}^2) \longrightarrow 0 \text{ as } \varepsilon \to 0.
\end{aligned} \tag{6.112}$$

The relations (6.108) and (6.112) imply that the following asymptotic relation takes place:

$$\theta_\varepsilon^{(w_\circ)} = \sum_{n=1}^{[u_\varepsilon]} \theta_{\varepsilon,n}^{(w_\circ)} \xrightarrow{\mathsf{P}} 1 \text{ as } \varepsilon \to 0. \tag{6.113}$$

Let us now consider the case, where the function $0 < w_\varepsilon \to \infty$ as $\varepsilon \to 0$ satisfies the following condition:

$\mathbf{D}_{19}$: $A_\varepsilon = \frac{\dot{u}_\varepsilon}{q_{\varepsilon,\bar{1}} w_\varepsilon} \to A \in (0, \infty)$ as $\varepsilon \to 0$.

Under the condition $\mathbf{B}_{13}$, the condition $\mathbf{D}_{19}$ is necessary and sufficient for the fulfilment of the condition $\mathbf{D}_1$, which in this case takes the form of the following asymptotic relation:

$$\theta_\varepsilon^{(w)} = \sum_{n=1}^{[u_\varepsilon]} \theta_{\varepsilon,n}^{(w)} \Rightarrow \theta_0^{(w)} \text{ as } \varepsilon \to 0, \tag{6.114}$$

where $\theta_0^{(w)}$ is a non-negative random variable with distribution function not concentrated at zero.

Moreover, in this case, the limiting random variable $\theta_0^{(w)} = A$ with probability 1.

Indeed, assume that the conditions $\mathbf{B}_{13}$ and $\mathbf{D}_{19}$ are satisfied.

In this case, the relation (6.113) and the asymptotic relation appearing in the condition $\mathbf{D}_{19}$ imply that

$$\theta_\varepsilon^{(w)} = \sum_{n=1}^{[u_\varepsilon]} \theta_{\varepsilon,n}^{(w)} = \sum_{n=1}^{[u_\varepsilon]} \theta_{\varepsilon,n}^{(w_\circ)} \frac{\dot{u}_\varepsilon}{q_{\varepsilon,\bar{1}} w_\varepsilon}$$
$$= \theta_\varepsilon^{(w_\circ)} \frac{\dot{u}_\varepsilon}{q_{\varepsilon,\bar{1}} w_\varepsilon} \xrightarrow{\mathsf{P}} A \text{ as } \varepsilon \to 0. \tag{6.115}$$

Thus, the above sufficiency statement is true.

Let us assume that the condition $\mathbf{B}_{13}$ and the relation (6.114) hold, but the condition $\mathbf{D}_{19}$ is not satisfied.

In this case, one can always select two subsequences $0 < \varepsilon'_k, \varepsilon''_k \leq 1$ such that: (a) $\varepsilon'_k, \varepsilon''_k \to 0$ as $n \to \infty$, (b) $\frac{\dot{u}_{\varepsilon'_k}}{q_{\varepsilon'_k,\bar{1}} w_{\varepsilon'_k}} \to A' \in [0, \infty]$ as $k \to \infty$, (c) $\frac{\dot{u}_{\varepsilon''_k}}{q_{\varepsilon''_k,\bar{1}} w_{\varepsilon''_k}} \to A'' \in [0, \infty]$ as $k \to \infty$, and (d) $A' \neq A''$.

In this case, the relation (6.113) implies that

$$\theta_{\varepsilon'_k}^{(w)} = \sum_{n=1}^{[u_{\varepsilon'_k}]} \theta_{\varepsilon'_k,n}^{(w)} = \sum_{n=1}^{[u_{\varepsilon'_k}]} \theta_{\varepsilon'_k,n}^{(w_\circ)} \frac{\dot{u}_{\varepsilon'_k}}{q_{\varepsilon'_k,\bar{1}} w_{\varepsilon'_k}}$$
$$= \theta_{\varepsilon'_k}^{(w_\circ)} \frac{\dot{u}_{\varepsilon'_k}}{q_{\varepsilon'_k,\bar{1}} w_{\varepsilon'_k}} \xrightarrow{\mathsf{P}} A' \text{ as } k \to \infty \tag{6.116}$$

and

$$\theta_{\varepsilon''_k}^{(w)} = \sum_{n=1}^{[u_{\varepsilon''_k}]} \theta_{\varepsilon''_k,n}^{(w)} = \sum_{n=1}^{[u_{\varepsilon''_k}]} \theta_{\varepsilon''_k,n}^{(w_\circ)} \frac{\dot{u}_{\varepsilon''_k}}{q_{\varepsilon''_k,\bar{1}} w_{\varepsilon''_k}}$$
$$= \theta_{\varepsilon''_k}^{(w_\circ)} \frac{\dot{u}_{\varepsilon''_k}}{q_{\varepsilon''_k,\bar{1}} w_{\varepsilon''_k}} \xrightarrow{\mathsf{P}} A'' \text{ as } k \to \infty. \tag{6.117}$$

The above two relations contradict to the relation (6.114).

Thus, the above necessity statement is also true, i.e., the condition $\mathbf{D}_{19}$ is satisfied.

In this case, due to the relation (6.115), the limiting random variable $\theta_0^{(w)}$ in relation (6.114) equals to the constant A with probability 1.

The relation appearing in the condition $\mathbf{D}_{19}$ obviously implies that the relation (6.96) holds and, thus, the condition $\mathbf{C}_8$ also holds.

The condition $\mathbf{D}_{19}$ plays, in this case, the role of the condition $\mathbf{D}_1$.

Finally, the relation (6.108) implies that

$$\mathsf{E}_{\bar{1}} \kappa_{\varepsilon,1}^2 \leq \frac{2 + 2\sqrt{B_\varepsilon} q_{\varepsilon,\bar{1}} + B_\varepsilon q_{\varepsilon,\bar{1}}^2}{(w_\varepsilon q_{\varepsilon,\bar{1}})^2}$$
$$= \frac{2 + 2\sqrt{B_\varepsilon} q_{\varepsilon,\bar{1}} + B_\varepsilon q_{\varepsilon,\bar{1}}^2}{q_{\varepsilon,\bar{1}} w_\varepsilon} \frac{\dot{u}_\varepsilon}{q_{\varepsilon,\bar{1}} w_\varepsilon} \frac{u_\varepsilon}{\dot{u}_\varepsilon} p_{\varepsilon,\bar{1}} = o_{\bar{1}}(\varepsilon) p_{\varepsilon,\bar{1}}, \tag{6.118}$$

where

$$o_{\bar{1}}(\varepsilon) = \frac{2 + 2\sqrt{B_\varepsilon} q_{\varepsilon,\bar{1}} + B_\varepsilon q^2_{\varepsilon,\bar{1}}}{q_{\varepsilon,\bar{1}} w_\varepsilon} \frac{\dot{u}_\varepsilon}{q_{\varepsilon,\bar{1}} w_\varepsilon} \frac{u_\varepsilon}{\dot{u}_\varepsilon} \to 0 \text{ as } \varepsilon \to 0. \tag{6.119}$$

The relations (6.118) and (6.119) imply that the condition $\mathbf{C}_5$ is satisfied (for $\alpha, \beta = 2$), and, thus, the condition $\mathbf{C}_1$ also holds for the Markov renewal processes $(\eta_{\varepsilon,n}, \kappa_{\varepsilon,n}, \rho_{\varepsilon,n})$.

Therefore, according to Lemma 4.6, all conditions of Theorem 4.4 are satisfied for the Markov renewal processes $(\check{\eta}_{\varepsilon,n}, \check{\kappa}_{\varepsilon,n}, \check{\rho}_{\varepsilon,n})$. In this case, Theorem 4.4 takes the following form.

Theorem 6.2 *Let the condition* $\mathbf{B}_{13}$ *be satisfied. Then:*

(i) *The condition* $\mathbf{D}_{19}$ *is necessary and sufficient for the fulfilment (for some or any initial distributions* $\bar{q}_\varepsilon$*, respectively, in the statements of necessity and sufficiency) of the asymptotic relation* $\xi_{\mathbb{S}_\varepsilon} \xrightarrow{\text{d}} \xi_0$ *as* $\varepsilon \to 0$*, where* ξ_0 *is a non-negative random variable with distribution function not concentrated at zero.*

(ii) *The limiting random variable* ξ_0 *has an exponential distribution with the parameter* A^{-1}.

In conclusion, we would like to note that typical sufficient conditions for convergence in distribution for random functionals, such as normalised first failure times $\xi_{\mathbb{S}_\varepsilon}$, usually require convergence of all intensities $\mu_{\varepsilon,i,+}, i = 1, \ldots, N$ (as functions of $\varepsilon \to 0$) to some limiting values.

The conditions $\mathbf{B}_{13}$ and $\mathbf{D}_{19}$ do not require the convergence of these intensities. They can fluctuate almost arbitrary in the interval $(0, \infty)$. Convergence of only one "leading" parameter A_ε appearing in the condition $\mathbf{D}_{19}$ is required. Moreover, this convergence is not only sufficient but also a necessary condition for convergence in distribution for the normalised first failure times $\xi_{\mathbb{S}_\varepsilon}$.

Chapter 7
First-Rare-Event Times for Perturbed M/M-Type Queuing Systems

In this chapter, the results obtained in Chaps. 2–4 are illustrated by applications to perturbed M/M-type queuing systems with bounded and unbounded buffers. We are especially interested in the applications of the results of Chap. 4 related to the first-rare-event times, based on Markov renewal processes with transition periods and extending phase spaces for modulating semi-Markov processes. We present necessary and sufficient conditions for convergence in distribution of first-rare-event times for a number of models of perturbed M/M-type queuing systems with bounded and unbounded buffers. This chapter includes two sections.

In Sect. 7.1, we present necessary and sufficient conditions for the convergence in the distribution for the first-rare-event times for perturbed M/M queuing systems with limited queue buffers and rare failures caused by losses of customers trying to enter the system with a full buffer. Two cases are considered when the buffer size is fixed and the input stream intensity tends to zero, and when the buffer size can tend to infinity. The ergodic asymptotics is given in Theorems 7.1 and 7.2.

In Sect. 7.2, we present necessary and sufficient conditions of convergence in distribution for the first-rare-event times for perturbed M/M-type queuing systems with infinite queue buffer and rare failure events caused by too large number of customers appearing in the queue. The corresponding asymptotic result is given in Theorem 7.3.

7.1 Rare Events for Queuing Systems with Bounded Buffers

In this section, we give necessary and sufficient conditions for the convergence in distribution for first failure event times for perturbed M/M-type queuing system with bounded queue buffers.

D. Silvestrov, *Perturbed Semi-Markov Type Processes I*,
https://doi.org/10.1007/978-3-030-92403-4_7

7.1.1 Queuing Systems with a Buffer of Fixed Size

Let $\mathbf{S}_\varepsilon$ for each $\varepsilon \in (0, 1]$ be a standard M/M queuing system with bounded queue buffer. This system has one server and a bounded queue buffer of a fixed size $N \geq 1$ (including the place in the server). It is assumed that: **(a)** The input flow of customers is a standard Poisson flow with parameter $\mu_{\varepsilon,+} > 0$; **(b)** customers are placed in the queue buffer according to the order they arrive in the system; **(c)** a new customer coming in the system is immediately served if the queue buffer is empty, is placed in the queue buffer if the number of customers in the system is less than N, or leaves the system if the queue buffer is full, i.e., the number of customers in the system equals to N; **(d)** the service times for different customers are independent random variables, which have an exponential distribution with parameter $\mu_{\varepsilon,-} > 0$; **(f)** the input flow and all service processes are mutually independent.

The state of the system at the moment t is represented by the random variable $\hat{\eta}_\varepsilon(t)$, which is the number of customers in the buffer at this moment. The process $\hat{\eta}_\varepsilon(t)$ is a birth–death-type semi-Markov process, which has a phase space $\hat{\mathbb{X}} = \{0, \ldots, N\}$ and transition probabilities,

$$\hat{Q}_{\varepsilon,ij}(t) = \hat{p}_{\varepsilon,ij} \hat{F}_{\varepsilon,i}(t),\ t \geq 0,\ i, j \in \hat{\mathbb{X}}. \tag{7.1}$$

The $(N + 1) \times (N + 1)$ matrix of transition probabilities for the corresponding embedded Markov chain $\hat{\eta}_{\varepsilon,n}$ has the following form:

$$\hat{\mathbf{P}}_\varepsilon = \|\hat{p}_{\varepsilon,ij}\| = \begin{bmatrix} 0 & 1 & 0 & 0 & \cdots & 0 \\ p_{\varepsilon,-} & 0 & p_{\varepsilon,+} & 0 & \cdots & 0 \\ \cdots & \cdots & \cdots & \cdots & \cdots & \cdots \\ 0 & \cdots & 0 & p_{\varepsilon,-} & 0 & p_{\varepsilon,+} \\ 0 & \cdots & 0 \cdots & 0 & p_{\varepsilon,-} & p_{\varepsilon,+} \end{bmatrix}, \tag{7.2}$$

where,

$$p_{\varepsilon,+} = \frac{\mu_{\varepsilon,+}}{\lambda_\varepsilon},\ p_{\varepsilon,-} = \frac{\mu_{\varepsilon,-}}{\lambda_\varepsilon},\ \lambda_\varepsilon = \mu_{\varepsilon,+} + \mu_{\varepsilon,-} \tag{7.3}$$

and the distribution functions of sojourn times have the following forms:

$$\hat{F}_{\varepsilon,i}(t) = \begin{cases} 1 - e^{-\mu_{\varepsilon,+} t} & \text{for } t \geq 0,\ i = 0, \\ 1 - e^{-\lambda_\varepsilon t} & \text{for } t \geq 0,\ i = 1, \ldots, N. \end{cases} \tag{7.4}$$

We also assume that the initial state of the system $\hat{\eta}_\varepsilon(0)$ is a random variable independent of service processes and $\bar{q}_\varepsilon = \langle q_{\varepsilon,0}, \ldots q_{\varepsilon,N} \rangle$ is its distribution.

It is useful to note that the semi-Markov process $\hat{\eta}_\varepsilon(t)$ is, in fact, a homogeneous Markov process. However, we do prefer to consider it as a semi-Markov process, since its trajectories can include jump transitions of the form $N \longrightarrow N$. In the standard model of continuous time Markov chains, jump transitions of the form $i \longrightarrow i$ are usually excluded.

Let us consider the case, where the following condition is satisfied:

$\mathbf{B}_{14}$: **(a)** $0 < \underline{\lim}_{\varepsilon \to 0} \mu_{\varepsilon,+} \leq \overline{\lim}_{\varepsilon \to 0} \mu_{\varepsilon,+} < \infty$, **(b)** $\mu_{\varepsilon,-} \to \infty$ as $\varepsilon \to 0$.

The jump transition of the form $N \to N$ occurs in the above system, when a new customer tries to enter the system when the buffer is full. According to the above description this customer will be lost for the system.

Such an event can be interpreted as rare due to the condition $\mathbf{B}_{14}$.

Let $0 = \hat{\tau}_{\varepsilon,0} < \hat{\tau}_{\varepsilon,1} < \cdots$ be successive moments of jumps for the semi-Markov process $\hat{\eta}_\varepsilon(t)$, $\hat{\eta}_{\varepsilon,n} = \hat{\eta}(\hat{\tau}_{\varepsilon,n})$, $n = 0, 1, \ldots$ be the corresponding discrete time embedded Markov chain, $\hat{\kappa}_{\varepsilon,n} = \hat{\tau}_{\varepsilon,n} - \hat{\tau}_{\varepsilon,n-1}$, $n = 1, \ldots$ be inter-jump times for the semi-Markov process $\hat{\eta}_\varepsilon(t)$, and $\hat{\nu}_\varepsilon(t) = \max(n \geq 1 : \hat{\tau}_{\varepsilon,n} \leq t)$ be the number of jumps for the semi-Markov process $\hat{\eta}(s)$ in the time interval $[0, t]$.

Let us also consider the Markov renewal process $(\check{\eta}_{\varepsilon,n}, \check{\kappa}_{\varepsilon,n}, \check{\rho}_{\varepsilon,n})$, $n = 0, 1, \ldots$ which has the phase space $\check{\mathbb{X}} \times [0, \infty) \times \{0, 1\}$, where $\check{\mathbb{X}} = \hat{\mathbb{X}}$, and components,

$$\begin{cases} \check{\eta}_{\varepsilon,n} &= \hat{\eta}_{\varepsilon,n}, \\ \check{\kappa}_{\varepsilon,n} &= \hat{\kappa}_{\varepsilon,n}/w_\varepsilon, \\ \check{\rho}_{\varepsilon,n} &= \mathrm{I}(\hat{\eta}_{\varepsilon,n-1} = N, \hat{\eta}_{\varepsilon,n} = N), \end{cases} \tag{7.5}$$

where $0 < w_\varepsilon \to \infty$ as $\varepsilon \to 0$ is some normalisation function, which we will define below.

It is natural to define the normalised first-rare-event time for the system $\mathbf{S}_\varepsilon$, which can also be called the first failure time, as follows:

$$\xi_{\mathbf{S}_\varepsilon} = \sum_{n=1}^{\check{\nu}_\varepsilon} \check{\kappa}_{\varepsilon,n} = \sum_{n=1}^{\hat{\nu}_\varepsilon} \hat{\kappa}_{\varepsilon,n}/w_\varepsilon, \tag{7.6}$$

where

$$\check{\nu}_\varepsilon = \min(n \geq 1 : \check{\eta}_{\varepsilon,n-1} = N, \check{\eta}_{\varepsilon,n} = N). \tag{7.7}$$

The Markov chain $\check{\eta}_{\varepsilon,n}$ has the matrix of transition probabilities,

$$\check{\mathbf{P}}_\varepsilon = \hat{\mathbf{P}}_\varepsilon. \tag{7.8}$$

The condition $\mathbf{B}_{14}$ obviously implies that there exists $\varepsilon' \in (0, 1]$ such that $p_{\varepsilon,\pm} > 0$ for $\varepsilon \in (0, \varepsilon']$ and,

$$p_{\varepsilon,+} \to 0 \text{ as } \varepsilon \to 0. \tag{7.9}$$

This relation implies that,

$$\check{\mathbf{P}}_\varepsilon \to \check{\mathbf{P}}_0 \text{ as } \varepsilon \to 0, \tag{7.10}$$

where the limiting matrix $\check{\mathbf{P}}_0$ has the following form:

$$\check{\mathbf{P}}_0 = \begin{bmatrix} 0 & 1 & 0 & 0 & \cdots & 0 \\ 1 & 0 & 0 & 0 & \cdots & 0 \\ \cdots & \cdots & \cdots & \cdots & \cdots & \cdots \\ 0 & \cdots & 0 & 1 & 0 & 0 \\ 0 & \cdots & 0 \cdots & 0 & 1 & 0 \end{bmatrix}. \tag{7.11}$$

Let $\check{\eta}_{0,n}, n = 0, 1, \ldots$ be a Markov chain with the phase space $\check{\mathbb{X}}$ and the matrix of transition probabilities $\check{\mathbf{P}}_0$.

If $N = 1$, the phase space $\check{\mathbb{X}}$ of the limiting Markov chain $\check{\eta}_{0,n}$ is one class of communicative states.

If $N > 1$, the phase space $\check{\mathbb{X}} = \mathbb{X} \cup \bar{\mathbb{X}}$, where $\mathbb{X} = \{0, 1\}$ is a closed class of communicative states and $\bar{\mathbb{X}} = \{2, \ldots, N\}$ is a class of transient states.

First, consider the simplest case, where $N = 1$.

In this case, we can directly apply Theorem 2.1 to the Markov renewal processes $(\check{\eta}_{\varepsilon,n}, \check{\kappa}_{\varepsilon,n}, \check{\rho}_{\varepsilon,n})$.

The matrices $\check{\mathbf{P}}_\varepsilon$ and $\check{\mathbf{P}}_0$ take the following forms:

$$\check{\mathbf{P}}_\varepsilon = \begin{bmatrix} 0 & 1 \\ p_{\varepsilon,-} & p_{\varepsilon,+} \end{bmatrix} \text{ and } \check{\mathbf{P}}_0 = \begin{bmatrix} 0 & 1 \\ 1 & 0 \end{bmatrix}. \tag{7.12}$$

The condition $\mathbf{B}_1$ is obviously satisfied, i.e., the Markov chains $\check{\eta}_{\varepsilon,n}$ are asymptotically uniformly ergodic.

The stationary probabilities for the Markov chain $\check{\eta}_{\varepsilon,n}$,

$$\pi_{\varepsilon,0} = \frac{p_{\varepsilon,-}}{p_{\varepsilon,-} + 1} = \frac{\mu_{\varepsilon,-}}{2\mu_{\varepsilon,-} + \mu_{\varepsilon,+}}, \ \pi_{\varepsilon,1} = \frac{1}{p_{\varepsilon,-} + 1} = \frac{\mu_{\varepsilon,-} + \mu_{\varepsilon,+}}{2\mu_{\varepsilon,-} + \mu_{\varepsilon,+}}. \tag{7.13}$$

The condition $\mathbf{B}_{14}$ implies that,

$$\pi_{\varepsilon,0}, \ \pi_{\varepsilon,1} \longrightarrow \frac{1}{2} \text{ as } \varepsilon \longrightarrow 0. \tag{7.14}$$

In this case, the probabilities,

$$\check{p}_{\varepsilon,i} = \mathsf{P}_i\{\check{\rho}_{\varepsilon,1} = 1\} = \begin{cases} 0 & \text{for } i = 0, \\ p_{\varepsilon,+} & \text{for } i = 1. \end{cases} \tag{7.15}$$

The condition $\mathbf{B}_{14}$ and the relation (7.14) imply that $\check{p}_{\varepsilon,1} > 0$, for $\varepsilon \in (0, \varepsilon']$, and, for $i = 0, 1$,

$$\check{p}_{\varepsilon,i} \longrightarrow 0 \text{ as } \varepsilon \longrightarrow 0. \tag{7.16}$$

Thus, the condition $\mathbf{A}_1$ is satisfied.

Also, the corresponding averaged probability and the function u_ε are given by the following relations:

$$\check{p}_\varepsilon = \pi_{\varepsilon,0}\check{p}_{\varepsilon,0} + \pi_{\varepsilon,1}\check{p}_{\varepsilon,1} = \frac{p_{\varepsilon,+}}{p_{\varepsilon,-} + 1}, \tag{7.17}$$

and

$$\begin{aligned} u_\varepsilon = \check{p}_\varepsilon^{-1} &= \frac{p_{\varepsilon,-} + 1}{p_{\varepsilon,+}} \\ &= \frac{2\mu_{\varepsilon,-} + \mu_{\varepsilon,+}}{\mu_{\varepsilon,+}} \sim \frac{2}{\mu_{\varepsilon,+}}\mu_{\varepsilon,-} \text{ as } \varepsilon \longrightarrow 0. \end{aligned} \tag{7.18}$$

Also, the distribution functions,

$$\check{F}_{\varepsilon,i}(u) = \mathsf{P}_i\{\check{\kappa}_{\varepsilon,1} \leq u\} = \begin{cases} 1 - e^{-\mu_{\varepsilon,+}w_\varepsilon u} & \text{for } u \geq 0, i = 0, \\ 1 - e^{-\lambda_\varepsilon w_\varepsilon u} & \text{for } u \geq 0, i = 1 \end{cases} \tag{7.19}$$

and, thus, the averaged distribution function,

$$\begin{aligned} \check{F}_\varepsilon(u) &= \pi_{\varepsilon,0}\check{F}_{\varepsilon,0}(u) + \pi_{\varepsilon,1}\check{F}_{\varepsilon,1}(u) \\ &= \frac{p_{\varepsilon,-}}{p_{\varepsilon,-}+1}(1 - e^{-\mu_{\varepsilon,+}w_\varepsilon u}) + \frac{1}{p_{\varepsilon,-}+1}(1 - e^{-\lambda_\varepsilon w_\varepsilon u}), \ u \geq 0. \end{aligned} \tag{7.20}$$

In this case, for $u \geq 0$,

$$\mathsf{P}_i\{\check{\kappa}_{\varepsilon,1} \leq u, \check{\rho}_{\varepsilon,1} = 1\} = \begin{cases} (1 - e^{-\mu_{\varepsilon,+}w_\varepsilon u})\mathrm{I}(j = 0) & \text{for } j = 0, 1, \ i = 0, \\ (1 - e^{-\lambda_\varepsilon w_\varepsilon u})p_{\varepsilon,-} & \text{for } j = 0, \ i = 1, \\ (1 - e^{-\lambda_\varepsilon w_\varepsilon u})p_{\varepsilon,+} & \text{for } j = 1, \ i = 1. \end{cases} \tag{7.21}$$

Thus, the condition $\mathbf{C}_2$ is satisfied, and, according to Remark 2.6, the condition $\mathbf{C}_1$ reduces to the assumption that the following asymptotic relation holds, for $i = 0, 1$ and $\delta > 0$,

$$\mathsf{P}_i\{\check{\kappa}_{\varepsilon,1} > \delta\} = 1 - \check{F}_{\varepsilon,i}(\delta) \to 0 \text{ as } \varepsilon \to 0. \tag{7.22}$$

The condition $\mathbf{B}_{14}$ and the relation (7.21) imply that the relation (7.22) holds for any function $0 < w_\varepsilon \to \infty$ as $\varepsilon \to 0$.

In this case the averaged Laplace transforms,

$$\check{\phi}_\varepsilon(s) = \frac{p_{\varepsilon,-}}{p_{\varepsilon,-}+1}\frac{1}{1 + s/\mu_{\varepsilon,+}w_\varepsilon} + \frac{1}{p_{\varepsilon,-}+1}\frac{1}{1 + s/\mu_{\varepsilon,+}w_\varepsilon}, \ s \geq 0. \tag{7.23}$$

Let, as in Sect. 5.2.2, $\frac{1}{1+s} = 1 - s + o(s)$ be the asymptotic Taylor expansion of the first order for the function $\frac{1}{1+s}$ at the point 0. The relation (7.23) and the condition $\mathbf{B}_{13}$ imply that the following asymptotic relation takes place, for $s > 0$,

$$\begin{aligned} u_\varepsilon(1 - \phi_\varepsilon(s)) &= u_\varepsilon\big(\pi_{\varepsilon,0}(1 - \frac{1}{1 + s/\mu_{\varepsilon,+}w_\varepsilon}) + \pi_{\varepsilon,1}(1 - \frac{1}{1 + s/\lambda_\varepsilon w_\varepsilon})\big) \\ &= u_\varepsilon\big(\pi_{\varepsilon,0}\left(s/\mu_{\varepsilon,+}w_\varepsilon + o(s/\mu_{\varepsilon,+}w_\varepsilon)\right) \\ &\quad + \pi_{\varepsilon,1}\left(s/\lambda_\varepsilon w_\varepsilon + o(s/\lambda_\varepsilon w_\varepsilon)\right)\big) \\ &\sim u_\varepsilon\pi_{\varepsilon,0}s/\mu_{\varepsilon,+}w_\varepsilon \sim s\frac{\mu_{\varepsilon,-} + \mu_{\varepsilon,+}}{\mu^2_{\varepsilon,+}w_\varepsilon} \\ &\sim s\frac{\mu_{\varepsilon,-}}{\mu^2_{\varepsilon,+}w_\varepsilon} \text{ as } \varepsilon \to 0. \end{aligned} \tag{7.24}$$

From the relation (7.24) it follows that condition $\mathbf{D}_2$, in this case, is equivalent to the following condition:

$\mathbf{D}_{20}$: $A_\varepsilon = \frac{\pi_{\varepsilon,0}u_\varepsilon}{\mu_{\varepsilon,+}w_\varepsilon} = \frac{\mu_{\varepsilon,-}}{\mu^2_{\varepsilon,+}w_\varepsilon} \to A \in (0, \infty)$ as $\varepsilon \to 0$.

Moreover, in this case, the corresponding limiting cumulant $A(s) = As$, and, thus, the corresponding limiting distribution $G(u)$ is exponential with parameter A^{-1}.

It is also useful to note that the condition $\mathbf{D}_{20}$ is automatically satisfied if to choose the normalisation function $w_\varepsilon = \mu_{\varepsilon,-}/\mu^2_{\varepsilon,+}$. In this case, $A = 1$.

The variant of Theorem 2.1, which corresponds to the statements **(i)** and **(ii)** of this theorem, takes the following form.

Lemma 7.1 *Let $N = 1$ and the condition $\mathbf{B}_{14}$ is satisfied. Then:*

(i) *The condition $\mathbf{D}_{20}$ is necessary and sufficient for the fulfilment (for some or any initial distributions $\bar{q}_\varepsilon$, respectively, in the statements of necessity and sufficiency) of the asymptotic relation, $\xi_{\mathbf{S}_\varepsilon} \stackrel{\mathsf{d}}{\longrightarrow} \xi_0$ as $\varepsilon \to 0$, where ξ_0 is a non-negative random variable with distribution function not concentrated at zero. In this case:*

(ii) *The limiting random variable ξ_0 has an exponential distribution with the parameter A^{-1}.*

Let us now consider the case, where $N > 1$.

In this case, the condition $\mathbf{B}_1$ is not satisfied for the Markov chains $\check{\eta}_{\varepsilon,n}$.

However, we can use the results presented in Sect. 4.1.4.2.

We can choose the phases spaces,

$$\tilde{\mathbb{X}} = \check{\mathbb{X}} = \hat{\mathbb{X}} \text{ and } \mathbb{X} = \{0, 1\}. \tag{7.25}$$

As mentioned above, the condition $\mathbf{B}_{14}$ implies that there exist $\varepsilon' \in (0, 1]$ such that, for every $\varepsilon \in (0, \varepsilon']$, the phase space $\check{\mathbb{X}}$ is one class of communicative states for the Markov chain $\check{\eta}_{\varepsilon,n}$. Obviously, we can assume that $\varepsilon' = 1$.

Therefore, the condition $\mathbf{B}_6$ is satisfied and, thus, for any subset $\emptyset \neq \mathbb{D} \subseteq \check{\mathbb{X}}$ and $i \in \check{\mathbb{X}}$,

$$\mathsf{P}_i\{\check{\nu}_{\varepsilon,\mathbb{D}} < \infty\} = 1, \tag{7.26}$$

where

$$\check{\nu}_{\varepsilon,\mathbb{D}} = \min(n \geq 1 : \check{\eta}_{\varepsilon,n} \in \mathbb{D}). \tag{7.27}$$

The relation (7.26) makes it possible to define the successive times of hitting the Markov chain $\check{\eta}_{\varepsilon,n}$ in the set $\mathbb{X}$,

$$\check{\nu}_{\varepsilon,\mathbb{X},n} = \min(k > \check{\nu}_{\varepsilon,\mathbb{X},n-1}, \check{\eta}_{\varepsilon,k} \in \mathbb{X}), n = 1, 2, \ldots, \text{ where } \check{\nu}_{\varepsilon,\mathbb{X},0} = 0. \tag{7.28}$$

The corresponding Markov renewal process with transition period $(\tilde{\eta}_{\varepsilon,n}, \tilde{\kappa}_{\varepsilon,n}, \tilde{\rho}_{\varepsilon,n})$, $n = 0, 1, \ldots$ (with the phase space $\tilde{\mathbb{X}}$, for the initial state $\tilde{\eta}_{\varepsilon,0}$, and the phase space $\mathbb{X}$, for the random variables $\tilde{\eta}_{\varepsilon,n}$, $n \geq 1$) can be constructed using the following recurrent relations:

$$\begin{cases} \tilde{\eta}_{\varepsilon,n} &= \check{\eta}_{\varepsilon,\check{\nu}_{\varepsilon,\mathbb{X},n}}, \ n = 1, 2, \ldots, \ \tilde{\eta}_{\varepsilon,0} = \check{\eta}_{\varepsilon,0}, \\ \tilde{\kappa}_{\varepsilon,n} &= \sum_{k=\check{\nu}_{\varepsilon,\mathbb{X},n-1}+1}^{\check{\nu}_{\varepsilon,\mathbb{X},n}} \check{\kappa}_{\varepsilon,k}, \ n = 1, 2, \ldots, \ \tilde{\kappa}_{\varepsilon,0} = \check{\kappa}_{\varepsilon,0}, \\ \tilde{\rho}_{\varepsilon,n} &= 1 - \prod_{k=\check{\nu}_{\varepsilon,\mathbb{X},n-1}+1}^{\check{\nu}_{\varepsilon,\mathbb{X},n}} (1 - \check{\rho}_{\varepsilon,k}), \ n = 1, 2, \ldots, \ \tilde{\rho}_{\varepsilon,0} = \check{\rho}_{\varepsilon,0}. \end{cases} \tag{7.29}$$

Also, as in Sect. 4.1.3.1, the corresponding shifted Markov renewal process can be defined by the following relation:

$$(\eta_{\varepsilon,n}, \kappa_{\varepsilon,n}, \rho_{\varepsilon,n}) = (\tilde{\eta}_{\varepsilon,n+1}, \tilde{\kappa}_{\varepsilon,n+1}, \tilde{\rho}_{\varepsilon,n+1}), n = 0, 1, \ldots . \tag{7.30}$$

Obviously, the following inequality holds, for every $\varepsilon \in (0, \varepsilon']$:

$$\tilde{\xi}_{\varepsilon,-} \leq \xi_{\mathbf{S}_\varepsilon} \leq \tilde{\xi}_{\varepsilon,+}. \tag{7.31}$$

where

$$\tilde{\xi}_{\varepsilon,-} = \sum_{k=1}^{\tilde{\nu}_\varepsilon - 1} \tilde{\kappa}_{\varepsilon,k}, \ \tilde{\xi}_{\varepsilon,+} = \sum_{k=1}^{\tilde{\nu}_\varepsilon} \tilde{\kappa}_{\varepsilon,k}, \tag{7.32}$$

and

$$\tilde{\nu}_\varepsilon = \min(n \geq 1 : \tilde{\rho}_{\varepsilon,n} = 1). \tag{7.33}$$

Let us show that the conditions $\mathbf{C}_1$ (for the Markov renewal processes $(\eta_{\varepsilon,n}, \kappa_{\varepsilon,n}, \rho_{\varepsilon,n})$) and $\mathbf{A}_5$, $\mathbf{B}_6$, $\mathbf{B}_7$, $\mathbf{C}_8$ (for the Markov renewal processes $(\check{\eta}_{\varepsilon,n}, \check{\kappa}_{\varepsilon,n}, \check{\rho}_{\varepsilon,n})$) are satisfied and, thus, Lemma 4.6 can be applied.

The relation (7.9) implies that the condition $\mathbf{B}_7$ holds.

Thus, the condition $\mathbf{B}_1$ holds for the Markov chains $\eta_{\varepsilon,n}$, which are asymptotically uniformly ergodic.

The Markov chain $\eta_{\varepsilon,n}$ has the phase space $\mathbb{X} = \{0, 1\}$ and the matrix of transition probabilities,

$$\mathbf{P}_\varepsilon = \begin{bmatrix} 0 & 1 \\ p_{\varepsilon,-} & p_{\varepsilon,+} \end{bmatrix}. \tag{7.34}$$

Obviously,

$$\mathbf{P}_\varepsilon \longrightarrow \mathbf{P}_0 = \begin{bmatrix} 0 & 1 \\ 1 & 0 \end{bmatrix} \text{ as } \varepsilon \longrightarrow 0. \tag{7.35}$$

The Markov chain $\eta_{\varepsilon,n}$ has the stationary probabilities,

$$\pi_{\varepsilon,0} = \frac{p_{\varepsilon,-}}{p_{\varepsilon,-} + 1} = \frac{\mu_{\varepsilon,-}}{2\mu_{\varepsilon,-} + \mu_{\varepsilon,+}}, \ \pi_{\varepsilon,1} = \frac{1}{p_{\varepsilon,-} + 1} = \frac{\mu_{\varepsilon,-} + \mu_{\varepsilon,+}}{2\mu_{\varepsilon,-} + \mu_{\varepsilon,+}}, \tag{7.36}$$

and, thus,

$$\pi_{\varepsilon,0}, \ \pi_{\varepsilon,1} \longrightarrow \frac{1}{2} \text{ as } \varepsilon \longrightarrow 0. \tag{7.37}$$

In this case, for $i \in \check{\mathbb{X}}$,

$$\check{p}_{\varepsilon,i} = \mathsf{P}_i\{\check{\rho}_{\varepsilon,1} = 1\} = \begin{cases} 0 & \text{for } i = 0, \ldots, N-1, \\ p_{\varepsilon,+} & \text{for } i = N. \end{cases} \tag{7.38}$$

The relations (7.9) and (7.38) imply that the condition $\mathbf{A}_5$ is satisfied.

In this case, for $u \geq 0$ and $i \in \tilde{\mathbb{X}}$,

$$\mathsf{P}_i\{\check{\kappa}_{\varepsilon,1} \leq u\} = \begin{cases} 1 - e^{-\mu_{\varepsilon,+} w_\varepsilon u} & \text{for } i = 0, \\ 1 - e^{-\lambda_\varepsilon w_\varepsilon u} & \text{for } i = 1, \ldots, N. \end{cases} \tag{7.39}$$

Therefore, the condition $\mathbf{B}_{14}$ implies that the condition $\mathbf{C}_8$ is satisfied, for any function $0 \leq w_\varepsilon \to 0$ as $\varepsilon \to 0$.

Let us introduce the hitting times, for $j \in \check{\mathbb{X}}$,

$$\check{\nu}_{\varepsilon,j} = \min(n \geq 1 : \check{\eta}_{\varepsilon,n} = j), \tag{7.40}$$

and

$$\dot{\nu}_{\varepsilon,N} = \min(n \geq 1 : \check{\eta}_{\varepsilon,n-1} = N, \check{\eta}_{\varepsilon,n} = N). \tag{7.41}$$

In this case, the probabilities $p_{\varepsilon,i} = \mathsf{P}_i\{\tilde{\rho}_{\varepsilon,1} = 1\}, i \in \tilde{\mathbb{X}}$ take the following form:

$$p_{\varepsilon,i} = \mathsf{P}_i\{\tilde{\rho}_{\varepsilon,1} = 1\} = \begin{cases} 0 & \text{for } i = 0, \\ p_{\varepsilon,+}\mathsf{P}_{\varepsilon,2,N} & \text{for } i = 1, \\ \mathsf{P}_{\varepsilon,i,N} & \text{for } i = 2, \ldots \tilde{m}, \end{cases} \tag{7.42}$$

where, for $i = 2, \ldots, N$,

$$\mathsf{P}_{\varepsilon,i,N} = \mathsf{P}_i\{\dot{\nu}_{\varepsilon,N} < \check{\nu}_{\varepsilon,1}\}. \tag{7.43}$$

The condition $\mathbf{B}_6$ implies that, for $\varepsilon \in (0, \varepsilon']$, the probabilities $\mathsf{P}_{\varepsilon,iN}, i = 2, \ldots, N$ are the only positive solution for the following system of linear equations:

$$\mathsf{P}_{\varepsilon,i,N} = p_{\varepsilon,-}\mathsf{P}_{\varepsilon,i-1,N} + p_{\varepsilon,+}\mathsf{P}_{\varepsilon,i+1,N}, \ 2 \leq i \leq N, \tag{7.44}$$

where $\mathsf{P}_{\varepsilon,1,N} = 0$, $\mathsf{P}_{\varepsilon,N+1,N} = 1$.

This system can be rewritten in the equivalent form of the recurrent relations,

$$\mathsf{P}_{\varepsilon,i+1,N} - \mathsf{P}_{\varepsilon,i,N} = (\mathsf{P}_{\varepsilon,i,N} - \mathsf{P}_{\varepsilon,i-1,N})\frac{p_{\varepsilon,-}}{p_{\varepsilon,+}}, \ 2 \leq i \leq N, \tag{7.45}$$

or

$$\mathsf{P}_{\varepsilon,i+1,N} - \mathsf{P}_{\varepsilon,i,N} = \mathsf{P}_{\varepsilon,2,N}(\frac{p_{\varepsilon,-}}{p_{\varepsilon,+}})^{i-1}, \ 2 \leq i \leq N. \tag{7.46}$$

Summing from $i = 2$ to $i = N$ in the relation (7.46), we get the following relation:

$$1 - \mathsf{P}_{\varepsilon,2,N} = \mathsf{P}_{\varepsilon,2,N} \sum_{i=2}^{N} (\frac{p_{\varepsilon,-}}{p_{\varepsilon,+}})^{i-1}, \tag{7.47}$$

which gives the following explicit formula:

$$\mathsf{P}_{\varepsilon,2,N} = \frac{1 - \frac{p_{\varepsilon,-}}{p_{\varepsilon,+}}}{1 - (\frac{p_{\varepsilon,-}}{p_{\varepsilon,+}})^N} = (\frac{p_{\varepsilon,+}}{p_{\varepsilon,-}})^{N-1} \frac{1 - \frac{p_{\varepsilon,+}}{p_{\varepsilon,-}}}{1 - (\frac{p_{\varepsilon,+}}{p_{\varepsilon,-}})^N}. \tag{7.48}$$

Also, summing from $i = 2$ to $i = j - 1$ in the relation (7.46), we obtain the following formula, for $2 \leq j \leq N$:

$$\mathsf{P}_{\varepsilon,j,N} = \mathsf{P}_{\varepsilon,2,N} \cdot \frac{1 - (\frac{p_{\varepsilon,-}}{p_{\varepsilon,+}})^{j-1}}{1 - \frac{p_{\varepsilon,-}}{p_{\varepsilon,+}}} = \mathsf{P}_{\varepsilon,2,N} \cdot (\frac{p_{\varepsilon,+}}{p_{\varepsilon,-}})^{2-j} \frac{1 - (\frac{p_{\varepsilon,+}}{p_{\varepsilon,-}})^{j-1}}{1 - \frac{p_{\varepsilon,+}}{p_{\varepsilon,-}}}. \tag{7.49}$$

Therefore, the averaged rare-event probability p_ε for the Markov chain $\eta_{\varepsilon,n}$, and the function $u_\varepsilon = p_\varepsilon^{-1}$ take the following forms:

$$p_\varepsilon = \pi_{\varepsilon,0} p_{\varepsilon,0} + \pi_{\varepsilon,1} p_{\varepsilon,1} = \pi_{\varepsilon,1} p_{\varepsilon,1} = \frac{1}{1 + p_{\varepsilon,-}} \cdot p_{\varepsilon,+} (\frac{p_{\varepsilon,+}}{p_{\varepsilon,-}})^{N-1} \frac{1 - \frac{p_{\varepsilon,+}}{p_{\varepsilon,-}}}{1 - (\frac{p_{\varepsilon,+}}{p_{\varepsilon,-}})^N}, \tag{7.50}$$

and

$$u_\varepsilon = p_\varepsilon^{-1} = (\pi_{\varepsilon,1} p_{\varepsilon,1})^{-1} \sim \frac{2}{\mu_{\varepsilon,+}^N} \mu_{\varepsilon,-}^N \to \infty \text{ as } \varepsilon \to 0. \tag{7.51}$$

Consider the random variables,

$$\tilde{\kappa}_{\varepsilon,1} = \sum_{k=1}^{\check{\nu}_{\varepsilon,\mathbb{X},1}} \check{\kappa}_{\varepsilon,k} = \sum_{k=1}^{\check{\nu}_{\varepsilon,\mathbb{X},1}} \hat{\kappa}_{\varepsilon,k} / w_\varepsilon \tag{7.52}$$

and the distribution functions, for $j \in \tilde{\mathbb{X}}$,

$$\hat{F}_{\varepsilon,j}(u) = \mathsf{P}_j \{ \sum_{k=1}^{\check{\nu}_{\varepsilon,\mathbb{X},1}} \hat{\kappa}_{\varepsilon,k} \leq u \}, \ u \geq 0. \tag{7.53}$$

It is useful to note that,

$$\mathsf{P}_0 \{ \check{\nu}_{\varepsilon,\mathbb{X},1} = 1 \} = 1. \tag{7.54}$$

In this case, the distribution functions,

$$F_{\varepsilon,i}(u) = \mathsf{P}_i \{ \kappa_{\varepsilon,1} \leq u \} = \begin{cases} 1 - e^{-\mu_{\varepsilon,+} w_\varepsilon u}, u \geq 0 & \text{for } u \geq 0, i = 0, \\ \hat{F}_{\varepsilon,1}(u w_\varepsilon) & \text{for } u \geq 0, i = 1. \end{cases}$$

The distribution function $F_{\varepsilon,0}(\cdot)$ is exponential with parameter $\mu_{\varepsilon,+} w_\varepsilon$, and, thus, its two first moments are,

$$\mathsf{E}_0 \kappa_{\varepsilon,1} = \frac{1}{\mu_{\varepsilon,+} w_\varepsilon}, \ \mathsf{E}_0 \kappa_{\varepsilon,1}^2 = \frac{2}{(\mu_{\varepsilon,+} w_\varepsilon)^2}. \tag{7.55}$$

In this case, the quantities $\check{e}_{\varepsilon,i,\alpha,\beta}$ appearing in the condition $\mathbf{C}_9$ and the relation (4.109) take the following form, for $i = 1$ and $\alpha = \beta = 2$,

$$\check{e}_{\varepsilon,1,2,2} = (\mathsf{E}_1 \check{\kappa}_{\varepsilon,1}^4 \vee \max_{j \in \tilde{\mathbb{X}} \setminus \mathbb{X}} \mathsf{E}_j \check{\kappa}_{\varepsilon,1}^4)^{\frac{1}{2}} = (\frac{4!}{\lambda_\varepsilon^4 w_\varepsilon^4})^{\frac{1}{2}} = \frac{\sqrt{4!}}{\lambda_\varepsilon^2 w_\varepsilon^2} \tag{7.56}$$

and, thus, by Lemma 4.7,

$$\mathsf{E}_1 \kappa_{\varepsilon,1}^2 \leq \check{e}_{\varepsilon,1,2,2} a_{p,N,2,2} \leq \sqrt{4!} a_{p,N,2,2} \frac{1}{\lambda_\varepsilon^2 w_\varepsilon^2} \tag{7.57}$$

and

$$\mathsf{E}_1 \kappa_{\varepsilon,1} \le \sqrt{\mathsf{E}_1 \kappa^2_{\varepsilon,1}} \le \sqrt{5 a_{p,N,2,2}} \frac{1}{\lambda_\varepsilon w_\varepsilon}, \tag{7.58}$$

where the constant $a_{p,N,2,2}$ is given by relation (4.108).

Let $\hat{\theta}_{\varepsilon,n}, n = 1, 2, \dots$ be a sequence of i.i.d. random variables with the distribution function $\hat{F}_\varepsilon(\cdot) = \pi_{\varepsilon,0}\hat{F}_{\varepsilon,0}(\cdot) + \pi_{\varepsilon,1}\hat{F}_{\varepsilon,1}(\cdot)$.

In this case, $\theta^{(w)}_{\varepsilon,n} = \hat{\theta}_{\varepsilon,n}/w_\varepsilon, n = 1, 2, \dots$ is also a sequence of i.i.d. random variables with the distribution function $F_\varepsilon(\cdot) = \pi_{\varepsilon,0}F_{\varepsilon,0}(\cdot) + \pi_{\varepsilon,1}F_{\varepsilon,1}(\cdot)$.

Using relations (7.36), (7.37), (7.55), and (7.58) and taking into account that, by condition $\mathbf{B}_{13}$, $\lambda_\varepsilon \to \infty$ as $\varepsilon \to 0$, we get,

$$\begin{aligned} \mathsf{E}\theta^{(w)}_{\varepsilon,1} &= \pi_{\varepsilon,0}\mathsf{E}_0\tilde{\kappa}_{\varepsilon,1} + \pi_{\varepsilon,1}\mathsf{E}_1\tilde{\kappa}_{\varepsilon,1} \\ &= \frac{p_{\varepsilon,-}}{p_{\varepsilon,-}+1}\frac{1}{\mu_{\varepsilon,+}w_\varepsilon} + \frac{1}{p_{\varepsilon,-}+1}\mathsf{E}_1\kappa_{\varepsilon,1} = \frac{C_\varepsilon}{w_\varepsilon}, \end{aligned} \tag{7.59}$$

where

$$\begin{aligned} C_\varepsilon &= \frac{p_{\varepsilon,-}}{p_{\varepsilon,-}+1}\frac{1}{\mu_{\varepsilon,+}} + \frac{1}{p_{\varepsilon,-}+1}\mathsf{E}_1\kappa_{\varepsilon,1}w_\varepsilon \\ &\sim C_{\varepsilon,\circ} = \frac{1}{2\mu_{\varepsilon,+}} \ \text{as } \varepsilon \to 0. \end{aligned} \tag{7.60}$$

The condition $\mathbf{B}_{14}$ implies that,

$$0 < \underline{\lim}_{\varepsilon\to 0} C_{\varepsilon,\circ} \le \overline{\lim}_{\varepsilon\to 0} C_{\varepsilon\circ} < \infty. \tag{7.61}$$

Also, using relations (7.36), (7.37), and (7.57), we get,

$$\begin{aligned} \mathsf{E}(\theta^{(w)}_{\varepsilon,1})^2 &= \pi_{\varepsilon,0}\mathsf{E}_0\kappa^2_{\varepsilon,1} + \pi_{\varepsilon,1}\mathsf{E}_1\kappa^2_{\varepsilon,1} \\ &= \frac{p_{\varepsilon,-}}{p_{\varepsilon,-}+1}\frac{2}{(\mu_{\varepsilon,+}w_\varepsilon)^2} + \frac{1}{p_{\varepsilon,-}+1}\mathsf{E}_1\kappa^2_{\varepsilon,1} = \frac{B_\varepsilon}{w^2_\varepsilon}, \end{aligned} \tag{7.62}$$

where

$$\begin{aligned} B_\varepsilon &= \frac{p_{\varepsilon,-}}{p_{\varepsilon,-}+1}\frac{2}{\mu^2_{\varepsilon,+}} + \frac{1}{p_{\varepsilon,-}+1}\mathsf{E}_1\kappa_{\varepsilon,1}w^2_\varepsilon \\ &\le B_{\varepsilon,\circ} = \frac{p_{\varepsilon,-}}{p_{\varepsilon,-}+1}\frac{2}{\mu^2_{\varepsilon,+}} + \frac{1}{p_{\varepsilon,-}+1}\sqrt{4!}a_{p,N,2,2}\frac{1}{\lambda^2_\varepsilon} \end{aligned} \tag{7.63}$$

The condition $\mathbf{B}_{14}$ implies that,

$$0 < \underline{\lim}_{\varepsilon\to 0} B_{\varepsilon,\circ} \le \overline{\lim}_{\varepsilon\to 0} B_{\varepsilon,\circ} < \infty. \tag{7.64}$$

Consider the random variables,

$$\theta_\varepsilon^{(w)} = \sum_{n=1}^{[u_\varepsilon]} \theta_{\varepsilon,n}^{(w)}. \tag{7.65}$$

Let us choose the normalisation function,

$$w_{\circ,\varepsilon} = C_{\varepsilon,\circ} u_\varepsilon. \tag{7.66}$$

The relations (7.59), (7.60), and (7.61) imply that,

$$\mathsf{E}\theta_\varepsilon^{(w_\circ)} = [u_\varepsilon]\mathsf{E}\theta_{\varepsilon,1}^{(w_\circ)} = \frac{[u_\varepsilon]C_\varepsilon}{C_{\varepsilon,\circ}u_\varepsilon} \longrightarrow 1 \text{ as } \varepsilon \to 0. \tag{7.67}$$

Also, the relations (7.61), (7.62), (7.63), and (7.64) imply that,

$$\begin{aligned} \mathrm{Var}\theta_\varepsilon^{(w_\circ)} &= [u_\varepsilon]\mathrm{Var}\theta_{\varepsilon,1}^{(w_\circ)} \\ &\leq [u_\varepsilon]\mathsf{E}(\theta_{\varepsilon,1}^{(w_\circ)})^2 \leq \frac{[u_\varepsilon]B_{\varepsilon,\circ}}{C_{\varepsilon,\circ}^2 u_\varepsilon^2} \longrightarrow 0 \text{ as } \varepsilon \to 0. \end{aligned} \tag{7.68}$$

The relations (7.67) and (7.68) imply that the following asymptotic relation takes place:

$$\theta_\varepsilon^{(w_\circ)} = \sum_{n=1}^{[u_\varepsilon]} \theta_{\varepsilon,n}^{(w_\circ)} \stackrel{\mathsf{P}}{\longrightarrow} 1 \text{ as } \varepsilon \to 0. \tag{7.69}$$

Let us now consider the case, where the function $0 < w_\varepsilon \to \infty$ as $\varepsilon \to 0$ satisfies the following condition:

$\mathbf{D}_{21}$: $A_\varepsilon = \frac{C_{\varepsilon,\circ}u_\varepsilon}{w_\varepsilon} = \frac{\mu_{\varepsilon,-}^N}{\mu_{\varepsilon,+}^{N+1} w_\varepsilon} \longrightarrow A \in (0, \infty)$ as $\varepsilon \to 0$.

Under the condition $\mathbf{B}_{14}$, the condition $\mathbf{D}_{21}$ is necessary and sufficient for the fulfilment of the condition $\mathbf{D}_1$, which in this case takes the form of the following asymptotic relation,

$$\theta_\varepsilon^{(w)} = \sum_{n=1}^{[u_\varepsilon]} \theta_{\varepsilon,n}^{(w)} \Rightarrow \theta_0^{(w)} \text{ as } \varepsilon \to 0, \tag{7.70}$$

where $\theta_0^{(w)}$ is a non-negative random variable with distribution function not concentrated at zero.

Moreover, in this case the limiting random variable $\theta_0^{(w)} = A$ with probability 1.

Indeed, let the conditions $\mathbf{B}_{14}$ and $\mathbf{D}_{21}$ be satisfied.

In this case, the relation (7.69) and the asymptotic relation appearing in the condition $\mathbf{D}_{21}$ imply that,

$$\begin{aligned} \theta_\varepsilon^{(w)} &= \sum_{n=1}^{[u_\varepsilon]} \theta_{\varepsilon,n}^{(w)} = \sum_{n=1}^{[u_\varepsilon]} \hat{\theta}_{\varepsilon,n}/w_\varepsilon \\ &= \sum_{n=1}^{[u_\varepsilon]} \theta_{\varepsilon,n}^{(w_\circ)} \frac{C_{\varepsilon,\circ}u_\varepsilon}{w_\varepsilon} = \theta_\varepsilon^{(w_\circ)} \frac{C_{\varepsilon,\circ}u_\varepsilon}{w_\varepsilon} \stackrel{\mathsf{P}}{\longrightarrow} A \text{ as } \varepsilon \to 0. \end{aligned} \tag{7.71}$$

Thus, the above sufficiency statement is true.

Let us assume that the condition $\mathbf{B}_{14}$ is satisfied and the relation (7.70) holds, but the condition $\mathbf{D}_{20}$ is not satisfied.

In this case, two subsequences $0 < \varepsilon'_k, \varepsilon''_k \leq 1$ can be selected such that: (a) $\varepsilon'_k, \varepsilon''_k \to 0$ as $n \to \infty$, (b) $\frac{C_{\varepsilon'_k,\circ} u_{\varepsilon'_k}}{w_{\varepsilon'_k}} \to A' \in [0, \infty]$ as $k \to \infty$, (c) $\frac{C_{\varepsilon''_k,\circ} u_{\varepsilon''_k}}{w_{\varepsilon''_k}} \to A'' \in [0, \infty]$ as $k \to \infty$, (d) $A' \neq A''$.

In this case, the relation (7.69) implies that

$$
\begin{aligned}
\theta^{(w)}_{\varepsilon'_k} &= \sum_{n=1}^{[u_{\varepsilon'_k}]} \theta^{(w)}_{\varepsilon'_k,n} = \sum_{n=1}^{[u_{\varepsilon'_k}]} \theta^{(w_\circ)}_{\varepsilon'_k,n} \frac{C_{\varepsilon'_k,\circ} u_{\varepsilon'_k}}{w_{\varepsilon'_k}} \\
&= \theta^{(w_\circ)}_{\varepsilon'_k} \frac{C_{\varepsilon'_k,\circ} u_{\varepsilon'_k}}{w_{\varepsilon'_k}} \xrightarrow{\mathsf{d}} A' \text{ as } k \to \infty
\end{aligned} \tag{7.72}
$$

and

$$
\begin{aligned}
\theta^{(w)}_{\varepsilon''_k} &= \sum_{n=1}^{[u_{\varepsilon''_k}]} \theta^{(w)}_{\varepsilon''_k,n} = \sum_{n=1}^{[u_{\varepsilon''_k}]} \theta^{(w_\circ)}_{\varepsilon''_k,n} \frac{A_{\varepsilon''_k,\circ} u_{\varepsilon''_k}}{w_{\varepsilon''_k}} \\
&= \theta^{(w_\circ)}_{\varepsilon''_k} \frac{C_{\varepsilon''_k,\circ} u_{\varepsilon''_k}}{w_{\varepsilon''_k}} \xrightarrow{\mathsf{d}} A'' \text{ as } k \to \infty.
\end{aligned} \tag{7.73}
$$

The above two relations contradict to the relation (7.70).

Thus, the above necessity statement is also true, i.e., the condition $\mathbf{D}_{20}$ is satisfied.

In this case, according to the relation (7.71), the limiting random variable $\theta^{(w)}_0 = A$ with probability 1.

The condition $\mathbf{D}_{21}$ plays, in this case, the role of the condition $\mathbf{D}_1$.

Finally, by Lemma 4.7, the following relation takes place, for $\varepsilon \in (0, \varepsilon''']$:

$$
\begin{aligned}
\mathsf{E}_1 \kappa^2_{\varepsilon,1} &\leq \check{e}_{\varepsilon,1,2,2} a_{p,N,2,2} = \Big(\frac{\sqrt{4!}}{\lambda^2_\varepsilon w^2_\varepsilon}\Big) a_{p,N,2,2} \\
&= \Big(\frac{\sqrt{4!}}{\lambda^2_\varepsilon w_\varepsilon}\Big) a_{p,N,2,2} \frac{u_\varepsilon}{w_\varepsilon} \pi_{\varepsilon,1} p_{\varepsilon,1} = o_1(\varepsilon) p_{\varepsilon,1},
\end{aligned} \tag{7.74}
$$

where

$$
o_1(\varepsilon) = \Big(\frac{\sqrt{4!}}{\lambda^2_\varepsilon w_\varepsilon}\Big) a_{p,N,2,2} \frac{u_\varepsilon}{w_\varepsilon} \pi_{\varepsilon,1} \to 0 \text{ as } \varepsilon \to 0. \tag{7.75}
$$

The relations (7.74) and (7.75) imply that the condition $\mathbf{C}_5$ is satisfied (for $\alpha, \beta = 2$), and, thus, condition $\mathbf{C}_1$ holds for the Markov renewal processes $(\eta_{\varepsilon,n}, \kappa_{\varepsilon,n}, \rho_{\varepsilon,n})$.

Therefore, according to Lemma 4.6, all conditions of Theorem 4.4 are satisfied for the Markov renewal processes $(\check{\eta}_{\varepsilon,n}, \check{\kappa}_{\varepsilon,n}, \check{\rho}_{\varepsilon,n})$. In this case, Theorem 4.4 takes the following form.

Theorem 7.1 *Let $N > 1$ and the condition $\mathbf{B}_{14}$ is satisfied. Then*:

(i) *Condition $\mathbf{D}_{21}$ is necessary and sufficient for holding (for some or any initial distributions $\bar{q}_\varepsilon$, respectively, in statements of necessity and sufficiency) of the asymptotic relation $\xi_{\mathbf{S}_\varepsilon} \stackrel{\mathsf{d}}{\longrightarrow} \xi_0$ as $\varepsilon \to 0$, where ξ_0 is a non-negative random variable with distribution not concentrated at zero.*

(ii) *The limiting random variable ξ_0 has an exponential distribution with the parameter A^{-1}.*

7.1.2 Queuing Systems with Asymptotically Unbounded Buffers

Let now $\mathbf{S}_\varepsilon$ be, for every $\varepsilon \in (0, 1]$, a standard M/M queuing system described in Sect. 4.3.1.1, but with a buffer size N_ε, which depend on ε in such way that $N_\varepsilon \to \infty$ as $\varepsilon \to 0$.

The state of system at the moment t is represented by the random variable $\hat{\eta}_\varepsilon(t)$, which is the number of customers in the buffer at this moment. The process $\hat{\eta}_\varepsilon(t)$ is a birth–death-type semi-Markov process, which has the phase space $\hat{\mathbb{X}}_\varepsilon = \{0, \ldots, N_\varepsilon\}$ and transition probabilities,

$$\hat{Q}_{\varepsilon,ij}(t) = \hat{p}_{\varepsilon,ij}\hat{F}_{\varepsilon,i}(t),\ t \geq 0,\ i, j \in \hat{\mathbb{X}}_\varepsilon, \tag{7.76}$$

The $(N_\varepsilon + 1) \times (N_\varepsilon + 1)$ matrix of transition probabilities for the corresponding embedded Markov chain $\hat{\eta}_{\varepsilon,n}$ has the following form:

$$\mathbf{P}_\varepsilon = \|\hat{p}_{\varepsilon,ij}\| = \begin{bmatrix} 0 & 1 & 0 & 0 & \cdots & 0 \\ p_{\varepsilon,-} & 0 & p_{\varepsilon,+} & 0 & \cdots & 0 \\ \cdots & \cdots & \cdots & \cdots & \cdots & \cdots \\ 0 & \cdots & 0 & p_{\varepsilon,-} & 0 & p_{\varepsilon,+} \\ 0 & \cdots & 0 \cdots & 0 & p_{\varepsilon,-} & p_{\varepsilon,+} \end{bmatrix}, \tag{7.77}$$

where,

$$p_{\varepsilon,+} = \frac{\mu_{\varepsilon,+}}{\lambda_\varepsilon},\ p_{\varepsilon,-} = \frac{\mu_{\varepsilon,-}}{\lambda_\varepsilon},\ \lambda_\varepsilon = \mu_{\varepsilon,+} + \mu_{\varepsilon,-} \tag{7.78}$$

and the distribution functions of sojourn times have the following form:

$$\hat{F}_{\varepsilon,i}(t) = \begin{cases} 1 - e^{-\mu_{\varepsilon,+}t} & \text{for } t \geq 0,\ i = 0, \\ 1 - e^{-\lambda_\varepsilon t} & \text{for } t \geq 0,\ i = 1, \ldots, N_\varepsilon. \end{cases} \tag{7.79}$$

It is useful to note that the semi-Markov process $\hat{\eta}_\varepsilon(t)$ is, in fact, a homogeneous Markov process. However, we do prefer consider it as a semi-Markov process, since its trajectories can include jump transitions of the form $N_\varepsilon \longrightarrow N_\varepsilon$. In the standard model of continuous time Markov chain jump transitions of the form $i \longrightarrow i$ are usually excluded.

Let us consider the case, where the following condition is satisfied:

B$_{15}$: **(a)** $0 < \underline{\lim}_{\varepsilon\to 0}\, \mu_{\varepsilon,\pm} \leq \overline{\lim}_{\varepsilon\to 0}\, \mu_{\varepsilon,\pm} < \infty$, **(b)** $\overline{\lim}_{\varepsilon\to 0} \frac{\mu_{\varepsilon,+}}{\mu_{\varepsilon,-}} < 1$, **(c)** $1 \leq N_\varepsilon \to \infty$ as $\varepsilon \to 0$.

The jump transition of the form $N_\varepsilon \longrightarrow N_\varepsilon$ occurs in the above system, when a new customer tries to enter the system in the situation where the buffer is full. According to the above description this customer will be lost for the system.

Such an event can be interpreted as rare due to the condition **B**$_{15}$.

Let $0 = \hat{\zeta}_{\varepsilon,0} < \hat{\zeta}_{\varepsilon,1} < \cdots$ be sequential moments of jumps for the semi-Markov process $\hat{\eta}_\varepsilon(t)$, $\hat{\eta}_{\varepsilon,n} = \hat{\eta}(\hat{\zeta}_{\varepsilon,n}), n = 0, 1, \ldots$ be the corresponding discrete time embedded Markov chain, $\hat{\kappa}_{\varepsilon,n} = \hat{\zeta}_{\varepsilon,n} - \hat{\zeta}_{\varepsilon,n-1}, n = 1, \ldots$ be inter-jump times for the semi-Markov process $\hat{\eta}_\varepsilon(t)$, and $\hat{\nu}_\varepsilon(t) = \max(n \geq 1 : \hat{\zeta}_{\varepsilon,n} \leq t)$ be the number of jumps for the semi-Markov process $\hat{\eta}(s)$ in the time interval $[0, t]$.

Let us consider the Markov renewal process $(\check{\eta}_{\varepsilon,n}, \check{\kappa}_{\varepsilon,n}, \check{\rho}_{\varepsilon,n}), n = 0, 1, \ldots$ which has the phase space $\check{\mathbb{X}}_\varepsilon \times [0, \infty) \times \{0, 1\}$, where $\check{\mathbb{X}}_\varepsilon = \hat{\mathbb{X}}_\varepsilon$, and components,

$$\begin{cases} \check{\eta}_{\varepsilon,n} = \hat{\eta}_{\varepsilon,n}, \\ \check{\kappa}_{\varepsilon,n} = \hat{\kappa}_{\varepsilon,n}/w_\varepsilon, \\ \check{\rho}_{\varepsilon,n} = \mathrm{I}(\hat{\eta}_{\varepsilon,n-1} = N_\varepsilon, \hat{\eta}_{\varepsilon,n} = N_\varepsilon), \end{cases} \tag{7.80}$$

where $0 < w_\varepsilon \longrightarrow \infty$ as $\varepsilon \longrightarrow 0$ is some normalisation function, which we will define below.

As in Sect. 7.1.1, let us define the normalised first-rare-event time for the system $\mathbf{S}_\varepsilon$, which can also be called the first failure time, as follows:

$$\xi_{\mathbf{S}_\varepsilon} = \sum_{n=1}^{\check{\nu}_\varepsilon} \check{\kappa}_{\varepsilon,n} = \sum_{n=1}^{\hat{\nu}_\varepsilon} \hat{\kappa}_{\varepsilon,n}/w_\varepsilon, \tag{7.81}$$

where

$$\check{\nu}_\varepsilon = \min(n \geq 1 : \check{\eta}_{\varepsilon,n-1} = N_\varepsilon, \check{\eta}_{\varepsilon,n} = N_\varepsilon). \tag{7.82}$$

In this case, we can choose the phases spaces,

$$\tilde{\mathbb{X}} = \{1, \ldots, \tilde{m}\}, \text{ where } \tilde{m} \geq 2 \text{ and } \mathbb{X} = \{0, 1\}. \tag{7.83}$$

Since, $N_\varepsilon \longrightarrow \infty$ as $\varepsilon \longrightarrow 0$, it can be assumed that $N_\varepsilon \geq \tilde{m}$ and, thus, $\tilde{\mathbb{X}} \subseteq \check{\mathbb{X}}_\varepsilon$, for $\varepsilon \in (0, 1]$.

Let us assume that the initial state $\check{\eta}_{\varepsilon,0} \in \tilde{\mathbb{X}}$.

The condition **B**$_{15}$ obviously implies that,

$$0 < \underline{\lim}_{\varepsilon\to 0}\, p_{\varepsilon,+} \leq \overline{\lim}_{\varepsilon\to 0}\, p_{\varepsilon,+} < \frac{1}{2}. \tag{7.84}$$

In this case, there exist $\varepsilon' \in (0, 1]$ such that, for every $\varepsilon \in (0, \varepsilon']$, the phase space $\check{\mathbb{X}}$ is one class of communicative states for the Markov chain $\check{\eta}_{\varepsilon,n}$. Obviously, we can assume that $\varepsilon' = 1$.

In this case, the condition $\mathbf{B}_5$ holds and, thus, for $i \in \check{\mathbb{X}}_\varepsilon$,

$$\mathsf{P}_i\{\check{\nu}_{\varepsilon,\mathbb{X}} < \infty\} = 1, \tag{7.85}$$

where

$$\check{\nu}_{\varepsilon,\mathbb{X}} = \min(n \geq 1 : \check{\eta}_{\varepsilon,n} \in \mathbb{X}). \tag{7.86}$$

The relation (7.85) makes it possible to define the successive times of hitting the Markov chain $\check{\eta}_{\varepsilon,n}$ in the set $\mathbb{X}$,

$$\check{\nu}_{\varepsilon,\mathbb{X},n} = \min(k > \check{\nu}_{\varepsilon,\mathbb{X},n-1}, \check{\eta}_{\varepsilon,k} \in \mathbb{X}), n = 1, 2, \dots, \text{ where } \check{\nu}_{\varepsilon,\mathbb{X},0} = 0. \tag{7.87}$$

The corresponding Markov renewal process with transition period $(\tilde{\eta}_{\varepsilon,n}, \tilde{\kappa}_{\varepsilon,n}, \tilde{\rho}_{\varepsilon,n})$, $n = 0, 1, \dots$ (with the phase space $\check{\mathbb{X}}$, for the initial state $\tilde{\eta}_{\varepsilon,0}$, and the phase space $\mathbb{X}$, for the random variables $\tilde{\eta}_{\varepsilon,n}$, $n \geq 1$) can be constructed using the following recurrent relations:

$$\begin{cases} \tilde{\eta}_{\varepsilon,n} &= \check{\eta}_{\varepsilon,\check{\nu}_{\varepsilon,\mathbb{X},n}}, \ n = 1, 2, \dots, \ \tilde{\eta}_{\varepsilon,0} = \check{\eta}_{\varepsilon,0}, \\ \tilde{\kappa}_{\varepsilon,n} &= \sum_{k=\check{\nu}_{\varepsilon,\mathbb{X},n-1}+1}^{\check{\nu}_{\varepsilon,\mathbb{X},n}} \check{\kappa}_{\varepsilon,k}, \ n = 1, 2, \dots, \ \tilde{\kappa}_{\varepsilon,0} = \check{\kappa}_{\varepsilon,0}, \\ \tilde{\rho}_{\varepsilon,n} &= 1 - \prod_{k=\check{\nu}_{\varepsilon,\mathbb{X},n-1}+1}^{\check{\nu}_{\varepsilon,\mathbb{X},n}} (1 - \check{\rho}_{\varepsilon,k}), \ n = 1, 2, \dots, \ \tilde{\rho}_{\varepsilon,0} = \check{\rho}_{\varepsilon,0}. \end{cases} \tag{7.88}$$

Also, as in Sect. 4.1.3.1, the corresponding shifted Markov renewal process can be defined by the following relation:

$$(\eta_{\varepsilon,n}, \kappa_{\varepsilon,n}, \rho_{\varepsilon,n}) = (\tilde{\eta}_{\varepsilon,n+1}, \tilde{\kappa}_{\varepsilon,n+1}, \tilde{\rho}_{\varepsilon,n+1}), n = 0, 1, \dots. \tag{7.89}$$

Obviously, the following inequality holds, for every $\varepsilon \in (0, 1]$:

$$\tilde{\xi}_{\varepsilon,-} \leq \xi_{\mathbf{S}_\varepsilon} \leq \tilde{\xi}_{\varepsilon,+}, \tag{7.90}$$

where

$$\tilde{\xi}_{\varepsilon,-} = \sum_{k=1}^{\tilde{\nu}_\varepsilon - 1} \tilde{\kappa}_{\varepsilon,k}, \ \tilde{\xi}_{\varepsilon,+} = \sum_{k=1}^{\tilde{\nu}_\varepsilon} \tilde{\kappa}_{\varepsilon,k}, \tag{7.91}$$

and

$$\tilde{\nu}_\varepsilon = \min(n \geq 1 : \tilde{\rho}_{\varepsilon,n} = 1). \tag{7.92}$$

Let us show that conditions of Theorem 4.4 are satisfied.

The Markov chain $\eta_{\varepsilon,n}$ has the phase space $\mathbb{X} = \{0, 1\}$ and the matrix of transition probabilities,

$$\mathbf{P}_\varepsilon = \begin{bmatrix} 0 & 1 \\ p_{\varepsilon,-} & p_{\varepsilon,+} \end{bmatrix}. \tag{7.93}$$

The relation (7.84) implies that condition $\mathbf{B}_1$ holds for Markov chains $\eta_{\varepsilon,n}$, which are asymptotically uniformly ergodic.

The stationary probabilities for the Markov chain $\eta_{\varepsilon,n}$,

$$\pi_{\varepsilon,0} = \frac{p_{\varepsilon,-}}{p_{\varepsilon,-}+1} = \frac{\mu_{\varepsilon,-}}{2\mu_{\varepsilon,-}+\mu_{\varepsilon,+}},\ \pi_{\varepsilon,1} = \frac{1}{p_{\varepsilon,-}+1} = \frac{\mu_{\varepsilon,-}+\mu_{\varepsilon,+}}{2\mu_{\varepsilon,-}+\mu_{\varepsilon,+}} \tag{7.94}$$

and, thus, for $i = 0, 1$,

$$0 < \lim_{\varepsilon\to 0} \pi_{\varepsilon,i} \leq \overline{\lim_{\varepsilon\to 0}}\, \pi_{\varepsilon,i} < 1. \tag{7.95}$$

In this case,

$$\check{p}_{\varepsilon,i} = \mathsf{P}_i\{\check{\rho}_{\varepsilon,1} = 1\} = \begin{cases} 0 & \text{for } i = 0, \dots, N_\varepsilon - 1, \\ p_{\varepsilon,+} & \text{for } i = N_\varepsilon. \end{cases} \tag{7.96}$$

It is easy to see that the probabilities $p_{\varepsilon,i} = \mathsf{P}_i\{\tilde{\rho}_{\varepsilon,1} = 1\}, i \in \tilde{\mathbb{X}}$ coincide for the above queuing system and the queue system (with a bounded buffer of size N_ε), which was the subject of research in Sect. 7.1.1.

Thus, the probabilities $p_{\varepsilon,i}, i \in \tilde{\mathbb{X}}$ take the following forms:

$$p_{\varepsilon,i} = \mathsf{P}_i\{\tilde{\rho}_{\varepsilon,1} = 1\} = \begin{cases} 0 & \text{for } i = 0, \\ p_{\varepsilon,+}\mathsf{P}_{\varepsilon,2,N_\varepsilon} & \text{for } i = 1, \\ \mathsf{P}_{\varepsilon,i,N_\varepsilon} & \text{for } i = 2, \dots, \bar{m}, \end{cases} \tag{7.97}$$

where, by the relation (7.48),

$$\mathsf{P}_{\varepsilon,2,N_\varepsilon} = \frac{1 - \frac{p_{\varepsilon,-}}{p_{\varepsilon,+}}}{1 - (\frac{p_{\varepsilon,-}}{p_{\varepsilon,+}})^{N_\varepsilon}} = (\frac{p_{\varepsilon,+}}{p_{\varepsilon,-}})^{N_\varepsilon - 1} \frac{1 - \frac{p_{\varepsilon,+}}{p_{\varepsilon,-}}}{1 - (\frac{p_{\varepsilon,+}}{p_{\varepsilon,-}})^{N_\varepsilon}}, \tag{7.98}$$

and, by the relation (7.49), for $j = 2, \dots, N_\varepsilon$,

$$\mathsf{P}_{\varepsilon,j,N_\varepsilon} = \mathsf{P}_{\varepsilon,2,N_\varepsilon} \cdot \frac{1 - (\frac{p_{\varepsilon,-}}{p_{\varepsilon,+}})^{j-1}}{1 - \frac{p_{\varepsilon,-}}{p_{\varepsilon,+}}} = \mathsf{P}_{\varepsilon,2,N_\varepsilon} \cdot (\frac{p_{\varepsilon,+}}{p_{\varepsilon,-}})^{2-j} \frac{1 - (\frac{p_{\varepsilon,+}}{p_{\varepsilon,-}})^{j-1}}{1 - \frac{p_{\varepsilon,+}}{p_{\varepsilon,-}}}. \tag{7.99}$$

The condition $\mathbf{B}_{15}$ and the relations (7.84), (7.98), (7.99) imply that $p_{\varepsilon,i} > 0, i \in \tilde{\mathbb{X}}$, for $\varepsilon \in (0, \varepsilon']$ and, for every $i \in \tilde{\mathbb{X}}$,

$$p_{\varepsilon,i} \to 0 \text{ as } \varepsilon \to 0. \tag{7.100}$$

Therefore the condition $\mathbf{A}_1$ (for the Markov renewal processes $(\eta_{\varepsilon,n}, \kappa_{\varepsilon,n}, \rho_{\varepsilon,n})$ and the condition $\mathbf{A}_3$ (for the Markov renewal processes $(\tilde{\eta}_{\varepsilon,n}, \tilde{\kappa}_{\varepsilon,n}, \tilde{\rho}_{\varepsilon,n})$) are satisfied.

In this case, the averaged rare-event probability p_ε for the Markov chain $\eta_{\varepsilon,n}$ and function $u_\varepsilon = p_\varepsilon^{-1}$ take the following forms:

$$
\begin{aligned}
p_\varepsilon &= \pi_{\varepsilon,0} p_{\varepsilon,0} + \pi_{\varepsilon,1} p_{\varepsilon,1} = \pi_{\varepsilon,1} p_{\varepsilon,1} \\
&= \frac{1}{1 + p_{\varepsilon,-}} \cdot p_{\varepsilon,+} (\frac{p_{\varepsilon,+}}{p_{\varepsilon,-}})^{N_\varepsilon - 1} \frac{1 - \frac{p_{\varepsilon,+}}{p_{\varepsilon,-}}}{1 - (\frac{p_{\varepsilon,+}}{p_{\varepsilon,-}})^{N_\varepsilon}},
\end{aligned} \tag{7.101}
$$

and

$$
u_\varepsilon = p_\varepsilon^{-1} \sim \frac{1 + p_{\varepsilon,-}}{p_{\varepsilon,-} - p_{\varepsilon,+}} (\frac{p_{\varepsilon,-}}{p_{\varepsilon,+}})^{N_\varepsilon} \to \infty \text{ as } \varepsilon \to 0. \tag{7.102}
$$

Let us consider the random variables,

$$
\tilde{\kappa}_{\varepsilon,1} = \sum_{k=1}^{\check{\nu}_{\varepsilon,\mathbb{X},1}} \check{\kappa}_{\varepsilon,k}. \tag{7.103}
$$

Let us also introduce the distribution functions, for $j \in \check{\mathbb{X}}_\varepsilon$,

$$
\hat{F}_{\varepsilon,j}(u) = \mathsf{P}_j\{ \sum_{k=1}^{\check{\nu}_{\varepsilon,\mathbb{X},1}} \hat{\kappa}_{\varepsilon,k} \leq u\}, \ u \geq 0. \tag{7.104}
$$

It is useful to note that,

$$
\mathsf{P}_0\{\check{\nu}_{\varepsilon,\mathbb{X}} = 1\} = 1. \tag{7.105}
$$

In this case, the distribution functions,

$$
F_{\varepsilon,i}(u) = \mathsf{P}_i\{\tilde{\kappa}_{\varepsilon,1} \leq u\} = \begin{cases} 1 - e^{-\mu_{\varepsilon,+} w_\varepsilon u} & \text{for } u \geq 0, \ j = 0, \\ \hat{F}_{\varepsilon,i}(u w_\varepsilon) & \text{for } u \geq 0, \ i = 1, \ldots, N_\varepsilon. \end{cases} \tag{7.106}
$$

In what follows, we use the following stochastic representation:

$$
\tau_{\varepsilon,i} \stackrel{\mathsf{d}}{=} \begin{cases} \zeta_{\varepsilon,+} & \text{for } i = 0, \\ \zeta_\varepsilon + \mathrm{I}(\iota_\varepsilon = +1)\tau_{\varepsilon,i+1} & \text{for } i = 1, 2, \\ \zeta_\varepsilon + \mathrm{I}(\iota_\varepsilon = -1)\tau_{\varepsilon,i-1} + \mathrm{I}(\iota_\varepsilon = +1)\tau_{\varepsilon,i+1} & \text{for } 2 < i \leq N_\varepsilon, \end{cases} \tag{7.107}
$$

where: (a) the random variable $\tau_{\varepsilon,i}$ has the distribution function $\mathsf{P}_i\{\tilde{\kappa}_{\varepsilon,1} \leq \cdot\}$, for $i = 0, 1, \ldots, N_\varepsilon$, (b) the random variable $\tau_{\varepsilon,N_\varepsilon+1} \stackrel{\mathsf{d}}{=} \tau_{\varepsilon,N_\varepsilon}$, (c) the random variable ζ_ε is exponentially distributed with parameter $\lambda_\varepsilon w_\varepsilon$, (d) the random variable $\zeta_{\varepsilon,+}$ is exponentially distributed with parameter $\mu_{\varepsilon,+} w_\varepsilon$, (d) the random variable ι_ε takes values -1 and $+1$ with probabilities, respectively, $p_{\varepsilon,-}$ and $p_{\varepsilon,+}$, (e) the random variables ζ_ε, ι_ε, $\tau_{\varepsilon,i-1}$, and $\tau_{\varepsilon,i+1}$ are mutually independent, for every $i = 1, \ldots, N_\varepsilon$.

The condition $\mathbf{B}_{15}$ implies that there exists $\varepsilon_n \in (0, \varepsilon']$ such that, for $\varepsilon \in (0, \varepsilon_n]$, moments $\mathsf{E}^{(n)}_{\varepsilon,i,N_\varepsilon} < \infty, i = 1, \ldots, N_\varepsilon$ for $n \geq 1$ (see, for example Silvestrov (1980a) or Gyllenberg and Silvestrov (2008)).

The representation (7.107) implies that,

$$
\mathsf{E}_0 \tilde{\kappa}_{\varepsilon,1} = \frac{n!}{(\mu_{\varepsilon,+} w_\varepsilon)^n}, \ n \geq 1. \tag{7.108}
$$

The representation (7.107) also implies that the following recurrent relations hold:

$$\mathsf{E}_1 \tilde{\kappa}^n_{\varepsilon,1} = \frac{n!}{(\lambda_\varepsilon w_\varepsilon)^n} + \sum_{k=0}^{n-1} \binom{n}{k} \frac{k!}{(\lambda_\varepsilon w_\varepsilon)^k} \cdot p_{\varepsilon,+} \mathsf{E}_2 \tilde{\kappa}^{n-k}_{\varepsilon,1},\ n \geq 1. \tag{7.109}$$

Let us denote,

$$\mathsf{E}^{(n)}_{\varepsilon,i,N_\varepsilon} = \mathsf{E}_i \tilde{\kappa}^n_{\varepsilon,1},\ n \geq 1,\ 2 \leq i \leq N_\varepsilon. \tag{7.110}$$

The recurrent relations (7.109) can be rewritten in the following form:

$$\begin{aligned} \mathsf{E}^{(n)}_{\varepsilon,i,N_\varepsilon} &= \mathsf{F}^{(n)}_{\varepsilon,i,N_\varepsilon} + p_{\varepsilon,-} \mathsf{E}^{(n)}_{\varepsilon,i-1,N_\varepsilon} \\ &\quad + p_{\varepsilon,+} \mathsf{E}^{(n)}_{\varepsilon,i+1,N_\varepsilon},\ 2 \leq i \leq N_\varepsilon, n \geq 1, \end{aligned} \tag{7.111}$$

where,

$$\mathsf{E}^{(k)}_{\varepsilon,1,N_\varepsilon} = 0,\ \mathsf{E}^{(k)}_{\varepsilon,N_\varepsilon+1,N_\varepsilon} = \mathsf{E}^{(k)}_{\varepsilon,N_\varepsilon,N_\varepsilon}, n \geq 1, \tag{7.112}$$

and

$$\begin{aligned} \mathsf{F}^{(n)}_{\varepsilon,i,N_\varepsilon} &= \sum_{k=1}^{n} \binom{n}{k} \mathsf{E} \rho^k_\varepsilon (p_{\varepsilon,-} \mathsf{E}^{(n-k)}_{\varepsilon,i-1,N_\varepsilon} + p_{\varepsilon,+} \mathsf{E}^{(n-k)}_{\varepsilon,i+1,N_\varepsilon}) \\ &= \sum_{k=1}^{n} \binom{n}{k} \frac{k!}{(\lambda_\varepsilon w_\varepsilon)^k} (\mathsf{E}^{(n-k)}_{\varepsilon,i,N_\varepsilon} - \mathsf{F}^{(n-k)}_{\varepsilon,i,N_\varepsilon}),\ 2 \leq i \leq N_\varepsilon,\ n \geq 1. \end{aligned} \tag{7.113}$$

In particular, the recurrent relations (7.111) for expectations ($n = 1$) take the following form:

$$\mathsf{E}^{(1)}_{\varepsilon,i,N_\varepsilon} = \frac{1}{\lambda_\varepsilon w_\varepsilon} + p_{\varepsilon,-} \mathsf{E}^{(1)}_{\varepsilon,i-1,N_\varepsilon} + p_{\varepsilon,+} \mathsf{E}^{(1)}_{\varepsilon,i+1,N_\varepsilon},\ 2 \leq i \leq N_\varepsilon, \tag{7.114}$$

where $\mathsf{E}^{(1)}_{\varepsilon,1,N_\varepsilon} = 0,\ \mathsf{E}^{(1)}_{\varepsilon,N_\varepsilon+1,N_\varepsilon} = \mathsf{E}^{(1)}_{\varepsilon,N_\varepsilon,N_\varepsilon}$.

These recurrent relations be rewritten in the following equivalent form:

$$\begin{aligned} \mathsf{E}^{(1)}_{\varepsilon,i+1,N_\varepsilon} - \mathsf{E}^{(1)}_{\varepsilon,i,N_\varepsilon} &= -\frac{1}{\lambda_\varepsilon w_\varepsilon p_{\varepsilon,+}} \\ &\quad + (\mathsf{E}^{(1)}_{\varepsilon,i,N_\varepsilon} - \mathsf{E}^{(1)}_{\varepsilon,i-1,N_\varepsilon}) \frac{p_{\varepsilon,-}}{p_{\varepsilon,+}},\ 2 \leq i \leq N_\varepsilon, \end{aligned} \tag{7.115}$$

or

$$\begin{aligned} \mathsf{E}^{(1)}_{\varepsilon,i+1,N_\varepsilon} - \mathsf{E}^{(1)}_{\varepsilon,i,N_\varepsilon} &= -\frac{1}{\lambda_\varepsilon w_\varepsilon p_{\varepsilon,+}} \sum_{k=1}^{i-1} (\frac{p_{\varepsilon,-}}{p_{\varepsilon,+}})^{k-1} \\ &\quad + \mathsf{E}^{(1)}_{\varepsilon,2,N_\varepsilon} (\frac{p_{\varepsilon,-}}{p_{\varepsilon,+}})^{i-1},\ 2 \leq i \leq N_\varepsilon. \end{aligned} \tag{7.116}$$

Taking $i = N_\varepsilon$ in the relation (7.116), we get the following relation:

$$0 = -\frac{1}{\lambda_\varepsilon w_\varepsilon p_{\varepsilon,+}} \frac{1 - (\frac{p_{\varepsilon,-}}{p_{\varepsilon,+}})^{N_\varepsilon - 1}}{1 - \frac{p_{\varepsilon,-}}{p_{\varepsilon,+}}} + \mathsf{E}^{(1)}_{\varepsilon,2,N_\varepsilon} (\frac{p_{\varepsilon,-}}{p_{\varepsilon,+}})^{N_\varepsilon - 1}, \tag{7.117}$$

which gives the following explicit formula:

$$\begin{aligned} \mathsf{E}_2 \tilde{\kappa}_{\varepsilon,1} &= \mathsf{E}^{(1)}_{\varepsilon,2,N_\varepsilon} \\ &= \frac{1}{\lambda_\varepsilon w_\varepsilon p_{\varepsilon,+}} \frac{1 - (\frac{p_{\varepsilon,-}}{p_{\varepsilon,+}})^{N_\varepsilon - 1}}{1 - \frac{p_{\varepsilon,-}}{p_{\varepsilon,+}}} \cdot (\frac{p_{\varepsilon,+}}{p_{\varepsilon,-}})^{N_\varepsilon - 1} \\ &= \frac{1}{\lambda_\varepsilon w_\varepsilon p_{\varepsilon,+}} \frac{1 - (\frac{p_{\varepsilon,+}}{p_{\varepsilon,-}})^{N_\varepsilon - 1}}{\frac{p_{\varepsilon,-}}{p_{\varepsilon,+}} - 1}. \end{aligned} \tag{7.118}$$

In addition, summing from $i = 2$ to $i = j - 1$ in the relation (7.116), we obtain the following formula, for $2 \leq j \leq N_\varepsilon$:

$$\begin{aligned} \mathsf{E}_j \tilde{\kappa}_{\varepsilon,1} &= \mathsf{E}^{(1)}_{\varepsilon,j,N_\varepsilon} \\ &= \mathsf{E}^{(1)}_{\varepsilon,2,N_\varepsilon} + \frac{1}{\lambda_\varepsilon w_\varepsilon p_{\varepsilon,+}} \sum_{i=2}^{j-1} \frac{1 - (\frac{p_{\varepsilon,-}}{p_{\varepsilon,+}})^{i-1}}{\frac{p_{\varepsilon,-}}{p_{\varepsilon,+}} - 1} + \mathsf{E}^{(1)}_{\varepsilon,2,N_\varepsilon} \cdot \Big(\sum_{i=2}^{j-1} (\frac{p_{\varepsilon,-}}{p_{\varepsilon,+}})^{i-1} \Big) \\ &= \frac{1}{\lambda_\varepsilon w_\varepsilon p_{\varepsilon,+}} \cdot \sum_{i=1}^{j-1} \frac{(1 - (\frac{p_{\varepsilon,+}}{p_{\varepsilon,-}})^{N_\varepsilon - i})}{\frac{p_{\varepsilon,-}}{p_{\varepsilon,+}} - 1}. \end{aligned} \tag{7.119}$$

Also, the relation (7.109), taken for $n = 1$, and the relation (7.118) imply that,

$$\begin{aligned} \mathsf{E}_1 \tilde{\kappa}_{\varepsilon,1} &= \mathsf{E}_1 \kappa_{\varepsilon,1} = \frac{1}{\lambda_\varepsilon w_\varepsilon} + p_{\varepsilon,+} \mathsf{E}_2 \tilde{\kappa}_{\varepsilon,1} \\ &= \frac{1}{\lambda_\varepsilon w_\varepsilon} (1 + \frac{1 - (\frac{p_{\varepsilon,+}}{p_{\varepsilon,-}})^{N_\varepsilon - 1}}{\frac{p_{\varepsilon,-}}{p_{\varepsilon,+}} - 1}). \end{aligned} \tag{7.120}$$

The condition $\mathbf{B}_{15}$ and the relations (7.108), (7.118), and (7.119), and (7.120) imply that, for $\delta > 0$ and $j \in \tilde{\mathbb{X}}$,

$$\mathsf{P}_j \{ \tilde{\kappa}_{\varepsilon,1} > \delta \} \leq \frac{\mathsf{E}_{\varepsilon,j,N_\varepsilon}}{\delta} \to 0 \text{ as } \varepsilon \to 0, \tag{7.121}$$

for any function $0 < w_\varepsilon \to 0$ as $\varepsilon \to 0$.

The relation (7.121) implies that the condition $\mathbf{C}_6$ (for the Markov renewal processes $(\tilde{\eta}_{\varepsilon,n}, \tilde{\kappa}_{\varepsilon,n}, \tilde{\rho}_{\varepsilon,n})$) is satisfied.

The recurrent relations (7.111) for second moments (for $n = 2$) take the following form:

$$\begin{aligned} \mathsf{E}^{(2)}_{\varepsilon,i,N_\varepsilon} &= \mathsf{F}^{(2)}_{\varepsilon,i,N_\varepsilon} + p_{\varepsilon,-} \mathsf{E}^{(2)}_{\varepsilon,i-1,N_\varepsilon} \\ &\quad + p_{\varepsilon,+} \mathsf{E}^{(2)}_{\varepsilon,i+1,N_\varepsilon}, \ 2 \leq i \leq N_\varepsilon, \end{aligned} \tag{7.122}$$

where $\mathsf{E}^{(2)}_{\varepsilon,1,N_\varepsilon} = 0$, $\mathsf{E}^{(2)}_{\varepsilon,N_\varepsilon+1,N_\varepsilon} = \mathsf{E}^{(2)}_{\varepsilon,N_\varepsilon,N_\varepsilon}$ and,

$$\begin{aligned}
\mathsf{F}^{(2)}_{\varepsilon,i,N_\varepsilon} &= \frac{2}{(\lambda_\varepsilon w_\varepsilon)^2} + 2\frac{1}{\lambda_\varepsilon w_\varepsilon}(p_{\varepsilon,-}\mathsf{E}^{(n-k)}_{\varepsilon,i-1,N_\varepsilon} + p_{\varepsilon,+}\mathsf{E}^{(1)}_{\varepsilon,i+1,N_\varepsilon}) \\
&= \frac{2}{(\lambda_\varepsilon w_\varepsilon)^2} + 2\frac{1}{\lambda_\varepsilon w_\varepsilon}(\mathsf{E}^{(1)}_{\varepsilon,i,N_\varepsilon} - \frac{1}{\lambda_\varepsilon w_\varepsilon}) \\
&= \frac{2}{\lambda_\varepsilon w_\varepsilon}\mathsf{E}^{(1)}_{\varepsilon,i,N_\varepsilon}, \ 1 \leq i \leq N_\varepsilon.
\end{aligned} \tag{7.123}$$

The relations (7.122) can be rewritten in the equivalent form,

$$\begin{aligned}
\mathsf{E}^{(2)}_{\varepsilon,i+1,N_\varepsilon} - \mathsf{E}^{(2)}_{\varepsilon,i,N_\varepsilon} &= -\frac{2}{\lambda_\varepsilon w_\varepsilon p_{\varepsilon,+}}\mathsf{E}^{(1)}_{\varepsilon,i,N_\varepsilon} \\
&\quad + (\mathsf{E}^{(2)}_{\varepsilon,i,N_\varepsilon} - \mathsf{E}^{(2)}_{\varepsilon,i-1,N_\varepsilon})\frac{p_{\varepsilon,-}}{p_{\varepsilon,+}}, \ 2 \leq i \leq N_\varepsilon,
\end{aligned} \tag{7.124}$$

or

$$\begin{aligned}
\mathsf{E}^{(2)}_{\varepsilon,i+1,N_\varepsilon} - \mathsf{E}^{(2)}_{\varepsilon,i,N_\varepsilon} &= -\frac{2}{\lambda_\varepsilon w_\varepsilon p_{\varepsilon,+}}\sum_{k=1}^{i-1}\mathsf{E}^{(1)}_{\varepsilon,i-k+1,N_\varepsilon}(\frac{p_{\varepsilon,-}}{p_{\varepsilon,+}})^{k-1} \\
&\quad + \mathsf{E}^{(2)}_{\varepsilon,2,N_\varepsilon}(\frac{p_{\varepsilon,-}}{p_{\varepsilon,+}})^{i-1}, \ 2 \leq i \leq N_\varepsilon.
\end{aligned} \tag{7.125}$$

Taking $i = N_\varepsilon$ in the relation (7.125), we get the following relation:

$$\begin{aligned}
0 = &-\frac{2}{\lambda_\varepsilon w_\varepsilon p_{\varepsilon,+}}\sum_{k=1}^{N_\varepsilon-1}\mathsf{E}^{(1)}_{\varepsilon,N_\varepsilon-k+1,N_\varepsilon}(\frac{p_{\varepsilon,-}}{p_{\varepsilon,+}})^{k-1} \\
&+ \mathsf{E}^{(2)}_{\varepsilon,2,N_\varepsilon}(\frac{p_{\varepsilon,-}}{p_{\varepsilon,+}})^{N_\varepsilon-1},
\end{aligned} \tag{7.126}$$

which gives the following explicit formula:

$$\begin{aligned}
\mathsf{E}_2\tilde{\kappa}^2_{\varepsilon,1} &= \mathsf{E}^{(2)}_{\varepsilon,2,N_\varepsilon} \\
&= \frac{2}{\lambda_\varepsilon w_\varepsilon p_{\varepsilon,+}}\sum_{k=1}^{N_\varepsilon-1}\mathsf{E}^{(1)}_{\varepsilon,N_\varepsilon-k+1,N_\varepsilon}(\frac{p_{\varepsilon,-}}{p_{\varepsilon,+}})^{k-1}\cdot(\frac{p_{\varepsilon,+}}{p_{\varepsilon,-}})^{N_\varepsilon-1} \\
&= \frac{2}{\lambda_\varepsilon w_\varepsilon p_{\varepsilon,+}}\sum_{r=2}^{N_\varepsilon}\mathsf{E}^{(1)}_{\varepsilon,r,N_\varepsilon}(\frac{p_{\varepsilon,+}}{p_{\varepsilon,-}})^{r-1} \\
&= \frac{2}{(\lambda_\varepsilon w_\varepsilon p_{\varepsilon,+})^2}\sum_{r=2}^{N_\varepsilon}\sum_{i=1}^{r-1}\frac{(1-(\frac{p_{\varepsilon,+}}{p_{\varepsilon,-}})^{N_\varepsilon-i})}{\frac{p_{\varepsilon,-}}{p_{\varepsilon,+}}-1}(\frac{p_{\varepsilon,+}}{p_{\varepsilon,-}})^{r-1}.
\end{aligned} \tag{7.127}$$

Also, the relation (7.109), taken for $n = 2$, and the relations (7.118) and (7.127) imply that,

$$
\begin{aligned}
\mathsf{E}_1\tilde{\kappa}^2_{\varepsilon,1} = \mathsf{E}_1\kappa^2_{\varepsilon,1} &= \frac{2}{(\lambda_\varepsilon w_\varepsilon)^2} + 2\frac{1}{\lambda_\varepsilon w_\varepsilon} p_{\varepsilon,+}\mathsf{E}_2\tilde{\kappa}_{\varepsilon,1} + p_{\varepsilon,+}\mathsf{E}_2\tilde{\kappa}^2_{\varepsilon,1} \\
&= \frac{2}{(\lambda_\varepsilon w_\varepsilon)^2}\Big(1 + \frac{1 - (\frac{p_{\varepsilon,+}}{p_{\varepsilon,-}})^{N_\varepsilon - 1}}{\frac{p_{\varepsilon,-}}{p_{\varepsilon,+}} - 1} \\
&\quad + \frac{1}{p_{\varepsilon,+}} \sum_{r=2}^{N_\varepsilon} \sum_{i=1}^{r-1} \frac{(1 - (\frac{p_{\varepsilon,+}}{p_{\varepsilon,-}})^{N_\varepsilon - i})}{\frac{p_{\varepsilon,-}}{p_{\varepsilon,+}} - 1} \Big(\frac{p_{\varepsilon,+}}{p_{\varepsilon,-}}\Big)^{r-1}.
\end{aligned} \tag{7.128}
$$

The relation (7.128) implies that,

$$
\mathsf{E}_1\tilde{\kappa}^2_{\varepsilon,1} \leq \frac{B_\varepsilon}{w_\varepsilon^2}. \tag{7.129}
$$

where

$$
B_\varepsilon = \frac{2}{\lambda_\varepsilon^2}\Big(1 + \frac{1}{\frac{p_{\varepsilon,-}}{p_{\varepsilon,+}} - 1} + \frac{1}{p_{\varepsilon,+}} \sum_{r=2}^{N_\varepsilon} \frac{(r-1)(\frac{p_{\varepsilon,+}}{p_{\varepsilon,-}})^r}{\frac{p_{\varepsilon,-}}{p_{\varepsilon,+}} - 1}\Big). \tag{7.130}
$$

The condition $\mathbf{B}_{15}$ obviously implies that,

$$
0 < \varliminf_{\varepsilon\to 0} B_\varepsilon \leq \varlimsup_{\varepsilon\to 0} B_\varepsilon < \infty. \tag{7.131}
$$

Let $\hat{\theta}_{\varepsilon,n}, n = 1, 2, \ldots$ be, for $\varepsilon \in (0, \varepsilon_2]$, i.i.d. random variables with the distribution function $\hat{F}_{\varepsilon,\mathbb{X}}(\cdot) = \pi_{\varepsilon,0}\hat{F}_{\varepsilon,0,\mathbb{X}}(\cdot) + \pi_{\varepsilon,1}\hat{F}_{\varepsilon,1,\mathbb{X}}(\cdot)$.

Consider also the sequence of i.i.d. random variables $\theta^{(w)}_{\varepsilon,n} = \hat{\theta}_{\varepsilon,n}/w_\varepsilon, n = 1, 2, \ldots$. Obviously, these random variables have the distribution function $F_\varepsilon(\cdot) = \pi_{\varepsilon,0}\hat{F}_{\varepsilon,0,\mathbb{X}}(\cdot w_\varepsilon) + \pi_{\varepsilon,1}\hat{F}_{\varepsilon,1,\mathbb{X}}(\cdot w_\varepsilon) = \pi_{\varepsilon,0}F_{\varepsilon,0}(\cdot) + \pi_{\varepsilon,1}F_{\varepsilon,1}(\cdot)$.

The relations (7.94), (7.109), and (7.120) imply that,

$$
\mathsf{E}\theta^{(w)}_{\varepsilon,1} = \pi_{\varepsilon,0}\mathsf{E}_0\tilde{\kappa}_{\varepsilon,1} + \pi_{\varepsilon,1}\mathsf{E}_1\tilde{\kappa}_{\varepsilon,1} = \frac{C_{\varepsilon,\circ}}{w_\varepsilon}, \tag{7.132}
$$

where

$$
C_{\varepsilon,\circ} = \frac{p_{\varepsilon,-}}{p_{\varepsilon,-} + 1}\frac{1}{\mu_{\varepsilon,+}} + \frac{1}{p_{\varepsilon,-} + 1}\frac{1}{\lambda_\varepsilon}\Big(1 + \frac{1 - (\frac{p_{\varepsilon,+}}{p_{\varepsilon,-}})^{N_\varepsilon - 1}}{\frac{p_{\varepsilon,-}}{p_{\varepsilon,+}} - 1}\Big). \tag{7.133}
$$

The condition $\mathbf{B}_{15}$ implies that,

$$
0 < \varliminf_{\varepsilon\to 0} C_{\varepsilon,\circ} \leq \varlimsup_{\varepsilon\to 0} C_{\varepsilon,\circ} < \infty. \tag{7.134}
$$

Also, the relations (7.94), (7.109), and (7.128) imply that,

$$
\mathsf{E}(\theta^{(w)}_{\varepsilon,1})^2 = \pi_{\varepsilon,0}\mathsf{E}_0\tilde{\kappa}^2_{\varepsilon,1} + \pi_{\varepsilon,1}\mathsf{E}_0\tilde{\kappa}^2_{\varepsilon,1} \leq \frac{B_{\varepsilon,\circ}}{w_\varepsilon^2}, \tag{7.135}
$$

where

$$B_{\varepsilon,\circ} = \frac{p_{\varepsilon,-}}{p_{\varepsilon,-}+1}\frac{2}{\mu^2_{\varepsilon,+}} + \frac{1}{p_{\varepsilon,-}+1}B_\varepsilon. \tag{7.136}$$

The relation (7.131) implies that,

$$0 < \varliminf_{\varepsilon\to 0} B_{\varepsilon,\circ} \leq \varlimsup_{\varepsilon\to 0} B_{\varepsilon,\circ} < \infty. \tag{7.137}$$

Let us consider the random variables,

$$\theta^{(w)}_\varepsilon = \sum_{n=1}^{[u_\varepsilon]} \theta^{(w)}_{\varepsilon,n} = \sum_{n=1}^{[u_\varepsilon]} \hat{\theta}_{\varepsilon,n}/w_\varepsilon. \tag{7.138}$$

First, we choose the normalisation function,

$$w_{\circ,\varepsilon} = C_{\varepsilon,\circ}u_\varepsilon. \tag{7.139}$$

The relations (7.132), (7.133), and (7.134) imply that,

$$\mathsf{E}\theta^{(w_\circ)}_\varepsilon = [u_\varepsilon]\mathsf{E}\theta^{(w_\circ)}_{\varepsilon,1} = [u_\varepsilon]\frac{C_{\varepsilon,\circ}}{C_{\varepsilon,\circ}u_\varepsilon} \longrightarrow 1 \text{ as } \varepsilon \to 0. \tag{7.140}$$

Also, the relations (7.134), (7.135), (7.136), and (7.137) imply that,

$$\begin{aligned} \mathrm{Var}\theta^{(w_\circ)}_\varepsilon &= [u_\varepsilon]\mathrm{Var}\theta^{(w_\circ)}_{\varepsilon,1} \\ &\leq [u_\varepsilon]\mathsf{E}(\theta^{(w_\circ)}_{\varepsilon,1})^2 \leq [u_\varepsilon]\frac{B_{\varepsilon,\circ}}{C^2_{\varepsilon,\circ}u^2_\varepsilon} \longrightarrow 0 \text{ as } \varepsilon \to 0. \end{aligned} \tag{7.141}$$

The relations (7.140) and (7.141) obviously imply that the following asymptotic relation takes place:

$$\theta^{(w_\circ)}_\varepsilon = \sum_{n=1}^{[u_\varepsilon]} \theta^{(w_\circ)}_{\varepsilon,n} \xrightarrow{\mathsf{P}} 1 \text{ as } \varepsilon \to 0. \tag{7.142}$$

Let us now consider the case, where the function $0 < w_\varepsilon \longrightarrow \infty$ as $\varepsilon \to 0$ satisfies the following condition:

$\mathbf{D}_{22}$: $A_\varepsilon = u_\varepsilon\mathsf{E}\theta_{\varepsilon,1} = C_{\varepsilon,\circ}\frac{u_\varepsilon}{w_\varepsilon} \longrightarrow A \in (0,\infty)$ as $\varepsilon \to 0$.

Under the condition $\mathbf{B}_{15}$, the condition $\mathbf{D}_{21}$ is necessary and sufficient for the fulfilment of the condition $\mathbf{D}_1$, which in this case takes the form of asymptotic relation,

$$\theta^{(w)}_\varepsilon = \sum_{n=1}^{[u_\varepsilon]} \theta^{(w)}_{\varepsilon,n} \Rightarrow \theta^{(w)}_0 \text{ as } \varepsilon \to 0, \tag{7.143}$$

where $\theta^{(w)}_0$ is a non-negative random variable with distribution function not concentrated at zero.

Moreover, in this case, the limiting random variable $\theta_0^{(w)} = A$ with probability 1. Indeed, let the conditions $\mathbf{B}_{15}$ and $\mathbf{D}_{22}$ be satisfied.

In this case, relations (7.142) and the asymptotic relation appearing in the condition $\mathbf{D}_{22}$ imply that,

$$\begin{aligned}\theta_\varepsilon^{(w)} &= \sum_{n=1}^{[u_\varepsilon]} \theta_{\varepsilon,n}^{(w)} = \sum_{n=1}^{[u_\varepsilon]} \theta_{\varepsilon,n}^{(w_\circ)} \frac{C_{\varepsilon,\circ} u_\varepsilon}{w_\varepsilon} \\ &= \theta_\varepsilon^{(w_\circ)} \frac{C_{\varepsilon,\circ} u_\varepsilon}{w_\varepsilon} \xrightarrow{\mathsf{P}} A \text{ as } \varepsilon \to 0. \end{aligned} \tag{7.144}$$

Thus, the above sufficiency statement is true.

Let us assume that the condition $\mathbf{B}_{15}$ is satisfied and the relation (7.143) holds, but the condition $\mathbf{D}_{22}$ is not satisfied.

In this case, two subsequences $0 < \varepsilon'_k, \varepsilon''_k \le 1$ can be selected such that: (a) $\varepsilon'_k, \varepsilon''_k \to 0$ as $n \to \infty$, (b) $\frac{A_{\varepsilon'_k} u_{\varepsilon'_k}}{w_{\varepsilon'_k}} \to A' \in [0, \infty]$ as $k \to \infty$, (c) $\frac{A_{\varepsilon''_k} u_{\varepsilon''_k}}{w_{\varepsilon''_k}} \to A'' \in [0, \infty]$ as $k \to \infty$, (d) $A' \ne A''$.

In this case, relation (7.142) implies that

$$\begin{aligned}\theta_{\varepsilon'_k}^{(w)} &= \sum_{n=1}^{[u_{\varepsilon'_k}]} \theta_{\varepsilon'_k,n}^{(w)} = \sum_{n=1}^{[u_{\varepsilon'_k}]} \theta_{\varepsilon'_k,n}^{(w_\circ)} \frac{C_{\varepsilon'_k,\circ} u_{\varepsilon'_k}}{w_{\varepsilon'_k}} \\ &= \theta_{\varepsilon'_k}^{(w_\circ)} \frac{C_{\varepsilon'_k,\circ} u_{\varepsilon'_k}}{w_{\varepsilon'_k}} \xrightarrow{\mathsf{d}} A' \text{ as } k \to \infty \end{aligned} \tag{7.145}$$

and

$$\begin{aligned}\theta_{\varepsilon''_k}^{(w)} &= \sum_{n=1}^{[u_{\varepsilon''_k}]} \theta_{\varepsilon''_k,n}^{(w)} = \sum_{n=1}^{[u_{\varepsilon''_k}]} \theta_{\varepsilon''_k,n}^{(w_\circ)} \frac{C_{\varepsilon''_k,\circ} u_{\varepsilon''_k}}{w_{\varepsilon''_k}} \\ &= \theta_{\varepsilon''_k}^{(w_\circ)} \frac{C_{\varepsilon''_k,\circ} u_{\varepsilon''_k}}{w_{\varepsilon''_k}} \xrightarrow{\mathsf{d}} A'' \text{ as } k \to \infty. \end{aligned} \tag{7.146}$$

The above two relations contradict to the relation noplutaop.

Thus, the above necessity statement is also true, i.e., the condition $\mathbf{D}_{22}$ is satisfied.

In this case, according to the relation (7.71), the limiting random variable $\theta_0^{(w)} = A$ with probability 1.

The condition $\mathbf{D}_{22}$ plays, in this case, the role of the condition $\mathbf{D}_1$.

Finally, the relations (7.129)–(7.131) imply, by Lemma 4.7, that the following relation takes place, for $\varepsilon \in (0, \varepsilon''']$:

$$\mathsf{E}_1 \kappa_{\varepsilon,1}^2 \le \frac{B_{\varepsilon,1}}{w_\varepsilon^2} = \frac{B_{\varepsilon,1}}{w_\varepsilon} \frac{u_\varepsilon}{w_\varepsilon} \pi_{\varepsilon,1} p_{\varepsilon,1} = o_1(\varepsilon) p_{\varepsilon,1}, \tag{7.147}$$

where

$$o_1(\varepsilon) = \frac{B_{\varepsilon,1}}{w_\varepsilon}\frac{u_\varepsilon}{w_\varepsilon}\pi_{\varepsilon,1} \to 0 \text{ as } \varepsilon \to 0. \tag{7.148}$$

The relation (7.147) implies that condition $\mathbf{C}_5$ is satisfied (for $\alpha, \beta = 2$), and, thus, the condition $\mathbf{C}_1$ is satisfied for the Markov renewal processes $(\eta_{\varepsilon,n}, \kappa_{\varepsilon,n}, \rho_{\varepsilon,n})$.

Therefore, all conditions of Theorem 4.4 hold for Markov renewal process $(\check{\eta}_{\varepsilon,n}, \check{\kappa}_{\varepsilon,n}, \check{\rho}_{\varepsilon,n})$. In this case, Theorem 4.4 takes the following form.

Theorem 7.2 *Let the condition* $\mathbf{B}_{15}$ *be satisfied. Then*:

(i) *The condition* $\mathbf{D}_{22}$ *is necessary and sufficient for the fulfilment (for some or any initial distributions* $\bar{q}_\varepsilon$ *concentrated on the space* $\tilde{\mathbb{X}}$*, respectively, in statements of necessity and sufficiency) of the asymptotic relation* $\xi_{\mathbf{S}_\varepsilon} \xrightarrow{\mathsf{d}} \xi_0$ *as* $\varepsilon \to 0$*, where* ξ_0 *is a non-negative random variable with distribution not concentrated in zero.*

(ii) *The limiting random variable* ξ_0 *has the exponential distribution with the parameter* A^{-1}.

It is useful to note that the role of $\tilde{\mathbb{X}}$ can be played by any finite subset $\tilde{\mathbb{X}} = \{1, \ldots, \tilde{m}\}$ of the space $\check{\mathbb{X}}_\varepsilon$.

7.2 Rare Events for Queuing Systems with Unbounded Buffers

In this section, we give necessary and sufficient conditions of convergence in distribution for the first failure event times for perturbed M/M queuing systems with an unbounded (infinite) queue buffer.

Let $\mathbf{S}_\varepsilon$ be, for every $\varepsilon \in (0, 1]$, a standard M/M queuing system with an unbounded queue buffer. This system has one server and an unbounded queue buffer (including the place in the server). It is assumed that: **(a)** the input flow of customers is a standard Poisson flow with parameter $\mu_{\varepsilon,+} > 0$; **(b)** customers are placed in the queue buffer according to the order they arrive in the system; **(c)** a new customer coming in the system is immediately served if the queue buffer is empty, otherwise it is placed in the queue buffer; **(d)** the service times for different customers are independent random variables, which have an exponential distribution with parameter $\mu_{\varepsilon,-} > 0$; **(e)** the input flow and all service processes are mutually independent.

The state of system at the moment t is represented by the random variable $\hat{\eta}_\varepsilon(t)$, which is the number of customers in the buffer at this instant. The process $\hat{\eta}_\varepsilon(t)$ is a birth–death-type continuous Markov chain. It has the phase space $\hat{\mathbb{X}} = \{0, 1, \ldots\}$ and transition probabilities of the corresponding embedded Markov chain,

$$\hat{p}_{\varepsilon,ij} = \begin{cases} 1 & \text{for } i = 0, j = 1, \\ p_{\varepsilon,\pm} & \text{for } i > 1, j = i + \frac{1 \pm 1}{2}, \\ 0 & \text{otherwise,} \end{cases} \tag{7.149}$$

where,

$$p_{\varepsilon,+} = \frac{\mu_{\varepsilon,+}}{\lambda_\varepsilon}, \ p_{\varepsilon,-} = \frac{\mu_{\varepsilon,-}}{\lambda_\varepsilon}, \ \lambda_\varepsilon = \mu_{\varepsilon,+} + \mu_{\varepsilon,-} \tag{7.150}$$

and distribution functions of sojourn times,

$$\hat{F}_{\varepsilon,i}(t) = \begin{cases} 1 - e^{-\mu_{\varepsilon,+}t} & \text{for } t \geq 0, \ i = 0, \\ 1 - e^{-\lambda_\varepsilon t} & \text{for } t \geq 0, \ i \geq 1. \end{cases} \tag{7.151}$$

We consider two cases: subcritical and critical.

In the subcritical case, the following condition is assumed to be satisfied:

B$_{16}$: **(a)** $0 < \underline{\lim}_{\varepsilon \to 0} \mu_{\varepsilon,\pm} \leq \overline{\lim}_{\varepsilon \to 0} \mu_{\varepsilon,\pm} < \infty$, **(b)** $\overline{\lim}_{\varepsilon \to 0} \frac{\mu_{\varepsilon,+}}{\mu_{\varepsilon,-}} < 1$.

In the critical case, the following condition is assumed to be satisfied:

B$_{17}$: **(a)** $0 < \underline{\lim}_{\varepsilon \to 0} \mu_{\varepsilon,\pm} \leq \overline{\lim}_{\varepsilon \to 0} \mu_{\varepsilon,\pm} < \infty$, **(b)** $\frac{\mu_{\varepsilon,+}}{\mu_{\varepsilon,-}} < 1$, for $\varepsilon \in (0, 1]$ and $\frac{\mu_{\varepsilon,+}}{\mu_{\varepsilon,-}} \to 1$ as $\varepsilon \to 0$.

Let N_ε be an integer threshold parameter such that,

$$1 \leq N_\varepsilon \to \infty \text{ as } \varepsilon \to 0. \tag{7.152}$$

The event $\{\eta_{\varepsilon,n} \geq N_\varepsilon + 1\}$ occurs in the above system, when the number of customers in the system exceeds the threshold N_ε.

Such event can be interpreted as a rare one, due to condition **B**$_{16}$, or **B**$_{17}$, and the assumption (7.152).

Let $0 = \hat{\tau}_{\varepsilon,0} < \hat{\tau}_{\varepsilon,1} < \cdots$ be successive moments of jumps for the semi-Markov process $\hat{\eta}_\varepsilon(t)$, $\hat{\eta}_{\varepsilon,n} = \hat{\eta}(\hat{\tau}_{\varepsilon,n})$, $n = 0, 1, \ldots$ be the corresponding discrete time embedded Markov chain, $\hat{\kappa}_{\varepsilon,n} = \hat{\tau}_{\varepsilon,n} - \hat{\tau}_{\varepsilon,n-1}$, $n = 1, \ldots$ be the inter-jump times for the semi-Markov process $\hat{\eta}_\varepsilon(t)$, and $\hat{\nu}_\varepsilon(t) = \max(n \geq 1 : \hat{\tau}_{\varepsilon,n} \leq t)$ be the number of jumps for the Markov chain $\hat{\eta}(s)$ in the time interval $[0, t]$.

Let us consider the Markov renewal process $(\check{\eta}_{\varepsilon,n}, \check{\kappa}_{\varepsilon,n}, \check{\rho}_{\varepsilon,n})$, $n = 0, 1, \ldots$ which has the phase space $\check{\mathbb{X}} \times [0, \infty) \times \{0, 1\}$, where $\check{\mathbb{X}} = \hat{\mathbb{X}}$ and components,

$$\begin{cases} \check{\eta}_{\varepsilon,n} = \hat{\eta}_{\varepsilon,n}, \\ \check{\kappa}_{\varepsilon,n} = \hat{\kappa}_{\varepsilon,n}/w_\varepsilon, \\ \check{\rho}_{\varepsilon,n} = \mathrm{I}(\hat{\eta}_{\varepsilon,n} \geq N_\varepsilon + 1), \end{cases} \tag{7.153}$$

where $0 < w_\varepsilon \to \infty$ as $\varepsilon \to 0$ is some normalisation function, which we will define below.

As in Sect. 7.1.2, let us define the normalised first-rare-event time for the system $\mathbf{S}_\varepsilon$, which can also be called the first failure time, as follows:

$$\xi_{\mathbf{S}_\varepsilon} = \sum_{n=1}^{\check{\nu}_\varepsilon} \check{\kappa}_{\varepsilon,n} = \sum_{n=1}^{\hat{\nu}_\varepsilon} \hat{\kappa}_{\varepsilon,n}/w_\varepsilon, \tag{7.154}$$

where

$$\check{\nu}_\varepsilon = \min(n \geq 1 : \check{\eta}_{\varepsilon,n} \geq N_\varepsilon + 1). \tag{7.155}$$

In this case, we choose the phases spaces,

$$\breve{\mathbb{X}} = \{1, \ldots, \tilde{m}\} \subset \breve{\mathbb{X}}, \text{ where } \tilde{m} \geq 2 \text{ and } \mathbb{X} = \{0, 1\}. \tag{7.156}$$

Since, $N_\varepsilon \to \infty$ as $\varepsilon \to 0$, it can be assumed that $N_\varepsilon \geq \tilde{m}$, for $\varepsilon \in (0, 1]$.
Let us assume that the initial state $\breve{\eta}_{\varepsilon,0} \in \breve{\mathbb{X}}$.
The condition $\mathbf{B}_{16}$ implies that,

$$0 < \varliminf_{\varepsilon \to 0} p_{\varepsilon,+} \leq \varlimsup_{\varepsilon \to 0} p_{\varepsilon,+} < \frac{1}{2}, \tag{7.157}$$

while the condition $\mathbf{B}_{17}$ implies that,

$$p_{\varepsilon,+} < \frac{1}{2}, \ \varepsilon \in (0, 1] \text{ and } \lim_{\varepsilon \to 0} p_{\varepsilon,+} = \frac{1}{2}. \tag{7.158}$$

Therefore, there exist $\varepsilon' \in (0, 1]$ such that, for every $\varepsilon \in (0, \varepsilon']$, the phase space $\breve{\mathbb{X}}$ is one class of communicative states for the Markov chain $\breve{\eta}_{\varepsilon,n}$. Obviously, we can assume that $\varepsilon' = 1$.

In this case, the condition $\mathbf{B}_5$ is satisfied and, thus, for $i \in \breve{\mathbb{X}}$,

$$\mathsf{P}_i\{\breve{\nu}_{\varepsilon,\mathbb{X}} < \infty\} = 1, \tag{7.159}$$

where

$$\breve{\nu}_{\varepsilon,\mathbb{X}} = \min(n \geq 1 : \breve{\eta}_{\varepsilon,n} \in \mathbb{X}). \tag{7.160}$$

The relation (7.159) makes it possible to define the successive times of hitting the Markov chain $\breve{\eta}_{\varepsilon,n}$ in the set $\mathbb{X}$,

$$\breve{\nu}_{\varepsilon,\mathbb{X},n} = \min(k > \breve{\nu}_{\varepsilon,\mathbb{X},n-1}, \breve{\eta}_{\varepsilon,k} \in \mathbb{X}), n = 1, 2, \ldots, \text{ where } \breve{\nu}_{\varepsilon,\mathbb{X},0} = 0. \tag{7.161}$$

The corresponding Markov renewal process with transition period $(\tilde{\eta}_{\varepsilon,n}, \tilde{\kappa}_{\varepsilon,n}, \tilde{\rho}_{\varepsilon,n}), n = 0, 1, \ldots$ (with the phase space $\breve{\mathbb{X}}$, for the initial state $\tilde{\eta}_{\varepsilon,0}$, and the phase space $\mathbb{X}$, for the random variables $\tilde{\eta}_{\varepsilon,n}, n \geq 1$) can be constructed using the following recurrent relations:

$$\begin{cases} \tilde{\eta}_{\varepsilon,n} = \breve{\eta}_{\varepsilon,\breve{\nu}_{\varepsilon,\mathbb{X},n}}, \ n = 1, 2, \ldots, \ \tilde{\eta}_{\varepsilon,0} = \breve{\eta}_{\varepsilon,0}, \\ \tilde{\kappa}_{\varepsilon,n} = \sum_{k=\breve{\nu}_{\varepsilon,\mathbb{X},n-1}+1}^{\breve{\nu}_{\varepsilon,\mathbb{X},n}} \breve{\kappa}_{\varepsilon,k}, \ n = 1, 2, \ldots, \ \tilde{\kappa}_{\varepsilon,0} = \breve{\kappa}_{\varepsilon,0}, \\ \tilde{\rho}_{\varepsilon,n} = 1 - \prod_{k=\breve{\nu}_{\varepsilon,\mathbb{X},n-1}+1}^{\breve{\nu}_{\varepsilon,\mathbb{X},n}} (1 - \breve{\rho}_{\varepsilon,k}), \ n = 1, 2, \ldots, \ \tilde{\rho}_{\varepsilon,0} = \breve{\rho}_{\varepsilon,0}. \end{cases} \tag{7.162}$$

Also, as in Sect. 4.1.3.1, the corresponding shifted Markov renewal process can be defined by the following relation:

$$(\eta_{\varepsilon,n}, \kappa_{\varepsilon,n}, \rho_{\varepsilon,n}) = (\tilde{\eta}_{\varepsilon,n+1}, \tilde{\kappa}_{\varepsilon,n+1}, \tilde{\rho}_{\varepsilon,n+1}), n = 0, 1, \ldots. \tag{7.163}$$

Obviously, the following inequality holds, for every $\varepsilon \in (0, 1]$:

$$\tilde{\xi}_{\varepsilon,-} \leq \xi_{\mathbf{S}_\varepsilon} \leq \tilde{\xi}_{\varepsilon,+}, \tag{7.164}$$

where

$$\tilde{\xi}_{\varepsilon,-} = \sum_{k=1}^{\tilde{\nu}_\varepsilon - 1} \tilde{\kappa}_{\varepsilon,k}, \ \tilde{\xi}_{\varepsilon,+} = \sum_{k=1}^{\tilde{\nu}_\varepsilon} \tilde{\kappa}_{\varepsilon,k}, \tag{7.165}$$

and

$$\tilde{\nu}_\varepsilon = \min(n \geq 1 : \tilde{\rho}_{\varepsilon,n} = 1). \tag{7.166}$$

Let us show that conditions of Theorem 4.4 hold.

In the subcritical case, the subsequent analysis and results are partially similar to those presented in Sect. 7.1.2, but in the critical case, the results are different.

The Markov chain $\eta_{\varepsilon,n}$ has the phase space $\mathbb{X} = \{0, 1\}$ and the matrix of transition probabilities,

$$\mathbf{P}_\varepsilon = \begin{bmatrix} 0 & 1 \\ p_{\varepsilon,-} & p_{\varepsilon,+} \end{bmatrix}. \tag{7.167}$$

The relation (7.157) implies that the condition $\mathbf{B}_1$ is satisfied for Markov chains $\eta_{\varepsilon,n}$, which are asymptotically uniformly ergodic.

The stationary probabilities for the Markov chain $\eta_{\varepsilon,n}$,

$$\pi_{\varepsilon,0} = \frac{p_{\varepsilon,-}}{p_{\varepsilon,-} + 1} = \frac{\mu_{\varepsilon,-}}{2\mu_{\varepsilon,-} + \mu_{\varepsilon,+}}, \ \pi_{\varepsilon,1} = \frac{1}{p_{\varepsilon,-} + 1} = \frac{\mu_{\varepsilon,-} + \mu_{\varepsilon,+}}{2\mu_{\varepsilon,-} + \mu_{\varepsilon,+}} \tag{7.168}$$

and, thus, for $i = 0, 1$,

$$0 < \varliminf_{\varepsilon \to 0} \pi_{\varepsilon,i} \leq \varlimsup_{\varepsilon \to 0} \pi_{\varepsilon,i} < 1. \tag{7.169}$$

In this case,

$$\check{p}_{\varepsilon,i} = \mathsf{P}_i\{\check{\rho}_{\varepsilon,1} = 1\} = \begin{cases} 0 & \text{for } i = 0, \dots, N_\varepsilon - 1, \\ p_{\varepsilon,+} & \text{for } i = N_\varepsilon, N_\varepsilon + 1, \\ 1 & \text{for } i > N_\varepsilon + 1, \end{cases} \tag{7.170}$$

and, for $u \geq 0$,

$$\mathsf{P}_i\{\check{\kappa}_{\varepsilon,1} \leq u\} = \begin{cases} 1 - e^{-\mu_{\varepsilon,+} w_\varepsilon u} & \text{for } i = 0, \\ 1 - e^{-\lambda_\varepsilon w_\varepsilon u} & \text{for } i \geq 1. \end{cases} \tag{7.171}$$

It is easy to see that the probabilities $p_{\varepsilon,i} = \mathsf{P}_i\{\tilde{\rho}_{\varepsilon,1} = 1\}, i \in \tilde{\mathbb{X}}$ coincides for the above model of M/M queuing system and the model of M/M queuing system (with bounded buffer of size N_ε), which was the subject of research in Sect. 7.1.2.

Therefore, according to the relations (7.172), (7.98), and (7.99) the probabilities $p_{\varepsilon,i} = \mathsf{P}_i\{\tilde{\rho}_{\varepsilon,1} = 1\}, i \in \tilde{\mathbb{X}}$ take the following form:

$$p_{\varepsilon,i} = \mathsf{P}_i\{\tilde{\rho}_{\varepsilon,1} = 1\} = \begin{cases} 0 & \text{for } i = 0, \\ p_{\varepsilon,+}\mathsf{P}_{\varepsilon,2,N_\varepsilon} & \text{for } i = 1, \\ \mathsf{P}_{\varepsilon,i,N_\varepsilon} & \text{for } i = 2, \dots, \tilde{m}, \end{cases} \tag{7.172}$$

where,

$$\mathsf{P}_{\varepsilon,2,N_\varepsilon} = \frac{1 - \frac{p_{\varepsilon,-}}{p_{\varepsilon,+}}}{1 - (\frac{p_{\varepsilon,-}}{p_{\varepsilon,+}})^{N_\varepsilon}} = (\frac{p_{\varepsilon,+}}{p_{\varepsilon,-}})^{N_\varepsilon - 1} \frac{1 - \frac{p_{\varepsilon,+}}{p_{\varepsilon,-}}}{1 - (\frac{p_{\varepsilon,+}}{p_{\varepsilon,-}})^{N_\varepsilon}}, \tag{7.173}$$

and, for $j = 2, \ldots, N_\varepsilon$,

$$\mathsf{P}_{\varepsilon,j,N_\varepsilon} = \mathsf{P}_{\varepsilon,2,N_\varepsilon} \cdot \frac{1 - (\frac{p_{\varepsilon,-}}{p_{\varepsilon,+}})^{j-1}}{1 - \frac{p_{\varepsilon,-}}{p_{\varepsilon,+}}} = \mathsf{P}_{\varepsilon,2,N_\varepsilon} \cdot (\frac{p_{\varepsilon,+}}{p_{\varepsilon,-}})^{2-j} \frac{1 - (\frac{p_{\varepsilon,+}}{p_{\varepsilon,-}})^{j-1}}{1 - \frac{p_{\varepsilon,+}}{p_{\varepsilon,-}}}. \tag{7.174}$$

The condition $\mathbf{B}_{16}$, or $\mathbf{B}_{16}$, and the relations (7.172)–(7.174) imply that $p_{\varepsilon,i} > 0, i \in \tilde{\mathbb{X}}$, for $\varepsilon \in (0, \varepsilon']$.

The condition $\mathbf{B}_{16}$ obviously implies that, for every $i \in \tilde{\mathbb{X}}$,

$$p_{\varepsilon,i} \to 0 \text{ as } \varepsilon \to 0. \tag{7.175}$$

This relation also holds in the case, where the condition $\mathbf{B}_{17}$ is assumed to be satisfied.

Indeed, the relation (7.173), the condition $\mathbf{B}_{17}$, and the assumption $N_\varepsilon \to \infty$ as $\varepsilon \to 0$, imply that, for $j = 2, \ldots, \bar{m}$ and $N \geq 1$,

$$\begin{aligned}
\overline{\lim_{\varepsilon\to 0}} \mathsf{P}_{\varepsilon,j,N_\varepsilon} &= \overline{\lim_{\varepsilon\to 0}} \frac{\frac{p_{\varepsilon,-}}{p_{\varepsilon,+}} - 1}{(\frac{p_{\varepsilon,-}}{p_{\varepsilon,+}})^{N_\varepsilon} - 1} \cdot \frac{1 - (\frac{p_{\varepsilon,-}}{p_{\varepsilon,+}})^{j-1}}{1 - \frac{p_{\varepsilon,-}}{p_{\varepsilon,+}}} \\
&\leq \overline{\lim_{\varepsilon\to 0}} \frac{\frac{p_{\varepsilon,-}}{p_{\varepsilon,+}} - 1}{(\frac{p_{\varepsilon,-}}{p_{\varepsilon,+}})^{N} - 1} \cdot \frac{1 - (\frac{p_{\varepsilon,-}}{p_{\varepsilon,+}})^{j-1}}{1 - \frac{p_{\varepsilon,-}}{p_{\varepsilon,+}}} = \frac{j-1}{N} \to 0 \text{ as } N \to \infty.
\end{aligned} \tag{7.176}$$

Thus, the condition $\mathbf{B}_{16}$, or $\mathbf{B}_{17}$, implies that the conditions $\mathbf{A}_1$ (for the Markov renewal processes $(\eta_{\varepsilon,n}, \kappa_{\varepsilon,n}, \rho_{\varepsilon,n})$) and $\mathbf{A}_3$ (for the Markov renewal processes $(\tilde{\eta}_{\varepsilon,n}, \tilde{\kappa}_{\varepsilon,n}, \tilde{\rho}_{\varepsilon,n})$) are satisfied.

In this case, the averaged rare-event probability p_ε for the Markov chain $\eta_{\varepsilon,n}$, and function $u_\varepsilon = p_\varepsilon^{-1}$ take the following forms:

$$\begin{aligned}
p_\varepsilon &= \pi_{\varepsilon,0} p_{\varepsilon,0} + \pi_{\varepsilon,1} p_{\varepsilon,1} = \pi_{\varepsilon,1} p_{\varepsilon,1} \\
&= \frac{1}{1 + p_{\varepsilon,-}} \cdot p_{\varepsilon,+} \frac{\frac{p_{\varepsilon,-}}{p_{\varepsilon,+}} - 1}{(\frac{p_{\varepsilon,-}}{p_{\varepsilon,+}})^{N_\varepsilon} - 1},
\end{aligned} \tag{7.177}$$

and

$$u_\varepsilon = p_\varepsilon^{-1} = \frac{1 + p_{\varepsilon,-}}{p_{\varepsilon,+}} \frac{(\frac{p_{\varepsilon,-}}{p_{\varepsilon,+}})^{N_\varepsilon} - 1}{\frac{p_{\varepsilon,-}}{p_{\varepsilon,+}} - 1} \to \infty \text{ as } \varepsilon \to 0. \tag{7.178}$$

Let us consider the random variables:

$$\tilde{\kappa}_{\varepsilon,1} = \sum_{k=1}^{\check{\nu}_{\varepsilon,\mathbb{X},1}} \check{\kappa}_{\varepsilon,k}. \tag{7.179}$$

Let us also introduce the distribution functions, for $j = 0, 1, \ldots,$

$$\hat{F}_{\varepsilon,j,\mathbb{X}}(u) = \mathsf{P}_j\{\sum_{k=1}^{\check{\nu}_{\varepsilon,\mathbb{X},1}} \hat{\kappa}_{\varepsilon,k} \leq u\},\ u \geq 0. \tag{7.180}$$

It is useful to note that,

$$\mathsf{P}_0\{\check{\nu}_{\varepsilon,\mathbb{X}} = 1\} = 1. \tag{7.181}$$

In this case, distribution functions,

$$\begin{aligned} F_{\varepsilon,j}(u) &= \mathsf{P}_i\{\tilde{\kappa}_{\varepsilon,1} \leq u\} \\ &= \begin{cases} 1 - e^{-\mu_{\varepsilon,+} w_\varepsilon u} & \text{for } u \geq 0,\ j = 0, \\ \hat{G}_{\varepsilon,1,\mathbb{X}}(u w_\varepsilon) & \text{for } u \geq 0,\ i = 1, 2, \ldots. \end{cases} \end{aligned} \tag{7.182}$$

The following stochastic representation takes place,

$$\tau_{\varepsilon,i} \overset{\mathsf{d}}{=} \begin{cases} \zeta'_\varepsilon & \text{for } i = 0, \\ \zeta_\varepsilon + \mathrm{I}(\iota_\varepsilon = +1)\tau'_{\varepsilon,2} & \text{for } i = 1, \\ \zeta_\varepsilon + \mathrm{I}(\iota_\varepsilon = +1)(\tau'_{\varepsilon,3} + \tau'_{\varepsilon,2}) & \text{for } i = 2, \\ \tau'_{\varepsilon,i} + \cdots + \tau'_{\varepsilon,2} & \text{for } i = 3, 4, \ldots, \end{cases} \tag{7.183}$$

where: (a) the random variable $\tau_{\varepsilon,i}$ has the distribution function $\mathsf{P}_i\{\tilde{\kappa}_{\varepsilon,1} \leq \cdot\}$, for $i = 0, 1, \ldots,$ (b) the random variable $\tau'_{\varepsilon,k} \overset{\mathsf{d}}{=} \tau_{\varepsilon,2}$, for $k = 2, 3, \ldots,$ (c) the random variable ζ_ε is exponentially distributed with parameter $\lambda_\varepsilon w_\varepsilon$, (d) the random variable ζ'_ε is exponentially distributed with parameter $\mu_{\varepsilon,+} w_\varepsilon$, (e) the random variable ι_ε takes values -1 and $+1$ with probabilities, respectively, $p_{\varepsilon,-}$ and $p_{\varepsilon,+}$, and (f) the random variables ζ_ε, ι_ε, $\tau'_{\varepsilon,k}$, $k = 2, 3, \ldots$ are mutually independent, for every $k = 2, 3, \ldots$.

The conditions $\mathbf{B}_{16}$, or $\mathbf{B}_{17}$, imply that there exists $\varepsilon'' \in (0, \varepsilon']$ such that $p_{\varepsilon,+} < p_{\varepsilon,-}$, for $\varepsilon \in (0, \varepsilon'']$. As follows from the well-known results related to random walks that, for $\varepsilon \in (0, \varepsilon'']$ and $n \geq 1, i = 0, 1, \ldots,$

$$\mathsf{E}^{(n)}_{\varepsilon,i} = \mathsf{E}_i \tilde{\kappa}^n_{\varepsilon,1} < \infty. \tag{7.184}$$

The stochastic representation (7.183) implies that the following relation holds for the expectations $\mathsf{E}^{(1)}_{\varepsilon,i} = \mathsf{E}_i \tilde{\kappa}_{\varepsilon,1}, i = 0, 1, \ldots,$

$$\mathsf{E}^{(1)}_{\varepsilon,i} = \begin{cases} \frac{1}{\mu_{\varepsilon,+} w_\varepsilon} & \text{for } i = 0, \\ \frac{1}{\lambda_\varepsilon w_\varepsilon} + p_{\varepsilon,+} \mathsf{E}^{(1)}_{\varepsilon,2} & \text{for } i = 1, \\ \frac{1}{\lambda_\varepsilon w_\varepsilon} + 2 p_{\varepsilon,+} \mathsf{E}^{(1)}_{\varepsilon,2} & \text{for } i = 2, \\ (i-1)\mathsf{E}^{(1)}_{\varepsilon,2} & \text{for } i = 3, 4, \ldots. \end{cases} \tag{7.185}$$

The relation (7.185) takes the form of the following relation, for $i = 2$:

$$\mathsf{E}^{(1)}_{\varepsilon,2} = \frac{1}{\lambda_\varepsilon w_\varepsilon} + 2p_{\varepsilon,+}\mathsf{E}^{(1)}_{\varepsilon,2}, \tag{7.186}$$

which gives the following explicit formula:

$$\mathsf{E}^{(1)}_{\varepsilon,2} = \frac{1}{\lambda_\varepsilon w_\varepsilon}\frac{1}{p_{\varepsilon,-} - p_{\varepsilon,+}}. \tag{7.187}$$

The relation (7.187) allows us to rewrite the relation (7.185) in the following explicit form:

$$\mathsf{E}^{(1)}_{\varepsilon,i} = \begin{cases} \frac{1}{\mu_{\varepsilon,+}w_\varepsilon} & \text{for } i = 0, \\ \frac{1}{\lambda_\varepsilon w_\varepsilon}\frac{p_{\varepsilon,-}}{p_{\varepsilon,-}-p_{\varepsilon,+}} & \text{for } i = 1, \\ \frac{1}{\lambda_\varepsilon w_\varepsilon}\frac{1}{p_{\varepsilon,-}-p_{\varepsilon,+}} & \text{for } i = 2, \\ \frac{1}{\lambda_\varepsilon w_\varepsilon}\frac{i-1}{p_{\varepsilon,-}-p_{\varepsilon,+}} & \text{for } i = 3, 4, \ldots. \end{cases} \tag{7.188}$$

The condition $\mathbf{B}_{16}$, or $\mathbf{B}_{17}$, and the relation (7.188) imply that, for $\delta > 0$ and $i \in \check{\mathbb{X}}$,

$$\mathsf{P}_i\{\tilde{\kappa}_{\varepsilon,1} > \delta\} \leq \frac{\mathsf{E}^{(1)}_{\varepsilon,i}}{\delta} \to 0 \text{ as } \varepsilon \to 0, \tag{7.189}$$

for any function $0 < w_\varepsilon \to 0$ as $\varepsilon \to 0$.

The relation (7.189) implies that the condition $\mathbf{C}_6$ (for the Markov renewal processes $(\tilde{\eta}_{\varepsilon,n}, \tilde{\kappa}_{\varepsilon,n}, \tilde{\rho}_{\varepsilon,n})$) is satisfied.

The stochastic representations (7.183) imply that the following relation holds for second moments $\mathsf{E}^{(2)}_{\varepsilon,i} = \mathsf{E}_i\tilde{\kappa}_{\varepsilon,1}, i = 0, 1, \ldots,$

$$\mathsf{E}^{(2)}_{\varepsilon,i} = \begin{cases} \frac{2}{(\mu_{\varepsilon,+}w_\varepsilon)^2} & \text{for } i = 0, \\ \frac{2}{(\lambda_\varepsilon w_\varepsilon)^2} + 2\frac{1}{\lambda_\varepsilon w_\varepsilon}p_{\varepsilon,+}\mathsf{E}^{(1)}_{\varepsilon,2} + p_{\varepsilon,+}\mathsf{E}^{(2)}_{\varepsilon,2} & \text{for } i = 1, \\ \frac{2}{(\lambda_\varepsilon w_\varepsilon)^2} + \frac{4p_{\varepsilon,+}}{\lambda_\varepsilon w_\varepsilon}\mathsf{E}^{(1)}_{\varepsilon,2} + 2p_{\varepsilon,+}\big(\mathsf{E}^{(2)}_{\varepsilon,2} + (\mathsf{E}^{(1)}_{\varepsilon,2})^2\big) & \text{for } i = 2, \\ (i-1)\mathsf{E}^{(2)}_{\varepsilon,2} + (i-1)(i-2)(\mathsf{E}^{(1)}_{\varepsilon,2})^2 & \text{for } i = 3, 4, \ldots. \end{cases} \tag{7.190}$$

Relation (7.190) takes the form of the following relation, for $i = 2$:

$$\begin{aligned} \mathsf{E}^{(2)}_{\varepsilon,2} = {} & \frac{2}{(\lambda_\varepsilon w_\varepsilon)^2} + \frac{4p_{\varepsilon,+}}{\lambda_\varepsilon w_\varepsilon}\frac{1}{\lambda_\varepsilon w_\varepsilon}\frac{1}{p_{\varepsilon,-} - p_{\varepsilon,+}} \\ & + 2p_{\varepsilon,+}\mathsf{E}^{(2)}_{\varepsilon,2} + 2p_{\varepsilon,+}\frac{1}{(\lambda_\varepsilon w_\varepsilon)^2}\frac{1}{(p_{\varepsilon,-} - p_{\varepsilon,+})^2} \end{aligned} \tag{7.191}$$

which gives the following explicit formula:

$$\begin{aligned} \mathsf{E}^{(2)}_{\varepsilon,2} = {} & \frac{2}{\lambda^2_\varepsilon}\frac{1}{w^2_\varepsilon(p_{\varepsilon,-} - p_{\varepsilon,+})} \\ & + \frac{4p_{\varepsilon,+}}{\lambda^2_\varepsilon}\frac{1}{w^2_\varepsilon(p_{\varepsilon,-} - p_{\varepsilon,+})^2} + \frac{2p_{\varepsilon,+}}{\lambda^2_\varepsilon}\frac{1}{w^2_\varepsilon(p_{\varepsilon,-} - p_{\varepsilon,+})^3}. \end{aligned} \tag{7.192}$$

Also, the relation (7.190), for $i = 1$, and the relation (7.192) imply that,

$$\begin{aligned} \mathsf{E}^{(2)}_{\varepsilon,1} &= \frac{2}{(\lambda_\varepsilon w_\varepsilon)^2} + 2\frac{1}{\lambda_\varepsilon w_\varepsilon} p_{\varepsilon,+}\mathsf{E}^{(1)}_{\varepsilon,2} + p_{\varepsilon,+}\mathsf{E}^{(2)}_{\varepsilon,2} \\ &= \frac{2}{\lambda^2_\varepsilon}\frac{1}{w^2_\varepsilon} + \frac{4p_{\varepsilon,+}}{\lambda^2_\varepsilon}\frac{1}{w^2_\varepsilon(p_{\varepsilon,-} - p_{\varepsilon,+})} \\ &+ \frac{4p^2_{\varepsilon,+}}{\lambda^2_\varepsilon}\frac{1}{w^2_\varepsilon(p_{\varepsilon,-} - p_{\varepsilon,+})^2} + \frac{2p^2_{\varepsilon,+}}{\lambda^2_\varepsilon}\frac{1}{w^2_\varepsilon(p_{\varepsilon,-} - p_{\varepsilon,+})^3}. \end{aligned} \tag{7.193}$$

The relation (7.193) implies that,

$$\mathsf{E}\tilde{\kappa}^2_{\varepsilon,1} = \mathsf{E}^{(2)}_{\varepsilon,1} = \frac{B_\varepsilon}{(p_{\varepsilon,-} - p_{\varepsilon,+})^3}\frac{1}{w^2_\varepsilon}, \tag{7.194}$$

where

$$\begin{aligned} B_\varepsilon &= \frac{2(p_{\varepsilon,-} - p_{\varepsilon,+})^3}{\lambda^2_\varepsilon} + \frac{4p_{\varepsilon,+}(p_{\varepsilon,-} - p_{\varepsilon,+})^2}{\lambda^2_\varepsilon} \\ &+ \frac{4p^2_{\varepsilon,+}(p_{\varepsilon,-} - p_{\varepsilon,+})}{\lambda^2_\varepsilon} + \frac{2p^2_{\varepsilon,+}}{\lambda^2_\varepsilon}. \end{aligned} \tag{7.195}$$

The condition $\mathbf{B}_{16}$, or $\mathbf{B}_{17}$, implies that,

$$0 < \varliminf_{\varepsilon\to 0} B_\varepsilon \leq \varlimsup_{\varepsilon\to 0} B_\varepsilon < \infty. \tag{7.196}$$

Let $\hat{\theta}_{\varepsilon,n}, n = 1, 2, \ldots$ be, for $\varepsilon \in (0, \varepsilon'']$, i.i.d. random variables with the distribution function $\hat{F}_{\varepsilon,\mathbb{X}}(\cdot) = \pi_{\varepsilon,0}\hat{F}_{\varepsilon,0,\mathbb{X}}(\cdot) + \pi_{\varepsilon,1}\hat{F}_{\varepsilon,1,\mathbb{X}}(\cdot)$.

Let us also consider the sequence of i.i.d. random variables $\theta^{(w)}_{\varepsilon,n} = \hat{\theta}_{\varepsilon,n}/w_\varepsilon, n = 1, 2, \ldots$. These random variables have the distribution function $F_\varepsilon(\cdot) = \pi_{\varepsilon,0}\hat{F}_{\varepsilon,0,\mathbb{X}}(\cdot w_\varepsilon) + \pi_{\varepsilon,1}\hat{F}_{\varepsilon,1,\mathbb{X}}(\cdot w_\varepsilon) = \pi_{\varepsilon,0}F_{\varepsilon,0}(\cdot) + \pi_{\varepsilon,1}F_{\varepsilon,1}(\cdot)$.

The relations (7.94), (7.109), and (7.120) imply that,

$$\mathsf{E}\theta^{(w)}_{\varepsilon,1} = \pi_{\varepsilon,0}\mathsf{E}_0\tilde{\kappa}_{\varepsilon,1} + \pi_{\varepsilon,1}\mathsf{E}_1\tilde{\kappa}_{\varepsilon,1} = \frac{A_{\varepsilon,\circ}}{(p_{\varepsilon,-} - p_{\varepsilon,+})}\frac{1}{w_\varepsilon}, \tag{7.197}$$

where

$$A_{\varepsilon,\circ} = \frac{p_{\varepsilon,-}}{p_{\varepsilon,-} + 1}\frac{1}{\mu_{\varepsilon,+}}(p_{\varepsilon,-} - p_{\varepsilon,+}) + \frac{1}{p_{\varepsilon,-} + 1}\frac{1}{\lambda_\varepsilon}p_{\varepsilon,-}. \tag{7.198}$$

The condition $\mathbf{B}_{16}$, or $\mathbf{B}_{17}$, implies that,

$$0 < \varliminf_{\varepsilon\to 0} A_{\varepsilon,\circ} \leq \varlimsup_{\varepsilon\to 0} A_{\varepsilon,\circ} < \infty. \tag{7.199}$$

Also, the relations (7.94), (7.190), and (7.193) imply that,

$$\mathsf{E}(\theta^{(w)}_{\varepsilon,1})^2 = \pi_{\varepsilon,0}\mathsf{E}_0\kappa^2_{\varepsilon,1} + \pi_{\varepsilon,1}\mathsf{E}_0\kappa^2_{\varepsilon,1} = \frac{B_{\varepsilon,\circ}}{(p_{\varepsilon,-} - p_{\varepsilon,+})^3}\frac{1}{w^2_\varepsilon}, \tag{7.200}$$

where

$$B_{\varepsilon,\circ} = \frac{p_{\varepsilon,-}}{p_{\varepsilon,-}+1}\frac{2(p_{\varepsilon,-} - p_{\varepsilon,+})^3}{\mu^2_{\varepsilon,+}} + \frac{1}{p_{\varepsilon,-}+1}B_\varepsilon. \tag{7.201}$$

The condition $\mathbf{B}_{16}$, or $\mathbf{B}_{17}$, implies that,

$$0 < \underline{\lim}_{\varepsilon\to 0} B_{\varepsilon,\circ} \leq \overline{\lim}_{\varepsilon\to 0} B_{\varepsilon,\circ} < \infty. \tag{7.202}$$

Let us consider the random variables,

$$\theta^{(w)}_\varepsilon = \sum_{n=1}^{[u_\varepsilon]} \theta^{(w)}_{\varepsilon,n} = \sum_{n=1}^{[u_\varepsilon]} \hat{\theta}_{\varepsilon,n}/w_\varepsilon. \tag{7.203}$$

Let us, first, choose the normalisation function,

$$w_{\circ,\varepsilon} = A_{\varepsilon,\circ}\frac{u_\varepsilon}{p_{\varepsilon,-} - p_{\varepsilon,+}}. \tag{7.204}$$

The relations (7.197), (7.198), and (7.199) imply that,

$$\mathsf{E}\theta^{(w_\circ)}_\varepsilon = [u_\varepsilon]\mathsf{E}\theta^{(w_\circ)}_{\varepsilon,1} = [u_\varepsilon]\frac{A_{\varepsilon,\circ}/(p_{\varepsilon,-} - p_{\varepsilon,+})}{A_{\varepsilon,\circ}u_\varepsilon/(p_{\varepsilon,-} - p_{\varepsilon,+})} \to 1 \text{ as } \varepsilon \to 0. \tag{7.205}$$

Also, the relations (7.197), (7.198), (7.200), and (7.201) imply that,

$$\begin{aligned}\mathrm{Var}\theta^{(w_\circ)}_\varepsilon &= [u_\varepsilon]\mathrm{Var}\theta^{(w_\circ)}_{\varepsilon,1} = [u_\varepsilon](\mathsf{E}(\theta^{(w_\circ)}_{\varepsilon,1})^2 - (\mathsf{E}\theta^{(w_\circ)}_{\varepsilon,1})^2) \\ &= [u_\varepsilon]\Big(\frac{B_{\varepsilon,\circ}/(p_{\varepsilon,-} - p_{\varepsilon,+})^3}{A^2_{\varepsilon,\circ}u^2_\varepsilon/(p_{\varepsilon,-} - p_{\varepsilon,+})^2} - \Big(\frac{A_{\varepsilon,\circ}/(p_{\varepsilon,-} - p_{\varepsilon,+})}{A_{\varepsilon,\circ}u_\varepsilon/(p_{\varepsilon,-} - p_{\varepsilon,+})}\Big)^2\Big) \\ &= \frac{[u_\varepsilon]}{u_\varepsilon}\Big(\frac{B_{\varepsilon,\circ}}{A^2_{\varepsilon,\circ}u_\varepsilon/(p_{\varepsilon,-} - p_{\varepsilon,+})} - 1\Big).\end{aligned} \tag{7.206}$$

At this stage, the subsequent analysis differs for the subcritical case, where the condition $\mathbf{B}_{16}$ is satisfied and the critical case, where the condition $\mathbf{B}_{17}$ is satisfied.

If the condition $\mathbf{B}_{16}$ is satisfied, $\underline{\lim}_{\varepsilon\to 0}(p_{\varepsilon,-} - p_{\varepsilon,+}) > 0$, and, thus, the relation (7.206) implies that,

$$\mathrm{Var}\,\theta^{(w_\circ)}_\varepsilon \to 0 \text{ as } \varepsilon \to 0. \tag{7.207}$$

If the condition $\mathbf{B}_{17}$ holds, $\lim_{\varepsilon\to 0}(p_{\varepsilon,-} - p_{\varepsilon,+}) = 0$. In this case, the relation (7.178) implies that,

$$u_\varepsilon(p_{\varepsilon,-} - p_{\varepsilon,+}) = \frac{1 + p_{\varepsilon,-}}{p_{\varepsilon,+}} \frac{(\frac{p_{\varepsilon,-}}{p_{\varepsilon,+}})^{N_\varepsilon - 1}}{\frac{p_{\varepsilon,-}}{p_{\varepsilon,+}} - 1}(p_{\varepsilon,-} - p_{\varepsilon,+})$$
$$= (1 + p_{\varepsilon,-})((\frac{p_{\varepsilon,-}}{p_{\varepsilon,+}})^{N_\varepsilon - 1} - 1). \tag{7.208}$$

The relation (7.208) implies that,

$$u_\varepsilon(p_{\varepsilon,-} - p_{\varepsilon,+}) \longrightarrow \infty \text{ as } \varepsilon \to 0, \tag{7.209}$$

and, thus, the relation (7.207) holds, if and only if the following condition is satisfied, which is in addition to the condition $\mathbf{B}_{17}$:

$\mathbf{B}_{18}$: $(\frac{\mu_{\varepsilon,-}}{\mu_{\varepsilon,+}} - 1)N_\varepsilon \longrightarrow \infty$ as $\varepsilon \to 0$.

According to the above remarks, if the condition $\mathbf{B}_{16}$ holds, or the conditions $\mathbf{B}_{17}$ and $\mathbf{B}_{18}$ hold, then the relations (7.205) and (7.207) imply that the following asymptotic relation takes place:

$$\theta_\varepsilon^{(w_\circ)} = \sum_{n=1}^{[u_\varepsilon]} \theta_{\varepsilon,n}^{(w_\circ)} \xrightarrow{\mathsf{P}} 1 \text{ as } \varepsilon \to 0. \tag{7.210}$$

Let us now consider the case, where the function $0 < w_\varepsilon \longrightarrow \infty$ as $\varepsilon \to 0$ satisfies the following condition:

$\mathbf{D}_{23}$: $A_\varepsilon = u_\varepsilon \mathsf{E}\theta_{\varepsilon,1} = A_{\varepsilon,\circ}\frac{u_\varepsilon}{w_\varepsilon(p_{\varepsilon,-} - p_{\varepsilon,+})} \longrightarrow A \in (0, \infty)$ as $\varepsilon \to 0$.

Under condition $\mathbf{B}_{16}$, or the conditions $\mathbf{B}_{17}$ and $\mathbf{B}_{18}$, the condition $\mathbf{D}_{23}$ is necessary and sufficient for holding of condition $\mathbf{D}_1$, which in this case takes the form of the following asymptotic relation:

$$\theta_\varepsilon^{(w)} = \sum_{n=1}^{[u_\varepsilon]} \theta_{\varepsilon,n}^{(w)} \Longrightarrow \theta_0^{(w)} \text{ as } \varepsilon \to 0, \tag{7.211}$$

where $\theta_0^{(w)}$ is a non-negative random variable with distribution function not concentrated at zero.

Moreover, in this case, the limiting random variable $\theta_0^{(w)} = A$ with probability 1.

Indeed, let the condition $\mathbf{B}_{16}$ and $\mathbf{D}_{23}$ or the conditions $\mathbf{B}_{17}$, $\mathbf{B}_{18}$ and $\mathbf{D}_{23}$ be satisfied.

In this case, the relation (7.210) and the asymptotic relation appearing in the condition $\mathbf{D}_{23}$ imply that,

$$\theta_\varepsilon^{(w)} = \sum_{n=1}^{[u_\varepsilon]} \theta_{\varepsilon,n}^{(w)} = \sum_{n=1}^{[u_\varepsilon]} \theta_{\varepsilon,n}^{(w_\circ)} \frac{A_{\varepsilon,\circ} u_\varepsilon}{(p_{\varepsilon,-} - p_{\varepsilon,+})w_\varepsilon}$$
$$= \theta_\varepsilon^{(w_\circ)} \frac{A_{\varepsilon,\circ} u_\varepsilon}{(p_{\varepsilon,-} - p_{\varepsilon,+})w_\varepsilon} \xrightarrow{\mathsf{P}} A \text{ as } \varepsilon \to 0. \tag{7.212}$$

Thus, the above sufficiency statement is true.

Let us assume that the condition $\mathbf{B}_{16}$ is satisfied and the relation (7.211) holds, or the conditions $\mathbf{B}_{17}$ and $\mathbf{B}_{18}$ are satisfied and the relation (7.211) holds, but the condition $\mathbf{D}_{23}$ is not satisfied.

In this case, two subsequences $0 < \varepsilon'_k, \varepsilon''_k \leq 1$ can be selected such that: (a) $\varepsilon'_k, \varepsilon''_k \to 0$ as $n \to \infty$, (b) $\frac{A_{\varepsilon'_k,\circ} u_{\varepsilon'_k}}{(p_{\varepsilon'_k,-} - p_{\varepsilon'_k,+}) w_{\varepsilon'_k}} \to A' \in [0, \infty]$ as $k \to \infty$, (c) $\frac{A_{\varepsilon''_k,\circ} u_{\varepsilon''_k}}{(p_{\varepsilon''_k,-} - p_{\varepsilon''_k,+}) w_{\varepsilon''_k}} \to A'' \in [0, \infty]$ as $k \to \infty$, (d) $A' \neq A''$.

In this case, relation (7.211) implies that

$$\begin{aligned} \theta^{(w)}_{\varepsilon'_k} &= \sum_{n=1}^{[u_{\varepsilon'_k}]} \theta^{(w)}_{\varepsilon'_k,n} = \sum_{n=1}^{[u_{\varepsilon'_k}]} \theta^{(w_\circ)}_{\varepsilon'_k,n} \frac{A_{\varepsilon'_k,\circ} u_{\varepsilon'_k}}{(p_{\varepsilon'_k,-} - p_{\varepsilon'_k,+}) w_{\varepsilon'_k}} \\ &= \theta^{(w_\circ)}_{\varepsilon'_k} \frac{A_{\varepsilon'_k,\circ} u_{\varepsilon'_k}}{(p_{\varepsilon'_k,-} - p_{\varepsilon'_k,+}) w_{\varepsilon'_k}} \xrightarrow{\mathsf{d}} A' \text{ as } k \to \infty \end{aligned} \tag{7.213}$$

and

$$\begin{aligned} \theta^{(w)}_{\varepsilon''_k} &= \sum_{n=1}^{[u_{\varepsilon''_k}]} \theta^{(w)}_{\varepsilon''_k,n} = \sum_{n=1}^{[u_{\varepsilon''_k}]} \theta^{(w_\circ)}_{\varepsilon''_k,n} \frac{A_{\varepsilon''_k,\circ} u_{\varepsilon''_k}}{(p_{\varepsilon''_k,-} - p_{\varepsilon''_k,+}) w_{\varepsilon''_k}} \\ &= \theta^{(w_\circ)}_{\varepsilon''_k} \frac{A_{\varepsilon''_k,\circ} u_{\varepsilon''_k}}{(p_{\varepsilon''_k,-} - p_{\varepsilon''_k,+}) w_{\varepsilon''_k}} \xrightarrow{\mathsf{d}} A'' \text{ as } k \to \infty. \end{aligned} \tag{7.214}$$

The above two relations contradict to the relation (7.213).

Thus, the above necessity statement is also true, i.e. the condition $\mathbf{D}_{23}$ is satisfied.

In this case, due to relation (7.212), the limiting random variable $\theta^{(w)}_0$ in relation (7.211) should be constant A.

Condition $\mathbf{D}_{23}$ plays, in this case, the role of condition $\mathbf{D}_1$.

Finally, by Lemma 4.7, relations (7.194)–(7.196) imply that the following relation takes place, for $\varepsilon \in (0, \varepsilon''']$:

$$\mathsf{E}_1 \kappa^2_{\varepsilon,1} \leq \frac{B_{\varepsilon,1}}{w^2_\varepsilon} = \frac{B_{\varepsilon,1}}{w_\varepsilon} \frac{u_\varepsilon}{w_\varepsilon} \pi_{\varepsilon,1} p_{\varepsilon,1} = o_1(\varepsilon) p_{\varepsilon,1}. \tag{7.215}$$

Since, $o_1(\varepsilon) = \frac{B_{\varepsilon,1}}{w_\varepsilon} \frac{u_\varepsilon}{w_\varepsilon} \pi_{\varepsilon,1} \to 0$ as $\varepsilon \to 0$, the relation (7.215) implies that the condition $\mathbf{C}_5$ holds, and, thus, the condition $\mathbf{C}_1$ is satisfied for Markov renewal processes $(\eta_{\varepsilon,n}, \kappa_{\varepsilon,n}, \rho_{\varepsilon,n})$.

Therefore, all conditions of Theorem 4.4 hold for Markov renewal process $(\check{\eta}_{\varepsilon,n}, \check{\kappa}_{\varepsilon,n}, \check{\rho}_{\varepsilon,n})$. In this case, Theorem 4.4 takes the following form.

Theorem 7.3 *Let the condition* $\mathbf{B}_{16}$ *be satisfied, or the conditions* $\mathbf{B}_{17}$ *and* $\mathbf{B}_{18}$ *be satisfied. Then*:

(i) *The condition* $\mathbf{D}_{23}$ *is necessary and sufficient for the fulfilment (for some or any initial distributions* $\bar{q}_\varepsilon$ *concentrated on the space* $\tilde{\mathbb{X}}$*, respectively, in the statements of necessity and sufficiency) of the asymptotic relation,* $\xi_{\mathbf{S}_\varepsilon} \xrightarrow{\mathsf{d}} \xi_0$ *as* $\varepsilon \to 0$*, where* ξ_0 *is a non-negative random variable with distribution function not concentrated at zero.*

(ii) *The limiting random variable* ξ_0 *has an exponential distribution with the parameter* A^{-1}.

It is useful to note that the role of $\tilde{\mathbb{X}}$ can be played by any finite subset $\tilde{\mathbb{X}} = \{1, \ldots, \tilde{m}\}$ of the space $\hat{\mathbb{X}}$.

Part II
Hitting Times and Phase Space Reduction for Perturbed Semi-Markov Processes

Chapter 8
Asymptotically Comparable Functions

In this chapter, we introduce the notion of a complete family of asymptotically comparable functions and present operating rules and formulas for such families. This concept plays a key role in perturbation conditions for semi-Markov type processes. We also consider typical examples of such families, namely, families of asymptotically comparable power type, power-exponential type, and power-logarithmic type functions. This chapter includes two sections.

In Sect. 8.1 we give a definition of complete family of asymptotically comparable functions. In Lemmas 8.2 and 8.3, we present operating rules (summation, multiplication, and division) for such families.

In Sect. 8.2, we consider the complete families of asymptotically comparable power-type, power-exponential-type, and power-logarithmic-type functions. Operating rules for these complete families are presented for asymptotically comparable power-type functions in Lemmas 8.4 and 8.5, for power-exponential-type functions in Lemmas 8.6 and 8.7, and for power-logarithmic-type functions in Lemmas 8.8 and 8.9.

8.1 Complete Families of Asymptotically Comparable Functions

In this section, the notion of a complete family of asymptotically comparable functions is introduced and some general properties of such families are described.

8.1.1 Definitions for Families of Asymptotically Comparable Functions

We consider functions $h(\varepsilon)$ defined on the interval $(0, 1]$ and taking values in the interval $(0, \infty)$. Let $\mathcal{H} = \{h(\varepsilon)\}$ be a non-empty family of such functions.

D. Silvestrov, *Perturbed Semi-Markov Type Processes I*,
https://doi.org/10.1007/978-3-030-92403-4_8

Definition 8.1 $\mathcal{H} = \{h(\cdot)\}$ is a complete family of asymptotically comparable functions if: **(1)** it is closed with respect to operations of summation, multiplication, and division, **(2)** there exists $\lim_{\varepsilon\to 0} h(\varepsilon) = a_{\mathcal{H}}[h(\cdot)] \in [0, \infty]$, for any function $h(\cdot) \in \mathcal{H}$.

Here, $a_{\mathcal{H}}[h(\cdot)]$ is some functional acting from the functional space $\mathcal{H}$ to the interval $[0, \infty]$.

We call $a_{\mathcal{H}}[h(\cdot)]$ the comparability limit for the function $h(\cdot)$.

Definition 8.2 $\mathcal{H}'$ is a family of asymptotically comparable functions if it is a subfamily of some complete family of asymptotically comparable functions $\mathcal{H}$.

Let functions $h_k(\varepsilon) \in \mathcal{H}', k = 1, 2, \ldots$, indices $j_k = \pm 1, k = 1, 2, \ldots$. Then, obviously, the following limit exists, for any $n \geq 1$:

$$\lim_{\varepsilon\to 0} \prod_{k=1}^{n} h_k(\varepsilon)^{j_k} = a_{\mathcal{H}'}[\prod_{k=1}^{n} h_k(\cdot)^{j_k}] \in [0, \infty]. \tag{8.1}$$

Indeed, according to the obvious induction, functions $\prod_{k=1}^{n} h_k(\varepsilon)^{j_k}$ belong to family $\mathcal{H}$, for $n \geq 1$.

Let $\mathcal{H}'$ be a family of functions defined on interval $(0, 1]$, taking values in the interval $(0, \infty)$, for which the asymptotic relation (8.1) holds.

Let us now define the family $[\mathcal{H}']$ of all functions given by the following relation:

$$h(\varepsilon) = \frac{\sum_{r'=1}^{n'} \prod_{k'=1}^{n'_{r'}} h'_{r',k'}(\varepsilon)^{j'_{r',k'}}}{\sum_{r''=1}^{n''} \prod_{k''=1}^{n''_{r''}} h''_{r'',k''}(\varepsilon)^{j''_{r'',k''}}}, \tag{8.2}$$

where, (a) functions $h'_{r',k'}(\varepsilon) \in \mathcal{H}$, for $1 \leq k' \leq n'_{r'} < \infty, 1 \leq r' \leq n' < \infty$; (b) functions $h''_{r'',k''}(\varepsilon) \in \mathcal{H}$, for $1 \leq k'' \leq n''_{r''} < \infty, 1 \leq r'' \leq n'' < \infty$; (c) indices $j'_{r',k'} = \pm 1$, for $1 \leq k' \leq n'_{r'} < \infty, 1 \leq r' \leq n' < \infty$, (d) indices $j''_{r'',k''} = \pm 1$, for $1 \leq k'' \leq n''_{r''} < \infty, 1 \leq r'' \leq n'' < \infty$.

Obviously, $\mathcal{H}' \subseteq [\mathcal{H}']$. Indeed, any function $h(\varepsilon) \in \mathcal{H}'$ can be represented in the form, $h(\varepsilon) = h(\varepsilon)^2/h(\varepsilon) \in [\mathcal{H}']$.

Also, the constant $1 = 1(\varepsilon), \varepsilon \in (0, 1]$ belongs to the family $[\mathcal{H}']$. Indeed, $1 = 1(\varepsilon)$ can be represented in the form, $1(\varepsilon) = h(\varepsilon)/h(\varepsilon) \in [\mathcal{H}']$.

We call the family $[\mathcal{H}']$ the closure of the family $\mathcal{H}'$.

The following lemma presents the basic properties of the family $[\mathcal{H}']$.

Lemma 8.1 *Let $\mathcal{H}'$ be a family of functions, for which the asymptotic relation* (8.1) *holds. Then its closure $[\mathcal{H}']$ is a complete family of asymptotically comparable functions.*

Proof The fulfilment of the condition **(1)** given in the Definition 8.1 for the family $[\mathcal{H}']$ and the corresponding operating formulas for calculating sums, products, and quotients for functions from the family $[\mathcal{H}']$ are obvious.

The following relation, which holds for any function $h(\varepsilon) \in [\mathcal{H}']$ given by the relation (8.2), implies that the family $[\mathcal{H}']$ satisfies the condition **(2)** given in the Definition 8.1,

$$\lim_{\varepsilon\to 0} h(\varepsilon) = \lim_{\varepsilon\to 0} \frac{\sum_{r'=1}^{n'} \prod_{k'=1}^{n'_{r'}} h'_{r',k'}(\varepsilon)^{j'_{r',k'}}}{\sum_{r''=1}^{n''} \prod_{k''=1}^{n''_{r''}} h''_{r'',k''}(\varepsilon)^{j''_{r'',k''}}}$$
$$= \lim_{\varepsilon\to 0} \sum_{r'=1}^{n'} \Big(\sum_{r''=1}^{n''} \prod_{k''=1}^{n''_{r''}} h''_{r'',k''}(\varepsilon)^{j''_{r'',k''}} \prod_{k'=1}^{n'_{r'}} h'_{r',k'}(\varepsilon)^{-j'_{r',k'}} \Big)^{-1}$$
$$= \sum_{r'=1}^{n'} \Big(\sum_{r''=1}^{n''} a_{\mathcal{H}'}\Big[\prod_{k''=1}^{n''_{r''}} h''_{r'',k''}(\cdot)^{j''_{r'',k''}} \prod_{k'=1}^{n'_{r'}} h'_{r',k'}(\cdot)^{-j'_{r',k'}}\Big]\Big)^{-1} \in [0, \infty]. \quad (8.3)$$

Thus, $[\mathcal{H}']$ is, indeed, a complete family of asymptotically comparable functions. □

Lemma 8.1 allows us to give alternative variants of Definitions 8.1 and 8.2.

Since $\mathcal{H}' \subseteq [\mathcal{H}']$, we can say that $\mathcal{H}'$ is a family of asymptotically comparable functions if the asymptotic relation (8.1) holds for functions from the family $\mathcal{H}'$.

Moreover, the family of asymptotically comparable functions $\mathcal{H}'$ is complete if $\mathcal{H}' = [\mathcal{H}']$.

The complete family of asymptotically comparable functions $\mathcal{H}$ can be called trivial if comparability limits $a_{\mathcal{H}}[h(\cdot)] \in (0, \infty)$ for all functions $h(\cdot) \in \mathcal{H}$.

In such case, there always exists a family of asymptotically comparable functions $\tilde{\mathcal{H}}$ such that: (a) $\mathcal{H} \subset \tilde{\mathcal{H}}$, (b) there exists a function $\tilde{h}(\cdot) \in \tilde{\mathcal{H}}$ such that its comparability limit $a_{\mathcal{H}}[\tilde{h}(\cdot)] = \infty$.

Indeed, take an arbitrary function $u(\varepsilon)$, defined on interval $(0, 1]$, taking values in interval $(0, \infty)$ and such that $\lim_{\varepsilon\to 0} u(\varepsilon) = \infty$.

Let us consider the family of functions $\mathcal{H}_{u(\cdot)} = \mathcal{H} \cup \{u(\cdot)\}$.

Obviously, the asymptotic relation (8.1) holds for functions from this family, and, moreover, for any sequence of functions $h_k(\varepsilon) \in \mathcal{H}_{u(\cdot)}, k = 1, 2, \ldots$, indices $j_k = \pm 1, k = 1, 2, \ldots$, and $n \geq 1$, the following limit exists, for any $n \geq 1$:

$$\lim_{\varepsilon\to 0} \prod_{k=1}^{n} h_k(\varepsilon)^{j_k} = a_{\mathcal{H}_{u(\cdot)}}\Big[\prod_{k=1}^{n} h_k(\cdot)^{j_k}\Big] = \begin{cases} 0 & \text{if } b^{(n)}_{u(\cdot)} < 0, \\ a^{(n)}_{\mathcal{H}} & \text{if } b^{(n)}_{u(\cdot)} = 0, \\ \infty & \text{if } b^{(n)}_{u(\cdot)} > 0, \end{cases} \quad (8.4)$$

where

$$a^{(n)}_{\mathcal{H}} = \prod_{1\leq k\leq n: h_k(\cdot)\in\mathcal{H}} a_{\mathcal{H}}[h_k(\cdot)^{j_k}], \; b^{(n)}_{u(\cdot)} = \sum_{1\leq k\leq n: h_k(\cdot)=u(\cdot)} j_k. \quad (8.5)$$

According to Lemma 8.1, we can define the complete family of asymptotically comparable $\tilde{\mathcal{H}} = [\mathcal{H}_{u(\cdot)}]$, as the closure of the family $\mathcal{H}_{u(\cdot)}$.

Since, $\mathcal{H}$ is itself a complete family of comparable functions, the family $[\mathcal{H}_{u(\cdot)}]$ consists of all functions of the form,

$$h(\varepsilon) = \frac{\sum_{r'=n'_-}^{n'_+} h'_{r'}(\varepsilon)u(\varepsilon)^{r'}}{\sum_{r''=n''_-}^{n''_+} h''_{r''}(\varepsilon)u(\varepsilon)^{r''}}, \quad (8.6)$$

where, (a) functions $h'_{r'}(\varepsilon) \in \mathcal{H}$, for $-\infty < n'_- \leq r' \leq n'_+ < \infty$; (b) functions $h''_{r''}(\varepsilon) \in \mathcal{H}$, for $-\infty < n''_- \leq r'' \leq n''_+ < \infty$.

Moreover, for any function $h(\cdot) \in [\mathcal{H}_{u(\cdot)}]$,

$$a_{[\mathcal{H}_{u(\cdot)}]}[h(\cdot)] = \lim_{\varepsilon \to 0} h(\varepsilon) = \begin{cases} 0 & \text{if } n'_+ < n''_+, \\ a_{\mathcal{H}}[h'_{n'_+}(\cdot)]/a_{\mathcal{H}}[h''_{n''_+}(\cdot)] & \text{if } n'_+ = n''_+, \\ \infty & \text{if } n'_+ > n''_+. \end{cases} \tag{8.7}$$

8.1.2 Operating Rules for Asymptotically Comparable Functions

The following lemmas, which are obvious consequences of the operating rules for limits of sums, products, and quotients of functions, present operating rules for functions from a complete family of asymptotically comparable functions.

Lemma 8.2 *Let $\mathcal{H}$ be a complete family of asymptotically comparable functions, $h_i(\cdot) \in \mathcal{H}$, $i = 1, 2$, and $\lim_{\varepsilon \to 0} h_i(\varepsilon) = a_{\mathcal{H}}[h_i(\cdot)]$, for $i = 1, 2$. Then, the following operating sum, product, and quotient formulas for calculating comparability limits take place*:

(a) *If $h'(\cdot) = h_1(\cdot) + h_2(\cdot)$. Then, $h'(\cdot) \in \mathcal{H}$ and,*

$$a_{\mathcal{H}}[h'(\cdot)] = \lim_{\varepsilon \to 0} h'(\varepsilon) = a_{\mathcal{H}}[h_1(\cdot)] + a_{\mathcal{H}}[h_2(\cdot)]. \tag{8.8}$$

(b) *If $h''(\cdot) = h_1(\cdot) \cdot h_2(\cdot)$. Then, $h''(\cdot) \in \mathcal{H}$ and,*

$$a_{\mathcal{H}}[h''(\cdot)] = \lim_{\varepsilon \to 0} h''(\varepsilon)$$

$$= \begin{cases} a_{\mathcal{H}}[h_1(\cdot) \cdot h_2(\cdot)] & \text{if } a_{\mathcal{H}}[h_1(\cdot)] = 0,\ a_{\mathcal{H}}[h_2(\cdot)] = \infty, \\ & \quad \text{or } a_{\mathcal{H}}[h_1(\cdot)] = \infty,\ a_{\mathcal{H}}[h_2(\cdot)] = 0, \\ a_{\mathcal{H}}[h_1(\cdot)] \cdot a_{\mathcal{H}}[h_2(\cdot)] & \text{otherwise.} \end{cases} \tag{8.9}$$

(c) *If $h'''(\cdot) = h_1(\cdot)/h_2(\cdot)$. Then, $h'''(\cdot) \in \mathcal{H}$ and,*

$$a_{\mathcal{H}}[h'''(\cdot)] = \lim_{\varepsilon \to 0} h'''(\varepsilon)$$

$$= \begin{cases} a_{\mathcal{H}}[h_1(\cdot)/h_2(\cdot)] & \text{if } a_{\mathcal{H}}[h_1(\cdot)],\ a_{\mathcal{H}}[h_2(\cdot)] = 0, \\ & \quad \text{or } a_{\mathcal{H}}[h_1(\cdot)],\ a_{\mathcal{H}}[h_2(\cdot)] = \infty, \\ a_{\mathcal{H}}[h_1(\cdot)]/a_{\mathcal{H}}[h_2(\cdot)] & \text{otherwise.} \end{cases} \tag{8.10}$$

The operating rules presented in Lemma 8.1 admit the following multivariate generalisations.

Lemma 8.3 *Let $\mathcal{H}$ be a complete family of asymptotically comparable functions, $h_i(\cdot) \in \mathcal{H}$, $i = 1, 2, \ldots$, and $\lim_{\varepsilon \to 0} h_i(\varepsilon) = a_{\mathcal{H}}[h_i(\cdot)]$, for $i = 1, 2, \ldots$. Then,* *the following operating multiple sum and multiple product/quotient formulas for computing comparability limits take place*:

(a) *If* $h^{(n)}(\cdot) = \sum_{i=1}^{n} h_i(\cdot)$, *for* $n \geq 1$. *Then,* $h^{(n)}(\cdot) \in \mathcal{H}$, *for* $n \geq 1$, *and,*

$$\begin{aligned} a_{\mathcal{H}}[h^{(n)}(\cdot)] &= \lim_{\varepsilon \to 0} h^{(n)}(\varepsilon) \\ &= \sum_{i=1}^{n} a_{\mathcal{H}}[h_i(\cdot)] \\ &= a_{\mathcal{H}}[h^{(n-1)}(\cdot)] + a_{\mathcal{H}}[h_n(\cdot)], \end{aligned} \tag{8.11}$$

where $a_{\mathcal{H}}[h^{(0)}(\cdot)] \equiv 0$.

(b) *If* $h_\circ^{(n)}(\cdot) = \prod_{i=1}^{n} h_i(\cdot)^{\jmath_i}$, *for* $n \geq 1$, *where* $\jmath_i = \pm 1, i = 1, 2, \ldots$. *Then,* $h_\circ^{(n)}(\cdot) \in \mathcal{H}$, *for* $n \geq 1$, *and,*

$$a_{\mathcal{H}}[h_\circ^{(n)}(\cdot)] = \lim_{\varepsilon \to 0} h_\circ^{(n)}(\varepsilon)$$

$$= \begin{cases} a_{\mathcal{H}}[\prod_{i=1}^{n} h_i(\cdot)^{\jmath_i}] & \text{if } \min_{1 \leq i \leq n} a_{\mathcal{H}}[h_i(\cdot)]^{\jmath_i} = 0, \\ & \quad \max_{1 \leq i \leq n} a_{\mathcal{H}}[h_i(\cdot)]^{\jmath_i} = \infty, \\ \prod_{i=1}^{n} a_{\mathcal{H}}[h_i(\cdot)^{\jmath_i}] & \text{otherwise}, \end{cases} \tag{8.12}$$

$$= \begin{cases} a_{\mathcal{H}}[h_\circ^{(n-1)}(\cdot) h_n(\cdot)^{\jmath_n}] & \text{if } a_{\mathcal{H}}[h_\circ^{(n-1)}(\cdot)] \wedge a_{\mathcal{H}}[h_n(\cdot)]^{\jmath_n} = 0, \\ & \quad a_{\mathcal{H}}[h_\circ^{(n-1)}(\cdot)] \vee a_{\mathcal{H}}[h_n(\cdot)]^{\jmath_i} = \infty, \\ a_{\mathcal{H}}[h_\circ^{(n-1)}(\cdot)] a_{\mathcal{H}}[h_n(\cdot)]^{\jmath_n} & \text{otherwise}, \end{cases} \tag{8.13}$$

where $a_{\mathcal{H}}[h_\circ^{(0)}(\cdot)] \equiv 1$.

The relations, given in Lemmas 8.2 and 8.3 allow expressing the comparability limits for asymptotically comparable functions, obtained as results of operations of summation, multiplication, or division, in terms of the corresponding comparability limits for functions involved in the above operations, except for the asymptotic cases of the forms $0 \cdot \infty$, $0/0$, or ∞/∞. In such cases, some additional information about the asymptotic behaviour of the corresponding functions is required so that the limits can be calculated more explicitly. The complete families of asymptotically comparable functions considered in the next section represent models, in which the corresponding parameters and limits for asymptotically comparable functions can be calculated explicitly.

8.2 Examples of Complete Families of Asymptotically Comparable Functions

In this section, we present three typical examples of complete families of asymptotically comparable functions, namely, families of asymptotically comparable functions of power, power exponential, and power-logarithmic types.

8.2.1 Asymptotically Comparable Power-Type Functions

Let us consider the family of asymptotically comparable power-type functions $\mathcal{H}_1 = \{h(\cdot)\}$, which includes all functions $h(\cdot)$ defined on the interval $(0, 1]$, taking values in the interval $(0, \infty)$ and such that, for any function $h(\cdot) \in \mathcal{H}_1$, there exist constants $a_h > 0$ and $b_h \in (-\infty, \infty)$ such that the following asymptotic relation holds:

$$\frac{h(\varepsilon)}{a_h \varepsilon^{b_h}} \to 1 \text{ as } \varepsilon \to 0. \tag{8.14}$$

We call a_h and b_h comparability parameters for the function $h(\cdot)$.

Lemma 8.4 *The following statements are true for the family of asymptotically comparable power-type functions* $\mathcal{H}_1$:

(i) *If the functions* $h_i(\cdot) \in \mathcal{H}_1, i = 1, 2$, *i.e.,* $h_i(\varepsilon)/a_i\varepsilon^{b_i} \to 1$ *as* $\varepsilon \to 0$, *for* $i = 1, 2$, *where* $a_i > 0, b_i \in (-\infty, \infty), i = 1, 2$, *then the following operating formulas for calculating the comparability parameters and limits for the sums, products, and quotients of these functions take place*:

(a) *The function* $h'(\varepsilon) = h_1(\varepsilon) + h_2(\varepsilon) \in \mathcal{H}_1$, *and the relation* (8.14) *holds for this function, with comparability parameters,*

$$a' = a_1 \mathrm{I}(b_1 < b_2) + (a_1 + a_2)\mathrm{I}(b_1 = b_2) + a_2 \mathrm{I}(b_1 > b_2), \ b' = b_1 \wedge b_2, \tag{8.15}$$

i.e.,

$$\frac{h'(\varepsilon)}{a'\varepsilon^{b'}} \to 1 \text{ as } \varepsilon \to 0. \tag{8.16}$$

(b) *The function* $h''(\varepsilon) = h_1(\varepsilon) \cdot h_2(\varepsilon) \in \mathcal{H}_1$, *and the relation* (8.14) *holds for this function, with comparability parameters,*

$$a'' = a_1 a_2, \ b'' = b_1 + b_2, \tag{8.17}$$

i.e.,

$$\frac{h''(\varepsilon)}{a''\varepsilon^{b''}} \to 1 \text{ as } \varepsilon \to 0. \tag{8.18}$$

(c) *The function* $h'''(\varepsilon) = h_1(\varepsilon) \cdot h_2(\varepsilon)^{-1} \in \mathcal{H}_1$, *and the relation* (8.14) *holds for this function, with comparability parameters*

$$a''' = a_1/a_2, \ b''' = b_1 - b_2, \tag{8.19}$$

i.e.,

$$\frac{h'''(\varepsilon)}{a'''\varepsilon^{b'''}} \to 1 \text{ as } \varepsilon \to 0. \tag{8.20}$$

(ii) *If the function $h(\cdot) \in \mathcal{H}_1$, i.e., $h(\varepsilon)/a_h\varepsilon^{b_h} \to 1$ as $\varepsilon \to 0$, for some $a_h > 0, b_h \in (-\infty, \infty)$, then there exists the corresponding comparability limit,*

$$\lim_{\varepsilon \to 0} h(\varepsilon) = a_{\mathcal{H}_1}[h(\cdot)] = \begin{cases} 0 & \text{if } b_h > 0, \\ a_h & \text{if } b_h = 0, \\ \infty & \text{if } b_h < 0. \end{cases} \tag{8.21}$$

(iii) *$\mathcal{H}_1$ is a complete family of asymptotically comparable functions.*

Proof First, suppose $b_1 < b_2$. Then,

$$\begin{aligned} \frac{h'(\varepsilon)}{a'\varepsilon^{b'}} &= \frac{h_1(\varepsilon)}{a_1\varepsilon^{b_1}}\frac{a_1}{a'}\varepsilon^{b_1-b'} + \frac{h_2(\varepsilon)}{a_2\varepsilon^{b_2}}\frac{a_2}{a}\varepsilon^{b_2-b'} \\ &= \frac{h_1(\varepsilon)}{a_1\varepsilon^{b_1}} + \frac{h_2(\varepsilon)}{a_2\varepsilon^{b_2}}\frac{a_2}{a_1}\varepsilon^{b_2-b_1} \to 1 \text{ as } \varepsilon \to 0. \end{aligned} \tag{8.22}$$

Second, suppose $b_1 = b_2$. Then,

$$\begin{aligned} \frac{h'(\varepsilon)}{a'\varepsilon^{b'}} &= \frac{h_1(\varepsilon)}{a_1\varepsilon^{b_1}}\frac{a_1}{a}\varepsilon^{b_1-b'} + \frac{h_2(\varepsilon)}{a_2\varepsilon^{b_2}}\frac{a_2}{a}\varepsilon^{b_2-b} \\ &= \frac{h_1(\varepsilon)}{a_1\varepsilon^{b_1}}\frac{a_1}{a_1+a_2} + \frac{h_2(\varepsilon)}{a_2\varepsilon^{b_2}}\frac{a_2}{a_1+a_2} \to 1 \text{ as } \varepsilon \to 0. \end{aligned} \tag{8.23}$$

The third case, where $b_1 > b_2$, is similar to the first considered above.

Thus, the family $\mathcal{H}_1$ is closed with respect to the operation of summation.

Also, the relation (8.14) holds for the function $h^{(j)}(\varepsilon) = h_1(\varepsilon)h_2(\varepsilon)^j$, with the corresponding comparability parameters $h_j = a_1a_2^j$ and $b_j = b_1 + b_2 j$, for $j = \pm 1$. Indeed,

$$\frac{h^{(j)}(\varepsilon)}{a_j\varepsilon^{b_j}} = \frac{h_1(\varepsilon)h_2(\varepsilon)^j}{a_1a_2^j\varepsilon^{b_1+b_2j}} = \frac{h_1(\varepsilon)}{a_1\varepsilon^{b_1}}\Big(\frac{h_2(\varepsilon)}{a_2\varepsilon^{b_2}}\Big)^j \to 1 \text{ as } \varepsilon \to 0. \tag{8.24}$$

Thus, the family $\mathcal{H}_1$ is closed with respect to the operations of multiplication and division.

Finally, the relation (8.14) obviously implies that, for any function $h(\cdot) \in \mathcal{H}_1$, for which the asymptotic relation (8.14) takes the form, $h(\varepsilon)/a_h\varepsilon^{b_h} \to 1$ as $\varepsilon \to 0$, for some $a_h > 0, b_h \in (-\infty, \infty)$, there exists the following limit:

$$\begin{aligned} \lim_{\varepsilon \to 0} h(\varepsilon) &= \lim_{\varepsilon \to 0} a_h\varepsilon^{b_h} \\ &= a[h(\cdot)] = \begin{cases} 0 & \text{if } b_h > 0, \\ a_h & \text{if } b_h = 0, \\ \infty & \text{if } b_h < 0. \end{cases} \end{aligned} \tag{8.25}$$

Therefore, $\mathcal{H}_1$ is a complete family of asymptotically comparable functions. □

The following lemma generalises Lemma 8.4 to the case of multiple operations of summation, multiplication and division.

Lemma 8.5 *Let $h_i(\cdot) \in \mathcal{H}_1$, $i = 1, 2, \ldots$ and the asymptotic relation* (8.14) *takes for these functions the following forms, $h_i(\varepsilon)/a_i\varepsilon^{b_i} \longrightarrow 1$ as $\varepsilon \longrightarrow 0$, for $i = 1, 2, \ldots$, where $a_i > 0$, $b_i \in (-\infty, \infty)$, $i = 1, 2, \ldots$. Then the following operating formulas for calculating comparability parameters and limits for the multiple sums and products/quotients of these functions take place*:

(a) *The function $h^{(n)}(\cdot) = \sum_{i=1}^{n} h_i(\cdot) \in \mathcal{H}_1$, for $n \geq 1$, and the relation* (8.14) *holds for this function with comparability parameters,*

$$a^{(n)} = \sum_{1 \leq i \leq n:\, b_i = b^{(n)}} a_i, \quad b^{(n)} = \min_{1 \leq i \leq n} b_i, \tag{8.26}$$

i.e.,

$$\frac{h^{(n)}(\varepsilon)}{a^{(n)}\varepsilon^{b^{(n)}}} \longrightarrow 1 \text{ as } \varepsilon \longrightarrow 0, \tag{8.27}$$

and, thus,

$$\lim_{\varepsilon \to 0} h^{(n)}(\varepsilon) = a_{\mathcal{H}_1}[h^{(n)}(\cdot)] = \begin{cases} 0 & \text{if } b^{(n)} > 0, \\ a^{(n)} & \text{if } b^{(n)} = 0, \\ \infty & \text{if } b^{(n)} < 0. \end{cases} \tag{8.28}$$

(b) *The function $h_\circ^{(n)}(\cdot) = \prod_{i=1}^{n} h_i(\cdot)^{j_i} \in \mathcal{H}_1$, for $j_i = \pm 1, i = 1, 2, \ldots, n$ and $n \geq 1$, and the relation* (8.14) *holds for this function with comparability parameters,*

$$a_\circ^{(n)} = \prod_{1 \leq i \leq n} a_i^{j_i}, \quad b_\circ^{(n)} = \sum_{1 \leq i \leq n} b_i j_i, \tag{8.29}$$

i.e.,

$$\frac{h_\circ^{(n)}(\varepsilon)}{a_\circ^{(n)}\varepsilon^{b_\circ^{(n)}}} \longrightarrow 1 \text{ as } \varepsilon \longrightarrow 0, \tag{8.30}$$

and, thus,

$$\lim_{\varepsilon \to 0} h_\circ^{(n)}(\varepsilon) = a_{\mathcal{H}_1}[h_\circ^{(n)}(\cdot)] = \begin{cases} 0 & \text{if } b_\circ^{(n)} > 0, \\ a_\circ^{(n)} & \text{if } b_\circ^{(n)} = 0, \\ \infty & \text{if } b_\circ^{(n)} < 0. \end{cases} \tag{8.31}$$

It is also worth mentioning the family of asymptotically comparable integer-power-type functions $\hat{\mathcal{H}}_1 = \{h(\cdot)\}$, which includes all functions $h(\cdot)$ defined on the interval $(0, 1]$, taking values in the interval $(0, \infty)$ and such that, for any function $h(\cdot) \in \hat{\mathcal{H}}_1$, there exist a constant $a_h > 0$ and an integer $b_h \in (-\infty, \infty)$ such that the following asymptotic relation holds:

$$\frac{h(\varepsilon)}{a_h \varepsilon^{b_h}} \to 1 \text{ as } \varepsilon \to 0. \tag{8.32}$$

Obviously, $\hat{\mathcal{H}}_1$ is a complete family of asymptotically comparable functions, and $\hat{\mathcal{H}}_1 \subset \mathcal{H}_1$.

All relations appearing in Lemmas 8.4 and 8.5 hold for functions from the family $\hat{\mathcal{H}}_1$. The only difference is that the parameters b', b'', b'', $b^{(n)}$, and $b^{(n)}_\circ$, which appears in the relations (8.15)–(8.31), are integers in this case.

In applications, characteristics of perturbed stochastic models $f(\varepsilon)$ are often linearly perturbed positive functions, which means that they admit the following asymptotic expansions, $f(\varepsilon) = f_0 + f_1\varepsilon + o_f(\varepsilon)$ as $\varepsilon \to 0$, where, (a) $f_0 > 0$ or $f_0 = 0, f_1 > 0$, (b) $o_f(\varepsilon)/\varepsilon \to 0$ as $\varepsilon \to 0$.

Any such function $f(\varepsilon)$ belongs to the family $\hat{\mathcal{H}}_1$. Indeed, the following asymptotic relation holds, $f(\varepsilon)/a_f\varepsilon^{b_f} \to 1$ as $\varepsilon \to 0$, where $a_f = f_0, b_f = 0$, if $f_0 > 0$, or $a_f = f_1, b_f = 1$, if $f_0 = 0, f_1 > 0$.

A case similar to the one above is where $f(\varepsilon)$ are assumed to be nonlinearly perturbed functions, which means that they admit the following asymptotic expansions, $f(\varepsilon) = f_0 + f_1\varepsilon + \cdots + f_k\varepsilon^k + o_f(\varepsilon)$ as $\varepsilon \to 0$, where: (a) $f_n = 0, n < r < k, f_r > 0$, for some $0 \leq r \leq k$, (b) $o_f(\varepsilon)/\varepsilon^k \to 0$ as $\varepsilon \to 0$.

Any such function $f(\varepsilon)$ belongs to the family $\hat{\mathcal{H}}_1$. Indeed, the following asymptotic relation holds, $f(\varepsilon)/a_f\varepsilon^{b_f} \to 1$ as $\varepsilon \to 0$, where $a_f = f_r, b_f = r$.

8.2.2 Asymptotically Comparable Power-Exponential-Type Functions

Let us consider the family of asymptotically comparable power-exponential-type functions $\mathcal{H}_2 = \{h(\cdot)\}$, which includes all functions $h(\cdot)$ defined on the interval $(0, 1]$, taking values in the interval $(0, \infty)$ and such that, for any function $h(\cdot) \in \mathcal{H}_2$, there exist constants $a_h > 0$ and $b_h, c_h \in (-\infty, \infty)$ such that the following asymptotic relation holds:

$$\frac{h(\varepsilon)}{a_h \varepsilon^{b_h} e^{-c_h \varepsilon^{-1}}} \to 1 \text{ as } \varepsilon \to 0. \tag{8.33}$$

We call a_h, b_h, and c_h comparability parameters for the function $h(\cdot)$.

Lemma 8.6 *The following statements are true for the family of asymptotically comparable power-exponential-type functions $\mathcal{H}_2$:*

(i) *If the functions $h_i(\cdot) \in \mathcal{H}_2, i = 1, 2$, i.e., $h_i(\varepsilon)/a_i\varepsilon^{b_i}e^{-c_i\varepsilon^{-1}} \to 1$ as $\varepsilon \to 0$, for $i = 1, 2$, where $a_i > 0, b_i, c_i \in (-\infty, \infty), i = 1, 2$, then the following operating formulas for calculating the comparability parameters and limits for the sums, products, and quotients of these functions take place:* **(a)** *The function*

$h'(\varepsilon) = h_1(\varepsilon) + h_2(\varepsilon) \in \mathcal{H}_2$, *and the relation* (8.33) *holds for this function, with comparability parameters,*

$$\begin{aligned} a' &= a_1(\mathrm{I}(c_1 < c_2) + \mathrm{I}(c_1 = c_2, b_1 < b_2)) \\ &\quad + (a_1 + a_2)\mathrm{I}(c_1 = c_2, b_1 = b_2) + a_2(\mathrm{I}(c_1 > c_2) + \mathrm{I}(c_1 = c_2, b_1 > b_2)), \\ b' &= b_1\mathrm{I}(c_1 < c_2) + (b_1 \wedge b_2)\mathrm{I}(c_1 = c_2) + b_2\mathrm{I}(c_1 > c_2), \\ c' &= c_1 \wedge c_2, \end{aligned} \tag{8.34}$$

i.e.,

$$\frac{h'(\varepsilon)}{a'\varepsilon^{b'}e^{-c'\varepsilon^{-1}}} \to 1 \text{ as } \varepsilon \to 0. \tag{8.35}$$

(b) *The function* $h''(\varepsilon) = h_1(\varepsilon)h_2(\varepsilon) \in \mathcal{H}_2$, *and the relation* (8.33) *holds for this function, with comparability parameters,*

$$a'' = a_1 a_2', \ b'' = b_1 + b_2, \ c'' = c_1 + c_2, \tag{8.36}$$

i.e.,

$$\frac{h''(\varepsilon)}{a''\varepsilon^{b''}e^{-c''\varepsilon^{-1}}} \to 1 \text{ as } \varepsilon \to 0. \tag{8.37}$$

(c) *The function* $h'''(\varepsilon) = h_1(\varepsilon)/h_2(\varepsilon) \in \mathcal{H}_2$, *and the relation* (8.33) *holds for this function, with comparability parameters,*

$$a''' = a_1/a_2', \ b''' = b_1 - b_2, \ c''' = c_1 - c_2, \tag{8.38}$$

i.e.,

$$\frac{h'''(\varepsilon)}{a'''\varepsilon^{b'''}e^{-c'''\varepsilon^{-1}}} \to 1 \text{ as } \varepsilon \to 0. \tag{8.39}$$

(ii) *If the function* $h(\cdot) \in \mathcal{H}_2$, *i.e.,* $h(\varepsilon)/a_h\varepsilon^{b_h}e^{-c_h\varepsilon^{-1}} \to 1$ *as* $\varepsilon \to 0$, *for some* $a_h > 0$, $b_h, c_h \in (-\infty, \infty)$, *then there exists the corresponding comparability limit,*

$$\lim_{\varepsilon\to 0} h(\varepsilon) = a_{\mathcal{H}_2}[h(\cdot)] = \begin{cases} 0 & \text{if } c_h > 0 \text{ or } c_h = 0, \ b_h > 0, \\ a_h & \text{if } c_h = 0, \ b_h = 0, \\ \infty & \text{if } c_h < 0 \text{ or } c_h = 0, \ b_h < 0. \end{cases} \tag{8.40}$$

(iii) $\mathcal{H}_2$ *is a complete family of asymptotically comparable functions.*

Proof First, suppose $c_1 < c_2$ or $c_1 = c_2, b_1 < b_2$. Then,

$$\frac{h'(\varepsilon)}{a'\varepsilon^{b'}e^{-c'\varepsilon^{-1}}} = \frac{h_1(\varepsilon)}{a_1\varepsilon^{b_1}e^{-c_1\varepsilon^{-1}}}\frac{a_1}{a'}\varepsilon^{b_1-b}e^{-(c_1-c')\varepsilon^{-1}}$$
$$+ \frac{h_2(\varepsilon)}{a_2\varepsilon^{b_2}e^{-c_2\varepsilon^{-1}}}\frac{a_2}{a'}\varepsilon^{b_2-b'}e^{-(c_2-c')\varepsilon^{-1}}$$
$$= \frac{h_1(\varepsilon)}{a_1\varepsilon^{b_1}e^{-c_1\varepsilon^{-1}}} + \frac{h_2(\varepsilon)}{a_2\varepsilon^{b_2}e^{-c_2\varepsilon^{-1}}}\frac{a_2}{a_1}\varepsilon^{b_2-b_1}e^{-(c_2-c_1)\varepsilon^{-1}}$$
$$\to 1 \text{ as } \varepsilon \to 0. \tag{8.41}$$

Second, suppose $c_1 = c_2$, $b_1 = b_2$. Then,

$$\frac{h'(\varepsilon)}{a'\varepsilon^{b}e^{-c'\varepsilon^{-1}}} = \frac{h_1(\varepsilon)}{a_1\varepsilon^{b_1}e^{-c_1\varepsilon^{-1}}}\frac{a_1}{a'}\varepsilon^{b_1-b'}e^{-(c_1-c')\varepsilon^{-1}}$$
$$+ \frac{h_2(\varepsilon)}{a_2\varepsilon^{b_2}e^{-c_2\varepsilon^{-1}}}\frac{a_2}{a'}\varepsilon^{b_2-b'}e^{-(c_2-c')\varepsilon^{-1}}$$
$$= \frac{h_1(\varepsilon)}{a_1\varepsilon^{b_1}e^{-c_1\varepsilon^{-1}}}\frac{a_1}{a_1+a_2} + \frac{h_2(\varepsilon)}{a_2\varepsilon^{b_2}e^{-c_2\varepsilon^{-1}}}\frac{a_2}{a_1+a_2}$$
$$\to 1 \text{ as } \varepsilon \to 0. \tag{8.42}$$

The third case, where $c_1 > c_2$ or $c_1 = c_2, b_1 > b_2$, is similar to the first case considered above.

Thus, the family $\mathcal{H}_2$ is closed with respect to the operation of summation.

Also, the relation (8.33) holds for the function $h^{(\jmath)}(\varepsilon) = h_1(\varepsilon)h_2(\varepsilon)^{\jmath}$, with the corresponding comparability parameters $h_\jmath = a_1a_2^\jmath$, $b_\jmath = b_1 + b_2\jmath$, and $c_\jmath = c_1 + c_2\jmath$, for $\jmath = \pm 1$. Indeed,

$$\frac{h^{(\jmath)}(\varepsilon)}{a_\jmath\varepsilon^{b_\jmath}e^{-c_\jmath\varepsilon^{-1}}} = \frac{h_1(\varepsilon)h_2(\varepsilon)^\jmath}{a_1a_2^\jmath\varepsilon^{b_1+b_2\jmath}e^{-(c_1+c_2\jmath)\varepsilon^{-1}}}$$
$$= \frac{h_1(\varepsilon)}{a_1\varepsilon^{b_1}e^{-c_1\varepsilon^{-1}}}\left(\frac{h_2(\varepsilon)}{a_2\varepsilon^{b_2}e^{-c_2\varepsilon^{-1}}}\right)^\jmath \to 1 \text{ as } \varepsilon \to 0. \tag{8.43}$$

Thus, the family $\mathcal{H}_2$ is closed with respect to the operations of multiplication and division.

Finally, the relation (8.33) obviously implies that, for any function $h(\cdot) \in \mathcal{H}_2$, for which the asymptotic relation (8.33) takes the form, $h(\varepsilon)/a_h\varepsilon^{b_h}e^{-c_h\varepsilon^{-1}} \to 1$ as $\varepsilon \to 0$, for some $a_h > 0$, $b_h, c_h \in (-\infty, \infty)$, there exists the following limit:

$$\lim_{\varepsilon\to 0} h(\varepsilon) = \lim_{\varepsilon\to 0} a_h\varepsilon^{b_h}e^{-c_h\varepsilon^{-1}}$$
$$= a_{\mathcal{H}_2}[h(\cdot)] = \begin{cases} 0 & \text{if } c_h > 0 \text{ or } c_h = 0,\ b_h > 0, \\ a_h & \text{if } c_h = 0,\ b_h = 0, \\ \infty & \text{if } c_h < 0 \text{ or } c_h = 0,\ b_h < 0. \end{cases} \tag{8.44}$$

Therefore, $\mathcal{H}_2$ is a complete family of asymptotically comparable functions. □

The following lemma generalises Lemma 8.6 to the case of multiple operations of summation, multiplication and division.

Lemma 8.7 *Let $h_i(\cdot) \in \mathcal{H}_2$, $i = 1, 2, \ldots$ and the asymptotic relation* (8.33) *takes for these functions the following forms, $h_i(\varepsilon)/a_i \varepsilon^{b_i} e^{-c_i \varepsilon^{-1}} \to 1$ as $\varepsilon \to 0$, for $i = 1, 2, \ldots$, where $a_i > 0$, $b_i, c_i \in (-\infty, \infty)$, $i = 1, 2, \ldots$. Then the following operating formulas for calculating comparability parameters and limits for the multiple sums and products/quotients of these functions take place*:

(a) *The function $h^{(n)}(\cdot) = \sum_{i=1}^{n} h_i(\cdot) \in \mathcal{H}_1$, for $n \geq 1$, and the relation* (8.14) *holds for this function with comparability parameters,*

$$a^{(n)} = \sum_{1 \leq i \leq n:\, b_i = b^{(n)}, c_i = c^{(n)}} a_i, \ b^{(n)} = \min_{1 \leq i \leq n: c_i = c^{(n)}} b_i, \ c^{(n)} = \min_{1 \leq i \leq n} c_i, \tag{8.45}$$

i.e.,

$$\frac{h^{(n)}(\varepsilon)}{a^{(n)} \varepsilon^{b^{(n)}} e^{-c^{(n)} \varepsilon^{-1}}} \to 1 \text{ as } \varepsilon \to 0, \tag{8.46}$$

and, thus,

$$\lim_{\varepsilon \to 0} h^{(n)}(\varepsilon) = a_{\mathcal{H}_2}[h^{(n)}(\cdot)] = \begin{cases} 0 & \text{if } c^{(n)} > 0 \text{ or } c^{(n)} = 0,\ b^{(n)} > 0, \\ a^{(n)} & \text{if } c^{(n)} = 0,\ b^{(n)} = 0, \\ \infty & \text{if } c^{(n)} < 0 \text{ or } c^{(n)} = 0,\ b^{(n)} < 0. \end{cases} \tag{8.47}$$

(b) *The function $h_{\circ}^{(n)}(\cdot) = \prod_{i=1}^{n} h_i(\cdot)^{j_i} \in \mathcal{H}_1$, for $j_i = \pm 1, i = 1, 2, \ldots, n$ and $n \geq 1$, and the relation* (8.33) *holds for this function with comparability parameters,*

$$a_{\circ}^{(n)} = \prod_{1 \leq i \leq n} a_i^{j_i}, \ b_{\circ}^{(n)} = \sum_{1 \leq i \leq n} b_i j_i, \ c_{\circ}^{(n)} = \sum_{1 \leq i \leq n} c_i j_i, \tag{8.48}$$

i.e.,

$$\frac{h_{\circ}^{(n)}(\varepsilon)}{a_{\circ}^{(n)} \varepsilon^{b_{\circ}^{(n)}} e^{-c_{\circ}^{(n)} \varepsilon^{-1}}} \to 1 \text{ as } \varepsilon \to 0, \tag{8.49}$$

and, thus,

$$\lim_{\varepsilon \to 0} h_{\circ}^{(n)}(\varepsilon) = a_{\mathcal{H}_2}[h_{\circ}^{(n)}(\cdot)] = \begin{cases} 0 & \text{if } c_{\circ}^{(n)} > 0 \text{ or } c_{\circ}^{(n)} = 0,\ b_{\circ}^{(n)} > 0, \\ a_{\circ}^{(n)} & \text{if } c_{\circ}^{(n)} = 0,\ b_{\circ}^{(n)} = 0, \\ \infty & \text{if } c_{\circ}^{(n)} < 0 \text{ or } c_{\circ}^{(n)} = 0,\ b_{\circ}^{(n)} < 0. \end{cases} \tag{8.50}$$

8.2.3 Asymptotically Comparable Power-Logarithmic-Type Functions

Let us consider the family of asymptotically comparable power-logarithmic-type functions $\mathcal{H}_3 = \{h(\cdot)\}$, which includes all functions $h(\cdot)$ defined on the interval $(0, 1]$, taking values in the interval $(0, \infty)$ and such that, for any function $h(\cdot) \in \mathcal{H}_2$, there exist constants $a_h > 0$ and $b_h, d_h \in (-\infty, \infty)$ such that the following asymptotic relation holds:

$$\frac{h(\varepsilon)}{a_h \varepsilon^{b_h}(1 + \ln \varepsilon^{-1})^{-d_h}} \to 1 \text{ as } \varepsilon \to 0. \tag{8.51}$$

We call a_h, b_h and d_h comparability parameters for the function $h(\cdot)$.

Lemma 8.8 *The following statements are true for the family of asymptotically comparable power-logarithmic-type functions $\mathcal{H}_3$:*

(i) *If the functions $h_i(\cdot) \in \mathcal{H}_3, i = 1, 2$, i.e., $h_i(\varepsilon)/a_i\varepsilon^{b_i}(1 + \ln \varepsilon^{-1})^{-d_i} \to 1$ as $\varepsilon \to 0$, for $i = 1, 2$, where $a_i > 0$, $b_i, d_i \in (-\infty, \infty), i = 1, 2$, then the following operating formulas for calculating the comparability parameters and limits for the sums, products, and quotients of these functions take place:*

(a) *The function $h'(\varepsilon) = h_1(\varepsilon) + h_2(\varepsilon) \in \mathcal{H}_3$, and the relation* (8.51) *holds for this function, with comparability parameters,*

$$\begin{aligned} a' &= a_1(\mathrm{I}(b_1 < b_2) + \mathrm{I}(b_1 = b_2, d_1 < d_2)) \\ &\quad + (a_1 + a_2)\mathrm{I}(b_1 = b_2, d_1 = d_2) + a_2(\mathrm{I}(b_1 > b_2) + \mathrm{I}(b_1 = b_2, d_1 > d_2)), \\ b' &= b_1 \wedge b_2, \\ d' &= d_1\mathrm{I}(b_1 < b_2) + (d_1 \wedge d_2)\mathrm{I}(b_1 = b_2) + d_2\mathrm{I}(b_1 > b_2). \end{aligned} \tag{8.52}$$

i.e.,

$$\frac{h'(\varepsilon)}{a'\varepsilon^{b'}(1 + \ln \varepsilon^{-1})^{-d'}} \to 1 \text{ as } \varepsilon \to 0. \tag{8.53}$$

(b) *The function $h''(\varepsilon) = h_1(\varepsilon)h_2(\varepsilon) \in \mathcal{H}_3$, and the relation* (8.51) *holds for this function, with comparability parameters,*

$$a'' = a_1a_2,\ b'' = b_1 + b_2,\ d'' = d_1 + d_2, \tag{8.54}$$

i.e.,

$$\frac{h''(\varepsilon)}{a''\varepsilon^{b''}(1 + \ln \varepsilon^{-1})^{-d''}} \to 1 \text{ as } \varepsilon \to 0. \tag{8.55}$$

(c) *The function $h'''(\varepsilon) = h_1(\varepsilon)/h_2(\varepsilon) \in \mathcal{H}_3$, and the relation* (8.33) *holds for this function, with comparability parameters,*

$$a''' = a_1/a_2,\ b''' = b_1 - b_2,\ d''' = d_1 - d_2, \tag{8.56}$$

i.e.,

$$\frac{h'''(\varepsilon)}{a'''\varepsilon^{b'''}(1 + \ln \varepsilon^{-1})^{-d'''}} \to 1 \text{ as } \varepsilon \to 0. \tag{8.57}$$

(ii) *If the function $h(\cdot) \in \mathcal{H}_3$, i.e., $h(\varepsilon)/a_h\varepsilon^{b_h}(1 + \ln \varepsilon^{-1})^{-d_h} \to 1$ as $\varepsilon \to 0$, for some $a_h > 0$, $b_h, d_h \in (-\infty, \infty)$, then there exists the following limit:*

$$\lim_{\varepsilon \to 0} h(\varepsilon) = a_{\mathcal{H}_3}[h(\cdot)] = \begin{cases} 0 & \text{if } b_h > 0 \text{ or } b_h = 0,\ d_h > 0, \\ a_h & \text{if } b_h = 0,\ d_h = 0, \\ \infty & \text{if } b_h < 0 \text{ or } b_h = 0,\ d_h < 0. \end{cases} \tag{8.58}$$

(iii) *$\mathcal{H}_3$ is a complete family of asymptotically comparable functions.*

Proof First, suppose $b_1 < b_2$ or $b_1 = b_2, d_1 < d_2$. Then,

$$\begin{aligned} \frac{h'(\varepsilon)}{a'\varepsilon^{b}(1 + \ln \varepsilon^{-1})^{-d'}} &= \frac{h_1(\varepsilon)}{a_1\varepsilon^{b_1}(1 + \ln \varepsilon^{-1})^{-d_1}} \frac{a_1}{a'}\varepsilon^{b_1-b'}(1 + \ln \varepsilon^{-1})^{-(d_1-d')} \\ &\quad + \frac{h_2(\varepsilon)}{a_2\varepsilon^{b_2}(1 + \ln \varepsilon^{-1})^{-d_2}} \frac{a_2}{a'}\varepsilon^{b_2-b'}(1 + \ln \varepsilon^{-1})^{-(d_2-d')} \\ &= \frac{h_1(\varepsilon)}{a_1\varepsilon^{b_1}(1 + \ln \varepsilon^{-1})^{-d_1}} \\ &\quad + \frac{h_2(\varepsilon)}{a_2\varepsilon^{b_2}(1 + \ln \varepsilon^{-1})^{-d_2}} \frac{a_2}{a_1}\varepsilon^{b_2-b_1}(1 + \ln \varepsilon^{-1})^{-(d_2-d_1)} \\ &\to 1 \text{ as } \varepsilon \to 0. \end{aligned} \tag{8.59}$$

Second, suppose $b_1 = b_2, d_1 = d_2$. Then,

$$\begin{aligned} \frac{h'(\varepsilon)}{a'\varepsilon^{b}(1 + \ln \varepsilon^{-1})^{-d'}} &= \frac{h_1(\varepsilon)}{a_1\varepsilon^{b_1}(1 + \ln \varepsilon^{-1})^{-d_1}} \frac{a_1}{a'}\varepsilon^{b_1-b'}(1 + \ln \varepsilon^{-1})^{-(d_1-d')} \\ &\quad + \frac{h_2(\varepsilon)}{a_2\varepsilon^{b_2}(1 + \ln \varepsilon^{-1})^{-d_2}} \frac{a_2}{a'}\varepsilon^{b_2-b'}(1 + \ln \varepsilon^{-1})^{-(d_2-d')} \\ &= \frac{h_1(\varepsilon)}{a_1\varepsilon^{b_1}(1 + \ln \varepsilon^{-1})^{-d_1}} \frac{a_1}{a_1 + a_2} \\ &\quad + \frac{h_2(\varepsilon)}{a_2\varepsilon^{b_2}(1 + \ln \varepsilon^{-1})^{-d_2}} \frac{a_2}{a_1 + a_2} \\ &\to 1 \text{ as } \varepsilon \to 0. \end{aligned} \tag{8.60}$$

Thus, the family $\mathcal{H}_3$ is closed with respect to the operation of summation.

Also, the relation (8.51) holds for the function $h^{(\jmath)}(\varepsilon) = h_1(\varepsilon)h_2(\varepsilon)^{\jmath}$, with the comparability parameters $h_\jmath = a_1 a_2^{\jmath}$, $b_\jmath = b_1 + b_2\jmath$, and $d_\jmath = d_1 + d_2\jmath$ for $\jmath = \pm 1$. Indeed,

$$\begin{aligned} \frac{h^{(\jmath)}(\varepsilon)}{a_\jmath\varepsilon^{b_\jmath}(1 + \ln \varepsilon^{-1})^{-d_\jmath}} &= \frac{h_1(\varepsilon)h_2(\varepsilon)^{\jmath}}{a_1 a_2^{\jmath}\varepsilon^{b_1+b_2\jmath}(1 + \ln \varepsilon^{-1})^{-(d_1+d_2\jmath)}} \\ &= \frac{h_1(\varepsilon)}{a_1\varepsilon^{b_1}(1 + \ln \varepsilon^{-1})^{-d_1}} \Big(\frac{h_2(\varepsilon)}{a_2\varepsilon^{b_2}(1 + \ln \varepsilon^{-1})^{-d_2}}\Big)^{\jmath} \\ &\to 1 \text{ as } \varepsilon \to 0. \end{aligned} \tag{8.61}$$

Thus, the family $\mathcal{H}_3$ is closed with respect to the operations of multiplication and division.

Finally, the relation (8.51) obviously implies that, for any function $h(\cdot) \in \mathcal{H}_3$, for which the asymptotic relation (8.51) holds, i.e., $h(\varepsilon)/a_h\varepsilon^{b_h}(1+\ln\varepsilon^{-1})^{-d_h} \to 1$ as $\varepsilon \to 0$, for some $a_h > 0$, $b_h, d_h \in (-\infty, \infty)$, there exists the following limit:

$$\lim_{\varepsilon\to 0} h(\varepsilon) = \lim_{\varepsilon\to 0} a_h\varepsilon^{b_h}(1+\ln\varepsilon^{-1})^{-d_h}$$

$$= a_{\mathcal{H}_3}[h(\cdot)] = \begin{cases} 0 & \text{if } b_h > 0 \text{ or } b_h = 0,\ d_h > 0, \\ a_h & \text{if } b_h = 0,\ d_h = 0, \\ \infty & \text{if } b_h < 0 \text{ or } b_h = 0,\ d_h < 0. \end{cases} \tag{8.62}$$

Therefore, $\mathcal{H}_3$ is a complete family of asymptotically comparable functions. □

The following lemma generalises Lemma 8.8 to the case of multiple operations of summation, multiplication, and division.

Lemma 8.9 *Let $h_i(\cdot) \in \mathcal{H}_3$, $i = 1, 2, \ldots$ and the asymptotic relation* (8.51) *takes for these functions the following forms, $h_i(\varepsilon)/a_i\varepsilon^{b_i}(1+\ln\varepsilon^{-1})^{-d_i} \to 1$ as $\varepsilon \to 0$, for $i = 1, 2, \ldots$, where $a_i > 0$, $b_i, d_i \in (-\infty, \infty)$, $i = 1, 2, \ldots$. Then the following operating formulas for calculating comparability parameters and limits for the multiple sums and products/quotients of these functions take place:*

(a) *The function $h^{(n)}(\cdot) = \sum_{i=1}^n h_i(\cdot) \in \mathcal{H}_3$, for $n \geq 1$, and the relation* (8.14) *holds for this function with comparability parameters,*

$$a^{(n)} = \sum_{1\leq i\leq n:\, d_i = d^{(n)}, b_i = b^{(n)}} a_i,\ b^{(n)} = \min_{1\leq i\leq n} b_i,\ d^{(n)} = \min_{1\leq i\leq n: b_i = b^{(n)}} d_i, \tag{8.63}$$

i.e.,

$$\frac{h^{(n)}(\varepsilon)}{a^{(n)}\varepsilon^{b^{(n)}}(1+\ln\varepsilon^{-1})^{-d^{(n)}}} \to 1 \text{ as } \varepsilon \to 0, \tag{8.64}$$

and, thus,

$$\lim_{\varepsilon\to 0} h^{(n)}(\varepsilon) = a_{\mathcal{H}_3}[h^{(n)}(\cdot)] = \begin{cases} 0 & \text{if } b^{(n)} > 0 \text{ or } b^{(n)} = 0,\ d^{(n)} > 0, \\ a^{(n)} & \text{if } b^{(n)} = 0,\ d^{(n)} = 0, \\ \infty & \text{if } b^{(n)} < 0 \text{ or } b^{(n)} = 0,\ d^{(n)} < 0. \end{cases} \tag{8.65}$$

(b) *The function $h_\circ^{(n)}(\cdot) = \prod_{i=1}^n h_i(\cdot)^{\jmath_i} \in \mathcal{H}_1$, for $\jmath_i = \pm 1, i = 1, 2, \ldots, n$ and $n \geq 1$, and the relation* (8.33) *holds for this function with comparability parameters,*

$$a_\circ^{(n)} = \prod_{1\leq i\leq n} a_i^{\jmath_i},\ b_\circ^{(n)} = \sum_{1\leq i\leq n} b_i\jmath_i,\ d_\circ^{(n)} = \sum_{1\leq i\leq n} d_i\jmath_i, \tag{8.66}$$

i.e.,

$$\frac{h_\circ^{(n)}(\varepsilon)}{a_\circ^{(n)}\varepsilon^{b_\circ^{(n)}}(1+\ln\varepsilon^{-1})^{-d_\circ^{(n)}}} \to 1 \text{ as } \varepsilon \to 0, \tag{8.67}$$

and, thus,

$$\lim_{\varepsilon\to 0} h_\circ^{(n)}(\varepsilon) = a_{\mathcal{H}_2}[h_\circ^{(n)}(\cdot)] = \begin{cases} 0 & \text{if } b_\circ^{(n)} > 0 \text{ or } b_\circ^{(n)} = 0,\ d_\circ^{(n)} > 0, \\ a_\circ^{(n)} & \text{if } b_\circ^{(n)} = 0,\ d_\circ^{(n)} = 0, \\ \infty & \text{if } b_\circ^{(n)} < 0 \text{ or } b_\circ^{(n)} = 0,\ d_\circ^{(n)} < 0. \end{cases} \tag{8.68}$$

In conclusion, we note that many other examples of complete families of asymptotically comparable functions based on power, exponential, logarithmic, iterated logarithmic, and other types of functions can be constructed in a similar way.

Chapter 9
Perturbed Semi-Markov Processes and Reduction of Phase Space

In this chapter, we present models of regularly and singularly perturbed semi-Markov processes and formulate basic perturbation conditions and asymptotic comparability conditions based on the notion of a complete family of asymptotically comparable functions. We also describe asymptotic procedures of removing virtual transitions and one-state reducing the phase space, and an asymptotic recurrent phase space reduction algorithm for perturbed semi-Markov processes. These procedures and algorithm are provided by explicit formulas for recalculating normalisation functions, Laplace transforms for limit distributions, and limit expectations appearing in perturbation conditions for semi-Markov processes with reduced phase space. This chapter includes four sections.

In Sect. 9.1, models of regularly and singularly perturbed semi-Markov processes are introduced and basic perturbation conditions are formulated.

In Sect. 9.2, the asymptotic procedure of removing virtual transition for perturbed semi-Markov processes is described. The corresponding recurrent formulas for re-calculating normalisation functions, limiting distributions and expectations in the corresponding perturbation conditions are given in Lemmas 9.3–9.13 and summarised in Theorem 9.1.

In Sect. 9.3, the asymptotic procedure of one-state reduction of phase space for perturbed semi-Markov processes is described. The corresponding recurrent formulas for re-calculating normalisation functions, limiting distributions and expectations in the corresponding perturbation conditions are given in Lemmas 9.14–9.25 and summarised in Theorem 9.2.

In Sect. 9.4, the asymptotic recurrent algorithm of phase space reduction for perturbed semi-Markov processes is described. The invariance properties of the perturbation conditions with respect to the above asymptotic recurrent phase space reduction algorithm are summarised in Theorem 9.3.

D. Silvestrov, *Perturbed Semi-Markov Type Processes I*,
https://doi.org/10.1007/978-3-030-92403-4_9

9.1 Perturbed Semi-Markov Processes

In Sect. 9.1, models of regularly and singularly perturbed semi-Markov processes are introduced and basic perturbation conditions are formulated.

9.1.1 Perturbed Semi-Markov Processes

Let $\mathbb{X} = \{1, \ldots, m\}$ be a finite set. Let also $(\eta_{\varepsilon,n}, \kappa_{\varepsilon,n}), n = 0, 1, \ldots$ be, for every $\varepsilon \in (0, 1]$, a Markov renewal process, i.e., a homogeneous Markov chain, with the phase space $\mathbb{X} \times [0, \infty)$ and transition probabilities, for $(i, s), (j, t) \in \mathbb{X} \times [0, \infty)$,

$$Q_{\varepsilon,ij}(t) = \mathsf{P}\{\eta_{\varepsilon,1} = j, \kappa_{\varepsilon,1} \leq t/\eta_{\varepsilon,0} = i, \kappa_{\varepsilon,0} = s\}. \tag{9.1}$$

The first component of the Markov renewal process $\eta_{\varepsilon,n}, n = 0, 1, \ldots$ is itself a homogeneous, so-called embedded, Markov chain, with the phase space $\mathbb{X}$ and transition probabilities, for $i, j \in \mathbb{X}$,

$$p_{\varepsilon,ij} = Q_{\varepsilon,ij}(\infty) = \mathsf{P}\{\eta_{\varepsilon,1} = j, /\eta_{\varepsilon,0} = i\}. \tag{9.2}$$

The above Markov renewal process is used to define a semi-Markov process,

$$\eta_\varepsilon(t) = \eta_{\varepsilon,\nu_\varepsilon(t)}, \ t \geq 0, \tag{9.3}$$

where $\zeta_{\varepsilon,n} = \kappa_{\varepsilon,1} + \cdots + \kappa_{\varepsilon,n}, n = 1, 2, \ldots, \zeta_{\varepsilon,0} = 0$, are the corresponding moments of jumps, and $\nu_\varepsilon(t) = \max(n \geq 1 : \zeta_{\varepsilon,n} \leq t)$ is the number of jumps in the interval $[0, t]$ for the above semi-Markov process.

Perturbed semi-Markov processes are the main objects of study in the book. The basic fact concerning these processes can be found in the books listed in Sect. B.2.7.

Here we would like to mention again two important cases. The semi-Markov process $\eta_\varepsilon(t)$ is a continuous time Markov chain, if its transition probabilities, $Q_{\varepsilon,ij}(t) = (1 - e^{-\lambda_{\varepsilon,i} t}) p_{\varepsilon,ij}, \ t \geq 0, \ i, j \in \mathbb{X}$ or a discrete time Markov chain embedded in continuous time, if its transition probabilities, $Q_{\varepsilon,ij}(t) = \mathrm{I}(t \geq 1) p_{\varepsilon,ij}, \ t \geq 0, \ i, j \in \mathbb{X}$.

Let us introduce the distribution function of sojourn time $\kappa_{\varepsilon,n}$ for the semi-Markov process $\eta_\varepsilon(t)$, for $i \in \mathbb{X}$,

$$F_{\varepsilon,i}(t) = \mathsf{P}_i\{\kappa_{\varepsilon,1} \leq t\} = \sum_{j \in \mathbb{X}} Q_{\varepsilon,ij}(t), \ t \geq 0. \tag{9.4}$$

We assume that the following regularity condition is satisfied:

G: **(a)** $p_{\varepsilon,ij} > 0, \varepsilon \in (0, 1]$ or $p_{\varepsilon,ij} = 0, \varepsilon \in (0, 1]$, for every $i, j \in \mathbb{X}$, **(b)** $F_{\varepsilon,i}(0) < 1$, for $i \in \mathbb{X}, \varepsilon \in (0, 1]$.

Let us introduce sets, for $i \in \mathbb{X}$ and $\varepsilon \in (0, 1]$,

$$\mathbb{Y}_{\varepsilon,i} = \{j \in \mathbb{X} : p_{\varepsilon,ij} > 0\}. \tag{9.5}$$

The following obvious lemma presents an equivalent form of the condition **G (a)**.

Lemma 9.1 *The condition* **G (a)** *is satisfied if and only if sets* $\mathbb{Y}_{\varepsilon,i} = \mathbb{Y}_{1,i}, \varepsilon \in (0, 1]$, *for* $i \in \mathbb{X}$.

The condition **G (b)** guarantees that $\mathsf{P}_i\{\lim_{n\to\infty} \zeta_{\varepsilon,n} = \infty\} = 1$, for any $i \in \mathbb{X}$ and $\varepsilon \in (0, 1]$. Therefore, the process $\eta_\varepsilon(t), t \geq 0$ is well defined on the time interval $[0, \infty)$, for each $\varepsilon \in (0, 1]$.

We also assume that the following ergodicity condition is satisfied:

H: For any $i, j \in \mathbb{X}$, there exist $n_{ij} \geq 1$ and a chain of states $i = r_0, r_1, \ldots, r_{n_{ij}} = j$ such that $\prod_{1\leq l\leq n_{ij}} p_{1,r_{l-1}r_l} > 0$.

The condition **G (a)** implies that the product $\prod_{1\leq l\leq n_{ij}} p_{\varepsilon,r_{l-1}r_l}$ is a positive number or equals zero simultaneously for all $\varepsilon \in (0, 1]$.

Therefore, the condition **H (b)** implies that the phase space $\mathbb{X}$ is one class of communicative states for the embedded Markov chain $\eta_{\varepsilon,n}$, and, thus, this Markov chain is ergodic, for every $\varepsilon \in (0, 1]$.

9.1.2 Perturbation Conditions

In what follows, we assume that the conditions **G** and **H** are satisfied.

We also use that the following perturbation condition:

I: $p_{\varepsilon,ij} \longrightarrow p_{0,ij}$ as $\varepsilon \longrightarrow 0$, for $i, j \in \mathbb{X}$.

Since the matrix $\mathbf{P}_\varepsilon = \|p_{\varepsilon,ij}\|$ is stochastic, for $\varepsilon \in (0, 1]$, the condition **I** implies that the matrix $\mathbf{P}_0 = \|p_{0,ij}\|$ is also stochastic.

Let $\eta_{0,n}, n = 0, 1, \ldots$ be a Markov chain with the phase space $\mathbb{X}$ and the matrix of transition probabilities $\mathbf{P}_0$.

Condition **I** allows us to interpret the Markov chain $\eta_{\varepsilon,n}$, for $\varepsilon \in (0, 1]$, as a perturbed version of the Markov chain $\eta_{0,n}$.

The case, where the phase space $\mathbb{X}$ is one class of communicative states plus, possibly, the class of transient states for the Markov chain $\eta_{0,n}$, refers to the model with **regular** perturbations.

The case, where the phase space $\mathbb{X}$ is split into several closed classes of communicative states plus, possibly, the class of transient states for the Markov chain $\eta_{0,n}$, refers to the model with **singular** perturbations.

Let us introduce sets, for $i \in \mathbb{X}$,

$$\mathbb{Y}_{0,i} = \{j \in \mathbb{X} : p_{0,ij} > 0\}. \tag{9.6}$$

Lemma 9.2 *The conditions* **G** *and* **I** *imply that* $\mathbb{Y}_{0,i} \subseteq \mathbb{Y}_{1,i}$, *for* $i \in \mathbb{X}$.

The transition probabilities $Q_{\varepsilon,ij}(t)$ can, for every $\varepsilon \in (0, 1]$, be represented in the following form, for $t \geq 0, i, j \in \mathbb{X}$:

$$Q_{\varepsilon,ij}(t) = F_{\varepsilon,ij}(t) p_{\varepsilon,ij}, \tag{9.7}$$

where

$$F_{\varepsilon,ij}(t) = \mathsf{P}\{\kappa_{\varepsilon,1} \leq t/\eta_{\varepsilon,0} = i, \eta_{\varepsilon,1} = j\}. \tag{9.8}$$

If $j \in \mathbb{Y}_{1,i}, i \in \mathbb{X}$, then,

$$F_{\varepsilon,ij}(t) = p_{\varepsilon,ij}^{-1} Q_{\varepsilon,ij}(t), t \geq 0. \tag{9.9}$$

Let us recall the distribution function of the sojourn time in state $i \in \mathbb{X}$ for the semi-Markov process $\eta_\varepsilon(t)$,

$$\begin{aligned} F_{\varepsilon,i}(t) &= \mathsf{P}\{\kappa_{\varepsilon,1} \leq t/\eta_{\varepsilon,0} = i\} \\ &= \sum_{j \in \mathbb{X}} Q_{\varepsilon,ij}(t) = \sum_{j \in \mathbb{Y}_{1,i}} Q_{\varepsilon,ij}(t) = \sum_{j \in \mathbb{Y}_{1,i}} F_{\varepsilon,ij}(t) p_{\varepsilon,ij}. \end{aligned} \tag{9.10}$$

If $j \in \bar{\mathbb{Y}}_{1,i}, i \in \mathbb{X}$, an arbitrary distribution function concentrated on $[0, \infty)$ can play the role of $F_{\varepsilon,ij}(t)$.

The choice of $F_{\varepsilon,ij}(t)$ for $j \in \bar{\mathbb{Y}}_{1,i}, i \in \mathbb{X}$ does not affect the transition probabilities $Q_{\varepsilon,ij}(t)$. In any case, the following relations hold, for $t \geq 0, i, j \in \mathbb{X}$ and $\varepsilon \in (0, 1]$:

$$Q_{\varepsilon,ij}(t) = F_{\varepsilon,ij}(t) p_{\varepsilon,ij}. \tag{9.11}$$

In what follows, we use the standard variant and choose,

$$F_{\varepsilon,ij}(t) = F_{\varepsilon,i}(t), \ t \geq 0. \tag{9.12}$$

In the case of continuous and discrete time Markov chains, $F_{\varepsilon,ij}(t) = F_{\varepsilon,i}(t) = 1 - e^{-\lambda_{\varepsilon,i} t}, t \geq 0, i \in \mathbb{X}$ and $F_{\varepsilon,ij}(t) = F_{\varepsilon,i}(t) = \mathrm{I}(t \geq 1), t \geq 0, i \in \mathbb{X}$, respectively.

We also use the following perturbation condition:

J: **(a)** $F_{\varepsilon,ij}(\cdot\, u_{\varepsilon,i}) \Rightarrow F_{0,ij}(\cdot)$ as $\varepsilon \to 0$, for $j \in \mathbb{Y}_{1,i}, i \in \mathbb{X}$, **(b)** $F_{0,ij}(\cdot), j \in \mathbb{Y}_{1,i}, i \in \mathbb{X}$ are proper distribution functions such that $F_{0,ij}(0) < 1, j \in \mathbb{Y}_{1,i}, i \in \mathbb{X}$, **(c)** $u_{\varepsilon,i} \in (0, \infty), \varepsilon \in (0, 1]$, for $i \in \mathbb{X}$.

We call the functions $u_{\varepsilon,i}$ "initial local" normalisation functions.

The standard case is, where all initial normalisation functions $u_{\varepsilon,i} = 1, \varepsilon \in (0, 1]$.

The conditions **G**, **I**, and **J** imply that the following relation of weak convergence holds, for $i \in \mathbb{X}$:

$$
\begin{aligned}
F_{\varepsilon,i}(\cdot u_{\varepsilon,i}) &= \sum_{j \in \mathbb{Y}_{1,i}} F_{\varepsilon,ij}(\cdot u_{\varepsilon,i}) p_{\varepsilon,ij} \\
&\Rightarrow \sum_{j \in \mathbb{Y}_{1,i}} F_{0,ij}(\cdot) p_{0,ij} \\
&= \sum_{j \in \mathbb{Y}_{0,i}} F_{0,ij}(\cdot) p_{0,ij} = F_{0,i}(\cdot) \text{ as } \varepsilon \to 0. \qquad (9.13)
\end{aligned}
$$

The condition **J** and Lemma 9.2 imply that $F_{0,i}(t)$ is a proper distribution function such that $F_{0,i}(0) < 1$, for $i \in \mathbb{X}$.

The relations (9.12) and (9.13) imply that, in fact, the weak convergence relation appearing in the condition **J** is satisfied for all $i, j \in \mathbb{X}$.

The condition **J (b)** guarantees that the initial local normalisation functions $u_{\varepsilon,i}$ do not deliver excessive levels of local normalisation.

Let us introduce Laplace transforms, for $i, j \in \mathbb{X}$ and $\varepsilon \in (0, 1]$,

$$
\phi_{\varepsilon,ij}(s) = \int_0^\infty e^{-st} F_{\varepsilon,ij}(dt), \ s \geq 0 \qquad (9.14)
$$

and

$$
\phi_{\varepsilon,i}(s) = \int_0^\infty e^{-st} F_{\varepsilon,i}(dt), \ s \geq 0. \qquad (9.15)
$$

According to the relation (9.12), for $j \in \bar{\mathbb{Y}}_{1,i}, i \in \mathbb{X}$ and $\varepsilon \in (0, 1]$,

$$
\phi_{\varepsilon,ij}(s) = \phi_{\varepsilon,i}(s), \ s \geq 0. \qquad (9.16)
$$

The condition **J** can also be re-formulated in the following equivalent form:

J$^\circ$: **(a)** $\phi_{\varepsilon,ij}(s/u_{\varepsilon,i}) \to \phi_{0,ij}(s)$ as $\varepsilon \to 0$, for $s \geq 0$ and $j \in \mathbb{Y}_{1,i}, i \in \mathbb{X}$, **(b)** $\phi_{0,ij}(s) = \int_0^\infty e^{-st} F_{0,ij}(dt), s \geq 0, j \in \mathbb{Y}_{1,i}, i \in \mathbb{X}$ are Laplace transforms of proper distribution functions such that $F_{0,ij}(0) < 1, j \in \mathbb{Y}_{1,i}, i \in \mathbb{X}$, **(c)** $u_{\varepsilon,i} \in (0, \infty), \varepsilon \in (0, 1]$, for $i \in \mathbb{X}$.

It is useful to note that the relation, $F_{0,ij}(0) < 1$ holds if and only if $\phi_{0,ij}(s) < 1, s > 0$. Note also that all these inequalities hold if $\phi_{0,ij}(s_0) < 1$, for some $s_0 > 0$.

The conditions **G**, **I**, and **J**$^\circ$ imply that the following convergence relation holds, for $i \in \mathbb{X}$ and $s \geq 0$:

$$
\begin{aligned}
\phi_{\varepsilon,i}(s u_{\varepsilon,i}) &= \sum_{j \in \mathbb{Y}_{1,i}} \phi_{\varepsilon,ij}(s u_{\varepsilon,i}) p_{\varepsilon,ij} \\
&\Rightarrow \sum_{j \in \mathbb{Y}_{1,i}} \phi_{0,ij}(s) p_{0,ij} \\
&= \sum_{j \in \mathbb{Y}_{0,i}} \phi_{0,ij}(s) p_{0,ij} = \phi_{0,i}(s) \text{ as } \varepsilon \to 0. \qquad (9.17)
\end{aligned}
$$

The relations (9.15) and (9.17) imply that, in fact, the convergence relation appearing in the condition **J**$^\circ$ is satisfied for all $i, j \in \mathbb{X}$.

Let us now introduce the semi-Markov transition probabilities, for $t \geq 0, i, j \in \mathbb{X}$,

$$Q_{0,ij}(t) = \begin{cases} F_{0,ij}(t)p_{0,ij} & \text{for } j \in \mathbb{Y}_{0,i}, \\ 0 & \text{for } j \in \bar{\mathbb{Y}}_{0,i}. \end{cases} \tag{9.18}$$

The probabilities $Q_{0,ij}(t), t \geq 0, i, j \in \mathbb{X}$ can serve as transition probabilities for some Markov renewal process $(\eta_{0,n}, \kappa_{0,n}), n = 0, 1, \ldots$, with the phase space $\mathbb{X} \times [0, \infty)$. We also can define the corresponding semi-Markov process $\eta_0(t) = \eta_{0,\nu_0(t)}, t \geq 0$, where $\zeta_{0,n} = \kappa_{0,1} + \cdots + \kappa_{0,n}, n = 1, 2, \ldots, \zeta_{0,0} = 0$, are the corresponding instants of jumps, and $\nu_0(t) = \max(n \geq 1 : \zeta_{0,n} \leq t)$ is the number of jumps in the interval $[0, t]$ for the above semi-Markov process.

As mentioned above, the probabilities $F_{0,i}(0) < 1$, for $i \in \mathbb{X}$. This implies that $\mathsf{P}_i\{\lim_{n\to\infty} \zeta_{0,n} = \infty\} = 1$, for $i \in \mathbb{X}$. Thus, the process $\eta_0(t)$ is well defined on the time interval $[0, \infty)$.

The conditions **I**, **J** and Lemma 9.2 imply that, for $i, j \in \mathbb{X}$,

$$Q_{\varepsilon,ij}(\cdot) \Rightarrow Q_{0,ij}(\cdot) \text{ as } \varepsilon \to 0. \tag{9.19}$$

Also, according to the condition **I**, for $i, j \in \mathbb{X}$,

$$Q_{\varepsilon,ij}(\infty) = p_{\varepsilon,ij} \to p_{0,ij} = Q_{0,ij}(\infty) \text{ as } \varepsilon \to 0. \tag{9.20}$$

The relations (9.19) and (9.20) allow us to interpret the semi-Markov process $\eta_\varepsilon(t)$, for $\varepsilon \in (0, 1]$ as a perturbed version of the semi-Markov process $\eta_0(t)$.

Let us also introduce the expectations of inter-jump times, for $i, j \in \mathbb{X}$ and $\varepsilon \in (0, 1]$,

$$f_{\varepsilon,ij} = \int_0^\infty t F_{\varepsilon,ij}(dt), \tag{9.21}$$

and

$$f_{\varepsilon,i} = \int_0^\infty t F_{\varepsilon,i}(dt). \tag{9.22}$$

According to the relation (9.12), for $j \in \bar{\mathbb{Y}}_{1,i}, i \in \mathbb{X}$ and $\varepsilon \in (0, 1]$,

$$f_{\varepsilon,ij} = f_{\varepsilon,i}. \tag{9.23}$$

We also use the following perturbation condition:

K: **(a)** $f_{\varepsilon,ij} < \infty, j \in \mathbb{Y}_{1,i}, i \in \mathbb{X}$, for $\varepsilon \in (0, 1]$, **(b)** $f_{\varepsilon,ij}/u_{\varepsilon,i} \to f_{0,ij} = \int_0^\infty t F_{0,ij}(dt) < \infty$ as $\varepsilon \to 0$, for $j \in \mathbb{Y}_{1,i}, i \in \mathbb{X}$.

Note that the perturbation conditions **J** and **K** assume not only weak convergence of the distribution functions $F_{\varepsilon,ij}(\cdot u_{\varepsilon,i})$ to the distribution function $F_{0,ij}(\cdot)$, as $\varepsilon \to 0$, but also the convergence of their first moments $f_{\varepsilon,ij}/u_{\varepsilon,i}$ as $\varepsilon \to 0$ to the corresponding first moments $f_{0,ij}$ of the distribution function $F_{0,ij}(\cdot)$, for $j \in \mathbb{Y}_{1,i}, i \in \mathbb{X}$.

Note also that the conditions **J** and **K** imply that, for $j \in \mathbb{Y}_{1,i}, i \in \mathbb{X}$,

$$f_{0,ij} \in (0, \infty). \tag{9.24}$$

The conditions **G**, **I**, **J**, and **K** also imply that the following asymptotic relation holds, for $i \in \mathbb{X}$:

$$\begin{aligned}\frac{f_{\varepsilon,i}}{u_{\varepsilon,i}} &= \sum_{j \in \mathbb{Y}_{1,i}} \frac{f_{\varepsilon,ij}}{u_{\varepsilon,i}} p_{\varepsilon,ij} \\ &\to \sum_{j \in \mathbb{Y}_{1,i}} f_{0,ij} p_{0,ij} = \sum_{j \in \mathbb{Y}_{0,i}} f_{0,ij} p_{0,ij} \\ &= f_{0,i} \text{ as } \varepsilon \to 0. \end{aligned} \tag{9.25}$$

It is useful to note that the expectation $f_{0,i} \in (0, \infty)$, for $i \in \mathbb{X}$.

The relation (9.25) implies that, in fact, the convergence relation given in the condition **K** holds for all $i, j \in \mathbb{X}$.

Let us denote, for $i, j \in \mathbb{X}$ and $\varepsilon \in [0, 1]$,

$$e_{\varepsilon,ij} = \int_0^\infty s Q_{\varepsilon,ij}(ds). \tag{9.26}$$

It also follows from the conditions **G**, **I**, **J**, and **K** that the following asymptotic relation complementing the relations (9.19) and (9.20) holds, for $i, j \in \mathbb{X}$:

$$u_{\varepsilon,i}^{-1} e_{\varepsilon,ij} \to e_{0,ij} \text{ as } \varepsilon \to 0. \tag{9.27}$$

In what follows, we also impose on the initial normalisation functions $u_{\varepsilon,i}$ the following natural asymptotic stability condition:

L: $u_{\varepsilon,i} \to u_{0,i} \in (0, \infty]$ as $\varepsilon \to 0$, for $i \in \mathbb{X}$.

In the standard case, where all initial normalisation functions $u_{\varepsilon,i} = 1, \varepsilon \in (0, 1]$, for $i \in \mathbb{X}$, the condition **L** automatically holds, with limits $u_{0,i} = 1, i \in \mathbb{X}$.

9.1.3 Additional Asymptotic Comparability Conditions

The recurrent asymptotic analysis of perturbed semi-Markov processes requires to impose some stronger conditions on transition probabilities $p_{\varepsilon,ij}$ and normalisation functions $v_{\varepsilon,i}$.

Let us introduce the following condition of asymptotic comparability for transition probabilities based on some complete family of asymptotically comparable functions $\mathcal{H}$:

$\mathbf{I}_{\mathcal{H}}$: The functions $p_{\cdot,ij}, j \in \mathbb{Y}_{1,i}, i \in \mathbb{X}$ belong to a complete family of asymptotically comparable functions $\mathcal{H}$.

Note that the condition $\mathbf{I}_{\mathcal{H}}$ implies the fulfilment of the condition **I**.

Since, the matrix $\mathbf{P}_\varepsilon = \|p_{\varepsilon,ij}\|$ is stochastic, for $\varepsilon \in [0, 1]$, the corresponding comparability limits $a_{\mathcal{H}}[p_{\cdot,ij}]$ satisfy the following relation:

$$a_{\mathcal{H}}[p_{\cdot,ij}] = p_{0,ij} \in [0, 1],\ i, j \in \mathbb{X},\ \sum_{j \in \mathbb{X}} a_{\mathcal{H}}[p_{\cdot,ij}] = 1, i \in \mathbb{X}. \tag{9.28}$$

The perturbation analysis and related recurrent algorithms of phase space reduction presented in this chapter also involve other probabilities such as, $\bar{p}_{\varepsilon,ii} = 1 - p_{\varepsilon,ii}$, $\tilde{p}_{\varepsilon,ij} = p_{\varepsilon,ij}/\bar{p}_{\varepsilon,ii}$, ${}_k p_{\varepsilon,ij} = \tilde{p}_{\varepsilon,ij} + \tilde{p}_{\varepsilon,ik}\tilde{p}_{\varepsilon,kj}$, ${}_k q_{\varepsilon,ij} = \tilde{p}_{\varepsilon,ij}/{}_k p_{\varepsilon,ij}$, ${}_k\bar{p}_{\varepsilon,ii} = 1 - {}_k\bar{p}_{\varepsilon,ii}$, etc.

In all cases, the condition $\mathbf{I}_{\mathcal{H}}$ and Lemmas 8.2 and 8.3 imply that the above functions belong to the family $\mathcal{H}$ and, thus, they converge to some limits, as $\varepsilon \to 0$.

Moreover, if the role of $\mathcal{H}$ is played by one of the families $\mathcal{H}_1$, $\mathcal{H}_2$, or $\mathcal{H}_3$, then the corresponding comparability parameters and limits can be calculated explicitly as rational functions of the comparability parameters and limits for transition probabilities $p_{\varepsilon,ij}$ using the operating rules and formulas given in Lemmas 8.4–8.9.

Since, the matrix $\mathbf{P}_\varepsilon = \|p_{\varepsilon,ij}\|$ is stochastic, for $\varepsilon \in [0, 1]$, the conditions $\mathbf{I}_{\mathcal{H}_1}$ take the following form:

$\mathbf{I}_{\mathcal{H}_1}$: $\frac{p_{\varepsilon,ij}}{a_{ij}\varepsilon^{b_{ij}}} \to 1$ as $\varepsilon \to 0$, where $a_{ij} > 0$, $b_{ij} \in [0, \infty)$, $j \in \mathbb{Y}_{1,i}, i \in \mathbb{X}$.

The condition $\mathbf{I}_{\mathcal{H}_1}$ implies that, for $j \in \mathbb{Y}_{1,i}, i \in \mathbb{X}$,

$$\lim_{\varepsilon \to 0} p_{\varepsilon,ij} = p_{0,ij} = \begin{cases} 0 & \text{if } b_{ij} > 0, \\ a_{ij} & \text{if } b_{ij} = 0. \end{cases} \tag{9.29}$$

Moreover, the following relation holds:

$$\sum_{j \in \mathbb{Y}_{1,i}:\, b_{ij}=0} a_{ij} = 1,\ i \in \mathbb{X}. \tag{9.30}$$

Similarly, the condition $\mathbf{I}_{\mathcal{H}_2}$ takes the following form:

$\mathbf{I}_{\mathcal{H}_2}$: $\frac{p_{\varepsilon,ij}}{a_{ij}\varepsilon^{b_{ij}} e^{-c_{ij}\varepsilon^{-1}}} \to 1$ as $\varepsilon \to 0$, where $a_{ij} > 0$, $b_{ij}, c_{ij} \in [0, \infty)$, $j \in \mathbb{Y}_{1,i}, i \in \mathbb{X}$.

The condition $\mathbf{I}_{\mathcal{H}_2}$ implies that, for $j \in \mathbb{Y}_{1,i}, i \in \mathbb{X}$,

$$\lim_{\varepsilon \to 0} p_{\varepsilon,ij} = p_{0,ij} = \begin{cases} 0 & \text{if } c_{ij} > 0 \text{ or } c_{ij} = 0, b_{ij} > 0, \\ a_{ij} & \text{if } b_{ij}, c_{ij} = 0. \end{cases} \tag{9.31}$$

Moreover, the following relation holds:

$$\sum_{j \in \mathbb{Y}_{1,i}:\, b_{ij}, c_{ij}=0} a_{ij} = 1,\ i \in \mathbb{X}. \tag{9.32}$$

Finally, the condition $\mathbf{I}_{\mathcal{H}_3}$ takes the following form:

$\mathbf{I}_{\mathcal{H}_3}$: $\frac{p_{\varepsilon,ij}}{a_{ij}\varepsilon^{b_{ij}}(1+\ln \varepsilon^{-1})^{-d_{ij}}} \to 1$ as $\varepsilon \to 0$, where $a_{ij} > 0$, $b_{ij}, d_{ij} \in [0, \infty)$, $j \in \mathbb{Y}_{1,i}, i \in \mathbb{X}$.

The condition $\mathbf{I}_{\mathcal{H}_3}$ implies that, for $j \in \mathbb{Y}_{1,i}, i \in \mathbb{X}$,

$$\lim_{\varepsilon\to 0} p_{\varepsilon,ij} = p_{0,ij} = \begin{cases} 0 & \text{if } b_{ij} > 0 \text{ or } b_{ij} = 0, d_{ij} > 0, \\ a_{ij} & \text{if } b_{ij}, d_{ij} = 0. \end{cases} \tag{9.33}$$

Moreover, the following relation holds,

$$\sum_{j\in\mathbb{Y}_{1,i}:\, b_{ij}, d_{ij}=0} a_{ij} = 1,\ i \in \mathbb{X}. \tag{9.34}$$

Let us also use the following condition of asymptotic comparability for normalisation functions based on the same complete family of asymptotically comparable functions $\mathcal{H}$ as in the condition $\mathbf{I}_{\mathcal{H}}$:

$\mathbf{L}_{\mathcal{H}}$: The functions $u_{\cdot,i}$, $i \in \mathbb{X}$ belong to the complete family of asymptotically comparable functions $\mathcal{H}$ appearing in the condition $\mathbf{I}_{\mathcal{H}}$.

The conditions $\mathbf{L}$ and $\mathbf{L}_{\mathcal{H}}$ imply that the corresponding comparability limits $a_{\mathcal{H}}[u_{\cdot,i}]$ satisfy the following relation:

$$a_{\mathcal{H}}[u_{\cdot,i}] = u_{0,i} \in (0,\infty],\ i \in \mathbb{X}. \tag{9.35}$$

Note also that in the case of standard initial normalisation functions $u_{\cdot,i} \equiv 1, i \in \mathbb{X}$, condition $\mathbf{L}_{\mathcal{H}}$ is automatically satisfied for any family of asymptotically comparable functions $\mathcal{H}$.

The perturbation analysis and related recurrent algorithms of phase space reduction presented in this chapter also involve other functions such as, $\tilde{u}_{\varepsilon,i} = \bar{p}_{\varepsilon,ii}^{-1} u_{\varepsilon,i}$, $\tilde{w}_{\varepsilon,ij} = \tilde{u}_{\varepsilon,i}/\tilde{u}_{\varepsilon,j}$, ${}_k\tilde{u}_{\varepsilon,i} = {}_k\bar{p}_{\varepsilon,ii}^{-1}\tilde{u}_{\varepsilon,i}$, etc.

In all cases, the conditions $\mathbf{I}_{\mathcal{H}}$, $\mathbf{L}_{\mathcal{H}}$ and Lemmas 8.2 and 8.3 imply that the above functions belong to the family $\mathcal{H}$ and, thus, they converge to some limits, as $\varepsilon \to 0$.

Moreover, if the role of $\mathcal{H}$ is played by one of the families $\mathcal{H}_1$, $\mathcal{H}_2$, or $\mathcal{H}_3$, then the corresponding comparability parameters and limits can be calculated explicitly as rational functions of the comparability parameters and limits for transition probabilities $p_{\varepsilon,ij}$ and the initial normalisation functions $u_{\varepsilon,i}$, using the operating rules and formulas given in Lemmas 8.4–8.9.

The condition $\mathbf{L}_{\mathcal{H}_1}$ takes the following form:

$\mathbf{L}_{\mathcal{H}_1}$: $\frac{u_{\varepsilon,i}}{a_i\varepsilon^{b_i}} \to 1$ as $\varepsilon \to 0$, where $a_i > 0$, $b_i \in (-\infty, 0]$, $i \in \mathbb{X}$.

The condition $\mathbf{L}_{\mathcal{H}_1}$ implies that, for $i \in \mathbb{X}$,

$$\lim_{\varepsilon\to 0} u_{\varepsilon,i} = u_{0,i} = \begin{cases} a_i & \text{if } b_i = 0, \\ \infty & \text{if } b_i < 0. \end{cases} \tag{9.36}$$

Similarly, the condition $\mathbf{L}_{\mathcal{H}_2}$ takes the following form:

$\mathbf{L}_{\mathcal{H}_2}$: $\frac{u_{\varepsilon,i}}{a_i\varepsilon^{b_i}e^{-c_i\varepsilon^{-1}}} \to 1$ as $\varepsilon \to 0$, where $a_i > 0$, $b_i, c_i \in (-\infty, 0]$, $i \in \mathbb{X}$.

The condition $\mathbf{L}_{\mathcal{H}_2}$ implies that, for $i \in \mathbb{X}$,

$$\lim_{\varepsilon\to 0} u_{\varepsilon,i} = u_{0,i} = \begin{cases} a_i & \text{if } b_{ij}, c_{ij} = 0, \\ \infty & \text{if } c_{ij} < 0 \text{ or } c_{ij} = 0, b_{ij} < 0. \end{cases} \tag{9.37}$$

Finally, the condition $\mathbf{L}_{\mathcal{H}_3}$ takes the following form:

$\mathbf{L}_{\mathcal{H}_3}$: $\frac{u_{\varepsilon,i}}{a_i \varepsilon^{b_i}(1+\ln \varepsilon^{-1})^{-d_i}} \longrightarrow 1$ as $\varepsilon \to 0$, where $a_i > 0$, $b_i, d_i \in (-\infty, 0]$, $i \in \mathbb{X}$.

The condition $\mathbf{L}_{\mathcal{H}_3}$ implies that, for $i \in \mathbb{X}$,

$$\lim_{\varepsilon\to 0} u_{\varepsilon,i} = u_{0,i} = \begin{cases} a_i & \text{if } b_i, d_i = 0, \\ \infty & \text{if } b_i < 0 \text{ or } b_i = 0, d_i < 0. \end{cases} \tag{9.38}$$

9.1.4 Perturbed Markov Chains

In the case of perturbed continuous and discrete Markov chains conditions **G**, **H**, **I**, $\mathbf{I}_{\mathcal{H}}$, **L**, and $\mathbf{L}_{\mathcal{H}}$ do not change.

In the case of continuous time Markov chains (see, the relation (1.47)), the condition **J** takes the form of the asymptotic relation $0 < \lambda_{\varepsilon,i} \longrightarrow [0, \infty)$ as $\varepsilon \to 0$, for $i \in \mathbb{X}$. In this case the role of initial normalisation functions is played by functions $u_{\cdot,i} = \lambda^{-1}_{\cdot,i}, i \in \mathbb{X}$. In this case the condition **J** is obviously satisfied and the corresponding limiting distribution functions $F_{0,ij}(t) = 1 - e^{-t}, t \geq 0, j \in \mathbb{Y}_{1,i}, i \in \mathbb{X}$ are exponential with parameter 1.

The condition **K** is also satisfied, with the corresponding limiting expectations $f_{0,ij} = 1, j \in \mathbb{Y}_{1,i}, i \in \mathbb{X}$.

In the case of discrete time Markov chains (see, the relation (1.48)), the condition **J** is automatically satisfied, with the corresponding initial normalisation functions $u_{\cdot,i} \equiv 1, i \in \mathbb{X}$ and the limiting distribution functions $F_{0,ij}(t) = \mathrm{I}(t \geq 1), t \geq 0, j \in \mathbb{Y}_{1,i}, i \in \mathbb{X}$.

The condition **K** is also satisfied, with the corresponding limiting expectations $f_{0,ij} = 1, j \in \mathbb{Y}_{1,i}, i \in \mathbb{X}$.

9.2 Removing of Virtual Transitions for Perturbed Semi-Markov Processes

In Sect. 9.2, we describe the procedure of removing virtual transitions of the form $i \to i$ from the trajectories of perturbed semi-Markov processes. It is shown that the basic perturbation conditions imposed on the original semi-Markov processes are also satisfied for the semi-Markov processes with removed virtual transitions. We also give explicit formulas for recalculating normalisation functions, limiting distributions and expectations in the corresponding perturbation conditions for semi-Markov processes with removed virtual transitions.

9.2.1 Semi-Markov Processes with Removed Virtual Transitions

9.2.1.1 Definition of Semi-Markov Processes with Removed Virtual Transitions. As above, we assume that conditions **G** and **H** are satisfied.

In what follows in this section, we also assume that $\varepsilon \in (0, 1]$.

Let us define the stopping times for the Markov chain $\eta_{\varepsilon,n}$, for $r = 0, 1, \ldots$,

$$\theta_\varepsilon[r] = \min(n > r : \eta_{\varepsilon,n} \neq \eta_{\varepsilon,r}). \tag{9.39}$$

By the definition, $\theta_\varepsilon[r]$ is the first moment after r, when the Markov chain $\eta_{\varepsilon,n}$ changes its state.

We also define the following successive stopping times:

$$\rho_{\varepsilon,n} = \theta_\varepsilon[\rho_{\varepsilon,n-1}], n = 1, 2, \ldots, \text{ where } \rho_{\varepsilon,0} = 0. \tag{9.40}$$

We now construct a new Markov renewal process $(\tilde{\eta}_{\varepsilon,n}, \tilde{\kappa}_{\varepsilon,n}), n = 0, 1, \ldots$ with the phase space $\mathbb{X} \times [0, \infty)$,

$$(\tilde{\eta}_{\varepsilon,n}, \tilde{\kappa}_{\varepsilon,n}) = \begin{cases} (\eta_{\varepsilon,0}, 0), & \text{for } n = 0, \\ (\eta_{\varepsilon,\rho_{\varepsilon,n}}, \sum_{l=\rho_{\varepsilon,n-1}+1}^{\rho_{\varepsilon,n}} \kappa_{\varepsilon,l}), & \text{for } n = 1, 2, \ldots. \end{cases} \tag{9.41}$$

We can also define the corresponding semi-Markov process,

$$\tilde{\eta}_\varepsilon(t) = \tilde{\eta}_{\varepsilon,\tilde{\nu}_\varepsilon(t)}, t \geq 0, \tag{9.42}$$

where

$$\tilde{\zeta}_{\varepsilon,n} = \tilde{\kappa}_{\varepsilon,1} + \cdots + \tilde{\kappa}_{\varepsilon,n}, n = 1, 2, \ldots, \tilde{\zeta}_{\varepsilon,0} = 0, \tag{9.43}$$

are the successive moments of jumps, and

$$\tilde{\nu}_\varepsilon(t) = \max(n \geq 1 : \tilde{\zeta}_{\varepsilon,n} \leq t), \tag{9.44}$$

is the number of jumps in the interval $[0, t]$ for the above semi-Markov process.

Below, we use the symbol $*$ and the superscript $^{(*n)}$ to denote convolution operation and n-fold convolution for a proper or improper distribution function.

The definition of stopping times $\rho_{\varepsilon,n}$ implies that the transition probabilities for the above Markov renewal process are determined by the following relation:

$$\begin{aligned} \tilde{Q}_{\varepsilon,ij}(t) &= \mathsf{P}\{\tilde{\eta}_{\varepsilon,1} = j, \tilde{\kappa}_{\varepsilon,1} \leq t/\tilde{\eta}_{\varepsilon,0} = i\} \\ &= \mathrm{I}(j \neq i) \sum_{n=0}^{\infty} Q_{\varepsilon,ii}^{(*n)}(t) * Q_{\varepsilon,ij}(t), t \geq 0, i, j \in \mathbb{X}, \end{aligned} \tag{9.45}$$

where, as usual, $Q_{\varepsilon,ii}^{(*0)}(t) = \mathrm{I}(t \geq 0), t \geq 0$.

Accordingly, the transition probabilities of the embedded Markov chain $\tilde{\eta}_{\varepsilon,n}$ are given by the following relation:

$$\tilde{p}_{\varepsilon,ij} = \mathsf{P}\{\tilde{\eta}_{\varepsilon,1} = j/\tilde{\eta}_{\varepsilon,0} = i\}$$
$$= \mathrm{I}(j \neq i)\frac{p_{\varepsilon,ij}}{\bar{p}_{\varepsilon,ii}},\ i, j \in \mathbb{X}, \tag{9.46}$$

where

$$\bar{p}_{\varepsilon,ii} = 1 - p_{\varepsilon,ii} = \sum_{j \neq i} p_{\varepsilon,ij},\ i \in \mathbb{X}. \tag{9.47}$$

Note that the conditions **G** and **H** guarantee that the probabilities $\bar{p}_{\varepsilon,ii} > 0, \varepsilon \in (0, 1]$, for $i \in \mathbb{X}$.

9.2.1.2 Condition $\tilde{\mathbf{G}}$. For $i \in \mathbb{X}$ and $\varepsilon \in (0, 1]$, we introduce the distribution functions of the sojourn times for the semi-Markov process $\tilde{\eta}_\varepsilon(t)$,

$$\tilde{F}_{\varepsilon,i}(t) = \mathsf{P}_i\{\tilde{\kappa}_{\varepsilon,i} \leq t\} = \sum_{j \in \mathbb{X}} \tilde{Q}_{\varepsilon,ij}(t),\ t \geq 0. \tag{9.48}$$

The following condition plays the role of the condition **G** for the semi-Markov processes $\tilde{\eta}_\varepsilon(t)$:

$\tilde{\mathbf{G}}$: **(a)** $\tilde{p}_{\varepsilon,ij} > 0, \varepsilon \in (0, 1]$ or $\tilde{p}_{\varepsilon,ij} = 0, \varepsilon \in (0, 1]$, for every $i, j \in \mathbb{X}$, **(b)** $\tilde{F}_{\varepsilon,i}(0) < 1, i \in \mathbb{X}, \varepsilon \in (0, 1]$.

Lemma 9.3 *The conditions* **G** *and* **H**, *assumed to be satisfied for the semi-Markov processes $\eta_\varepsilon(t)$, entail the fulfilment of the condition* **G** *for the semi-Markov processes $\tilde{\eta}_\varepsilon(t)$ in the form of condition* $\tilde{\mathbf{G}}$.

Proof The condition **G (a)** and the relation (9.46) obviously imply that the condition $\tilde{\mathbf{G}}$ **(a)** is satisfied.

Since, $Q_{\varepsilon,ii}(0) \leq F_{\varepsilon,i}(0) < 1$, for $i \in \mathbb{X}$, the condition **G (b)** and the relations (9.45) imply that, for $i \in \mathbb{X}$ and $\varepsilon \in (0, 1]$,

$$\tilde{F}_{\varepsilon,i}(0) = \sum_{n=0}^{\infty} Q^n_{\varepsilon,ii}(0) \sum_{j \neq i} Q_{\varepsilon,ij}(0) = \frac{F_{\varepsilon,i}(0) - Q_{\varepsilon,ii}(0)}{1 - Q_{\varepsilon,ii}(0)} < 1. \tag{9.49}$$

Thus, the regularity condition **G (b)** is also satisfied for the semi-Markov processes $\tilde{\eta}_\varepsilon(t)$. □

The condition $\tilde{\mathbf{G}}$ implies that the process $\tilde{\eta}_\varepsilon(t)$ is well defined on the time interval $[0, \infty)$, for each $\varepsilon \in (0, 1]$.

Let us introduce sets, for $i \in \mathbb{X}$ and $\varepsilon \in (0, 1$,

$$\tilde{\mathbb{Y}}_{\varepsilon,i} = \{j \in \mathbb{X} : \tilde{p}_{\varepsilon,ij} > 0\}. \tag{9.50}$$

The following obvious lemma takes place.

Lemma 9.4 *The condition* $\tilde{\mathbf{G}}$ **(a)** *is satisfied if and only if the sets* $\tilde{\mathbb{Y}}_{\varepsilon,i} = \tilde{\mathbb{Y}}_{1,i}, \varepsilon \in (0, 1]$, *for* $i \in \mathbb{X}$, *and the sets,*

$$\tilde{\mathbb{Y}}_{1,i} = \mathbb{Y}_{1,i} \setminus \{i\},\ i \in \mathbb{X}. \tag{9.51}$$

9.2.1.3 Condition $\tilde{\mathbf{H}}$. The following condition plays the role of the condition **H** for the semi-Markov processes $\tilde{\eta}_\varepsilon(t)$:

$\tilde{\mathbf{H}}$: For any $i, j \in \mathbb{X}$, there exist $\tilde{n}_{ij} \geq 1$ a chain of states $i = \tilde{r}_0, \tilde{r}_1, \ldots, \tilde{r}_{\tilde{n}_{ij}} = j$ such that $\prod_{1 \leq l \leq \tilde{n}_{ij}} \tilde{p}_{1, \tilde{r}_{l-1} \tilde{r}_l} > 0$.

Lemma 9.5 *The conditions* **G** *and* **H**, *assumed to be satisfied for the semi-Markov processes $\eta_\varepsilon(t)$, entail the fulfilment of the condition* **H** *for the semi-Markov processes $\tilde{\eta}_\varepsilon(t)$ in the form of condition* $\tilde{\mathbf{H}}$.

Proof It can always be assumed that a chain of states $r_0, \ldots, r_{n_{ij}}$ appearing in the condition **H** satisfies the additional assumption that $r_{l-1} \neq r_l, l = 1, \ldots, n_{ij}$. For any such chain, $\prod_{l=1}^{n_{ij}} \tilde{p}_{\varepsilon, r_{l-1} r_l} \geq \prod_{l=1}^{n_{ij}} p_{\varepsilon, r_{l-1} r_l} > 0$. Therefore, the condition **H** is satisfied for the semi-Markov processes $\tilde{\eta}_\varepsilon(t)$. □

9.2.2 Perturbation Conditions for Semi-Markov Processes with Removed Virtual Transitions

We are going to describe assumptions, under which basic conditions **I**–**L** and $\mathbf{I}_{\mathcal{H}}$, $\mathbf{L}_{\mathcal{H}}$ hold for the semi-Markov processes $\tilde{\eta}_\varepsilon(t)$ constructed with the use of the described above procedure of removing virtual transitions for the semi-Markov processes $\eta_\varepsilon(t)$.

9.2.2.1 Conditions $\tilde{\mathbf{I}}$ and $\tilde{\mathbf{I}}_{\mathcal{H}}$. The following conditions play the roles of the conditions **I** and $\mathbf{I}_{\mathcal{H}}$ for the semi-Markov processes $\tilde{\eta}_\varepsilon(t)$:

$\tilde{\mathbf{I}}$: $\tilde{p}_{\varepsilon, ij} = \mathrm{I}(j \neq i) \frac{p_{\varepsilon, ij}}{\bar{p}_{\varepsilon, ii}} \to \tilde{p}_{0, ij} \in [0, 1]$ as $\varepsilon \to 0$, for $i, j \in \mathbb{X}$

and

$\tilde{\mathbf{I}}_{\mathcal{H}}$: Functions $\tilde{p}_{\cdot, ij}, j \in \tilde{\mathbb{Y}}_{1,i}, i \in \mathbb{X}$ belong to the complete family of asymptotically comparable functions $\mathcal{H}$ appearing in the condition $\mathbf{I}_{\mathcal{H}}$.

Recall that the condition $\tilde{\mathbf{H}}$ implies that the probabilities $\bar{p}_{\varepsilon, ii} > 0, i \in \mathbb{X}$, for every $\varepsilon \in (0, 1]$.

Since, in this case, the matrix $\tilde{\mathbf{P}}_\varepsilon = \|\tilde{p}_{\varepsilon, ij}\|$ is stochastic, for $\varepsilon \in (0, 1]$, the condition $\tilde{\mathbf{I}}$ implies that the matrix $\tilde{\mathbf{P}}_0 = \|\tilde{p}_{0, ij}\|$ is also stochastic.

Let $\tilde{\eta}_{0,n}, n = 0, 1, \ldots$ be a Markov chain with the phase space $\mathbb{X}$ and the matrix of transition probabilities $\tilde{\mathbf{P}}_0$. The condition $\tilde{\mathbf{I}}$ allows us to interpret the Markov chain $\tilde{\eta}_{\varepsilon,n}$, for $\varepsilon \in (0, 1]$, as a perturbed version of the Markov chain $\tilde{\eta}_{0,n}$.

Let us introduce, for $i \in \mathbb{X}$, sets,

$$\tilde{\mathbb{Y}}_{0,i} = \{j \in \mathbb{X} : \tilde{p}_{0, ij} > 0\}. \tag{9.52}$$

Lemma 9.6 *The conditions $\tilde{\mathbf{G}}$ and $\tilde{\mathbf{I}}$ imply that $\tilde{\mathbb{Y}}_{0,i} \subseteq \tilde{\mathbb{Y}}_{1,i}$, for $i \in \mathbb{X}$.*

Lemma 9.7 *The conditions* **G**, **H**, **I**, $\mathbf{I}_{\mathcal{H}}$, *assumed to be satisfied for the semi-Markov processes $\eta_\varepsilon(t)$, entail the fulfilment of the conditions* **I**, $\mathbf{I}_{\mathcal{H}}$ *for the semi-Markov processes $\tilde{\eta}_\varepsilon(t)$ in the form of conditions* $\tilde{\mathbf{I}}$, $\tilde{\mathbf{I}}_{\mathcal{H}}$.

Proof It follows from the relation (9.47) and Lemma 8.2 according to which the functions $\tilde{p}_{\cdot,ij}, j \in \tilde{\mathbb{Y}}_{1,i}, i \in \mathbb{X}$ belong to the family $\mathcal{H}$. □

The following two lemmas present operating formulas which express the comparability parameters and limits for the functions $\bar{p}_{\cdot,ii} = \sum_{j \neq i} p_{\cdot,ij}, i \in \mathbb{X}$ and $\tilde{p}_{\cdot,ij} = p_{\cdot,ij}/\bar{p}_{\cdot,ii}, j \in \tilde{\mathbb{Y}}_{1,i}, i \in \mathbb{X}$ appearing in the conditions $\tilde{\mathbf{I}}$ and $\tilde{\mathbf{I}}_{\mathcal{H}}$ in terms of the comparability parameters and limits for the functions appearing in the conditions **I** and $\mathbf{I}_{\mathcal{H}}$.

Lemma 9.8 *Let the conditions* **G**, **H**, **I**, $\mathbf{I}_{\mathcal{H}}$ *be satisfied. Then*:

(i) *The functions* $\bar{p}_{\cdot,ii} = \sum_{j \neq i} p_{\cdot,ij}, i \in \mathbb{X}$ *belong to the family* $\mathcal{H}$ *and the corresponding comparability limits can be calculated by applying to these functions the operating multiple sum formula given in Lemma 8.3, which, in this case, gives the relation* $\lim_{\varepsilon \to 0} \bar{p}_{\varepsilon,ii} = \lim_{\varepsilon \to 0} \sum_{j \neq i} p_{\varepsilon,ij} = \sum_{j \neq i} p_{0,ij} = \bar{p}_{0,ii}$, *for* $i \in \mathbb{X}$.

(ii) *If* $\mathcal{H}$ *is one of the families,* $\mathcal{H}_1$, $\mathcal{H}_2$, *or* $\mathcal{H}_3$, *then the corresponding comparability parameters and limits can be calculated by applying the operating multiple sum formulas given, respectively, in Lemmas 8.5, 8.7, or 8.9 to the functions* $\bar{p}_{\cdot,ii} = \sum_{j \neq i} p_{\cdot,ij}, i \in \mathbb{X}$.

Lemma 9.9 *Let the conditions* **G**, **H**, **I**, $\mathbf{I}_{\mathcal{H}}$ *hold. Then*:

(i) *The functions* $\tilde{p}_{\cdot,ij} = p_{\cdot,ij}/\bar{p}_{\cdot,ii}, j \in \tilde{\mathbb{Y}}_{1,i}, i \in \mathbb{X}$ *belong to the family* $\mathcal{H}$ *and the corresponding comparability limits can be calculated by applying to these functions the operating quotient formula given in Lemma 8.2 to these functions.*

(ii) *If* $\mathcal{H}$ *is one of the families,* $\mathcal{H}_1$, $\mathcal{H}_2$, *or* $\mathcal{H}_3$, *then the corresponding comparability parameters and limits can be calculated by applying the operating quotient formulas given, respectively, in Lemmas 8.4, 8.6, or 8.8 to the functions* $\tilde{p}_{\cdot,ij} = p_{\cdot,ij}/\bar{p}_{\cdot,ii}, j \in \tilde{\mathbb{Y}}_{1,i}, i \in \mathbb{X}$.

9.2.2.2 Condition $\tilde{\mathbf{J}}$. Let us assume that the conditions **G**, **H**, **I**, $\mathbf{I}_{\mathcal{H}}$, $\mathbf{J}^\circ$ (equivalent to the condition **J**) and **K** are satisfied.

The distribution function $\tilde{F}_{\varepsilon,ij}(t), t \geq 0$, which is an analogue of the distribution function $F_{\varepsilon,ij}(t), t \geq 0$ is defined (for $\varepsilon \in (0,1]$) by the following relation, for $j \in \tilde{\mathbb{Y}}_{1,i}, i \in \mathbb{X}$,

$$\begin{aligned} \tilde{F}_{\varepsilon,ij}(t) &= \mathsf{P}\{\tilde{\kappa}_{\varepsilon,1} \leq t / \tilde{\eta}_{\varepsilon,0} = i, \tilde{\eta}_{\varepsilon,1} = j\} \\ &= \frac{1}{\tilde{p}_{\varepsilon,ij}} \sum_{n=0}^{\infty} F^{(*n)}_{\varepsilon,ii}(t) * F_{\varepsilon,ij}(t) p^n_{\varepsilon,ii} p_{\varepsilon,ij}, \ t \geq 0, \end{aligned} \tag{9.53}$$

and, for $j \notin \tilde{\mathbb{Y}}_{1,i}, i \in \mathbb{X}$,

$$\tilde{F}_{\varepsilon,ij}(t) = \tilde{F}_{\varepsilon,i}(t), \ t \geq 0, \tag{9.54}$$

where, for $i \in \mathbb{X}$,

$$\begin{aligned}\tilde{F}_{\varepsilon,i}(t) &= \mathsf{P}\{\tilde{\kappa}_{\varepsilon,1} \leq t/\tilde{\eta}_{\varepsilon,0} = i\} \\ &= \sum_{j \in \mathbb{X}} \tilde{Q}_{\varepsilon,ij}(t) = \sum_{j \in \tilde{\mathbb{Y}}_{1,i}} \tilde{F}_{\varepsilon,ij}(t)\tilde{p}_{\varepsilon,ij}, t \geq 0.\end{aligned} \tag{9.55}$$

The corresponding Laplace transform $\tilde{\phi}_{\varepsilon,ij}(s)$ takes the following form, for $j \in \tilde{\mathbb{Y}}_{1,i}, i \in \mathbb{X}$:

$$\begin{aligned}\tilde{\phi}_{\varepsilon,ij}(s) &= \mathsf{E}\{e^{-s\tilde{\kappa}_{\varepsilon,1}}/\tilde{\eta}_{\varepsilon,0} = i, \tilde{\eta}_{\varepsilon,1} = j\} = \int_0^\infty e^{-st}\tilde{F}_{\varepsilon,ij}(dt) \\ &= \frac{1}{\tilde{p}_{\varepsilon,ij}} \cdot \frac{\phi_{\varepsilon,ij}(s)p_{\varepsilon,ij}}{1 - \phi_{\varepsilon,ii}(s)p_{\varepsilon,ii}} = \frac{\phi_{\varepsilon,ij}(s)\bar{p}_{\varepsilon,ii}}{1 - \phi_{\varepsilon,ii}(s)p_{\varepsilon,ii}} \\ &= \frac{\phi_{\varepsilon,ij}(s)}{1 + p_{\varepsilon,ii}\bar{p}_{\varepsilon,ii}^{-1}(1 - \phi_{\varepsilon,ii}(s))}, \ s \geq 0,\end{aligned} \tag{9.56}$$

and, for $j \notin \tilde{\mathbb{Y}}_{1,i}, i \in \mathbb{X}$,

$$\tilde{\phi}_{\varepsilon,ij}(s) = \tilde{\phi}_{\varepsilon,i}(s), s \geq 0, \tag{9.57}$$

where

$$\tilde{\phi}_{\varepsilon,i}(s) = \mathsf{E}\{e^{-s\tilde{\kappa}_{\varepsilon,1}}/\tilde{\eta}_{\varepsilon,0} = i\} = \sum_{j \in \tilde{\mathbb{Y}}_{1,i}} \tilde{\phi}_{\varepsilon,ij}(s)\tilde{p}_{\varepsilon,ij}, \ s \geq 0. \tag{9.58}$$

Note that the above choice of $\tilde{F}_{\varepsilon,ij}(t)$ for $j \notin \tilde{\mathbb{Y}}_{1,i}, i \in \mathbb{X}$ has no effect on the transition probabilities $\tilde{Q}_{\varepsilon,ij}(t)$. The following relations hold for $t \geq 0, i, j \in \mathbb{X}$ and $\varepsilon \in (0, 1]$,

$$\tilde{Q}_{\varepsilon,ij}(t) = \tilde{F}_{\varepsilon,ij}(t)\tilde{p}_{\varepsilon,ij}. \tag{9.59}$$

The following condition plays the role of condition **J** for the semi-Markov processes $\tilde{\eta}_\varepsilon(t)$:

$\tilde{\mathbf{J}}$: **(a)** $\tilde{F}_{\varepsilon,ij}(\cdot\, \tilde{u}_{\varepsilon,i}) \Rightarrow \tilde{F}_{0,ij}(\cdot)$ as $\varepsilon \to 0$, for $j \in \tilde{\mathbb{Y}}_{1,i}, i \in \mathbb{X}$, **(b)** $\tilde{F}_{0,ij}(\cdot), j \in \tilde{\mathbb{Y}}_{1,i}, i \in \mathbb{X}$ are proper distribution functions such that $\tilde{F}_{0,ij}(0) < 1, j \in \tilde{\mathbb{Y}}_{1,i}, i \in \mathbb{X}$, **(c)** $\tilde{u}_{\varepsilon,i} \in (0, \infty), \varepsilon \in (0, 1]$, for $i \in \mathbb{X}$.

The role of condition $\tilde{\mathbf{J}}^\circ$ (equivalent to condition **J**), is played by the following condition equivalent to the condition $\tilde{\mathbf{J}}$:

$\tilde{\mathbf{J}}^\circ$: **(a)** $\tilde{\phi}_{\varepsilon,ij}(s/\tilde{u}_{\varepsilon,i}) \to \tilde{\phi}_{0,ij}(s)$ as $\varepsilon \to 0$, for $s \geq 0$ and $j \in \tilde{\mathbb{Y}}_{1,i}, i \in \mathbb{X}$, **(b)** $\tilde{\phi}_{0,ij}(s) = \int_0^\infty e^{-st}\tilde{F}_{0,ij}(dt), s \geq 0, j \in \tilde{\mathbb{Y}}_{1,i}, i \in \mathbb{X}$ are Laplace transforms of proper distribution functions such that $\tilde{F}_{0,ij}(0) < 1, j \in \tilde{\mathbb{Y}}_{1,i}, i \in \mathbb{X}$, **(c)** $\tilde{u}_{\varepsilon,i} \in (0, \infty), \varepsilon \in (0, 1]$, for $i \in \mathbb{X}$,

The operation of removing virtual transitions aggregates the sojourn times for the semi-Markov process $\eta_\varepsilon(t)$. Obviously, $\mathsf{P}_i\{\theta_\varepsilon[0] = n\} = p_{\varepsilon,ii}^{n-1}(1 - p_{\varepsilon,ii}), n = 1, 2, \ldots$ and, thus, $\mathsf{E}_i\theta_\varepsilon[0] = \bar{p}_{\varepsilon,ii}^{-1}$, for $i \in \mathbb{X}$ and $\varepsilon \in (0, 1]$.

This note provides a hint for compensating the aggregation of sojourn times using the new local normalisation functions $\tilde{u}_{\varepsilon,i} \in (0, \infty), \varepsilon \in (0, 1]$, defined by the following relation, for $i \in \mathbb{X}$,

$$\tilde{u}_{\varepsilon,i} = \bar{p}_{\varepsilon,ii}^{-1} u_{\varepsilon,i}. \tag{9.60}$$

Recall that the conditions **G** and **H** imply that the probability $\bar{p}_{\varepsilon,ii} \in (0,1], \varepsilon \in (0,1]$, and thus $\bar{p}_{\varepsilon,ii}^{-1} \in [1,\infty), \varepsilon \in (0,1]$, for $i \in \mathbb{X}$. Therefore, the normalisation functions $\tilde{u}_{\varepsilon,i} \in (0,\infty), \varepsilon \in (0,1]$, for $i \in \mathbb{X}$.

Let us define the set of asymptotically absorbing states,

$$\mathbb{Y}_0 = \{i \in \mathbb{X} : p_{0,ii} = 1\}. \tag{9.61}$$

The asymptotic relations and the corresponding limiting Laplace transforms and distributions appearing in the condition $\tilde{\mathbf{J}}^\circ$ ($\tilde{\mathbf{J}}$) take two different forms for asymptotically non-absorbing states $i \in \bar{\mathbb{Y}}_0$ and absorbing states $i \in \mathbb{Y}_0$.

First, let us consider the non-absorbing case, where $i \in \bar{\mathbb{Y}}_0$.

Using the conditions **I**, $\mathbf{I}_{\mathcal{H}}$, $\mathbf{J}^\circ$ and the relations (9.17) and (9.56), we get the following relation, for $j \in \tilde{\mathbb{Y}}_{1,i}, i \in \bar{\mathbb{Y}}_0$:

$$\begin{aligned}\tilde{\phi}_{\varepsilon,ij}(s/\tilde{u}_{\varepsilon,i}) &= \frac{\phi_{\varepsilon,ij}(\bar{p}_{\varepsilon,ii}s/u_{\varepsilon,i})\bar{p}_{\varepsilon,ii}}{1 - p_{\varepsilon,ii}\phi_{\varepsilon,ii}(\bar{p}_{\varepsilon,ii}s/u_{\varepsilon,i})} \\ &\to \tilde{\phi}_{0,ij}(s) \text{ as } \varepsilon \to 0, \text{ for } s \geq 0,\end{aligned} \tag{9.62}$$

where

$$\tilde{\phi}_{0,ij}(s) = \frac{\phi_{0,ij}(\bar{p}_{0,ii}s)\bar{p}_{0,ii}}{1 - p_{0,ii}\phi_{0,ii}(\bar{p}_{0,ii}s)}, s \geq 0. \tag{9.63}$$

Obviously, $\tilde{\phi}_{0,ij}(s) = \int_0^\infty e^{-st}\tilde{F}_{0,ij}(dt) = \mathsf{E}e^{-s\tilde{\kappa}_{0,ij}}, s \geq 0$, is the Laplace transform for the distribution function $\tilde{F}_{0,ij}(\cdot)$ of some non-negative random variable $\tilde{\kappa}_{0,ij}$. In fact, the random variable $\tilde{\kappa}_{0,ij}$ is a geometric type sum of random variables, that is,

$$\tilde{\kappa}_{0,ij} = \sum_{n=1}^{\rho_{0,i}-1} \bar{p}_{0,ii}\kappa_{0,ii,n} + \bar{p}_{0,ii}\kappa_{0,ij}, \tag{9.64}$$

where: (a) $\rho_{0,i}$ is a geometrically distributed random variable with parameter $\bar{p}_{0,ii} = 1 - p_{0,ii}$ (which takes value n with probability $\bar{p}_{0,ii}p_{0,ii}^{n-1}$, for $n = 1, 2, \ldots$); (b) $\kappa_{0,ii,n}, n = 1, 2, \ldots$ are i.i.d. random variables with the Laplace transform $\phi_{0,ii}(s)$, (d) $\kappa_{0,ij}$ is a random variable with the Laplace transform $\phi_{0,ij}(s)$; (c) the random variables $\rho_{0,i}, \kappa_{0,ij}, \kappa_{0,ii,n}, n = 1, 2, \ldots$ are mutually independent;

Obviously, $\tilde{F}_{0,ij}(0) < 1$, for $j \in \tilde{\mathbb{Y}}_{1,i}, i \in \bar{\mathbb{Y}}_0$, i.e., the part of condition $\tilde{\mathbf{J}}^\circ$ related to states $j \in \tilde{\mathbb{Y}}_{1,i}, i \in \bar{\mathbb{Y}}_0$ is satisfied.

Second, let us consider the absorbing case, where $i \in \mathbb{Y}_0$.

Let: (a) $\hat{\kappa}_{\varepsilon,ij,n}, n = 1, 2, \ldots$ be i.i.d. random variables with distribution function $\hat{F}_{\varepsilon,ij}(\cdot) = F_{\varepsilon,ij}(\cdot\, u_{\varepsilon,i})$, for every $j \in \mathbb{Y}_{1,i}, i \in \mathbb{Y}_0$ and $\varepsilon \in (0,1]$; (b) $\hat{\kappa}_{0,ij}$ be a random variable with the distribution function $\hat{F}_{0,ij}(\cdot) = F_{0,ij}(\cdot)$, for every $j \in \mathbb{Y}_{1,i}, i \in \mathbb{Y}_0$; (c) $\hat{f}_{\varepsilon,ij} = f_{\varepsilon,ij} = \int_0^\infty t\hat{F}_{\varepsilon,ij}(dt)$, for $j \in \mathbb{Y}_{1,i}, i \in \mathbb{X}, \varepsilon \in (0,1]$, and $\hat{f}_{0,ij} = f_{0,ij} = \int_0^\infty t\hat{F}_{0,ij}(dt)$, for $j \in \mathbb{Y}_{1,i}, i \in \mathbb{X}$.

By the central criterium of convergence for sums of i.i.d random variables (see, for example, Loève (1977)), the conditions **J** and **K** imply that the following weak LLN (law of large numbers) type relation holds, for any $0 < u_\varepsilon \to \infty$ as $\varepsilon \to 0$ and $j \in \mathbb{Y}_{1,i}, i \in \mathbb{Y}_0$,

$$\frac{\sum_{n \leq u_\varepsilon} \hat{\kappa}_{\varepsilon,ij,n}}{u_\varepsilon} \xrightarrow{\text{d}} \hat{f}_{0,ij} \text{ as } \varepsilon \to 0. \tag{9.65}$$

Indeed, the above conditions imply that (a) $\hat{F}_{\varepsilon,ij}(\cdot) \Rightarrow \hat{F}_{0,ij}(\cdot)$ as $\varepsilon \to 0$, and (b) $\hat{f}_{\varepsilon,ij} \to \hat{f}_{0,ij} = f_{0,ij} < \infty$ as $\varepsilon \to 0$, for $j \in \mathbb{Y}_{1,i}, i \in \mathbb{Y}_0$.

Let $0 < s_k \to \infty$ as $k \to \infty$ be a sequence of continuity points for the distribution function $\hat{F}_{0,ij}(\cdot)$. The above asymptotic relations (a) and (b) obviously imply that, for any $t > 0$,

$$\begin{aligned}
\overline{\lim_{\varepsilon \to 0}} \int_{tu_\varepsilon}^{\infty} s\hat{F}_{\varepsilon,ij}(ds) &\leq \overline{\lim_{\varepsilon \to 0}} \int_{s_k}^{\infty} s\hat{F}_{\varepsilon,ij}(ds) \\
&= \overline{\lim_{\varepsilon \to 0}} (\hat{f}_{\varepsilon,ij} - \int_0^{s_k} s\hat{F}_{\varepsilon,ij}(ds)) \\
&= \hat{f}_{0,ij} - \int_0^{s_k} s\hat{F}_{0,ij}(ds) \to 0 \text{ as } k \to \infty,
\end{aligned} \tag{9.66}$$

and, thus, the following relation holds, for any $t > 0$:

$$\overline{\lim_{\varepsilon \to 0}} \int_{tu_\varepsilon}^{\infty} s\hat{F}_{\varepsilon,ij}(ds) = 0. \tag{9.67}$$

The relation (9.67) implies that, for any $t > 0$,

$$\begin{aligned}
u_\varepsilon \mathsf{P}_1\{u_\varepsilon^{-1}\hat{\kappa}_{\varepsilon,ij,1} > t\} &= u_\varepsilon(1 - \hat{F}_{\varepsilon,ij}(tu_\varepsilon)) \\
&\leq t^{-1} \int_{tu_\varepsilon}^{\infty} s\hat{F}_{\varepsilon,ij}(ds) \to 0 \text{ as } \varepsilon \to 0.
\end{aligned} \tag{9.68}$$

Also, the relation (9.67) implies that, for any $t > 0$,

$$\begin{aligned}
u_\varepsilon \mathsf{E}_1 u_\varepsilon^{-1}\hat{\kappa}_{\varepsilon,ij,1}\mathrm{I}(u_\varepsilon^{-1}\hat{\kappa}_{\varepsilon,ij,1} \leq t) &= \int_0^{tu_\varepsilon} s\hat{F}_{\varepsilon,ij}(ds) \\
&\to \hat{f}_{0,ij} \text{ as } \varepsilon \to 0.
\end{aligned} \tag{9.69}$$

The relations (9.68) and (9.69) imply, by the above-mentioned central criterion of convergence, that the relation (9.65) holds.

Let us introduce Laplace transforms, for $j \in \mathbb{Y}_{1,i}, i \in \mathbb{X}$ and $\varepsilon \in (0, 1]$,

$$\hat{\phi}_{\varepsilon,ij}(s) = \int_0^{\infty} e^{-st}\hat{F}_{\varepsilon,ij}(dt) = \phi_{\varepsilon,ij}(s/u_{\varepsilon,i}), s \geq 0. \tag{9.70}$$

As well known (see, for example, Feller (1971)), the relation (9.65) is equivalent to the following relation:

$$u_\varepsilon(1 - \hat{\phi}_{\varepsilon,ij}(s/u_\varepsilon)) = u_\varepsilon(1 - \phi_{\varepsilon,ij}(s/u_\varepsilon u_{\varepsilon,i})) \\ \longrightarrow f_{0,ij}s \text{ as } \varepsilon \longrightarrow 0, \text{ for } s \geq 0. \tag{9.71}$$

Since, $0 < \bar{p}_{\varepsilon,ii} \longrightarrow 0$ as $\varepsilon \longrightarrow 0$, for $i \in \mathbb{Y}_0$, we can, in this case, choose,

$$u_\varepsilon = \bar{p}_{\varepsilon,ii}^{-1}. \tag{9.72}$$

Using the conditions **I**, $\mathbf{I}_{\mathcal{H}}$, $\mathbf{J}^\circ$, **K** and the relations (9.17), (9.25), (9.56), (9.71), (9.72) and taking into account that the state $i \in \mathbb{Y}_{1,i}$ if $i \in \mathbb{Y}_0$, we get the following relation, for $j \in \tilde{\mathbb{Y}}_{1,i}, i \in \mathbb{Y}_0$:

$$\tilde{\phi}_{\varepsilon,ij}(s/\tilde{u}_{\varepsilon,i}) = \frac{\phi_{\varepsilon,ij}(s/\bar{p}_{\varepsilon,ii}^{-1}u_{\varepsilon,i})}{1 + p_{\varepsilon,ii}\bar{p}_{\varepsilon,ii}^{-1}(1 - \phi_{\varepsilon,ii}(s/\bar{p}_{\varepsilon,ii}^{-1}u_{\varepsilon,i}))} \\ \longrightarrow \frac{1}{1 + f_{0,ii}s} = \tilde{\phi}_{0,ij}(s) \text{ as } \varepsilon \longrightarrow 0, \text{ for } s \geq 0. \tag{9.73}$$

In this case, $\tilde{\phi}_{0,ij}(s) = \tilde{\phi}_{0,i}(s) = \int_0^\infty e^{-st}\tilde{F}_{0,i}(dt)$ is the Laplace transform of an exponentially distributed random variable $\tilde{\kappa}_{0,ij} = \kappa_{0,i}$, with parameter $f_{0,ii}^{-1}$.

Note that $f_{0,ii} > 0$ and the distribution function $\tilde{F}_{0,i}(t) = 1 - \exp\{-f_{0,ii}^{-1}t\}$ has no atom in zero.

Therefore, the part of condition $\tilde{\mathbf{J}}^\circ$ related to states $j \in \tilde{\mathbb{Y}}_{1,i}, i \in \mathbb{Y}_0$ is also satisfied.

Also, the conditions $\tilde{\mathbf{I}}$, $\tilde{\mathbf{I}}_{\mathcal{H}}$ and the relations (9.58), (9.62) and (9.73) imply that the following relation takes place, for $i \in \mathbb{X}$:

$$\tilde{\phi}_{\varepsilon,i}(s) = \sum_{j \in \tilde{\mathbb{Y}}_{1,i}} \tilde{\phi}_{\varepsilon,ij}(s)\tilde{p}_{\varepsilon,ij} \\ \longrightarrow \sum_{j \in \tilde{\mathbb{Y}}_{1,i}} \tilde{\phi}_{0,ij}(s)\tilde{p}_{0,ij} = \sum_{j \in \tilde{\mathbb{Y}}_{1,i}} \tilde{\phi}_{0,ij}(s)\tilde{p}_{0,ij} \\ = \sum_{j \in \tilde{\mathbb{Y}}_{0,i}} \tilde{\phi}_{0,ij}(s)\tilde{p}_{0,ij} = \tilde{\phi}_{0,i}(s) \text{ as } \varepsilon \longrightarrow 0, \text{ for } s \geq 0. \tag{9.74}$$

The above remarks can be summarised in the following lemma.

Lemma 9.10 *The conditions* **G**, **H**, **I**, $\mathbf{I}_{\mathcal{H}}$, $\mathbf{J}^\circ$ (**J**), **K**, *assumed to be satisfied for the semi-Markov processes* $\eta_\varepsilon(t)$, *entail the fulfilment of the condition* $\mathbf{J}^\circ$ (**J**) *for the semi-Markov processes* $\tilde{\eta}_\varepsilon(t)$ *in the form of condition* $\tilde{\mathbf{J}}^\circ$ ($\tilde{\mathbf{J}}$), *where:*

(i) *The asymptotic relation appearing in the condition* $\mathbf{J}^\circ$ *takes two different forms, given in the relations (9.62) or (9.73), for the cases where the initial state i is, respectively, asymptotically non-absorbing or asymptotically absorbing state, that is,* $i \in \overline{\mathbb{Y}}_0$ *or* $i \in \mathbb{Y}_0$.

(ii) *The normalisation functions* $\tilde{u}_{\cdot,i}, i \in \mathbb{X}$ *appearing in the condition* $\mathbf{J}^\circ$ *are given by the relation (9.60).*

9.2.2.3 Condition $\tilde{\mathbf{K}}$. Let us assume that the conditions **G**, **H**, **I**, $\mathbf{I}_{\mathcal{H}}$, **J**, and **K** hold.

The expectation $\tilde{f}_{\varepsilon,ij}$, which is an analogue of the expectation $f_{\varepsilon,ij}$, is defined (for $\varepsilon \in (0,1]$) by the following relation, for $j \in \tilde{\mathbb{Y}}_{1,i}, i \in \mathbb{X}$:

$$\begin{aligned}\tilde{f}_{\varepsilon,ij} &= \mathsf{E}\{\tilde{\kappa}_{\varepsilon,1}/\tilde{\eta}_{\varepsilon,0} = i, \tilde{\eta}_{\varepsilon,1} = j\} \\ &= f_{\varepsilon,ij} + p_{\varepsilon,ii}\bar{p}_{\varepsilon,ii}^{-1} f_{\varepsilon,ii}.\end{aligned} \tag{9.75}$$

Also, for $i \in \mathbb{X}$,

$$\tilde{f}_{\varepsilon,i} = \mathsf{E}\{\tilde{\kappa}_{\varepsilon,1}/\tilde{\eta}_{\varepsilon,0} = i\} = \sum_{j \in \tilde{\mathbb{Y}}_{1,i}} \tilde{f}_{\varepsilon,ij}\tilde{p}_{\varepsilon,ij}. \tag{9.76}$$

The following condition plays the role of condition **K** for the semi-Markov processes $\tilde{\eta}_\varepsilon(t)$:

$\tilde{\mathbf{K}}$: **(a)** $\tilde{f}_{\varepsilon,ij} < \infty, j \in \mathbb{Y}_{1,i}, i \in \mathbb{X}$, for every $\varepsilon \in (0,1]$, **(b)** $\tilde{f}_{\varepsilon,ij}/\tilde{u}_{\varepsilon,i} \to \tilde{f}_{0,ij} = \int_0^\infty t\tilde{F}_{0,ij}(dt) < \infty$ as $\varepsilon \to 0$, for $j \in \mathbb{Y}_{1,i}, i \in \mathbb{X}$.

The conditions **G**, **H**, $\mathbf{I}_{\mathcal{H}}$, **J**, **K** and the relation (9.75) imply that, for $j \in \tilde{\mathbb{Y}}_{1,i}, i \in \mathbb{X}$,

$$\begin{aligned}\frac{\tilde{f}_{\varepsilon,ij}}{\tilde{u}_{\varepsilon,i}} &= \bar{p}_{\varepsilon,ii}\frac{f_{\varepsilon,ij}}{u_{\varepsilon,i}} + p_{\varepsilon,ii}\frac{f_{\varepsilon,ii}}{u_{\varepsilon,i}} \\ &\to \bar{p}_{0,ii}f_{0,ij} + p_{0,ii}f_{0,ii} = \tilde{f}_{0,ij} \text{ as } \varepsilon \to 0.\end{aligned} \tag{9.77}$$

Note that the above relation works for both cases, where $j \in \tilde{\mathbb{Y}}_{1,i}, i \in \bar{\mathbb{Y}}_0$ or $j \in \tilde{\mathbb{Y}}_{1,i}, i \in \mathbb{Y}_0$.

Also, the following relation takes place, for $i \in \mathbb{X}$:

$$\begin{aligned}\frac{\tilde{f}_{\varepsilon,i}}{\tilde{u}_{\varepsilon,i}} &= \sum_{j \in \tilde{\mathbb{Y}}_{1,i}} \frac{\tilde{f}_{\varepsilon,ij}}{u_{\varepsilon,i}}\tilde{p}_{\varepsilon,ij} \\ &\to \sum_{j \in \tilde{\mathbb{Y}}_{1,i}} \tilde{f}_{0,ij}\tilde{p}_{0,ij} = \sum_{j \in \tilde{\mathbb{Y}}_{0,i}} \tilde{f}_{0,ij}\tilde{p}_{0,ij} \\ &= \tilde{f}_{0,i} = \int_0^\infty t\tilde{F}_{0,i}(dt) \text{ as } \varepsilon \to 0.\end{aligned} \tag{9.78}$$

The above remarks can be summarised in the following lemma.

Lemma 9.11 *The conditions* **G**, **H**, **I**, $\mathbf{I}_{\mathcal{H}}$, **J**, **K** *assumed to be satisfied for the semi-Markov processes $\eta_\varepsilon(t)$, entail the fulfilment of the condition* **K** *for the semi-Markov processes $\tilde{\eta}_\varepsilon(t)$ in the form of condition* $\tilde{\mathbf{K}}$, *where*:

(i) *The asymptotic relation appearing in condition* **K** *takes the form of relation (9.77).*

(ii) *The normalisation functions $\tilde{u}_{\cdot,i}, i \in \mathbb{X}$ appearing in the condition* **K** *are given by relation (9.60).*

Remark 9.1 It is useful noting that the conditions **G**, **H**, and **K (a)**, assumed to be satisfied for the semi-Markov processes $\eta_\varepsilon(t)$, entail the fulfilment of the condition **K (a)** for the semi-Markov processes $\tilde{\eta}_\varepsilon(t)$ in the form of condition $\tilde{\mathbf{K}}$ **(a)**. This follows from the relation (9.75).

9.2.2.4 Conditions $\tilde{\mathbf{L}}$ and $\tilde{\mathbf{L}}_{\mathcal{H}}$. The following conditions play the roles of condition **L** and $\mathbf{L}_{\mathcal{H}}$ for the semi-Markov processes $\tilde{\eta}_\varepsilon(t)$:

$\tilde{\mathbf{L}}$: $\tilde{u}_{\varepsilon,i} \to \tilde{u}_{0,i} \in (0, \infty]$ as $\varepsilon \to 0$, for $i \in \mathbb{X}$

and

$\tilde{\mathbf{L}}_{\mathcal{H}}$: The functions $\tilde{u}_{\cdot,i}, i \in \mathbb{X}$ belong to the complete family of asymptotically comparable functions $\mathcal{H}$ appearing in the condition $\mathbf{L}_{\mathcal{H}}$.

Lemma 9.12 *The conditions* **G**, **H** *and* **I**, $\mathbf{I}_{\mathcal{H}}$, **L**, $\mathbf{L}_{\mathcal{H}}$, *assumed to be satisfied for the semi-Markov processes $\eta_\varepsilon(t)$, entail the fulfilment of the conditions* **L**, $\mathbf{L}_{\mathcal{H}}$ *for the semi-Markov processes $\tilde{\eta}_\varepsilon(t)$ in the form of conditions* $\tilde{\mathbf{L}}$, $\tilde{\mathbf{L}}_{\mathcal{H}}$.

Proof It follows from the relation (9.60) and Lemma 8.2 that the functions $\tilde{u}_{\cdot,i} = \bar{p}_{\cdot,ii}^{-1} u_{\cdot,i}, i \in \mathbb{X}$ belong to the family $\mathcal{H}$. Moreover, the limit $\tilde{u}_{0,i} = \bar{p}_{0,ii}^{-1} u_{0,i} \in (0, \infty]$, for $i \in \mathbb{X}$. □

The following lemma presents operating formulas which express the comparability parameters and limits for the functions appearing in conditions $\tilde{\mathbf{L}}$ and $\tilde{\mathbf{L}}_{\mathcal{H}}$ in terms of the comparability parameters and limits for the functions appearing in the conditions **I**, **L** and $\mathbf{I}_{\mathcal{H}}$, $\mathbf{L}_{\mathcal{H}}$.

Lemma 9.13 *Let the conditions* **G**, **H**, **I**, $\mathbf{I}_{\mathcal{H}}$, **L**, $\mathbf{L}_{\mathcal{H}}$ *be satisfied. Then*:

(i) *The functions $\tilde{u}_{\cdot,i} = \bar{p}_{\cdot,ii}^{-1} u_{\cdot,i}, i \in \mathbb{X}$ belong to the family $\mathcal{H}$ and the corresponding comparability limits can be calculated by applying to these functions the operating quotient formula given in Lemma 8.2.*

(ii) *If $\mathcal{H}$ is one of the families, $\mathcal{H}_1$, $\mathcal{H}_2$, or $\mathcal{H}_3$, then the corresponding comparability parameters and limits can be calculated by applying to the functions $\tilde{u}_{\cdot,i} = \bar{p}_{\cdot,ii}^{-1} u_{\cdot,i}, i \in \mathbb{X}$ the operating quotient formulas given, respectively, in Lemmas 8.5, 8.7, or 8.9.*

9.2.2.5 Summary of Perturbation Conditions for Semi-Markov Process $\tilde{\eta}_\varepsilon(t)$. The following theorem summarises the remarks made in Sect. 9.2.

Theorem 9.1 *The conditions* **G**, **H**, **I**, $\mathbf{I}_{\mathcal{H}}$, **L**, $\mathbf{L}_{\mathcal{H}}$, **J**, **K**, *assumed to be satisfied for the semi-Markov processes $\eta_\varepsilon(t)$, entail the fulfilment of these conditions for the semi-Markov processes $\tilde{\eta}_\varepsilon(t)$ in the form of conditions* $\tilde{\mathbf{G}}$, $\tilde{\mathbf{H}}$, $\tilde{\mathbf{I}}$, $\tilde{\mathbf{I}}_{\mathcal{H}}$, $\tilde{\mathbf{L}}$, $\tilde{\mathbf{L}}_{\mathcal{H}}$, $\tilde{\mathbf{J}}$, $\tilde{\mathbf{K}}$ *presented in Lemmas 9.1–9.13.*

9.3 One-State Reduction of Phase Space for Perturbed Semi-Markov Processes

In Sect. 9.3, we describe the procedure of one-state phase space reduction for perturbed semi-Markov processes. It is shown that the basic perturbation conditions imposed on the original semi-Markov processes are also satisfied for the semi-Markov processes with the reduced phase space. We also give explicit formulas for recalculating normalisation functions, limiting distributions and expectations in the corresponding perturbation conditions for semi-Markov processes with the reduced phase space.

9.3.1 A Procedure of One-Step Phase Space Reduction for Perturbed Semi-Markov Processes

9.3.1.1 Definition of Semi-Markov Process with Reduced Phase Space. We assume now that the number of states m in the phase space $\mathbb{X}$ is greater than 1.

We also assume that the conditions **G** and **H** are satisfied for the semi-Markov processes $\eta_\varepsilon(t)$ and, therefore, these condition are also satisfied for the semi-Markov processes $\tilde{\eta}_\varepsilon(t)$ in the form of conditions $\tilde{\mathbf{G}}$ and $\tilde{\mathbf{H}}$.

In what follows in this section, we assume that $\varepsilon \in (0, 1]$.

Let us choose some state $k \in \mathbb{X}$ and consider the reduced phase space ${}_k\mathbb{X} = \mathbb{X} \setminus \{k\}$.

Let us define the stopping times for the Markov chain $\tilde{\eta}_{\varepsilon,n}$, for $r = 0, 1, \ldots$,

$$ {}_k\alpha_\varepsilon[r] = \min(n > r : \tilde{\eta}_{\varepsilon,n} \in {}_k\mathbb{X}). \tag{9.79} $$

By the definition, ${}_k\alpha_\varepsilon[r]$ is the first after r time of hitting in the reduced phase space ${}_k\mathbb{X}$ by the Markov chain $\tilde{\eta}_{\varepsilon,n}$.

Since the Markov chain $\tilde{\eta}_{\varepsilon,n}$ does not make virtual transitions (of the form $i \to i$), the following relation takes place, for $r = 0, 1, \ldots$,

$$ {}_k\alpha_\varepsilon[r] = \begin{cases} r + 1 \text{ if } \tilde{\eta}_{\varepsilon,r+1} \in {}_k\mathbb{X}, \\ r + 2 \text{ if } \tilde{\eta}_{\varepsilon,r+1} = k. \end{cases} \tag{9.80} $$

We also define the following successive stopping times:

$$ {}_k\beta_{\varepsilon,n} = {}_k\alpha_\varepsilon[{}_k\beta_{\varepsilon,n-1}], n = 1, 2, \ldots, \text{ where } {}_k\beta_{\varepsilon,0} = \mathrm{I}(\tilde{\eta}_{\varepsilon,0} = k). \tag{9.81} $$

Let us now construct a new Markov renewal process $({}_k\eta_{\varepsilon,n}, {}_k\kappa_{\varepsilon,n}), n = 0, 1, \ldots$, with the phase space ${}_k\mathbb{X} \times [0, \infty)$,

$$ ({}_k\eta_{\varepsilon,n}, {}_k\kappa_{\varepsilon,n}) = \begin{cases} (\tilde{\eta}_{\varepsilon, {}_k\beta_{\varepsilon,0}}, 0) & \text{for } n = 0, \\ (\tilde{\eta}_{\varepsilon, {}_k\beta_{\varepsilon,n}}, \sum_{l={}_k\beta_{\varepsilon,n-1}+1}^{{}_k\beta_{\varepsilon,n}} \tilde{\kappa}_{\varepsilon,l}) & \text{for } n = 1, 2, \ldots. \end{cases} \tag{9.82} $$

We also can define the corresponding reduced semi-Markov process,

$$ {}_k\eta_\varepsilon(t) = {}_k\eta_{\varepsilon,\, {}_k\nu_\varepsilon(t)}, t \geq 0, \tag{9.83} $$

where,

$$ {}_k\zeta_{\varepsilon,n} = {}_k\kappa_{\varepsilon,1} + \cdots + {}_k\kappa_{\varepsilon,n}, n = 1, 2, \ldots, {}_k\zeta_{\varepsilon,0} = 0, \tag{9.84} $$

are the successive moments of jumps, and,

$$ {}_k\nu_\varepsilon(t) = \max(n \geq 1 : {}_k\zeta_{\varepsilon,n} \leq t), \tag{9.85} $$

is the number of jumps in the interval $[0, t]$ for the above semi-Markov process.

The definition of stopping times ${}_k\beta_{\varepsilon,n}$ implies that the transition probabilities of the above Markov renewal process are determined by the following relation, for $t \geq 0, i, j \in {}_k\mathbb{X}$:

$$ \begin{aligned} {}_kQ_{\varepsilon,ij}(t) &= \mathsf{P}\{{}_k\eta_{\varepsilon,1} = j, {}_k\kappa_{\varepsilon,1} \leq t / {}_k\eta_{\varepsilon,0} = i\} \\ &= \tilde{Q}_{\varepsilon,ij}(t) + \tilde{Q}_{\varepsilon,ik}(t) * \tilde{Q}_{\varepsilon,kj}(t). \end{aligned} \tag{9.86} $$

Accordingly, the transition probabilities of the embedded Markov chain ${}_k\eta_{\varepsilon,n}$ are given by the following relation, for $i, j \in {}_k\mathbb{X}$:

$$ \begin{aligned} {}_kp_{\varepsilon,ij} &= \mathsf{P}\{{}_k\eta_{\varepsilon,1} = j / {}_k\eta_{\varepsilon,0} = i\} \\ &= \tilde{p}_{\varepsilon,ij} + \tilde{p}_{\varepsilon,ik}\, \tilde{p}_{\varepsilon,kj}. \end{aligned} \tag{9.87} $$

Note that the transition probabilities $\tilde{Q}_{\varepsilon,ii}(t) = 0, t \geq 0, i \in \mathbb{X}$ and $\tilde{p}_{\varepsilon,ii} = 0, i \in \mathbb{X}$, since the semi-Markov process $\tilde{\eta}_\varepsilon(t)$ does not make virtual transitions of the form $i \to i$.

The semi-Markov process ${}_k\eta_\varepsilon(t)$ can be considered as the result of some two-stage procedure applied to the original semi-Markov process $\eta_\varepsilon(t)$. At the first stage, the semi-Markov process $\eta_\varepsilon(t)$ is transformed in the semi-Markov process $\tilde{\eta}_\varepsilon(t)$, with the use of the procedure of removing virtual transitions. At the second stage, the semi-Markov process $\tilde{\eta}_\varepsilon(t)$ is transformed in the semi-Markov ${}_k\eta_\varepsilon(t)$, with the use of the described above procedure of phase space reduction.

The following relation holds, for $i \in {}_k\mathbb{X}$:

$$ \mathsf{P}_i\{{}_k\eta(t) \in {}_k\mathbb{X}, t \geq 0\} = 1. \tag{9.88} $$

The relation (9.88) allows us to consider ${}_k\eta(t)$, as a semi-Markov process with the reduced phase space ${}_k\mathbb{X}$. We call ${}_k\eta(t)$ a reduced semi-Markov process.

9.3.1.2 Condition ${}_\mathbf{k}$G. Let us assume that the conditions **G** and **H** are satisfied.

Let us introduce the distribution functions of sojourn times for the reduced semi-Markov process ${}_k\eta_\varepsilon(t)$, for $i \in {}_k\mathbb{X}$ and $\varepsilon \in (0, 1]$,

$$\begin{aligned} {}_kF_{\varepsilon,i}(t) &= \mathsf{P}\{{}_k\kappa_{\varepsilon,i} \leq t /\ {}_k\eta_{\varepsilon,0} = i\} \\ &= \sum_{j \in {}_k\mathbb{X}} {}_kQ_{\varepsilon,ij}(t),\ t \geq 0. \end{aligned} \tag{9.89}$$

The following condition plays the role of the condition **G** for the reduced semi-Markov process ${}_k\eta_\varepsilon(t)$:

${}_\mathbf{k}\mathbf{G}$: **(a)** ${}_kp_{\varepsilon,ij} > 0, \varepsilon \in (0,1]$ or ${}_kp_{\varepsilon,ij} = 0, \varepsilon \in (0,1]$, for $i, j \in {}_k\mathbb{X}$, **(b)** ${}_kF_{\varepsilon,i}(0) < 1, i \in {}_k\mathbb{X}, \varepsilon \in (0,1]$.

Lemma 9.14 *The conditions* **G** *and* **H**, *assumed to be satisfied for the semi-Markov processes* $\eta_\varepsilon(t)$, *entail the fulfilment of the condition* **G** *for the reduced semi-Markov processes* ${}_k\eta_\varepsilon(t)$ *in the form of condition* ${}_\mathbf{k}\mathbf{G}$.

Proof Lemma 9.3 implies that the condition $\tilde{\mathbf{G}}$ is satisfied. The condition $\tilde{\mathbf{G}}$ **(a)** and the relation (9.87) obviously imply that the condition ${}_k\mathbf{G}$ **(a)** is satisfied.

The condition $\tilde{\mathbf{G}}$ **(b)** and the relation (9.89) obviously imply, for $i \in {}_k\mathbb{X}$ and $\varepsilon \in (0,1]$,

$$\begin{aligned} {}_kF_{\varepsilon,i}(0) &= \sum_{j \neq i,k} \tilde{Q}_{\varepsilon,ij}(0) + \tilde{Q}_{\varepsilon,ik}(0)\tilde{F}_{\varepsilon,k}(0) \\ &\leq \sum_{j \neq i,k} \tilde{Q}_{\varepsilon,ij}(0) + \tilde{Q}_{\varepsilon,ik}(0) = \tilde{F}_{\varepsilon,i}(0) < 1. \end{aligned} \tag{9.90}$$

Thus, the regularity condition ${}_k\mathbf{G}$ is satisfied for the semi-Markov process ${}_k\eta_\varepsilon(t)$. □

Therefore, the reduced semi-Markov process ${}_k\eta_\varepsilon(t)$ is well defined on the time interval $[0,\infty)$, for every $\varepsilon \in (0,1]$.

Let us introduce sets, for $i \in {}_k\mathbb{X}$ and $\varepsilon \in (0,1]$,

$${}_k\mathbb{Y}_{\varepsilon,i} = \{j \in {}_k\mathbb{X} : {}_kp_{\varepsilon,ij} > 0\}. \tag{9.91}$$

The following lemma takes place.

Lemma 9.15 *The condition* ${}_\mathbf{k}\mathbf{G}$ **(a)** *is satisfied if and only if the sets* ${}_k\mathbb{Y}_{\varepsilon,i} = {}_k\mathbb{Y}_{1,i}, \varepsilon \in (0,1]$, *for* $i \in {}_k\mathbb{X}$, *and sets,*

$${}_k\mathbb{Y}_{1,i} = \begin{cases} (\tilde{\mathbb{Y}}_{1,i} \cup \tilde{\mathbb{Y}}_{1,k}) \setminus \{k\} & \text{if } k \in \tilde{\mathbb{Y}}_{1,i}, i \in {}_k\mathbb{X}, \\ \tilde{\mathbb{Y}}_{1,i} & \text{if } k \notin \tilde{\mathbb{Y}}_{1,i}, i \in {}_k\mathbb{X}. \end{cases} \tag{9.92}$$

Proof The proof obviously follows from the relation (9.87). □

9.3.1.3 Condition ${}_\mathbf{k}\mathbf{H}$. The following conditions plays the role of condition **H** for the reduced semi-Markov processes ${}_k\eta_\varepsilon(t)$:

${}_\mathbf{k}\mathbf{H}$: For any $i, j \in {}_k\mathbb{X}$, there exist ${}_kn_{ij} \geq 1$ and a chain of states $i = {}_kr_0, {}_kr_1, \ldots, {}_kr_{{}_kn_{ij}} = j$ such that $\prod_{1 \leq l \leq {}_kn_{ij}} {}_kp_{1,{}_kr_{l-1}\,{}_kr_l} > 0$.

Lemma 9.16 *The conditions* **G** *and* **H**, *assumed to be satisfied for the semi-Markov processes* $\eta_\varepsilon(t)$, *entail the fulfilment of the condition* **H** *for the reduced semi-Markov processes* ${}_k\eta_\varepsilon(t)$ *in the form of condition* $_{\mathbf{k}}\mathbf{H}$.

Proof The condition **H** for the semi-Markov processes $\tilde{\eta}_\varepsilon(t)$, is, under the condition $\tilde{\mathbf{G}}$ **(a)**, equivalent to the assumption that, for any states $i, j \in \mathbb{X}$, there exist an integer number $\tilde{n}_{ij}$ and a chain of states $i = \tilde{r}_0, \tilde{r}_1, \ldots, \tilde{r}_{\tilde{n}_{ij}} = j$ such that $\tilde{r}_{l-1} \neq \tilde{r}_l, l = 1, \ldots, \tilde{n}_{ij}$ and $\prod_{l=1}^{\tilde{n}_{ij}} \tilde{p}_{1,\tilde{r}_{l-1}\tilde{r}_l} > 0$. Suppose that the states $i = \tilde{r}_0, j = \tilde{r}_{\tilde{n}_{ij}} \in {}_k\mathbb{X}$. If the states $\tilde{r}_{l-1}, \tilde{r}_l \in {}_k\mathbb{X}$, for some $1 \leq l \leq n_i$, then the probability ${}_kp_{1,\tilde{r}_{l-1}\tilde{r}_l} \geq \tilde{p}_{1,\tilde{r}_{l-1}\tilde{r}_l} > 0$. If the state $\tilde{r}_l = k$, for some $1 \leq l \leq n$, then the states $\tilde{r}_{l-1}, \tilde{r}_{l+1} \in {}_k\mathbb{X}$ and the probability ${}_kp_{1,\tilde{r}_{l-1}\tilde{r}_{l+1}} \geq \tilde{p}_{1,\tilde{r}_{l-1}k}\tilde{p}_{\varepsilon,k\tilde{r}_{l+1}} > 0$. Let $i = {}_kr_0, {}_kr_1, \ldots, {}_kr_{{}_kn_{ij}} = j$ be the new chain of states obtained from the chain $i = \tilde{r}_0, \tilde{r}_1, \ldots, \tilde{r}_{\tilde{n}_{ij}}$ by excluding all states $\tilde{r}_l, 1 \leq l \leq \tilde{n}_{ij} - 1$ such that $\tilde{r}_l = k$. The above inequalities imply that $\prod_{l=1}^{{}_kn_{ij}} {}_kp_{1,{}_kr_{l-1}{}_kr_l} > 0$ for this new chain. Therefore, the condition $_{\mathbf{k}}\mathbf{H}$ is satisfied for the reduced semi-Markov processes ${}_k\eta_\varepsilon(t)$. □

9.3.2 Perturbation Conditions for Reduced Semi-Markov Processes

We are going to describe the assumptions, under which the basic conditions **I–L** and $\mathbf{I}_{\mathcal{H}}$, $\mathbf{L}_{\mathcal{H}}$ are valid for the reduced semi-Markov processes ${}_k\eta_\varepsilon(t)$, constructed by using the described above procedure of one-state reduction of the phase space for semi-Markov processes $\tilde{\eta}_\varepsilon(t)$.

9.3.2.1 Conditions $_{\mathbf{k}}\mathbf{I}$ **and** $\mathbf{I}_{\mathbf{k}\mathcal{H}}$. The following conditions play the roles of the conditions **I** and $\mathbf{I}_{\mathcal{H}}$ for the reduced semi-Markov processes ${}_k\eta_\varepsilon(t)$:

$_{\mathbf{k}}\mathbf{I}$: ${}_kp_{\varepsilon,ij} = \tilde{p}_{\varepsilon,ij} + \tilde{p}_{\varepsilon,ik}\tilde{p}_{\varepsilon,kj} \longrightarrow {}_kp_{0,ij} = \tilde{p}_{0,ij} + \tilde{p}_{0,ik}\tilde{p}_{0,kj}$ as $\varepsilon \longrightarrow 0$, for $i, j \in {}_k\mathbb{X}$,

and

$_{\mathbf{k}}\mathbf{I}_{\mathcal{H}}$: The functions ${}_kp_{\cdot,ij}, j \in {}_k\mathbb{Y}_{1,i}, i \in {}_k\mathbb{X}$ belong to the complete family of asymptotically comparable functions $\mathcal{H}$ appearing in the condition $\mathbf{I}_{\mathcal{H}}$.

Recall that the condition $_{\mathbf{k}}\mathbf{H}$ implies that the probabilities ${}_kp_{\varepsilon,ii} > 0, i \in {}_k\mathbb{X}$, for $\varepsilon \in (0, 1]$.

Since, the matrix ${}_k\mathbf{P}_\varepsilon = \|{}_kp_{\varepsilon,ij}\|$ is stochastic, for $\varepsilon \in (0, 1]$, the condition ${}_k\mathbf{I}$ implies that the matrix ${}_k\mathbf{P}_0 = \|{}_kp_{0,ij}\|$ is also stochastic.

Let ${}_k\eta_{0,n}, n = 0, 1, \ldots$ be a Markov chain with the phase space ${}_k\mathbb{X}$ and the matrix of transition probabilities ${}_k\mathbf{P}_0$. The condition $_{\mathbf{k}}\mathbf{I}$ allows us to interpret the Markov chains ${}_k\eta_{\varepsilon,n}$, for $\varepsilon \in (0, 1]$, as a perturbed version of the Markov chain ${}_k\eta_{0,n}$.

Lemma 9.17 *The conditions* **G**, **H**, **I**, $\mathbf{I}_{\mathcal{H}}$, *assumed to be satisfied for the semi-Markov processes* $\eta_\varepsilon(t)$, *entail the fulfilment of the conditions* **I**, $\mathbf{I}_{\mathcal{H}}$ *for the reduced semi-Markov processes* ${}_k\eta_\varepsilon(t)$ *in the form of conditions* $_{\mathbf{k}}\mathbf{I}$, $_{\mathbf{k}}\mathbf{I}_{\mathcal{H}}$.

Proof The conditions **G**, **H**, **I**, $\mathbf{I}_{\mathcal{H}}$ assumed to be satisfied for the semi-Markov processes $\eta_\varepsilon(t)$ imply that these conditions are also satisfied for the semi-Markov processes $\tilde{\eta}_\varepsilon(t)$ in the form of conditions $\tilde{\mathbf{G}}$, $\tilde{\mathbf{H}}$, $\tilde{\mathbf{I}}$, $\tilde{\mathbf{I}}_{\mathcal{H}}$. Due to this fact, the statement of Lemma 9.17 follows from the relation (9.87) and Lemmas 8.2 and 9.7, according to which, the functions ${}_kp_{\cdot,ij}, j \in {}_k\mathbb{Y}_{1,i}, i \in {}_k\mathbb{X}$ belong to the family $\mathcal{H}$. □

The following lemma presents operating formulas, which express the comparability parameters and limits for functions appearing in the conditions ${}_k\mathbf{I}$ and ${}_k\mathbf{I}_{\mathcal{H}}$ in terms of the comparability parameters and limits for the functions appearing in the conditions $\tilde{\mathbf{I}}$ and $\tilde{\mathbf{I}}_{\mathcal{H}}$.

Lemma 9.18 *Let conditions* **G**, **H**, **I**, $\mathbf{I}_{\mathcal{H}}$ *be satisfied. Then*:

(i) *The functions* ${}_kp_{\cdot,ij} = \tilde{p}_{\cdot,ij} + \tilde{p}_{\cdot,ik}\tilde{p}_{\cdot,kj}, j \in {}_k\mathbb{Y}_{1,i}, i \in {}_k\mathbb{X}$ *belong to the family* $\mathcal{H}$ *and the corresponding comparability limits can be computed by applying the operating product formula given in Lemma 8.2 to the functions* $\tilde{p}_{\cdot,ik}\tilde{p}_{\cdot,kj}$ *and then the operating sum formula given in Lemma 8.2 to the functions* ${}_kp_{\cdot,ij} = \tilde{p}_{\cdot,ij} + \tilde{p}_{\cdot,ik}\tilde{p}_{\cdot,kj}$, $j \in {}_k\mathbb{Y}_{1,i}, i \in {}_k\mathbb{X}$.

(ii) *If* $\mathcal{H}$ *is one of the families*, $\mathcal{H}_1$, $\mathcal{H}_2$, *or* $\mathcal{H}_3$, *then the corresponding comparability parameters and limits can be computed by applying the operating product formulas given, respectively, in Lemmas 8.4, 8.6, or 8.8 to the functions* $\tilde{p}_{\cdot,ik}\tilde{p}_{\cdot,kj}$ *and then the operating sum formulas given, respectively, in Lemmas 8.4, 8.6, or 8.8 to the functions* ${}_kp_{\cdot,ij} = \tilde{p}_{\cdot,ij} + \tilde{p}_{\cdot,ik}\tilde{p}_{\cdot,kj}, j \in {}_k\mathbb{Y}_{1,i}, i \in {}_k\mathbb{X}$.

9.3.2.2 Condition ${}_k$J. Let us assume that the conditions **G**, **H**, **I**, $\mathbf{I}_{\mathcal{H}}$, **L**, $\mathbf{L}_{\mathcal{H}}$, $\mathbf{J}^\circ$ (equivalent to the condition **J**), and **K**, are satisfied for the semi-Markov processes $\eta_\varepsilon(t)$. Then, according to Lemmas 9.3, 9.5, 9.10, and 9.11, these conditions are also satisfied for the semi-Markov processes $\tilde{\eta}_\varepsilon(t)$ in the form of conditions $\tilde{\mathbf{G}}$, $\tilde{\mathbf{H}}$, $\tilde{\mathbf{I}}$, $\tilde{\mathbf{I}}_{\mathcal{H}}$, $\tilde{\mathbf{L}}$, $\tilde{\mathbf{L}}_{\mathcal{H}}$, $\tilde{\mathbf{J}}^\circ$ (equivalent to the condition $\tilde{\mathbf{J}}$), and $\tilde{\mathbf{K}}$.

The distribution function ${}_kF_{\varepsilon,ij}(t), t \geq 0$, which is an analogue of the distribution function $\tilde{F}_{\varepsilon,ij}(t), t \geq 0$ is defined (for $\varepsilon \in (0, 1]$) by the following relation, for $j \in {}_k\mathbb{Y}_{1,i}, i \in {}_k\mathbb{X}$:

$$\begin{aligned} {}_kF_{\varepsilon,ij}(t) &= \mathsf{P}\{{}_k\eta_{\varepsilon,1} = j,\ {}_k\kappa_{\varepsilon,1} \leq t /\ {}_k\eta_{\varepsilon,0} = i,\ {}_k\eta_{\varepsilon,1} = j\} \\ &= \tilde{F}_{\varepsilon,ij}(t)\frac{\tilde{p}_{\varepsilon,ij}}{{}_kp_{\varepsilon,ij}} + \tilde{F}_{\varepsilon,ik}(t) * \tilde{F}_{\varepsilon,kj}(t)\frac{\tilde{p}_{\varepsilon,ik}\tilde{p}_{\varepsilon,kj}}{{}_kp_{\varepsilon,ij}}, t \geq 0 \end{aligned} \tag{9.93}$$

and, for $j \notin {}_k\mathbb{Y}_{1,i}, i \in {}_k\mathbb{X}$,

$${}_kF_{\varepsilon,ij}(t) = {}_kF_{\varepsilon,i}(t),\ t \geq 0, \tag{9.94}$$

where, for $i \in {}_k\mathbb{X}$,

$$\begin{aligned} {}_kF_{\varepsilon,i}(t) &= \mathsf{P}\{{}_k\kappa_{\varepsilon,1} \leq t /\ {}_k\eta_{\varepsilon,0} = i\} \\ &= \sum_{j \in {}_k\mathbb{X}} {}_kQ_{\varepsilon,ij}(t) = \sum_{j \in {}_k\mathbb{Y}_{1,i}} {}_kF_{\varepsilon,ij}(t)\ {}_kp_{\varepsilon,ij}, t \geq 0. \end{aligned} \tag{9.95}$$

The relations (9.93)–(9.95) imply that the corresponding Laplace transforms ${}_k\phi_{\varepsilon,ij}(s), s \geq 0$ take the following form, for $j \in {}_k\mathbb{Y}_{1,i}, i \in {}_k\mathbb{X}$:

$$\begin{aligned} {}_k\phi_{\varepsilon,ij}(s) &= \mathsf{E}\{e^{-s\, {}_k\kappa_{\varepsilon,1}} / {}_k\eta_{\varepsilon,0} = i,\ {}_k\eta_{\varepsilon,1} = j\} \\ &= \tilde{\phi}_{\varepsilon,ij}(s) \frac{\tilde{p}_{\varepsilon,ij}}{{}_kp_{\varepsilon,ij}} + \tilde{\phi}_{\varepsilon,ik}(s)\tilde{\phi}_{\varepsilon,kj}(s) \frac{\tilde{p}_{\varepsilon,ik}\tilde{p}_{\varepsilon,kj}}{{}_kp_{\varepsilon,ij}}, \ s \geq 0, \end{aligned} \tag{9.96}$$

and, for $j \notin {}_k\mathbb{Y}_{1,i}, i \in {}_k\mathbb{X}$,

$${}_k\phi_{\varepsilon,ij}(a) = {}_k\phi_{\varepsilon,i}(s), s \geq 0, \tag{9.97}$$

where, for $i \in {}_k\mathbb{X}$,

$${}_k\phi_{\varepsilon,i}(s) = \mathsf{E}\{e^{-s\, {}_k\kappa_{\varepsilon,1}} / {}_k\eta_{\varepsilon,0} = i\} = \sum_{j \in {}_k\mathbb{Y}_{1,i}} {}_k\phi_{\varepsilon,ij}(s)\, {}_kp_{\varepsilon,ij}, \ s \geq 0. \tag{9.98}$$

The following condition plays the role of the condition **J** for the reduced semi-Markov processes ${}_k\eta_\varepsilon(t)$:

$_{\mathbf{k}}$J: **(a)** ${}_kF_{\varepsilon,ij}(\cdot\, {}_ku_{\varepsilon,i}) = \mathsf{P}\{ {}_k\kappa_{\varepsilon,1}/ {}_ku_{\varepsilon,i} \leq \cdot / \eta_{\varepsilon,0} = i, \eta_{\varepsilon,1} = j\} \Rightarrow {}_kF_{0,ij}(\cdot)$ as $\varepsilon \to 0$, for $j \in {}_k\mathbb{Y}_{1,i}, i \in {}_k\mathbb{X}$, **(b)** ${}_kF_{0,ij}(\cdot), j \in {}_k\mathbb{Y}_{1,i}, i \in {}_k\mathbb{X}$ are proper distribution functions such that ${}_kF_{0,ij}(0) < 1, j \in {}_k\mathbb{Y}_{1,i}, i \in {}_k\mathbb{X}$, **(c)** ${}_ku_{\varepsilon,i} \in (0, \infty), \varepsilon \in (0, 1]$, for $i \in {}_k\mathbb{X}$,

The role of the condition **J**° (equivalent to the condition **J**) is played by the following condition equivalent to the condition **$_{\mathbf{k}}$J**:

$_{\mathbf{k}}$J°: **(a)** ${}_k\phi_{\varepsilon,ij}(s/ {}_ku_{\varepsilon,i}) \to {}_k\phi_{0,ij}(s)$ as $\varepsilon \to 0$, for $s \geq 0$ and $j \in {}_k\mathbb{Y}_{1,i}, i \in {}_k\mathbb{X}$, **(b)** ${}_k\phi_{0,ij}(s) = \int_0^\infty e^{-st}\, {}_kF_{0,ij}(dt), s \geq 0, j \in {}_k\mathbb{Y}_{1,i}, i \in {}_k\mathbb{X}$ are Laplace transforms of proper distribution functions such that ${}_kF_{0,ij}(0) < 1, j \in {}_k\mathbb{Y}_{1,i}, i \in {}_k\mathbb{X}$, **(c)** ${}_ku_{\varepsilon,i} \in (0, \infty), \varepsilon \in (0, 1]$, for $i \in {}_k\mathbb{X}$,

The above operation of excluding the state k does not aggregate the sojourn times for the semi-Markov process $\tilde{\eta}_\varepsilon(t)$. That is why, the local normalisation functions ${}_ku_{\varepsilon,i} \in [1, \infty), \varepsilon \in (0, 1]$ are defined by the following relation, for $i \in {}_k\mathbb{X}$:

$${}_ku_{\varepsilon,i} = \tilde{u}_{\varepsilon,i} = \bar{p}_{\varepsilon,ii}^{-1}\, u_{\varepsilon,i}. \tag{9.99}$$

Recall that the conditions **G** and **H** imply that the probability $\bar{p}_{\varepsilon,ii} \in (0, 1], \varepsilon \in (0, 1]$, and, thus, $\bar{p}_{\varepsilon,ii}^{-1} \in [1, \infty), \varepsilon \in (0, 1]$, for $i \in \mathbb{X}$. That is why, the normalisation functions ${}_ku_{\varepsilon,i} \in (0, \infty), \varepsilon \in (0, 1]$, for $i \in {}_k\mathbb{X}$.

The relation (9.99) and Lemma 8.2 imply that the functions ${}_ku_{\cdot,i}, i \in \mathbb{X}$ belong to the family $\mathcal{H}$. Moreover, the limit $\tilde{u}_{0,i} = \bar{p}_{0,ii}^{-1} u_{0,i} \in (0, \infty]$, for $i \in \mathbb{X}$.

To obtain the corresponding asymptotic relations and limiting Laplace transforms appearing in the condition **$_{\mathbf{k}}$J°** (**$_{\mathbf{k}}$J**), it is necessary to use some additional conditions imposed on the transition probabilities of the semi-Markov processes $\tilde{\eta}_\varepsilon(t)$.

The first is the following condition:

$_{\mathbf{k}}\hat{\mathbf{I}}$: ${}_kq_{\varepsilon,ij} = \frac{\tilde{p}_{\varepsilon,ij}}{\tilde{p}_{\varepsilon,ij} + \tilde{p}_{\varepsilon,ik}\, \tilde{p}_{\varepsilon,kj}} = \frac{\tilde{p}_{\varepsilon,ij}}{{}_kp_{\varepsilon,ij}} \to {}_kq_{0,ij} \in [0, 1]$ as $\varepsilon \to 0$, for $j \in {}_k\mathbb{Y}_{1,i}, i \in {}_k\mathbb{X}$.

Lemma 9.19 *Let the conditions* **G**, **H**, **I**, *and* $\mathbf{I}_{\mathcal{H}}$ *be satisfied for the semi-Markov processes* $\eta_\varepsilon(t)$. *Then:*

(i) *The condition* ${}_k\hat{\mathbf{I}}$ *is satisfied.*

(ii) *If* $j \notin \tilde{\mathbb{Y}}_{1,i}$, $j \in {}_k\mathbb{Y}_{1,i}, i \in {}_k\mathbb{X}$, *then* ${}_kq_{\varepsilon,ij} = 0, \varepsilon \in (0, 1]$, *and thus, the asymptotic relation appearing in the condition* ${}_k\hat{\mathbf{I}}$ *holds, with the limit* ${}_kq_{0,ij} = 0$.

(iii) *If* $j \in \tilde{\mathbb{Y}}_{1,i}$, $j \in {}_k\mathbb{Y}_{1,i}, i \in {}_k\mathbb{X}$, *then the function* ${}_kq_{\varepsilon,ij} \in \mathcal{H}$ *and the corresponding comparability limits can be calculated by applying the operating quotient formula given in Lemma 8.2 to the functions* $\tilde{p}_{\cdot,ij}$ *and* ${}_kp_{\cdot,ij}$.

(iv) *If* $\mathcal{H}$ *is one of the families,* $\mathcal{H}_1$, $\mathcal{H}_2$, *or* $\mathcal{H}_3$, *then the corresponding comparability parameters and limits can be calculated by applying the quotient formulas given, respectively, in Lemmas 8.4, 8.6, or 8.8 to the functions* $\tilde{p}_{\cdot,ij}$ *and* ${}_kp_{\cdot,ij}$.

Proof Conditions **G**, **H**, **I**, $\mathbf{I}_{\mathcal{H}}$, assumed to be satisfied for the semi-Markov processes $\eta_\varepsilon(t)$, entail the fulfilment of these conditions for the semi-Markov processes $\tilde{\eta}_\varepsilon(t)$ in the form of conditions $\tilde{\mathbf{G}}$, $\tilde{\mathbf{H}}$, $\tilde{\mathbf{I}}$, $\tilde{\mathbf{I}}_{\mathcal{H}}$.

If $j \notin \tilde{\mathbb{Y}}_{1,i}$, $j \in {}_k\mathbb{Y}_{1,i}, i \in {}_k\mathbb{X}$, then $\tilde{p}_{\varepsilon,ij} = 0$, ${}_kp_{\varepsilon,ij} > 0, \varepsilon \in (0, 1]$. Thus, ${}_kq_{\varepsilon,ij} = 0, \varepsilon \in (0, 1]$. Therefore, ${}_kq_{\varepsilon,ij} \to {}_kq_{0,ij} = 0$ as $\varepsilon \to 0$.

If $j \in \tilde{\mathbb{Y}}_{1,i}$, $j \in {}_k\mathbb{Y}_{1,i}, i \in {}_k\mathbb{X}$, then $\tilde{p}_{\varepsilon,ij}, {}_kp_{\varepsilon,ij} > 0, \varepsilon \in (0, 1]$ and, thus, the function ${}_kq_{\cdot,ij} \in \mathcal{H}$, as the quotient of the functions $\tilde{p}_{\cdot,ij}, {}_kp_{\cdot,ij} \in \mathcal{H}$. Therefore, there exists $\lim_{\varepsilon\to 0} {}_kq_{\varepsilon,ij} = {}_kq_{0,ij}$. Since, ${}_kq_{\varepsilon,ij} \in (0, 1], \varepsilon \in (0, 1]$, the limit ${}_kq_{0,ij} \in [0, 1]$. □

The second is the following condition of asymptotic comparability for normalisation functions:

$\hat{\mathbf{L}}$: $w_{\varepsilon,j,i} = \frac{\tilde{u}_{\varepsilon,j}}{\tilde{u}_{\varepsilon,i}} \to w_{0,j,i} \in [0, \infty]$ as $\varepsilon \to 0$, for $i, j \in \mathbb{X}$.

Lemma 9.20 *Let the conditions* **G**, **H**, **I**, $\mathbf{I}_{\mathcal{H}}$, **L**, $\mathbf{L}_{\mathcal{H}}$ *be satisfied for the semi-Markov processes* $\eta_\varepsilon(t)$. *Then:*

(i) *The condition* $\hat{\mathbf{L}}$ *is satisfied.*

(ii) *The functions* $w_{\cdot,j,i}, i, j \in \mathbb{X}$ *belong to the family* $\mathcal{H}$, *and the corresponding comparability limits can be calculated by applying the operating quotient formula given in Lemma 8.2 to the functions* $u_{\cdot,j}$ *and* $u_{\cdot,i}$, *for* $i, j \in \mathbb{X}$.

(iii) *If* $\mathcal{H}$ *is one of the families,* $\mathcal{H}_1$, $\mathcal{H}_2$, *or* $\mathcal{H}_3$, *then the corresponding comparability parameters and limits can be computed by applying the operating quotient formulas given, respectively, in Lemmas 8.4, 8.6, or 8.8 to the functions* $\tilde{u}_{\cdot,j}$ *and* $\tilde{u}_{\cdot,i}$, *for* $i, j \in \mathbb{X}$.

Proof Conditions **G**, **H**, **I**, $\mathbf{I}_{\mathcal{H}}$, **L**, $\mathbf{L}_{\mathcal{H}}$ assumed to be satisfied for the semi-Markov processes $\eta_\varepsilon(t)$ imply that these conditions are also satisfied for the semi-Markov processes $\tilde{\eta}_\varepsilon(t)$ in the form of conditions $\tilde{\mathbf{G}}$, $\tilde{\mathbf{H}}$, $\tilde{\mathbf{I}}$, $\tilde{\mathbf{I}}_{\mathcal{H}}$, $\tilde{\mathbf{L}}$, $\tilde{\mathbf{L}}_{\mathcal{H}}$.

Since $\tilde{u}_{\varepsilon,i} \in (0, \infty), \varepsilon \in (0, 1]$, for $i \in \mathbb{X}$, the functions $w_{\varepsilon,j,i} \in (0, \infty), \varepsilon \in (0, 1]$, for $i, j \in \mathbb{X}$. Moreover, the functions $\tilde{w}_{\cdot,ji} \in \mathcal{H}$ as the quotients of the functions $\tilde{u}_{\cdot,j}, \tilde{u}_{\cdot,i} \in \mathcal{H}$. Therefore, there exists $\lim_{\varepsilon\to 0} w_{\varepsilon,j,i} = w_{0,j,i} \in [0, \infty]$, for $i, j \in \mathbb{X}$. □

From the condition $\hat{\mathbf{L}}$ it follows that there exist sets $\mathbb{W}_l, l = 0, \dots, L$ such that,

(a) $\mathbb{W}_{l'} \cap \mathbb{W}_{l''} = \emptyset, l' \neq l''$.

(b) $\cup_{l=0}^{L} \mathbb{W}_l = \mathbb{X}$.

(c) $w_{0,j,i} \in (0, \infty), j, i \in \mathbb{W}_l$, for $0 \leq l \leq L$.

(d) $w_{0,j,i} = 0, j \in \mathbb{W}_{l'}, i \in \mathbb{W}_{l''}$, for $0 \leq l' < l'' \leq L$. (9.100)

We are especially interested in the set $\mathbb{W}_0$, which includes the so-called least absorbing states in the phase space $\mathbb{X}$.

In Sect. 9.3.2.4, we present a simple algorithm for finding the set $\mathbb{W}_0$.

The third is the following absorbing rate condition:

${}_{\mathbf{k}}\mathbf{M}$: The state $k \in \mathbb{W}_0$, i.e., $w_{0,k,i} \in [0, \infty)$, for $i \in \mathbb{X}$.

We assume that the conditions $\mathbf{G}, \mathbf{H}, \mathbf{I}, \mathbf{I}_{\mathcal{H}}, \mathbf{L}, \mathbf{L}_{\mathcal{H}}, \mathbf{J}^\circ$ (equivalent to the condition $\mathbf{J}$), and $\mathbf{K}$ are satisfied for the semi-Markov processes $\eta_\varepsilon(t)$. Thus, according to Lemmas 9.3, 9.5, 9.10, 9.11, 9.19, and 9.20, the conditions $\tilde{\mathbf{G}}, \tilde{\mathbf{H}}, \tilde{\mathbf{I}}, \tilde{\mathbf{I}}_{\mathcal{H}}, \tilde{\mathbf{L}}, \tilde{\mathbf{L}}_{\mathcal{H}}\ \tilde{\mathbf{J}}^\circ$ (equivalent to condition $\tilde{\mathbf{J}}$), and ${}_{\mathbf{k}}\hat{\mathbf{I}}$, $\hat{\mathbf{L}}$ are satisfied for the semi-Markov processes $\tilde{\eta}_\varepsilon(t)$.

Suppose also that the condition ${}_{\mathbf{k}}\mathbf{M}$ is satisfied.

Using the above conditions and the relation (9.96), we obtain the following relation, which plays the role of asymptotic relation appearing in the condition ${}_{\mathbf{k}}\mathbf{J}^\circ$, for $j \in {}_k\mathbb{Y}_{1,i}, i \in {}_k\mathbb{X}$,

$$\begin{aligned} {}_k\phi_{\varepsilon,ij}(s/{}_ku_{\varepsilon,i}) &= \tilde{\phi}_{\varepsilon,ij}(s/\tilde{u}_{\varepsilon,i})\frac{\tilde{p}_{\varepsilon,ij}}{{}_kp_{\varepsilon,ij}} \\ &\quad + \tilde{\phi}_{\varepsilon,ik}(s/\tilde{u}_{\varepsilon,i})\tilde{\phi}_{\varepsilon,kj}(\frac{\tilde{u}_{\varepsilon,k}}{\tilde{u}_{\varepsilon,i}}s/\tilde{u}_{\varepsilon,k})\frac{\tilde{p}_{\varepsilon,ik}\tilde{p}_{\varepsilon,kj}}{{}_kp_{\varepsilon,ij}} \\ &\quad \longrightarrow {}_k\phi_{0,ij}(s) \text{ as } \varepsilon \to 0, \text{ for } s \geq 0, \end{aligned} \tag{9.101}$$

where

$$\begin{aligned} {}_k\phi_{0,ij}(s) &= \tilde{\phi}_{0,ij}(s)\,{}_kq_{0,ij} \\ &\quad + \tilde{\phi}_{0,ik}(s)\tilde{\phi}_{0,kj}(w_{0,k,i}s)(1 - {}_kq_{0,ij}),\ s \geq 0. \end{aligned} \tag{9.102}$$

Also, the conditions ${}_{\mathbf{k}}\mathbf{I}_{\mathcal{H}}$ and the relation (9.101) imply that the following asymptotic relation holds, for $i \in {}_k\mathbb{X}$:

$$\begin{aligned} {}_k\phi_{\varepsilon,i}(s/{}_ku_{\varepsilon,i}) &= \sum_{j \in {}_k\mathbb{Y}_{1,i}} {}_k\phi_{\varepsilon,ij}(s/{}_ku_{\varepsilon,i})\,{}_kp_{\varepsilon,ij} \\ &\longrightarrow \sum_{j \in {}_k\mathbb{Y}_{1,i}} {}_k\phi_{0,ij}(s)\,{}_kp_{0,ij} = \sum_{j \in {}_k\mathbb{Y}_{0,i}} {}_k\phi_{0,ij}(s)\,{}_kp_{0,ij} \\ &= {}_k\phi_{0,i}(s) \text{ as } \varepsilon \to 0, \text{ for } s \geq 0. \end{aligned} \tag{9.103}$$

In the case, where the condition $_{\mathbf{k}}\mathbf{M}$ is satisfied, the condition $_{\mathbf{k}}\mathbf{J}^{\circ}$ **(b)** is satisfied for any $_kq_{0,ij} \in [0, 1]$, even, if $_kq_{0,ij} = 0$ or $_kq_{0,ij} = 1$.

Indeed, if $j \in {}_k\mathbb{Y}_{1,i}$ and $_kq_{0,ij} = 0$, then $\tilde{p}_{0,ik}, \tilde{p}_{0,kj} > 0$, and, thus, $k \in \tilde{\mathbb{Y}}_{1,i}$ and $j \in \tilde{\mathbb{Y}}_{1,k}$. In this case, the limiting functions $\tilde{\phi}_{0,ik}(s)$ and $\tilde{\phi}_{0,kj}(w_{0,k,i}s)$ are the Laplace transforms of some proper distribution functions which are not concentrated at zero, and, thus, the limiting function $_k\phi_{0,ij}(s) = \tilde{\phi}_{0,ik}(s)\tilde{\phi}_{0,kj}(w_{0,k,i}s)$ is the Laplace transform of some proper distribution function which is not concentrated in 0.

If $j \in {}_k\mathbb{Y}_{1,i}$ and $_kq_{0,ij} \in (0, 1)$, then $\tilde{p}_{0,ij}, \tilde{p}_{0,ik}, \tilde{p}_{0,kj} > 0$ and, thus, $j, k \in \tilde{\mathbb{Y}}_{1,i}$ and $j \in \tilde{\mathbb{Y}}_{1,k}$. In this case, the limiting functions $\tilde{\phi}_{0,ij}(s)$, $\tilde{\phi}_{0,ik}(s)$ and $\tilde{\phi}_{0,kj}(w_{0,k,i}s)$ are the Laplace transforms of some proper distribution functions which are not concentrated at zero, and, thus, the limiting function $_k\phi_{0,ij}(s) = \tilde{\phi}_{0,ij}(s)\,{}_kq_{0,ij} + \tilde{\phi}_{0,ik}(s)\tilde{\phi}_{0,kj}(w_{0,k,i}s)(1 - {}_kq_{0,ij})$ is the Laplace transform of some proper distribution function which is not concentrated at zero.

Finally, if $j \in {}_k\mathbb{Y}_{1,i}$ and $_kq_{0,ij} = 1$, then $\tilde{p}_{0,ij} > 0$ and, thus, $j \in \tilde{\mathbb{Y}}_{1,i}$. In this case, the limiting function $_k\phi_{0,ij}(s) = \tilde{\phi}_{0,ij}(s)$ is the Laplace transform of some proper distribution function which is not concentrated at zero.

The above remarks can be summarised in the following lemma.

Lemma 9.21 *The conditions* **G**, **H**, **I**, $\mathbf{I}_{\mathcal{H}}$, **L**, $\mathbf{L}_{\mathcal{H}}$, $\mathbf{J}^{\circ}$ **(J)**, **K**, *and* $_{\mathbf{k}}\mathbf{M}$, *assumed to be satisfied for the semi-Markov processes* $\eta_\varepsilon(t)$, *entail the fulfilment of the condition* $\mathbf{J}^{\circ}$ **(J)** *for the reduced semi-Markov processes* $_k\eta_\varepsilon(t)$ *in the form of condition* $_{\mathbf{k}}\mathbf{J}^{\circ}$ ($_{\mathbf{k}}\mathbf{J}$), *where*:

(i) *The asymptotic relation appearing in the condition* $_{\mathbf{k}}\mathbf{J}^{\circ}$ *takes the form given in relation (9.101).*

(ii) *The normalisation functions* $_ku_{\cdot,i}, i \in {}_k\mathbb{X}$ *appearing in the condition* $_{\mathbf{k}}\mathbf{J}^{\circ}$ *are given by the relation (9.99).*

It is worth commenting on the need to assume the fulfilment of the condition $_{\mathbf{k}}\mathbf{M}$.

If the condition $_{\mathbf{k}}\mathbf{M}$ is not satisfied, it is possible that the coefficient $w_{0,ki} = \infty$ for some $i \in {}_k\mathbb{X}$. In this case, it follows from the relation (9.96) that, for $j \in {}_k\mathbb{Y}_{1,i}$ and $s \geq 0$,

$$\overline{\lim_{\varepsilon\to 0}}\ {}_k\phi_{\varepsilon,ij}(s/{}_ku_{\varepsilon,i}) \leq \tilde{\phi}_{0,ij}(s)\,{}_kq_{0,ij} + \tilde{\phi}_{0,ik}(s)\mathrm{I}(s = 0) + \tilde{F}_{0,kj}(0)\mathrm{I}(s > 0))(1 - {}_kq_{0,ij}). \tag{9.104}$$

In this case, the relation (9.104) implies that the condition $_{\mathbf{k}}\mathbf{J}_1$ **(b)** is not satisfied, if also $_kq_{0,ij} < 1$ and $j \in \tilde{\mathbb{Y}}_{1,k}$.

Indeed, $\tilde{F}_{0,kj}(0) < 1$ if $j \in \tilde{\mathbb{Y}}_{1,k}$. Therefore, the function on the right-hand side of the relation (9.104) converges to limit $\tilde{F}_{0,kj}(0)(1 - {}_kq_{0,ij}) < 1$ as $s \to 0$. This fact and the relation (9.104) imply that the functions $_k\phi_{\varepsilon,ij}(s/{}_ku_{\varepsilon,i})$ cannot converge pointwise to the Laplace transform of some proper distribution function concentrated on the interval $[0, \infty)$.

9.3.2.3 Condition $_{\mathbf{k}}\mathbf{K}$. We assume that the conditions **G**, **H**, **I**, $\mathbf{I}_{\mathcal{H}}$, **L**, $\mathbf{L}_{\mathcal{H}}$, $\mathbf{J}^{\circ}$ (equivalent to the condition **J**), **K** are satisfied for the semi-Markov processes $\eta_\varepsilon(t)$.

As mentioned above, this implies that conditions $\tilde{\mathbf{G}}, \tilde{\mathbf{H}}, \tilde{\mathbf{I}}, \tilde{\mathbf{I}}_{\mathcal{H}}, \tilde{\mathbf{L}}, \tilde{\mathbf{L}}_{\mathcal{H}}, \tilde{\mathbf{J}}^\circ$ (equivalent to condition $\tilde{\mathbf{J}}$), ${}_{\mathbf{k}}\hat{\mathbf{I}}$, and $\hat{\mathbf{L}}$ are satisfied for the semi-Markov processes $\tilde{\eta}_\varepsilon(t)$.

We also assume that the condition ${}_{\mathbf{k}}\mathbf{M}$ is satisfied.

The expectation ${}_k f_{\varepsilon,ij}$, which is an analogue of the expectation $f_{\varepsilon,ij}$ is defined, for $\varepsilon \in (0, 1]$ by the following relation, for $j \in {}_k\mathbb{Y}_{1,i}, i \in {}_k\mathbb{X}$,

$$\begin{aligned} {}_k f_{\varepsilon,ij} &= \mathsf{E}\{{}_k\kappa_{\varepsilon,1} / {}_k\eta_{\varepsilon,0} = i, \ {}_k\eta_{\varepsilon,1} = j\} \\ &= \tilde{f}_{\varepsilon,ij}\tilde{p}_{\varepsilon,ij} + (\tilde{f}_{\varepsilon,ik} + \tilde{f}_{\varepsilon,kj})\tilde{p}_{\varepsilon,ik}\tilde{p}_{\varepsilon,kj}. \end{aligned} \tag{9.105}$$

Also, for $i \in {}_k\mathbb{X}$,

$${}_k f_{\varepsilon,i} = \mathsf{E}\{{}_k\kappa_{\varepsilon,1} / {}_k\eta_{\varepsilon,0} = i\} = \sum_{j \in {}_k\mathbb{Y}_{1,i}} {}_k f_{\varepsilon,ij} \, {}_k p_{\varepsilon,ij}. \tag{9.106}$$

The following condition plays the role of condition $\mathbf{K}$ for the reduced semi-Markov processes ${}_k\eta_\varepsilon(t)$:

${}_{\mathbf{k}}\mathbf{K}$: **(a)** ${}_k f_{\varepsilon,ij} < \infty, j \in {}_k\mathbb{Y}_{1,i}, i \in {}_k\mathbb{X}$, for every $\varepsilon \in (0, 1]$, **(b)** ${}_k f_{\varepsilon,ij} / {}_k u_{\varepsilon,i} \to {}_k f_{0,ij} = \int_0^\infty s \, {}_k F_{0,ij}(ds) < \infty$ as $\varepsilon \to 0$, for $j \in {}_k\mathbb{Y}_{1,i}, \ i \in {}_k\mathbb{X}$.

The conditions $\mathbf{G}, \mathbf{H}, \mathbf{I}, \mathbf{I}_{\mathcal{H}}, \mathbf{L}, \mathbf{L}_{\mathcal{H}}, \mathbf{J}, \mathbf{K}, {}_{\mathbf{k}}\mathbf{M}$ and the relation (9.75) imply that, for $j \in {}_k\mathbb{Y}_{1,i}, i \in {}_k\mathbb{X}$,

$$\begin{aligned} \frac{{}_k f_{\varepsilon,ij}}{{}_k u_{\varepsilon,i}} &= \frac{\tilde{f}_{\varepsilon,ij}}{\tilde{u}_{\varepsilon,i}} \frac{\tilde{p}_{\varepsilon,ij}}{{}_k p_{\varepsilon,ij}} + \Big(\frac{\tilde{f}_{\varepsilon,ik}}{\tilde{u}_{\varepsilon,i}} + \frac{\tilde{f}_{\varepsilon,kj}}{\tilde{u}_{\varepsilon,k}} \frac{\tilde{u}_{\varepsilon,k}}{\tilde{u}_{\varepsilon,i}}\Big) \frac{\tilde{p}_{\varepsilon,ik}\tilde{p}_{\varepsilon,kj}}{{}_k p_{\varepsilon,ij}} \\ &\to {}_k f_{0,ij} = \tilde{f}_{0,ij} \, {}_k q_{0,ij} + (\tilde{f}_{0,ik} + \tilde{f}_{0,kj} w_{0,k,i})(1 - {}_k q_{0,ij}) \\ &= \int_0^\infty t \, {}_k F_{0,ij}(dt) \text{ as } \varepsilon \to 0, \ \text{for } s \ge 0. \end{aligned} \tag{9.107}$$

Also, the following relation takes place, for $i \in {}_k\mathbb{X}$:

$$\begin{aligned} \frac{{}_k f_{\varepsilon,i}}{{}_k u_{\varepsilon,i}} &= \sum_{j \in \tilde{\mathbb{Y}}_{1,i}} \frac{{}_k f_{\varepsilon,ij}}{{}_k u_{\varepsilon,i}} \, {}_k p_{\varepsilon,ij} \\ &\to \sum_{j \in \tilde{\mathbb{Y}}_{1,i}} {}_k f_{0,ij} \, {}_k p_{0,ij} = \sum_{j \in \tilde{\mathbb{Y}}_{0,i}} {}_k f_{0,ij} \, {}_k p_{0,ij} \\ &= {}_k f_{0,i} = \int_0^\infty t \, {}_k F_{0,i}(dt) \text{ as } \varepsilon \to 0. \end{aligned} \tag{9.108}$$

The above remarks can be summarised in the following lemma.

Lemma 9.22 *The conditions* $\mathbf{G}, \mathbf{H}, \mathbf{I}, \mathbf{I}_{\mathcal{H}}, \mathbf{L}, \mathbf{L}_{\mathcal{H}}, \mathbf{J}$ $(\mathbf{J}^\circ)$, $\mathbf{K}$, *and* ${}_{\mathbf{k}}\mathbf{M}$, *assumed to be satisfied for the semi-Markov processes* $\eta_\varepsilon(t)$, *entail the fulfilment of the condition* $\mathbf{K}$ *for the reduced semi-Markov processes* ${}_k\eta_\varepsilon(t)$ *in the form of condition* ${}_{\mathbf{k}}\mathbf{K}$, *where:*

(i) *The asymptotic relation appearing in the condition* ${}_{\mathbf{k}}\mathbf{K}$ *takes the form of the relation (9.107).*

(ii) *The normalisation functions* ${}_k u_{\cdot,i}, i \in {}_k\mathbb{X}$ *appearing in the condition* ${}_{\mathbf{k}}\mathbf{K}$ *are given by the relation (9.99).*

Remark 9.2 It is useful noting that the conditions **G**, **H**, and **K (a)**, assumed to be satisfied for the semi-Markov processes $\eta_\varepsilon(t)$, entail that, for any state $k \in \mathbb{X}$, the condition **K (a)** is satisfied for the reduced semi-Markov processes ${}_k\eta_\varepsilon(t)$ in the form of condition ${}_{\mathbf{k}}\mathbf{K}$ **(a)**. This follows from the relation (9.105).

9.3.2.4 Set $\mathbb{W}_0$ and condition ${}_{\mathbf{k}}\mathbf{M}$. The condition $\hat{\mathbf{L}}$ implies that the set $\mathbb{W}_0 \neq \emptyset$. Obviously, $w_{0,k,i} \in (0, \infty)$, for $k, i \in \mathbb{W}_0$, while $w_{0,k,i} = 0$, for $k \in \mathbb{W}_0, i \in \mathbb{X} \setminus \mathbb{W}_0$. Also, $\mathbb{W}_0 = \mathbb{X}$, if $w_{0,k,i} \in (0, \infty)$, for any $k, i \in \mathbb{X}$.

The set $\mathbb{W}_0$ can be found using the following simple algorithm. Let us order by an arbitrary way states in space $\mathbb{X}$, i.e., represent this space in the form $\mathbb{X} = \{i_1, \dots, i_m\}$. Let us now define the states $i^*_n, n = 1, \dots, m$ using the following recurrent procedure:

$$\begin{aligned} i^*_1 &= i_1, \\ i^*_2 &= i^*_1 \mathrm{I}(w_{0,i^*_1,i_2} = 0) + i_2 \mathrm{I}(w_{0,i^*_1,i_2} \in (0, \infty]), \\ &\dots \\ i^*_m &= i^*_{m-1} \mathrm{I}(w_{0,i^*_{m-1},i_m} = 0) + i_m \mathrm{I}(w_{0,i^*_{m-1},i_m} \in (0, \infty]). \end{aligned} \tag{9.109}$$

The following simple lemma takes place.

Lemma 9.23 *Let the condition* $\hat{\mathbf{L}}$ *be satisfied. Then, the state* $i^*_m \in \mathbb{W}_0$ *and the set,*

$$\mathbb{W}_0 = \{i_n \in \mathbb{X} : w_{0,i^*_m,i_n} \in (0, \infty)\}. \tag{9.110}$$

9.3.2.5 Conditions L and $\mathbf{L}_{\mathcal{H}}$. The following conditions play the roles of the conditions **L** and $\mathbf{L}_{\mathcal{H}}$ for the reduced semi-Markov processes ${}_k\eta_\varepsilon(t)$:

${}_{\mathbf{k}}\mathbf{L}$: ${}_k u_{\varepsilon,i} \longrightarrow {}_k u_{0,i} \in (0, \infty]$ as $\varepsilon \longrightarrow 0$, for $i \in {}_k\mathbb{X}$

and

${}_{\mathbf{k}}\mathbf{L}_{\mathcal{H}}$: The functions ${}_k u_{\cdot,i}, i \in {}_k\mathbb{X}$ belong to the complete family of asymptotically comparable functions $\mathcal{H}$ appearing in the condition $\mathbf{L}_{\mathcal{H}}$.

The following two lemmas are direct corollaries of Lemmas 9.12 and 9.13, since the normalisation functions ${}_k u_{\cdot,i} = \tilde{u}_{\cdot,i}, i \in {}_k\mathbb{X}$.

Lemma 9.24 *The conditions* **G**, **H**, **I**, $\mathbf{I}_{\mathcal{H}}$, **L**, $\mathbf{L}_{\mathcal{H}}$, *assumed to be satisfied for the semi-Markov processes* $\eta_\varepsilon(t)$, *entail the fulfilment of the conditions* **L**, $\mathbf{L}_{\mathcal{H}}$ *for the reduced semi-Markov processes* ${}_k\eta_\varepsilon(t)$ *in the form of conditions* ${}_{\mathbf{k}}\mathbf{L}$, ${}_{\mathbf{k}}\mathbf{L}_{\mathcal{H}}$.

Lemma 9.25 *Let conditions* **G**, **H**, **I**, $\mathbf{I}_{\mathcal{H}}$, **L**, $\mathbf{L}_{\mathcal{H}}$ *be satisfied. Then*:

(i) *The functions* ${}_k u_{\cdot,i} = \tilde{u}_{\cdot,i}, i \in {}_k\mathbb{X}$ *belong to the family* $\mathcal{H}$ *and the corresponding comparability limits can be calculated by applying to these functions the operating quotient formula given in Lemma 9.13.*

(ii) *If* $\mathcal{H}$ *is one of the families,* $\mathcal{H}_1$, $\mathcal{H}_2$, *or* $\mathcal{H}_3$, *then the corresponding comparability parameters and limits can be computed by applying the operating quotient formulas given in Lemma 9.13 to the functions* ${}_k u_{\cdot,i} = \tilde{u}_{\cdot,i}, i \in {}_k\mathbb{X}$.

9.3.2.6 Summary of Perturbation Conditions for Semi-Markov Process ${}_k\eta_\varepsilon(t)$. The following theorem summarises remarks made in Sect. 9.3.2.

Theorem 9.2 *The conditions* **G, H, I,** $\mathbf{I}_{\mathcal{H}}$, **L,** $\mathbf{L}_{\mathcal{H}}$, **J** (**J**°), **K**, *and* ${}_{\mathbf{k}}\mathbf{M}$, *assumed to be satisfied for the semi-Markov processes* $\eta_\varepsilon(t)$, *entail the fulfilment of the conditions* **G, H, I,** $\mathbf{I}_{\mathcal{H}}$, **L,** $\mathbf{L}_{\mathcal{H}}$, **J** (**J**°), **K**, *and hold for reduced semi-Markov processes* ${}_k\eta_\varepsilon(t)$ *in the form of conditions* ${}_{\mathbf{k}}\mathbf{G}$, ${}_{\mathbf{k}}\mathbf{H}$, ${}_{\mathbf{k}}\mathbf{I}$, ${}_{\mathbf{k}}\mathbf{I}_{\mathcal{H}}$, ${}_{\mathbf{k}}\mathbf{L}$, ${}_{\mathbf{k}}\mathbf{L}_{\mathcal{H}}$, ${}_{\mathbf{k}}\mathbf{J}$ (${}_{\mathbf{k}}\mathbf{J}^\circ$), ${}_{\mathbf{k}}\mathbf{K}$, *presented in Lemmas 9.14–9.25.*

Remark 9.3 The relations (9.101) and (9.107) imply that condition ${}_{\mathbf{k}}\mathbf{M}$ can be weakened in Theorem 9.2 and replaced by the following local absorbing rate condition:

${}_{\mathbf{k}}\hat{\mathbf{M}}$: $w_{0,k,i} \in [0, \infty)$, for $i \in \mathbb{X}$ such that $k \in \mathbb{Y}_{1,i}$.

9.4 Recurrent Reduction of Phase Space for Perturbed Semi-Markov Processes

In this section, we describe an asymptotic multi-step algorithm for recurrent phase space reduction for perturbed semi-Markov processes.

9.4.1 The Recurrent Phase Space Reduction Algorithm

9.4.1.1 Notation for Semi-Markov Processes with Reduced Phase Spaces. An important element of the recurrent phase space reduction algorithm presented in this section is the procedure of successive exclusion of a sequence of states $\bar{k}_n = \langle k_1, k_2, \ldots, k_h \rangle$, for some $1 \leq h \leq m-1$ from the phase space $\mathbb{X}$ of the semi-Markov process $\eta_\varepsilon(t)$. This procedure results in reduction of the phase space $\mathbb{X}$ to the space $\mathbb{X} \setminus \{k_1, k_2, \ldots, k_h\}$.

We use the abbreviated notation ${}_{\bar{k}_h}\eta_\varepsilon(t)$ for the reduced semi-Markov process obtained by successive exclusion of the sequence of states $\bar{k}_n = \langle k_1, k_2, \ldots, k_h \rangle$ from the phase space $\mathbb{X}$ of semi-Markov processes $\eta_\varepsilon(t)$.

We also use the abbreviated notation ${}_{\bar{k}_h}\tilde{\eta}_\varepsilon(t)$ for the semi-Markov process obtained by the removal of virtual transitions from trajectories of the reduced semi-Markov process ${}_{\bar{k}_n}\eta_\varepsilon(t)$.

The semi-Markov processes ${}_{\bar{k}_h}\eta_\varepsilon(t)$ and ${}_{\bar{k}_h}\tilde{\eta}_\varepsilon(t)$ have the phase space ${}_{\bar{k}_h}\mathbb{X} = \mathbb{X} \setminus \{k_1, k_2, \ldots, k_h\}$.

In particular, we change the notation from ${}_{k_1}\eta_\varepsilon(t)$ and ${}_{k_1}\tilde{\eta}_\varepsilon(t)$, respectively, to ${}_{\bar{k}_1}\eta_\varepsilon(t)$ and ${}_{\bar{k}_1}\tilde{\eta}_\varepsilon(t)$, for the reduced semi-Markov process ${}_{k_1}\eta_\varepsilon(t)$ obtained by the exclusion of the state $k = k_1$ from the initial phase space $\mathbb{X}$ and the reduced semi-Markov process ${}_{k_1}\tilde{\eta}_\varepsilon(t)$ by removal of virtual transitions from trajectories of the reduced semi-Markov process ${}_{k_1}\eta_\varepsilon(t)$.

In the similar way, we index random variables, in particular hitting times, phase spaces and other related sets, conditions, probabilities, expectations and other quantities and objects related to the semi-Markov processes ${}_{\bar{k}_h}\eta_\varepsilon(t)$ and ${}_{\bar{k}_h}\tilde{\eta}_\varepsilon(t)$ by the left lower index ${}_{\bar{k}_h}$ (in such a way as ${}_{\bar{k}_h}f$). This is done in order to distinguish these random variables, phase spaces and other sets, probabilities, expectations and other quantities and objects, for $h = 1, \dots, m-1$.

To obtain similar forms for recurrent relations connecting random variables, probabilities and expectations and other quantities and objects resulted by excluding the sequences of states $\bar{k}_{h-1}$ and $\bar{k}_h$ from the phase space $\mathbb{X}$, we also use the left lower index ${}_{\bar{k}_0}$ (for the "empty" sequence $\bar{k}_0 = \langle\rangle$) for $h = 1$. Thus, we change the notation from $\eta_\varepsilon(t)$ and $\tilde{\eta}_\varepsilon(t)$, respectively, to ${}_{\bar{k}_0}\eta_\varepsilon(t)$ and ${}_{\bar{k}_0}\tilde{\eta}_\varepsilon(t)$.

In similar way, we index ${}_{\bar{k}_0}\mathbb{X} = \mathbb{X}$ and other related sets, conditions, probabilities, expectations and other quantities and objects related to the semi-Markov processes ${}_{\bar{k}_0}\eta_\varepsilon(t) = \eta_\varepsilon(t)$ and ${}_{\bar{k}_0}\tilde{\eta}_\varepsilon(t) = \tilde{\eta}_\varepsilon(t)$.

Let us describe the recurrent algorithm described above in more detail.

9.4.1.2 The First Step of the Phase Space Reduction Algorithm. Initially, it should be assumed that the conditions ${}_{\bar{\mathbf{k}}_0}\mathbf{G} = \mathbf{G}$ and ${}_{\bar{\mathbf{k}}_0}\mathbf{H} = \mathbf{H}$ are satisfied for the semi-Markov processes ${}_{\bar{k}_0}\eta_\varepsilon(t) = \eta_\varepsilon(t)$. Other conditions are indicated below.

The first step of the phase space reduction algorithm includes two sub-steps.

At the first sub-step, the original semi-Markov process ${}_{\bar{k}_0}\eta_\varepsilon(t) = \eta_\varepsilon(t)$ is transformed in the semi-Markov process ${}_{\bar{k}_0}\tilde{\eta}_\varepsilon(t) = \tilde{\eta}_\varepsilon(t)$ by applying the procedure of removal of virtual transitions described in Sect. 9.2. The resulting semi-Markov process ${}_{\bar{k}_0}\tilde{\eta}_\varepsilon(t)$ has the phase space ${}_{\bar{k}_0}\mathbb{X} = \mathbb{X}$.

At the second sub-step, the semi-Markov process ${}_{\bar{k}_0}\tilde{\eta}_\varepsilon(t) = \tilde{\eta}_\varepsilon(t)$ is transformed in the reduced semi-Markov process ${}_{\bar{k}_1}\eta_\varepsilon(t) = {}_{k_1}\eta_\varepsilon(t)$ by applying the procedure of one-state reduction of phase space described in Sect. 9.3. The resulting reduced semi-Markov process ${}_{\bar{k}_1}\eta_\varepsilon(t)$ has the reduced phase space ${}_{\bar{k}_1}\mathbb{X} = {}_{k_1}\mathbb{X}$.

There are five variations of the procedures described above that use different sets of conditions.

In the variant **(i)**, it is initially assumed that the only conditions ${}_{\bar{\mathbf{k}}_0}\mathbf{G}$ and ${}_{\bar{\mathbf{k}}_0}\mathbf{H}$ are satisfied.

In this case, Lemmas 9.3–9.5 imply that the conditions ${}_{\bar{\mathbf{k}}_0}\mathbf{G}$ and ${}_{\bar{\mathbf{k}}_0}\mathbf{H}$ are satisfied for the semi-Markov processes ${}_{\bar{k}_0}\tilde{\eta}_\varepsilon(t)$ in the form of conditions ${}_{\bar{\mathbf{k}}_0}\tilde{\mathbf{G}} = \tilde{\mathbf{G}}$ and ${}_{\bar{\mathbf{k}}_0}\tilde{\mathbf{H}} = \tilde{\mathbf{H}}$. Also, Lemmas 9.14–9.16 imply that the conditions ${}_{\bar{\mathbf{k}}_0}\mathbf{G}$ and ${}_{\bar{\mathbf{k}}_0}\mathbf{H}$ are satisfied for the reduced semi-Markov processes ${}_{\bar{k}_1}\eta_\varepsilon(t)$ in the form of conditions ${}_{\bar{\mathbf{k}}_1}\mathbf{G} = {}_{\mathbf{k}_1}\mathbf{G}$ and ${}_{\bar{\mathbf{k}}_1}\mathbf{H} = {}_{\mathbf{k}_1}\mathbf{H}$.

In the variant **(ii)**, it is initially assumed that the conditions ${}_{\bar{\mathbf{k}}_0}\mathbf{G}$, ${}_{\bar{\mathbf{k}}_0}\mathbf{H}$, and ${}_{\bar{\mathbf{k}}_0}\mathbf{K}$ **(a)** $= {}_{\mathbf{k}_0}\mathbf{K}$ **(a)** (the assumption about finiteness of the expectations $f_{\varepsilon,ij}, j \in \mathbb{Y}_{1,i}, i \in \mathbb{X}$, for $\varepsilon \in (0, 1]$) are satisfied.

In this case, Lemmas 9.3–9.5 and Remark 9.1 imply that the conditions ${}_{\bar{k}_0}\mathbf{G}$, ${}_{\bar{\mathbf{k}}_0}\mathbf{H}$, and ${}_{\bar{\mathbf{k}}_0}\mathbf{K}$ **(a)** are satisfied for the semi-Markov processes ${}_{\bar{k}_0}\tilde{\eta}_\varepsilon(t)$ in the form of conditions ${}_{\bar{\mathbf{k}}_0}\tilde{\mathbf{G}}$, ${}_{\bar{\mathbf{k}}_0}\tilde{\mathbf{H}}$, and ${}_{\bar{\mathbf{k}}_0}\tilde{\mathbf{K}}$ **(a)** $= {}_{\mathbf{k}_0}\tilde{\mathbf{K}}$ **(a)**. Also, Lemmas 9.14–9.16 and Remark 9.1 imply that the conditions ${}_{\bar{\mathbf{k}}_0}\mathbf{G}$, ${}_{\bar{\mathbf{k}}_0}\mathbf{H}$, and ${}_{\bar{\mathbf{k}}_0}\mathbf{K}$ **(a)** are satisfied for the reduced semi-

Markov processes ${}_{\bar{k}_1}\eta_\varepsilon(t)$ in the form of conditions ${}_{\bar{\mathbf{k}}_1}\mathbf{G}$, ${}_{\bar{\mathbf{k}}_1}\mathbf{H}$ and ${}_{\bar{\mathbf{k}}_1}\mathbf{K}$ **(a)** = ${}_{\mathbf{k}_1}\mathbf{K}$ **(a)**.

In the variant **(iii)**, it is initially assumed that the conditions ${}_{\bar{\mathbf{k}}_0}\mathbf{G}$, ${}_{\bar{\mathbf{k}}_0}\mathbf{H}$, ${}_{\bar{\mathbf{k}}_0}\mathbf{I} = {}_{\mathbf{k}_0}\mathbf{I}$, and ${}_{\bar{\mathbf{k}}_0}\mathbf{I}_{\mathcal{H}} = {}_{\mathbf{k}_0}\mathbf{I}_{\mathcal{H}}$ are satisfied.

In this case, Lemmas 9.6–9.9 imply that the conditions ${}_{\bar{\mathbf{k}}_0}\mathbf{G}$, ${}_{\bar{\mathbf{k}}_0}\mathbf{H}$, ${}_{\bar{\mathbf{k}}_0}\mathbf{I}$, and ${}_{\bar{\mathbf{k}}_0}\mathbf{I}_{\mathcal{H}}$ are satisfied for the semi-Markov processes ${}_{\bar{k}_0}\tilde{\eta}_\varepsilon(t)$ in the form of conditions ${}_{\bar{\mathbf{k}}_0}\tilde{\mathbf{G}}$, ${}_{\bar{\mathbf{k}}_0}\tilde{\mathbf{H}}$, ${}_{\bar{\mathbf{k}}_0}\tilde{\mathbf{I}} = {}_{\mathbf{k}_0}\tilde{\mathbf{I}}$, and ${}_{\bar{\mathbf{k}}_0}\tilde{\mathbf{I}}_{\mathcal{H}} = {}_{\mathbf{k}_0}\tilde{\mathbf{I}}_{\mathcal{H}}$. Also, Lemmas 9.17–9.18 imply that the conditions ${}_{\bar{\mathbf{k}}_0}\mathbf{G}$, ${}_{\bar{\mathbf{k}}_0}\mathbf{H}$, ${}_{\bar{\mathbf{k}}_0}\mathbf{I}$, and ${}_{\bar{\mathbf{k}}_0}\mathbf{I}_{\mathcal{H}}$ are satisfied for the reduced semi-Markov processes ${}_{\bar{k}_1}\eta_\varepsilon(t)$ in the form of conditions ${}_{\bar{\mathbf{k}}_1}\mathbf{G}$, ${}_{\bar{\mathbf{k}}_1}\mathbf{H}$, ${}_{\bar{\mathbf{k}}_1}\mathbf{I} = {}_{\mathbf{k}_1}\mathbf{I}$, and ${}_{\bar{\mathbf{k}}_1}\mathbf{I}_{\mathcal{H}} = {}_{\mathbf{k}_1}\mathbf{I}_{\mathcal{H}}$.

In the variant **(iv)**, it is initially assumed that the conditions ${}_{\bar{\mathbf{k}}_0}\mathbf{G}$, ${}_{\bar{\mathbf{k}}_0}\mathbf{H}$, ${}_{\bar{\mathbf{k}}_0}\mathbf{I}$, ${}_{\bar{\mathbf{k}}_0}\mathbf{I}_{\mathcal{H}}$, ${}_{\bar{\mathbf{k}}_0}\mathbf{L} = {}_{\mathbf{k}_0}\mathbf{L}$, and ${}_{\bar{\mathbf{k}}_0}\mathbf{L}_{\mathcal{H}} = {}_{\mathbf{k}_0}\mathbf{L}_{\mathcal{H}}$ are satisfied.

In this case, Lemmas 9.12–9.13 imply that the conditions ${}_{\bar{\mathbf{k}}_0}\mathbf{G}$, ${}_{\bar{\mathbf{k}}_0}\mathbf{H}$, ${}_{\bar{\mathbf{k}}_0}\mathbf{I}$, ${}_{\bar{\mathbf{k}}_0}\mathbf{I}_{\mathcal{H}}$, ${}_{\bar{\mathbf{k}}_0}\mathbf{L}$, and ${}_{\bar{\mathbf{k}}_0}\mathbf{L}_{\mathcal{H}}$ are satisfied for the semi-Markov processes ${}_{\bar{k}_0}\tilde{\eta}_\varepsilon(t)$ in the form of conditions ${}_{\bar{\mathbf{k}}_0}\tilde{\mathbf{G}}$, ${}_{\bar{\mathbf{k}}_0}\tilde{\mathbf{H}}$, ${}_{\bar{\mathbf{k}}_0}\tilde{\mathbf{I}}$, ${}_{\bar{\mathbf{k}}_0}\tilde{\mathbf{I}}_{\mathcal{H}}$, ${}_{\bar{\mathbf{k}}_0}\tilde{\mathbf{L}} = {}_{\mathbf{k}_0}\tilde{\mathbf{L}}$, and ${}_{\bar{\mathbf{k}}_0}\tilde{\mathbf{L}}_{\mathcal{H}} = {}_{\mathbf{k}_0}\tilde{\mathbf{L}}_{\mathcal{H}}$. Also, Lemmas 9.24–9.25 imply that the conditions ${}_{\bar{\mathbf{k}}_0}\mathbf{G}$, ${}_{\bar{\mathbf{k}}_0}\mathbf{H}$, ${}_{\bar{\mathbf{k}}_0}\mathbf{I}$, ${}_{\bar{\mathbf{k}}_0}\mathbf{I}_{\mathcal{H}}$, ${}_{\bar{\mathbf{k}}_0}\mathbf{L}$, and ${}_{\bar{\mathbf{k}}_0}\mathbf{L}_{\mathcal{H}}$ are satisfied for the reduced semi-Markov processes ${}_{\bar{k}_1}\eta_\varepsilon(t)$ in the form of conditions ${}_{\bar{\mathbf{k}}_1}\mathbf{G}$, ${}_{\bar{\mathbf{k}}_1}\mathbf{H}$, ${}_{\bar{\mathbf{k}}_1}\mathbf{I}$, ${}_{\bar{\mathbf{k}}_1}\mathbf{I}_{\mathcal{H}}$, ${}_{\bar{\mathbf{k}}_0}\mathbf{L} = {}_{\mathbf{k}_0}\mathbf{L}$, and ${}_{\bar{\mathbf{k}}_1}\mathbf{L}_{\mathcal{H}} = {}_{\mathbf{k}_1}\mathbf{L}_{\mathcal{H}}$.

Finally, in the variant **(v)**, it is initially assumed that the conditions ${}_{\bar{\mathbf{k}}_0}\mathbf{G}$, ${}_{\bar{\mathbf{k}}_0}\mathbf{H}$, ${}_{\bar{\mathbf{k}}_0}\mathbf{I}$, ${}_{\bar{\mathbf{k}}_0}\mathbf{I}_{\mathcal{H}}$, ${}_{\bar{\mathbf{k}}_0}\mathbf{L}$, ${}_{\bar{\mathbf{k}}_0}\mathbf{L}_{\mathcal{H}}$, ${}_{\bar{\mathbf{k}}_0}\mathbf{J} = {}_{\mathbf{k}_0}\mathbf{J}$, ${}_{\bar{\mathbf{k}}_0}\mathbf{K} = {}_{\mathbf{k}_0}\mathbf{K}$ are satisfied. In addition, it should be assumed that the condition ${}_{\bar{\mathbf{k}}_1}\mathbf{M} = {}_{\mathbf{k}_1}\mathbf{M}$ (state $k_1 \in {}_{\bar{k}_0}\mathbb{W}_0 = \mathbb{W}_0$), which is required to implement the second sub-algorithm of transforming the semi-Markov process ${}_{\bar{k}_0}\tilde{\eta}_\varepsilon(t)$ in the reduced semi-Markov process ${}_{\bar{k}_1}\eta_\varepsilon(t)$, is satisfied.

In this case, Lemmas 9.10–9.11 imply that the conditions ${}_{\bar{\mathbf{k}}_0}\mathbf{G}$, ${}_{\bar{\mathbf{k}}_0}\mathbf{H}$, ${}_{\bar{\mathbf{k}}_0}\mathbf{I}$, ${}_{\bar{\mathbf{k}}_0}\mathbf{I}_{\mathcal{H}}$, ${}_{\bar{\mathbf{k}}_0}\mathbf{L}$, ${}_{\bar{\mathbf{k}}_0}\mathbf{L}_{\mathcal{H}}$, ${}_{\bar{\mathbf{k}}_0}\mathbf{J}$, and ${}_{\bar{\mathbf{k}}_0}\mathbf{K}$ are satisfied for the semi-Markov processes ${}_{\bar{k}_0}\tilde{\eta}_\varepsilon(t)$ in the form of conditions ${}_{\bar{\mathbf{k}}_0}\tilde{\mathbf{G}}$, ${}_{\bar{\mathbf{k}}_0}\tilde{\mathbf{H}}$, ${}_{\bar{\mathbf{k}}_0}\tilde{\mathbf{I}}$, ${}_{\bar{\mathbf{k}}_0}\tilde{\mathbf{I}}_{\mathcal{H}}$, ${}_{\bar{\mathbf{k}}_0}\tilde{\mathbf{L}}$, ${}_{\bar{\mathbf{k}}_0}\tilde{\mathbf{L}}_{\mathcal{H}}$, ${}_{\bar{\mathbf{k}}_0}\tilde{\mathbf{J}} = {}_{\mathbf{k}_0}\tilde{\mathbf{J}}$, and ${}_{\bar{\mathbf{k}}_0}\tilde{\mathbf{K}} = {}_{\mathbf{k}_0}\tilde{\mathbf{K}}$. Also, Lemmas 9.21–9.23 imply that the conditions ${}_{\bar{\mathbf{k}}_0}\mathbf{G}$, ${}_{\bar{\mathbf{k}}_0}\mathbf{H}$, ${}_{\bar{\mathbf{k}}_0}\mathbf{I}$, ${}_{\bar{\mathbf{k}}_0}\mathbf{I}_{\mathcal{H}}$, ${}_{\bar{\mathbf{k}}_0}\mathbf{L}$, ${}_{\bar{\mathbf{k}}_0}\mathbf{L}_{\mathcal{H}}$, ${}_{\bar{\mathbf{k}}_0}\mathbf{J}$, and ${}_{\bar{\mathbf{k}}_0}\mathbf{K}$ are satisfied for the reduced semi-Markov processes ${}_{\bar{k}_1}\eta_\varepsilon(t)$ in the form of conditions ${}_{\bar{\mathbf{k}}_1}\mathbf{G}$, ${}_{\bar{\mathbf{k}}_1}\mathbf{H}$, ${}_{\bar{\mathbf{k}}_1}\mathbf{I}$, ${}_{\bar{\mathbf{k}}_1}\mathbf{I}_{\mathcal{H}}$, ${}_{\bar{\mathbf{k}}_1}\mathbf{L}$, ${}_{\bar{\mathbf{k}}_1}\mathbf{L}_{\mathcal{H}}$, ${}_{\bar{\mathbf{k}}_1}\mathbf{J} = {}_{\mathbf{k}_1}\mathbf{J}$, and ${}_{\bar{\mathbf{k}}_1}\mathbf{K} = {}_{\mathbf{k}_1}\mathbf{K}$.

What is important is that in the variants **(i)**–**(iv)**, the state k_1 can be an arbitrary state in the phase space $\mathbb{X}$, while, in the variant **(v)**, the state k_1 should be chosen such that $k_1 \in {}_{\bar{k}_0}\mathbb{W}_0 = \mathbb{W}_0$, that is, the condition ${}_{\bar{\mathbf{k}}_1}\mathbf{M} = {}_{\mathbf{k}_1}\mathbf{M}$ is assumed to be satisfied.

9.4.1.3 The Recurrent Phase Space Reduction Algorithm. As mentioned above, initially it should assumed that the conditions ${}_{\bar{\mathbf{k}}_0}\mathbf{G} = \mathbf{G}$ and ${}_{\bar{\mathbf{k}}_0}\mathbf{H} = \mathbf{H}$ are satisfied for the semi-Markov processes ${}_{\bar{k}_0}\eta_\varepsilon(t) = \eta_\varepsilon(t)$.

The first step of phase space reduction algorithm described in Sect. 9.4.1.2 can be recurrently repeated h times for some $1 \leq h \leq m - 1$.

The h-th step of the phase space reduction algorithm includes two sub-steps similar to those described above for the first step. The only difference is that the corresponding sub-algorithms must be applied to the reduced semi-Markov process ${}_{\bar{k}_{h-1}}\eta_\varepsilon(t)$ instead of the semi-Markov process ${}_{\bar{k}_0}\eta_\varepsilon(t)$.

First, the reduced semi-Markov process ${}_{\bar{k}_{h-1}}\eta_\varepsilon(t)$ is transformed in the semi-Markov process ${}_{\bar{k}_{h-1}}\tilde{\eta}_\varepsilon(t)$ by applying the procedure of removal of virtual transitions described above in Sects. 9.2 and 9.4.1.2. The resulting semi-Markov process ${}_{\bar{k}_{h-1}}\tilde{\eta}_\varepsilon(t)$ has the phase space ${}_{\bar{k}_{h-1}}\mathbb{X} = \mathbb{X} \setminus \{k_1, \ldots, k_{h-1}\}$.

Second, the semi-Markov process ${}_{\bar{k}_{h-1}}\tilde{\eta}_\varepsilon(t)$ is transformed in the reduced semi-Markov process ${}_{\bar{k}_h}\eta_\varepsilon(t)$ by applying the procedure of one-state reduction of phase space described in Sects. 9.3 and 9.4.1.2. The resulting semi-Markov process ${}_{\bar{k}_h}\eta_\varepsilon(t)$ has the reduced phase space ${}_{\bar{k}_h}\mathbb{X} = {}_{\bar{k}_{h-1}}\mathbb{X} \setminus \{k_h\} = \mathbb{X} \setminus \{k_1, \ldots, k_h\}$.

Again, there are five variants of the procedures described above that use different sets of conditions.

In the variant **(i)**, the conditions ${}_{\bar{\mathbf{k}}_{h-1}}\mathbf{G}$ and ${}_{\bar{\mathbf{k}}_{h-1}}\mathbf{H}$ replace the initial conditions ${}_{\bar{\mathbf{k}}_0}\mathbf{G}$ and ${}_{\bar{\mathbf{k}}_0}\mathbf{H}$. The fulfilment of the conditions ${}_{\bar{\mathbf{k}}_{h-1}}\mathbf{G}$ and ${}_{\bar{\mathbf{k}}_{h-1}}\mathbf{H}$ was the result of the $h-1$ steps in the variant **(i)** of the recurrent phase space reduction algorithm.

In this case, Lemmas 9.3–9.5 imply that the conditions ${}_{\bar{\mathbf{k}}_{h-1}}\mathbf{G}$ and ${}_{\bar{\mathbf{k}}_{h-1}}\mathbf{H}$ are satisfied for the semi-Markov processes ${}_{\bar{k}_{h-1}}\tilde{\eta}_\varepsilon(t)$ in the form of conditions ${}_{\bar{\mathbf{k}}_{h-1}}\tilde{\mathbf{G}}$ and ${}_{\bar{\mathbf{k}}_{h-1}}\tilde{\mathbf{H}}$. Also, Lemmas 9.14–9.16 imply that the conditions ${}_{\bar{k}_{h-1}}\mathbf{G}$ and ${}_{\bar{\mathbf{k}}_{h-1}}\mathbf{H}$ are satisfied for the reduced semi-Markov processes ${}_{\bar{k}_h}\eta_\varepsilon(t)$ in the form of conditions ${}_{\bar{\mathbf{k}}_h}\mathbf{G}$ and ${}_{\bar{\mathbf{k}}_h}\mathbf{H}$.

In the variant **(ii)**, the conditions ${}_{\bar{\mathbf{k}}_{h-1}}\mathbf{G}$, ${}_{\bar{\mathbf{k}}_{h-1}}\mathbf{H}$, and ${}_{\bar{\mathbf{k}}_{h-1}}\mathbf{K}$ **(a)** replace the initial conditions ${}_{\bar{\mathbf{k}}_0}\mathbf{G}$, ${}_{\bar{\mathbf{k}}_0}\mathbf{H}$, and ${}_{\bar{\mathbf{k}}_0}\mathbf{K}$ **(a)**. The fulfilment of the conditions ${}_{\bar{\mathbf{k}}_{h-1}}\mathbf{G}$, ${}_{\bar{\mathbf{k}}_{h-1}}\mathbf{H}$, and ${}_{\bar{\mathbf{k}}_{h-1}}\mathbf{K}$ **(a)** was the result of the $h-1$ steps in the variant **(ii)** of the recurrent phase space reduction algorithm.

In this case, Lemmas 9.3–9.5 and Remark 9.1 imply that the conditions ${}_{\bar{\mathbf{k}}_{h-1}}\mathbf{G}$, ${}_{\bar{\mathbf{k}}_{h-1}}\mathbf{H}$, and ${}_{\bar{\mathbf{k}}_{h-1}}\mathbf{K}$ **(a)** are satisfied for the semi-Markov processes ${}_{\bar{k}_{h-1}}\tilde{\eta}_\varepsilon(t)$ in the form of conditions ${}_{\bar{\mathbf{k}}_{h-1}}\tilde{\mathbf{G}}$, ${}_{\bar{\mathbf{k}}_{h-1}}\tilde{\mathbf{H}}$, and ${}_{\bar{\mathbf{k}}_{h-1}}\tilde{\mathbf{K}}$ **(a)**. Also, Lemmas 9.14–9.16 and Remark 9.1 imply that the conditions ${}_{\bar{\mathbf{k}}_{h-1}}\mathbf{G}$, ${}_{\bar{\mathbf{k}}_{h-1}}\mathbf{H}$, and ${}_{\bar{\mathbf{k}}_{h-1}}\mathbf{K}$ are satisfied for the reduced semi-Markov processes ${}_{\bar{k}_h}\eta_\varepsilon(t)$ in the form of conditions ${}_{\bar{\mathbf{k}}_h}\mathbf{G}$, ${}_{\bar{\mathbf{k}}_h}\mathbf{H}$ and ${}_{\bar{\mathbf{k}}_h}\mathbf{K}$ **(a)**.

In the variant **(iii)**, the conditions ${}_{\bar{\mathbf{k}}_{h-1}}\mathbf{G}$, ${}_{\bar{\mathbf{k}}_{h-1}}\mathbf{H}$, ${}_{\bar{\mathbf{k}}_{h-1}}\mathbf{I}$, and ${}_{\bar{\mathbf{k}}_{h-1}}\mathbf{I}_{\mathcal{H}}$ replace the initial conditions ${}_{\bar{\mathbf{k}}_0}\mathbf{G}$, ${}_{\bar{\mathbf{k}}_0}\mathbf{H}$, ${}_{\bar{\mathbf{k}}_0}\mathbf{I}$, and ${}_{\bar{\mathbf{k}}_0}\mathbf{I}_{\mathcal{H}}$, The fulfilment of the conditions ${}_{\bar{\mathbf{k}}_{h-1}}\mathbf{G}$, ${}_{\bar{\mathbf{k}}_{h-1}}\mathbf{H}$, ${}_{\bar{\mathbf{k}}_{h-1}}\mathbf{I}$, and ${}_{\bar{\mathbf{k}}_{h-1}}\mathbf{I}_{\mathcal{H}}$ was the result of the $h-1$ steps in the variant **(iii)** of the recurrent phase space reduction algorithm.

In this case, Lemmas 9.6–9.9 imply that the conditions ${}_{\bar{k}_{h-1}}\mathbf{G}$, ${}_{\bar{\mathbf{k}}_{h-1}}\mathbf{H}$, ${}_{\bar{\mathbf{k}}_{h-1}}\mathbf{I}$, and ${}_{\bar{\mathbf{k}}_{h-1}}\mathbf{I}_{\mathcal{H}}$ are satisfied for the semi-Markov processes ${}_{\bar{k}_{h-1}}\tilde{\eta}_\varepsilon(t)$ in the form of conditions ${}_{\bar{\mathbf{k}}_{h-1}}\tilde{\mathbf{G}}$, ${}_{\bar{\mathbf{k}}_{h-1}}\tilde{\mathbf{H}}$, ${}_{\bar{\mathbf{k}}_{h-1}}\tilde{\mathbf{I}}$, and ${}_{\bar{\mathbf{k}}_{h-1}}\tilde{\mathbf{I}}_{\mathcal{H}}$. Also, Lemmas 9.17–9.18 imply that the conditions ${}_{\bar{k}_{h-1}}\mathbf{G}$, ${}_{\bar{\mathbf{k}}_{h-1}}\mathbf{H}$, ${}_{\bar{\mathbf{k}}_{h-1}}\mathbf{I}$, and ${}_{\bar{\mathbf{k}}_{h-1}}\mathbf{I}_{\mathcal{H}}$ are satisfied for the reduced semi-Markov processes ${}_{\bar{k}_h}\eta_\varepsilon(t)$ in the form of conditions ${}_{\bar{\mathbf{k}}_h}\mathbf{G}$, ${}_{\bar{\mathbf{k}}_h}\mathbf{H}$, ${}_{\bar{\mathbf{k}}_h}\mathbf{I}$, and ${}_{\bar{\mathbf{k}}_h}\mathbf{I}_{\mathcal{H}}$.

In the variant **(iv)**, the conditions ${}_{\bar{k}_{h-1}}\mathbf{G}$, ${}_{\bar{\mathbf{k}}_{h-1}}\mathbf{H}$, ${}_{\bar{\mathbf{k}}_{h-1}}\mathbf{I}$, ${}_{\bar{\mathbf{k}}_{h-1}}\mathbf{I}_{\mathcal{H}}$, ${}_{\bar{\mathbf{k}}_{h-1}}\mathbf{L}$, and ${}_{\bar{\mathbf{k}}_{h-1}}\mathbf{L}_{\mathcal{H}}$ replace the initial conditions ${}_{\bar{k}_0}\mathbf{G}$, ${}_{\bar{\mathbf{k}}_0}\mathbf{H}$, ${}_{\bar{\mathbf{k}}_0}\mathbf{I}$, ${}_{\bar{\mathbf{k}}_0}\mathbf{I}_{\mathcal{H}}$, ${}_{\bar{\mathbf{k}}_0}\mathbf{L}$, and ${}_{\bar{\mathbf{k}}_0}\mathbf{L}_{\mathcal{H}}$. The fulfilment of the conditions ${}_{\bar{k}_{h-1}}\mathbf{G}$, ${}_{\bar{\mathbf{k}}_{h-1}}\mathbf{H}$, ${}_{\bar{\mathbf{k}}_{h-1}}\mathbf{I}$, ${}_{\bar{\mathbf{k}}_{h-1}}\mathbf{I}_{\mathcal{H}}$, ${}_{\bar{\mathbf{k}}_{h-1}}\mathbf{L}$, and ${}_{\bar{\mathbf{k}}_{h-1}}\mathbf{L}_{\mathcal{H}}$ was the result of the $h-1$ step in the variant **(iv)** of the recurrent phase space reduction algorithm.

In this case, Lemmas 9.12–9.13 imply that the conditions ${}_{\bar{k}_{h-1}}\mathbf{G}$, ${}_{\bar{\mathbf{k}}_{h-1}}\mathbf{H}$, ${}_{\bar{\mathbf{k}}_{h-1}}\mathbf{I}$, ${}_{\bar{\mathbf{k}}_{h-1}}\mathbf{I}_{\mathcal{H}}$, ${}_{\bar{\mathbf{k}}_{h-1}}\mathbf{L}$, and ${}_{\bar{\mathbf{k}}_{h-1}}\mathbf{L}_{\mathcal{H}}$ are satisfied for the semi-Markov processes ${}_{\bar{k}_{h-1}}\tilde{\eta}_\varepsilon(t)$ in the form of conditions ${}_{\bar{\mathbf{k}}_{h-1}}\tilde{\mathbf{G}}$, ${}_{\bar{\mathbf{k}}_{h-1}}\tilde{\mathbf{H}}$, ${}_{\bar{\mathbf{k}}_{h-1}}\tilde{\mathbf{I}}$, ${}_{\bar{\mathbf{k}}_{h-1}}\tilde{\mathbf{I}}_{\mathcal{H}}$, ${}_{\bar{\mathbf{k}}_{h-1}}\tilde{\mathbf{L}}$, and ${}_{\bar{\mathbf{k}}_{h-1}}\tilde{\mathbf{L}}_{\mathcal{H}}$. Also, Lemmas 9.24–9.25 imply that the conditions ${}_{\bar{k}_{h-1}}\mathbf{G}$, ${}_{\bar{\mathbf{k}}_{h-1}}\mathbf{H}$, ${}_{\bar{\mathbf{k}}_{h-1}}\mathbf{I}$, ${}_{\bar{\mathbf{k}}_{h-1}}\mathbf{I}_{\mathcal{H}}$, ${}_{\bar{\mathbf{k}}_{h-1}}\mathbf{L}$, and ${}_{\bar{\mathbf{k}}_{h-1}}\mathbf{L}_{\mathcal{H}}$ hold for the reduced semi-Markov processes ${}_{\bar{k}_h}\eta_\varepsilon(t)$ in the form of conditions ${}_{\bar{\mathbf{k}}_h}\mathbf{G}$, ${}_{\bar{\mathbf{k}}_h}\mathbf{H}$, ${}_{\bar{\mathbf{k}}_h}\mathbf{I}$, ${}_{\bar{\mathbf{k}}_h}\mathbf{I}_{\mathcal{H}}$, ${}_{\bar{\mathbf{k}}_h}\mathbf{L}$, and ${}_{\bar{\mathbf{k}}_h}\mathbf{L}_{\mathcal{H}}$.

Finally, in the variant **(v)**, the conditions ${}_{\bar{k}_{h-1}}\mathbf{G}$, ${}_{\bar{\mathbf{k}}_{h-1}}\mathbf{H}$, ${}_{\bar{\mathbf{k}}_{h-1}}\mathbf{I}$, ${}_{\bar{\mathbf{k}}_{h-1}}\mathbf{I}_{\mathcal{H}}$, ${}_{\bar{\mathbf{k}}_{h-1}}\mathbf{L}$, ${}_{\bar{\mathbf{k}}_{h-1}}\mathbf{L}_{\mathcal{H}}$, ${}_{\bar{\mathbf{k}}_{h-1}}\mathbf{J}$, ${}_{\bar{\mathbf{k}}_{h-1}}\mathbf{K}$, and ${}_{\bar{\mathbf{k}}_h}\mathbf{M}$ (state $k_l \in {}_{\bar{k}_{l-1}}\mathbb{W}_0$, for $l = 1, \ldots, h$) replace the conditions ${}_{\bar{k}_0}\mathbf{G}$, ${}_{\bar{\mathbf{k}}_0}\mathbf{H}$, ${}_{\bar{\mathbf{k}}_0}\mathbf{I}$, ${}_{\bar{\mathbf{k}}_0}\mathbf{I}_{\mathcal{H}}$, ${}_{\bar{\mathbf{k}}_0}\mathbf{L}$, ${}_{\bar{\mathbf{k}}_0}\mathbf{L}_{\mathcal{H}}$, ${}_{\bar{\mathbf{k}}_0}\mathbf{J}$, ${}_{\bar{\mathbf{k}}_0}\mathbf{K}$ and ${}_{\bar{\mathbf{k}}_1}\mathbf{M}$ (state $k_1 \in {}_{\bar{k}_0}\mathbb{W}_0 = \mathbb{W}_0$). The fulfilment of the conditions ${}_{\bar{k}_{h-1}}\mathbf{G}$, ${}_{\bar{\mathbf{k}}_{h-1}}\mathbf{H}$, ${}_{\bar{\mathbf{k}}_{h-1}}\mathbf{I}$, ${}_{\bar{\mathbf{k}}_{h-1}}\mathbf{I}_{\mathcal{H}}$, ${}_{\bar{\mathbf{k}}_{h-1}}\mathbf{L}$, ${}_{\bar{\mathbf{k}}_{h-1}}\mathbf{L}_{\mathcal{H}}$, ${}_{\bar{\mathbf{k}}_{h-1}}\mathbf{J}$, ${}_{\bar{\mathbf{k}}_{h-1}}\mathbf{K}$ was the result of the $h - 1$ step in the variant **(v)** of the recurrent phase space reduction algorithm. At the $(h - 1)$-th step, the fulfilment of the condition ${}_{\bar{\mathbf{k}}_{h-1}}\mathbf{M}$ was additionally assumed. At the h-th step, this assumption should be extended to the assumption that the condition ${}_{\bar{\mathbf{k}}_h}\mathbf{M}$, which is required to implement the second sub-algorithm of transforming the semi-Markov process ${}_{\bar{k}_{h-1}}\tilde{\eta}_\varepsilon(t)$ in the reduced semi-Markov process ${}_{\bar{k}_h}\eta_\varepsilon(t)$, is satisfied.

In this case, Lemmas 9.10–9.11 imply that the conditions ${}_{\bar{k}_{h-1}}\mathbf{G}$, ${}_{\bar{\mathbf{k}}_{h-1}}\mathbf{H}$, ${}_{\bar{\mathbf{k}}_{h-1}}\mathbf{I}$, ${}_{\bar{\mathbf{k}}_{h-1}}\mathbf{I}_{\mathcal{H}}$, ${}_{\bar{\mathbf{k}}_{h-1}}\mathbf{L}$, ${}_{\bar{\mathbf{k}}_{h-1}}\mathbf{L}$, ${}_{\bar{\mathbf{k}}_{h-1}}\mathbf{J}$, and ${}_{\bar{\mathbf{k}}_{h-1}}\mathbf{K}$ are satisfied for the semi-Markov processes ${}_{\bar{k}_{h-1}}\tilde{\eta}_\varepsilon(t)$ in the form of conditions ${}_{\bar{\mathbf{k}}_{h-1}}\tilde{\mathbf{G}}$, ${}_{\bar{\mathbf{k}}_{h-1}}\tilde{\mathbf{H}}$, ${}_{\bar{\mathbf{k}}_{h-1}}\tilde{\mathbf{I}}$, ${}_{\bar{\mathbf{k}}_{h-1}}\tilde{\mathbf{I}}_{\mathcal{H}}$, ${}_{\bar{\mathbf{k}}_{h-1}}\tilde{\mathbf{L}}$, ${}_{\bar{\mathbf{k}}_{h-1}}\tilde{\mathbf{L}}_{\mathcal{H}}$, ${}_{\bar{\mathbf{k}}_{h-1}}\tilde{\mathbf{J}}$, and ${}_{\bar{\mathbf{k}}_{h-1}}\tilde{\mathbf{K}}$. Also, Lemmas 9.21–9.23 imply that the conditions ${}_{\bar{k}_{h-1}}\mathbf{G}$, ${}_{\bar{\mathbf{k}}_{h-1}}\mathbf{H}$, ${}_{\bar{\mathbf{k}}_{h-1}}\mathbf{I}$, ${}_{\bar{\mathbf{k}}_{h-1}}\mathbf{I}_{\mathcal{H}}$, ${}_{\bar{\mathbf{k}}_{h-1}}\mathbf{L}$, ${}_{\bar{\mathbf{k}}_{h-1}}\mathbf{L}_{\mathcal{H}}$, ${}_{\bar{\mathbf{k}}_{h-1}}\mathbf{J}$, ${}_{\bar{\mathbf{k}}_{h-1}}\mathbf{K}$ are satisfied for the reduced semi-Markov processes ${}_{\bar{k}_h}\eta_\varepsilon(t)$ in the form of conditions ${}_{\bar{\mathbf{k}}_h}\mathbf{G}$, ${}_{\bar{\mathbf{k}}_h}\mathbf{H}$, ${}_{\bar{\mathbf{k}}_h}\mathbf{I}$, ${}_{\bar{\mathbf{k}}_h}\mathbf{I}_{\mathcal{H}}$, ${}_{\bar{\mathbf{k}}_h}\mathbf{L}$, ${}_{\bar{\mathbf{k}}_h}\mathbf{L}_{\mathcal{H}}$, ${}_{\bar{\mathbf{k}}_h}\mathbf{J}$, and ${}_{\bar{\mathbf{k}}_h}\mathbf{K}$.

What is important is that in the variants **(i)**–**(iv)**, the states $k_1, \ldots, k_h$ can be arbitrary different states in the phase space $\mathbb{X}$. In the variant **(v)**, the state k_1 should be chosen such that $k_1 \in {}_{\bar{k}_0}\mathbb{W}_0, \ldots$, the state k_h should be chosen such that $k_h \in {}_{\bar{k}_{n-1}}\mathbb{W}_0$, that is, the condition ${}_{\bar{\mathbf{k}}_h}\mathbf{M}$ is assumed to be satisfied.

Also, the algorithm described in Sects. 9.2 and 9.4.1.2 can additionally be applied to the reduced semi-Markov process ${}_{\bar{k}_h}\eta_\varepsilon(t)$ and this semi-Markov process be transformed in the reduced semi-Markov process ${}_{\bar{k}_h}\tilde{\eta}_\varepsilon(t)$. By Theorems 9.1 and 9.2, the conditions ${}_{\bar{\mathbf{k}}_h}\tilde{\mathbf{G}}$, ${}_{\bar{\mathbf{k}}_h}\tilde{\mathbf{H}}$, ${}_{\bar{\mathbf{k}}_h}\tilde{\mathbf{I}}$, ${}_{\bar{\mathbf{k}}_h}\tilde{\mathbf{I}}_{\mathcal{H}}$, ${}_{\bar{\mathbf{k}}_h}\tilde{\mathbf{L}}$, ${}_{\bar{\mathbf{k}}_h}\tilde{\mathbf{L}}_{\mathcal{H}}$, ${}_{\bar{\mathbf{k}}_h}\tilde{\mathbf{J}}$, and ${}_{\bar{\mathbf{k}}_h}\tilde{\mathbf{K}}$ are satisfied for the semi-Markov processes ${}_{\bar{k}_h}\tilde{\eta}_\varepsilon(t)$.

It is useful to note that the normalisation functions ${}_{\bar{k}_{h-1}}\tilde{u}_{\cdot,i}, i \in {}_{\bar{k}_{h-1}}\mathbb{X}, h = 1, \ldots, m$ used for the perturbed reduced semi-Markov processes ${}_{\bar{k}_{h-1}}\tilde{\eta}_\varepsilon(t)$ take the following forms:

$$
{}_{\bar{k}_{h-1}}\tilde{u}_{\varepsilon,i} = \prod_{r=0}^{h-1} {}_{\bar{k}_r}\bar{p}_{\varepsilon,ii}^{-1}\, u_{\varepsilon,i} \tag{9.111}
$$

and the normalisation functions ${}_{\bar{k}_h}u_{\cdot,i}, i \in {}_{\bar{k}_h}\mathbb{X}, n = 1, \ldots, m - 1$ used for the perturbed reduced semi-Markov processes ${}_{\bar{k}_h}\eta_\varepsilon(t)$ take the following forms:

$$
{}_{\bar{k}_h}u_{\varepsilon,i} = {}_{\bar{k}_{h-1}}\tilde{u}_{\varepsilon,i}. \tag{9.112}
$$

It is also worth noting that the normalisation functions ${}_{\bar{k}_{h-1}}\tilde{u}_{\cdot,i}, i \in {}_{\bar{k}_{h-1}}\mathbb{X}, h = 1, \ldots, m$ and ${}_{\bar{k}_h}u_{\cdot,i}, i \in {}_{\bar{k}_h}\mathbb{X}, h = 1, \ldots, m - 1$ belong to the space of asymptotically comparable functions $\mathcal{H}$.

The absorbing rates condition ${}_{\bar{\mathbf{k}}_h}\mathbf{M}$ mentioned above takes the following form, for $h = 1, \ldots, m - 1$:

${}_{\bar{\mathbf{k}}_h}\mathbf{M}$: $k_l \in {}_{\bar{k}_{l-1}}\mathbb{W}_0$, i.e., ${}_{\bar{k}_{l-1}}w_{0,k_l,i} = \lim_{\varepsilon\to 0} \frac{{}_{\bar{k}_{l-1}}\tilde{u}_{\varepsilon,k_l}}{{}_{\bar{k}_{l-1}}\tilde{u}_{\varepsilon,i}} \in [0, \infty)$, $i \in {}_{\bar{k}_{l-1}}\mathbb{X}$, for $l = 1, \ldots, h$.

This condition means that at each step of successive exclusion of the states $k_1, \ldots, k_h$ from the phase space $\mathbb{X}$, the state k_l is chosen from the set of the least absorbing states ${}_{\bar{k}_{l-1}}\mathbb{W}_0$ (this set is a subset of the phase space ${}_{\bar{k}_{l-1}}\mathbb{X}$) for the semi-Markov process ${}_{\bar{k}_{l-1}}\tilde{\eta}_\varepsilon(t)$, sequentially for $l = 1, \ldots, h$.

The conditions $\mathbf{G}$, $\mathbf{H}$, $\mathbf{I}_{\mathcal{H}}$, and $\mathbf{L}_{\mathcal{H}}$ imply that the limits in the asymptotic relations appearing in the condition ${}_{\bar{\mathbf{k}}_\mathbf{h}}\mathbf{M}$ exist and can be recurrently calculated using Lemmas 8.2–8.9.

We denote by $\mathcal{M}_\mathbf{h}$ the set of sequences of states $\bar{k}_h$, such that all asymptotic relations appearing in the condition ${}_{\bar{\mathbf{k}}_\mathbf{h}}\mathbf{M}$ hold (with the limits indicated in this condition).

According to Lemmas 9.20 and 9.23, the conditions $\mathbf{G}$, $\mathbf{H}$, $\mathbf{I}_{\mathcal{H}}$, and $\mathbf{L}_{\mathcal{H}}$ imply that the set $\mathcal{M}_\mathbf{h}$ is not empty.

Thus, the condition ${}_{\bar{\mathbf{k}}_\mathbf{h}}\mathbf{M}$, in fact, means only that the sequence $\bar{k}_h$ is selected from the set $\mathcal{M}_\mathbf{h}$.

9.4.2 Summary of Recurrent Phase Space Reduction Algorithm

The above phase space reduction algorithm is realised in sequential recurrent steps of construction the sequence of semi-Markov processes with removed virtual transitions and reduced phase spaces, for some $0 \leq h \leq m-1$. In addition, virtual transitions can be removed from trajectories of the resulting semi-Markov process. This sequence of recurrent transformations can be represented by the following symbolic diagram:

$$\begin{aligned} \eta_\varepsilon(t) &= {}_{\bar{k}_0}\eta_\varepsilon(t) \\ &\longrightarrow {}_{\bar{k}_0}\tilde{\eta}_\varepsilon(t) \longrightarrow {}_{\bar{k}_1}\eta_\varepsilon(t) \\ &\ldots \\ &\longrightarrow {}_{\bar{k}_{h-1}}\tilde{\eta}_\varepsilon(t) \longrightarrow {}_{\bar{k}_h}\eta_\varepsilon(t) \\ &\longrightarrow {}_{\bar{k}_h}\tilde{\eta}_\varepsilon(t). \end{aligned} \tag{9.113}$$

The corresponding recurrent formulas for finding transition characteristics of the above perturbed semi-Markov process with removed virtual transitions and reduced phase spaces are given in Sects. 9.2–9.4.

As mentioned above, five variants of recurrent phase space reduction algorithm can be realised.

In the variant **(i)**, the initial set of conditions $\langle \mathbf{G}, \mathbf{H} \rangle$ is imposed on the semi-Markov processes $\eta_\varepsilon(t)$. This variant of the recurrent phase space reduction algorithm is described in Sect. 9.4.3.3. It insures the fulfilment of the similar set of conditions for the corresponding transformed semi-Markov processes. This can be represented by the following symbolic diagram:

$$\begin{aligned} \langle \mathbf{G}, \mathbf{H} \rangle &= \langle {}_{\bar{\mathbf{k}}_0}\mathbf{G}, {}_{\bar{\mathbf{k}}_0}\mathbf{H} \rangle \\ &\Downarrow \end{aligned}$$

$$
\begin{array}{c}
\langle {}_{\bar{\mathbf{k}}_0}\tilde{\mathbf{G}},\ {}_{\bar{\mathbf{k}}_0}\tilde{\mathbf{H}}\rangle \\
\Downarrow \\
\langle {}_{\bar{\mathbf{k}}_1}\mathbf{G},\ {}_{\bar{\mathbf{k}}_1}\mathbf{H}\rangle \\
\Downarrow \\
\cdots \\
\Downarrow \\
\langle {}_{\bar{\mathbf{k}}_{h-1}}\tilde{\mathbf{G}},\ {}_{\bar{\mathbf{k}}_{h-1}}\tilde{\mathbf{H}}\rangle \\
\Downarrow \\
\langle {}_{\bar{\mathbf{k}}_h}\mathbf{G},\ {}_{\bar{\mathbf{k}}_h}\mathbf{H}\rangle \\
\Downarrow \\
\langle {}_{\bar{\mathbf{k}}_h}\tilde{\mathbf{G}},\ {}_{\bar{\mathbf{k}}_h}\tilde{\mathbf{H}}\rangle.
\end{array} \tag{9.114}
$$

The corresponding recurrent relations for "re-calculation" of the above sets of conditions are given in Lemmas 9.3–9.5 and 9.14–9.16.

In the variant **(ii)**, the initial set of condition $\langle \mathbf{G}, \mathbf{H}, \mathbf{K}(\mathbf{a})\rangle$ is imposed on the semi-Markov processes $\eta_\varepsilon(t)$. This variant of the recurrent phase space reduction algorithm is described in Sect. 9.4.1. It insures the fulfilment of the similar set of conditions for the corresponding transformed semi-Markov processes. This can be represented by the following symbolic diagram:

$$
\begin{array}{rl}
\langle \mathbf{G},\ \mathbf{H},\ \mathbf{K}(\mathbf{a})\rangle = & \langle {}_{\bar{\mathbf{k}}_0}\mathbf{G},\ {}_{\bar{\mathbf{k}}_0}\mathbf{H},\ {}_{\bar{\mathbf{k}}_0}\mathbf{K}(\mathbf{a})\rangle \\
& \Downarrow \\
& \langle {}_{\bar{\mathbf{k}}_0}\tilde{\mathbf{G}},\ {}_{\bar{\mathbf{k}}_0}\tilde{\mathbf{H}},\ {}_{\bar{\mathbf{k}}_0}\tilde{\mathbf{K}}(\mathbf{a})\rangle \\
& \Downarrow \\
& \langle {}_{\bar{\mathbf{k}}_1}\mathbf{G},\ {}_{\bar{\mathbf{k}}_1}\mathbf{H},\ {}_{\bar{\mathbf{k}}_1}\mathbf{K}(\mathbf{a})\rangle \\
& \Downarrow \\
& \cdots \\
& \Downarrow \\
& \langle {}_{\bar{\mathbf{k}}_{h-1}}\tilde{\mathbf{G}},\ {}_{\bar{\mathbf{k}}_{h-1}}\tilde{\mathbf{H}},\ {}_{\bar{\mathbf{k}}_{h-1}}\tilde{\mathbf{K}}(\mathbf{a})\rangle \\
& \Downarrow \\
& \langle {}_{\bar{\mathbf{k}}_h}\mathbf{G},\ {}_{\bar{\mathbf{k}}_h}\mathbf{H},\ {}_{\bar{\mathbf{k}}_h}\mathbf{K}(\mathbf{a})\rangle \\
& \Downarrow \\
& \langle {}_{\bar{\mathbf{k}}_h}\tilde{\mathbf{G}},\ {}_{\bar{\mathbf{k}}_h}\tilde{\mathbf{H}},\ {}_{\bar{\mathbf{k}}_h}\tilde{\mathbf{K}}(\mathbf{a})\rangle.
\end{array} \tag{9.115}
$$

The corresponding recurrent relations for "re-calculation" of the above set of conditions are given in Lemmas 9.3–9.5 and 9.14–9.16 and Remarks 9.1 and 9.2.

In the variant **(iii)**, the initial set of conditions $\langle \mathbf{G}, \mathbf{H}, \mathbf{I}, \mathbf{I}_{\mathcal{H}}\rangle$ are imposed on the semi-Markov processes $\eta_\varepsilon(t)$. This variant of the recurrent phase space reduction algorithm is also described in Sect. 9.4.1. It insures the fulfilment of the similar set of conditions for the corresponding transformed semi-Markov processes. This can be represented by the following symbolic diagram:

$$
\begin{aligned}
\langle \mathbf{G}, \mathbf{H}, \mathbf{I}, \mathbf{I}_{\mathcal{H}} \rangle &= \langle_{\bar{\mathbf{k}}_0}\mathbf{G}, {}_{\bar{\mathbf{k}}_0}\mathbf{H}, {}_{\bar{\mathbf{k}}_0}\mathbf{I}, {}_{\bar{\mathbf{k}}_0}\mathbf{I}_{\mathcal{H}} \rangle \\
&\Downarrow \\
&\langle_{\bar{\mathbf{k}}_0}\tilde{\mathbf{G}}, {}_{\bar{\mathbf{k}}_0}\tilde{\mathbf{H}}, {}_{\bar{\mathbf{k}}_0}\tilde{\mathbf{I}}, {}_{\bar{\mathbf{k}}_0}\tilde{\mathbf{I}}_{\mathcal{H}} \rangle \\
&\Downarrow \\
&\langle_{\bar{\mathbf{k}}_1}\mathbf{G}, {}_{\bar{\mathbf{k}}_1}\mathbf{H}, {}_{\bar{\mathbf{k}}_1}\mathbf{I}, {}_{\bar{\mathbf{k}}_1}\mathbf{I}_{\mathcal{H}} \rangle \\
&\Downarrow \\
&\cdots \\
&\Downarrow \\
&\langle_{\bar{\mathbf{k}}_{h-1}}\tilde{\mathbf{G}}, {}_{\bar{\mathbf{k}}_{h-1}}\tilde{\mathbf{H}}, {}_{\bar{\mathbf{k}}_{h-1}}\tilde{\mathbf{I}}, {}_{\bar{\mathbf{k}}_{h-1}}\tilde{\mathbf{I}}_{\mathcal{H}} \rangle \\
&\Downarrow \\
&\langle_{\bar{\mathbf{k}}_h}\mathbf{G}, {}_{\bar{\mathbf{k}}_h}\mathbf{H}, {}_{\bar{\mathbf{k}}_h}\mathbf{I}, {}_{\bar{\mathbf{k}}_h}\mathbf{I}_{\mathcal{H}} \rangle \\
&\Downarrow \\
&\langle_{\bar{\mathbf{k}}_h}\tilde{\mathbf{G}}, {}_{\bar{\mathbf{k}}_h}\tilde{\mathbf{H}}, {}_{\bar{\mathbf{k}}_h}\tilde{\mathbf{I}}, {}_{\bar{\mathbf{k}}_h}\tilde{\mathbf{I}}_{\mathcal{H}} \rangle.
\end{aligned}
\tag{9.116}
$$

The corresponding recurrent relations for "re-calculation" of the above set of conditions are given in Lemmas 9.6–9.9 and 9.17–9.18.

In the variant **(iv)**, the initial set of conditions $\langle \mathbf{G}, \mathbf{H}, \mathbf{I}, \mathbf{I}_{\mathcal{H}}, \mathbf{L}, \mathbf{L}_{\mathcal{H}}$ are imposed on the semi-Markov processes $\eta_\varepsilon(t)$. This variant of the recurrent phase space reduction algorithm is also described in Sect. 9.4.1. It insures the fulfilment of the similar set of conditions for the corresponding transformed semi-Markov processes. This can be represented by the following symbolic diagram:

$$
\begin{aligned}
\langle \mathbf{G}, \mathbf{H}, \mathbf{I}, \mathbf{I}_{\mathcal{H}}, \mathbf{L}, \mathbf{L}_{\mathcal{H}} \rangle &= \langle_{\bar{\mathbf{k}}_0}\mathbf{G}, {}_{\bar{\mathbf{k}}_0}\mathbf{H}, {}_{\bar{\mathbf{k}}_0}\mathbf{I}, {}_{\bar{\mathbf{k}}_0}\mathbf{I}_{\mathcal{H}}, {}_{\bar{\mathbf{k}}_0}\mathbf{L}, {}_{\bar{\mathbf{k}}_0}\mathbf{L}_{\mathcal{H}} \rangle \\
&\Downarrow \\
&\langle_{\bar{\mathbf{k}}_0}\tilde{\mathbf{G}}, {}_{\bar{\mathbf{k}}_0}\tilde{\mathbf{H}}, {}_{\bar{\mathbf{k}}_0}\tilde{\mathbf{I}}, {}_{\bar{\mathbf{k}}_0}\tilde{\mathbf{I}}_{\mathcal{H}}, {}_{\bar{\mathbf{k}}_0}\tilde{\mathbf{L}}, {}_{\bar{\mathbf{k}}_0}\tilde{\mathbf{L}}_{\mathcal{H}} \rangle \\
&\Downarrow \\
&\langle_{\bar{\mathbf{k}}_1}\mathbf{G}, {}_{\bar{\mathbf{k}}_1}\mathbf{H}, {}_{\bar{\mathbf{k}}_1}\mathbf{I}, {}_{\bar{\mathbf{k}}_1}\mathbf{I}_{\mathcal{H}}, {}_{\bar{\mathbf{k}}_1}\mathbf{L}, {}_{\bar{\mathbf{k}}_1}\mathbf{L}_{\mathcal{H}} \rangle \\
&\Downarrow \\
&\cdots \\
&\Downarrow \\
&\langle_{\bar{\mathbf{k}}_{h-1}}\tilde{\mathbf{G}}, {}_{\bar{\mathbf{k}}_{h-1}}\tilde{\mathbf{H}}, {}_{\bar{\mathbf{k}}_{h-1}}\tilde{\mathbf{I}}, {}_{\bar{\mathbf{k}}_{h-1}}\tilde{\mathbf{I}}_{\mathcal{H}}, {}_{\bar{\mathbf{k}}_{h-1}}\tilde{\mathbf{L}}, {}_{\bar{\mathbf{k}}_{h-1}}\tilde{\mathbf{L}}_{\mathcal{H}} \rangle \\
&\Downarrow \\
&\langle_{\bar{\mathbf{k}}_h}\mathbf{G}, {}_{\bar{\mathbf{k}}_h}\mathbf{H}, {}_{\bar{\mathbf{k}}_h}\mathbf{I}, {}_{\bar{\mathbf{k}}_h}\mathbf{I}_{\mathcal{H}}, {}_{\bar{\mathbf{k}}_h}\mathbf{L}, {}_{\bar{\mathbf{k}}_h}\mathbf{L}_{\mathcal{H}} \rangle \\
&\Downarrow \\
&\langle_{\bar{\mathbf{k}}_h}\tilde{\mathbf{G}}, {}_{\bar{\mathbf{k}}_h}\tilde{\mathbf{H}}, {}_{\bar{\mathbf{k}}_h}\tilde{\mathbf{I}}, {}_{\bar{\mathbf{k}}_h}\tilde{\mathbf{I}}_{\mathcal{H}}, {}_{\bar{\mathbf{k}}_h}\tilde{\mathbf{L}}, {}_{\bar{\mathbf{k}}_h}\tilde{\mathbf{L}}_{\mathcal{H}} \rangle.
\end{aligned}
\tag{9.117}
$$

The corresponding recurrent relations for "re-calculation" of the above set of conditions are given in Lemmas 9.12–9.13 and 9.24–9.25.

In the variant (**v**), the initial set of conditions $\langle \mathbf{G}, \mathbf{H}, \mathbf{I}, \mathbf{I}_{\mathcal{H}}, \mathbf{L}, \mathbf{L}_{\mathcal{H}}, \mathbf{J}, \mathbf{K} \rangle$ and the condition ${}_{\bar{\mathbf{k}}_h}\mathbf{M}$ are imposed on the semi-Markov processes $\eta_\varepsilon(t)$ (at the second sub-step of each step of the algorithm). This variant of the recurrent phase space reduction algorithm is also described in Sect. 9.4.1. It insures the fulfilment of the similar set of conditions for the corresponding transformed semi-Markov processes. This can be presented by the following symbolic diagram:

$$
\begin{array}{rl}
\langle \mathbf{G}, \mathbf{H}, \mathbf{I}, \mathbf{I}_{\mathcal{H}}, \mathbf{L}, \mathbf{L}_{\mathcal{H}}, \mathbf{J}, \mathbf{K} \rangle & = \langle {}_{\bar{\mathbf{k}}_0}\mathbf{G}, {}_{\bar{\mathbf{k}}_0}\mathbf{H}, {}_{\bar{\mathbf{k}}_0}\mathbf{I}, {}_{\bar{\mathbf{k}}_0}\mathbf{I}_{\mathcal{H}}, {}_{\bar{\mathbf{k}}_0}\mathbf{L}, {}_{\bar{\mathbf{k}}_0}\mathbf{L}_{\mathcal{H}}, {}_{\bar{\mathbf{k}}_0}\mathbf{J}, {}_{\bar{\mathbf{k}}_0}\mathbf{K} \rangle \\
& \Downarrow \\
& \langle {}_{\bar{\mathbf{k}}_0}\tilde{\mathbf{G}}, {}_{\bar{\mathbf{k}}_0}\tilde{\mathbf{H}}, {}_{\bar{\mathbf{k}}_0}\tilde{\mathbf{I}}, {}_{\bar{\mathbf{k}}_0}\tilde{\mathbf{I}}_{\mathcal{H}}, {}_{\bar{\mathbf{k}}_0}\tilde{\mathbf{L}}, {}_{\bar{\mathbf{k}}_0}\tilde{\mathbf{L}}_{\mathcal{H}}, {}_{\bar{\mathbf{k}}_0}\tilde{\mathbf{J}}, {}_{\bar{\mathbf{k}}_0}\tilde{\mathbf{K}} \rangle \\
& \Downarrow \\
{}_{\bar{\mathbf{k}}_1}\mathbf{M} \Rightarrow & \langle {}_{\bar{\mathbf{k}}_1}\mathbf{G}, {}_{\bar{\mathbf{k}}_1}\mathbf{H}, {}_{\bar{\mathbf{k}}_1}\mathbf{I}, {}_{\bar{\mathbf{k}}_1}\mathbf{I}_{\mathcal{H}}, {}_{\bar{\mathbf{k}}_1}\mathbf{L}, {}_{\bar{\mathbf{k}}_1}\mathbf{L}_{\mathcal{H}}, {}_{\bar{\mathbf{k}}_1}\mathbf{J}, {}_{\bar{\mathbf{k}}_1}\mathbf{K} \rangle \\
& \Downarrow \\
& \cdots \\
& \Downarrow \\
& \langle {}_{\bar{\mathbf{k}}_{h-1}}\tilde{\mathbf{G}}, {}_{\bar{\mathbf{k}}_{h-1}}\tilde{\mathbf{H}}, {}_{\bar{\mathbf{k}}_{h-1}}\tilde{\mathbf{I}}, {}_{\bar{\mathbf{k}}_{h-1}}\tilde{\mathbf{I}}_{\mathcal{H}}, {}_{\bar{\mathbf{k}}_{h-1}}\tilde{\mathbf{L}}, {}_{\bar{\mathbf{k}}_{h-1}}\tilde{\mathbf{L}}_{\mathcal{H}}, {}_{\bar{\mathbf{k}}_{h-1}}\tilde{\mathbf{J}}, {}_{\bar{\mathbf{k}}_{h-1}}\tilde{\mathbf{K}} \rangle \\
& \Downarrow \\
{}_{\bar{\mathbf{k}}_h}\mathbf{M} \Rightarrow & \langle {}_{\bar{\mathbf{k}}_h}\mathbf{G}, {}_{\bar{\mathbf{k}}_h}\mathbf{H}, {}_{\bar{\mathbf{k}}_h}\mathbf{I}, {}_{\bar{\mathbf{k}}_h}\mathbf{I}_{\mathcal{H}}, {}_{\bar{\mathbf{k}}_h}\mathbf{L}, {}_{\bar{\mathbf{k}}_h}\mathbf{L}_{\mathcal{H}}, {}_{\bar{\mathbf{k}}_h}\mathbf{J}, {}_{\bar{\mathbf{k}}_h}\mathbf{K} \rangle \\
& \Downarrow \\
& \langle {}_{\bar{\mathbf{k}}_h}\tilde{\mathbf{G}}, {}_{\bar{\mathbf{k}}_h}\tilde{\mathbf{H}}, {}_{\bar{\mathbf{k}}_h}\tilde{\mathbf{I}}, {}_{\bar{\mathbf{k}}_h}\tilde{\mathbf{I}}_{\mathcal{H}}, {}_{\bar{\mathbf{k}}_h}\tilde{\mathbf{L}}, {}_{\bar{\mathbf{k}}_h}\tilde{\mathbf{L}}_{\mathcal{H}}, {}_{\bar{\mathbf{k}}_h}\tilde{\mathbf{J}}, {}_{\bar{\mathbf{k}}_h}\tilde{\mathbf{K}} \rangle .
\end{array}
\tag{9.118}
$$

The corresponding recurrent relations for "re-calculation" of the above set of conditions are given in Lemmas 9.10–9.11 and 9.21–9.23.

What is important is that, in the variants (**i**)–(**iv**), the sequence $\bar{k}_h = \langle k_1, \dots, k_h \rangle$ can (for every $1 \leq h \leq m-1$) be an arbitrary sequence of different states from the phase space $\mathbb{X}$.

In the variant (**v**), the sequence $\bar{k}_h = \langle k_1, \dots, k_h \rangle$ should (for every $1 \leq h \leq m-1$) satisfy the condition ${}_{\bar{\mathbf{k}}_h}\mathbf{M}$ (at the second sub-step of each step of the algorithm).

Note also that ${}_{\bar{\mathbf{k}}_0}\mathbf{M}$ is an "empty" condition, which is always satisfied. This explains the absence of its symbol in the first line of the above diagram.

The following theorem summarises the remarks made above.

Theorem 9.3 *Let the conditions* **G**, **H** *be satisfied for the semi-Markov processes* $\eta_\varepsilon(t)$. *Then, for* $h = 0, \dots, m-1$:

(i) *The conditions* **G**, **H** *are satisfied for the reduced semi-Markov processes* ${}_{\bar{k}_n}\eta_\varepsilon(t)$ *and* ${}_{\bar{k}_n}\tilde{\eta}_\varepsilon(t)$ *in the form of conditions, respectively,* ${}_{\bar{\mathbf{k}}_h}\mathbf{G}$, ${}_{\bar{\mathbf{k}}_h}\mathbf{H}$ *and* ${}_{\bar{\mathbf{k}}_h}\tilde{\mathbf{G}}$, ${}_{\bar{\mathbf{k}}_h}\tilde{\mathbf{H}}$.

(ii) *If, in addition to the conditions* **G**, **H**, *the condition* **K (a)** *is satisfied for the semi-Markov processes* $\eta_\varepsilon(t)$, *then the condition* **K (a)** *is satisfied for the reduced* *semi-Markov processes* ${}_{\bar{k}_h}\eta_\varepsilon(t)$ *and* ${}_{\bar{k}_h}\tilde{\eta}_\varepsilon(t)$ *in the form of conditions, respectively,* ${}_{\bar{\mathbf{k}}_\mathbf{h}}\mathbf{K}$ **(a)** *and* ${}_{\bar{\mathbf{k}}_\mathbf{h}}\tilde{\mathbf{K}}$, **(a)**.

(iii) *If, in addition to the conditions* **G**, **H**, *the conditions* **I**, $\mathbf{I}_{\mathcal{H}}$ *are satisfied for the semi-Markov processes* $\eta_\varepsilon(t)$, *then the conditions* **I**, $\mathbf{I}_{\mathcal{H}}$ *are satisfied for the reduced semi-Markov processes* ${}_{\bar{k}_h}\eta_\varepsilon(t)$ *and* ${}_{\bar{k}_h}\tilde{\eta}_\varepsilon(t)$ *in the form of conditions, respectively,* ${}_{\bar{\mathbf{k}}_\mathbf{h}}\mathbf{I}$, ${}_{\bar{\mathbf{k}}_\mathbf{h}}\mathbf{I}_{\mathcal{H}}$, *and* ${}_{\bar{\mathbf{k}}_\mathbf{h}}\tilde{\mathbf{I}}$, ${}_{\bar{\mathbf{k}}_\mathbf{h}}\tilde{\mathbf{I}}_{\mathcal{H}}$.

(iv) *If, in addition to the conditions* **G**, **H**, *the conditions* **I**, $\mathbf{I}_{\mathcal{H}}$, **L**, $\mathbf{L}_{\mathcal{H}}$ *are satisfied for the semi-Markov processes* $\eta_\varepsilon(t)$, *then the conditions* **I**, $\mathbf{I}_{\mathcal{H}}$, **L**, $\mathbf{L}_{\mathcal{H}}$ *are satisfied for the reduced semi-Markov processes* ${}_{\bar{k}_h}\eta_\varepsilon(t)$ *and* ${}_{\bar{k}_h}\tilde{\eta}_\varepsilon(t)$ *in the form of conditions, respectively,* ${}_{\bar{\mathbf{k}}_\mathbf{h}}\mathbf{I}$, ${}_{\bar{\mathbf{k}}_\mathbf{h}}\mathbf{I}_{\mathcal{H}}$, ${}_{\bar{\mathbf{k}}_\mathbf{h}}\mathbf{L}$, ${}_{\bar{\mathbf{k}}_\mathbf{h}}\mathbf{L}_{\mathcal{H}}$, *and* ${}_{\bar{\mathbf{k}}_\mathbf{h}}\tilde{\mathbf{I}}$, ${}_{\bar{\mathbf{k}}_\mathbf{h}}\tilde{\mathbf{I}}_{\mathcal{H}}$, ${}_{\bar{\mathbf{k}}_\mathbf{h}}\tilde{\mathbf{L}}$, ${}_{\bar{\mathbf{k}}_\mathbf{h}}\tilde{\mathbf{L}}_{\mathcal{H}}$.

(v) *If, in addition to the conditions* **G**, **H**, *the conditions* **I**, $\mathbf{I}_{\mathcal{H}}$, **L**, $\mathbf{L}_{\mathcal{H}}$, **J**, **K** *and* ${}_{\bar{\mathbf{k}}_\mathbf{h}}\mathbf{M}$ *are satisfied for the semi-Markov processes* $\eta_\varepsilon(t)$, *then the conditions* **J**, **K** *are satisfied for the reduced semi-Markov processes* ${}_{\bar{k}_h}\eta_\varepsilon(t)$ *and* ${}_{\bar{k}_h}\tilde{\eta}_\varepsilon(t)$ *in the forms of conditions, respectively,* ${}_{\bar{\mathbf{k}}_\mathbf{h}}\mathbf{I}$, ${}_{\bar{\mathbf{k}}_\mathbf{h}}\mathbf{I}_{\mathcal{H}}$, ${}_{\bar{\mathbf{k}}_\mathbf{h}}\mathbf{L}$, ${}_{\bar{\mathbf{k}}_\mathbf{h}}\mathbf{L}_{\mathcal{H}}$, ${}_{\bar{\mathbf{k}}_\mathbf{h}}\mathbf{J}$, ${}_{\bar{\mathbf{k}}_\mathbf{h}}\mathbf{K}$ *and* ${}_{\bar{\mathbf{k}}_\mathbf{h}}\tilde{\mathbf{I}}$, ${}_{\bar{\mathbf{k}}_\mathbf{h}}\tilde{\mathbf{I}}_{\mathcal{H}}$, ${}_{\bar{\mathbf{k}}_\mathbf{h}}\tilde{\mathbf{L}}$, ${}_{\bar{\mathbf{k}}_\mathbf{h}}\tilde{\mathbf{L}}_{\mathcal{H}}$, ${}_{\bar{\mathbf{k}}_\mathbf{h}}\tilde{\mathbf{J}}$, ${}_{\bar{\mathbf{k}}_\mathbf{h}}\tilde{\mathbf{K}}$.

Chapter 10
Asymptotics of Hitting Times for Perturbed Semi-Markov Processes

This chapter plays a key role in Part II. Here we apply asymptotic recurrent phase space reduction algorithms for regularly and singularly perturbed finite semi-Markov processes to an asymptotic analysis of distributions of hitting times. It is important that the hitting times are asymptotically invariant in distribution with respect to the proposed procedures for reducing the phase space. We formulate conditions that ensure that the basic perturbation conditions imposed on the original semi-Markov processes are also satisfied for semi-Markov processes with reduced phase spaces. We also give recurrent formulas for recalculating normalisation functions, limiting distributions and expectations appearing in the corresponding perturbation conditions for semi-Markov processes with reduced phase spaces. Acting in this way, we recursively reduce the asymptotic analysis to the case, where the hitting time coincides with the first transition time for the corresponding reduced semi-Markov processes and obtain theorems on weak convergence of hitting times for regularly and singularly perturbed finite semi-Markov processes. This chapter includes three sections.

In Sect. 10.1, we introduce hitting times, present variants of asymptotic recurrent algorithms of phase space reduction suitable for asymptotic analysis of hitting times for perturbed semi-Markov processes, and summarise the corresponding results in Theorem 10.1. In Lemmas 10.1–10.12, we describe the invariance properties of hitting times with respect to the above algorithms of phase space reduction and present recurrent formulas connecting distributions and Laplace transforms of hitting times for reduced semi-Markov processes.

In Sect. 10.2, we present the results of an asymptotic analysis of distributions of hitting times for regularly and singularly perturbed semi-Markov processes. Theorems 10.2–10.4 present results on the weak convergence of the distributions of hitting times $\tau_{\varepsilon,\mathbb{D}}$ in a domain $\mathbb{D}$ for the case, where the initial state of the indicated semi-Markov processes belongs to the domain $\bar{\mathbb{D}}$.

In Sect. 10.3, we present results of an asymptotic analysis of distributions of return times for regularly and singularly perturbed semi-Markov processes. Theorems 10.5 and 10.6 present results on the weak convergence of the distributions of return

D. Silvestrov, *Perturbed Semi-Markov Type Processes I*,
https://doi.org/10.1007/978-3-030-92403-4_10

times, that is, hitting times $\tau_{\varepsilon,\mathbb{D}}$ for the case, where the initial state of the indicated semi-Markov processes belongs to the domain $\mathbb{D}$.

10.1 Hitting Times for Perturbed Semi-Markov Processes

In this section, we introduce hitting times and present variants of asymptotic recurrent algorithms of phase space reduction adapted to the asymptotic analysis of hitting times for perturbed semi-Markov processes.

10.1.1 Hitting Times and Related Asymptotic Problems

10.1.1.1 Hitting Times. Let $\mathbb{D}$ be a non-empty subset of $\mathbb{X}$ and $\varepsilon \in (0, 1]$.

The object of our interest are random functionals,

$$\tau_{\varepsilon,\mathbb{D}} = \sum_{n=1}^{\nu_{\varepsilon,\mathbb{D}}} \kappa_{\varepsilon,n}, \tag{10.1}$$

where

$$\nu_{\varepsilon,\mathbb{D}} = \min(n \geq 1 : \eta_{\varepsilon,n} \in \mathbb{D}). \tag{10.2}$$

The random functionals $\nu_{\varepsilon,\mathbb{D}}$ and $\tau_{\varepsilon,\mathbb{D}}$ are the first hitting times in the domain $\mathbb{D}$, respectively, for the embedded Markov chain $\eta_{\varepsilon,n}$ and the semi-Markov process $\eta_\varepsilon(t)$.

We assume that the basic regularity conditions **G** and **H** are satisfied for the semi-Markov processes $\eta_\varepsilon(t)$.

Let us introduce, for $\varepsilon \in (0, 1]$ and $i \in \mathbb{X}$, $j \in \mathbb{D}$, distributions,

$$G_{\varepsilon,\mathbb{D},ij}(t) = \mathsf{P}_i\{\tau_{\varepsilon,\mathbb{D}} \leq t, \eta_\varepsilon(\tau_{\varepsilon,\mathbb{D}}) = j\},\ t \geq 0, \text{ for } j \in \mathbb{D}, \tag{10.3}$$

Laplace transforms,

$$\Psi_{\varepsilon.\mathbb{D},ij}(s) = \int_0^\infty e^{-st} G_{\varepsilon,\mathbb{D},ij}(dt),\ s \geq 0, \tag{10.4}$$

expectations,

$$E_{\varepsilon,\mathbb{D},ij} = \mathsf{E}_i \tau_{\varepsilon,\mathbb{D}} \mathrm{I}(\eta_\varepsilon(\tau_{\varepsilon,\mathbb{D}}) = j), \tag{10.5}$$

and hitting probabilities,

$$P_{\varepsilon,\mathbb{D},ij} = \mathsf{P}_i\{\eta_\varepsilon(\tau_{\varepsilon,\mathbb{D}}) = j\} = G_{\varepsilon,\mathbb{D},ij}(\infty). \tag{10.6}$$

Let us also introduce conditional distribution functions,

$$F_{\varepsilon,\mathbb{D},ij}(t) = \mathsf{P}_i\{\tau_{\varepsilon,\mathbb{D}} \leq t/\eta_\varepsilon(\tau_{\varepsilon,\mathbb{D}}) = j\},\ t \geq 0. \tag{10.7}$$

We define sets,

$$\mathbb{Y}_{\varepsilon,\mathbb{D},i} = \{j \in \mathbb{D} : P_{\varepsilon,\mathbb{D},ij} > 0\}. \tag{10.8}$$

As usual,

$$F_{\varepsilon,\mathbb{D},ij}(\cdot) = \begin{cases} P^{-1}_{\varepsilon,\mathbb{D},ij} G_{\varepsilon,\mathbb{D},ij}(\cdot) & \text{if } j \in \mathbb{Y}_{\varepsilon,\mathbb{D},i}, \\ F_{\varepsilon,\mathbb{D},i}(\cdot) & \text{if } j \notin \mathbb{Y}_{\varepsilon,\mathbb{D},i}, \end{cases} \tag{10.9}$$

where,

$$\begin{aligned} F_{\varepsilon,\mathbb{D},i}(\cdot) &= \sum_{j \in \mathbb{D}} G_{\varepsilon,\mathbb{D},ij}(\cdot) \\ &= \sum_{r \in \mathbb{Y}_{\varepsilon,\mathbb{D},i}} F_{\varepsilon,\mathbb{D},ir}(\cdot) P_{\varepsilon,\mathbb{D},ir},\ t \geq 0. \end{aligned} \tag{10.10}$$

Let us also introduce Laplace transforms,

$$\Phi_{\varepsilon,\mathbb{D},ij}(s) = \int_0^\infty e^{-st} F_{\varepsilon,\mathbb{D},ij}(dt),\ s \geq 0, \text{ for } j \in \mathbb{D}. \tag{10.11}$$

Obviously,

$$\Phi_{\varepsilon,\mathbb{D},ij}(\cdot) = \begin{cases} P^{-1}_{\varepsilon,\mathbb{D},ij} \Psi_{\varepsilon,\mathbb{D},ij}(\cdot) & \text{if } j \in \mathbb{Y}_{\varepsilon,\mathbb{D},i}, \\ \Phi_{\varepsilon,\mathbb{D},i}(\cdot) & \text{if } j \notin \mathbb{Y}_{\varepsilon,\mathbb{D},i}, \end{cases} \tag{10.12}$$

where,

$$\begin{aligned} \Phi_{\varepsilon,\mathbb{D},i}(\cdot) &= \sum_{j \in \mathbb{D}} \Psi_{\varepsilon,\mathbb{D},ij}(\cdot) \\ &= \sum_{r \in \mathbb{Y}_{\varepsilon,\mathbb{D},i}} \Phi_{\varepsilon,\mathbb{D},ir}(\cdot) P_{\varepsilon,\mathbb{D},ir},\ s \geq 0. \end{aligned} \tag{10.13}$$

Finally, let us introduce conditional expectations,

$$F_{\varepsilon,\mathbb{D},ij} = \int_0^\infty t F_{\varepsilon,\mathbb{D},ij}(dt),\ s \geq 0, \text{ for } j \in \mathbb{D}. \tag{10.14}$$

Obviously,

$$F_{\varepsilon,\mathbb{D},ij} = \begin{cases} P^{-1}_{\varepsilon,\mathbb{D},ij} E_{\varepsilon,\mathbb{D},ij} & \text{if } j \in \mathbb{Y}_{\varepsilon,\mathbb{D},i}, \\ F_{\varepsilon,\mathbb{D},i} & \text{if } j \notin \mathbb{Y}_{\varepsilon,\mathbb{D},i}, \end{cases} \tag{10.15}$$

where,

$$F_{\varepsilon,\mathbb{D},i} = \sum_{j \in \mathbb{D}} E_{\varepsilon,\mathbb{D},ij} = \sum_{r \in \mathbb{Y}_{\varepsilon,\mathbb{D},i}} F_{\varepsilon,\mathbb{D},ir} P_{\varepsilon,\mathbb{D},ir}. \tag{10.16}$$

10.1.1.2 Asymptotic Problems for Hitting Times and Related Functionals. We are interested in finding conditions that entail the fulfilment of the following asymptotic relations, for $i \in \mathbb{X}$, $j \in \mathbb{D}$,

$$\begin{aligned} P_{\varepsilon,\mathbb{D},ij} &= G_{\varepsilon,\mathbb{D},ij}(\infty) \\ &\to G_{0,\mathbb{D},ij}(\infty) = P_{0,\mathbb{D},ij} \text{ as } \varepsilon \to 0 \end{aligned} \tag{10.17}$$

and

$$\begin{aligned} G_{\varepsilon,\mathbb{D},ij}(\cdot\, \check{u}_{\varepsilon,i}) &= F_{\varepsilon,\mathbb{D},ij}(\cdot\, \check{u}_{\varepsilon,i}) P_{\varepsilon,\mathbb{D},ij} \\ &\Rightarrow G_{0,\mathbb{D},ij}(\cdot) = F_{0,\mathbb{D},ij}(\cdot) P_{0,\mathbb{D},ij} \text{ as } \varepsilon \to 0, \end{aligned} \tag{10.18}$$

where: (a) $P_{0,\mathbb{D},ij}, j \in \mathbb{D}$ is a discrete distribution, i.e. $P_{0,\mathbb{D},ij} \geq 0, j \in \mathbb{D}$ and $\sum_{j \in \mathbb{D}} P_{0,\mathbb{D},ij} = 1$, for $i \in \mathbb{X}$, (b) $F_{0,\mathbb{D},ij}(\cdot)$ is a proper distribution function, i.e., $F_{0,\mathbb{D},ij}(\infty) = 1$, for $i \in \mathbb{X}, j \in \mathbb{D}$, (c) $\check{u}_{\cdot,i} >, i \in \mathbb{X}$ are normalisation functions such that $\check{u}_{\varepsilon,i} > 0, \varepsilon \in (0, 1]$, for $i \in \mathbb{X}$.

The relations (10.17) and (10.18) imply that the following weak convergence relation holds, for $i \in \mathbb{X}$:

$$\begin{aligned} F_{\varepsilon,\mathbb{D},i}(\cdot\, \check{u}_{\varepsilon,i}) &= \sum_{j \in \mathbb{D}} G_{\varepsilon,\mathbb{D},ij}(\cdot\, \check{u}_{\varepsilon,i}) \\ &\Rightarrow \sum_{j \in \mathbb{D}} G_{0,\mathbb{D},ij}(\cdot) = F_{0,\mathbb{D},i}(\cdot) \text{ as } \varepsilon \to 0, \end{aligned} \tag{10.19}$$

where

$$F_{0,\mathbb{D},i}(\cdot) = \sum_{j \in \mathbb{D}} F_{0,\mathbb{D},ij}(\cdot) P_{0,\mathbb{D},ij}. \tag{10.20}$$

We are especially interested to clarify conditions, under which the corresponding limiting distributions $F_{0,\mathbb{D},i}(\cdot)$, $i \in \mathbb{X}$ are not concentrated at zero, i.e., for $i \in \mathbb{X}$,

$$F_{0,\mathbb{D},i}(0) < 1. \tag{10.21}$$

The relation (10.21) means, in fact, that the normalisation functions $\check{u}_{\cdot,i}, i \in \mathbb{X}$ are chosen correctly in the sense that the hitting times normalised by these functions do not converge in probability to 0.

It is useful noting that the relations (10.17) and (10.18) are equivalent to the following relation for Laplace transforms, for $i \in \mathbb{X}$, $j \in \mathbb{D}$:

$$\begin{aligned} \Psi_{\varepsilon,\mathbb{D},ij}(s/\check{u}_{\varepsilon,i}) &\to \Psi_{0,\mathbb{D},ij}(s) \\ &= \Phi_{0,\mathbb{D},ij}(s) P_{0,\mathbb{D},ij} \text{ as } \varepsilon \to 0, \text{ for } s \geq 0, \end{aligned} \tag{10.22}$$

where: (a) $\Phi_{0,\mathbb{D},ij}(s) = \int_0^\infty e^{-st} F_{0,\mathbb{D},ij}(dt), s \geq 0$ are Laplace transforms of proper distribution functions $F_{0,\mathbb{D},ij}(\cdot)$ (such that $F_{0,\mathbb{D},ij}(\infty) = 1$), for $i \in \mathbb{X}, j \in \mathbb{D}$, (b) $P_{0,\mathbb{D},ij}, j \in \mathbb{D}$ is a discrete distribution, i.e., $P_{0,\mathbb{D},ij} \geq 0, j \in \mathbb{D}$ and $\sum_{j \in \mathbb{D}} P_{0,\mathbb{D},ij} = 1$, for $i \in \mathbb{X}$.

The relation (10.22) implies that the following relation holds, for $i \in \mathbb{X}$:

$$\begin{aligned}\Phi_{\varepsilon,\mathbb{D},i}(s/\check{u}_{\varepsilon,i}) &= \sum_{j\in\mathbb{D}} \Psi_{\varepsilon,\mathbb{D},ij}(s/\check{u}_{\varepsilon,i}) \\ &\to \Phi_{0,\mathbb{D},i}(s) = \sum_{j\in\mathbb{D}} \Psi_{0,\mathbb{D},ij}(s) \\ &= \sum_{j\in\mathbb{D}} \Phi_{0,\mathbb{D},ij}(s)P_{0,\mathbb{D},ij} \text{ as } \varepsilon \to 0, \text{ for } s \geq 0. \end{aligned} \tag{10.23}$$

It is also useful noting that the relation (10.21) can be expressed in the equivalent form in terms of the corresponding Laplace transforms, for $i \in \mathbb{X}$,

$$\Phi_{0,\mathbb{D},i}(s) < 1, \text{ for } s > 0. \tag{10.24}$$

We also would like to find conditions, which would imply that there exist normalisation functions $\bar{u}_{\cdot,i} > 0, i \in \mathbb{X}$ such that the following convergence relation holds, for $i \in \mathbb{X}, j \in \mathbb{D}$:

$$\bar{u}_{\varepsilon,i}^{-1} E_{\varepsilon,\mathbb{D},ij} \to \bar{E}_{0,\mathbb{D},ij} < \infty \text{ as } \varepsilon \to 0. \tag{10.25}$$

We are going to present conditions insuring holding of the asymptotic relations (10.18), (10.22), and (10.25) and to construct effective recurrent algorithms, which would allow us to find the above normalisation functions, the Laplace transforms $\Psi_{0,\mathbb{D},ij}(s) = \int_0^\infty e^{-st} G_{0,\mathbb{D},ij}(dt), s \geq 0$, the limits of expectations $\bar{E}_{0,\mathbb{D},ij}$, and the limiting hitting probabilities $P_{0,\mathbb{D},ij}$, for $i \in \mathbb{X}, j \in \mathbb{D}$.

We are also interested to find additional conditions, under which the following asymptotic relation holds, for $i \in \mathbb{X}, j \in \mathbb{D}$:

$$\bar{u}_{\varepsilon,i}/\check{u}_{\varepsilon,i} \to C_i \in (0, \infty) \text{ as } \varepsilon \to 0, \tag{10.26}$$

and the limits for expectations satisfy the following equalities, for $i \in \mathbb{X}, j \in \mathbb{D}$:

$$C_i \bar{E}_{0,\mathbb{D},ij} = E_{0,\mathbb{D},ij} = \int_0^\infty t G_{0,\mathbb{D},ij}(dt). \tag{10.27}$$

The relations (10.25), (10.26), (10.27) imply that, for $i \in \mathbb{X}, j \in \mathbb{D}$,

$$\check{u}_{\varepsilon,i}^{-1} E_{\varepsilon,\mathbb{D},ij} \to E_{0,\mathbb{D},ij} < \infty \text{ as } \varepsilon \to 0. \tag{10.28}$$

The relation (10.28) means that the expectations $E_{\varepsilon,\mathbb{D},ij}$, normalised by the function $\check{u}_{\varepsilon,i}$ (which appears in the weak convergence relation (10.18) for the distributions $G_{\varepsilon,\mathbb{D},ij}(\cdot\, \check{u}_{\varepsilon,k_{\bar{m}}}))$, converge to the first moment $E_{0,\mathbb{D},ij}$ of the corresponding limiting distribution $G_{0,\mathbb{D},ij}(\cdot)$, as $\varepsilon \to 0$, for $i \in \mathbb{X}, j \in \mathbb{D}$.

There exist cases, where the weak convergence relation (10.18) and the convergence relation (10.25) hold, while the asymptotic relations (10.26) and (10.28) do not hold, and, for example, $\check{u}_{\varepsilon,i}^{-1} E_{\varepsilon,\mathbb{D},ij} \to \infty$ as $\varepsilon \to 0$.

The conditions **G** and **H** guarantee that, for any non-empty domain $\mathbb{D} \subseteq \mathbb{X}$, the initial state $i \in \mathbb{X}$ and $\varepsilon \in (0, 1]$,

$$\mathsf{P}_i\{\nu_{\varepsilon,\mathbb{D}} < \infty\} = \mathsf{P}_i\{\tau_{\varepsilon,\mathbb{D}} < \infty\} = 1. \tag{10.29}$$

The case where $\mathbb{D} = \mathbb{X}$ is trivial, since $\nu_{\varepsilon,\mathbb{X}} = 1$ and $\tau_{\varepsilon,\mathbb{X}} = \kappa_{\varepsilon,1}$. In this case, the asymptotic relations given in the conditions formulated in Sect. 9.1 solve the asymptotic problems formulated above.

Thus, we assume that $\mathbb{D} \subset \mathbb{X}$, and, thus, $m \geq 2$ and $1 \leq \bar{m} < m$, where $\bar{m}$ is the number of states in the domain $\bar{\mathbb{D}}$.

10.1.2 Hitting Times and Reduced Perturbation Conditions for Semi-Markov Processes

We study the asymptotics of hitting times for some non-empty domain $\mathbb{D} \subset \mathbb{X}$. The asymptotic problems formulated above have different solutions for two cases where the initial state $i \in \bar{\mathbb{D}}$ or $i \in \mathbb{D}$. The main is the former case.

The following obvious lemma plays an important role in what follows.

Lemma 10.1 *The distributions $G_{\varepsilon,\mathbb{D},ij}(t), t \geq 0, j \in \mathbb{D}, i \in \bar{\mathbb{D}}$ are, for every $\varepsilon \in (0, 1]$, completely determined by the transition probabilities $Q_{\varepsilon,kr}(t), t \geq 0, r \in \mathbb{X}, k \in \bar{\mathbb{D}}$ and, thus, the distributions $G_{\varepsilon,\mathbb{D},ij}(t), t \geq 0, j \in \mathbb{D}, i \in \bar{\mathbb{D}}$ coincide for semi-Markov processes with any form of transition probabilities $Q_{\varepsilon,kr}(t), t \geq 0, r \in \mathbb{X}, k \in \mathbb{D}$.*

This lemma lets us simplify the model in the case, when we are interested in studying the asymptotics of hitting times for some fixed domain $\mathbb{D}$ and only for the initial states $i \in \bar{\mathbb{D}}$. This case is the subject of our study in Sects. 10.1 and 10.2

In what follows, we assume that the transition probabilities of the semi-Markov processes $\eta_\varepsilon(t)$ satisfy the following condition:

$\mathbf{N}_{\mathbb{D}}$: $Q_{\varepsilon,kr}(t) = \frac{1}{m}\mathrm{I}(t \geq u_\varepsilon)$, for $t \geq 0, r \in \mathbb{X}, k \in \mathbb{D}$ and $\varepsilon \in (0, 1]$.

Here $u_\varepsilon, \varepsilon \in (0, 1]$ is some function taking values in interval $(0, \infty)$ such that $u_\varepsilon \longrightarrow \infty$ as $\varepsilon \longrightarrow 0$. We specify this function later.

The condition $\mathbf{N}_{\mathbb{D}}$ implies that the transition probabilities $p_{\varepsilon,ij} \equiv \frac{1}{m}, j \in \mathbb{X}$, for $i \in \mathbb{D}$ and, thus, the set $\mathbb{Y}_{1,i} = \mathbb{X}$, for $i \in \mathbb{D}$. Also, the distribution function $F_{\varepsilon,ij}(t) = F_{\varepsilon,i}(t) = \mathrm{I}(t \geq u_\varepsilon), t \geq 0$, for $j \in \mathbb{X}, i \in \mathbb{D}$.

Obviously, the relations appearing in the condition **G** hold for the above transition probabilities $p_{\varepsilon,ij}$ and distribution functions $F_{\varepsilon,i}(t)$, for $i \in \mathbb{D}$. The condition **G** can, in fact, be reduced to the following form:

$\mathbf{G}_{\mathbb{D}}$: **(a)** $p_{\varepsilon,ij} > 0, \varepsilon \in (0, 1]$ or $p_{\varepsilon,ij} = 0, \varepsilon \in (0, 1]$, for every $j \in \mathbb{X}, i \in \bar{\mathbb{D}}$, **(b)** $F_{\varepsilon,i}(0) < 1, i \in \bar{\mathbb{D}}, \varepsilon \in (0, 1]$.

Thus, the conditions $\mathbf{G}_{\mathbb{D}}$ and $\mathbf{N}_{\mathbb{D}}$ imply that the condition **G** is satisfied.

The conditions $\mathbf{G}_{\mathbb{D}}$ and $\mathbf{N}_{\mathbb{D}}$ make it possible to reduce the condition **H** to the following form:

$\mathbf{H}_{\mathbb{D}}$: For any $i \in \bar{\mathbb{D}}$, there exists a chain of states $i = j_0, j_1 \in \bar{\mathbb{D}}, \ldots, j_{n_i-1} \in \bar{\mathbb{D}}, j_{n_i} \in \mathbb{D}$ such that $\prod_{1 \leq l \leq n_i} p_{1,j_{l-1}j_l} > 0$.

Indeed, conditions $\mathbf{G}_{\mathbb{D}}$ and $\mathbf{N}_{\mathbb{D}}$ imply that, $p_{1,ij} = \frac{1}{m}$, for any $i \in \mathbb{D}, j \in \mathbb{X}$, and $\prod_{1 \leq l \leq n_i} p_{1,j_{l-1}j_l} p_{1,j_{n_i},j} = \prod_{1 \leq l \leq n_i} p_{1,j_{l-1}j_l} \frac{1}{m} > 0$ for any $i \in \bar{\mathbb{D}}, j \in \mathbb{X}$ and the chains of states appearing in the condition $\mathbf{H}_{\mathbb{D}}$.

Thus, the condition $\mathbf{H}$ is satisfied.

The transition probabilities $p_{\varepsilon,ij} \equiv \frac{1}{m}, j \in \mathbb{X}, i \in \mathbb{D}$ belong to any family of asymptotically comparable functions. The conditions $\mathbf{I}$ and $\mathbf{I}_{\mathcal{H}}$ can be reduced to the following forms:

$\mathbf{I}_{\mathbb{D}}$: $p_{\varepsilon,ij} \to p_{0,ij}$ as $\varepsilon \to 0$, for $j \in \mathbb{X}, i \in \bar{\mathbb{D}}$

and

$\mathbf{I}_{\mathbb{D},\mathcal{H}}$: The functions $p_{\cdot,ij}, j \in \mathbb{Y}_{1,i}, i \in \bar{\mathbb{D}}$ belong to a complete family of asymptotically comparable functions $\mathcal{H}$.

Obviously, the distribution functions $F_{\varepsilon,ij}(tu_\varepsilon) = \mathrm{I}(t \geq 1), t \geq 0$, for $\varepsilon \in (0, 1]$, $j \in \mathbb{X}, i \in \mathbb{D}$ and, thus, the asymptotic relation appearing in the condition $\mathbf{J}$ holds, with the normalisation function $u_{\varepsilon,i} = u_\varepsilon, \varepsilon \in (0, 1]$, for $i \in \mathbb{D}$ and the limiting distribution function $F_{0,ij}(t) = \mathrm{I}(t \geq 1), t \geq 0$, for $j \in \mathbb{X}, i \in \mathbb{D}$. The condition $\mathbf{J}$ can be reduced to the following form:

$\mathbf{J}_{\mathbb{D}}$: **(a)** $F_{\varepsilon,ij}(\cdot\, u_{\varepsilon,i}) = \mathsf{P}\{\kappa_{\varepsilon,1}/u_{\varepsilon,i} \leq \cdot/\eta_{\varepsilon,0} = i, \eta_{\varepsilon,1} = j\} \Rightarrow F_{0,ij}(\cdot)$ as $\varepsilon \to 0$, for $j \in \mathbb{Y}_{1,i}, i \in \bar{\mathbb{D}}$, **(b)** $F_{0,ij}(\cdot), j \in \mathbb{Y}_{1,i}, i \in \bar{\mathbb{D}}$ are proper distribution functions such that $F_{0,ij}(0) < 1, j \in \mathbb{Y}_{1,i}, i \in \bar{\mathbb{D}}$, **(c)** $u_{\varepsilon,i} \in (0, \infty), \varepsilon \in (0, 1]$, for $i \in \bar{\mathbb{D}}$

or equivalently:

$\mathbf{J}^{\circ}_{\mathbb{D}}$: **(a)** $\phi_{\varepsilon,ij}(s/u_{\varepsilon,i}) \to \phi_{0,ij}(s)$ as $\varepsilon \to 0$, for $s \geq 0$ and $j \in \mathbb{Y}_{1,i},\ i \in \bar{\mathbb{D}}$, **(b)** $\phi_{0,ij}(s) = \int_0^\infty e^{-st} F_{0,ij}(dt),\ s \geq 0,\ j \in \mathbb{Y}_{1,i},\ i \in \bar{\mathbb{D}}$ are Laplace transforms of proper distribution functions such that $F_{0,ij}(0) < 1, j \in \mathbb{Y}_{1,i}, i \in \bar{\mathbb{D}}$, **(c)** $u_{\varepsilon,i} \in (0, \infty), \varepsilon \in (0, 1]$, for $i \in \bar{\mathbb{D}}$.

Also, the expectations $f_{\varepsilon,ij} = u_\varepsilon$, for $\varepsilon \in (0, 1],\ j \in \mathbb{X}, i \in \mathbb{D}$ and, thus, the asymptotic relation appearing in the condition $\mathbf{K}$ holds, with the normalisation function $u_{\varepsilon,i} = u_\varepsilon, \varepsilon \in (0, 1]$, for $i \in \mathbb{D}$ and the limiting expectation $f_{0,ij} = 1$, for $j \in \mathbb{X}, i \in \mathbb{D}$. The condition $\mathbf{K}$ can be reduced to the following form:

$\mathbf{K}_{\mathbb{D}}$: **(a)** $f_{\varepsilon,ij} < \infty,\ j \in \mathbb{Y}_{1,i},\ i \in \bar{\mathbb{D}}$, for every $\varepsilon \in (0, 1]$, **(b)** $f_{\varepsilon,ij}/u_{\varepsilon,i} \to f_{0,ij} = \int_0^\infty t F_{0,ij}(dt) < \infty$ as $\varepsilon \to 0$, for $j \in \mathbb{Y}_{1,i}, i \in \bar{\mathbb{D}}$.

Finally, let us formulate the reduced forms of conditions $\mathbf{L}$ and $\mathbf{L}_{\mathcal{H}}$:

$\mathbf{L}_{\mathbb{D}}$: $u_{\varepsilon,i} \to u_{0,i} \in (0, \infty]$ as $\varepsilon \to 0$, for $i \in \bar{\mathbb{D}}$

and

$\mathbf{L}_{\mathbb{D},\mathcal{H}}$: Functions $u_{\cdot,i},\ i \in \bar{\mathbb{D}}$ belong to the complete family of asymptotically comparable functions $\mathcal{H}$ appearing in the condition $\mathbf{I}_{\mathbb{D},\mathcal{H}}$.

Recall that we denoted by $\bar{m}$ the number of states in domain $\bar{\mathbb{D}} = \{k_1, \ldots, k_m\}$.

We would like to realise the algorithm of sequential reduction of phase space $\mathbb{X}$ for the semi-Markov process $\eta_\varepsilon(t)$ by excluding states $k_1, \ldots, k_{\bar{m}-1} \in \bar{\mathbb{D}}$, so in the end we would get the semi-Markov process ${}_{\bar{k}_{\bar{m}-}}\eta_\varepsilon(t)$ and ${}_{\bar{k}_{\bar{m}-1}}\tilde{\eta}_\varepsilon(t)$, with the phase space ${}_{\bar{k}_{\bar{m}-1}}\mathbb{X} = \mathbb{D} \cup \{k_{\bar{m}}\}$.

The normalisation function u_ε appearing in the condition $\mathbf{N}_{\mathbb{D}}$ can always be chosen in such a way that, for $i \in {}_{\bar{k}_{r-1}}\bar{\mathbb{D}}$, $r = 1, \ldots, \bar{m}$,

$$\lim_{\varepsilon \to 0} \frac{{}_{\bar{k}_{r-1}}\tilde{u}_{\varepsilon,i}}{u_\varepsilon} = 0, \tag{10.30}$$

where

$${}_{\bar{k}_{r-1}}\bar{\mathbb{D}} = \bar{\mathbb{D}} \setminus \{k_1, \ldots, k_{r-1}\}. \tag{10.31}$$

In this case, the sets ${}_{\bar{k}_{l-1}}\mathbb{W}_0 \subseteq \bar{\mathbb{D}}$, for $l = 1, \ldots, \bar{m}$ and, thus, the conditions ${}_{\bar{\mathbf{k}}_\mathbf{n}}\mathbf{M}$, $n = 1, \ldots, \bar{m} - 1$ will take the following reduced form:

${}_{\bar{\mathbf{k}}_\mathbf{n}}\mathbf{M}_{\mathbb{D}}$: $k_l \in {}_{\bar{k}_{l-1}}\mathbb{W}_{\mathbb{D},0}$, i.e., ${}_{\bar{k}_{l-1}}w_{0,k_l,i} = \lim_{\varepsilon \to 0} \frac{{}_{\bar{k}_{l-1}}\tilde{u}_{\varepsilon,k_l}}{{}_{\bar{k}_{l-1}}\tilde{u}_{\varepsilon,i}} \in [0, \infty)$, $i \in {}_{\bar{k}_{l-1}}\bar{\mathbb{D}}$, for $l = 1, \ldots, n$.

In the above condition, we use the sets of the least absorbing states,

$${}_{\bar{k}_{l-1}}\mathbb{W}_{\mathbb{D},0} = \{k \in {}_{\bar{k}_{l-1}}\bar{\mathbb{D}} : {}_{\bar{k}_{l-1}}w_{0,k_l,i} \in [0, \infty), \, i \in {}_{\bar{k}_{l-1}}\bar{\mathbb{D}}\}, \ l = 1, \ldots, n. \tag{10.32}$$

Three cases should be considered, in order to prove the statement formulated in the relation (10.30).

(1) The function $\tilde{u}_\varepsilon = \sum_{r=1}^{\bar{m}} \sum_{i \in {}_{\bar{k}_{r-1}}\mathbb{X}} {}_{\bar{k}_{r-1}}\tilde{u}_{\varepsilon,i} \to \tilde{u}_0 = \infty$ as $\varepsilon \to 0$. In this case, we can choose $u_\varepsilon = \tilde{u}_\varepsilon^2$. In this case, the relation (10.30) obviously holds.

(2) The function $\tilde{u}_\varepsilon \to \tilde{u}_0 \in (0, \infty)$ as $\varepsilon \to 0$, but the family $\mathcal{H}$ is non-trivial, i.e., it includes a function $h(\varepsilon)$ such that the corresponding comparability limit $\lim_{\varepsilon \to 0} h(\varepsilon) = a_{\mathcal{H}}[h(\cdot)] = \infty$. In this case, we can choose $u_\varepsilon = h(\cdot)$. Again, the relation (10.30) holds.

(3) The family $\mathcal{H}$ is trivial, i.e., the corresponding comparability limit, $a_{\mathcal{H}}[h(\cdot)] = \lim_{\varepsilon \to 0} h(\varepsilon) \in (0, \infty)$, for any function $h(\cdot) \in \mathcal{H}$. In this case, we can choose any function $u_\varepsilon \to \infty$ as $\varepsilon \to 0$, and then replace the family $\mathcal{H}$ appearing in the conditions $\mathbf{I}_{\mathbb{D}}$ and $\mathbf{L}_{\mathbb{D}}$ by the new family of asymptotically comparable functions $\tilde{\mathcal{H}} = [\mathcal{H} \cup \{u_\cdot\}]$ (see Sect. 8.1.1). In this case, the relation (10.30) also holds.

The semi-Markov processes ${}_{\bar{k}_r}\eta_\varepsilon(t)$ and ${}_{\bar{k}_r}\tilde{\eta}_\varepsilon(t)$ have the phase space ${}_{\bar{k}_r}\mathbb{X} = \mathbb{D} \cup {}_{\bar{k}_r}\bar{\mathbb{D}}$, for $r = 0, \ldots, \bar{m} - 1$.

Importantly, their transition probabilities ${}_{\bar{k}_r}Q_{\varepsilon,ij}(t), t \geq 0, j \in {}_{\bar{k}_r}\mathbb{X}, i \in {}_{\bar{k}_r}\bar{\mathbb{D}}$ and ${}_{\bar{k}_r}\tilde{Q}_{\varepsilon,ij}(t), t \geq 0, j \in {}_{\bar{k}_r}\mathbb{X}, i \in_{\bar{k}_r} \bar{\mathbb{D}}$, for $r = 0, \ldots, \bar{m}$, are determined by the initial transition probabilities $Q_{\varepsilon,ij}(t), t \geq 0, j \in \mathbb{X}, i \in \bar{\mathbb{D}}$.

Therefore, the only transition probabilities $Q_{\varepsilon,ij}(t), t \geq 0, j \in \mathbb{X}, i \in \bar{\mathbb{D}}$ and the normalisation functions $u_{\varepsilon,i}, i \in \bar{\mathbb{D}}$ are used in the relations appearing in the conditions ${}_{\bar{\mathbf{k}}_\mathbf{r}}\mathbf{G}_{\mathbb{D}}$, ${}_{\bar{\mathbf{k}}_\mathbf{r}}\mathbf{H}_{\mathbb{D}}$, ${}_{\bar{\mathbf{k}}_\mathbf{r}}\mathbf{I}_{\mathbb{D},\mathcal{H}}$, ${}_{\bar{\mathbf{k}}_\mathbf{r}}\mathbf{I}_{\mathbb{D}}$, ${}_{\bar{\mathbf{k}}_\mathbf{r}}\mathbf{J}_{\mathbb{D}}$, ${}_{\bar{\mathbf{k}}_\mathbf{r}}\mathbf{K}_{\mathbb{D}}$, ${}_{\bar{\mathbf{k}}_\mathbf{r}}\mathbf{L}_{\mathbb{D},\mathcal{H}}$, ${}_{\bar{\mathbf{k}}_\mathbf{r}}\mathbf{L}_{\mathbb{D}}$, ${}_{\bar{\mathbf{k}}_\mathbf{r}}\tilde{\mathbf{G}}_{\mathbb{D}}$, ${}_{\bar{\mathbf{k}}_\mathbf{r}}\tilde{\mathbf{H}}_{\mathbb{D}}$, ${}_{\bar{\mathbf{k}}_\mathbf{r}}\tilde{\mathbf{I}}_{\mathbb{D},\mathcal{H}}$, ${}_{\bar{\mathbf{k}}_\mathbf{r}}\tilde{\mathbf{I}}_{\mathbb{D}}$, ${}_{\bar{\mathbf{k}}_\mathbf{r}}\tilde{\mathbf{J}}_{\mathbb{D}}$, ${}_{\bar{\mathbf{k}}_\mathbf{r}}\tilde{\mathbf{K}}_{\mathbb{D}}$, ${}_{\bar{\mathbf{k}}_\mathbf{r}}\tilde{\mathbf{L}}_{\mathbb{D},\mathcal{H}}$, ${}_{\bar{\mathbf{k}}_\mathbf{r}}\tilde{\mathbf{L}}_{\mathbb{D}}$, and ${}_{\bar{\mathbf{k}}_\mathbf{r}}\mathbf{M}_{\mathbb{D}}$, for $r = 0, 1, \ldots, \bar{m} - 1$.

Therefore, it is possible to use the family of asymptotically comparable functions $\mathcal{H}$ instead of $\tilde{\mathcal{H}}$ in the formulations of the above conditions, in particular, in the relations appearing in these conditions and the formulas for the quantities appearing in these relations, for the initial states $i \in \bar{\mathbb{D}}$. Also, for the initial states $i \in \bar{\mathbb{D}}$, it is possible to use the corresponding analogues of Lemmas 9.1–9.25.

As far as the initial states $i \in \mathbb{D}$ are concerned, the family of asymptotically comparable functions $\tilde{\mathcal{H}}$ and the normalisation function u_ε should be used. As it is pointed above, the condition $\mathbf{N}_{\mathbb{D}}$ implies that the corresponding relations appearing in the conditions **G**, **H**, $\mathbf{I}_{\mathbb{D}}$, **I**, **J**, **K**, **L**, and $\mathbf{L}_{\mathbb{D}}$ are automatically satisfied, for $i \in \mathbb{D}$. Therefore, the corresponding relations appearing in Lemmas 9.1–9.25 hold, for $i \in \mathbb{D}$. However, these relations are not used, in the case, when we are interested in the asymptotics of the hitting times, for the initial states $i \in \bar{\mathbb{D}}$.

The following theorem, which is an analogue of Theorem 9.3, summarises the remarks made in Sects. 9.4 and 10.1.2.

Theorem 10.1 *Let the conditions* $\mathbf{G}_{\mathbb{D}}$, $\mathbf{H}_{\mathbb{D}}$ *be satisfied for the semi-Markov processes* $\eta_\varepsilon(t)$. *Then, for* $n = 0, \ldots, \bar{m} - 1$:

(i) *The conditions* $\mathbf{G}_{\mathbb{D}}$, $\mathbf{H}_{\mathbb{D}}$ *are satisfied for the reduced semi-Markov processes* ${}_{\bar{k}_n}\eta_\varepsilon(t)$ *and* ${}_{\bar{k}_n}\tilde{\eta}_\varepsilon(t)$ *in the form of conditions, respectively,* ${}_{\bar{\mathbf{k}}_n}\mathbf{G}_{\mathbb{D}}$, ${}_{\bar{\mathbf{k}}_n}\mathbf{H}_{\mathbb{D}}$ *and* ${}_{\bar{\mathbf{k}}_n}\tilde{\mathbf{G}}_{\mathbb{D}}$, ${}_{\bar{\mathbf{k}}_n}\tilde{\mathbf{H}}_{\mathbb{D}}$.

(ii) *If, in addition to the conditions* $\mathbf{G}_{\mathbb{D}}$, $\mathbf{H}_{\mathbb{D}}$, *the condition* $\mathbf{K}_{\mathbb{D}}$ **(a)** *is satisfied for the semi-Markov processes* $\eta_\varepsilon(t)$, *then the condition* $\mathbf{K}_{\mathbb{D}}$ **(a)** *is satisfied for the reduced semi-Markov processes* ${}_{\bar{k}_n}\eta_\varepsilon(t)$ *and* ${}_{\bar{k}_n}\tilde{\eta}_\varepsilon(t)$ *in the form of conditions, respectively,* ${}_{\bar{\mathbf{k}}_n}\mathbf{K}_{\mathbb{D}}$ **(a)** *and* ${}_{\bar{\mathbf{k}}_n}\tilde{\mathbf{K}}_{\mathbb{D}}$ **(a)**.

(iii) *If, in addition to the conditions* $\mathbf{G}_{\mathbb{D}}$, $\mathbf{H}_{\mathbb{D}}$, *the conditions* $\mathbf{I}_{\mathbb{D}}$, $\mathbf{I}_{\mathbb{D},\mathcal{H}}$ *are satisfied for the semi-Markov processes* $\eta_\varepsilon(t)$, *then the conditions* $\mathbf{I}_{\mathbb{D}}$, $\mathbf{I}_{\mathbb{D},\mathcal{H}}$ *are satisfied for the reduced semi-Markov processes* ${}_{\bar{k}_n}\eta_\varepsilon(t)$ *and* ${}_{\bar{k}_n}\tilde{\eta}_\varepsilon(t)$ *in the form of conditions, respectively,* ${}_{\bar{\mathbf{k}}_n}\mathbf{I}_{\mathbb{D}}$, ${}_{\bar{\mathbf{k}}_n}\mathbf{I}_{\mathbb{D},\mathcal{H}}$ *and* ${}_{\bar{\mathbf{k}}_n}\tilde{\mathbf{I}}_{\mathbb{D}}$, ${}_{\bar{\mathbf{k}}_n}\tilde{\mathbf{I}}_{\mathbb{D},\mathcal{H}}$.

(iv) *If, in addition to the conditions* $\mathbf{G}_{\mathbb{D}}$, $\mathbf{H}_{\mathbb{D}}$, *the conditions* $\mathbf{I}_{\mathbb{D}}$, $\mathbf{I}_{\mathbb{D},\mathcal{H}}$, $\mathbf{L}_{\mathbb{D}}$, $\mathbf{L}_{\mathbb{D},\mathcal{H}}$ *are satisfied for the semi-Markov processes* $\eta_\varepsilon(t)$, *then conditions* $\mathbf{I}_{\mathbb{D},\mathcal{H}}$, $\mathbf{I}_{\mathbb{D}}$, $\mathbf{L}_{\mathbb{D},\mathcal{H}}$, $\mathbf{L}_{\mathbb{D}}$ *are satisfied for the reduced semi-Markov processes* ${}_{\bar{k}_n}\eta_\varepsilon(t)$ *and* ${}_{\bar{k}_n}\tilde{\eta}_\varepsilon(t)$ *in the form of conditions, respectively,* ${}_{\bar{\mathbf{k}}_n}\mathbf{I}_{\mathbb{D}}$, ${}_{\bar{\mathbf{k}}_n}\mathbf{I}_{\mathbb{D},\mathcal{H}}$, ${}_{\bar{\mathbf{k}}_n}\mathbf{L}_{\mathbb{D}}$, ${}_{\bar{\mathbf{k}}_n}\mathbf{L}_{\mathbb{D},\mathcal{H}}$ *and* ${}_{\bar{\mathbf{k}}_n}\tilde{\mathbf{I}}_{\mathbb{D}}$, ${}_{\bar{\mathbf{k}}_n}\tilde{\mathbf{I}}_{\mathbb{D},\mathcal{H}}$, ${}_{\bar{\mathbf{k}}_n}\tilde{\mathbf{L}}_{\mathbb{D}}$, ${}_{\bar{\mathbf{k}}_n}\tilde{\mathbf{L}}_{\mathbb{D},\mathcal{H}}$.

(v) *If, in addition to the conditions* $\mathbf{G}_{\mathbb{D}}$, $\mathbf{H}_{\mathbb{D}}$, *the conditions* $\mathbf{I}_{\mathbb{D}}$, $\mathbf{I}_{\mathbb{D},\mathcal{H}}$, $\mathbf{L}_{\mathbb{D}}$, $\mathbf{L}_{\mathbb{D},\mathcal{H}}$, $\mathbf{J}_{\mathbb{D}}$, $\mathbf{K}_{\mathbb{D}}$ *and* ${}_{\bar{\mathbf{k}}_n}\mathbf{M}_{\mathbb{D}}$ *are satisfied for the semi-Markov processes* $\eta_\varepsilon(t)$, *then the conditions* $\mathbf{I}_{\mathbb{D}}$, $\mathbf{I}_{\mathbb{D},\mathcal{H}}$, $\mathbf{L}_{\mathbb{D}}$, $\mathbf{L}_{\mathbb{D},\mathcal{H}}$, $\mathbf{J}_{\mathbb{D}}$, $\mathbf{K}_{\mathbb{D}}$ *are satisfied for the reduced semi-Markov processes* ${}_{\bar{k}_n}\eta_\varepsilon(t)$ *and* ${}_{\bar{k}_n}\tilde{\eta}_\varepsilon(t)$ *in the form of conditions, respectively,* ${}_{\bar{\mathbf{k}}_n}\mathbf{I}_{\mathbb{D}}$, ${}_{\bar{\mathbf{k}}_n}\mathbf{I}_{\mathbb{D},\mathcal{H}}$, ${}_{\bar{\mathbf{k}}_n}\mathbf{L}_{\mathbb{D}}$, ${}_{\bar{\mathbf{k}}_n}\mathbf{L}_{\mathbb{D},\mathcal{H}}$, ${}_{\bar{\mathbf{k}}_n}\mathbf{J}_{\mathbb{D}}$, ${}_{\bar{\mathbf{k}}_n}\mathbf{K}_{\mathbb{D}}$ *and* ${}_{\bar{\mathbf{k}}_n}\tilde{\mathbf{I}}_{\mathbb{D}}$, ${}_{\bar{\mathbf{k}}_n}\tilde{\mathbf{I}}_{\mathbb{D},\mathcal{H}}$, ${}_{\bar{\mathbf{k}}_n}\tilde{\mathbf{L}}_{\mathbb{D}}$, ${}_{\bar{\mathbf{k}}_n}\tilde{\mathbf{L}}_{\mathbb{D},\mathcal{H}}$, ${}_{\bar{\mathbf{k}}_n}\tilde{\mathbf{J}}_{\mathbb{D}}$, ${}_{\bar{\mathbf{k}}_n}\tilde{\mathbf{K}}_{\mathbb{D}}$.

10.1.3 Hitting Times for Semi-Markov Processes with Reduced Phase Spaces

In this subsection, we present forward and backward recurrent relations for hitting times, their distributions, and Laplace transforms. These relations create the base for asymptotic recurrent algorithms of phase space reductions for regularly and singularly perturbed semi-Markov processes used for obtaining limit theorems on weak convergence of distributions and convergence of expectations for hitting times.

Let us choose an arbitrary state $k \in \bar{\mathbb{D}}$ and introduce the hitting times for the semi-Markov processes $\tilde{\eta}_\varepsilon(t)$ and ${}_k\eta_\varepsilon(t)$, for $\varepsilon \in (0, 1]$,

$$\tilde{\tau}_{\varepsilon,\mathbb{D}} = \sum_{n=1}^{\tilde{\nu}_{\varepsilon,\mathbb{D}}} \tilde{\kappa}_{\varepsilon,n}, \text{ where } \tilde{\nu}_{\varepsilon,\mathbb{D}} = \min(n \geq 1 : \tilde{\eta}_{\varepsilon,n} \in \mathbb{D}) \tag{10.33}$$

and

$$ {}_k\tau_{\varepsilon,\mathbb{D}} = \sum_{n=1}^{{}_k\nu_{\varepsilon,\mathbb{D}}} {}_k\kappa_{\varepsilon,n}, \text{ where } {}_k\nu_{\varepsilon,\mathbb{D}} = \min(n \geq 1 : {}_k\eta_{\varepsilon,n} \in \mathbb{D}). \tag{10.34}$$

The definitions of the semi-Markov processes $\eta_\varepsilon(t)$, $\tilde{\eta}_\varepsilon(t)$, and the reduced semi-Markov processes ${}_k\eta_\varepsilon(t)$ imply that the following lemma takes place.

Lemma 10.2 *Let the conditions* $\mathbf{G}_\mathbb{D}$, $\mathbf{H}_\mathbb{D}$ *be satisfied. Then the conditions* $\tilde{\mathbf{G}}_\mathbb{D}$, $\tilde{\mathbf{H}}_\mathbb{D}$, ${}_\mathbf{k}\mathbf{G}_\mathbb{D}$, ${}_\mathbf{k}\mathbf{H}_\mathbb{D}$ *are also satisfied, and the following relations take place, for* $\varepsilon \in (0, 1]$*:*

$$\begin{aligned} \mathsf{P}_i\{\tau_{\varepsilon,\mathbb{D}} = \tilde{\tau}_{\varepsilon,\mathbb{D}} = {}_k\tau_{\varepsilon,\mathbb{D}}, \\ \eta_\varepsilon(\tau_{\varepsilon\mathbb{D}}) = \tilde{\eta}_\varepsilon(\tilde{\tau}_{\varepsilon\mathbb{D}}) = {}_k\eta_\varepsilon({}_k\tau_{\varepsilon,\mathbb{D}})\} = 1, \ i \in {}_k\bar{\mathbb{D}}, \end{aligned} \tag{10.35}$$

and

$$\begin{aligned} \mathsf{P}_k\{\tau_{\varepsilon,\mathbb{D}} = \tilde{\tau}_{\varepsilon,\mathbb{D}} = \tilde{\kappa}_{\varepsilon,1}\mathrm{I}(\tilde{\eta}_{\varepsilon,1} \in \mathbb{D}) + (\tilde{\kappa}_{\varepsilon,1} + {}_k\tau_{\varepsilon,\mathbb{D}})\mathrm{I}(\tilde{\eta}_{\varepsilon,1} \in {}_k\bar{\mathbb{D}}), \\ \eta_\varepsilon(\tau_{\varepsilon,\mathbb{D}}) = \tilde{\eta}_\varepsilon(\tilde{\tau}_{\varepsilon,\mathbb{D}}) = \tilde{\eta}_{\varepsilon,1}\mathrm{I}(\tilde{\eta}_{\varepsilon,1} \in \mathbb{D}) \\ + {}_k\eta_\varepsilon({}_k\tau_{\varepsilon,D})\mathrm{I}(\tilde{\eta}_{\varepsilon,1} \in {}_k\bar{\mathbb{D}})\} = 1. \end{aligned} \tag{10.36}$$

Proof The following obvious relation takes place for the semi-Markov processes $\eta_\varepsilon(t), t \geq 0$ and $\tilde{\eta}_\varepsilon(t), t \geq 0$, for $i \in \mathbb{X}$ and $\varepsilon \in (0, 1]$:

$$\mathsf{P}_i\{\eta_\varepsilon(t) = \tilde{\eta}_\varepsilon(t), \ t \geq 0\} = 1. \tag{10.37}$$

Let us define the following variants of the hitting times, for $s \geq 0$:

$$\tau_\varepsilon[s] = \inf(t \geq s : \eta_\varepsilon(t) \in \mathbb{D}), \tag{10.38}$$

and

$$\tilde{\tau}_\varepsilon[s] = \inf(t \geq s : \tilde{\eta}_\varepsilon(t) \in \mathbb{D}). \tag{10.39}$$

The relation (10.37) obviously implies that, for $s \geq 0, i \in \bar{\mathbb{D}}$ and $\varepsilon \in (0, 1]$,

$$\mathsf{P}_i\{\tau_\varepsilon[s] = \tilde{\tau}_\varepsilon[s], \ \eta_\varepsilon(\tau_\varepsilon[s]) = \tilde{\eta}_\varepsilon(\tilde{\tau}_\varepsilon[s])\} = 1. \tag{10.40}$$

Also, the relations (10.37) and (10.40) imply that, for $i \in \bar{\mathbb{D}}$ and $\varepsilon \in (0, 1]$,

$$\begin{aligned}\mathsf{P}_i\{\tau_{\varepsilon,\mathbb{D}} &= \tau_\varepsilon[0] = \tilde{\tau}_\varepsilon[0] = \tilde{\tau}_{\varepsilon,\mathbb{D}},\\ &\eta_\varepsilon(\tau_{\varepsilon,\mathbb{D}}) = \eta_\varepsilon(\tau_\varepsilon[0]) = \tilde{\eta}_\varepsilon(\tilde{\tau}_\varepsilon[0]) = \tilde{\eta}_\varepsilon(\tilde{\tau}_{\varepsilon,\mathbb{D}})\} = 1. \qquad (10.41)\end{aligned}$$

The relations (9.39)–(9.41) imply that: (a) if $\eta_{\varepsilon,0} = i \in \bar{\mathbb{D}}$ and $\eta_{\varepsilon,1} \in \mathbb{D}$, then the random functional $\theta_\varepsilon[0] = \nu_{\varepsilon,\mathbb{D}} = \tilde{\nu}_{\varepsilon,\mathbb{D}} = 1$, and, thus, $\tau_{\varepsilon,\mathbb{D}} = \tilde{\tau}_{\varepsilon,\mathbb{D}} = \kappa_{\varepsilon,1}$, and $\eta_\varepsilon(\tau_{\varepsilon,\mathbb{D}}) = \tilde{\eta}_\varepsilon(\tilde{\tau}_{\varepsilon,\mathbb{D}}) = \eta_{\varepsilon,1}$; (b) if $\eta_{\varepsilon,0} = i \in \bar{\mathbb{D}}$ and $\eta_{\varepsilon,1} \in \bar{\mathbb{D}}$, then $\tau_{\varepsilon,\mathbb{D}} = \kappa_{\varepsilon,1} + \tau_\varepsilon[\kappa_{\varepsilon,1}]$ and $\tilde{\tau}_{\varepsilon,\mathbb{D}} = \kappa_{\varepsilon,1} + \tilde{\tau}_\varepsilon[\kappa_{\varepsilon,1}]$.

Using the above remarks, the Markov property of the Markov renewal process $(\eta_{\varepsilon,n}, \kappa_{\varepsilon,n})$, and the relations (10.40), (10.41) we obtain, for $i \in \mathbb{D}$ and $\varepsilon \in (0, 1]$,

$$\begin{aligned}\mathsf{P}_i&\{\tau_{\varepsilon,\mathbb{D}} = \tilde{\tau}_{\varepsilon,\mathbb{D}}, \eta_\varepsilon(\tau_{\varepsilon,\mathbb{D}}) = \tilde{\eta}_\varepsilon(\tilde{\tau}_{\varepsilon,\mathbb{D}})\}\\ &= \mathsf{P}_i\{\tau_{\varepsilon,\mathbb{D}} = \tilde{\tau}_{\varepsilon,\mathbb{D}}, \eta_\varepsilon(\tau_{\varepsilon,\mathbb{D}}) = \tilde{\eta}_\varepsilon(\tilde{\tau}_{\varepsilon,\mathbb{D}}), \eta_{\varepsilon,1} \in \mathbb{D}\}\\ &\quad + \mathsf{P}_i\{\tau_{\varepsilon,\mathbb{D}} = \tilde{\tau}_{\varepsilon,\mathbb{D}}, \eta_\varepsilon(\tau_{\varepsilon,\mathbb{D}}) = \tilde{\eta}_\varepsilon(\tilde{\tau}_{\varepsilon,\mathbb{D}}), \eta_{\varepsilon,1} \in \bar{\mathbb{D}}\}\\ &= \mathsf{P}_i\{\eta_{\varepsilon,1} \in \mathbb{D}\}\\ &\quad + \sum_{k \in \bar{\mathbb{D}}} \int_0^\infty \mathsf{P}_k\{s + \tau_\varepsilon[s] = s + \tilde{\tau}_\varepsilon[s], \eta_\varepsilon(\tau_\varepsilon[s]) = \tilde{\eta}_\varepsilon(\tilde{\tau}_\varepsilon[s])\}Q_{\varepsilon,ik}(ds)\\ &= \sum_{r \in \mathbb{D}} p_{\varepsilon,ir} + \sum_{k \in \bar{\mathbb{D}}} \int_0^\infty Q_{\varepsilon,ik}(ds) = 1. \qquad (10.42)\end{aligned}$$

Recall that the state $k \in \bar{\mathbb{D}}$ and ${}_k\bar{\mathbb{D}} = \bar{\mathbb{D}} \setminus \{k\}$, ${}_k\bar{\mathbb{X}} = {}_k\bar{\mathbb{D}} \cup \mathbb{D}$.

The relations (9.79)–(9.82) imply that ${}_k\beta_{\varepsilon,0} = 0$, if $\tilde{\eta}_{\varepsilon,0} = i \in {}_k\mathbb{X}$. In this case, the hitting times $\tilde{\nu}_{\varepsilon,\mathbb{D}}$ and ${}_k\nu_{\varepsilon,\mathbb{D}}$ are connected by the following relation, for $\varepsilon \in (0, 1]$:

$$\mathsf{P}_i\{\tilde{\nu}_{\varepsilon,\mathbb{D}} = {}_k\beta_{\varepsilon,\, {}_k\nu_{\varepsilon,\mathbb{D}}}\} = 1, \ i \in {}_k\mathbb{X}. \qquad (10.43)$$

Thus, the hitting times $\tilde{\tau}_{\varepsilon,\mathbb{D}}$ and ${}_k\tau_{\varepsilon,\mathbb{D}}$ are connected by the following relation, for $\varepsilon \in (0, 1]$:

$$\begin{aligned}\mathsf{P}_i\{\tilde{\tau}_{\varepsilon,\mathbb{D}} &= \sum_{n=1}^{\tilde{\nu}_{\varepsilon,\mathbb{D}}} \tilde{\kappa}_{\varepsilon,n} = \sum_{n=1}^{{}_k\beta_{\varepsilon,\,{}_k\nu_{\varepsilon,\mathbb{D}}}} \tilde{\kappa}_{\varepsilon,n}\\ &= \sum_{n=1}^{{}_k\nu_{\varepsilon,\mathbb{D}}} {}_k\kappa_{\varepsilon,n} = {}_k\tau_{\varepsilon,\mathbb{D}}\} = 1, \ i \in {}_k\bar{\mathbb{D}}, \qquad (10.44)\end{aligned}$$

and the random variables $\tilde{\eta}_\varepsilon(\tilde{\tau}_{\varepsilon,\mathbb{D}})$ and ${}_k\eta_\varepsilon({}_k\tau_{\varepsilon,\mathbb{D}})$ are connected by the following relation, for $\varepsilon \in (0, 1]$:

$$\begin{aligned}\mathsf{P}_i\{\tilde{\eta}_\varepsilon(\tilde{\tau}_{\varepsilon,\mathbb{D}}) &= \tilde{\eta}_{\varepsilon,\tilde{\nu}_{\varepsilon,\mathbb{D}}} = \tilde{\eta}_{\varepsilon,\,{}_k\beta_{\varepsilon,\,{}_k\nu_{\varepsilon,\mathbb{D}}}}\\ &= {}_k\eta_{\varepsilon,\,{}_k\nu_{\varepsilon,\mathbb{D}}} = {}_k\eta_\varepsilon({}_k\tau_{\varepsilon,\mathbb{D}})\} = 1, \ i \in {}_k\mathbb{X}. \qquad (10.45)\end{aligned}$$

The relations (10.42), (10.44), and (10.45) imply that the relation (10.35) holds.

The relations (9.79)–(9.82) also imply that $\beta_{\varepsilon,0} = 1$ and $\tilde{\eta}_{\varepsilon,1} \in {}_k\mathbb{X}$, if $\tilde{\eta}_{\varepsilon,0} = k$. In this case, the following relation similar to (10.43) takes place, for $\varepsilon \in (0, 1]$:

$$\mathsf{P}_k\{\tilde{\nu}_{\varepsilon,\mathbb{D}} = \mathrm{I}(\tilde{\eta}_{\varepsilon,1} \in \mathbb{D}) + {}_k\beta_{\varepsilon,\, {}_k\nu_{\varepsilon,\mathbb{D}}}\mathrm{I}(\tilde{\eta}_{\varepsilon,1} \in {}_k\bar{\mathbb{D}})\} = 1. \tag{10.46}$$

Thus, the hitting times $\tilde{\tau}_{\varepsilon,\mathbb{D}}$ and ${}_k\tau_{\varepsilon,\mathbb{D}}$ are connected by the following relation, for $\varepsilon \in (0, 1]$:

$$\begin{aligned}\mathsf{P}_k\{\tilde{\tau}_{\varepsilon,\mathbb{D}} &= \sum_{n=1}^{\tilde{\nu}_{\varepsilon,\mathbb{D}}} \tilde{\kappa}_{\varepsilon,n} \\ &= \tilde{\kappa}_{\varepsilon,1}\mathrm{I}(\tilde{\eta}_{\varepsilon,1} \in \mathbb{D}) + (\tilde{\kappa}_{\varepsilon,1} + \sum_{n=2}^{{}_k\beta_{\varepsilon,\, {}_k\nu_{\varepsilon,\mathbb{D}}}} \tilde{\kappa}_{\varepsilon,n})\mathrm{I}(\tilde{\eta}_{\varepsilon,1} \in {}_k\bar{\mathbb{D}}) \\ &= \tilde{\kappa}_{\varepsilon,1}\mathrm{I}(\tilde{\eta}_{\varepsilon,1} \in \mathbb{D}) + (\tilde{\kappa}_{\varepsilon,1} + \sum_{n=1}^{{}_k\nu_{\varepsilon,\mathbb{D}}} {}_k\kappa_{\varepsilon,n})\mathrm{I}(\tilde{\eta}_{\varepsilon,1} \in {}_k\bar{\mathbb{D}}) \\ &= \tilde{\kappa}_{\varepsilon,1}\mathrm{I}(\tilde{\eta}_{\varepsilon,1} \in \mathbb{D}) + (\tilde{\kappa}_{\varepsilon,1} + {}_k\tau_{\varepsilon,\mathbb{D}})\mathrm{I}(\tilde{\eta}_{\varepsilon,1} \in {}_k\bar{\mathbb{D}})\} = 1,\end{aligned} \tag{10.47}$$

and the random variables $\tilde{\eta}_\varepsilon(\tilde{\tau}_{\varepsilon,\mathbb{D}})$ and ${}_k\eta_\varepsilon({}_k\tau_{\varepsilon,\mathbb{D}})$ are connected by the following relation, for $\varepsilon \in (0, 1]$:

$$\mathsf{P}_k\{\tilde{\eta}_\varepsilon(\tilde{\tau}_{\varepsilon,\mathbb{D}}) = \tilde{\eta}_{\varepsilon,1}\mathrm{I}(\tilde{\eta}_{\varepsilon,1} \in \mathbb{D}) + {}_k\eta_\varepsilon({}_k\tau_{\varepsilon,D})\mathrm{I}(\tilde{\eta}_{\varepsilon,1} \in {}_k\bar{\mathbb{D}})\} = 1. \tag{10.48}$$

The relations (10.47) and (10.48) imply that the relation (10.36) holds. □

Let us also introduce the following distributions, for $i \in {}_k\mathbb{X}$, $k \in \bar{\mathbb{D}}$ and $\varepsilon \in (0, 1]$:

$$\tilde{G}_{\varepsilon,\mathbb{D},ij}(t) = \mathsf{P}_i\{\tilde{\tau}_{\varepsilon,\mathbb{D}} \le t,\ \tilde{\eta}_\varepsilon(\tilde{\tau}_{\varepsilon\mathbb{D}}) = j\},\ t \ge 0, j \in \mathbb{D}, \tag{10.49}$$

and

$${}_kG_{\varepsilon,\mathbb{D},ij}(t) = \mathsf{P}_i\{{}_k\tau_{\varepsilon,\mathbb{D}} \le t,\ {}_k\eta_\varepsilon({}_k\tau_{\varepsilon\mathbb{D}}) = j\},\ t \ge 0, j \in \mathbb{D}. \tag{10.50}$$

The following lemma is a corollary of Lemma 10.2.

Lemma 10.3 *Let the conditions* $\mathbf{G}_\mathbb{D}$, $\mathbf{H}_\mathbb{D}$ *be satisfied. Then the conditions* $\tilde{\mathbf{G}}_\mathbb{D}$, $\tilde{\mathbf{H}}_\mathbb{D}$, ${}_\mathbf{k}\mathbf{G}_\mathbb{D}$, ${}_\mathbf{k}\mathbf{H}_\mathbb{D}$ *are also satisfied, and the following relations take place, for* $\varepsilon \in (0, 1]$:

$$G_{\varepsilon,\mathbb{D},ij}(t) = \tilde{G}_{\varepsilon,\mathbb{D},ij}(t) = {}_kG_{\varepsilon,\mathbb{D},ij}(t),\ t \ge 0, j \in \mathbb{D},\ i \in {}_k\mathbb{X}, \tag{10.51}$$

and

$$\begin{aligned}G_{\varepsilon,\mathbb{D},kj}(t) &= \tilde{G}_{\varepsilon,\mathbb{D},kj}(t) \\ &= \tilde{F}_{\varepsilon,kj}(t)\tilde{p}_{\varepsilon,kj} + \sum_{r \in {}_k\bar{\mathbb{D}}} (\tilde{F}_{\varepsilon,kr}(t) * {}_kG_{\varepsilon,\mathbb{D},rj}(t)\tilde{p}_{\varepsilon,kr} \\ &= \tilde{F}_{\varepsilon,kj}(t)\tilde{p}_{\varepsilon,kj} + \sum_{r \in {}_k\bar{\mathbb{D}}} (\tilde{F}_{\varepsilon,kr}(t) * G_{\varepsilon,\mathbb{D},rj}(t)\tilde{p}_{\varepsilon,kr},\ t \ge 0, j \in \mathbb{D}.\end{aligned} \tag{10.52}$$

Proof The equalities given in the relations (10.51) are obvious corollaries of the relation (10.35).

The first equality given in the relation (10.52) follows from the relation (10.36).

The relations (9.79)–(9.82) imply that, in the case where $\tilde{\eta}_{\varepsilon,0} = k$, the random variable $\beta_{\varepsilon,0} = 1$ and, thus, the random functional $({}_k\tau_{\varepsilon,\mathbb{D}}, {}_k\eta_\varepsilon({}_k\tau_{\varepsilon,\mathbb{D}}))$ is determined by the trajectory of the Markov renewal process $(\tilde{\eta}_{\varepsilon,n}, \tilde{\kappa}_{\varepsilon,n})$ for $n \geq 1$. This allows us to use the relation (10.36) and the Markov property of Markov renewal process $(\tilde{\eta}_{\varepsilon,n}, \tilde{\kappa}_{\varepsilon,n})$, and to obtain the following relation, for $t \geq 0, j \in \mathbb{D}$ and $\varepsilon \in (0, 1]$:

$$\begin{aligned}\tilde{G}_{\varepsilon,\mathbb{D},kj}(t) &= \mathsf{P}_k\{\tilde{\kappa}_{\varepsilon,1} \leq t, \tilde{\eta}_{\varepsilon,1} = j\} \\ &\quad + \sum_{r \in {}_k\bar{\mathbb{D}}} \mathsf{P}_k\{\tilde{\kappa}_{\varepsilon,1} + {}_k\tau_{\varepsilon,\mathbb{D}} \leq t, \ {}_k\eta_\varepsilon({}_k\tau_{\varepsilon,\mathbb{D}}) = j, \tilde{\eta}_{\varepsilon,1} = r\} \\ &= \tilde{F}_{\varepsilon,kj}(t)\tilde{p}_{\varepsilon,kj} + \sum_{r \in {}_k\bar{\mathbb{D}}} (\tilde{F}_{\varepsilon,kr}(t) * {}_kG_{\varepsilon,\mathbb{D},rj}(t)\tilde{p}_{\varepsilon,kr}. \end{aligned} \tag{10.53}$$

Thus, the second equality given in the relation (10.52) holds.

The third equality given in the relation (10.52) follows from the second equality given in this relation and the relation (10.51). □

Let us also introduce the following Laplace transforms, for $i \in {}_k\mathbb{X}$ and $\varepsilon \in (0, 1]$:

$$\tilde{\Psi}_{\varepsilon,\mathbb{D},ij}(s) = \mathsf{E}_i \exp\{-s\tilde{\tau}_{\varepsilon,\mathbb{D}}\}\mathrm{I}(\tilde{\eta}_\varepsilon(\tilde{\tau}_{\varepsilon,\mathbb{D}}) = j)\}, \ s \geq 0, j \in \mathbb{D}, \tag{10.54}$$

and

$${}_k\Psi_{\varepsilon,\mathbb{D},ij}(s) = \mathsf{E}_i \exp\{-s\,{}_k\tau_{\varepsilon,\mathbb{D}}\}\mathrm{I}({}_k\eta_\varepsilon({}_k\tau_{\varepsilon,\mathbb{D}}) = j)\}, \ s \geq 0, j \in \mathbb{D}. \tag{10.55}$$

The next lemma reformulates the statements of Lemma 9.3 in an equivalent form of relations for the Laplace transforms of the hitting times.

Lemma 10.4 *Let the conditions* $\mathbf{G}_\mathbb{D}$, $\mathbf{H}_\mathbb{D}$ *be satisfied. Then the conditions* $\tilde{\mathbf{G}}_\mathbb{D}$, $\tilde{\mathbf{H}}_\mathbb{D}$, ${}_\mathbf{k}\mathbf{G}_\mathbb{D}$, ${}_\mathbf{k}\mathbf{H}_\mathbb{D}$ *are also satisfied, and the following relations take place, for* $\varepsilon \in (0, 1]$*:*

$$\Psi_{\varepsilon,\mathbb{D},ij}(s) = \tilde{\Psi}_{\varepsilon,\mathbb{D},ij}(s) = {}_k\Psi_{\varepsilon,\mathbb{D},ij}(s), \ s \geq 0, j \in \mathbb{D}, \ i \in {}_k\mathbb{X} \tag{10.56}$$

and

$$\begin{aligned}\Psi_{\varepsilon,\mathbb{D},kj}(s) &= \tilde{\Psi}_{\varepsilon,\mathbb{D},kj}(s) \\ &= \tilde{\phi}_{\varepsilon,kj}(s)\tilde{p}_{\varepsilon,kj} + \sum_{r \in {}_k\bar{\mathbb{D}}} {}_k\Psi_{\varepsilon,\mathbb{D},rj}(s)\tilde{\phi}_{\varepsilon,kr}(s)\tilde{p}_{\varepsilon,kr} \\ &= \tilde{\phi}_{\varepsilon,kj}(s)\tilde{p}_{\varepsilon,kj} + \sum_{r \in {}_k\bar{\mathbb{D}}} \Psi_{\varepsilon,\mathbb{D},rj}(s)\tilde{\phi}_{\varepsilon,kr}(s)\tilde{p}_{\varepsilon,kr}, \ s \geq 0, j \in \mathbb{D}. \end{aligned} \tag{10.57}$$

Lemmas 10.2–10.4 allow us to reduce the study of the asymptotics for distributions of the hitting times from the case of semi-Markov processes $\eta_\varepsilon(t)$ to the simpler case of reduced semi-Markov processes $\tilde{\eta}_\varepsilon(t)$ and ${}_k\eta_\varepsilon(t)$.

10.1.4 Recurrent Relations for Distributions and Laplace Transforms of Hitting Times

10.1.4.1 Recurrent Relations for Distributions and Laplace Transforms. Let $\bar{k}_n = \langle k_1, \ldots, k_n \rangle$ be, for $1 \leq n \leq \bar{m}$, an arbitrary sequence of different states from the domain $\bar{\mathbb{D}}$. As is Sect. 9.4.1, we use the notation $\bar{k}_0 = \langle\rangle$ for an "empty" sequence of states.

Let us introduce the hitting times for the semi-Markov processes ${}_{\bar{k}_n}\tilde{\eta}_\varepsilon(t)$ and ${}_{\bar{k}_n}\eta_\varepsilon(t)$, for $n = 0, \ldots, \bar{m} - 1$ and $\varepsilon \in (0, 1]$,

$$
{}_{\bar{k}_n}\tilde{\tau}_{\varepsilon,\mathbb{D}} = \sum_{n=1}^{{}_{\bar{k}_n}\tilde{\nu}_{\varepsilon,\mathbb{D}}} {}_{\bar{k}_n}\tilde{\kappa}_{\varepsilon,n}, \text{ where } {}_{\bar{k}_n}\tilde{\nu}_{\varepsilon,\mathbb{D}} = \min(n \geq 1 : {}_{\bar{k}_n}\tilde{\eta}_{\varepsilon,n} \in \mathbb{D}) \tag{10.58}
$$

and

$$
{}_{\bar{k}_n}\tau_{\varepsilon,\mathbb{D}} = \sum_{n=1}^{{}_{\bar{k}_n}\nu_{\varepsilon,\mathbb{D}}} {}_{\bar{k}_n}\kappa_{\varepsilon,n}, \text{ where } {}_{\bar{k}_n}\nu_{\varepsilon,\mathbb{D}} = \min(n \geq 1 : {}_{\bar{k}_n}\eta_{\varepsilon,n} \in \mathbb{D}). \tag{10.59}
$$

Let us also introduce the following distributions, for $i \in {}_{\bar{k}_n}\mathbb{X}$, $n = 0, \ldots, \bar{m} - 1$ and $\varepsilon \in (0, 1]$:

$$
{}_{\bar{k}_n}\tilde{G}_{\varepsilon,\mathbb{D},ij}(t) = \mathsf{P}_i\{{}_{\bar{k}_n}\tilde{\tau}_{\varepsilon,\mathbb{D}} \leq t,\ {}_{\bar{k}_n}\tilde{\eta}_\varepsilon({}_{\bar{k}_n}\tilde{\tau}_{\varepsilon,\mathbb{D}}) = j\},\ t \geq 0, j \in \mathbb{D}, \tag{10.60}
$$

and

$$
{}_{\bar{k}_n}G_{\varepsilon,\mathbb{D},ij}(t) = \mathsf{P}_i\{{}_{\bar{k}_n}\tau_{\varepsilon,\mathbb{D}} \leq t,\ {}_{\bar{k}_n}\eta_\varepsilon({}_{\bar{k}_n}\tau_{\varepsilon,\mathbb{D}}) = j\},\ t \geq 0, j \in \mathbb{D}. \tag{10.61}
$$

Finally, let us introduce the following Laplace transforms, for $i \in {}_{\bar{k}_n}\mathbb{X}$, $n = 0, \ldots, \bar{m} - 1$ and $\varepsilon \in (0, 1]$:

$$
{}_{\bar{k}_n}\tilde{\Psi}_{\varepsilon,\mathbb{D},ij}(s) = \mathsf{E}_i \exp\{-s\, {}_{\bar{k}_n}\tilde{\tau}_{\varepsilon,\mathbb{D}}\}\mathrm{I}({}_{\bar{k}_n}\tilde{\eta}_\varepsilon({}_{\bar{k}_n}\tilde{\tau}_{\varepsilon,\mathbb{D}}) = j)\},\ s \geq 0, j \in \mathbb{D}, \tag{10.62}
$$

and

$$
{}_{\bar{k}_n}\Psi_{\varepsilon,\mathbb{D},ij}(s) = \mathsf{E}_i \exp\{-s\, {}_{\bar{k}_n}\tau_{\varepsilon,\mathbb{D}}\}\mathrm{I}({}_{\bar{k}_n}\eta_\varepsilon({}_{\bar{k}_n}\tau_{\varepsilon,\mathbb{D}}) = j)\},\ s \geq 0, j \in \mathbb{D}. \tag{10.63}
$$

Lemmas 10.5–10.10 are similar to Lemmas 10.2–10.4. They play a key role in phase space reduction algorithms and obtaining recurrent weak convergence relations for hitting times.

Lemma 10.5 *Let conditions* $\mathbf{G}_\mathbb{D} = {}_{\bar{k}_0}\mathbf{G}_\mathbb{D}$, $\mathbf{H}_\mathbb{D} = {}_{\bar{k}_0}\mathbf{H}_\mathbb{D}$ *hold. Then, for every* $n = 1, \ldots, \bar{m} - 1$, *the conditions* ${}_{\bar{k}_{n-1}}\tilde{\mathbf{G}}_\mathbb{D}$, ${}_{\bar{k}_{n-1}}\tilde{\mathbf{H}}_\mathbb{D}$, ${}_{\bar{k}_n}\mathbf{G}_\mathbb{D}$, ${}_{\bar{k}_n}\mathbf{H}_\mathbb{D}$ *are also satisfied, and the following relations take place, for* $\varepsilon \in (0, 1]$:

$$
\begin{aligned}
&\mathsf{P}_i\{\tau_{\varepsilon,\mathbb{D}} = {}_{\bar{k}_{n-1}}\tilde{\tau}_{\varepsilon,\mathbb{D}} = {}_{\bar{k}_n}\tau_{\varepsilon,\mathbb{D}}, \\
&\quad \eta_\varepsilon(\tau_{\varepsilon,\mathbb{D}}) = {}_{\bar{k}_{n-1}}\tilde{\eta}_\varepsilon({}_{\bar{k}_{n-1}}\tilde{\tau}_{\varepsilon,\mathbb{D}}) = {}_{\bar{k}_n}\eta_\varepsilon({}_{\bar{k}_n}\tau_{\varepsilon,\mathbb{D}})\} = 1, \ j \in \mathbb{D}, i \in {}_{\bar{k}_n}\bar{\mathbb{D}}, \qquad (10.64)
\end{aligned}
$$

and

$$
\begin{aligned}
\mathsf{P}_{k_n}\{\tau_{\varepsilon,\mathbb{D}} &= {}_{\bar{k}_{n-1}}\tilde{\tau}_{\varepsilon,\mathbb{D}} = {}_{\bar{k}_{n-1}}\tilde{\kappa}_{\varepsilon,1}\mathrm{I}({}_{\bar{k}_{n-1}}\tilde{\eta}_{\varepsilon,1} \in \mathbb{D}) \\
&\quad + ({}_{\bar{k}_{n-1}}\tilde{\kappa}_{\varepsilon,1} + {}_{\bar{k}_n}\tau_{\varepsilon,\mathbb{D}})\mathrm{I}({}_{\bar{k}_{n-1}}\tilde{\eta}_{\varepsilon,1} \in {}_{\bar{k}_n}\bar{\mathbb{D}}), \\
\eta_\varepsilon(\tau_{\varepsilon,\mathbb{D}}) &= {}_{\bar{k}_{n-1}}\tilde{\eta}_\varepsilon({}_{\bar{k}_{n-1}}\tilde{\tau}_{\varepsilon,\mathbb{D}}) = {}_{\bar{k}_{n-1}}\tilde{\eta}_{\varepsilon,1}\mathrm{I}({}_{\bar{k}_{n-1}}\tilde{\eta}_{\varepsilon,1} \in \mathbb{D}) \\
&\quad + {}_{\bar{k}_n}\eta_\varepsilon({}_{\bar{k}_n}\tau_{\varepsilon,\mathbb{D}})\mathrm{I}({}_{\bar{k}_{n-1}}\tilde{\eta}_{\varepsilon,1} \in {}_{\bar{k}_n}\bar{\mathbb{D}})\} = 1, \ j \in \mathbb{D}. \qquad (10.65)
\end{aligned}
$$

Lemma 10.6 *Let the conditions* $\mathbf{G}_\mathbb{D} = {}_{\bar{k}_0}\mathbf{G}_\mathbb{D}$, $\mathbf{H}_\mathbb{D} = {}_{\bar{k}_0}\mathbf{H}_\mathbb{D}$ *be satisfied. Then, for every* $n = 1, \dots, \bar{m}-1$, *the conditions* ${}_{\bar{k}_{n-1}}\tilde{\mathbf{G}}_\mathbb{D}$, ${}_{\bar{k}_{n-1}}\tilde{\mathbf{H}}_\mathbb{D}$, ${}_{\bar{k}_n}\mathbf{G}_\mathbb{D}$, ${}_{\bar{k}_n}\mathbf{H}_\mathbb{D}$ *are also satisfied, and the following relations take place, for* $\varepsilon \in (0, 1]$:

$$
\begin{aligned}
G_{\varepsilon,\mathbb{D},ij}(t) &= {}_{\bar{k}_{n-1}}\tilde{G}_{\varepsilon,\mathbb{D},ij}(t) \\
&= {}_{\bar{k}_n}G_{\varepsilon,\mathbb{D},ij}(t), \ t \ge 0, j \in \mathbb{D}, i \in {}_{\bar{k}_n}\bar{\mathbb{D}}, \qquad (10.66)
\end{aligned}
$$

and

$$
\begin{aligned}
G_{\varepsilon,\mathbb{D},k_nj}(t) &= {}_{\bar{k}_{n-1}}\tilde{G}_{\varepsilon,\mathbb{D},k_nj}(t) = {}_{\bar{k}_{n-1}}\tilde{F}_{\varepsilon,k_nj}(t)\,{}_{\bar{k}_{n-1}}\tilde{p}_{\varepsilon,k_nj} \\
&\quad + \sum_{r \in {}_{\bar{k}_n}\bar{\mathbb{D}}} {}_{\bar{k}_{n-1}}\tilde{F}_{\varepsilon,k_nr}(t) * {}_{\bar{k}_n}G_{\varepsilon,\mathbb{D},rj}(t)\,{}_{\bar{k}_{n-1}}\tilde{p}_{\varepsilon,k_nr} \\
&= {}_{\bar{k}_{n-1}}\tilde{F}_{\varepsilon,k_nj}(t)\,{}_{\bar{k}_{n-1}}\tilde{p}_{\varepsilon,k_nj} \\
&\quad + \sum_{r \in {}_{\bar{k}_n}\bar{\mathbb{D}}} {}_{\bar{k}_{n-1}}\tilde{F}_{\varepsilon,k_nr}(t) * G_{\varepsilon,\mathbb{D},rj}(t)\,{}_{\bar{k}_{n-1}}\tilde{p}_{\varepsilon,k_nr}, \ t \ge 0, j \in \mathbb{D}. \qquad (10.67)
\end{aligned}
$$

Lemma 10.7 *Let the conditions* $\mathbf{G}_\mathbb{D} = {}_{\bar{k}_0}\mathbf{G}_\mathbb{D}$, $\mathbf{H}_\mathbb{D} = {}_{\bar{k}_0}\mathbf{H}_\mathbb{D}$ *be satisfied. Then, for every* $n = 1, \dots, \bar{m}-1$, *the conditions* ${}_{\bar{k}_{n-1}}\tilde{\mathbf{G}}_\mathbb{D}$, ${}_{\bar{k}_{n-1}}\tilde{\mathbf{H}}_\mathbb{D}$, ${}_{\bar{k}_n}\mathbf{G}_\mathbb{D}$, ${}_{\bar{k}_n}\mathbf{H}_\mathbb{D}$ *are also satisfied, and the following relations take place, for* $\varepsilon \in (0, 1]$:

$$
\begin{aligned}
\Psi_{\varepsilon,\mathbb{D},ij}(s) &= {}_{\bar{k}_{n-1}}\tilde{\Psi}_{\varepsilon,\mathbb{D},ij}(s) \\
&= {}_{\bar{k}_n}\Psi_{\varepsilon,\mathbb{D},ij}(s), \ s \ge 0, j \in \mathbb{D}, i \in {}_{\bar{k}_n}\mathbb{X}, \qquad (10.68)
\end{aligned}
$$

and

$$
\begin{aligned}
\Psi_{\varepsilon,\mathbb{D},k_nj}(s) &= {}_{\bar{k}_{n-1}}\tilde{\Psi}_{\varepsilon,\mathbb{D},k_nj}(s) = {}_{\bar{k}_{n-1}}\tilde{\phi}_{\varepsilon,k_nj}(s)\,{}_{\bar{k}_{n-1}}\tilde{p}_{\varepsilon,k_nj} \\
&\quad + \sum_{r \in {}_{\bar{k}_n}\bar{\mathbb{D}}} {}_{\bar{k}_n}\Psi_{\varepsilon,\mathbb{D},rj}(s)\,{}_{\bar{k}_{n-1}}\tilde{\phi}_{\varepsilon,k_nr}(s)\,{}_{\bar{k}_{n-1}}\tilde{p}_{\varepsilon,k_nr}
\end{aligned}
$$

$$= {}_{\bar{k}_{n-1}}\tilde{\phi}_{\varepsilon,k_nj}(s)\, {}_{\bar{k}_{n-1}}\tilde{p}_{\varepsilon,k_nj} + \sum_{r \in {}_{\bar{k}_n}\bar{\mathbb{D}}} \Psi_{\varepsilon,\mathbb{D},rj}(s)\, {}_{\bar{k}_{n-1}}\tilde{\phi}_{\varepsilon,k_nr}(s)\, {}_{\bar{k}_{n-1}}\tilde{p}_{\varepsilon,k_nr}, \ s \geq 0, j \in \mathbb{D}. \tag{10.69}$$

Lemma 10.8 *Let the conditions* $\mathbf{G}_{\mathbb{D}} = {}_{\bar{k}_0}\mathbf{G}_{\mathbb{D}}$, $\mathbf{H}_{\mathbb{D}} = {}_{\bar{k}_0}\mathbf{H}_{\mathbb{D}}$ *be satisfied. Then, for every* $n = 0, \ldots, \bar{m} - 1$, *the conditions* ${}_{\bar{k}_n}\mathbf{G}_{\mathbb{D}}$, ${}_{\bar{k}_n}\mathbf{H}_{\mathbb{D}}$, ${}_{\bar{k}_n}\tilde{\mathbf{G}}_{\mathbb{D}}$, ${}_{\bar{k}_n}\tilde{\mathbf{H}}_{\mathbb{D}}$ *are also satisfied, and the following relations take place, for* $\varepsilon \in (0, 1]$:

$$\mathsf{P}_i\{\tau_{\varepsilon,\mathbb{D}} = {}_{\bar{k}_n}\tau_{\varepsilon,\mathbb{D}} = {}_{\bar{k}_n}\tilde{\tau}_{\varepsilon,\mathbb{D}}, \ \eta_\varepsilon(\tau_{\varepsilon,\mathbb{D}}) = {}_{\bar{k}_n}\eta_\varepsilon({}_{\bar{k}_n}\tau_{\varepsilon,\mathbb{D}}) = {}_{\bar{k}_n}\tilde{\eta}_\varepsilon({}_{\bar{k}_n}\tilde{\tau}_{\varepsilon,\mathbb{D}})\} = 1, i \in {}_{\bar{k}_n}\mathbb{X}. \tag{10.70}$$

Lemma 10.9 *Let the conditions* $\mathbf{G}_{\mathbb{D}} = {}_{\bar{k}_0}\mathbf{G}_{\mathbb{D}}$, $\mathbf{H}_{\mathbb{D}} = {}_{\bar{k}_0}\mathbf{H}_{\mathbb{D}}$ *hold. Then, for every* $n = 0, \ldots, \bar{m} - 1$, *the conditions* ${}_{\bar{k}_n}\mathbf{G}_{\mathbb{D}}$, ${}_{\bar{k}_n}\mathbf{H}_{\mathbb{D}}$, ${}_{\bar{k}_n}\tilde{\mathbf{G}}_{\mathbb{D}}$, ${}_{\bar{k}_n}\tilde{\mathbf{H}}_{\mathbb{D}}$ *are also satisfied, and the following relations take place, for* $\varepsilon \in (0, 1]$:

$$G_{\varepsilon,\mathbb{D},ij}(t) = {}_{\bar{k}_n}G_{\varepsilon,\mathbb{D},ij}(t) = {}_{\bar{k}_n}\tilde{G}_{\varepsilon,\mathbb{D},ij}(t), \ t \geq 0, j \in \mathbb{D}, i \in {}_{\bar{k}_n}\bar{\mathbb{D}}. \tag{10.71}$$

Lemma 10.10 *Let the conditions* $\mathbf{G}_{\mathbb{D}} = {}_{\bar{k}_0}\mathbf{G}_{\mathbb{D}}$, $\mathbf{H}_{\mathbb{D}} = {}_{\bar{k}_0}\mathbf{H}_{\mathbb{D}}$ *be satisfied. Then, for every* $n = 0, \ldots, \bar{m} - 1$, *the conditions* ${}_{\bar{k}_n}\mathbf{G}_{\mathbb{D}}$, ${}_{\bar{k}_n}\mathbf{H}_{\mathbb{D}}$, ${}_{\bar{k}_n}\tilde{\mathbf{G}}_{\mathbb{D}}$, ${}_{\bar{k}_n}\tilde{\mathbf{H}}_{\mathbb{D}}$ *are also satisfied, and the following relations take place, for* $\varepsilon \in (0, 1]$:

$$\Psi_{\varepsilon,\mathbb{D},ij}(s) = {}_{\bar{k}_n}\Psi_{\varepsilon,\mathbb{D},ij}(s) = {}_{\bar{k}_n}\tilde{\Psi}_{\varepsilon,\mathbb{D},ij}(s), \ s \geq 0, j \in \mathbb{D}, i \in {}_{\bar{k}_n}\bar{\mathbb{D}}. \tag{10.72}$$

Proof Lemmas 10.5–10.7 can be proved by repeated application of Lemmas 10.2–10.4. For $r = 1$, the statements of Lemmas 10.5–10.7 coincide with the statements of Lemmas 10.2–10.4, as applied to the semi-Markov processes ${}_{\bar{k}_0}\tilde{\eta}_\varepsilon(t)$ and ${}_{\bar{k}_1}\eta_\varepsilon(t)$, for $k = k_1$. For $r = 2$, the statements of Lemmas 10.5–10.7 coincide with the statements of Lemmas 10.2–10.4, as applied to the semi-Markov processes ${}_{\bar{k}_1}\tilde{\eta}_\varepsilon(t)$ and ${}_{\bar{k}_2}\eta_\varepsilon(t)$, for $k = k_2$, etc. Finally, for $r = n$, the statements of Lemmas 10.5–10.7 coincide with the statements of Lemmas 10.2–10.4, as applied to the semi-Markov processes ${}_{\bar{k}_{n-1}}\tilde{\eta}_\varepsilon(t)$ and ${}_{\bar{k}_n}\eta_\varepsilon(t)$, for $k = k_n$.

For example, the recurrent procedure described above gives the following relation, which hold, for $r = 1, \ldots, n$, $1 \leq n \leq \bar{m} - 1$:

$$\mathsf{P}_i\{{}_{\bar{k}_{r-1}}\tau_{\varepsilon,\mathbb{D}} = {}_{\bar{k}_{r-1}}\tilde{\tau}_{\varepsilon,\mathbb{D}} = {}_{\bar{k}_r}\tau_{\varepsilon,\mathbb{D}}, \ {}_{\bar{k}_{r-1}}\eta_\varepsilon({}_{\bar{k}_{r-1}}\tau_{\varepsilon,\mathbb{D}}) = {}_{\bar{k}_{r-1}}\tilde{\eta}_\varepsilon({}_{\bar{k}_{r-1}}\tilde{\tau}_{\varepsilon,\mathbb{D}}) = {}_{\bar{k}_r}\eta_\varepsilon({}_{\bar{k}_r}\tau_{\varepsilon,\mathbb{D}})\} = 1, \ j \in \mathbb{D}, i \in {}_{\bar{k}_r}\bar{\mathbb{D}}. \tag{10.73}$$

The relation (10.73) obviously entails the fulfilment of the relation (10.64) given in Lemma 10.5.

Lemmas 10.8–10.10 can be proved by additional application of Lemmas 10.2–10.4 to the semi-Markov processes ${}_{\bar{k}_n}\eta_\varepsilon(t)$ and ${}_{\bar{k}_n}\tilde{\eta}_\varepsilon(t)$. □

10.1.4.2 Distributions of Hitting Times and Permutation of Excluding Sequences of States. We assume now that the number of states m in the phase space $\mathbb{X}$ is larger than 1.

Let the domain $\bar{\mathbb{D}} = \{k_1, \ldots, k_{\bar{m}}\}$. Also, let $\bar{k}_n = \langle k_1, \ldots, k_n \rangle$ be a subsequence of the sequence $\bar{k}_{\bar{m}} = \langle k_1, \ldots, k_{\bar{m}} \rangle$ and $\bar{k}'_{\bar{m}} = \langle k'_1, \ldots, k'_n \rangle$ be some permutation of sequence $\bar{k}_n$, for $n = 0, 1, \ldots, \bar{m} - 1$.

Here, as in Sect. 9.4.1.1, $\bar{k}_0 = \langle \rangle$ is the "empty" subsequence.

The next two lemmas are direct corollaries of Lemmas 9.8 and 9.9.

Lemma 10.11 *Let the domain $\bar{\mathbb{D}} = \{k_1, \ldots, k_{\bar{m}}\}$ and the conditions* **G** *and* **H** *be satisfied for the semi-Markov processes $\eta_\varepsilon(t)$. Then, for $n = 0, 1, \ldots, \bar{m} - 1$ and any permutation $\bar{k}'_n = \langle k'_1, \ldots, k'_n \rangle$ of the subsequence $\bar{k}_n = \langle k_1, \ldots, k_n \rangle$, the following relation takes place, for $\varepsilon \in (0, 1]$:*

$$\mathsf{P}_i\{ {}_{\bar{k}_n}\tau_{\varepsilon,\mathbb{D}} = {}_{\bar{k}'_n}\tau_{\varepsilon,\mathbb{D}}, \ \eta_\varepsilon({}_{\bar{k}_n}\tau_{\varepsilon,\mathbb{D}}) = \eta_\varepsilon({}_{\bar{k}'_n}\tau_{\varepsilon,\mathbb{D}})\} = 1, \ i \in {}_{\bar{k}_n}\bar{\mathbb{D}}. \tag{10.74}$$

Lemma 10.12 *Let the domain $\bar{\mathbb{D}} = \{k_1, \ldots, k_{\bar{m}}\}$ and the conditions* **G** *and* **H** *be satisfied for the semi-Markov processes $\eta_\varepsilon(t)$. Then, for $n = 0, 1, \ldots, \bar{m} - 1$ and any permutation $\bar{k}'_n = \langle k'_1, \ldots, k'_n \rangle$ of the subsequence $\bar{k}_n = \langle k_1, \ldots, k_n \rangle$, the following relation takes place, for $\varepsilon \in (0, 1]$:*

$${}_{\bar{k}_n}G_{\varepsilon,ij}(t) = {}_{\bar{k}'_n}G_{\varepsilon,ij}(t), \ t \geq 0, j \in \mathbb{D}, i \in {}_{\bar{k}_n}\bar{\mathbb{D}}. \tag{10.75}$$

10.2 Weak Asymptotics for Distributions of Hitting Times

In this section, we present limit theorems on the weak convergence of distributions of hitting times for regularly and singularly perturbed finite semi-Markov processes.

10.2.1 Asymptotics for Hitting Probabilities

Further, we assume that the domain $\bar{\mathbb{D}} = \{k_1, \ldots, k_{\bar{m}}\}$ and $\bar{k}_n = \langle k_1, \ldots, k_n \rangle, 1 \leq n \leq \bar{m}$ are subsequences of the sequence $\bar{k}_{\bar{m}} = \langle k_1, \ldots, k_{\bar{m}} \rangle$.

Obviously, for any $i \in \bar{\mathbb{D}}$, there exists the unique $1 \leq N(\bar{k}_m, i) \leq \bar{m}$ such that,

$$k_{N(\bar{k}_m, i)} = i. \tag{10.76}$$

Note that, $N(\bar{k}_m, k_n) = n$, and, thus, $k_{N(\bar{k}_m, k_n)} = k_n$, for $n = 1, \ldots, \bar{m}$.

First, let us give conditions of convergence for the hitting probabilities defined, for $i \in \bar{\mathbb{D}}, j \in \mathbb{D}$ and $\varepsilon \in [0, 1]$,

$$\begin{aligned} P_{\varepsilon,\mathbb{D},ij} &= \mathsf{P}_i\{\eta_\varepsilon(\tau_{\varepsilon,\mathbb{D}}) = j\} \\ &= G_{\varepsilon,\mathbb{D},ij}(\infty) = \Psi_{\varepsilon,\mathbb{D},ij}(0). \end{aligned} \tag{10.77}$$

Note that the conditions $\mathbf{G}_{\mathbb{D}}$ and $\mathbf{H}_{\mathbb{D}}$ imply that, for any $i \in \bar{\mathbb{D}}$ and $\varepsilon \in (0, 1]$,

$$P_{\varepsilon,\mathbb{D},ij} \geq 0, j \in \mathbb{D}, \sum_{j \in \mathbb{D}} P_{\varepsilon,\mathbb{D},ij} = 1. \tag{10.78}$$

Taking $s = 0$ in the relation (10.69), we get the following backward recurrent relations for the hitting probabilities $P_{\varepsilon,\mathbb{D},k_n j}, n = \bar{m}, \ldots, 1$, for every $j \in \mathbb{D}$ and $\varepsilon \in (0, 1]$:

$$P_{\varepsilon,\mathbb{D},k_n j} = {}_{\bar{k}_{n-1}}\tilde{p}_{\varepsilon,k_n j} + \sum_{n+1 \leq l \leq \bar{m}} P_{\varepsilon,\mathbb{D},k_l j}\, {}_{\bar{k}_{n-1}}\tilde{p}_{\varepsilon,k_n k_l}, \ n = \bar{m}, \ldots, 1. \tag{10.79}$$

Let us introduce the following sets, for $i \in \bar{\mathbb{D}}$ and $\varepsilon \in [0, 1]$:

$$\mathbb{Y}_{\varepsilon,\mathbb{D},i} = \{j \in \mathbb{D} : P_{\varepsilon,\mathbb{D},ij} > 0\}. \tag{10.80}$$

Lemma 10.13 *Let the conditions* $\mathbf{G}_{\mathbb{D}}$ *and* $\mathbf{H}_{\mathbb{D}}$ *be satisfied. Then, the sets,* $\mathbb{Y}_{\varepsilon,\mathbb{D},i}, i \in \bar{\mathbb{D}}$ *do not depend on* $\varepsilon \in (0, 1]$, *i.e., for* $i \in \bar{\mathbb{D}}$,

$$\mathbb{Y}_{\varepsilon,\mathbb{D},i} = \mathbb{Y}_{1,\mathbb{D},i}, \ \varepsilon \in (0, 1]. \tag{10.81}$$

Proof According to the statement **(i)** of Theorem 10.1, either ${}_{\bar{k}_{n-1}}\tilde{p}_{\varepsilon,k_n j} = 0, \varepsilon \in (0, 1]$, or ${}_{\bar{k}_{n-1}}\tilde{p}_{\varepsilon,k_n j} > 0, \varepsilon \in (0, 1]$, for $j \in \mathbb{X}, n = 1, \ldots, \bar{m}$.

The relation (10.79) taken for $n = \bar{m}$, implies that $P_{\varepsilon,\mathbb{D},k_{\bar{m}},j} = {}_{\bar{k}_{\bar{m}-1}}\tilde{p}_{\varepsilon,k_{\bar{m}} j}$, for $j \in \mathbb{X}$. The above remarks imply that the sets $\mathbb{Y}_{\varepsilon,\mathbb{D},k_{\bar{m}}} = \mathbb{Y}_{1,\mathbb{D},k_{\bar{m}}}, \varepsilon \in (0, 1]$.

Then, the above remarks and the relation (10.79) taken for $n = \bar{m} - 1$, imply that the sets $\mathbb{Y}_{\varepsilon,\mathbb{D},k_{\bar{m}-1}} = \mathbb{Y}_{1,\mathbb{D},k_{\bar{m}-1}}, \varepsilon \in (0, 1]$.

Continuing in this way, we get that, $\mathbb{Y}_{\varepsilon,\mathbb{D},k_n} = \mathbb{Y}_{1,\mathbb{D},k_n}$, for any $n = \bar{m}, \ldots, 1$ and $\varepsilon \in (0, 1]$.

It remains noting that the functions $P_{\cdot,\mathbb{D},ij} = P_{\cdot,\mathbb{D},k_{N(\bar{k}_m,i)},j}$, for $i \in \bar{\mathbb{D}}, j \in \mathbb{D}$ and the sets $\mathbb{Y}_{\varepsilon,\mathbb{D},i} = \mathbb{Y}_{\varepsilon,\mathbb{D},k_{N(\bar{k}_m,i)}}$, for $i \in \bar{\mathbb{D}}$ and $\varepsilon \in (0, 1]$. □

The following useful lemma takes place.

Lemma 10.14 *Let the conditions* $\mathbf{G}_{\mathbb{D}}$, $\mathbf{H}_{\mathbb{D}}$, *and* $\mathbf{I}_{\mathbb{D}}$ *be satisfied. Then:*

(i) *The functions* $P_{\cdot,\mathbb{D},k_n j} \in \mathcal{H}$, *for* $j \in \mathbb{D}, n = 1, \ldots, \bar{m}$, *and the following relation takes place, for* $j \in \mathbb{D}, n = 1, \ldots, \bar{m}$:

$$P_{\varepsilon,\mathbb{D},k_n j} \longrightarrow P_{0,\mathbb{D},k_n j} \text{ as } \varepsilon \longrightarrow 0, \tag{10.82}$$

where limits $P_{0,\mathbb{D},k_n j}$ *satisfy the following relations,* $P_{0,\mathbb{D},k_n j} \geq 0, j \in \mathbb{D}, \sum_{j \in \mathbb{D}} P_{0,\mathbb{D},k_n j} = 1$, *for* $n = \bar{m}, \ldots, 1$,

(ii) *The limits* $P_{0,\mathbb{D},k_n j}, j \in \mathbb{D}, n = 1, \ldots, \bar{m}$ *are given by the relations (10.84), (10.85), and (10.86).*

Proof The statement **(ii)** of Theorem 10.1 implies that, for every $j \in {}_{\bar{k}_{n-1}}\mathbb{X} = {}_{\bar{k}_{n-1}}\bar{\mathbb{D}} \cup \mathbb{D}, n = 1, \ldots, \bar{m}$,

$${}_{\bar{k}_{n-1}}\tilde{p}_{\varepsilon,k_n j} \longrightarrow {}_{\bar{k}_{n-1}}\tilde{p}_{0,k_n j} \text{ as } \varepsilon \longrightarrow 0. \tag{10.83}$$

The above remarks and the relation (10.79), taken for $n = \bar{m}$, imply that the functions $P_{\cdot,\mathbb{D},k_{\bar{m}}j} \in \mathcal{H}, j \in \mathbb{D}$ and that the following relation takes place, for $j \in \mathbb{D}$:

$$\begin{aligned} P_{\varepsilon,\mathbb{D},k_{\bar{m}}j} &= {}_{\bar{k}_{\bar{m}-1}}\tilde{p}_{\varepsilon,k_{\bar{m}}j} \\ &\to {}_{\bar{k}_{\bar{m}-1}}\tilde{p}_{0,k_{\bar{m}}j} = P_{0,\mathbb{D},k_{\bar{m}}j} \text{ as } \varepsilon \to 0. \end{aligned} \tag{10.84}$$

Then, the above remarks, the relation (10.79), taken for $n = \bar{m} - 1$, and the relation (10.84) imply that the functions $P_{\cdot,\mathbb{D},k_{\bar{m}-1}j} \in \mathcal{H}, j \in \mathbb{D}$ and that the following relation takes place, for $j \in \mathbb{D}$:

$$\begin{aligned} P_{\varepsilon,\mathbb{D},k_{\bar{m}-1}j} &= {}_{\bar{k}_{\bar{m}-1}}\tilde{p}_{\varepsilon,k_{\bar{m}-1}j} + P_{\varepsilon,\mathbb{D},k_m j}\, {}_{\bar{k}_{\bar{m}-1}}\tilde{p}_{\varepsilon,k_{\bar{m}-1}k_m} \\ &\to {}_{\bar{k}_{\bar{m}-1}}\tilde{p}_{0,k_{\bar{m}-1}j} + P_{0,\mathbb{D},k_m j}\, {}_{\bar{k}_{\bar{m}-1}}\tilde{p}_{0,k_{\bar{m}-1}k_m} \\ &= P_{0,\mathbb{D},k_{\bar{m}-1},j} \text{ as } \varepsilon \to 0. \end{aligned} \tag{10.85}$$

Then, the above remarks and the relation (10.79), taken recurrently for $n = \bar{m}, \ldots, 1$ imply that, the functions $P_{\cdot,\mathbb{D},k_n j} \in \mathcal{H}, j \in \mathbb{D}$ and that the following relation takes place, for $j \in \mathbb{D}$:

$$\begin{aligned} P_{\varepsilon,\mathbb{D},k_n j} &= {}_{\bar{k}_{n-1}}\tilde{p}_{\varepsilon,k_n j} + \sum_{n+1 \leq l \leq \bar{m}} P_{\varepsilon,\mathbb{D},k_l j}\, {}_{\bar{k}_{n-1}}\tilde{p}_{\varepsilon,k_n k_l} \\ &\to {}_{\bar{k}_{n-1}}\tilde{p}_{0,k_n j} + \sum_{n+1 \leq l \leq \bar{m}} P_{0,\mathbb{D},k_l j}\, {}_{\bar{k}_{n-1}}\tilde{p}_{0,k_n k_l} \\ &= P_{0,\mathbb{D},k_n j} \text{ as } \varepsilon \to 0. \end{aligned} \tag{10.86}$$

The relation (10.78) obviously implies that the limits $P_{0,\mathbb{D},k_n j}$ satisfy the additional relations $P_{0,\mathbb{D},k_n j} \geq 0, j \in \mathbb{D}, \sum_{j \in \mathbb{D}} P_{0,\mathbb{D},k_n j} = 1$, for $n = \bar{m}, \ldots, 1$.

The proof is complete. □

The relation, $P_{\cdot,\mathbb{D},ij} = P_{\cdot,\mathbb{D},k_{N(\bar{k}_m,i)},j}$, for $i \in \bar{\mathbb{D}}, j \in \mathbb{D}$, makes it possible to rewrite relation (10.82) in an alternative equivalent form, for $j \in \mathbb{D}, i \in \bar{\mathbb{D}}$,

$$P_{\varepsilon,\mathbb{D},ij} \to P_{0,\mathbb{D},ij} \text{ as } \varepsilon \to 0. \tag{10.87}$$

Note that relations (10.81) and (10.82) imply that, for $i \in \bar{\mathbb{D}}$,

$$\mathbb{Y}_{0,\mathbb{D},i} = \{j \in \mathbb{D} : P_{0,\mathbb{D},ij} > 0\} \subseteq \mathbb{Y}_{1,\mathbb{D},i}. \tag{10.88}$$

It should be also mentioned that the limits $P_{0,\mathbb{D},ij}, i \in \bar{\mathbb{D}}, j \in \mathbb{D}$ may not coincide with the corresponding hitting probabilities for the Markov chain $\eta_{0,n}$, with the matrix of transition probabilities $\mathbf{P}_{0,\mathbb{D}} = \|p_{0,ij}\|_{i \in \bar{\mathbb{D}}, j \in \mathbb{D}}$ given by the basic perturbation condition $\mathbf{I}_{\mathbb{D}}$. In fact, it is possible that the random variable $\nu_{0,\mathbb{D}} = \min(n \geq 1, \eta_{0,n} \in \mathbb{D})$ is improper, and, thus, the probability $\mathsf{P}_i\{\nu_{0,\mathbb{D}} < \infty\} = \sum_{j \in \mathbb{D}} \mathsf{P}_i\{\nu_{0,\mathbb{D}} < \infty, \eta_{0,\nu_{0,\mathbb{D}}} = j\}$ takes a value less than 1, for some $i \in \bar{\mathbb{D}}$.

10.2.2 Weak Asymptotic for Distributions of Hitting Times for the Case Where the Initial State Belongs to Domain $\bar{\mathbb{D}}$

In what follows, we assume that the domain $\bar{\mathbb{D}} = \{k_1, \ldots, k_{\bar{m}}\}$ and $\bar{k}_n = \langle k_1, \ldots, k_n \rangle$, for $1 \leq n \leq \bar{m}$ are subsequences of the sequence $\bar{k}_{\bar{m}} = \langle k_1, \ldots, k_{\bar{m}} \rangle$.

In this section, we consider the case, where the conditions $\mathbf{G}_{\mathbb{D}}$, $\mathbf{H}_{\mathbb{D}}$, $\mathbf{I}_{\mathbb{D},\mathcal{H}}$, $\mathbf{J}_{\mathbb{D}}$, $\mathbf{K}_{\mathbb{D}}$, $\mathbf{L}_{\mathbb{D},\mathcal{H}}$ (and, thus, also the conditions $\mathbf{I}_{\mathbb{D}}$ and $\mathbf{L}_{\mathbb{D}}$), and ${}_{\bar{\mathbf{k}}_{\bar{m}-1}}\mathbf{M}_{\mathbb{D}}$ are satisfied for the semi-Markov processes $\eta_\varepsilon(t)$.

10.2.2.1 The Case of the Most Absorbing Initial State in the Domain $\bar{\mathbb{D}}$. First, let us consider the case, where the initial state is $k_{\bar{m}}$, which is the most absorbing state in domain $\bar{\mathbb{D}}$.

In this case, domain ${}_{\bar{k}_{\bar{m}-1}}\bar{\mathbb{D}} = \bar{\mathbb{D}} \setminus \{k_1, \ldots, k_{\bar{m}-1}\} = \{k_{\bar{m}}\}$ is a one-state set.

In what follows, we use the following normalisation function:

$$\check{u}_{\varepsilon,k_{\bar{m}}} = {}_{\bar{k}_{\bar{m}-1}}\tilde{u}_{\varepsilon,k_{\bar{m}}} = \prod_{l=0}^{\bar{m}-1} (1 - {}_{\bar{k}_l}p_{\varepsilon,k_{\bar{m}}k_{\bar{m}}})^{-1} u_{\varepsilon,k_{\bar{m}}}. \tag{10.89}$$

The following theorem takes place.

Theorem 10.2 *Let the conditions* $\mathbf{G}_{\mathbb{D}}$, $\mathbf{H}_{\mathbb{D}}$, $\mathbf{I}_{\mathbb{D}}$, $\mathbf{I}_{\mathbb{D},\mathcal{H}}$, $\mathbf{L}_{\mathbb{D}}$, $\mathbf{L}_{\mathbb{D},\mathcal{H}}$, $\mathbf{J}_{\mathbb{D}}$, $\mathbf{K}_{\mathbb{D}}$, *and* ${}_{\bar{\mathbf{k}}_{\bar{m}-1}}\mathbf{M}_{\mathbb{D}}$ *be satisfied for the semi-Markov processes* $\eta_\varepsilon(t)$. *Then:*

(i) *The following asymptotic relation takes place, for* $j \in \mathbb{D}$:

$$\begin{aligned} G_{\varepsilon,\mathbb{D},k_{\bar{m}}j}(\cdot\, \check{u}_{\varepsilon,k_{\bar{m}}}) &= F_{\varepsilon,\mathbb{D},k_{\bar{m}}j}(\cdot) P_{\varepsilon,\mathbb{D},k_{\bar{m}}j} \\ &\Rightarrow G_{0,\mathbb{D},k_{\bar{m}}j}(\cdot) = F_{0,\mathbb{D},k_{\bar{m}}j}(\cdot) P_{0,\mathbb{D},k_{\bar{m}}j} \ \text{as}\ \varepsilon \to 0, \end{aligned} \tag{10.90}$$

where: (a) $\check{u}_{\cdot,k_{\bar{m}}}$ *is a normalisation function, which belongs to the family of asymptotically comparable functions* $\mathcal{H}$, *(b)* $F_{0,\mathbb{D},k_{\bar{m}}j}(\cdot)$, $j \in \mathbb{D}$ *is a proper distribution function, i.e.,* $F_{0,\mathbb{D},k_{\bar{m}}j}(\infty) = 1$, *for* $j \in \mathbb{D}$, *(c)* $P_{0,\mathbb{D},k_{\bar{m}}j}$, $j \in \mathbb{D}$ *is a discrete distribution, i.e.* $P_{0,\mathbb{D},k_{\bar{m}}j} \geq 0$, $j \in \mathbb{D}$ *and* $\sum_{j\in\mathbb{D}} P_{0,\mathbb{D},k_{\bar{m}}j} = 1$.

(ii) *The normalisation function* $\check{u}_{\cdot,k_{\bar{m}}}$ *is given by the relation (10.89).*

(iii) *The Laplace transforms* $\Psi_{0,\mathbb{D},k_{\bar{m}}j}(\cdot)$, $j \in \mathbb{D}$ *of distributions* $G_{0,\mathbb{D},k_{\bar{m}}j}(\cdot)$, $j \in \mathbb{D}$ *are given by the relations (10.95).*

(iv) *The limiting distribution functions* $F_{0,\mathbb{D},k_{\bar{m}}j}(\cdot) = {}_{\bar{k}_{\bar{m}-1}}\tilde{F}_{0,k_{\bar{m}}j}(\cdot)$, $j \in \mathbb{D}$ *are not concentrated at zero, i.e.,* $F_{0,\mathbb{D},k_{\bar{m}}j}(0) < 1$, $j \in \mathbb{D}$ *and, thus, the distribution function* $F_{0,\mathbb{D},k_{\bar{m}}}(\cdot) = \sum_{j\in\mathbb{D}} F_{0,\mathbb{D},k_{\bar{m}}j}(\cdot) P_{0,\mathbb{D},k_{\bar{m}}j}$ *is also not concentrated at zero, i.e.,* $F_{0,\mathbb{D},k_{\bar{m}}}(0) < 1$.

(v) *The limits of hitting probabilities* $P_{0,\mathbb{D},k_{\bar{m}}j} = {}_{\bar{k}_{\bar{m}-1}}\tilde{p}_{0,k_{\bar{m}}j}$, $j \in \mathbb{D}$ *are given by the relation (10.84).*

Proof The reduced semi-Markov processes ${}_{\bar{k}_{\bar{m}-1}}\eta_\varepsilon(t)$ and ${}_{\bar{k}_{\bar{m}-1}}\tilde{\eta}_\varepsilon(t)$ have the phase space ${}_{\bar{k}_{\bar{m}-1}}\mathbb{X} = \mathbb{X} \setminus \{k_1, \ldots, k_{\bar{m}-1}\} = \mathbb{D} \cup \{k_{\bar{m}}\}$. Therefore,

$$\mathsf{P}_{k_{\bar{m}}}\{ {}_{\bar{k}_{\bar{m}-1}}\tilde{\tau}_{\varepsilon,\mathbb{D}} = {}_{\bar{k}_{\bar{m}-1}}\tilde{\kappa}_{\varepsilon,1},\ {}_{\bar{k}_{\bar{m}-1}}\tilde{\eta}_\varepsilon({}_{\bar{k}_{\bar{m}-1}}\tilde{\tau}_{\varepsilon,\mathbb{D}}) = {}_{\bar{k}_{\bar{m}-1}}\tilde{\eta}_{\varepsilon,1}\} = 1. \tag{10.91}$$

Lemma 10.10 and the relation (10.91) imply that the following relation takes place, $t \geq 0, j \in \mathbb{D}$ and $\varepsilon \in (0, 1]$:

$$\begin{aligned} G_{\varepsilon,\mathbb{D},k_{\bar{m}}j}(t) &= F_{\varepsilon,\mathbb{D},k_{\bar{m}}j}(t)P_{\varepsilon,\mathbb{D},k_{\bar{m}}j} \\ &= {}_{\bar{k}_{\bar{m}-1}}\tilde{Q}_{\varepsilon,k_{\bar{m}}j}(t) = {}_{\bar{k}_{\bar{m}-1}}\tilde{F}_{\varepsilon,k_{\bar{m}}j}(t)\,{}_{\bar{k}_{\bar{m}-1}}\tilde{p}_{\varepsilon,k_{\bar{m}}j}. \end{aligned} \tag{10.92}$$

where as usual, for $t \geq 0, j \in \mathbb{D}$ such that $P_{\varepsilon,\mathbb{D},k_{\bar{m}}j} = 0$,

$$\begin{aligned} G_{\varepsilon,\mathbb{D},k_{\bar{m}}j}(t) &= F_{\varepsilon,\mathbb{D},k_{\bar{m}}}(t) = \sum_{j \in \mathbb{D}} F_{\varepsilon,\mathbb{D},k_{\bar{m}}j}(t)P_{\varepsilon,\mathbb{D},k_{\bar{m}}j} \\ &= \sum_{j \in \mathbb{D}} {}_{\bar{k}_{\bar{m}-1}}\tilde{F}_{\varepsilon,k_{\bar{m}}j}(t)\,{}_{\bar{k}_{\bar{m}-1}}\tilde{p}_{\varepsilon,k_{\bar{m}}j} = F_{\varepsilon,k_{\bar{m}}}(t). \end{aligned} \tag{10.93}$$

According to Theorem 10.1, the conditions ${}_{\bar{\mathbf{k}}_{\bar{m}-1}}\tilde{\mathbf{G}}_{\mathbb{D}}$, ${}_{\bar{\mathbf{k}}_{\bar{m}-1}}\tilde{\mathbf{H}}_{\mathbb{D}}$, ${}_{\bar{\mathbf{k}}_{\bar{m}-1}}\tilde{\mathbf{I}}_{\mathbb{D}}$, ${}_{\bar{\mathbf{k}}_{\bar{m}-1}}\tilde{\mathbf{I}}_{\mathbb{D},\mathcal{H}}$, ${}_{\bar{\mathbf{k}}_{\bar{m}-1}}\tilde{\mathbf{L}}_{\mathbb{D}}$, ${}_{\bar{\mathbf{k}}_{\bar{m}-1}}\tilde{\mathbf{L}}_{\mathbb{D},\mathcal{H}}$, ${}_{\bar{\mathbf{k}}_{\bar{m}-1}}\tilde{\mathbf{J}}_{\mathbb{D}}$, ${}_{\bar{\mathbf{k}}_{\bar{m}-1}}\tilde{\mathbf{K}}_{\mathbb{D}}$ are satisfied for semi-Markov processes ${}_{\bar{k}_{\bar{m}-1}}\tilde{\eta}_{\varepsilon}(t)$.

According to the conditions ${}_{\bar{\mathbf{k}}_{n-1}}\mathbf{H}_{\mathbb{D}}$, $n = 1, \ldots, \bar{m}$, the probabilities ${}_{\bar{k}_{n-1}}\tilde{p}_{\varepsilon,k_{\bar{m}}k_{\bar{m}}} \in [0, 1)$, for $n = 1, \ldots, \bar{m}$. Also, according to the conditions $\mathbf{J}_{\mathbb{D}}$ the initial normalisation function $u_{\varepsilon,k_m} \in (0, \infty), \varepsilon \in (0, 1]$. Therefore, the function $\check{u}_{\varepsilon,k_{\bar{m}}} \in (0, \infty), \varepsilon \in (0, 1]$. The conditions $\mathbf{I}_{\mathbb{D}}$, $\mathbf{I}_{\mathbb{D},\mathcal{H}}$, $\mathbf{L}_{\mathbb{D}}$, $\mathbf{L}_{\mathbb{D},\mathcal{H}}$ and Lemma 8.3 imply the function $\check{u}_{\cdot,k_{\bar{m}}} \in \mathcal{H}$. Moreover, the corresponding comparability limit, $\lim_{\varepsilon \to 0} \check{u}_{\varepsilon,k_{\bar{m}}} \in (0, \infty]$.

Lemma 10.12 implies that the asymptotic relation (10.84) holds.

Theorem 10.1 implies that the following asymptotic relation, which coincides with the asymptotic relation appearing in the condition ${}_{\bar{\mathbf{k}}_{\bar{m}-1}}\tilde{\mathbf{J}}_{\mathbb{D}}$, holds, for $j \in \mathbb{D}$,

$${}_{\bar{k}_{\bar{m}-1}}F_{\varepsilon,\mathbb{D},k_{\bar{m}}j}(\cdot\,\check{u}_{\varepsilon,k_{\bar{m}}}) \Rightarrow {}_{\bar{k}_{\bar{m}-1}}F_{0,\mathbb{D},k_{\bar{m}}j}(\cdot) \text{ as } \varepsilon \to 0. \tag{10.94}$$

The relations (10.84) and (10.94) obviously imply that the asymptotic relation (10.90) holds, for $j \in \mathbb{D}$.

In this case, the limiting distribution $G_{0,\mathbb{D},k_{\bar{m}}j}(\cdot)$ has the following Laplace transform, for $j \in \mathbb{D}$:

$$\begin{aligned} \Psi_{0,\mathbb{D},k_{\bar{m}}j}(s) &= \int_0^\infty e^{-st} G_{0,\mathbb{D},k_{\bar{m}}j}(dt) \\ &= {}_{\bar{k}_{\bar{m}-1}}\tilde{\phi}_{0,k_{\bar{m}}j}(s)\,{}_{\bar{k}_{\bar{m}-1}}\tilde{p}_{0,k_{\bar{m}}j}, \ s \geq 0. \end{aligned} \tag{10.95}$$

According to the condition ${}_{\bar{\mathbf{k}}_{\bar{m}-1}}\tilde{\mathbf{J}}_{\mathbb{D}}$, the limiting distribution functions ${}_{\bar{k}_{\bar{m}-1}}\tilde{F}_{0,k_{\bar{m}}j}(\cdot)$, $j \in \mathbb{D}$ are not concentrated at zero, i.e., ${}_{\bar{k}_{\bar{m}-1}}\tilde{F}_{0,k_{\bar{m}}j}(0) < 1, j \in \mathbb{D}$.

The relation (10.90) entails the fulfilment of the following asymptotic relation for distribution of hitting times, for $i \in \bar{\mathbb{D}}$:

$$\begin{aligned} F_{\varepsilon,\mathbb{D},k_{\bar{m}}}(\cdot\,\check{u}_{\varepsilon,k_{\bar{m}}}) &= \sum_{j \in \mathbb{D}} G_{\varepsilon,\mathbb{D},k_{\bar{m}}j}(\cdot\,\check{u}_{\varepsilon,k_{\bar{m}}}) \\ &\Rightarrow F_{0,\mathbb{D},k_{\bar{m}}}(\cdot) = \sum_{j \in \mathbb{D}} F_{0,\mathbb{D},k_{\bar{m}}j}(\cdot)P_{0,\mathbb{D},k_{\bar{m}}j} \end{aligned}$$

$$= \sum_{j \in \mathbb{D}} {}_{\bar{k}_{\bar{m}-1}}\tilde{F}_{0,k_{\bar{m}}j}(\cdot)\, {}_{\bar{k}_{\bar{m}-1}}\tilde{p}_{0,k_{\bar{m}}j} \text{ as } \varepsilon \to 0. \tag{10.96}$$

The above remarks and the relation (10.96) obviously imply that,

$$F_{0,\mathbb{D},k_{\bar{m}}}(0) < 1. \tag{10.97}$$

The proof is complete. □

10.2.2.2 The Case of the Second Most Absorbing Initial State in Domain $\bar{\mathbb{D}}$. Let us now consider the case, where the initial state is $k_{\bar{m}-1}$, which can be referred as the second most absorbing state in the domain $\bar{\mathbb{D}}$.

The reduced semi-Markov processes ${}_{\bar{k}_{\bar{m}-2}}\eta_\varepsilon(t)$ and ${}_{\bar{k}_{\bar{m}-2}}\tilde{\eta}_\varepsilon(t)$ have the phase space ${}_{\bar{k}_{\bar{m}-2}}\mathbb{X} = \mathbb{X} \setminus \{k_1, \ldots, k_{\bar{m}-2}\} = \mathbb{D} \cup \{k_{\bar{m}-1}, k_{\bar{m}}\}$, while domain ${}_{\bar{k}_{\bar{m}-2}}\bar{\mathbb{D}} = \bar{\mathbb{D}} \setminus \{k_1, \ldots, k_{\bar{m}-2}\} = \{k_{\bar{m}-1}, k_{\bar{m}}\}$ is a two-state set.

In what follows, we use the following normalisation function:

$$\check{u}_{\varepsilon,k_{\bar{m}-1}} = \begin{cases} {}_{\bar{k}_{\bar{m}-1}}\tilde{u}_{\varepsilon,k_{\bar{m}}} & \text{if } {}_{\bar{k}_{\bar{m}-2}}\tilde{p}_{0,k_{\bar{m}-1}k_{\bar{m}}} > 0, \\ {}_{\bar{k}_{\bar{m}-2}}\tilde{u}_{\varepsilon,k_{\bar{m}-1}} & \text{if } {}_{\bar{k}_{\bar{m}-2}}\tilde{p}_{0,k_{\bar{m}-1}k_{\bar{m}}} = 0, \end{cases}$$

$$= \begin{cases} \prod_{l=0}^{\bar{m}-1}(1 - {}_{\bar{k}_l}p_{\varepsilon,k_{\bar{m}}k_{\bar{m}}})^{-1}u_{\varepsilon,k_{\bar{m}}} & \text{if } {}_{\bar{k}_{\bar{m}-2}}\tilde{p}_{0,k_{\bar{m}-1}k_{\bar{m}}} > 0, \\ \prod_{l=0}^{\bar{m}-2}(1 - {}_{\bar{k}_l}p_{\varepsilon,k_{\bar{m}-1}k_{\bar{m}-1}})^{-1}u_{\varepsilon,k_{\bar{m}-1}} & \text{if } {}_{\bar{k}_{\bar{m}-2}}\tilde{p}_{0,k_{\bar{m}-1}k_{\bar{m}}} = 0. \end{cases} \tag{10.98}$$

The following theorem takes place.

Theorem 10.3 *Let the conditions* $\mathbf{G}_\mathbb{D}$, $\mathbf{H}_\mathbb{D}$, $\mathbf{I}_\mathbb{D}$, $\mathbf{I}_{\mathbb{D},\mathcal{H}}$, $\mathbf{L}_\mathbb{D}$, $\mathbf{L}_{\mathbb{D},\mathcal{H}}$, $\mathbf{J}_\mathbb{D}$, $\mathbf{K}_\mathbb{D}$, *and* ${}_{\bar{\mathbf{k}}_{\bar{\mathbf{m}}-1}}\mathbf{M}_\mathbb{D}$ *be satisfied for the semi-Markov processes* $\eta_\varepsilon(t)$. *Then*:

(i) *The following asymptotic relation takes place, for* $j \in \mathbb{D}$,

$$\begin{aligned} G_{\varepsilon,\mathbb{D},k_{\bar{m}-1}j}(\cdot\, \check{u}_{\varepsilon,k_{\bar{m}-1}}) &\Rightarrow G_{0,\mathbb{D},k_{\bar{m}-1}j}(\cdot) \\ &= F_{0,\mathbb{D},k_{\bar{m}-1}j}(\cdot)P_{0,\mathbb{D},k_{\bar{m}-1}j} \text{ as } \varepsilon \to 0, \end{aligned} \tag{10.99}$$

where: (a) $\check{u}_{\cdot,k_{\bar{m}-1}}$ *is a normalisation function, which belongs to the family of asymptotically comparable functions* $\mathcal{H}$, *(b)* $F_{0,\mathbb{D},k_{\bar{m}-1}j}(\cdot)$, $j \in \mathbb{D}$ *is a proper distribution function, i.e.,* $F_{0,\mathbb{D},k_{\bar{m}}j}(\infty) = 1$, *for* $j \in \mathbb{D}$, *(c)* $P_{0,\mathbb{D},k_{\bar{m}}j}$, $j \in \mathbb{D}$ *is a discrete distribution, i.e.* $P_{0,\mathbb{D},k_{\bar{m}}j} \geq 0$, $j \in \mathbb{D}$ *and* $\sum_{j\in\mathbb{D}} P_{0,\mathbb{D},k_{\bar{m}}j} = 1$.

(ii) *The normalisation function* $\check{u}_{\cdot,k_{\bar{m}-1}}$ *is given by relation (10.98).*

(iii) *The Laplace transforms* $\Psi_{0,\mathbb{D},k_{\bar{m}-1}j}(\cdot)$, $j \in \mathbb{D}$ *of distributions* $G_{0,\mathbb{D},k_{\bar{m}-1}j}(\cdot)$, $j \in \mathbb{D}$ *are given by the relation (10.103), in the case where* ${}_{\bar{k}_{\bar{m}-2}}\tilde{p}_{0,k_{\bar{m}-1}k_{\bar{m}}} > 0$, *or by the relation (10.112), in the case where* ${}_{\bar{k}_{\bar{m}-2}}\tilde{p}_{0,k_{\bar{m}-1}k_{\bar{m}}} = 0$.

(iv) *The limiting distribution function* $G_{0,\mathbb{D},k_{\bar{m}-1}}(\cdot) = \sum_{j\in\mathbb{D}} F_{0,\mathbb{D},k_{\bar{m}-1}j}(\cdot)P_{0,\mathbb{D},k_{\bar{m}-1}j}$ *is not concentrated at zero, i.e.,* $G_{0,\mathbb{D},k_{\bar{m}-1}}(0) < 1$.

(v) *The limits of hitting probabilities* $P_{0,\mathbb{D},k_{\bar{m}-1}j}$, $j \in \mathbb{D}$ *are given by the relation (10.85).*

Proof Lemmas 10.9 and 10.10 imply that the following relation takes place, for $s \geq 0$, $j \in \mathbb{D}$ and $\varepsilon \in (0, 1]$:

$$\begin{aligned}\Psi_{\varepsilon,\mathbb{D},k_{\bar{m}-1}j}(s) = {}& {}_{\bar{k}_{\bar{m}-2}}\tilde{\phi}_{\varepsilon,k_{\bar{m}-1}j}(s)\, {}_{\bar{k}_{\bar{m}-2}}\tilde{p}_{\varepsilon,k_{\bar{m}-1}j} \\ &+ \Psi_{\varepsilon,\mathbb{D},k_{\bar{m}}j}(s)\, {}_{\bar{k}_{\bar{m}-2}}\tilde{\phi}_{\varepsilon,k_{\bar{m}-1}k_{\bar{m}}}(s)\, {}_{\bar{k}_{\bar{m}-2}}\tilde{p}_{\varepsilon,k_{\bar{m}-1}k_{\bar{m}}}. \end{aligned} \tag{10.100}$$

Theorems 10.1 imply that the conditions ${}_{\bar{\mathbf{k}}_{\bar{m}-2}}\mathbf{G}_{\mathbb{D}}$, ${}_{\bar{\mathbf{k}}_{\bar{m}-2}}\mathbf{H}_{\mathbb{D}}$, ${}_{\bar{\mathbf{k}}_{\bar{m}-2}}\mathbf{I}_{\mathbb{D}}$, ${}_{\bar{\mathbf{k}}_{\bar{m}-2}}\mathbf{I}_{\mathbb{D},\mathcal{H}}$, ${}_{\bar{\mathbf{k}}_{\bar{m}-2}}\mathbf{L}_{\mathbb{D}}$, ${}_{\bar{\mathbf{k}}_{\bar{m}-2}}\mathbf{L}_{\mathbb{D},\mathcal{H}}$, ${}_{\bar{\mathbf{k}}_{\bar{m}-2}}\mathbf{J}_{\mathbb{D}}$, ${}_{\bar{\mathbf{k}}_{\bar{m}-2}}\mathbf{K}_{\mathbb{D}}$ are satisfied for the semi-Markov processes ${}_{\bar{k}_{\bar{m}-2}}\eta_\varepsilon(t)$, and conditions ${}_{\bar{\mathbf{k}}_{\bar{m}-1}}\tilde{\mathbf{G}}_{\mathbb{D}}$, ${}_{\bar{\mathbf{k}}_{\bar{m}-1}}\tilde{\mathbf{H}}_{\mathbb{D}}$, ${}_{\bar{\mathbf{k}}_{\bar{m}-1}}\tilde{\mathbf{I}}_{\mathbb{D}}$, ${}_{\bar{\mathbf{k}}_{\bar{m}-1}}\tilde{\mathbf{I}}_{\mathbb{D},\mathcal{H}}$, ${}_{\bar{\mathbf{k}}_{\bar{m}-1}}\tilde{\mathbf{L}}_{\mathbb{D}}$, ${}_{\bar{\mathbf{k}}_{\bar{m}-1}}\tilde{\mathbf{L}}_{\mathbb{D},\mathcal{H}}$, ${}_{\bar{\mathbf{k}}_{\bar{m}-1}}\tilde{\mathbf{J}}_{\mathbb{D}}$, ${}_{\bar{\mathbf{k}}_{\bar{m}-1}}\tilde{\mathbf{K}}_{\mathbb{D}}$ hold for the semi-Markov processes ${}_{\bar{k}_{\bar{m}-1}}\tilde{\eta}_\varepsilon(t)$.

According to the conditions ${}_{\bar{k}_{n-1}}\mathbf{H}_{\mathbb{D}}$, $n = 1, \ldots, \bar{m}$, the probabilities ${}_{\bar{k}_{n-1}}p_{\varepsilon,k_{\bar{m}}k_{\bar{m}}}$, $n = 1, \ldots, \bar{m}$ and ${}_{\bar{k}_{n-1}}p_{\varepsilon,k_{\bar{m}-1}k_{\bar{m}-1}}$, $n = 1, \ldots, \bar{m} - 1$ take values in the interval $[0, 1)$. Also, according to the condition $\mathbf{J}_{\mathbb{D}}$, the initial normalisation functions $u_{\varepsilon,k_{\bar{m}}}$, $u_{\varepsilon,k_{\bar{m}-1}}$ take values in the interval $(0, \infty)$. Thus, the normalisation function $\check{u}_{\varepsilon,k_{\bar{m}-1}}$ given by the relation (10.98) also takes values in the interval $(0, \infty)$. The conditions $\mathbf{I}_{\mathbb{D}}$, $\mathbf{I}_{\mathbb{D},\mathcal{H}}$, $\mathbf{L}_{\mathbb{D}}$, $\mathbf{L}_{\mathbb{D},\mathcal{H}}$, and Lemma 8.3 imply that the function $\check{u}_{\varepsilon,k_{\bar{m}-1}} \in \mathcal{H}$. Moreover, the corresponding comparability limit, $\lim_{\varepsilon\to 0}\check{u}_{\varepsilon,k_{\bar{m}-1}} \in (0, \infty]$.

First, let us consider the case, where the limiting probability,

$${}_{\bar{k}_{\bar{m}-2}}\tilde{p}_{0,k_{\bar{m}-1}k_{\bar{m}}} > 0. \tag{10.101}$$

In this case, the normalisation function $\check{u}_{\varepsilon,k_{\bar{m}-1}}$ takes the following form:

$$\check{u}_{\varepsilon,k_{\bar{m}-1}} = {}_{\bar{k}_{\bar{m}-1}}\tilde{u}_{\varepsilon,k_{\bar{m}}}. \tag{10.102}$$

The above remarks, Theorem 10.1, the relation (10.90) given in Theorem 10.2, and the relation (10.100) imply that the following relation takes place, for $j \in \mathbb{D}$:

$$\begin{aligned}
&\Psi_{\varepsilon,\mathbb{D},k_{\bar{m}-1}j}(s/\check{u}_{\varepsilon,k_{\bar{m}-1}}) \\
&\quad = {}_{\bar{k}_{\bar{m}-2}}\tilde{\phi}_{\varepsilon,k_{\bar{m}-1}j}\Big((1 - {}_{\bar{k}_{\bar{m}-1}}p_{\varepsilon,k_{\bar{m}}k_{\bar{m}}})\frac{{}_{\bar{k}_{\bar{m}-2}}\tilde{u}_{\varepsilon,k_{\bar{m}-1}}}{{}_{\bar{k}_{\bar{m}-2}}\tilde{u}_{\varepsilon,k_{\bar{m}}}}s/\,{}_{\bar{k}_{\bar{m}-2}}\tilde{u}_{\varepsilon,k_{\bar{m}-1}}\Big) \\
&\qquad \times {}_{\bar{k}_{\bar{m}-2}}\tilde{p}_{\varepsilon,k_{\bar{m}-1}j} \\
&\qquad + \Psi_{\varepsilon,\mathbb{D},{}_{\bar{k}_{\bar{m}}}j}(s/\,{}_{\bar{k}_{\bar{m}-1}}\tilde{u}_{\varepsilon,k_{\bar{m}}}) \\
&\qquad \times {}_{\bar{k}_{\bar{m}-2}}\tilde{\phi}_{\varepsilon,k_{\bar{m}-1}k_{\bar{m}}}\Big((1 - {}_{\bar{k}_{\bar{m}-1}}p_{\varepsilon,k_{\bar{m}}k_{\bar{m}}})\frac{{}_{\bar{k}_{\bar{m}-2}}\tilde{u}_{\varepsilon,k_{\bar{m}-1}}}{{}_{\bar{k}_{\bar{m}-2}}\tilde{u}_{\varepsilon,k_{\bar{m}}}}s/\,{}_{\bar{k}_{\bar{m}-2}}\tilde{u}_{\varepsilon,k_{\bar{m}-1}}\Big) \\
&\qquad \times {}_{\bar{k}_{\bar{m}-2}}\tilde{p}_{\varepsilon,k_{\bar{m}-1}k_{\bar{m}}} \\
&\quad \to {}_{\bar{k}_{\bar{m}-2}}\tilde{\phi}_{0,k_{\bar{m}-1}j}((1 - {}_{\bar{k}_{\bar{m}-1}}p_{0,k_{\bar{m}}k_{\bar{m}}})\, {}_{k_{\bar{m}-2}}w_{0,k_{\bar{m}-1},k_{\bar{m}}}s)\, {}_{\bar{k}_{\bar{m}-2}}\tilde{p}_{0,k_{\bar{m}-1}j} \\
&\qquad + \Psi_{0,\mathbb{D},k_{\bar{m}}j}(s) \\
&\qquad \times {}_{\bar{k}_{\bar{m}-2}}\tilde{\phi}_{0,k_{\bar{m}-1}k_{\bar{m}}}((1 - {}_{\bar{k}_{\bar{m}-1}}p_{0,k_{\bar{m}}k_{\bar{m}}})\, {}_{k_{\bar{m}-2}}w_{0,k_{\bar{m}-1},k_{\bar{m}}}s) \\
&\qquad \times {}_{\bar{k}_{\bar{m}-2}}\tilde{p}_{0,k_{\bar{m}-1}k_{\bar{m}}}
\end{aligned}$$

$$= {}_{\bar{k}_{\bar{m}-2}}\tilde{\phi}_{0,k_{\bar{m}-1}j}({}_{k_{\bar{m}-2}}\check{w}_{0,k_{\bar{m}-1},k_{\bar{m}}}s)\, {}_{\bar{k}_{\bar{m}-2}}\tilde{p}_{0,k_{\bar{m}-1}j}$$
$$+ \Psi_{0,\mathbb{D},k_{\bar{m}}j}(s)\, {}_{\bar{k}_{\bar{m}-2}}\tilde{\phi}_{0,k_{\bar{m}-1}k_{\bar{m}}}({}_{k_{\bar{m}-2}}\check{w}_{0,k_{\bar{m}-1},k_{\bar{m}}}s)\, {}_{\bar{k}_{\bar{m}-2}}\tilde{p}_{0,k_{\bar{m}-1}k_{\bar{m}}}$$
$$= \Psi_{0,\mathbb{D},k_{\bar{m}-1}j}(s) \text{ as } \varepsilon \to 0, \text{ for } s \geq 0, \tag{10.103}$$

where

$$_{k_{\bar{m}-2}}\check{w}_{0,k_{\bar{m}-1},k_{\bar{m}}} = (1 - {}_{\bar{k}_{\bar{m}-1}}p_{0,k_{\bar{m}}k_{\bar{m}}})\, {}_{k_{\bar{m}-2}}w_{0,k_{\bar{m}-1},k_{\bar{m}}}. \tag{10.104}$$

In this case, the limiting distribution $G_{0,\mathbb{D},k_{\bar{m}-1}j}(\cdot)$ has, for every $j \in \mathbb{D}$, the following Laplace transform $\Psi_{0,\mathbb{D},k_{\bar{m}-1}j}(s) = \int_0^\infty e^{-st} G_{0,\mathbb{D},k_{\bar{m}-1}j}(dt),\ s \geq 0$ given by the relation (10.103).

The relation (10.103) entails the fulfilment of the following asymptotic relation for the Laplace transforms of hitting times:

$$\Psi_{\varepsilon,\mathbb{D},k_{\bar{m}-1}}(s/\check{u}_{\varepsilon,k_{\bar{m}-1}}) = \sum_{j\in\mathbb{D}} \Psi_{\varepsilon,\mathbb{D},k_{\bar{m}-1}}(s/\check{u}_{\varepsilon,k_{\bar{m}-1}j})$$
$$\to \sum_{j\in\mathbb{D}} {}_{\bar{k}_{\bar{m}-2}}\tilde{\phi}_{0,k_{\bar{m}-1}j}({}_{k_{\bar{m}-2}}\check{w}_{0,k_{\bar{m}-1},k_{\bar{m}}}s)\, {}_{\bar{k}_{\bar{m}-2}}\tilde{p}_{0,k_{\bar{m}-1}j}$$
$$+ \Psi_{0,\mathbb{D},k_{\bar{m}}}(s)\, {}_{\bar{k}_{\bar{m}-2}}\tilde{\phi}_{0,k_{\bar{m}-1}k_{\bar{m}}}({}_{k_{\bar{m}-2}}\check{w}_{0,k_{\bar{m}-1},k_{\bar{m}}}s)\, {}_{\bar{k}_{\bar{m}-2}}\tilde{p}_{0,k_{\bar{m}-1}k_{\bar{m}}}$$
$$= \Psi_{0,\mathbb{D},k_{\bar{m}-1}}(s) \text{ as } \varepsilon \to 0, \text{ for } s \geq 0. \tag{10.105}$$

According to Theorem 10.2, the probability $G_{0,k_{\bar{m}}}(0) < 1$, and, thus, the Laplace transform,

$$\Psi_{0,\mathbb{D},\bar{k}_{\bar{m}}}(s) < 1, \text{ for } s > 0. \tag{10.106}$$

The above assumption (10.101) and the relations (10.105) and (10.106) imply that the following inequality holds, for $s > 0$:

$$\Psi_{0,\mathbb{D},k_{\bar{m}-1}}(s) \leq \sum_{j\in\mathbb{D}} {}_{\bar{k}_{\bar{m}-2}}\tilde{p}_{0,k_{\bar{m}-1}j} + \Psi_{0,\mathbb{D},\bar{k}_{\bar{m}}}(s) \cdot {}_{\bar{k}_{\bar{m}-2}}\tilde{p}_{0,k_{\bar{m}-1}k_{\bar{m}}}$$
$$< \sum_{j\in\mathbb{D}} {}_{\bar{k}_{\bar{m}-2}}\tilde{p}_{0,k_{\bar{m}-1}j} + \tilde{p}_{0,k_{\bar{m}-1}k_{\bar{m}}} = 1. \tag{10.107}$$

The relation (10.107) implies that

$$G_{0,\mathbb{D},k_{\bar{m}-1}}(0) < 1. \tag{10.108}$$

Second, let us consider the case, where the limiting probability,

$$_{\bar{k}_{\bar{m}-2}}\tilde{p}_{0,k_{\bar{m}-1}k_{\bar{m}}} = 0. \tag{10.109}$$

In this case, the normalisation function $\check{u}_{\varepsilon,k_{\bar{m}-1}}$ takes the following form:

$$\check{u}_{\varepsilon,k_{\bar{m}-1}} = {}_{\bar{k}_{\bar{m}-2}}\tilde{u}_{\varepsilon,k_{\bar{m}-1}}. \tag{10.110}$$

By Theorem 10.1, the probabilities,

$$ {}_{\bar{k}_{\bar{m}-2}}\tilde{p}_{\varepsilon,k_{\bar{m}-1}k_{\bar{m}}} \longrightarrow {}_{\bar{k}_{\bar{m}-2}}\tilde{p}_{0,k_{\bar{m}-1}k_{\bar{m}}} = 0 \text{ as } \varepsilon \to 0. \tag{10.111} $$

The above remarks, Theorem 10.1, the relation (10.90) given in Theorem 10.2, and the relations (10.100) and (10.111) imply that the following relation takes place, for $j \in \mathbb{D}$:

$$ \begin{aligned} &\Psi_{\varepsilon,\mathbb{D},k_{\bar{m}-1}j}(s/\check{u}_{\varepsilon,k_{\bar{m}-1}}) \\ &\quad = {}_{\bar{k}_{\bar{m}-2}}\tilde{\phi}_{\varepsilon,k_{\bar{m}-1}j}(s/\, {}_{\bar{k}_{\bar{m}-2}}\tilde{u}_{\varepsilon,k_{\bar{m}-1}})\, {}_{\bar{k}_{\bar{m}-2}}\tilde{p}_{\varepsilon,k_{\bar{m}-1}j} \\ &\qquad + \Psi_{\varepsilon,\mathbb{D},\,k_{\bar{m}}j}(s/\, {}_{\bar{k}_{\bar{m}-2}}\tilde{u}_{\varepsilon,k_{\bar{m}-1}}) \\ &\qquad\quad \times {}_{\bar{k}_{\bar{m}-2}}\tilde{\phi}_{\varepsilon,k_{\bar{m}-1}k_{\bar{m}}}(s/\, {}_{\bar{k}_{\bar{m}-2}}\tilde{u}_{\varepsilon,k_{\bar{m}-1}})\, {}_{\bar{k}_{\bar{m}-2}}\tilde{p}_{\varepsilon,k_{\bar{m}-1}k_{\bar{m}}} \\ &\quad \longrightarrow {}_{\bar{k}_{\bar{m}-2}}\tilde{\phi}_{0,k_{\bar{m}-1}j}(s)\, {}_{\bar{k}_{\bar{m}-2}}\tilde{p}_{0,k_{\bar{m}-1}j} \\ &\quad = \Psi_{0,\mathbb{D},k_{\bar{m}-1}j}(s) \text{ as } \varepsilon \to 0, \ \text{for } s \geq 0. \end{aligned} \tag{10.112} $$

In this case, the corresponding limiting distribution $G_{0,\mathbb{D},k_{\bar{m}-1}j}(\cdot)$ has, for every $j \in \mathbb{D}$, the Laplace transform $\Psi_{0,\mathbb{D},k_{\bar{m}-1}j}(s) = \int_0^\infty e^{-st} G_{0,\mathbb{D},k_{\bar{m}-1}j}(dt)$, $s \geq 0$ given by the relation (10.112).

The relation (10.112) entails the fulfilment of the following asymptotic relation for the Laplace transforms of hitting times:

$$ \begin{aligned} \Psi_{\varepsilon,\mathbb{D},k_{\bar{m}-1}}(s/\check{u}_{\varepsilon,k_{\bar{m}-1}}) &= \sum_{j \in \mathbb{D}} \Psi_{\varepsilon,\mathbb{D},k_{\bar{m}-1}j}(s/\check{u}_{\varepsilon,k_{\bar{m}-1}}) \\ &\longrightarrow \sum_{j \in \mathbb{D}} {}_{\bar{k}_{\bar{m}-2}}\tilde{\phi}_{0,k_{\bar{m}-1}j}(s)\, {}_{\bar{k}_{\bar{m}-2}}\tilde{p}_{0,k_{\bar{m}-1}j} \\ &= \Psi_{0,\mathbb{D},k_{\bar{m}-1}}(s) \text{ as } \varepsilon \to 0, \ \text{for } s \geq 0. \end{aligned} \tag{10.113} $$

By Theorem 10.1, the limiting distribution functions ${}_{\bar{k}_{\bar{m}-2}}\tilde{F}_{0,k_{\bar{m}-1}j}(\cdot)$, $j \in \mathbb{D}$ are not concentrated at zero, i.e. ${}_{\bar{k}_{\bar{m}-2}}\tilde{F}_{0,k_{\bar{m}-1}j}(0) < 1$, $j \in \mathbb{D}$ and, thus, for $j \in \mathbb{D}$,

$$ {}_{\bar{k}_{\bar{m}-2}}\tilde{\phi}_{0,k_{\bar{m}-1}j}(s) < 1, \ \text{for } s > 0. \tag{10.114} $$

The assumption (10.109) and the relations (10.113) and (10.114) imply that the following inequality holds, for $s > 0$:

$$ \begin{aligned} \Psi_{0,\mathbb{D},k_{\bar{m}-1}}(s) &\leq \sum_{j \in \mathbb{D}} {}_{\bar{k}_{\bar{m}-2}}\tilde{\phi}_{0,k_{\bar{m}-1}j}(s)\, {}_{\bar{k}_{\bar{m}-2}}\tilde{p}_{0,k_{\bar{m}-1}j} \\ &< \sum_{j \in \mathbb{D}} {}_{\bar{k}_{\bar{m}-2}}\tilde{p}_{0,k_{\bar{m}-1}j} = 1. \end{aligned} \tag{10.115} $$

The relation (10.115) implies that

$$ G_{0,\mathbb{D},k_{\bar{m}-1}}(0) < 1. \tag{10.116} $$

The proof is complete. □

10.2.2.3 The Case of an Arbitrary Initial State in the Domain $\bar{\mathbb{D}}$. The asymptotic relations (10.90) and (10.99) given, respectively, in Theorems 10.2 and 10.3 can be considered as the results of the first and the second steps in some backward asymptotic recurrent algorithm for computing the weak limits for distributions of hitting times $G_{\varepsilon,\mathbb{D},k_n j}(\cdot\, \check{u}_{\varepsilon,k_n})$, with some normalisation functions $\check{u}_{\varepsilon,k_n}$, for $n = \bar{m}, \ldots, 1$.

In what follows, we assume that $\bar{m} \geq 3$ and consider the general case, where the initial state is k_n, for some $1 \leq n \leq \bar{m}$.

The reduced semi-Markov processes ${}_{\bar{k}_{n-1}}\eta_\varepsilon(t)$ and ${}_{\bar{k}_{n-1}}\tilde{\eta}_\varepsilon(t)$ have the phase space ${}_{\bar{k}_{n-1}}\mathbb{X} = \mathbb{X} \setminus \{k_1, \ldots, k_{n-1}\} = \mathbb{D} \cup \{k_n, \ldots, k_{\bar{m}}\}$, while domain ${}_{\bar{k}_{n-1}}\bar{\mathbb{D}} = \bar{\mathbb{D}} \setminus \{k_1, \ldots, k_{n-1}\} = \{k_n, \ldots, k_{\bar{m}}\}$. The state k_n can be referred as the absorbing state of the order $\bar{m} - n + 1$ in the domain $\bar{\mathbb{D}}$.

In what follows, we use the following normalisation function, for $n = \bar{m}, \ldots, 1$:

$$\begin{aligned} \check{u}_{\varepsilon,k_n} &= {}_{\bar{k}_{\bar{n}(\bar{k}_{\bar{m}},n)-1}}\tilde{u}_{\varepsilon,k_{\bar{n}(\bar{k}_{\bar{m}},n)}} \\ &= \prod_{q=0}^{\bar{n}(\bar{k}_{\bar{m}},n)-1} (1 - {}_{\bar{k}_q}p_{\varepsilon,k_{\bar{n}(\bar{k}_{\bar{m}},n)}k_{\bar{n}(\bar{k}_{\bar{m}},n)}})^{-1} u_{\varepsilon,k_{\bar{n}(\bar{k}_{\bar{m}},n)}}, \end{aligned} \tag{10.117}$$

where

$$\bar{n}(\bar{k}_{\bar{m}}, n) = \begin{cases} \bar{n} \ \text{if there exists } n+1 \leq \bar{n} \leq \bar{m} \text{ such that} \\ \quad {}_{\bar{k}_{n-1}}\tilde{p}_{0,k_n k_l} = 0, \bar{n}+1 \leq l \leq \bar{m}, \text{ and } {}_{\bar{k}_{n-1}}\tilde{p}_{0,k_n k_{\bar{n}}} > 0, \\ n \ \text{if } {}_{\bar{k}_{n-1}}\tilde{p}_{0,k_n k_l} = 0, \ n+1 \leq l \leq \bar{m}. \end{cases} \tag{10.118}$$

Let us denote, for $1 \leq n', n'' \leq \bar{m}$,

$$\bar{n}' = \bar{n}(\bar{k}_{\bar{m}}, n'), \ \bar{n}'' = \bar{n}(\bar{k}_{\bar{m}}, n''). \tag{10.119}$$

In what follows, we use the following lemma.

Lemma 10.15 *Let the conditions* $\mathbf{G}_{\mathbb{D}}$, $\mathbf{H}_{\mathbb{D}}$, $\mathbf{I}_{\mathbb{D},\mathcal{H}}$, $\dot{\mathbf{L}}_{\mathbb{D},\mathcal{H}}$, *and* ${}_{\bar{\mathbf{k}}_{\bar{m}-1}}\mathbf{M}_{\mathbb{D}}$ *be satisfied for the semi-Markov processes* $\eta_\varepsilon(t)$. *Then, the following asymptotic relation takes place, for any* $1 \leq n', n'' \leq \bar{m}$:

$$\begin{aligned} \frac{\check{u}_{\varepsilon,k_{n'}}}{\check{u}_{\varepsilon,k_{n''}}} &\to {}_{\bar{k}_{\bar{m}}}\check{w}_{0,k_{n'},k_{n''}} \\ &= \begin{cases} \prod_{q=\bar{n}'}^{\bar{n}''-1}(1 - {}_{\bar{k}_q}p_{0,k_{\bar{n}''}k_{\bar{n}''}})\, {}_{\bar{k}_{\bar{n}'-1}}w_{0,\bar{k}_{\bar{n}'},\bar{k}_{\bar{n}''}} & \text{if } \bar{n}' < \bar{n}'', \\ 1 & \text{if } \bar{n}' = \bar{n}'', \\ \prod_{q=\bar{n}''}^{\bar{n}'-1}(1 - {}_{\bar{k}_q}p_{0,k_{\bar{n}'}k_{\bar{n}'}})\, {}_{\bar{k}_{\bar{n}''-1}}w_{0,\bar{k}_{\bar{n}''},\bar{k}_{\bar{n}'}} & \text{if } \bar{n}'' < \bar{n}'. \end{cases} \end{aligned} \tag{10.120}$$

Proof The relation (10.117), (10.118), and the conditions of the lemma imply that the following relation holds, for $1 \leq \bar{n}' < \bar{n}'' \leq \bar{m}$:

$$\frac{\check{u}_{\varepsilon,k_{n'}}}{\check{u}_{\varepsilon,k_{n''}}} = \prod_{q=\bar{n}'}^{\bar{n}''-1}(1 - {}_{\bar{k}_q}p_{\varepsilon,k_{\bar{n}''}k_{\bar{n}''}})\frac{{}_{\bar{k}_{\bar{n}'-1}}\tilde{u}_{\varepsilon,\bar{k}_{\bar{n}'}}}{{}_{\bar{k}_{\bar{n}'-1}}\tilde{u}_{\varepsilon,\bar{k}_{\bar{n}''}}}$$
$$\to \prod_{q=\bar{n}'}^{\bar{n}''-1}(1 - {}_{\bar{k}_q}p_{0,k_{\bar{n}''}k_{\bar{n}''}})\,{}_{\bar{k}_{\bar{n}'-1}}w_{0,\bar{k}_{\bar{n}'},\bar{k}_{\bar{n}'}} \text{ as } \varepsilon \to 0. \tag{10.121}$$

The proof for the case, $1 \leq \bar{n}' < \bar{n}'' \leq \bar{m}$, is similar. The case, $1 \leq \bar{n}' = \bar{n}'' \leq \bar{m}$ is trivial. □

The following theorem takes place.

Theorem 10.4 *Let the conditions* $\mathbf{G}_{\mathbb{D}}$, $\mathbf{H}_{\mathbb{D}}$, $\mathbf{I}_{\mathbb{D},\mathcal{H}}$, $\mathbf{J}_{\mathbb{D}}$, $\mathbf{K}_{\mathbb{D}}$, $\mathbf{L}_{\mathbb{D},\mathcal{H}}$ *(and, thus, also conditions* $\mathbf{I}_{\mathbb{D}}$ *and* $\mathbf{L}_{\mathbb{D}}$*) and* ${}_{\bar{\mathbf{k}}_{\bar{m}-1}}\mathbf{M}_{\mathbb{D}}$ *be satisfied for the semi-Markov processes* $\eta_\varepsilon(t)$. *Then*:

(i) *The following asymptotic relation takes place, for* $j \in \mathbb{D}$, $n = \bar{m}, \ldots, 1$:

$$G_{\varepsilon,\mathbb{D},k_nj}(\cdot\,\check{u}_{\varepsilon,k_n}) \Rightarrow G_{0,\mathbb{D},k_nj}(\cdot)$$
$$= F_{0,\mathbb{D},k_nj}(\cdot)P_{0,\mathbb{D},k_nj} \text{ as } \varepsilon \to 0, \tag{10.122}$$

where: (a) $\check{u}_{\cdot,k_n}, n = \bar{m}, \ldots, 1$ *are normalisation functions, which belong to the family of asymptotically comparable functions* $\mathcal{H}$, *(b)* $F_{0,\mathbb{D},k_nj}(\cdot), j \in \mathbb{D}, n = \bar{m}, \ldots, 1$ *are proper distribution functions, i.e.,* $F_{0,\mathbb{D},k_nj}(\infty) = 1$, *for* $j \in \mathbb{D}, n = \bar{m}, \ldots, 1$, *(c)* $P_{0,\mathbb{D},k_nj}, j \in \mathbb{D}, n = \bar{m}, \ldots, 1$ *are discrete distributions, i.e.,* $P_{0,\mathbb{D},k_nj} \geq 0, j \in \mathbb{D}$ *and* $\sum_{j\in\mathbb{D}} P_{0,\mathbb{D},k_{\bar{m}}j} = 1$, *for* $n = \bar{m}, \ldots, 1$.

(ii) *The normalisation functions* $\check{u}_{\varepsilon,k_n}, n = \bar{m}, \ldots, 1$ *are given by the relations (10.117) and (10.118).*

(iii) *The Laplace transforms* $\Psi_{0,\mathbb{D},k_nj}(\cdot), j \in \mathbb{D}, n = \bar{m}, \ldots, 1$ *of distributions* $G_{0,\mathbb{D},k_nj}(\cdot), j \in \mathbb{D}, n = \bar{m}, \ldots, 1$ *are given by the relations (10.127), if* $\bar{n}(\bar{k}_{\bar{m}}, n) = \bar{n}$, *where* $n + 1 \leq \bar{n} \leq \bar{m}$, *or by the relation (10.133), if* $\bar{n}(\bar{k}_{\bar{m}}, n) = n$.

(iv) *The limiting distribution functions* $G_{0,\mathbb{D},k_n}(\cdot) = \sum_{j\in\mathbb{D}} F_{0,\mathbb{D},k_nj}(\cdot)P_{0,\mathbb{D},k_nj}, n = \bar{m}, \ldots, 1$ *are not concentrated at zero, i.e.,* $G_{0,\mathbb{D},k_n}(0) < 1, n = \bar{m}, \ldots, 1$.

(v) *The limits of hitting probabilities* $P_{0,\mathbb{D},k_nj}, j \in \mathbb{D}$ *are given by the relation (10.86).*

Proof Lemma 10.7 implies that the following relation takes place, for $s \geq 0, j \in \mathbb{D}$ and $\varepsilon \in (0, 1]$:

$$\Psi_{\varepsilon,\mathbb{D},k_nj}(s) = {}_{\bar{k}_{n-1}}\tilde{\phi}_{\varepsilon,k_nj}(s)\,{}_{\bar{k}_{n-1}}\tilde{p}_{\varepsilon,k_nj} + \sum_{r\in {}_{\bar{k}_n}\bar{\mathbb{D}}} \Psi_{\varepsilon,\mathbb{D},rj}(s)$$
$$\times\, {}_{\bar{k}_{n-1}}\tilde{\phi}_{\varepsilon,k_nr}(s)\,{}_{\bar{k}_{n-1}}\tilde{p}_{\varepsilon,k_nr}$$
$$= {}_{\bar{k}_{n-1}}\tilde{\phi}_{\varepsilon,k_nj}(s)\,{}_{\bar{k}_{n-1}}\tilde{p}_{\varepsilon,k_nj} + \sum_{n+1\leq l\leq \bar{m}} \Psi_{\varepsilon,\mathbb{D},k_lj}(s)$$
$$\times\, {}_{\bar{k}_{n-1}}\tilde{\phi}_{\varepsilon,k_nk_l}(s)\,{}_{\bar{k}_{n-1}}\tilde{p}_{\varepsilon,k_nk_l}. \tag{10.123}$$

Theorems 10.1 implies that the conditions ${}_{\bar{\mathbf{k}}_n}\mathbf{G}_{\mathbb{D}}$, ${}_{\bar{\mathbf{k}}_n}\mathbf{H}_{\mathbb{D}}$, ${}_{\bar{\mathbf{k}}_n}\mathbf{I}_{\mathbb{D},\mathcal{H}}$, ${}_{\bar{\mathbf{k}}_n}\mathbf{J}_{\mathbb{D}}$, ${}_{\bar{\mathbf{k}}_n}\mathbf{K}_{\mathbb{D}}$, ${}_{\bar{\mathbf{k}}_n}\mathbf{L}_{\mathbb{D},\mathcal{H}}$ (and, thus, the conditions ${}_{\bar{\mathbf{k}}_n}\mathbf{I}_{\mathbb{D}}$ and ${}_{\bar{\mathbf{k}}_n}\mathbf{L}_{\mathbb{D}}$) are satisfied for the semi-Markov processes ${}_{\bar{k}_{\bar{m}-2}}\eta_\varepsilon(t)$, and the conditions ${}_{\bar{\mathbf{k}}_{\bar{m}-1}}\tilde{\mathbf{G}}_{\mathbb{D}}$, ${}_{\bar{\mathbf{k}}_{\bar{m}-1}}\tilde{\mathbf{H}}_{\mathbb{D}}$, ${}_{\bar{\mathbf{k}}_{\bar{m}-1}}\tilde{\mathbf{I}}_{\mathbb{D},\mathcal{H}}$, ${}_{\bar{\mathbf{k}}_{\bar{m}-1}}\tilde{\mathbf{J}}_{\mathbb{D}}$, ${}_{\bar{\mathbf{k}}_{\bar{m}-1}}\tilde{\mathbf{K}}_{\mathbb{D}}$, ${}_{\bar{\mathbf{k}}_{\bar{m}-1}}\tilde{\mathbf{L}}_{\mathbb{D},\mathcal{H}}$ (and, thus, the conditions ${}_{\bar{\mathbf{k}}_{\bar{m}-1}}\tilde{\mathbf{I}}_{\mathbb{D}}$ and ${}_{\bar{\mathbf{k}}_{\bar{m}-1}}\tilde{\mathbf{L}}_{\mathbb{D}}$) are satisfied for the semi-Markov processes ${}_{\bar{k}_{\bar{m}-1}}\tilde{\eta}_\varepsilon(t)$.

The probabilities ${}_{\bar{k}_q}p_{\varepsilon,k_{\bar{n}(\bar{k}_{\bar{m}},n)}k_{\bar{n}(\bar{k}_{\bar{m}},n)}}$, $q = 0, \ldots, \bar{n}(\bar{k}_{\bar{m}}, n) - 1$, $n = \bar{m}, \ldots, 1$ take values in the interval $[0, 1)$, since all conditions ${}_{\bar{k}_{n-1}}\mathbf{H}_{\mathbb{D}}$, $n = 1, \ldots, \bar{m}$ are satisfied. The condition $\mathbf{J}_{\mathbb{D}}$ implies that the initial normalisation functions u_{ε,k_n}, $n = \bar{m}, \ldots, 1$ take values in the interval $(0, \infty)$. Therefore, the normalisation functions $\check{u}_{\varepsilon,k_n}$, $n = \bar{m}, \ldots, 1$ given by the relations (10.117) and (10.118) also take values in the interval $(0, \infty)$. The functions $\check{u}_{\cdot,k_n}$, $n = \bar{m}, \ldots, 1$ belong to the family $\mathcal{H}$ due to the conditions $\mathbf{I}_{\mathbb{D}}$, $\mathbf{I}_{\mathbb{D},\mathcal{H}}$, $\mathbf{L}_{\mathbb{D}}$, $\mathbf{L}_{\mathbb{D},\mathcal{H}}$ and Lemma 8.3. Moreover, $\lim_{\varepsilon\to 0}\check{u}_{\varepsilon,k_n} \in (0, \infty]$, for $n = \bar{m}, \ldots, 1$.

Let us assume that, we have already realised $\bar{m} - n$ steps in the announced above asymptotic backward algorithm and, thus: (a) the relation (10.122) holds, for $j \in \mathbb{D}$, $l = \bar{m}, \ldots, n+1$, (b) the statements **(ii)**–**(v)** are true, for $j \in \mathbb{D}$, $l = \bar{m}, \ldots, n+1$, in particular, the distribution function $G_{0,\mathbb{D},k_l}(\cdot) = \sum_{j\in\mathbb{D}} F_{0,\mathbb{D},k_n j}(\cdot)P_{0,\mathbb{D},k_n j}$ is not concentrated at zero, i.e., $G_{0,\mathbb{D},k_l}(0) < 1$, for $j \in \mathbb{D}$, $l = \bar{m}, \ldots, n + 1$.

The above assumption (a) which assumes that the relation (10.122) holds can be also expressed in the equivalent form of the following relation for the corresponding Laplace transforms, for $j \in \mathbb{D}$, $l = \bar{m}, \ldots, n + 1$:

$$\Psi_{\varepsilon,\mathbb{D},k_l j}(s/\check{u}_{\varepsilon,k_l}) \to \Psi_{0,\mathbb{D},k_l j}(s) \text{ as } \varepsilon \to 0, \text{ for } s \geq 0. \tag{10.124}$$

According to Theorem 10.1, the probabilities,

$${}_{\bar{k}_{n-1}}\tilde{p}_{\varepsilon,k_n k_l} \to 0 \text{ as } \varepsilon \to 0, \text{ if } {}_{\bar{k}_{n-1}}\tilde{p}_{0,k_n k_l} = 0. \tag{10.125}$$

First, suppose that,

$$\bar{n}(\bar{k}_{\bar{m}}, n) = \bar{n}, \text{ where } n + 1 \leq \bar{n} \leq \bar{m}. \tag{10.126}$$

The above remarks, Theorems 10.2–10.3, the relations (10.124)–(10.125), and Lemma 10.15 imply that the following relation takes place, for $j \in \mathbb{D}$:

$$\begin{aligned}
&\Psi_{\varepsilon,\mathbb{D},k_n j}(s/\check{u}_{\varepsilon,k_n}) \\
&\quad = {}_{\bar{k}_{n-1}}\tilde{\phi}_{\varepsilon,k_n j}\Big(\prod_{q=n}^{\bar{n}-1}(1 - {}_{\bar{k}_q}p_{\varepsilon,k_{\bar{n}}k_{\bar{n}}})\frac{{}_{\bar{k}_{n-1}}\tilde{u}_{\varepsilon,k_n}}{{}_{\bar{k}_{n-1}}\tilde{u}_{\varepsilon,k_{\bar{n}}}} s/{}_{\bar{k}_{n-1}}\tilde{u}_{\varepsilon,k_n}\Big)\, {}_{\bar{k}_{n-1}}\tilde{p}_{\varepsilon,k_n j} \\
&\qquad + \sum_{\bar{n}+1\leq l\leq \bar{m}} \Psi_{\varepsilon,\mathbb{D},k_l j}(s/\,{}_{\bar{k}_{\bar{m}}}\check{u}_{\varepsilon,k_n})\, {}_{\bar{k}_{n-1}}\tilde{\phi}_{\varepsilon,k_n k_l}(s/\,{}_{\bar{k}_{\bar{m}}}\check{u}_{\varepsilon,k_n})\, {}_{\bar{k}_{n-1}}\tilde{p}_{\varepsilon,k_n k_l} \cdot \\
&\qquad + \Psi_{\varepsilon,\mathbb{D},k_{\bar{n}} j}(s/{}_{\bar{k}_{\bar{n}-1}}\tilde{u}_{\varepsilon,k_{\bar{n}}})
\end{aligned}$$

$$
\begin{aligned}
& \times {}_{\bar{k}_{n-1}}\tilde{\phi}_{\varepsilon,k_n k_{\bar{n}}}\Big(\prod_{q=n}^{\bar{n}-1}(1 - {}_{\bar{k}_q}p_{\varepsilon,k_{\bar{n}}k_{\bar{n}}})\frac{{}_{\bar{k}_{n-1}}\tilde{u}_{\varepsilon,k_n}}{{}_{\bar{k}_{n-1}}\tilde{u}_{\varepsilon,k_{\bar{n}}}} s/{}_{\bar{k}_{n-1}}\tilde{u}_{\varepsilon,k_n}\Big)\, {}_{\bar{k}_{n-1}}\tilde{p}_{\varepsilon,k_n k_{\bar{n}}} \cdot \\
& + \sum_{n+1 \leq l < \bar{n}} \Psi_{\varepsilon,\mathbb{D},k_l j}\Big(\prod_{q=l}^{\bar{n}-1}(1 - {}_{\bar{k}_q}p_{\varepsilon,k_{\bar{n}}k_{\bar{n}}})\frac{{}_{\bar{k}_{l-1}}\tilde{u}_{\varepsilon,k_l}}{{}_{\bar{k}_{l-1}}\tilde{u}_{\varepsilon,k_{\bar{n}}}} s/{}_{\bar{k}_{\bar{m}-l}}\tilde{u}_{\varepsilon,k_l}\Big) \\
& \times {}_{\bar{k}_{n-1}}\tilde{\phi}_{\varepsilon,k_n k_l}\Big(\prod_{q=n}^{\bar{n}-1}(1 - {}_{\bar{k}_q}p_{\varepsilon,k_{\bar{n}}k_{\bar{n}}})\frac{{}_{\bar{k}_{n-1}}\tilde{u}_{\varepsilon,k_n}}{{}_{\bar{k}_{n-1}}\tilde{u}_{\varepsilon,k_{\bar{n}}}} s/{}_{\bar{k}_{n-1}}\tilde{u}_{\varepsilon,k_n}\Big)\, {}_{\bar{k}_{n-1}}\tilde{p}_{\varepsilon,k_n k_l} \\
& \longrightarrow {}_{\bar{k}_{n-1}}\tilde{\phi}_{0,k_n j}\Big(\prod_{q=n}^{\bar{n}-1}(1 - {}_{\bar{k}_q}p_{0,k_{\bar{n}}k_{\bar{n}}})\, {}_{\bar{k}_{n-1}}w_{0,k_n,k_{\bar{n}}} s\Big)\, {}_{\bar{k}_{n-1}}\tilde{p}_{0,k_n j} \\
& + \Psi_{0,\mathbb{D},k_{\bar{n}} j}(s) \\
& \times {}_{\bar{k}_{n-1}}\tilde{\phi}_{0,k_n k_{\bar{n}}}\Big(\prod_{q=n}^{\bar{n}-1}(1 - {}_{\bar{k}_q}p_{0,k_{\bar{n}}k_{\bar{n}}})\, {}_{\bar{k}_{n-1}}w_{0,k_n,k_{\bar{n}}} s\Big)\, {}_{\bar{k}_{n-1}}\tilde{p}_{0,k_n k_{\bar{n}}} \\
& + \sum_{n+1 \leq l < \bar{n}} \Psi_{0,\mathbb{D},k_l j}\Big(\prod_{q=l}^{\bar{n}-1}(1 - {}_{\bar{k}_q}p_{0,k_{\bar{n}}k_{\bar{n}}})\, {}_{\bar{k}_{l-1}}w_{0,k_l,k_{\bar{n}}} s\Big) \\
& \times {}_{\bar{k}_{n-1}}\tilde{\phi}_{0,k_n k_l}\Big(\prod_{q=n}^{\bar{n}-1}(1 - {}_{\bar{k}_q}p_{0,k_{\bar{n}}k_{\bar{n}}})\, {}_{\bar{k}_{n-1}}w_{0,k_n,k_{\bar{n}}} s\Big)\, {}_{\bar{k}_{n-1}}\tilde{p}_{0,k_n k_l} \\
& = {}_{\bar{k}_{n-1}}\tilde{\phi}_{0,k_n j}({}_{\bar{k}_{n-1}}\check{w}_{0,k_n,k_{\bar{n}}} s)\, {}_{\bar{k}_{n-1}}\tilde{p}_{0,k_n j} \\
& + \Psi_{0,\mathbb{D},k_{\bar{n}} j}(s)\, {}_{\bar{k}_{n-1}}\tilde{\phi}_{0,k_n k_{\bar{n}}}({}_{\bar{k}_{n-1}}\check{w}_{0,k_n,k_{\bar{n}}} s)\, {}_{\bar{k}_{n-1}}\tilde{p}_{0,k_n k_{\bar{n}}} \\
& + \sum_{n+1 \leq l < \bar{n}} \Psi_{0,\mathbb{D},k_l j}({}_{\bar{k}_{l-1}}\check{w}_{0,k_l,k_{\bar{n}}} s)\, {}_{\bar{k}_{n-1}}\tilde{\phi}_{0,k_n k_l}({}_{\bar{k}_{n-1}}\check{w}_{0,k_n,k_{\bar{n}}} s)\, {}_{\bar{k}_{n-1}}\tilde{p}_{0,k_n k_l} \\
& = \Psi_{0,\mathbb{D},k_n j}(s) \text{ as } \varepsilon \to 0, \text{ for } s \geq 0.
\end{aligned} \tag{10.127}
$$

In this case, the limiting distribution $G_{0,\mathbb{D},k_n j}(\cdot)$ has, for every $j \in \mathbb{D}$, the Laplace transform $\Psi_{0,\mathbb{D},k_n j}(s) = \int_0^\infty e^{-st} G_{0,\mathbb{D},k_n j}(dt),\ s \geq 0$ given by the relation (10.127).

The relation (10.127) entails the fulfilment of the following asymptotic relation for the Laplace transforms of hitting times:

$$
\begin{aligned}
\Psi_{\varepsilon,\mathbb{D},k_n}(s/\check{u}_{\varepsilon,k_n}) & = \sum_{j \in \mathbb{D}} \Psi_{\varepsilon,\mathbb{D},k_n,j}(s/\check{u}_{\varepsilon,k_n}) \\
& \longrightarrow \sum_{j \in \mathbb{D}} {}_{\bar{k}_{n-1}}\tilde{\phi}_{0,k_n j}({}_{\bar{k}_{n-1}}\check{w}_{0,k_n,k_{\bar{n}}} s)\, {}_{\bar{k}_{n-1}}\tilde{p}_{0,k_n j} \\
& + \Psi_{0,\mathbb{D},k_{\bar{n}}}(s)\, {}_{\bar{k}_{n-1}}\tilde{\phi}_{0,k_n k_{\bar{n}}}({}_{\bar{k}_{n-1}}\check{w}_{0,k_n,k_{\bar{n}}} s)\, {}_{\bar{k}_{n-1}}\tilde{p}_{0,k_n k_{\bar{n}}} \\
& + \sum_{n+1 \leq l < \bar{n}} \Psi_{0,\mathbb{D},k_l}({}_{\bar{k}_{l-1}}\check{w}_{0,k_l,k_{\bar{n}}} s) \\
& \times {}_{\bar{k}_{n-1}}\tilde{\phi}_{0,k_n k_l}({}_{\bar{k}_{n-1}}\check{w}_{0,k_n,k_{\bar{n}}} s)\, {}_{\bar{k}_{n-1}}\tilde{p}_{0,k_n k_l} \\
& = \Psi_{0,\mathbb{D},k_n}(s) \text{ as } \varepsilon \to 0, \text{ for } s \geq 0.
\end{aligned} \tag{10.128}
$$

According to Theorem 10.2 and the above induction assumptions, the probability $G_{0,k_l}(0) < 1, l = \bar{m}, \ldots, n+1$, and, thus, for $l = \bar{m}, \ldots, n+1$, the Laplace transforms,

$$\Psi_{0,\mathbb{D},\bar{k}_l}(s) < 1, \text{ for } s > 0. \tag{10.129}$$

The assumption (10.126) and the relations (10.128) and (10.129) imply that the following inequality holds, for $s > 0$:

$$\begin{aligned}\Psi_{0,\mathbb{D},k_n}(s) &\leq \sum_{j\in\mathbb{D}} {}_{\bar{k}_{n-1}}\tilde{p}_{0,k_nj} + \Psi_{0,\mathbb{D},k_{\bar{n}}}(s)\, {}_{\bar{k}_{n-1}}\tilde{p}_{0,k_nk_{\bar{n}}} \\ &\quad + \sum_{n+1\leq l<\bar{n}} {}_{\bar{k}_{n-1}}\tilde{p}_{0,k_nk_l} \\ &< \sum_{j\in\mathbb{D}} {}_{\bar{k}_{n-1}}\tilde{p}_{0,k_nj} + {}_{\bar{k}_{n-1}}\tilde{p}_{0,k_nk_{\bar{n}}} \\ &\quad + \sum_{n+1\leq l<\bar{n}} {}_{\bar{k}_{n-1}}\tilde{p}_{0,k_nk_l} = 1.\end{aligned} \tag{10.130}$$

The relation (10.130) implies that,

$$G_{0,\mathbb{D},k_n}(0) < 1. \tag{10.131}$$

Second, suppose that,

$$\bar{n}(\bar{k}_{\bar{m}}, n) = n. \tag{10.132}$$

In this case, the above remarks, Theorems 10.3 and 10.4, and the relations (10.124) and (10.123) imply that the following relation takes place, for $j \in \mathbb{D}$:

$$\begin{aligned}&\Psi_{\varepsilon,\mathbb{D},k_nj}(s/\check{u}_{\varepsilon,k_n}) \\ &\quad = {}_{\bar{k}_{n-1}}\tilde{\phi}_{\varepsilon,k_nj}(s/{}_{\bar{k}_{n-1}}\check{u}_{\varepsilon,k_n})\, {}_{\bar{k}_{n-1}}\tilde{p}_{\varepsilon,k_nj} \\ &\quad\quad + \sum_{n+1\leq l\leq\bar{m}} \Psi_{\varepsilon,\mathbb{D},k_lj}(s/{}_{\bar{k}_{\bar{m}}}\check{u}_{\varepsilon,k_n})\, {}_{\bar{k}_{n-1}}\tilde{\phi}_{\varepsilon,k_nk_l}(s/{}_{\bar{k}_{\bar{m}}}\check{u}_{\varepsilon,k_n})\, {}_{\bar{k}_{n-1}}\tilde{p}_{\varepsilon,k_nk_l} \\ &\quad \to {}_{\bar{k}_{n-1}}\tilde{\phi}_{0,k_nj}(s)\, {}_{\bar{k}_{n-1}}\tilde{p}_{0,k_nj} \\ &\quad = \Psi_{0,\mathbb{D},k_nj}(s) \text{ as } \varepsilon \to 0, \text{ for } s \geq 0.\end{aligned} \tag{10.133}$$

In this case, the limiting distribution $G_{0,\mathbb{D},k_nj}(\cdot)$ has, for every $j \in \mathbb{D}$, the Laplace transform $\Psi_{0,\mathbb{D},k_nj}(s) = \int_0^\infty e^{-st} G_{0,\mathbb{D},k_nj}(dt), s \geq 0$ given by the relation (10.133).

The relation (10.133) entails the fulfilment of the following asymptotic relation for the Laplace transforms of hitting times:

$$\begin{aligned}\Psi_{\varepsilon,\mathbb{D},k_n}(s/\check{u}_{\varepsilon,k_n}) &= \sum_{j\in\mathbb{D}} \Psi_{\varepsilon,\mathbb{D},k_n}(s/\check{u}_{\varepsilon,k_nj}) \\ &\to \sum_{j\in\mathbb{D}} {}_{\bar{k}_{n-1}}\tilde{\phi}_{0,k_nj}(s)\, {}_{\bar{k}_{n-1}}\tilde{p}_{0,k_nj} \\ &= \Psi_{0,\mathbb{D},k_n}(s) \text{ as } \varepsilon \to 0, \text{ for } s \geq 0.\end{aligned} \tag{10.134}$$

By Theorem 10.1, the limiting distribution functions ${}_{\bar{k}_{n-1}}\tilde{F}_{0,k_nj}(\cdot), j \in \mathbb{D}$ are not concentrated in 0, i.e. ${}_{\bar{k}_{n-1}}\tilde{F}_{0,k_nj}(0) < 1, j \in \mathbb{D}$ and, thus, for $j \in \mathbb{D}$,

$$ {}_{\bar{k}_{n-1}}\tilde{\phi}_{0,k_nj}(s) < 1, \ \text{for } s > 0. \tag{10.135} $$

The assumption (10.132) and the relations (10.134) and (10.135) imply that the following inequality holds, for $s > 0$:

$$ \begin{aligned} \Psi_{0,\mathbb{D},k_n}(s) &= \sum_{j \in \mathbb{D}} {}_{\bar{k}_{n-1}}\tilde{\phi}_{0,k_nj}(s)\, {}_{\bar{k}_{n-1}}\tilde{p}_{0,k_nj} \\ &< \sum_{j \in \mathbb{D}} {}_{\bar{k}_{n-1}}\tilde{p}_{0,k_nj} = 1. \end{aligned} \tag{10.136} $$

The relation (10.136) implies that

$$ G_{0,\mathbb{D},k_n}(0) < 1. \tag{10.137} $$

As it was mentioned above, the above asymptotic relations (10.127) or (10.133) hold, due to Theorems 10.2 and 10.3, for $n = \bar{m}$ and $n = \bar{m}-1$. Sequential application of the asymptotic relations (10.127) and (10.133) makes it possible to obtain similar relations for $n = \bar{m} - 2, \ldots, 1$.

The formulas for the limiting Laplace transforms $\Psi_{0,\mathbb{D},k_nj}(s), n = 1, \ldots, \bar{m}$ given by the relations (10.127) and (10.133) have the recurrent character. In the case, $j \in \mathbb{D}, n = \bar{m}$, the Laplace transform $\Psi_{0,\mathbb{D},k_{\bar{m}}j}(s)$ given by the relation (10.133) coincides with the Laplace transform given, according to Theorem 10.2, by the relation (10.95). In the case, $j \in \mathbb{D}, n = \bar{m} - 1$, the Laplace transform $\Psi_{0,\mathbb{D},k_{\bar{m}-1}j}(s)$ given by the relations (10.127) and (10.133) coincide with the Laplace transform given, according to Theorem 10.3, by the relations (10.103) and (10.112). The sequential application of formulas given by the relations (10.127) and (10.133), for $j \in \mathbb{D}, n = \bar{m} - 2, \ldots, 1$ lets us compute the limiting Laplace transforms $\Psi_{0,\mathbb{D},k_nj}(s)$, for $j \in \mathbb{D}, n = \bar{m} - 2, \ldots, 1$. □

Remark 10.1 According to the statement **(i)** of Theorem 10.4, the normalisation function $\check{u}_{\varepsilon,k_n}$ belongs to the complete family of comparable functions $\mathcal{H}$. Let us assume, for example, $\mathcal{H} = \mathcal{H}_1$ is the complete family of asymptotically comparable power-type functions. The relation (10.117), recurrent formulas for finding of the transition probabilities $1 - {}_{\bar{k}_l}p_{\varepsilon,ii} = \sum_{j \in {}_{\bar{k}_l}\mathbb{X}, j \neq i} {}_{\bar{k}_l}p_{\varepsilon,ij}, i \in {}_{\bar{k}_l}\mathbb{X}, l = 0, \ldots, \bar{m} - 1$ given in Chap. 9, and the operating rules for asymptotically comparable functions given in Sect. 8.2.1 make it possible to obtain for the normalisation function $\check{u}_{\varepsilon,k_n}$ the asymptotic relation, $\check{u}_{\varepsilon,k_n} / {}_{\bar{k}_m}a_{k_n}\varepsilon^{{}_{\bar{k}_m}b_{k_n}} \to 1$ as $\varepsilon \to 0$, and recurrently calculate the parameters ${}_{\bar{k}_m}a_{k_n}, {}_{\bar{k}_m}b_{k_n}$.

Similar comments can be made in the cases, where $\mathcal{H} = \mathcal{H}_2$ is the complete family of asymptotically comparable power-exponential functions, or $\mathcal{H} = \mathcal{H}_3$ is the complete family of asymptotically comparable power-logarithmic functions.

It is also useful to reformulate the asymptotic relations (10.90), (10.99), and (10.122) given in Theorems 10.2–10.4 in the equivalent form, where i is an arbitrary state in the domain $\bar{\mathbb{D}}$.

The following relations obviously hold, for $i \in \bar{\mathbb{D}}$, $j \in \mathbb{D}$ and $\varepsilon \in [0, 1]$:

$$
\begin{aligned}
&\check{u}_{\varepsilon,i} = \check{u}_{\varepsilon,k_{N(\bar{k}_m,i)}},\\
&G_{\varepsilon,\mathbb{D},ij}(t) = G_{\varepsilon,\mathbb{D},k_{N(\bar{k}_m,i)},j}(t),\ t \geq 0,\\
&F_{\varepsilon,\mathbb{D},ij}(t) = F_{\varepsilon,\mathbb{D},k_{N(\bar{k}_m,i)},j}(t),\ t \geq 0,\\
&P_{\varepsilon,\mathbb{D},ij} = P_{\varepsilon,\mathbb{D},k_{N(\bar{k}_m,i)},j},\\
&\Psi_{\varepsilon,\mathbb{D},ij}(s) = \Psi_{0\varepsilon,\mathbb{D},k_{N(\bar{k}_m,i)},j}(s),\ s \geq 0. \qquad (10.138)
\end{aligned}
$$

The above remarks make it possible to reformulate the asymptotic relation (10.122) given in Theorem 10.4 in the following alternative equivalent forms, for $i \in \bar{\mathbb{D}}$, $j \in \mathbb{D}$:

$$
\begin{aligned}
G_{\varepsilon,\mathbb{D},ij}(\cdot\,\check{u}_{\varepsilon,i}) &\Rightarrow G_{0,\mathbb{D},ij}(\cdot)\\
&= F_{0,\mathbb{D},ij}(\cdot)P_{0,\mathbb{D},ij} \text{ as } \varepsilon \to 0, \qquad (10.139)
\end{aligned}
$$

or

$$
\begin{aligned}
\Psi_{\varepsilon,\mathbb{D},ij}(s/\check{u}_{\varepsilon,i}) &\Rightarrow \Psi_{0,\mathbb{D},ij}(\cdot)\\
&= \Phi_{0,\mathbb{D},ij}(s)P_{0,\mathbb{D},ij} \text{ as } \varepsilon \to 0, \text{ for } s \geq 0. \qquad (10.140)
\end{aligned}
$$

10.2.3 Admissible Normalisation Functions and Phase Types of Limiting Distributions

10.2.3.1 Admissible Normalisation Functions. In principle, the limiting distributions $G_{0,\mathbb{D},ij}(\cdot)$ appearing in the relation (10.139) depend on the sequence $\bar{k}_{\bar{m}}$ via the normalisation functions $\check{u}_{\varepsilon,i}$ given by the relations (10.117) and (10.138).

Let us denote by $\mathcal{K}_{\mathbb{D}}$ the class of all "admissible" sequences $\bar{k}_{\bar{m}} = \langle k_1, \ldots, k_{\bar{m}}\rangle$, for which the condition ${}_{\bar{\mathbf{k}}_{\bar{m}-1}}\mathbf{M}_{\mathbb{D}}$ is satisfied.

Let also $\mathcal{U}_{\mathbb{D}}$ be the class of all "admissible" normalisation functions $\check{u}_{\cdot,i}, i \in \bar{\mathbb{D}}$ generated by the sequences $\bar{k}_{\bar{m}} \in \mathcal{K}_{\mathbb{D}}$. In this context we can use the notation $\check{u}_{\cdot,i} = {}_{\bar{k}_{\bar{m}}}\check{u}_{\cdot,i}, i \in \bar{\mathbb{D}}$.

Under the conditions $\mathbf{G}_{\mathbb{D}}$, $\mathbf{H}_{\mathbb{D}}$, $\mathbf{I}_{\mathbb{D},\mathcal{H}}$, $\mathbf{L}_{\mathbb{D},\mathcal{H}}$, the normalisation functions ${}_{\bar{k}_{\bar{m}}}\check{u}_{\cdot,i}, i \in \bar{\mathbb{D}}$ from the class $\mathcal{U}_{\mathbb{D}}$ belong to the class of asymptotically comparable functions $\mathcal{H}$.

Therefore, for any two sequences $\bar{k}'_{\bar{m}} = \langle k'_1, \ldots, k'_{\bar{m}}\rangle$ and $\bar{k}''_{\bar{m}} = \langle k''_1, \ldots, k''_{\bar{m}}\rangle$ from class $\mathcal{K}_{\mathbb{D}}$ and the corresponding normalisation functions, ${}_{\bar{k}'_{\bar{m}}}\check{u}'_{\cdot,i}, i \in \mathbb{D}$ and ${}_{\bar{k}'_{\bar{m}}}\check{u}''_{\cdot,i}, i \in \mathbb{D}$ generated, respectively, by the sequence $\bar{k}'_{\bar{m}}$ and $\bar{k}''_{\bar{m}}$ and given by the relations (10.117) and (10.138), there exist limits,

$$\lim_{\varepsilon\to 0} \frac{{}_{\bar{k}'_{\bar{m}}}\check{u}'_{\cdot,i}}{{}_{\bar{k}''_{\bar{m}}}\check{u}''_{\cdot,i}} = w_{0,i}[\bar{k}'_{\bar{m}}, \bar{k}''_{\bar{m}}] \in [0, \infty], \ \text{for } i \in \bar{\mathbb{D}}. \tag{10.141}$$

Since the functions ${}_{\bar{k}'_{\bar{m}}}\check{u}'_{\cdot,i}$, ${}_{\bar{k}''_{\bar{m}}}\check{u}''_{\cdot,i}, n = 1, \ldots, \bar{m}$ belong to the class of asymptotically comparable functions $\mathcal{H}$, the limits appearing in the relation (10.141) can be computed with the use of the quotient formulas given in Lemma 8.2. In the case, where the family $\mathcal{H}$ is one of the families, $\mathcal{H}_1$, $\mathcal{H}_2$, or $\mathcal{H}_3$, the quotient formulas given, respectively, in Lemma 8.4, 8.6, or 8.8 should be used.

In the case, where conditions $\mathbf{G}_{\mathbb{D}}$, $\mathbf{H}_{\mathbb{D}}$, $\mathbf{I}_{\mathbb{D},\mathcal{H}}$, $\mathbf{L}_{\mathbb{D},\mathcal{H}}$, $\mathbf{J}_{\mathbb{D}}$, $\mathbf{K}_{\mathbb{D}}$ and ${}_{\bar{\mathbf{k}}_{\bar{m}-1}}\mathbf{M}_{\mathbb{D}}$ hold, the limiting distribution functions $F_{0,\mathbb{D},i}(\cdot) = \sum_{j\in\mathbb{D}} G_{0,\mathbb{D},ij}(\cdot), i \in \bar{\mathbb{D}}$ are proper and are not concentrated at zero.

This implies that the limits appearing in (10.141) satisfy the following relation, for any two sequences $\bar{k}'_{\bar{m}} = \langle k'_1, \ldots, k'_{\bar{m}}\rangle, \bar{k}''_{\bar{m}} = \langle k''_1, \ldots, k''_{\bar{m}}\rangle \in \mathcal{K}_{\mathbb{D}}$:

$$w_{0,i}[\bar{k}'_{\bar{m}}, \bar{k}''_{\bar{m}}] \in (0, \infty), \ i \in \bar{\mathbb{D}}. \tag{10.142}$$

The limiting distributions $G'_{0,\mathbb{D},ij}(\cdot)$ and $G''_{0,\mathbb{D},ij}(\cdot)$ appearing in the weak asymptotic relation (10.139) in the cases, where, respectively, the normalisation functions ${}_{\bar{k}'_{\bar{m}}}\check{u}'_{\cdot,i}$ or ${}_{\bar{k}''_{\bar{m}}}\check{u}''_{\cdot,i}$ used in the relation (10.139), differ just by the scaling factors $w_{0,i}[\bar{k}'_{\bar{m}}, \bar{k}''_{\bar{m}}]$, that is, for $i \in \bar{\mathbb{D}}, j \in \mathbb{D}$,

$$G'_{0,\mathbb{D},ij}(t) = G''_{0,\mathbb{D},ij}(t\, w_{0,i}[\bar{k}'_{\bar{m}}, \bar{k}''_{\bar{m}}]), \ t \geq 0. \tag{10.143}$$

The relations (10.141)–(10.143) allow not indexing the normalisation functions $\check{u}_{\cdot,k_n}$ and the limiting distributions $G_{0,\mathbb{D},k_n,j}(\cdot)$ (appearing in Theorems 10.2–10.4) and the normalisation functions $\check{u}_{\cdot,i}$ and the distributions $G_{0,\mathbb{D},i,j}(\cdot)$ (appearing in the asymptotic relation (10.139)) with index $\bar{k}_{\bar{m}}$.

10.2.3.2 Phase Types of Limiting Distributions for Hitting Times. The limiting conditional distributions $F_{0,\mathbb{D},ij}(\cdot)$, appearing in Theorems 10.2–10.4, have semi-Markov phase type, i.e., themselves are conditional distributions of hitting types for semi-Markov processes, possibly, with a smaller phase space than the phase space of the original semi-Markov processes.

In particular, it is possible that the limiting conditional distributions for hitting times belong to the well-known class of Markov phase type distributions (conditional distributions of hitting times for continuous time Markov chains). We refer to the works Asmussen (1997, 2003) and Asmussen and Albrecher (2010), where a detailed description of phase-type distributions can be found.

In particular, if, for some $1 \leq n \leq \bar{m}$, all distributions ${}_{\bar{k}_{n-1}}F_{0,ij}(\cdot), i \in {}_{\bar{k}_{n-1}}\bar{\mathbb{D}}, j \in {}_{\bar{k}_{n-1}}\mathbb{X}$ are exponential with parameters not depending on j, i.e., ${}_{\bar{k}_{n-1}}F_{0,ij}(t) = 1 - \exp\{-{}_{\bar{k}_{n-1}}\lambda_i t\}, t \geq 0, i \in {}_{\bar{k}_{n-1}}\bar{\mathbb{D}}, j \in {}_{\bar{k}_{n-1}}\mathbb{X}$ (here, the parameters ${}_{\bar{k}_{n-1}}\lambda_i \in (0, \infty)$), then the conditional limiting distributions $F_{0,\mathbb{D},ij}(\cdot), i \in {}_{\bar{k}_{n-1}}\bar{\mathbb{D}}, j \in {}_{\bar{k}_{n-1}}\mathbb{X}$ are of the Markov phase type.

10.3 Weak Asymptotics for Distributions of Return Times

In this section, we transfer the results presented in Sect. 10.2 to a model of perturbed semi-Markov processes with an initial state belonging to the domain $\mathbb{D}$.

10.3.1 Hitting and Return Times

10.3.1.1 Hitting and Return Times. Two variants of hitting type functionals can be considered for the case, where the initial state $\eta_{\varepsilon,0} \in \mathbb{D}$.

The first variant is the standard hitting time to the domain $\mathbb{D}$,

$$\tau_{\varepsilon,\mathbb{D}} = \sum_{n=1}^{\nu_{\varepsilon,\mathbb{D}}} \kappa_{\varepsilon,n}, \text{ where } \nu_{\varepsilon,\mathbb{D}} = \min(n \geq 1 : \eta_{\varepsilon,n} \in \mathbb{D}). \tag{10.144}$$

The second variant is the hitting type functional, which can be interpreted as a return time to the domain $\mathbb{D}$ (the hitting time to the domain $\mathbb{D}$ after the hitting in the domain $\bar{\mathbb{D}}$),

$$\dot{\tau}_{\varepsilon,\mathbb{D}} = \sum_{n=1}^{\dot{\nu}_{\varepsilon,\mathbb{D}}} \kappa_{\varepsilon,n}, \text{ where } \dot{\nu}_{\varepsilon,\mathbb{D}} = \min(n \geq \nu_{\varepsilon,\bar{\mathbb{D}}} : \eta_{\varepsilon,n} \in \mathbb{D}). \tag{10.145}$$

In the first case, the key role is played by the following relation, which takes place, for every $\varepsilon \in (0, 1]$ and $r \in \mathbb{D}$:

$$\begin{aligned} G_{\varepsilon,\mathbb{D},rj}(t) &= F_{\varepsilon,rj}(t)p_{\varepsilon,rj} \\ &\quad + \sum_{i \in \bar{\mathbb{D}}} G_{\varepsilon,\mathbb{D},ij}(t) * F_{\varepsilon,ri}(t)p_{\varepsilon,ri}, \text{ for } t \geq 0, j \in \mathbb{D}, \end{aligned} \tag{10.146}$$

which can also be rewritten in terms of the corresponding Laplace transforms, for every $\varepsilon \in (0, 1]$ and $r \in \mathbb{D}$,

$$\begin{aligned} \Psi_{\varepsilon,\mathbb{D},rj}(s) &= \phi_{\varepsilon,rj}(s)p_{\varepsilon,rj} \\ &\quad + \sum_{i \in \bar{\mathbb{D}}} \Psi_{\varepsilon,\mathbb{D},ij}(s)\phi_{\varepsilon,ri}(s)p_{\varepsilon,ri}, \text{ for } s \geq 0, j \in \mathbb{D}. \end{aligned} \tag{10.147}$$

In the second case, the key role is played by the following relation, which takes place, for every $\varepsilon \in (0, 1]$ and $r \in \mathbb{D}$:

$$\begin{aligned} \dot{G}_{\varepsilon,\mathbb{D},rj}(t) &= \mathsf{P}_r\{\dot{\tau}_{\varepsilon,\mathbb{D}} \leq t, \eta_\varepsilon(\dot{\tau}_{\varepsilon,\mathbb{D}}) = j\} \\ &= \sum_{i \in \bar{\mathbb{D}}} G_{\varepsilon,\mathbb{D},ij}(t) * G_{\varepsilon,\bar{\mathbb{D}},ri}(t), \text{ for } t \geq 0, j \in \mathbb{D}, \end{aligned} \tag{10.148}$$

which can also be rewritten in terms of the corresponding Laplace transforms, for every $\varepsilon \in (0, 1]$ and $r \in \mathbb{D}$,

$$\dot{\Psi}_{\varepsilon,\mathbb{D},rj}(s) = \mathsf{E}e^{-s\dot{\tau}_{\varepsilon,\mathbb{D}}}\mathrm{I}(\eta_\varepsilon(\dot{\tau}_{\varepsilon,\mathbb{D}}) = j) = \sum_{i \in \bar{\mathbb{D}}} \Psi_{\varepsilon,\mathbb{D},ij}(s)\Psi_{\varepsilon,\bar{\mathbb{D}},ri}(s), \text{ for } s \geq 0, j \in \mathbb{D}. \quad (10.149)$$

10.3.1.2 Regularity and Perturbation Conditions. The relations (10.144)–(10.149) tell us that the following conditions (similar to the basic conditions $\mathbf{G}_{\mathbb{D}}$–$\mathbf{K}_{\mathbb{D}}$ (for the case where the domains $\mathbb{D}$ and $\bar{\mathbb{D}}$ are swapped) should be used.

An analogue of the condition $\mathbf{G}_{\mathbb{D}}$ takes the following form:

$\dot{\mathbf{G}}_{\mathbb{D}}$: **(a)** $p_{\varepsilon,ij} > 0, \varepsilon \in (0, 1]$ or $p_{\varepsilon,ij} = 0, \varepsilon \in (0, 1]$, for every $j \in \mathbb{X}, i \in \mathbb{D}$, **(b)** $F_{\varepsilon,i}(0) < 1, i \in \mathbb{D}, \varepsilon \in (0, 1]$.

Recall the sets $\mathbb{Y}_{\varepsilon,i} = \{j \in \mathbb{X} : p_{\varepsilon,ij} > 0\}$ defined in the relation (9.5).

Lemma 9.1 implies that, under the condition $\mathbf{G}_{\mathbb{D}}$, the set $\mathbb{Y}_{\varepsilon,i} = \mathbb{Y}_{1,i}, \varepsilon \in (0, 1]$, for $i \in \bar{\mathbb{D}}$. The condition $\dot{\mathbf{G}}_{\mathbb{D}}$ implies that the set $\mathbb{Y}_{\varepsilon,i} = \mathbb{Y}_{1,i}, \varepsilon \in (0, 1]$, for $i \in \mathbb{D}$.

The following condition is an analogue of the condition $\mathbf{H}_{\mathbb{D}}$. It is required only in the case, where the return times $\dot{\tau}_{\varepsilon,\mathbb{D}}$ are objects of interest:

$\dot{\mathbf{H}}_{\mathbb{D}}$: For any $i \in \mathbb{D}$, there exists a chain of states $i = j_0, j_1 \in \mathbb{D}, \ldots, j_{\dot{n}_i - 1} \in \mathbb{D}, j_{\dot{n}_i} \in \bar{\mathbb{D}}$ such that $\prod_{1 \leq l \leq \dot{n}_i} p_{1,j_{l-1}j_l} > 0$.

Analogues of the conditions $\mathbf{I}_{\mathbb{D}}$ and $\mathbf{I}_{\mathbb{D},\mathcal{H}}$ take the following forms:

$\dot{\mathbf{I}}_{\mathbb{D}}$: $p_{\varepsilon,ij} \to p_{0,ij}$ as $\varepsilon \to 0$, for $i \in \mathbb{D}, j \in \mathbb{X}$.

and

$\dot{\mathbf{I}}_{\mathbb{D},\mathcal{H}}$: The functions $p_{\cdot,ij}, j \in \mathbb{Y}_{1,i}, i \in \mathbb{D}$ belong to the complete family of asymptotically comparable functions $\mathcal{H}$ appearing in the condition $\mathbf{I}_{\mathbb{D},\mathcal{H}}$.

Analogues of the equivalent conditions $\mathbf{J}_{\mathbb{D}}$ and $\mathbf{J}_{\mathbb{D},1}$ take the following forms:

$\dot{\mathbf{J}}_{\mathbb{D}}$: **(a)** $F_{\varepsilon,ij}(\cdot\, u_{\varepsilon,i}) \Rightarrow F_{0,ij}(\cdot)$ as $\varepsilon \to 0$, for $j \in \mathbb{Y}_{1,i}, i \in \mathbb{D}$, **(b)** $F_{0,ij}(\cdot), j \in \mathbb{Y}_{1,i}, i \in \mathbb{D}$ are proper distribution functions such that $F_{0,ij}(0) < 1, j \in \mathbb{Y}_{1,i}, i \in \mathbb{D}$, **(c)** $u_{\varepsilon,i} \in (0, \infty), \varepsilon \in (0, 1]$, for $i \in \mathbb{D}$.

and

$\dot{\mathbf{J}}^{\circ}_{\mathbb{D}}$: **(a)** $\phi_{\varepsilon,ij}(s/u_{\varepsilon,i}) \to \phi_{0,ij}(s)$ as $\varepsilon \to 0$, for $s \geq 0$ and $j \in \mathbb{Y}_{1,i}, i \in \mathbb{D}$, **(b)** $\phi_{0,ij}(s) = \int_0^\infty e^{-st} F_{0,ij}(dt), s \geq 0, j \in \mathbb{Y}_{1,i}, i \in \mathbb{D}$ are Laplace transforms of proper distribution functions such that $F_{0,ij}(0) < 1, j \in \mathbb{Y}_{1,i}, i \in \mathbb{D}$, **(c)** $u_{\varepsilon,i} \in (0, \infty), \varepsilon \in (0, 1]$, for $i \in \mathbb{D}$.

The following condition is an analogue of the condition $\mathbf{K}_{\mathbb{D}}$. It is required only in the case where the return times $\dot{\tau}_{\varepsilon,\mathbb{D}}$ are objects of interest:

$\dot{\mathbf{K}}_{\mathbb{D}}$: **(a)** $f_{\varepsilon,ij} < \infty, j \in \mathbb{Y}_{1,i}, i \in \mathbb{D}$, for every $\varepsilon \in (0, 1]$, **(b)** $f_{\varepsilon,ij}/u_{\varepsilon,i} \to f_{0,ij} = \int_0^\infty t F_{0,ij}(dt < \infty$ as $\varepsilon \to 0$, for $j \in \mathbb{Y}_{1,i}, i \in \mathbb{D}$.

Analogues of the conditions **L** and $\mathbf{L}_{\mathcal{H}}$ takes the following forms:

$\dot{\mathbf{L}}_{\mathbb{D}}$: $u_{\varepsilon,i} \longrightarrow u_{0,i} \in (0, \infty]$ as $\varepsilon \longrightarrow 0$, for $i \in \mathbb{D}$.

and

$\dot{\mathbf{L}}_{\mathbb{D},\mathcal{H}}$: Functions $u_{\cdot,i}$, $i \in \mathbb{D}$ belong to the complete family of asymptotically comparable functions $\mathcal{H}$ appearing in the condition $\mathbf{L}_{\mathbb{D},\mathcal{H}}$.

Let $\dot{m}$ be the number of states in the domain $\mathbb{D}$, and let $\bar{r}_{\dot{m}} = \langle r_1, \ldots, r_{\dot{m}} \rangle$ be a sequence of different states from the domain $\mathbb{D}$. Also, let $\bar{r}_n = \langle r_1, \ldots, r_n \rangle$, $n = 0, 1, \ldots, \dot{m}$ be the corresponding subsequences of the sequence $\bar{r}_n = \langle r_1, \ldots, r_{\dot{m}} \rangle$, As in Sect. 9.4.1, $\bar{r}_0 = \langle\rangle$ is an "empty" sequence.

Analogues of conditions ${}_{\bar{\mathbf{k}}_n}\mathbf{M}_{\mathbb{D}}$, $n = 1, \ldots, \bar{m} - 1$ take the following forms, for $n = 1, \ldots, \dot{m} - 1$

${}_{\bar{\mathbf{r}}_n}\dot{\mathbf{M}}_{\mathbb{D}}$: $r_l \in {}_{\bar{r}_{l-1}}\dot{\mathbb{W}}_{\mathbb{D},0}$, i.e., ${}_{\bar{r}_{l-1}}w_{0,r_l i} = \lim_{\varepsilon \to 0} \frac{{}_{\bar{r}_{l-1}}\tilde{u}_{\varepsilon,r_l}}{{}_{\bar{r}_{l-1}}\tilde{u}_{\varepsilon,i}} \in [0, \infty)$, $i \in {}_{\bar{r}_{l-1}}\mathbb{D}$, for $l = 1, \ldots, n$.

Here, the corresponding normalisation functions ${}_{\bar{r}_{n-1}}\tilde{u}_{\varepsilon,i}$ take the following forms, for $i \in {}_{\bar{r}_{l-1}}\mathbb{D} = \mathbb{D} \setminus \{r_1, \ldots, r_{l-1}\}$, $l = 1, \ldots, \dot{m} - 1$,

$$
{}_{\bar{r}_{l-1}}\tilde{u}_{\varepsilon,i} = \prod_{l=0}^{n-1} {}_{\bar{r}_l}\bar{p}_{\varepsilon,ii}^{-1}\, u_{\varepsilon,i}. \tag{10.150}
$$

and the sets ${}_{\bar{r}_{l-1}}\dot{\mathbb{W}}_{\mathbb{D},0}$ take the following forms, for $l = 1, \ldots, \dot{m} - 1$,

$$
{}_{\bar{r}_{l-1}}\dot{\mathbb{W}}_{\mathbb{D},0} = \{k \in {}_{\bar{r}_{l-1}}\mathbb{D} : {}_{\bar{r}_{l-1}}w_{0,r_l,i} \in [0, \infty),\ i \in {}_{\bar{r}_{l-1}}\mathbb{D}\}. \tag{10.151}
$$

It worth noting that the pair of conditions $\mathbf{G}_{\mathbb{D}}$ and $\dot{\mathbf{G}}_{\mathbb{D}}$ is equivalent to the condition **G**, the pair of conditions $\mathbf{I}_{\mathbb{D}}$ and $\dot{\mathbf{I}}_{\mathbb{D}}$ is equivalent to the condition **I**, the pair of conditions $\mathbf{I}_{\mathbb{D},\mathcal{H}}$ and $\dot{\mathbf{I}}_{\mathbb{D},\mathcal{H}}$ is equivalent to condition $\mathbf{I}_{\mathcal{H}}$, the pair of conditions $\mathbf{J}_{\mathbb{D}}$ and $\dot{\mathbf{J}}_{\mathbb{D}}$ is equivalent to the condition **J**, the pair of conditions $\mathbf{K}_{\mathbb{D}}$ and $\dot{\mathbf{K}}_{\mathbb{D}}$ is equivalent to the condition **K**, the pair of conditions $\mathbf{L}_{\mathbb{D}}$ and $\dot{\mathbf{L}}_{\mathbb{D}}$ is equivalent to the condition **L**, and the pair of conditions $\mathbf{L}_{\mathbb{D},\mathcal{H}}$ and $\dot{\mathbf{L}}_{\mathbb{D},\mathcal{H}}$ is equivalent to the condition $\mathbf{L}_{\mathcal{H}}$. However, the pair of conditions $\mathbf{H}_{\mathbb{D}}$ and $\dot{\mathbf{H}}_{\mathbb{D}}$ is weaker than the condition **H**.

10.3.2 Weak Asymptotics for Distributions of Hitting and Return Times

10.3.2.1 Convergence of Hitting and Return Probabilities. The relation (10.147) implies that the following relation holds, for $r, j \in \mathbb{D}$ and $\varepsilon \in (0, 1]$:

$$
\begin{aligned}
P_{\varepsilon,\mathbb{D},rj} &= \mathsf{P}_r\{\eta_\varepsilon(\tau_{\varepsilon,\mathbb{D}}) = j\} = \Psi_{\varepsilon,\mathbb{D},rj}(0) \\
&= p_{\varepsilon,rj} + \sum_{i \in \bar{\mathbb{D}}} P_{\varepsilon,\mathbb{D},ij} p_{\varepsilon,ri}.
\end{aligned} \tag{10.152}
$$

Note that the condition $\mathbf{H}_{\mathbb{D}}$ and the relation (10.152) imply that, for $r \in \mathbb{D}$ and $\varepsilon \in (0, 1]$,

$$P_{\varepsilon,\mathbb{D},rj} \geq 0, j \in \mathbb{D}, \sum_{j\in\mathbb{D}} P_{\varepsilon,\mathbb{D},rj} = 1. \tag{10.153}$$

The conditions $\mathbf{G}_{\mathbb{D}}$, $\mathbf{H}_{\mathbb{D}}$ and $\dot{\mathbf{G}}_{\mathbb{D}}$, Lemma 10.11, and the relation (10.152) also imply that the sets $\mathbb{Y}_{\varepsilon,\mathbb{D},r} = \{j \in \mathbb{D} : P_{\varepsilon,\mathbb{D},rj} > 0\}, r \in \mathbb{D}$ do not depend on $\varepsilon \in (0, 1]$, i.e., for $r \in \mathbb{D}$,

$$\mathbb{Y}_{\varepsilon,\mathbb{D},r} = \mathbb{Y}_{1,\mathbb{D},r}, \ \varepsilon \in (0, 1]. \tag{10.154}$$

The conditions $\mathbf{G}_{\mathbb{D}}$, $\mathbf{H}_{\mathbb{D}}$, $\mathbf{I}_{\mathbb{D}}$, $\mathbf{I}_{\mathbb{D},\mathcal{H}}$ and $\dot{\mathbf{G}}_{\mathbb{D}}$, $\dot{\mathbf{I}}_{\mathbb{D}}$, $\dot{\mathbf{I}}_{\mathbb{D},\mathcal{H}}$, Lemma 10.12, and the relation (10.152) imply that the functions $P_{\cdot,\mathbb{D},rj} \in \mathcal{H}, r, j \in \mathbb{D}$ and the following relation holds, for $r, j \in \mathbb{D}$:

$$\begin{aligned} P_{\varepsilon,\mathbb{D},rj} &= p_{\varepsilon,rj} + \sum_{i\in\bar{\mathbb{D}}} P_{\varepsilon,\mathbb{D},ij} p_{\varepsilon,ri} \\ &\to p_{0,rj} + \sum_{i\in\bar{\mathbb{D}}} P_{0,\mathbb{D},ij} p_{0,ri} = P_{0,\mathbb{D},rj} \text{ as } \varepsilon \to 0, \end{aligned} \tag{10.155}$$

where the limits appearing in the above relation satisfy the following relations, $P_{0,\mathbb{D},rj} \geq 0, r, j \in \mathbb{D}, \sum_{j\in\mathbb{D}} P_{0,\mathbb{D},rj} = 1$, for $r \in \mathbb{D}$.

The relation (10.149) implies that the following relation holds for the return probabilities, for $r, j \in \mathbb{D}$ and $\varepsilon \in (0, 1]$:

$$\begin{aligned} \dot{P}_{\varepsilon,\mathbb{D},rj} &= \mathsf{P}_r\{\eta_\varepsilon(\dot{\tau}_{\varepsilon,\mathbb{D}}) = j\} = \dot{\Psi}_{\varepsilon,\mathbb{D},rj}(0) \\ &= \sum_{i\in\bar{\mathbb{D}}} P_{\varepsilon,\mathbb{D},ij} P_{\varepsilon,\bar{\mathbb{D}},ri}. \end{aligned} \tag{10.156}$$

The conditions $\mathbf{G}_{\mathbb{D}}$, $\mathbf{H}_{\mathbb{D}}$, $\mathbf{I}_{\mathbb{D}}$, $\mathbf{I}_{\mathbb{D},\mathcal{H}}$ and $\dot{\mathbf{G}}_{\mathbb{D}}$, $\dot{\mathbf{H}}_{\mathbb{D}}$, $\dot{\mathbf{I}}_{\mathbb{D}}$, $\dot{\mathbf{I}}_{\mathbb{D},\mathcal{H}}$, Lemma 10.14, and the relation (10.156) imply that the functions $\dot{P}_{\cdot,\mathbb{D},rj} \in \mathcal{H}, r, j \in \mathbb{D}$ and the following relation holds, for $r, j \in \mathbb{D}$:

$$\begin{aligned} \dot{P}_{\varepsilon,\mathbb{D},rj} &= \sum_{i\in\bar{\mathbb{D}}} P_{\varepsilon,\mathbb{D},ij} P_{\varepsilon,\bar{\mathbb{D}},ri} \\ &\to \sum_{i\in\bar{\mathbb{D}}} P_{0,\mathbb{D},ij} P_{0,\bar{\mathbb{D}},ri} = \dot{P}_{0,\mathbb{D},rj} \text{ as } \varepsilon \to 0, \end{aligned} \tag{10.157}$$

where the limits appearing in the above relation satisfy the following relations:

$$\dot{P}_{0,\mathbb{D},rj} \geq 0, \ r, j \in \mathbb{D}, \sum_{j\in\mathbb{D}} \dot{P}_{0,\mathbb{D},rj} = 1, \ r \in \mathbb{D}. \tag{10.158}$$

10.3.2.2 Weak Asymptotics for Distributions of Hitting Times. Let us assume that the conditions $\mathbf{G}_{\mathbb{D}}$, $\mathbf{H}_{\mathbb{D}}$, $\mathbf{I}_{\mathbb{D}}$, $\mathbf{I}_{\mathbb{D},\mathcal{H}}$, $\mathbf{L}_{\mathbb{D}}$, $\mathbf{L}_{\mathbb{D},\mathcal{H}}$, $\mathbf{J}_{\mathbb{D}}$, $\mathbf{K}_{\mathbb{D}}$, ${}_{\bar{k}_{\bar{m}-1}}\mathbf{M}_{\mathbb{D}}$ and the conditions $\dot{\mathbf{G}}_{\mathbb{D}}$, $\dot{\mathbf{I}}_{\mathbb{D}}$, $\dot{\mathbf{I}}_{\mathbb{D},\mathcal{H}}$, $\dot{\mathbf{L}}_{\mathbb{D}}$, $\dot{\mathbf{L}}_{\mathbb{D},\mathcal{H}}$, $\dot{\mathbf{J}}_{\mathbb{D}}$ are satisfied for the semi-Markov processes $\eta_\varepsilon(t)$.

The conditions $\mathbf{G}_{\mathbb{D}}$, $\mathbf{H}_{\mathbb{D}}$, $\mathbf{I}_{\mathbb{D},\mathcal{H}}$, $\mathbf{L}_{\mathbb{D},\mathcal{H}}$, ${}_{\bar{\mathbf{k}}_{\bar{m}-1}}\mathbf{M}_{\mathbb{D}}$ and $\dot{\mathbf{I}}_{\mathbb{D},\mathcal{H}}$, $\dot{\mathbf{L}}_{\mathbb{D},\mathcal{H}}$ imply that the following condition holds:

$\hat{\mathbf{L}}_{\mathbb{D},1}$: $\frac{\check{u}_{\varepsilon,k_n}}{u_{\varepsilon,r}} \to {}_{\bar{k}_{\bar{m}}}\dot{w}_{0,k_n,r} \in [0,\infty]$ as $\varepsilon \to 0$, for $r \in \mathbb{D}$, $k_n \in \bar{\mathbb{D}}$.

Since the functions $\check{u}_{\cdot,k_n}, k_n \in \bar{\mathbb{D}}$ and $u_{\cdot,r}, r \in \mathbb{D}$ belong to the class of asymptotically comparable functions $\mathcal{H}$, the limits appearing in the condition $\hat{\mathbf{L}}_{\mathbb{D}}$ can be calculated using the quotient operating formulas given in Lemma 8.2. In the case, where the family $\mathcal{H}$ is one of the families, $\mathcal{H}_1$, $\mathcal{H}_2$, or $\mathcal{H}_3$, the quotient operating formulas given in Lemma 8.4, 8.6, or 8.8 should be used.

Let us introduce sets, for $r \in \mathbb{D}$,

$$ {}_{\bar{k}_{\bar{m}}}\bar{\mathbb{D}}_r = \{k_n \in \bar{\mathbb{D}} : p_{0,rk_n} > 0\} \tag{10.159} $$

and denote,

$$ \dot{n}(\bar{k}_{\bar{m}}, r) = \max_{k_n \in {}_{\bar{k}_{\bar{m}}}\bar{\mathbb{D}}_r} \bar{n}(\bar{k}_{\bar{m}}, n), \tag{10.160} $$

where $\bar{n}(\bar{k}_{\bar{m}}, n)$ are the numbers defined in the relation (10.118).

Let us define the normalisation functions, for $r \in \mathbb{D}$,

$$ \check{u}_{\varepsilon,r} = \begin{cases} u_{\varepsilon,r} & \text{if } {}_{\bar{k}_{\bar{m}}}\dot{w}_{0,k_{\dot{n}(\bar{k}_{\bar{m}},r)},r} \in [0,\infty), \\ \check{u}_{\varepsilon,k_{\dot{n}(\bar{k}_{\bar{m}},r)}} & \text{if } {}_{\bar{k}_{\bar{m}}}\dot{w}_{0,k_{\dot{n}(\bar{k}_{\bar{m}},r)},r} = \infty. \end{cases} \tag{10.161} $$

The following theorem takes place.

Theorem 10.5 *Let the conditions* $\mathbf{G}_{\mathbb{D}}$, $\mathbf{H}_{\mathbb{D}}$, $\mathbf{I}_{\mathbb{D}}$, $\mathbf{I}_{\mathbb{D},\mathcal{H}}$, $\mathbf{L}_{\mathbb{D}}$, $\mathbf{L}_{\mathbb{D},\mathcal{H}}$, $\mathbf{J}_{\mathbb{D}}$, $\mathbf{K}_{\mathbb{D}}$, ${}_{\bar{\mathbf{k}}_{\bar{m}-1}}\mathbf{M}_{\mathbb{D}}$ *and* $\dot{\mathbf{G}}_{\mathbb{D}}$, $\dot{\mathbf{I}}_{\mathbb{D}}$, $\dot{\mathbf{I}}_{\mathbb{D},\mathcal{H}}$, $\dot{\mathbf{L}}_{\mathbb{D}}$, $\dot{\mathbf{L}}_{\mathbb{D},\mathcal{H}}$, $\dot{\mathbf{J}}_{\mathbb{D}}$ *be satisfied for the semi-Markov processes* $\eta_\varepsilon(t)$. *Then*:

(i) *The following asymptotic relation takes place, for* $j, r \in \mathbb{D}$,

$$ \begin{aligned} G_{\varepsilon,\mathbb{D},rj}(\cdot\, \check{u}_{\varepsilon,r}) &\Rightarrow G_{0,\mathbb{D},rj}(\cdot) \\ &= F_{0,\mathbb{D},rj}(\cdot)P_{0,\mathbb{D},rj} \text{ as } \varepsilon \to 0, \end{aligned} \tag{10.162} $$

where: (a) $\check{u}_{\cdot,r}, r \in \mathbb{D}$ *are normalisation functions, which belong to the family of asymptotically comparable functions* $\mathcal{H}$, *(b)* $F_{0,\mathbb{D},rj}(\cdot), j, r \in \mathbb{D}$ *are proper distribution functions, i.e.,* $F_{0,\mathbb{D},rj}(\infty) = 1$, *for* $j, r \in \mathbb{D}$,, *(c)* $P_{0,\mathbb{D},rj}, j, r \in \mathbb{D}$ *are discrete distributions, i.e.,* $P_{0,\mathbb{D},rj} \geq 0, j, r \in \mathbb{D}$ *and* $\sum_{j\in\mathbb{D}} P_{0,\mathbb{D},rj} = 1$, *for* $r \in \mathbb{D}$.

(ii) *The normalisation functions* $\check{u}_{\cdot,r}, r \in \mathbb{D}$ *are given by the relation (10.161).*

(iii) *The Laplace transforms* $\Psi_{0,\mathbb{D},rj}(\cdot), j, r \in \mathbb{D}$ *of distributions* $G_{0,\mathbb{D},rj}(\cdot), j, r \in \mathbb{D}$ *are given by the relations (10.165) and (10.171).*

(iv) *The limiting distribution functions* $G_{0,\mathbb{D},r}(\cdot) = \sum_{j\in\mathbb{D}} F_{0,\mathbb{D},rj}(\cdot)P_{0,\mathbb{D},rj}, r \in \mathbb{D}$ *are not concentrated at zero, i.e.,* $G_{0,\mathbb{D},r}(0) < 1, r \in \mathbb{D}$.

(v) *The limits of hitting probabilities* $P_{0,\mathbb{D},rj}, j, r \in \mathbb{D}$ *are given by the relation (10.155).*

Proof Let us choose an arbitrary state $r \in \mathbb{D}$.

First, let us consider case, where the following assumption holds true:

$$
{}_{\bar{k}_{\bar{m}}}\dot{w}_{0,k_{\dot{n}(\bar{k}_{\bar{m}},r)},r} \in [0, \infty). \tag{10.163}
$$

In this case, the normalisation function,

$$
\check{u}_{\varepsilon,r} = u_{\varepsilon,r}. \tag{10.164}
$$

The conditions of Theorem 10.5, the relation (10.147), and Theorem 10.4 imply that the following relation holds, for $j \in \mathbb{D}$:

$$
\begin{aligned}
&\Psi_{\varepsilon,\mathbb{D},rj}(s/\check{u}_{\varepsilon,r}) \\
&\quad = \phi_{\varepsilon,rj}(s/u_{\varepsilon,r})p_{\varepsilon,rj} \\
&\qquad + \Psi_{\varepsilon,\mathbb{D},k_{\dot{n}(\bar{k}_{\bar{m}},r)},j}\Big(s\frac{\check{u}_{\varepsilon,k_{\dot{n}(\bar{k}_{\bar{m}},r)}}}{u_{\varepsilon,r}}/\check{u}_{\varepsilon,k_{\dot{n}(\bar{k}_{\bar{m}},r)}}\Big) \\
&\qquad \times \phi_{\varepsilon,rk_{\dot{n}(\bar{k}_{\bar{m}},r)}}(s/u_{\varepsilon,r})p_{\varepsilon,rk_{\dot{n}(\bar{k}_{\bar{m}},r)}} \\
&\qquad + \sum_{k_n \in {}_{\bar{k}_{\bar{m}}}\bar{\mathbb{D}}_r, n \neq \dot{n}(\bar{k}_{\bar{m}},r)} \Psi_{\varepsilon,\mathbb{D},k_nj}\Big(s\frac{\check{u}_{\varepsilon,k_{\dot{n}(\bar{k}_{\bar{m}},r)}}}{u_{\varepsilon,r}}\frac{\check{u}_{\varepsilon,k_n}}{\check{u}_{\varepsilon,k_{\dot{n}(\bar{k}_{\bar{m}},r)}}}/\check{u}_{\varepsilon,k_n}\Big) \\
&\qquad \times \phi_{\varepsilon,r,k_n}(s/u_{\varepsilon,r})p_{\varepsilon,rk_n} \\
&\qquad + \sum_{k_n \in \bar{\mathbb{D}} \setminus {}_{\bar{k}_{\bar{m}}}\bar{\mathbb{D}}_r} \Psi_{\varepsilon,\mathbb{D},k_nj}\Big(s\frac{\check{u}_{\varepsilon,k_{\dot{n}(\bar{k}_{\bar{m}},r)}}}{u_{\varepsilon,r}}\frac{\check{u}_{\varepsilon,k_n}}{\check{u}_{\varepsilon,k_{\dot{n}(\bar{k}_{\bar{m}},r)}}}/\check{u}_{\varepsilon,k_n}\Big) \\
&\qquad \times \phi_{\varepsilon,rk_n}(s/u_{\varepsilon,r})p_{\varepsilon,rk_n} \\
&\quad \longrightarrow \phi_{0,rj}(s)p_{0,rj} \\
&\qquad + \Psi_{0,\mathbb{D},k_{\dot{n}(\bar{k}_{\bar{m}},r)},j}(s\, {}_{\bar{k}_{\bar{m}}}\dot{w}_{0,k_{\dot{n}(\bar{k}_{\bar{m}},r)},r}) \\
&\qquad \times \phi_{0,r\,k_{\dot{n}(\bar{k}_{\bar{m}},r)}}(s)p_{0,r\,k_{\dot{n}(\bar{k}_{\bar{m}},r)}} \\
&\qquad + \sum_{k_n \in {}_{\bar{k}_{\bar{m}}}\bar{\mathbb{D}}_r, n \neq \dot{n}(\bar{k}_{\bar{m}},r)} \Psi_{\varepsilon,\mathbb{D},k_nj}(s\, {}_{\bar{k}_{\bar{m}}}\dot{w}_{0,k_{\dot{n}(\bar{k}_{\bar{m}},r)},r}\, {}_{\bar{k}_{\bar{m}}}\check{w}_{0,k_n,k_{\dot{n}(\bar{k}_{\bar{m}},r)}}) \\
&\qquad \times \phi_{0,r,k_n}(s)p_{0,rk_n} \\
&\quad = \Psi_{0,\mathbb{D},rj}(s) \text{ as } \varepsilon \longrightarrow 0, \text{ for } s \geq 0.
\end{aligned} \tag{10.165}
$$

In this case, the limiting distribution $G_{0,\mathbb{D},rj}(\cdot)$ has, for every $j \in \mathbb{D}$, the Laplace transform $\Psi_{0,\mathbb{D},rj}(s) = \int_0^\infty e^{-st}G_{0,\mathbb{D},rj}(dt), s \geq 0$ given by the relation (10.165).

Theorem 10.4 and the relation (10.165) also imply that the following relation holds:

$$
\begin{aligned}
\Psi_{\varepsilon,\mathbb{D},r}(s/\check{u}_{\varepsilon,r}) &= \sum_{j\in\mathbb{D}} \Psi_{\varepsilon,\mathbb{D},rj}(s/\check{u}_{\varepsilon,r}) \\
&\to \sum_{j\in\mathbb{D}} \phi_{0,rj}(s)p_{0,rj} \\
&\quad + \Psi_{0,\mathbb{D},k_{\dot{n}(\bar{k}_{\bar{m}},r)}}(s\, {}_{\bar{k}_{\bar{m}}}\dot{w}_{0,k_{\dot{n}(\bar{k}_{\bar{m}},r)},r}) \\
&\quad \times \phi_{0,r\,k_{\dot{n}(\bar{k}_{\bar{m}},r)}}(s)p_{0,r\,k_{\dot{n}(\bar{k}_{\bar{m}},r)}} \\
&\quad + \sum_{k_n\in {}_{\bar{k}_{\bar{m}}}\bar{\mathbb{D}}_r, n\neq \dot{n}(\bar{k}_{\bar{m}},r)} \Psi_{\varepsilon,\mathbb{D},k_n}(s\, {}_{\bar{k}_{\bar{m}}}\dot{w}_{0,k_{\dot{n}(\bar{k}_{\bar{m}},r)},r}\, {}_{\bar{k}_{\bar{m}}}\check{w}_{0,k_n,k_{\dot{n}(\bar{k}_{\bar{m}},r)}}) \\
&\quad \times \phi_{0,r,k_n}(s)p_{0,rk_n} \\
&= \Psi_{0,\mathbb{D},r}(s) \text{ as } \varepsilon\to 0, \text{ for } s\geq 0.
\end{aligned} \tag{10.166}
$$

The conditions $\mathbf{J}_{\mathbb{D},1}$ and $\dot{\mathbf{J}}_{\mathbb{D},1}$ imply that, $\phi_{0,rj}(s) < 1, j \in \mathbb{D}$ and $\phi_{0,k_nj}(s) < 1, k_n \in \bar{\mathbb{D}}$, for $s > 0$. These inequalities and the relation (10.166) imply that the following inequalities holds, for $s > 0$:

$$
\begin{aligned}
\Psi_{0,\mathbb{D},r}(s) &\leq \sum_{j\in\mathbb{D}} \phi_{0,rj}(s)p_{0,rj} \\
&\quad + \phi_{0,r\,k_{\dot{n}(\bar{k}_{\bar{m}},r)}}(s)p_{0,r\,k_{\dot{n}(\bar{k}_{\bar{m}},r)}} \\
&\quad + \sum_{k_n\in {}_{\bar{k}_{\bar{m}}}\bar{\mathbb{D}}_r, n\neq \dot{n}(\bar{k}_{\bar{m}},r)} \phi_{0,r,k_n}(s)p_{0,rk_n} \\
&< \sum_{j\in\mathbb{D}} p_{0,rj} + p_{0,r\,k_{\dot{n}(\bar{k}_{\bar{m}},r)}} \\
&\quad + \sum_{k_n\in {}_{\bar{k}_{\bar{m}}}\bar{\mathbb{D}}_r, n\neq \dot{n}(\bar{k}_{\bar{m}},r)} p_{0,rk_n} = 1.
\end{aligned} \tag{10.167}
$$

The relation (10.167) implies holding of the following relation:

$$
G_{0,\mathbb{D},r}(0) < 1. \tag{10.168}
$$

Second, let us consider case, where the following assumption holds true:

$$
{}_{\bar{k}_{\bar{m}}}\dot{w}_{0,k_{\dot{n}(\bar{k}_{\bar{m}},r)},r} = \infty. \tag{10.169}
$$

In this case, the normalisation function,

$$
\check{u}_{\varepsilon,r} = u_{\varepsilon,\check{u}_{\varepsilon,k_{\dot{n}(k_{\bar{m}},r)}}}. \tag{10.170}
$$

The conditions of Theorem 10.5, the relation (10.147) and Theorem 10.4 imply that the following relation holds, for $j \in \mathbb{D}$,

$$
\begin{aligned}
&\Psi_{\varepsilon,\mathbb{D},rj}(s/\breve{u}_{\varepsilon,r}) \\
&\quad = \phi_{\varepsilon,rj}(s\frac{u_{\varepsilon,r}}{\breve{u}_{\varepsilon,k_{\dot{n}(k_{\bar{m}},r)}}}/u_{\varepsilon,r})p_{\varepsilon,rj} \\
&\qquad + \Psi_{\varepsilon,\mathbb{D},k_{\dot{n}(\bar{k}_{\bar{m}},r)},j}(s/\breve{u}_{\varepsilon,k_{\dot{n}(\bar{k}_{\bar{m}},r)}}) \\
&\qquad \times \phi_{\varepsilon,rk_{\dot{n}(\bar{k}_{\bar{m}},r)}}(s\frac{u_{\varepsilon,r}}{\breve{u}_{\varepsilon,k_{\dot{n}(k_{\bar{m}},r)}}}/u_{\varepsilon,r})p_{\varepsilon,rk_{\dot{n}(\bar{k}_{\bar{m}},r)}} \\
&\qquad + \sum_{k_n \in {}_{\bar{k}_{\bar{m}}}\bar{\mathbb{D}}_r, n \neq \dot{n}(\bar{k}_{\bar{m}},r)} \Psi_{\varepsilon,\mathbb{D},k_n j}(s\frac{\breve{u}_{\varepsilon,k_n}}{\breve{u}_{\varepsilon,k_{\dot{n}(\bar{k}_{\bar{m}},r)}}}/\breve{u}_{\varepsilon,k_n}) \\
&\qquad \times \phi_{\varepsilon,r,k_n}(s/u_{\varepsilon,r})p_{\varepsilon,rk_n} \\
&\qquad + \sum_{k_n \in \bar{\mathbb{D}} \setminus {}_{\bar{k}_{\bar{m}}}\bar{\mathbb{D}}_r} \Psi_{\varepsilon,\mathbb{D},k_n j}(s\frac{\breve{u}_{\varepsilon,k_n}}{\breve{u}_{\varepsilon,k_{\dot{n}(\bar{k}_{\bar{m}},r)}}}/\breve{u}_{\varepsilon,k_n}) \\
&\qquad \times \phi_{\varepsilon,rk_n}(s\frac{u_{\varepsilon,r}}{\breve{u}_{\varepsilon,k_{\dot{n}(k_{\bar{m}},r)}}}/u_{\varepsilon,r})p_{\varepsilon,rk_n} \\
&\quad \longrightarrow p_{0,rj} + \Psi_{0,\mathbb{D},k_{\dot{n}(\bar{k}_{\bar{m}},r)},j}(s)p_{0,r\,k_{\dot{n}(\bar{k}_{\bar{m}},r)}} \\
&\qquad + \sum_{k_n \in {}_{\bar{k}_{\bar{m}}}\bar{\mathbb{D}}_r, n \neq \dot{n}(\bar{k}_{\bar{m}},r)} \Psi_{\varepsilon,\mathbb{D},k_n j}(s\,{}_{\bar{k}_{\bar{m}}}\breve{w}_{0,k_n,k_{\dot{n}(\bar{k}_{\bar{m}},r)}})p_{0,rk_n} \\
&\quad = \Psi_{0,\mathbb{D},rj}(s) \text{ as } \varepsilon \to 0, \text{ for } s \geq 0.
\end{aligned} \tag{10.171}
$$

In this case, the limiting distribution $G_{0,\mathbb{D},rj}(\cdot)$ has, for every $j \in \mathbb{D}$, the Laplace transform $\Psi_{0,\mathbb{D},rj}(s) = \int_0^\infty e^{-st}G_{0,\mathbb{D},rj}(dt), s \geq 0$ given by the relation (10.171).

The Relation (10.171) implies that the following relation holds,

$$
\begin{aligned}
&\Psi_{\varepsilon,\mathbb{D},r}(s/\breve{u}_{\varepsilon,r}) = \sum_{j \in \mathbb{D}} \Psi_{\varepsilon,\mathbb{D},rj}(s/\breve{u}_{\varepsilon,r}) \\
&\quad \longrightarrow p_{0,rj} + \Psi_{0,\mathbb{D},k_{\dot{n}(\bar{k}_{\bar{m}},r)}}(s)p_{0,r\,k_{\dot{n}(\bar{k}_{\bar{m}},r)}} \\
&\qquad + \sum_{k_n \in {}_{\bar{k}_{\bar{m}}}\bar{\mathbb{D}}_r, n \neq \dot{n}(\bar{k}_{\bar{m}},r)} \Psi_{\varepsilon,\mathbb{D},k_n}(s\,{}_{\bar{k}_{\bar{m}}}\breve{w}_{0,k_n,k_{\dot{n}(\bar{k}_{\bar{m}},r)}})p_{0,rk_n} \\
&\quad = \Psi_{0,\mathbb{D},r}(s) \text{ as } \varepsilon \to 0, \text{ for } s \geq 0.
\end{aligned} \tag{10.172}
$$

Theorem 10.4 implies that, $\Psi_{0,\mathbb{D},k_n,j}(s) < 1, k_n \in \bar{\mathbb{D}}$, for $s > 0$. These inequalities and relation (10.172) imply that the following inequality holds, for $s > 0$:

$$\begin{aligned}\Psi_{0,\mathbb{D},r}(s) &\leq p_{0,rj} + \Psi_{0,\mathbb{D},k_{\dot{n}(\bar{k}_{\bar{m}},r)}}(s)p_{0,r\,k_{\dot{n}(\bar{k}_{\bar{m}},r)}} \\ &\quad + \sum_{k_n \in {}_{\bar{k}_{\bar{m}}}\bar{\mathbb{D}}_r, n \neq \dot{n}(\bar{k}_{\bar{m}},r)} p_{0,rk_n} \\ &< \sum_{j \in \mathbb{D}} p_{0,rj} + p_{0,r\,k_{\dot{n}(\bar{k}_{\bar{m}},r)}} \\ &\quad + \sum_{k_n \in {}_{\bar{k}_{\bar{m}}}\bar{\mathbb{D}}_r, n \neq \dot{n}(\bar{k}_{\bar{m}},r)} p_{0,rk_n} = 1. \end{aligned} \tag{10.173}$$

Relation (10.167) implies holding of the following relation:

$$G_{0,\mathbb{D},r}(0) < 1. \tag{10.174}$$

The proof is complete. □

10.3.2.3 Weak Asymptotics for Distributions of Return Times. Let us assume that the conditions $\mathbf{G}_{\mathbb{D}}$, $\mathbf{H}_{\mathbb{D}}$, $\mathbf{I}_{\mathbb{D}}$, $\mathbf{I}_{\mathbb{D},\mathcal{H}}$, $\mathbf{L}_{\mathbb{D}}$, $\mathbf{L}_{\mathbb{D},\mathcal{H}}$, $\mathbf{J}_{\mathbb{D}}$, $\mathbf{K}_{\mathbb{D}}$, ${}_{\bar{\mathbf{k}}_{\bar{m}-1}}\mathbf{M}_{\mathbb{D}}$ and $\dot{\mathbf{G}}_{\mathbb{D}}$, $\dot{\mathbf{H}}_{\mathbb{D}}$, $\dot{\mathbf{I}}_{\mathbb{D}}$, $\dot{\mathbf{I}}_{\mathbb{D},\mathcal{H}}$, $\dot{\mathbf{L}}_{\mathbb{D}}$, $\dot{\mathbf{L}}_{\mathbb{D},\mathcal{H}}$, $\dot{\mathbf{J}}_{\mathbb{D}}$, $\dot{\mathbf{K}}_{\mathbb{D}}$ and ${}_{\bar{\mathbf{r}}_{\dot{m}-1}}\dot{\mathbf{M}}_{\mathbb{D}}$ are satisfied for the semi-Markov processes $\eta_\varepsilon(t)$.

The conditions $\mathbf{G}_{\mathbb{D}}$, $\mathbf{H}_{\mathbb{D}}$, $\mathbf{I}_{\mathbb{D},\mathcal{H}}$, $\mathbf{L}_{\mathbb{D},\mathcal{H}}$, ${}_{\bar{\mathbf{k}}_{\bar{m}-1}}\mathbf{M}_{\mathbb{D}}$ and $\dot{\mathbf{G}}_{\mathbb{D}}$, $\dot{\mathbf{H}}_{\mathbb{D}}$, $\dot{\mathbf{I}}_{\mathbb{D},\mathcal{H}}$, $\dot{\mathbf{L}}_{\mathbb{D},\mathcal{H}}$, ${}_{\bar{\mathbf{r}}_{\dot{m}-1}}\dot{\mathbf{M}}_{\mathbb{D}}$ imply that the following condition is satisfied:

$\hat{\mathbf{L}}_{\mathbb{D},2}$: $\frac{\check{u}_{\varepsilon,k_n}}{\check{u}_{\varepsilon,r_l}} \to {}_{\langle \bar{k}_{\bar{m}}, \bar{r}_{\dot{m}}\rangle}\dot{w}_{0,k_n,r_l} \in [0,\infty]$ as $\varepsilon \to 0$, for $r_l \in \mathbb{D}$, $k_n \in \bar{\mathbb{D}}$.

Since functions $\check{u}_{\cdot,k_n}$, $k_n \in \bar{\mathbb{D}}$ and $\check{u}_{\cdot,r_l}$, $r_l \in \mathbb{D}$ belong to the class of asymptotically comparable functions $\mathcal{H}$, the limits appearing in condition $\dot{\mathbf{L}}_{\mathbb{D},2}$ can be calculated using the quotient formulas given in Lemma 8.2. In the case, where the family $\mathcal{H}$ is one of the families, $\mathcal{H}_1$, $\mathcal{H}_2$, or $\mathcal{H}_3$, the quotient formulas given in Lemma 8.8, 8.6, or 8.8 should be used.

We introduce sets, for $r_l \in \mathbb{D}$,

$${}_{\langle \bar{k}_{\bar{m}}, \bar{r}_{\dot{m}}\rangle}\bar{\mathbb{D}}_{r_l} = \{k_n \in \bar{\mathbb{D}} : P_{0,\bar{\mathbb{D}},r_l k_n} > 0\} \tag{10.175}$$

and denote,

$$\dot{n}(\bar{k}_{\bar{m}}, \bar{r}_{\dot{m}}, r_l) = \max_{k_n \in {}_{\langle \bar{k}_{\bar{m}}, \bar{r}_{\dot{m}}\rangle}\bar{\mathbb{D}}_{r_l}} \bar{n}(\bar{k}_{\bar{m}}, n). \tag{10.176}$$

Let us define the normalisation functions, for $r_l \in \mathbb{D}$,

$$\dot{u}_{\varepsilon,r_l} = \begin{cases} \check{u}_{\varepsilon,r_l} & \text{if } {}_{\langle \bar{k}_{\bar{m}}, \bar{r}_{\dot{m}}\rangle}\dot{w}_{0,k_{\dot{n}(\bar{k}_{\bar{m}},\bar{r}_{\dot{m}},r_l)},r_l} \in [0,\infty), \\ \check{u}_{\varepsilon,k_{\dot{n}(k_{\bar{m}},r)}} & \text{if } {}_{\langle \bar{k}_{\bar{m}}, \bar{r}_{\dot{m}}\rangle}\dot{w}_{0,k_{\dot{n}(\bar{k}_{\bar{m}},\bar{r}_{\dot{m}},r_l)},r_l} = \infty. \end{cases} \tag{10.177}$$

The following theorem takes place.

Theorem 10.6 *Let the conditions* $\mathbf{G}$, $\mathbf{H}_{\mathbb{D}}$, $\dot{\mathbf{H}}_{\mathbb{D}}$, $\mathbf{I}_{\mathcal{H}}$, $\mathbf{L}_{\mathcal{H}}$ *(and, thus, also conditions* $\mathbf{I}$ *and* $\mathbf{L}$*),* $\mathbf{J}$, $\mathbf{K}$, *and* ${}_{\bar{\mathbf{k}}_{\bar{m}-1}}\mathbf{M}_{\mathbb{D}}$, ${}_{\bar{\mathbf{r}}_{\dot{m}-1}}\dot{\mathbf{M}}_{\mathbb{D}}$ *be satisfied for the semi-Markov processes* $\eta_\varepsilon(t)$. *Then*:

(i) *The following asymptotic relation takes place, for $j, r_l \in \mathbb{D}$:*

$$
\begin{aligned}
\dot{G}_{\varepsilon,\mathbb{D},r_l j}(\cdot\, \dot{u}_{\varepsilon,r}) &\Rightarrow \dot{G}_{0,\mathbb{D},r_l j}(\cdot) \\
&= \dot{F}_{0,\mathbb{D},r_l j}(\cdot)\dot{P}_{0,\mathbb{D},r_l j} \text{ as } \varepsilon \to 0,
\end{aligned} \tag{10.178}
$$

where: (a) $\dot{u}_{\cdot,r_l}, r_l \in \mathbb{D}$ are normalisation functions, which belong to the family of asymptotically comparable functions $\mathcal{H}$, (b) $\dot{F}_{0,\mathbb{D},r_l j}(\cdot), j, r_l \in \mathbb{D}$ are proper distribution functions, i.e., $\dot{F}_{0,\mathbb{D},r_l j}(\infty) = 1$, for $j, r_l \in \mathbb{D}$, (c) $\dot{P}_{0,\mathbb{D},r_l j}, j, r \in \mathbb{D}$ are discrete distributions, i.e., $\dot{P}_{0,\mathbb{D},r_l j} \geq 0, j, r \in \mathbb{D}$ and $\sum_{j\in\mathbb{D}} \dot{P}_{0,\mathbb{D},r_l j} = 1$, for $r_l \in \mathbb{D}$.

(ii) *The normalisation functions $\dot{u}_{\varepsilon,r}, r \in \mathbb{D}$ are given by the relation (10.177).*

(iii) *The Laplace transforms $\dot{\Psi}_{0,\mathbb{D},r_l j}(\cdot), j, r_l \in \mathbb{D}$ of distributions $\dot{G}_{0,\mathbb{D},r_l j}(\cdot), j, r_l \in \mathbb{D}$ are given by the relations (10.181) and (10.187).*

(iv) *The limiting distribution functions $\dot{G}_{0,\mathbb{D},r_l}(\cdot) = \sum_{j\in\mathbb{D}} \dot{F}_{0,\mathbb{D},r_l j}(\cdot)\dot{P}_{0,\mathbb{D},r_l j}, r_l \in \mathbb{D}$ are not concentrated at zero, i.e., $\dot{G}_{0,\mathbb{D},r_l}(0) < 1, r_l \in \mathbb{D}$.*

(v) *The limits of the return probabilities $\dot{P}_{0,\mathbb{D},r_l j}, j, r_l \in \mathbb{D}$ are given by the relations (10.155) and (10.157).*

Proof Let us now choose an arbitrary state $r_l \in \mathbb{D}$.

First, let us consider case, where the following assumption holds true:

$$
{}_{\langle \bar{k}_{\bar{m}},\bar{r}_{\bar{m}}\rangle}\dot{w}_{0,k_n,r_l} \in [0,\infty). \tag{10.179}
$$

In this case, the normalisation function,

$$
\dot{u}_{\varepsilon,r_l} = \check{u}_{\varepsilon,r_l}. \tag{10.180}
$$

The conditions of Theorem 10.6, the relation (10.149), and Theorem 10.4 imply that the following relation holds, for $j \in \mathbb{D}$:

$$
\begin{aligned}
&\dot{\Psi}_{\varepsilon,\mathbb{D},r_l j}(s/\dot{u}_{\varepsilon,r_l}) \\
&\quad = \Psi_{\varepsilon,\mathbb{D},k_{\dot{n}(\bar{k}_{\bar{m}},\bar{r}_{\bar{m}},r_l)},j}\Big(s\frac{\check{u}_{\varepsilon,k_{\dot{n}(\bar{k}_{\bar{m}},\bar{r}_{\bar{m}},r_l)}}}{\check{u}_{\varepsilon,r_l}}/\check{u}_{\varepsilon,k_{\dot{n}(\bar{k}_{\bar{m}},\bar{r}_{\bar{m}},r_l)}}\Big) \\
&\qquad \times \Psi_{\varepsilon,\bar{\mathbb{D}},r_l k_{\dot{n}(\bar{k}_{\bar{m}},\bar{r}_{\bar{m}},r_l)}}(s/\check{u}_{\varepsilon,r_l}) \\
&\qquad + \sum_{k_n \in {}_{\langle \bar{k}_{\bar{m}},\bar{r}_{\bar{m}}\rangle}\bar{\mathbb{D}}_{r_l}, n\neq\dot{n}(\bar{k}_{\bar{m}},\bar{r}_{\bar{m}},r_l)} \Psi_{\varepsilon,\mathbb{D},k_n j}\Big(s\frac{\check{u}_{\varepsilon,k_{\dot{n}(\bar{k}_{\bar{m}},\bar{r}_{\bar{m}},r_l)}}}{\check{u}_{\varepsilon,r_l}} \\
&\qquad \times \frac{\check{u}_{\varepsilon,k_n}}{\check{u}_{\varepsilon,k_{\dot{n}(\bar{k}_{\bar{m}},\bar{r}_{\bar{m}},r_l)}}}/\check{u}_{\varepsilon,k_n}\Big)\Psi_{\varepsilon,\bar{\mathbb{D}},r_l,k_n}(s/\check{u}_{\varepsilon,r_l}) \\
&\qquad + \sum_{k_n \in \bar{\mathbb{D}}\setminus {}_{\langle \bar{k}_{\bar{m}},\bar{r}_{\bar{m}}\rangle}\bar{\mathbb{D}}_{r_l}} \Psi_{\varepsilon,\mathbb{D},k_n j}\Big(s\frac{\check{u}_{\varepsilon,k_{\dot{n}(\bar{k}_{\bar{m}},\bar{r}_{\bar{m}},r_l)}}}{u_{\varepsilon,r_l}}
\end{aligned}
$$

$$\times \frac{\check{u}_{\varepsilon,k_n}}{\check{u}_{\varepsilon,k_{\dot{n}(\bar{k}_{\bar{m}},\bar{r}_{\dot{m}},r_l)}}}/\check{u}_{\varepsilon,k_n})\Psi_{\varepsilon,\bar{\mathbb{D}},r_l k_n}(s/\check{u}_{\varepsilon,r_l})$$
$$\longrightarrow \Psi_{0,\mathbb{D},k_{\dot{n}(\bar{k}_{\bar{m}},\bar{r}_{\dot{m}},r_l)},j}(s\ {}_{\langle \bar{k}_{\bar{m}},\bar{r}_{\dot{m}}\rangle}\dot{w}_{0,k_{\dot{n}(\bar{k}_{\bar{m}},r)},\bar{r}_{\dot{m}},r_l})$$
$$\times \Psi_{0,\bar{\mathbb{D}},r_l\,k_{\dot{n}(\bar{k}_{\bar{m}},\bar{r}_{\dot{m}},r_l)}}(s)$$
$$+ \sum_{k_n \in {}_{\langle \bar{k}_{\bar{m}},\bar{r}_{\dot{m}}\rangle}\bar{\mathbb{D}}_{r_l},n\neq \dot{n}(\bar{k}_{\bar{m}},\bar{r}_{\dot{m}},r_l)} \Psi_{\varepsilon,\mathbb{D},k_n j}(s\ {}_{\langle \bar{k}_{\bar{m}},\bar{r}_{\dot{m}}\rangle}\dot{w}_{0,k_{\dot{n}(\bar{k}_{\bar{m}},r_l)},r_l}$$
$$\times\ {}_{\bar{k}_{\bar{m}}}\check{w}_{0,k_n,k_{\dot{n}(\bar{k}_{\bar{m}},r_l)}})\Psi_{0,\bar{\mathbb{D}},r_l,k_n}(s)$$
$$= \dot{\Psi}_{0,\mathbb{D},r_l j}(s) \text{ as } \varepsilon \to 0, \text{ for } s \geq 0. \tag{10.181}$$

In this case, the limiting distribution $\dot{G}_{0,\mathbb{D},r_l j}(\cdot)$ has, for every $j \in \mathbb{D}$, the Laplace transform $\dot{\Psi}_{0,\mathbb{D},r_l j}(s) = \int_0^\infty e^{-st}\dot{G}_{0,\mathbb{D},r_l j}(dt), s \geq 0$ given by the relation (10.181).

Theorem 10.4 and relation (10.181) also imply that the following relation holds:

$$\dot{\Psi}_{\varepsilon,\mathbb{D},r_l}(s/\dot{u}_{\varepsilon,r_l}) = \sum_{j\in\mathbb{D}} \dot{\Psi}_{\varepsilon,\mathbb{D},r_l j}(s/\dot{u}_{\varepsilon,r_l})$$
$$\longrightarrow \Psi_{0,\mathbb{D},k_{\dot{n}(\bar{k}_{\bar{m}},\bar{r}_{\dot{m}},r_l)}}(s\ {}_{\langle \bar{k}_{\bar{m}},\bar{r}_{\dot{m}}\rangle}\dot{w}_{0,k_{\dot{n}(\bar{k}_{\bar{m}},r)},\bar{r}_{\dot{m}},r_l})$$
$$\times \Psi_{0,\bar{\mathbb{D}},r_l\,k_{\dot{n}(\bar{k}_{\bar{m}},\bar{r}_{\dot{m}},r_l)}}(s)$$
$$+ \sum_{k_n \in {}_{\langle \bar{k}_{\bar{m}},\bar{r}_{\dot{m}}\rangle}\bar{\mathbb{D}}_{r_l},n\neq \dot{n}(\bar{k}_{\bar{m}},\bar{r}_{\dot{m}},r_l)} \Psi_{\varepsilon,\mathbb{D},k_n}(s\ {}_{\langle \bar{k}_{\bar{m}},\bar{r}_{\dot{m}}\rangle}\dot{w}_{0,k_{\dot{n}(\bar{k}_{\bar{m}},r_l)},r_l}$$
$$\times\ {}_{\bar{k}_{\bar{m}}}\check{w}_{0,k_n,k_{\dot{n}(\bar{k}_{\bar{m}},r_l)}})\Psi_{0,\bar{\mathbb{D}},r_l,k_n}(s)$$
$$= \dot{\Psi}_{0,\mathbb{D},r_l}(s) \text{ as } \varepsilon \to 0, \text{ for } s \geq 0. \tag{10.182}$$

Theorem 10.4 implies that, $\Psi_{0,\bar{\mathbb{D}},r_l}(s) < 1$, for $s > 0$. These inequalities and the relation (10.172) imply that the following inequality holds, for $s > 0$:

$$\dot{\Psi}_{0,\mathbb{D},r_l}(s) \leq \Psi_{0,\bar{\mathbb{D}},r_l\,k_{\dot{n}(\bar{k}_{\bar{m}},\bar{r}_{\dot{m}},r_l)}}(s)$$
$$+ \sum_{k_n \in {}_{\langle \bar{k}_{\bar{m}},\bar{r}_{\dot{m}}\rangle}\bar{\mathbb{D}}_{r_l},n\neq \dot{n}(\bar{k}_{\bar{m}},\bar{r}_{\dot{m}},r_l)} \Psi_{0,\bar{\mathbb{D}},r_l,k_n}(s)$$
$$= \Psi_{0,\bar{\mathbb{D}},r_l}(s) < 1, \text{ for } s > 0. \tag{10.183}$$

The relation (10.183) implies that the following relation holds:

$$\dot{G}_{0,\mathbb{D},r_l}(0) < 1. \tag{10.184}$$

Second, let us consider case, where the following assumption holds true:

$${}_{\langle \bar{k}_{\bar{m}},\bar{r}_{\dot{m}}\rangle}\dot{w}_{0,k_n,r_l} = \infty. \tag{10.185}$$

In this case, the normalisation function,

$$\dot{u}_{\varepsilon,r_l} = \check{u}_{\varepsilon,k_{\dot{n}(k_{\bar{m}},r)}}. \tag{10.186}$$

The conditions of Theorem 10.6, the relation (10.149) and Theorem 10.4 imply that the following relation holds, for $j \in \mathbb{D}$:

$$\begin{aligned}
&\dot{\Psi}_{\varepsilon,\mathbb{D},r_l j}(s/\dot{u}_{\varepsilon,r_l}) \\
&\quad = \Psi_{\varepsilon,\mathbb{D},k_{\dot{n}(\bar{k}_{\bar{m}},\bar{r}_{\bar{m}},r_l)},j}(s/\check{u}_{\varepsilon,k_{\dot{n}(\bar{k}_{\bar{m}},\bar{r}_{\bar{m}},r_l)}}) \\
&\qquad \times \Psi_{\varepsilon,\bar{\mathbb{D}},r_l k_{\dot{n}(\bar{k}_{\bar{m}},\bar{r}_{\bar{m}},r_l)}}(s\frac{\check{u}_{\varepsilon,r_l}}{\check{u}_{\varepsilon,k_{\dot{n}(\bar{k}_{\bar{m}},\bar{r}_{\bar{m}},r_l)}}}/\check{u}_{\varepsilon,r_l}) \\
&\qquad + \sum_{k_n \in {}_{\langle \bar{k}_{\bar{m}},\bar{r}_{\bar{m}}\rangle}\bar{\mathbb{D}}_{r_l}, n \neq \dot{n}(\bar{k}_{\bar{m}},\bar{r}_{\bar{m}},r_l)} \Psi_{\varepsilon,\mathbb{D},k_n j}(s\frac{\check{u}_{\varepsilon,k_n}}{\check{u}_{\varepsilon,k_{\dot{n}(\bar{k}_{\bar{m}},\bar{r}_{\bar{m}},r_l)}}}/\check{u}_{\varepsilon,k_n}) \\
&\qquad \times \Psi_{\varepsilon,\bar{\mathbb{D}},r_l,k_n}(s\frac{\check{u}_{\varepsilon,r_l}}{\check{u}_{\varepsilon,k_{\dot{n}(\bar{k}_{\bar{m}},\bar{r}_{\bar{m}},r_l)}}}/\check{u}_{\varepsilon,r_l}) \\
&\qquad + \sum_{k_n \in \bar{\mathbb{D}} \setminus {}_{\langle \bar{k}_{\bar{m}},\bar{r}_{\bar{m}}\rangle}\bar{\mathbb{D}}_{r_l}} \Psi_{\varepsilon,\mathbb{D},k_n j}(s\frac{\check{u}_{\varepsilon,k_n}}{\check{u}_{\varepsilon,k_{\dot{n}(\bar{k}_{\bar{m}},\bar{r}_{\bar{m}},r_l)}}}/\check{u}_{\varepsilon,k_n}) \\
&\qquad \times \Psi_{\varepsilon,\bar{\mathbb{D}},r_l k_n}(s\frac{\check{u}_{\varepsilon,r_l}}{\check{u}_{\varepsilon,k_{\dot{n}(\bar{k}_{\bar{m}},\bar{r}_{\bar{m}},r_l)}}}/\check{u}_{\varepsilon,r_l}) \\
&\quad \to \Psi_{0,\mathbb{D},k_{\dot{n}(\bar{k}_{\bar{m}},\bar{r}_{\bar{m}},r_l)},j}(s) \\
&\qquad + \sum_{k_n \in {}_{\langle \bar{k}_{\bar{m}},\bar{r}_{\bar{m}}\rangle}\bar{\mathbb{D}}_{r_l}, n \neq \dot{n}(\bar{k}_{\bar{m}},\bar{r}_{\bar{m}},r_l)} \Psi_{0,\mathbb{D},k_n j}(s\ {}_{\bar{k}_{\bar{m}}}\check{w}_{0,k_n,k_{\dot{n}(\bar{k}_{\bar{m}},r)}}) \\
&\qquad \times P_{0,\bar{\mathbb{D}},r_l,k_n} \\
&\quad = \dot{\Psi}_{0,\mathbb{D},r_l j}(s) \text{ as } \varepsilon \to 0, \text{ for } s \geq 0.
\end{aligned} \tag{10.187}$$

In this case, the limiting distribution $\dot{G}_{0,\mathbb{D},r_l j}(\cdot)$ has, for every $j \in \mathbb{D}$, the Laplace transform $\dot{\Psi}_{0,\mathbb{D},r_l j}(s) = \int_0^\infty e^{-st}\dot{G}_{0,\mathbb{D},r_l j}(dt), s \geq 0$ given by the relation (10.187).

Theorem 10.4 and the relation (10.181) also imply that the following relation holds:

$$\begin{aligned}
\dot{\Psi}_{\varepsilon,\mathbb{D},r_l}(s/\dot{u}_{\varepsilon,r_l}) &= \sum_{j \in \mathbb{D}} \dot{\Psi}_{\varepsilon,\mathbb{D},r_l j}(s/\dot{u}_{\varepsilon,r_l}) \\
&\to \Psi_{0,\mathbb{D},k_{\dot{n}(\bar{k}_{\bar{m}},\bar{r}_{\bar{m}},r_l)}}(s) P_{0,\bar{\mathbb{D}},r_l\, k_{\dot{n}(\bar{k}_{\bar{m}},\bar{r}_{\bar{m}},r_l)}} \\
&\quad + \sum_{k_n \in {}_{\langle \bar{k}_{\bar{m}},\bar{r}_{\bar{m}}\rangle}\bar{\mathbb{D}}_{r_l}, n \neq \dot{n}(\bar{k}_{\bar{m}},\bar{r}_{\bar{m}},r_l)} \Psi_{0,\mathbb{D},k_n}(s\ {}_{\bar{k}_{\bar{m}}}\check{w}_{0,k_n,k_{\dot{n}(\bar{k}_{\bar{m}},r)}}) P_{0,\bar{\mathbb{D}},r_l,k_n} \\
&= \dot{\Psi}_{0,\mathbb{D},r_l}(s) \text{ as } \varepsilon \to 0, \text{ for } s \geq 0.
\end{aligned} \tag{10.188}$$

Theorem 10.4 implies that, $\Psi_{0,\mathbb{D},k_n}(s) < 1$, $k_n \in \bar{\mathbb{D}}$, for $s > 0$. These inequalities and the relation (10.172) imply that the following inequality holds, for $s > 0$:

$$\begin{aligned}
\dot{\Psi}_{0,\mathbb{D},r_l}(s) \leq \Psi_{0,\mathbb{D},k_{\dot{n}(\bar{k}_{\dot{m}},\bar{r}_{\dot{m}},r_l)}}(s) P_{0,\bar{\mathbb{D}},r_l\, k_{\dot{n}(\bar{k}_{\dot{m}},\bar{r}_{\dot{m}},r_l)}} \\
+ \sum_{k_n \in \,_{\langle \bar{k}_{\dot{m}},\bar{r}_{\dot{m}}\rangle}\bar{\mathbb{D}}_{r_l}, n \neq \dot{n}(\bar{k}_{\dot{m}},\bar{r}_{\dot{m}},r_l)} P_{0,\bar{\mathbb{D}},r_l,k_n} \\
< P_{0,\bar{\mathbb{D}},r_l\, k_{\dot{n}(\bar{k}_{\dot{m}},\bar{r}_{\dot{m}},r_l)}} \\
+ \sum_{k_n \in \,_{\langle \bar{k}_{\dot{m}},\bar{r}_{\dot{m}}\rangle}\bar{\mathbb{D}}_{r_l}, n \neq \dot{n}(\bar{k}_{\dot{m}},\bar{r}_{\dot{m}},r_l)} P_{0,\bar{\mathbb{D}},r_l,k_n} = 1.
\end{aligned} \tag{10.189}$$

The relation (10.189) implies that the following relation holds:

$$\dot{G}_{0,\mathbb{D},r_l}(0) < 1. \tag{10.190}$$

The proof is complete. □

Chapter 11
Asymptotics for Expectations of Hitting Times for Perturbed Semi-Markov Processes

In this chapter, we apply asymptotic recurrent phase space reduction algorithms for regularly and singularly perturbed finite semi-Markov processes to an asymptotic analysis of expectations of hitting times and obtain theorems about convergence of expectations of hitting times for regularly and singularly perturbed finite semi-Markov processes. Special attention is paid to the conditions under which the expectations of hitting times, normalised by the functions used in the weak limit theorems for the distributions of the hitting times, converge to the first moments of the corresponding limiting distributions. This may not be the case for singularly perturbed semi-Markov processes. Such simultaneous convergence plays an important role in ergodic theorems for perturbed Markov type processes. This chapter includes three sections.

In Sect. 11.1, we present variants of the asymptotic recurrent algorithms of phase space reduction suitable for asymptotic analysis of expectations of hitting times for regularly and singularly perturbed semi-Markov processes. The corresponding recurrent relations for expectations of hitting times are presented, in Lemmas 11.1–11.3.

In Sect. 11.2, the asymptotics for expectations of hitting times is presented in Theorems 11.1–11.3. In addition, in Theorem 11.4, conditions are given that guarantee the simultaneous convergence of distributions and expectations of the hit time for regularly and singularly perturbed semi-Markov processes.

In Sect. 11.3, limit theorems for expectations of return times and related asymptotic results are given in Theorems 11.5 and 11.6.

11.1 Expectations of Hitting Times

In this section, we present results concerning the convergence of expected response times for regularly and singularly perturbed semi-Markov processes. These results supplement the results on the weak asymptotics of the distributions of the impact moments for regularly and singularly perturbed semi-Markov processes.

D. Silvestrov, *Perturbed Semi-Markov Type Processes I*,
https://doi.org/10.1007/978-3-030-92403-4_11

11.1.1 Recurrent Relations for Expectations of Hitting Times

Let us introduce the expectations of hitting times for the original semi-Markov processes $\eta_\varepsilon(t)$, for $i \in \mathbb{X}$, $j \in \mathbb{D}$ and $\varepsilon \in [0, 1]$,

$$\begin{aligned} E_{\varepsilon,\mathbb{D},ij} &= \mathsf{E}_i \tau_{\varepsilon,\mathbb{D}} \mathrm{I}(\eta_\varepsilon(\tau_{\varepsilon,\mathbb{D}}) = j) \\ &= \int_0^\infty t G_{\varepsilon,\mathbb{D},ij}(dt), \end{aligned} \tag{11.1}$$

and

$$\begin{aligned} E_{\varepsilon,\mathbb{D},i} &= \mathsf{E}_i \tau_{\varepsilon,\mathbb{D}} \\ &= \sum_{j \in \mathbb{D}} E_{\varepsilon,\mathbb{D},ij} = \int_0^\infty t G_{\varepsilon,\mathbb{D},i}(dt). \end{aligned} \tag{11.2}$$

Let us, also, introduce the expectations, of hitting times for the reduced semi-Markov processes ${}_{\bar{k}_n}\eta_\varepsilon(t)$ and ${}_{\bar{k}_n}\tilde{\eta}_\varepsilon(t)$, for $i \in \bar{\mathbb{D}}, j \in \mathbb{D}, n = 0, \ldots, \bar{m} - 1$ and $\varepsilon \in [0, 1]$,

$$\begin{aligned} {}_{\bar{k}_n}E_{\varepsilon,\mathbb{D},ij} &= \mathsf{E}_i \, {}_{\bar{k}_n}\tau_{\varepsilon,\mathbb{D}} \mathrm{I}({}_{\bar{k}_n}\eta_\varepsilon({}_{\bar{k}_n}\tau_{\varepsilon,\mathbb{D}} = j) \\ &= \int_0^\infty t \, {}_{\bar{k}_n}G_{\varepsilon,\mathbb{D},ij}(dt) \end{aligned} \tag{11.3}$$

and

$$\begin{aligned} {}_{\bar{k}_n}\tilde{E}_{\varepsilon,\mathbb{D},ij} &= \mathsf{E}_i \, {}_{\bar{k}_n}\tilde{\tau}_{\varepsilon,\mathbb{D}} \mathrm{I}({}_{\bar{k}_n}\tilde{\eta}_\varepsilon({}_{\bar{k}_n}\tilde{\tau}_{\varepsilon,\mathbb{D}} = j) \\ &= \int_0^\infty t \, {}_{\bar{k}_n}\tilde{G}_{\varepsilon,\mathbb{D},ij}(dt). \end{aligned} \tag{11.4}$$

The following lemmas supplement Lemmas 10.5–10.10. As in these lemmas, let $\bar{k}_{\bar{m}} = \langle k_1, k_2, \ldots, k_{\bar{m}} \rangle$ be an arbitrary sequence of different states from the domain $\bar{\mathbb{D}}$ and $\bar{k}_n = \langle k_1, k_2, \ldots, k_n \rangle$, $n = 0, 1, \ldots, \bar{m}$ be the subsequences of the sequence $\bar{k}_{\bar{m}}$. As in Sect. 9.4.1, $\bar{r}_0 = \langle \rangle$ is an "empty" sequence.

Lemma 11.1 *Let the conditions* $\mathbf{G}_\mathbb{D} = {}_{\bar{k}_0}\mathbf{G}_\mathbb{D}$, $\mathbf{H}_\mathbb{D} = {}_{\bar{k}_0}\mathbf{H}_\mathbb{D}$, *and* $\mathbf{K}_\mathbb{D}(\mathbf{a}) = {}_{\bar{k}_0}\mathbf{K}_\mathbb{D}\,(\mathbf{a})$ *be satisfied. Then, for every* $n = 1, \ldots, \bar{m} - 1$, *the conditions* ${}_{\bar{k}_{n-1}}\tilde{\mathbf{G}}_\mathbb{D}$, ${}_{\bar{k}_{n-1}}\tilde{\mathbf{H}}_\mathbb{D}$, ${}_{\bar{k}_n}\mathbf{G}_\mathbb{D}$, ${}_{\bar{k}_n}\mathbf{H}_\mathbb{D}$ *and* ${}_{\bar{k}_{n-1}}\tilde{\mathbf{K}}_\mathbb{D}$, ${}_{\bar{k}_n}\mathbf{K}_\mathbb{D}$ *are satisfied, and the following relations take place, for* $\varepsilon \in (0, 1]$:

$$E_{\varepsilon,\mathbb{D},ij} = {}_{\bar{k}_{n-1}}\tilde{E}_{\varepsilon,\mathbb{D},ij} = {}_{\bar{k}_n}E_{\varepsilon,\mathbb{D},ij} < \infty, \ j \in \mathbb{D}, i \in {}_{\bar{k}_n}\bar{\mathbb{D}}, \tag{11.5}$$

and

$$E_{\varepsilon,\mathbb{D},k_n j} = {}_{\bar{k}_{n-1}}\tilde{E}_{\varepsilon,\mathbb{D},k_n j} = {}_{\bar{k}_{n-1}}\tilde{f}_{\varepsilon,k_n j} \, {}_{\bar{k}_{n-1}}\tilde{p}_{\varepsilon,k_n j}$$

$$+ \sum_{r \in \bar{k}_n \bar{\mathbb{D}}} ({}_{\bar{k}_{n-1}}\tilde{f}_{\varepsilon,k_n r}\, {}_{\bar{k}_{n-1}}P_{\varepsilon,\mathbb{D},rj} + {}_{\bar{k}_n}E_{\varepsilon,\mathbb{D},rj})\, {}_{\bar{k}_{n-1}}\tilde{p}_{\varepsilon,k_n r}$$

$$= {}_{\bar{k}_{n-1}}\tilde{f}_{\varepsilon,k_n j}\, {}_{\bar{k}_{n-1}}\tilde{p}_{\varepsilon,k_n j} + \sum_{r \in \bar{k}_n \bar{\mathbb{D}}} ({}_{\bar{k}_{n-1}}\tilde{f}_{\varepsilon,k_n r}\, P_{\varepsilon,\mathbb{D},rj} + E_{\varepsilon,\mathbb{D},rj})\, {}_{\bar{k}_{n-1}}\tilde{p}_{\varepsilon,k_n r} < \infty, \ j \in \mathbb{D}. \quad (11.6)$$

Lemma 11.2 *Let the conditions* $\mathbf{G}_{\mathbb{D}} = {}_{\bar{k}_0}\mathbf{G}_{\mathbb{D}}$, $\mathbf{H}_{\mathbb{D}} = {}_{\bar{k}_0}\mathbf{H}_{\mathbb{D}}$ *and* $\mathbf{K}_{\mathbb{D}} = {}_{\bar{k}_0}\mathbf{K}_{\mathbb{D}}$ **(a)** *be satisfied. Then, for every* $n = 0, \ldots, \bar{m} - 1$, *the conditions* ${}_{\bar{k}_n}\mathbf{G}_{\mathbb{D}}$, ${}_{\bar{k}_n}\mathbf{H}_{\mathbb{D}}$, ${}_{\bar{k}_n}\tilde{\mathbf{G}}_{\mathbb{D}}$, ${}_{\bar{k}_n}\tilde{\mathbf{H}}_{\mathbb{D}}$, *and* ${}_{\bar{k}_n}\mathbf{K}_{\mathbb{D}}$ *are satisfied, and the following relation takes place, for* $\varepsilon \in (0, 1]$:

$$E_{\varepsilon,\mathbb{D},ij} = {}_{\bar{k}_n}E_{\varepsilon,\mathbb{D},ij} = {}_{\bar{k}_n}\tilde{E}_{\varepsilon,\mathbb{D},ij} < \infty, \ j \in \mathbb{D}, i \in {}_{\bar{k}_n}\bar{\mathbb{D}}. \quad (11.7)$$

Proof The equalities given in the relations (11.5), (11.6), and (11.7) are direct corollaries of the equalities given, respectively, in the relations (10.66), (10.67), and (10.71), for both cases, where the corresponding expectations $E_{\varepsilon,\mathbb{D},ij}$ are finite or infinite. In the latter case, the product $E_{\varepsilon,\mathbb{D},rj}\, {}_{\bar{k}_{n-1}}\tilde{p}_{\varepsilon,k_n r}$ should be counted as 0, if the probability $\tilde{p}_{\varepsilon,k_n r} = 0$.

However, the statement **(ii)** of Theorem 10.1 implies that the condition ${}_{\bar{\mathbf{k}}_{\bar{m}-1}}\mathbf{K}$ is satisfied, and, thus, for $j \in \mathbb{D}$ and $\varepsilon \in (0, 1]$,

$$E_{\varepsilon,\mathbb{D},k_{\bar{m}}j} = {}_{\bar{k}_{\bar{m}-1}}\tilde{f}_{\varepsilon,k_{\bar{m}}j}\, {}_{\bar{k}_{\bar{m}-1}}\tilde{p}_{\varepsilon,k_{\bar{m}}j} < \infty. \quad (11.8)$$

The relation (11.6), taken for $n = \bar{m} - 1$, implies that, for $j \in \mathbb{D}$ and $\varepsilon \in (0, 1]$,

$$E_{\varepsilon,\mathbb{D},k_{\bar{m}-1}j} = {}_{\bar{k}_{\bar{m}-2}}\tilde{f}_{\varepsilon,k_{\bar{m}-1}j}\, {}_{\bar{k}_{\bar{m}-2}}\tilde{p}_{\varepsilon,k_{\bar{m}-1}j} + ({}_{\bar{k}_{\bar{m}-2}}\tilde{f}_{\varepsilon,k_{\bar{m}-1}k_{\bar{m}}}\, P_{\varepsilon,\mathbb{D},k_{\bar{m}}j} + E_{\varepsilon,\mathbb{D},k_{\bar{m}}j})\, {}_{\bar{k}_{\bar{m}-2}}\tilde{p}_{\varepsilon,k_{\bar{m}-1}k_{\bar{m}}} < \infty. \quad (11.9)$$

Continuing the recurrent use of the relation (11.6), for $n = \bar{m} - 2, \ldots, 1$, we prove that, for $n = \bar{m}, \ldots, 1, j \in \mathbb{D}$ and $\varepsilon \in (0, 1]$,

$$E_{\varepsilon,\mathbb{D},k_n j} < \infty. \quad (11.10)$$

The proof of Lemmas 11.1 and 11.2 is complete. □

11.1.2 Recurrent Relations for Limits of Expectations of Hitting Times

In what follows, we assume that the conditions of Theorem 10.4, $\mathbf{G}_{\mathbb{D}}$, $\mathbf{H}_{\mathbb{D}}$, $\mathbf{I}_{\mathbb{D}}$, $\mathbf{I}_{\mathbb{D},\mathcal{H}}$, $\mathbf{L}_{\mathbb{D}}$, $\mathbf{L}_{\mathbb{D},\mathcal{H}}$, $\mathbf{J}_{\mathbb{D}}$, $\mathbf{K}_{\mathbb{D}}$, and ${}_{\bar{\mathbf{k}}_{\bar{m}-1}}\mathbf{M}_{\mathbb{D}}$ are satisfied for the semi-Markov processes $\eta_\varepsilon(t)$.

According to Theorem 10.4, the weak convergence relation $G_{\varepsilon,\mathbb{D},ij}(\cdot \breve{u}_{\varepsilon,i}) \Rightarrow G_{0,\mathbb{D},k_n j}(\cdot)$ as $\varepsilon \to 0$, takes place, for $i \in \bar{\mathbb{D}}, j \in \mathbb{D}$. This relation implies that, for $i \in \bar{\mathbb{D}}, j \in \mathbb{D}$ and any sequence of $0 < T_n \to \infty$ as $n \to \infty$, which are continuity points for the distribution function $G_{0,\mathbb{D},ij}(\cdot)$,

$$\begin{aligned}\lim_{\varepsilon\to 0} E_{\varepsilon,\mathbb{D},k_nj}/\check{u}_{\varepsilon,i} &= \lim_{\varepsilon\to 0}\int_0^\infty t dG_{\varepsilon,\mathbb{D},ij}(t\check{u}_{\varepsilon,i}) \\ &\geq \lim_{\varepsilon\to 0}\int_0^{T_n} t dG_{\varepsilon,\mathbb{D},ij}(t\check{u}_{\varepsilon,i}) = \int_0^{T_n} t dG_{0,\mathbb{D},ij}(t). \qquad (11.11)\end{aligned}$$

By passing $n \longrightarrow \infty$ in the relation (11.11), we get the following relation:

$$\lim_{\varepsilon\to 0} E_{\varepsilon,\mathbb{D},ij}/\check{u}_{\varepsilon,i} \geq E_{0,\mathbb{D},ij} = \int_0^\infty t dG_{0,\mathbb{D},ij}(t). \qquad (11.12)$$

We are going to show that lower limit can be replaced by the usual limit for normalised expectations $E_{\varepsilon,\mathbb{D},ij}/\check{u}_{\varepsilon,i}$ on the left hand side of the relation (11.12) and clarify conditions under which this limit coincides with the expectation $E_{0,\mathbb{D},ij}$. We also show that, in some cases, $E_{\varepsilon,\mathbb{D},ij}/\check{u}_{\varepsilon,i}$ can converge to finite limits exceeding $E_{0,\mathbb{D},ij}$, or converge to ∞, as $\varepsilon \longrightarrow 0$.

We find normalisation functions $\bar{u}_{\varepsilon,i}$ such that the normalised expectations $E_{\varepsilon,\mathbb{D},ij}/\bar{u}_{\varepsilon,i} \longrightarrow \bar{E}_{\varepsilon,\mathbb{D},ij} \in (0,\infty)$ as $\varepsilon \longrightarrow 0$ and show that these functions are connected with normalisation functions $\check{u}_{\varepsilon,i}$ by asymptotic relations $\bar{u}_{\varepsilon,i}/\check{u}_{\varepsilon,i} \longrightarrow C_i \in (0,\infty]$ as $\varepsilon \longrightarrow 0$. From the two above asymptotic relations it follows that $E_{\varepsilon,\mathbb{D},ij}/\check{u}_{\varepsilon,i} \longrightarrow C_i\bar{E}_{0,\mathbb{D},ij} \in (0,\infty]$ as $\varepsilon \longrightarrow 0$. Finally, we clarify conditions under which the limits $C_i\bar{E}_{0,\mathbb{D},ij} = E_{0,\mathbb{D},ij}$.

The following lemma takes place.

Lemma 11.3 *Let the conditions of Theorem 10.4 be satisfied for the semi-Markov processes $\eta_\varepsilon(t)$. Then:*

(i) *The expectation $E_{0,\mathbb{D},ij} = \int_0^\infty tG_{0,\mathbb{D},ij}(dt) < \infty$, for $i \in \bar{\mathbb{D}}$, $j \in \mathbb{D}$ and given by the relation (11.13)*

(ii) *The expectation $E_{0,\mathbb{D},i} = \sum_{j\in\mathbb{D}} E_{0,\mathbb{D},ij} = \int_0^\infty tG_{0,\mathbb{D},i}(dt) > 0$, for $i \in \bar{\mathbb{D}}$.*

Proof The recurrent relations (10.127) for the Laplace transforms $\Psi_{0,\mathbb{D},k_nj}(s)$ of the limiting distributions $G_{0,\mathbb{D},k_nj}(t)$ let us (by computing the values of derivatives at zero for the corresponding Laplace transforms) get the following explicit recurrent relations for the expectations $E_{0,\mathbb{D},k_nj}$, for $n = \bar{m}, \ldots, 1$, $j \in \mathbb{D}$:

$$\begin{aligned}E_{0,\mathbb{D},k_nj} = \; & {}_{\bar{k}_{n-1}}\check{w}_{0,k_n,k_{\bar{n}}}\, {}_{\bar{k}_{n-1}}\tilde{f}_{0,k_nj}\, {}_{\bar{k}_{n-1}}\tilde{p}_{0,k_nj} \\ & + {}_{\bar{k}_{n-1}}\check{w}_{0,k_n,k_{\bar{n}}}\, {}_{\bar{k}_{n-1}}\tilde{f}_{0,k_nk_{\bar{n}}}\, P_{0,\mathbb{D},k_{\bar{n}}j}\, {}_{\bar{k}_{n-1}}\tilde{p}_{0,k_nk_{\bar{n}}} \\ & + E_{0,\mathbb{D},k_{\bar{n}}j}\, {}_{\bar{k}_{n-1}}\tilde{p}_{0,k_nk_{\bar{n}}} \\ & + \sum_{n+1\leq l<\bar{n}} {}_{\bar{k}_{n-1}}\check{w}_{0,k_n,k_{\bar{n}}}\, {}_{\bar{k}_{n-1}}\tilde{f}_{0,k_nk_l} P_{0,\mathbb{D},k_lj}\, {}_{\bar{k}_{n-1}}\tilde{p}_{0,k_nk_l} \\ & + \sum_{n+1\leq l<\bar{n}} {}_{\bar{k}_{l-1}}\check{w}_{0,k_l,k_{\bar{n}}}\, E_{0,\mathbb{D},k_lj}\, {}_{\bar{k}_{n-1}}\tilde{p}_{0,k_nk_l}, \qquad (11.13)\end{aligned}$$

where the parameter $n \leq \bar{n} \leq \bar{m}$ is defined by the relation (9.28).

The corresponding expectations $E_{0,\mathbb{D},k_nj}$ are finite or infinite. In the latter case, the product $E_{\varepsilon,\mathbb{D},k_nj}\, {}_{\bar{k}_{n-1}}\tilde{p}_{\varepsilon,k_nr}$ should be counted as 0, if the probability $\tilde{p}_{\varepsilon,k_nr} = 0$.

However, the statement **(ii)** of Theorem 10.1 implies that the condition ${}_{\bar{\mathbf{k}}_{\bar{m}-1}}\mathbf{K}$ is satisfied.

Thus, the relation (11.13) taken, for $n = \bar{m}, j \in \mathbb{D}$, gives the following relation:

$$E_{0,\mathbb{D},k_{\bar{m}}j} = {}_{\bar{k}_{\bar{m}-1}}\tilde{f}_{0,k_{\bar{m}}j}\ {}_{\bar{k}_{\bar{m}-1}}\tilde{p}_{0,k_{\bar{m}}j} < \infty. \tag{11.14}$$

Then, the relations (11.13) taken, for $n = \bar{m} - 1, j \in \mathbb{D}$ and the relation (11.14) give the following relation:

$$\begin{aligned} E_{0,\mathbb{D},k_{\bar{m}-1}j} &= {}_{k_{\bar{m}-2}}\check{w}_{0,k_{\bar{m}-1},k_{\bar{m}}}\ {}_{\bar{k}_{\bar{m}-2}}\tilde{f}_{0,k_{\bar{m}-1}j}\ {}_{\bar{k}_{\bar{m}-2}}\tilde{p}_{0,k_{\bar{m}-1}j} \\ &+ {}_{k_{\bar{m}-2}}\check{w}_{0,k_{\bar{m}-1},k_{\bar{m}}}\ {}_{\bar{k}_{\bar{m}-2}}\tilde{f}_{0,k_{\bar{m}-1}k_{\bar{m}}} P_{0,\mathbb{D},k_{m}j}\ {}_{\bar{k}_{\bar{m}-2}}\tilde{p}_{0,k_{\bar{m}-1}k_{\bar{m}}} \\ &+ E_{0,\mathbb{D},\,k_{\bar{m}}j}\ {}_{\bar{k}_{\bar{m}-2}}\tilde{p}_{0,k_{\bar{m}-1}k_{\bar{m}}} < \infty. \end{aligned} \tag{11.15}$$

Continuing the recurrent use of the relation (11.13), for $n = \bar{m} - 2, \ldots, 1$, we prove that, for $n = \bar{m}, \ldots, 1, j \in \mathbb{D}$,

$$E_{0,\mathbb{D},k_{n}j} < \infty. \tag{11.16}$$

The relation (11.16) completes the proof of the statement **(i)**.

The statement **(ii)** is obviously implied by the statement **(iv)** of Theorem 10.4 according to which $G_{0,\mathbb{D},i}(0) < 1$, for $i \in \bar{\mathbb{D}}$. □

11.2 Asymptotics of Expectations of Hitting Times

In this section, we present theorems on the asymptotics of expectations of hitting times, in the case where the initial state belongs to domain $\bar{\mathbb{D}}$.

11.2.1 Asymptotics for Expectations of Hitting Times for the Case Where an Initial State Belongs to Domain $\bar{\mathbb{D}}$

11.2.1.1 The Case with the Most Absorbing Initial State. Let us consider the case, where the condition ${}_{\bar{\mathbf{k}}_{\bar{m}-1}}\mathbf{M}_{\mathbb{D}}$ and the initial state is the most absorbing state k_m.

In this case, we can use the normalisation function, given in the relation (10.89), i.e., the same as in the corresponding weak convergence relation for hitting times given in Theorem 10.2,

$$\bar{u}_{\varepsilon,k_{\bar{m}}} = \check{u}_{\varepsilon,k_{\bar{m}}} = {}_{\bar{k}_{\bar{m}-1}}\tilde{u}_{\varepsilon,k_{\bar{m}}}. \tag{11.17}$$

The following theorem takes place.

Theorem 11.1 *Let the conditions* $\mathbf{G}_{\mathbb{D}}$, $\mathbf{H}_{\mathbb{D}}$, $\mathbf{I}_{\mathbb{D}}$, $\mathbf{I}_{\mathbb{D},\mathcal{H}}$, $\mathbf{L}_{\mathbb{D}}$, $\mathbf{L}_{\mathbb{D},\mathcal{H}}$, $\mathbf{J}_{\mathbb{D}}$, $\mathbf{K}_{\mathbb{D}}$, *and* ${}_{\bar{\mathbf{k}}_{\bar{m}-1}}\mathbf{M}_{\mathbb{D}}$ *be satisfied for the semi-Markov processes* $\eta_\varepsilon(t)$. *Then:*

(i) *The following asymptotic relation takes place, for* $j \in \mathbb{D}$,

$$\begin{aligned} E_{\varepsilon,\mathbb{D},k_{\bar{m}}j}/\bar{u}_{\varepsilon,k_{\bar{m}}} &\to \bar{E}_{0,\mathbb{D},k_{\bar{m}}j} = E_{0,\mathbb{D},k_{\bar{m}}j} \\ &= {}_{\bar{k}_{\bar{m}-1}}\tilde{f}_{0,k_{\bar{m}}j}\, {}_{\bar{k}_{\bar{m}-1}}\tilde{p}_{0,k_{\bar{m}}j} < \infty \text{ as } \varepsilon \to 0. \end{aligned} \tag{11.18}$$

(ii) *The normalisation function* $\bar{u}_{\cdot,k_{\bar{m}}} \in (0, \infty)$ *is given by the relation (11.17) and belongs to the family of asymptotically comparable functions* $\mathcal{H}$.

(iii) *The limit* $\bar{E}_{0,\mathbb{D},k_{\bar{m}}j} = E_{0,\mathbb{D},k_{\bar{m}}j}$ *is, for* $j \in \mathbb{D}$, *the first moment for the limiting distribution* $G_{0,\mathbb{D},k_{\bar{m}}j}(\cdot)$ *given in Theorem 9.2.*

(iv) $\bar{E}_{0,\mathbb{D},k_{\bar{m}}} = \sum_{j\in\mathbb{D}} \bar{E}_{0,\mathbb{D},k_{\bar{m}},j} \in (0, \infty)$.

Proof Lemma 10.14 and the relation (10.91), given in the proof of Theorem 10.2, imply, that the following relation takes place, for $j \in \mathbb{D}$,

$$\begin{aligned} E_{\varepsilon,\mathbb{D},k_{\bar{m}}j} &= {}_{\bar{k}_{\bar{m}-1}}E_{\varepsilon,\mathbb{D},k_{\bar{m}}j} \\ &= \mathsf{E}_{i}\, {}_{\bar{k}_n}\tilde{\tau}_{\varepsilon,\mathbb{D}} \mathrm{I}({}_{\bar{k}_n}\tilde{\eta}_{\varepsilon}({}_{\bar{k}_n}\tilde{\tau}_{\varepsilon,\mathbb{D}} = j) \\ &= \mathsf{E}_{i}\, {}_{\bar{k}_n}\kappa_{\varepsilon,1} \mathrm{I}({}_{\bar{k}_n}\eta_{\varepsilon,1} = j) \\ &= {}_{\bar{k}_{\bar{m}-1}}\tilde{f}_{\varepsilon,k_{\bar{m}}j}\, {}_{\bar{k}_{\bar{m}-1}}\tilde{p}_{\varepsilon,k_{\bar{m}}j}. \end{aligned} \tag{11.19}$$

According to Theorem 10.1, the conditions ${}_{\bar{\mathbf{k}}_{\bar{m}-1}}\mathbf{J}$ and ${}_{\bar{\mathbf{k}}_{\bar{m}-1}}\mathbf{K}$ are satisfied and, thus, the following relation holds, for $j \in \mathbb{D}$:

$$\begin{aligned} \frac{E_{\varepsilon,\mathbb{D},k_{\bar{m}}j}}{\bar{u}_{\varepsilon,k_{\bar{m}}}} &= \frac{{}_{\bar{k}_{\bar{m}-1}}\tilde{f}_{\varepsilon,k_{\bar{m}}j}}{{}_{\bar{k}_{\bar{m}-1}}\tilde{u}_{\varepsilon,k_{\bar{m}}}}\, {}_{\bar{k}_{\bar{m}-1}}\tilde{p}_{\varepsilon,k_{\bar{m}}j} \\ &\to {}_{\bar{k}_{\bar{m}-1}}\tilde{f}_{0,k_{\bar{m}}j}\, {}_{\bar{k}_{\bar{m}-1}}\tilde{p}_{0,k_{\bar{m}}j} \\ &= E_{0,\mathbb{D},k_{\bar{m}}j} = \bar{E}_{0,\mathbb{D},k_{\bar{m}}j} < \infty \text{ as } \varepsilon \to 0. \end{aligned} \tag{11.20}$$

The statements **(i)**–**(iii)** follow from the relation (11.20).

The relation (11.20) also implies that the following relation holds:

$$\begin{aligned} \frac{E_{\varepsilon,\mathbb{D},k_{\bar{m}}}}{\bar{u}_{\varepsilon,k_{\bar{m}}}} &= \sum_{j\in\mathbb{D}} \frac{{}_{\bar{k}_{\bar{m}-1}}\tilde{f}_{\varepsilon,k_{\bar{m}}j}}{{}_{\bar{k}_{\bar{m}-1}}\tilde{u}_{\varepsilon,k_{\bar{m}}}}\, {}_{\bar{k}_{\bar{m}-1}}\tilde{p}_{\varepsilon,k_{\bar{m}}j} \\ &\to \sum_{j\in\mathbb{D}} {}_{\bar{k}_{\bar{m}-1}}\tilde{f}_{0,k_{\bar{m}}j}\, {}_{\bar{k}_{\bar{m}-1}}\tilde{p}_{0,k_{\bar{m}}j} \\ &= E_{0,\mathbb{D},k_{\bar{m}}} = \bar{E}_{0,\mathbb{D},k_{\bar{m}}} < \infty \text{ as } \varepsilon \to 0. \end{aligned} \tag{11.21}$$

According to the statement **(v)** of Theorem 10.2, the distribution function $G_{0,\mathbb{D},k_{\bar{m}}}(\cdot)$ is not concentrated at zero, and, thus, its first moment,

$$E_{0,\mathbb{D},k_{\bar{m}}} = \bar{E}_{0,\mathbb{D},k_{\bar{m}}} > 0, \tag{11.22}$$

i.e., the statement **(iv)** is also true. □

11.2.1.2 The Case with the Second Most Absorbing Initial State. In what follows in this section, we assume that the conditions of Theorem 11.1 are satisfied.

Let us consider the case, where the initial state is the second most absorbing state k_{m-1}.

Let us define the following normalisation function:

$$\begin{aligned}\bar{u}_{\varepsilon,k_{\bar{m}-1}} &= {}_{\bar{k}_{\bar{m}-2}}\tilde{u}_{\varepsilon,k_{\bar{m}-1}} + \bar{u}_{\varepsilon,k_{\bar{m}}}\ {}_{\bar{k}_{\bar{m}-2}}\tilde{p}_{\varepsilon,k_{\bar{m}-1}k_{\bar{m}}} \\ &= {}_{\bar{k}_{\bar{m}-2}}\tilde{u}_{\varepsilon,k_{\bar{m}-1}} + {}_{\bar{k}_{\bar{m}-1}}\tilde{u}_{\varepsilon,k_{\bar{m}}}\ {}_{\bar{k}_{\bar{m}-2}}\tilde{p}_{\varepsilon,k_{\bar{m}-1}k_{\bar{m}}}. \end{aligned} \tag{11.23}$$

Obviously, $\bar{u}_{\varepsilon,k_{\bar{m}-1}} \in (0,\infty)$, $\varepsilon \in (0,1]$.

The normalisation function $\bar{u}_{\varepsilon,k_{\bar{m}-1}} \in \mathcal{H}$ since the functions ${}_{\bar{k}_{\bar{m}-2}}\tilde{u}_{\varepsilon,k_{\bar{m}-1}}$, $\bar{u}_{\varepsilon,k_{\bar{m}}}$, and ${}_{\bar{k}_{\bar{m}-2}}\tilde{p}_{\varepsilon,k_{\bar{m}-1}k_{\bar{m}}}$ belong to the family of asymptotically comparable functions $\mathcal{H}$.

The conditions $\mathbf{G}_{\mathbb{D}}$, $\mathbf{H}_{\mathbb{D}}$, $\mathbf{I}_{\mathbb{D}}$, $\mathbf{I}_{\mathbb{D},\mathcal{H}}$, $\mathbf{L}_{\mathbb{D}}$, $\mathbf{L}_{\mathbb{D},\mathcal{H}}$, and ${}_{\bar{\mathbf{k}}_{\bar{\mathbf{m}}-1}}\mathbf{M}_{\mathbb{D}}$ entail the fulfilment of the following condition:

$\bar{\mathbf{L}}_{\mathbb{D},1}$: $\dfrac{\bar{u}_{\varepsilon,k_{\bar{m}}}\ {}_{\bar{k}_{\bar{m}-2}}\tilde{p}_{\varepsilon,k_{\bar{m}-1}k_{\bar{m}}}}{\bar{u}_{\varepsilon,k_{\bar{m}-1}}} \to {}_{\bar{k}_{\bar{m}}}\bar{w}_{0,k_{\bar{m}-1},k_{\bar{m}}} \in [0,1]$ as $\varepsilon \to 0$.

Since the functions $\bar{u}_{\varepsilon,k_{\bar{m}}}$, ${}_{\bar{k}_{\bar{m}-2}}\tilde{p}_{\varepsilon,k_{\bar{m}-1}k_{\bar{m}}}$, $\bar{u}_{\varepsilon,k_{\bar{m}-1}}$, and $\bar{u}_{\varepsilon,k_{\bar{m}-1}}$ belong to the class of asymptotically comparable functions $\mathcal{H}$, the limits appearing in the condition $\bar{\mathbf{L}}_{\mathbb{D},1}$ can be calculated using the operating formulas given in Lemma 8.2. In the case, where the family $\mathcal{H}$ is one of the families, $\mathcal{H}_1$, $\mathcal{H}_2$, or $\mathcal{H}_3$, the operating formulas given in Lemma 8.4, 8.6, or 8.8 should be used.

The following theorem takes place.

Theorem 11.2 *Let the conditions* $\mathbf{G}_{\mathbb{D}}$, $\mathbf{H}_{\mathbb{D}}$, $\mathbf{I}_{\mathbb{D}}$, $\mathbf{I}_{\mathbb{D},\mathcal{H}}$, $\mathbf{L}_{\mathbb{D}}$, $\mathbf{L}_{\mathbb{D},\mathcal{H}}$, $\mathbf{J}_{\mathbb{D}}$, $\mathbf{K}_{\mathbb{D}}$, *and* ${}_{\bar{\mathbf{k}}_{\bar{\mathbf{m}}-1}}\mathbf{M}_{\mathbb{D}}$ *be satisfied for the semi-Markov processes* $\eta_\varepsilon(t)$. *Then*:

(i) *The following asymptotic relation takes place, for* $j \in \mathbb{D}$,

$$E_{\varepsilon,\mathbb{D},k_{\bar{m}-1}j}/\bar{u}_{\varepsilon,k_{\bar{m}-1}} \to \bar{E}_{0,\mathbb{D},k_{\bar{m}-1}j} < \infty \text{ as } \varepsilon \to 0. \tag{11.24}$$

(ii) *The normalisation function* $\bar{u}_{\varepsilon,k_{\bar{m}-1}} \in (0,\infty)$ *is given by the relation (11.23) and belongs to the family of asymptotically comparable functions* $\mathcal{H}$.

(iii) *The limits* $\bar{E}_{0,\mathbb{D},k_{\bar{m}-1}j}$, $j \in \mathbb{D}$ *are given by the relation (11.25).*

(iv) $\bar{E}_{0,\mathbb{D},k_{\bar{m}-1}} = \sum_{j\in\mathbb{D}} \bar{E}_{0,\mathbb{D},k_{\bar{m}-1}j} \in (0,\infty)$.

Proof The relation (11.9), (11.18), and the condition $\mathbf{L}_{\mathbb{D},1}$ imply that the following relation takes place, for $j \in \mathbb{D}$:

$$\begin{aligned}\frac{E_{\varepsilon,\mathbb{D},k_{\bar{m}-1}j}}{\bar{u}_{\varepsilon,k_{\bar{m}-1}}} &= \frac{{}_{\bar{k}_{\bar{m}-2}}\tilde{u}_{\varepsilon,k_{\bar{m}-1}}}{{}_{\bar{k}_{\bar{m}}}\bar{u}_{\varepsilon,k_{\bar{m}-1}}}\frac{{}_{\bar{k}_{\bar{m}-2}}\tilde{f}_{\varepsilon,k_{\bar{m}-1}j}}{{}_{\bar{k}_{\bar{m}-2}}\tilde{u}_{\varepsilon,k_{\bar{m}-1}}}\ {}_{\bar{k}_{\bar{m}-2}}\tilde{p}_{\varepsilon,k_{\bar{m}-1}j} \\ &\quad + \frac{{}_{\bar{k}_{\bar{m}-2}}\tilde{u}_{\varepsilon,k_{\bar{m}-1}}}{{}_{\bar{k}_{\bar{m}}}\bar{u}_{\varepsilon,k_{\bar{m}-1}}}\frac{{}_{\bar{k}_{\bar{m}-2}}\tilde{f}_{\varepsilon,k_{\bar{m}-1}k_{\bar{m}}}}{{}_{\bar{k}_{\bar{m}-2}}\tilde{u}_{\varepsilon,k_{\bar{m}-1}}}\ P_{\varepsilon,\mathbb{D},k_{\bar{m}}j}\ {}_{\bar{k}_{\bar{m}-2}}\tilde{p}_{\varepsilon,k_{\bar{m}-1}k_{\bar{m}}}\end{aligned}$$

$$
\begin{aligned}
&+ \frac{E_{\varepsilon,\mathbb{D},k_{\bar m}j}}{{}_{\bar k_{\bar m}}\bar u_{\varepsilon,k_{\bar m}}} \frac{{}_{\bar k_{\bar m}}\bar u_{\varepsilon,k_{\bar m}}\ {}_{\bar k_{\bar m-2}}\tilde p_{\varepsilon,k_{\bar m-1}k_{\bar m}}}{{}_{\bar k_{\bar m}}\bar u_{\varepsilon,k_{\bar m-1}}} \\
&\longrightarrow (1 - {}_{\bar k_{\bar m}}\bar w_{0,k_{\bar m-1},k_{\bar m}})\ {}_{\bar k_{\bar m-2}}\tilde f_{0,k_{\bar m-1}j}\ {}_{\bar k_{\bar m-2}}\tilde p_{0,k_{\bar m-1}j} \\
&\quad + (1 - {}_{\bar k_{\bar m}}\bar w_{0,k_{\bar m-1},k_{\bar m}})\ {}_{\bar k_{\bar m-2}}\tilde f_{0,k_{\bar m-1}k_{\bar m}} P_{0,\mathbb{D},k_{\bar m}j}\ {}_{\bar k_{\bar m-2}}\tilde p_{0,k_{\bar m-1}k_{\bar m}} \\
&\quad + {}_{\bar k_{\bar m}}\bar w_{0,k_{\bar m-1},k_{\bar m}} \bar E_{0,\mathbb{D},k_{\bar m}j} \\
&= \bar E_{0,\mathbb{D},k_{\bar m-1}j} < \infty \text{ as } \varepsilon \to 0. \qquad (11.25)
\end{aligned}
$$

The relation (11.25) also implies that the following relation holds:

$$
\begin{aligned}
\frac{E_{\varepsilon,\mathbb{D},k_{\bar m-1}}}{\bar u_{\varepsilon,k_{\bar m-1}}} &= \sum_{j\in\mathbb{D}} \frac{E_{\varepsilon,\mathbb{D},k_{\bar m-1}j}}{\bar u_{\varepsilon,k_{\bar m-1}}} \\
&\longrightarrow (1 - {}_{\bar k_{\bar m}}\bar w_{0,k_{\bar m-1},k_{\bar m}}) \sum_{j\in\mathbb{D}} {}_{\bar k_{\bar m-2}}\tilde f_{0,k_{\bar m-1}j}\ {}_{\bar k_{\bar m-2}}\tilde p_{0,k_{\bar m-1}j} \\
&\quad + (1 - {}_{\bar k_{\bar m}}\bar w_{0,k_{\bar m-1},k_{\bar m}})\ {}_{\bar k_{\bar m-2}}\tilde f_{0,k_{\bar m-1}k_{\bar m}}\ {}_{\bar k_{\bar m-2}}\tilde p_{0,k_{\bar m-1}k_{\bar m}} \\
&\quad + {}_{\bar k_{\bar m}}\bar w_0[k_{\bar m-1}, k_{\bar m}] \bar E_{0,\mathbb{D},k_{\bar m}} \\
&= (1 - {}_{\bar k_{\bar m}}\bar w_{0,k_{\bar m-1},k_{\bar m}})\Big(\sum_{j\in\mathbb{D}} {}_{\bar k_{\bar m-2}}\tilde f_{0,k_{\bar m-1}j}\ {}_{\bar k_{\bar m-2}}\tilde p_{0,k_{\bar m-1}j} \\
&\quad + {}_{\bar k_{\bar m-2}}\tilde f_{0,k_{\bar m-1}k_{\bar m}}\ {}_{\bar k_{\bar m-2}}\tilde p_{0,k_{\bar m-1}k_{\bar m}}\Big) + {}_{\bar k_{\bar m}}\bar w_{0,k_{\bar m-1},k_{\bar m}} \bar E_{0,\mathbb{D},k_{\bar m}} \\
&= \bar E_{0,\mathbb{D},k_{\bar m-1}} < \infty \text{ as } \varepsilon \to 0. \qquad (11.26)
\end{aligned}
$$

Theorem 10.1 implies that the conditions ${}_{\bar{\mathbf{k}}_{\bar{\mathbf{m}}-1}}\mathbf{J}$ and ${}_{\bar{\mathbf{k}}_{\bar{\mathbf{m}}-1}}\mathbf{K}$ are satisfied and, thus,

$$
{}_{\bar k_{\bar m}}\tilde f = \wedge_{j\in\mathbb{D}}\ {}_{\bar k_{\bar m-2}}\tilde f_{0,k_{\bar m-1}j} \wedge {}_{\bar k_{\bar m-2}}\tilde f_{0,k_{\bar m-1}k_{\bar m}} \in (0, \infty). \qquad (11.27)
$$

Also, according to Theorem 10.1, $\bar E_{0,\mathbb{D},k_{\bar m}} > 0$.

The above remarks and the relation (11.26) imply that the following inequality holds:

$$
\begin{aligned}
\bar E_{0,\mathbb{D},k_{\bar m-1}} &= (1 - {}_{\bar k_{\bar m}}\bar w_{0,k_{\bar m-1},k_{\bar m}})\Big(\sum_{j\in\mathbb{D}} {}_{\bar k_{\bar m-2}}\tilde f_{0,k_{\bar m-1}j}\ {}_{\bar k_{\bar m-2}}\tilde p_{0,k_{\bar m-1}j} \\
&\quad + {}_{\bar k_{\bar m-2}}\tilde f_{0,k_{\bar m-1}k_{\bar m}}\ {}_{\bar k_{\bar m-2}}\tilde p_{0,k_{\bar m-1}k_{\bar m}}\Big) + {}_{\bar k_{\bar m}}\bar w_{0,k_{\bar m-1},k_{\bar m}} \bar E_{0,\mathbb{D},k_{\bar m}} \\
&\geq {}_{\bar k_{\bar m}}\tilde f \cdot (1 - {}_{\bar k_{\bar m}}\bar w_0[k_{\bar m-1}, k_{\bar m}])\Big(\sum_{j\in\mathbb{D}} {}_{\bar k_{\bar m-2}}\tilde p_{0,k_{\bar m-1}j} + {}_{\bar k_{\bar m-2}}\tilde p_{0,k_{\bar m-1}k_{\bar m}}\Big) \\
&\quad + {}_{\bar k_{\bar m}}\bar w_{0,k_{\bar m-1},k_{\bar m}} \bar E_{0,\mathbb{D},k_{\bar m}} \\
&= {}_{\bar k_{\bar m}}\tilde f \cdot (1 - {}_{\bar k_{\bar m}}\bar w_{0,k_{\bar m-1},k_{\bar m}}) + {}_{\bar k_{\bar m}}\bar w_{0,k_{\bar m-1},k_{\bar m}} \bar E_{0,\mathbb{D},k_{\bar m}} > 0. \qquad (11.28)
\end{aligned}
$$

The proof is complete. □

11.2.1.3 The Case with an Arbitrary Initial State from the Domain $\bar{\mathbb{D}} \setminus \{k_m\}$. Let us now consider the general case, with the initial state being k_n, for some $n = 1, \ldots, \bar m - 1$.

We can use, in this case, the relation (11.6) given in Lemma 11.1. This relation can be written in the following form, for $j \in \mathbb{D}$ and $n = 1, \ldots, \bar{m} - 1$:

$$\begin{aligned} E_{\varepsilon,\mathbb{D},k_n j} = {}& {}_{\bar{k}_{n-1}}\tilde{f}_{\varepsilon,k_n j}\, {}_{\bar{k}_{n-1}}\tilde{p}_{\varepsilon,k_n j} \\ &+ \sum_{n+1 \leq l \leq \bar{m}} \big({}_{\bar{k}_{n-1}}\tilde{f}_{\varepsilon,k_n k_l} P_{\varepsilon,\mathbb{D},k_l j} \\ &+ E_{\varepsilon,\mathbb{D},k_l j}\big)\, {}_{\bar{k}_{n-1}}\tilde{p}_{\varepsilon,k_n k_l}. \end{aligned} \tag{11.29}$$

Let us define recurrently the following normalisation functions:

$$\bar{u}_{\varepsilon,k_n} = {}_{\bar{k}_{\bar{n}-1}}\tilde{u}_{\varepsilon,k_n} + \sum_{n+1 \leq l \leq \bar{m}} \bar{u}_{\varepsilon,k_l}\, {}_{\bar{k}_{n-1}}\tilde{p}_{\varepsilon,k_n k_l}, \ n = 1, \ldots, \bar{m} - 1. \tag{11.30}$$

Obviously, ${}_{\bar{k}_{\bar{m}}}\bar{u}_{\varepsilon,k_n} \in (0, \infty)$, $\varepsilon \in (0, 1]$, for $n = 1, \ldots, \bar{m} - 1$.

The normalisation function $\bar{u}_{\varepsilon,k_n}$ can be expressed in terms of the normalisation functions ${}_{\bar{k}_{l-1}}\tilde{u}_{\varepsilon,k_l}$, $l = n, \ldots, \bar{m}$, for $n = 1, \ldots, \bar{m}$.

In particular, according to the relations (11.17) and (11.23), the normalisation functions ${}_{\bar{k}_{\bar{m}}}\bar{u}_{\varepsilon,k_{\bar{m}}} = {}_{\bar{k}_{\bar{m}-1}}\tilde{u}_{\varepsilon,k_{\bar{m}}}$ and ${}_{\bar{k}_{\bar{m}}}\bar{u}_{\varepsilon,k_{\bar{m}-1}} = {}_{\bar{k}_{\bar{m}-2}}\tilde{u}_{\varepsilon,k_{\bar{m}-1}} + {}_{\bar{k}_{\bar{m}-1}}\tilde{u}_{\varepsilon,k_{\bar{m}}}\, {}_{\bar{k}_{\bar{m}-2}}\tilde{p}_{\varepsilon,k_{\bar{m}-1}k_{\bar{m}}}$.

Continuing the recurrent substitution in the relation (11.30), we get,

$$\begin{aligned} {}_{\bar{k}_{\bar{m}}}\bar{u}_{\varepsilon,k_{\bar{m}-2}} = {}& {}_{\bar{k}_{\bar{m}-3}}\tilde{u}_{\varepsilon,k_{\bar{m}-2}} \\ &+ \big({}_{\bar{k}_{\bar{m}-2}}\tilde{u}_{\varepsilon,k_{\bar{m}-1}} + {}_{\bar{k}_{\bar{m}-1}}\tilde{u}_{\varepsilon,k_{\bar{m}}}\, {}_{\bar{k}_{\bar{m}-2}}\tilde{p}_{\varepsilon,k_{\bar{m}-1}k_{\bar{m}}}\big)\, {}_{\bar{k}_{\bar{m}-3}}\tilde{p}_{\varepsilon,k_{\bar{m}-2}k_{\bar{m}-1}} \\ &+ {}_{\bar{k}_{\bar{m}-1}}\tilde{u}_{\varepsilon,k_{\bar{m}}}\, {}_{\bar{k}_{\bar{m}-3}}\tilde{p}_{\varepsilon,k_{\bar{m}-2}k_{\bar{m}}} \\ = {}& {}_{\bar{k}_{\bar{m}-3}}\tilde{u}_{\varepsilon,k_{\bar{m}-2}} \\ &+ {}_{\bar{k}_{\bar{m}-2}}\tilde{u}_{\varepsilon,k_{\bar{m}-1}}\, {}_{\bar{k}_{\bar{m}-2}}\tilde{p}_{\varepsilon,k_{\bar{m}-1}k_{\bar{m}}} \\ &+ {}_{\bar{k}_{\bar{m}-1}}\tilde{u}_{\varepsilon,k_{\bar{m}}}\big({}_{\bar{k}_{\bar{m}-3}}\tilde{p}_{\varepsilon,k_{\bar{m}-2}k_{\bar{m}-1}}\, {}_{\bar{k}_{\bar{m}-2}}\tilde{p}_{\varepsilon,k_{\bar{m}-1}k_{\bar{m}}} + {}_{\bar{k}_{\bar{m}-3}}\tilde{p}_{\varepsilon,k_{\bar{m}-2}k_{\bar{m}}}\big). \end{aligned} \tag{11.31}$$

Similar relations can be obtained for $n = \bar{m} - 3, \ldots, 1$.

The conditions $\mathbf{G}_{\mathbb{D}}$, $\mathbf{H}_{\mathbb{D}}$, $\mathbf{I}_{\mathbb{D}}$, $\mathbf{I}_{\mathbb{D},\mathcal{H}}$, $\mathbf{L}_{\mathbb{D}}$, $\mathbf{L}_{\mathbb{D},\mathcal{H}}$, and ${}_{\bar{\mathbf{k}}_{\bar{m}-1}}\mathbf{M}_{\mathbb{D}}$ imply the fulfilment of the following condition:

$\bar{\mathbf{L}}_{\mathbb{D},2}$: $\dfrac{\bar{u}_{\varepsilon,k_l}\, {}_{\bar{k}_{n-1}}\tilde{p}_{\varepsilon,k_n k_l}}{\bar{u}_{\varepsilon,k_n}} \to {}_{\bar{k}_{\bar{m}}}w_{0,k_n,k_l} \in [0, 1]$ as $\varepsilon \to 0$, for $l = n + 1, \ldots, \bar{m}$, $n = 1, \ldots, \bar{m} - 1$.

Since the functions $\bar{u}_{\cdot,k_l}$, ${}_{\bar{k}_{\bar{m}-2}}\tilde{p}_{\cdot,k_n k_l}$, $\bar{u}_{\cdot,k_{\bar{m}-1}}$, $\bar{u}_{\cdot,k_n}$, $l = n+1, \ldots, \bar{m}$, $n = 1, \ldots, \bar{m}-1$ belong to the class of asymptotically comparable functions $\mathcal{H}$, the limits appearing in the condition $\bar{\mathbf{L}}_{\mathbb{D},2}$ can be calculated using the operating formulas given in Lemma 8.2. In the case, where the family $\mathcal{H}$ is one of the families, $\mathcal{H}_1$, $\mathcal{H}_2$, or $\mathcal{H}_3$, the operating formulas given in Lemma 8.4, 8.6, or 8.8 should be used.

The following theorem takes place.

Theorem 11.3 *Let the conditions* $\mathbf{G}_{\mathbb{D}}$, $\mathbf{H}_{\mathbb{D}}$, $\mathbf{I}_{\mathbb{D}}$, $\mathbf{I}_{\mathbb{D},\mathcal{H}}$, $\mathbf{L}_{\mathbb{D}}$, $\mathbf{L}_{\mathbb{D},\mathcal{H}}$, $\mathbf{J}_{\mathbb{D}}$, $\mathbf{K}_{\mathbb{D}}$, *and* ${}_{\bar{\mathbf{k}}_{\bar{m}-1}}\mathbf{M}_{\mathbb{D}}$ *be satisfied for the semi-Markov processes* $\eta_\varepsilon(t)$. *Then*:

(i) *The following asymptotic relation takes place, for* $j \in \mathbb{D}$, $n = 1, \ldots, \bar{m} - 1$*:*

$$E_{\varepsilon,\mathbb{D},k_n j}/\bar{u}_{\varepsilon,k_n} \to \bar{E}_{0,\mathbb{D},k_n j} < \infty \text{ as } \varepsilon \to 0. \tag{11.32}$$

(ii) *The normalisation functions $\bar{u}_{\cdot,k_n} \in (0,\infty)$, $n = 1,\ldots,\bar{m}-1$ are given by the relation (11.30) and belong to the family of asymptotically comparable functions $\mathcal{H}$.*

(iii) *The limits $\bar{E}_{0,\mathbb{D},k_n j}$, $j \in \mathbb{D}$, $n = 1,\ldots,\bar{m}-1$ are given by the recurrent relation (11.35).*

(iv) $\bar{E}_{0,\mathbb{D},k_n} = \sum_{j\in\mathbb{D}} \bar{E}_{0,\mathbb{D},k_n j} \in (0,\infty)$, *for* $n = 1,\ldots,\bar{m}-1$.

Proof The condition ${}_{\bar{\mathbf{k}}_{\bar{m}}}\bar{\mathbf{L}}_{\mathbb{D},2}$ obviously implies that, for $n = 1,\ldots,\bar{m}-1$,

$$\begin{aligned}\frac{{}_{\bar{k}_{\bar{n}-1}}\tilde{u}_{\varepsilon,k_n}}{\bar{u}_{\varepsilon,k_n}} &\to {}_{\bar{k}_{\bar{m}}}\bar{w}_{0,k_n,k_n}\\ &= 1 - \sum_{l=n+1}^{\bar{m}} {}_{\bar{k}_{\bar{m}}}\bar{w}_{0,k_n,k_l} \in [0,1] \text{ as } \varepsilon \to 0.\end{aligned} \tag{11.33}$$

The asymptotic relation (11.24) given in Theorem 11.2 can be considered as a result of the first step in some backward asymptotic recurrent algorithm for computing limits for expectations of hitting times $E_{\varepsilon,\mathbb{D},k_n j}/\bar{u}_{\varepsilon,k_n}$, for $n = \bar{m},\ldots,1$, where the corresponding normalisation functions are given by the relations (11.23) and (11.30).

Suppose we have already implemented $\bar{m}-n$ steps in this backward algorithm, which gave the following asymptotic relations, for $j \in \mathbb{D}$, $l = \bar{m},\ldots,n+1$, for some $1 \le n \le \bar{m}-1$:

$$E_{\varepsilon,\mathbb{D},k_l,j}/\bar{u}_{\varepsilon,k_l} \to \bar{E}_{0,\mathbb{D},k_l j} < \infty \text{ as } \varepsilon \to 0 \tag{11.34}$$

with some limits $\bar{E}_{0,\mathbb{D},k_l j} < \infty$, $j \in \mathbb{D}$, $l = \bar{m},\ldots,n+1$.

Then, using the relation (11.29), we obtain the following relations, for $j \in \mathbb{D}$:

$$\begin{aligned}E_{\varepsilon,\mathbb{D},k_n j}/\bar{u}_{\varepsilon,k_n} &= \frac{{}_{\bar{k}_{\bar{n}-1}}\tilde{u}_{\varepsilon,k_n}}{\bar{u}_{\varepsilon,k_n}}\frac{{}_{\bar{k}_{n-1}}\tilde{f}_{\varepsilon,k_n j}}{{}_{\bar{k}_{\bar{n}-1}}\tilde{u}_{\varepsilon,k_n}}\,{}_{\bar{k}_{n-1}}\tilde{p}_{\varepsilon,k_n j}\\ &\quad + \sum_{n+1\le l\le \bar{m}} \frac{{}_{\bar{k}_{\bar{n}-1}}\tilde{u}_{\varepsilon,k_n}}{\bar{u}_{\varepsilon,k_n}}\frac{{}_{\bar{k}_{n-1}}\tilde{f}_{\varepsilon,k_n l}}{{}_{\bar{k}_{\bar{n}-1}}\tilde{u}_{\varepsilon,k_n}} P_{\varepsilon,\mathbb{D},k_l j}\,{}_{\bar{k}_{n-1}}\tilde{p}_{\varepsilon,k_n k_l}\\ &\quad + \sum_{n+1\le l\le \bar{m}} \frac{\bar{u}_{\varepsilon,k_l}\,{}_{\bar{k}_{n-1}}\tilde{p}_{\varepsilon,k_n k_l}}{\bar{u}_{\varepsilon,k_n}}\frac{E_{\varepsilon,\mathbb{D},k_l j}}{\bar{u}_{\varepsilon,k_l}}\\ &\to {}_{\bar{k}_{\bar{m}}}\bar{w}_{0,k_n,k_n}\big({}_{\bar{k}_{n-1}}\tilde{f}_{0,k_n j}\,{}_{\bar{k}_{n-1}}\tilde{p}_{0,k_n j}\\ &\quad + \sum_{n+1\le l\le \bar{m}} {}_{\bar{k}_{n-1}}\tilde{f}_{0,k_n l} P_{0,\mathbb{D},k_l j}\,{}_{\bar{k}_{n-1}}\tilde{p}_{0,k_n l}\big)\\ &\quad + \sum_{n+1\le l\le \bar{m}} {}_{\bar{k}_{\bar{m}}}\bar{w}_{0,k_n,k_l}\bar{E}_{0,\mathbb{D},k_l j}\\ &= \bar{E}_{0,\mathbb{D},k_n j} < \infty \text{ as } \varepsilon \to 0.\end{aligned} \tag{11.35}$$

By induction, the relation (11.35) holds for any $j \in \mathbb{D}, n = \bar{m} - 1, \ldots, 1$. In addition, (11.35) gives the explicit recurrent formulas for calculating the limits $\bar{E}_{0,\mathbb{D},k_n j} < \infty, j \in \mathbb{D}, n = \bar{m} - 1, \ldots, 1$. □

11.2.2 Conditions of Simultaneous Convergence for Distributions and Expectations of Hitting Times

11.2.2.1 Conditions of Simultaneous Convergence for Distributions and Expectations of Hitting Times. Theorem 11.2 implies that the expectations $E_{\varepsilon,\mathbb{D},k_{\bar{m}}j}$, normalised by the function $\check{u}_{\varepsilon,k_{\bar{m}}}$ (which is used in Theorem 11.2 as the normalisation function in the weak convergence relation (11.18) for the distributions $G_{\varepsilon,\mathbb{D},k_{\bar{m}}j}(\cdot\, \check{u}_{\varepsilon,k_{\bar{m}}}))$ converge to the first moment of the corresponding limiting distribution $G_{0,\mathbb{D},k_{\bar{m}}j}(\cdot)$, as $\varepsilon \to 0$.

The question arises about conditions, under which expectations $E_{\varepsilon,\mathbb{D},k_{\bar{m}-1}j}$, normalised by the function $\check{u}_{\varepsilon,k_{\bar{m}-1}}$, which is used in Theorem 11.3 as the normalisation function in the weak convergence relation for distributions $G_{\varepsilon,\mathbb{D},k_{\bar{m}-1}j}(\cdot\, \check{u}_{\varepsilon,k_{\bar{m}-1}})$, converge to the first moment of the corresponding limiting distribution $G_{0,\mathbb{D},k_{\bar{m}-1}j}(\cdot)$, as $\varepsilon \to 0$.

The following theorem gives the answer to this question.

Theorem 11.4 *Let the conditions* $\mathbf{G}_{\mathbb{D}}$, $\mathbf{H}_{\mathbb{D}}$, $\mathbf{I}_{\mathbb{D}}$, $\mathbf{I}_{\mathbb{D},\mathcal{H}}$, $\mathbf{L}_{\mathbb{D}}$, $\mathbf{L}_{\mathbb{D},\mathcal{H}}$, $\mathbf{J}_{\mathbb{D}}$, $\mathbf{K}_{\mathbb{D}}$, *and* ${}_{\bar{\mathbf{k}}_{\bar{m}-1}}\mathbf{M}_{\mathbb{D}}$ *be satisfied for the semi-Markov processes* $\eta_\varepsilon(t)$. *Then*:

(i) *If one of the following is true:* **(a)** $k_{\bar{m}} \in {}_{\bar{k}_{\bar{m}-2}}\tilde{\mathbb{Y}}_{1,k_{\bar{m}-1}}$ *and* ${}_{\bar{k}_{\bar{m}-2}}\tilde{p}_{0,k_{\bar{m}-1}k_{\bar{m}}} > 0$, *or,* **(b)** $k_{\bar{m}} \notin {}_{\bar{k}_{\bar{m}-2}}\tilde{\mathbb{Y}}_{1,k_{\bar{m}-1}}$, *or* **(c)** $k_{\bar{m}} \in {}_{\bar{k}_{\bar{m}-2}}\tilde{\mathbb{Y}}_{1,k_{\bar{m}-1}}$, ${}_{\bar{k}_{\bar{m}-2}}\tilde{p}_{0,k_{\bar{m}-1}k_{\bar{m}}} = 0$, *and* ${}_{\bar{k}_{\bar{m}}}\bar{w}_{0,k_{\bar{m}-1},k_{\bar{m}}} = 0$, *then the following asymptotic relation takes place, for* $j \in \mathbb{D}$:

$$E_{\varepsilon,\mathbb{D},k_{\bar{m}-1}j}/\check{u}_{\varepsilon,k_{\bar{m}-1}} \to E_{0,\mathbb{D},k_{\bar{m}-1}j} \text{ as } \varepsilon \to 0. \tag{11.36}$$

(ii) *If the following is true:* **(d)** $k_{\bar{m}} \in {}_{\bar{k}_{\bar{m}-2}}\tilde{\mathbb{Y}}_{1,k_{\bar{m}-1}}$, ${}_{\bar{k}_{\bar{m}-2}}\tilde{p}_{0,k_{\bar{m}-1}k_{\bar{m}}} = 0$, *and* ${}_{\bar{k}_{\bar{m}}}\bar{w}_{0,k_{\bar{m}-1},k_{\bar{m}}} \in (0, 1)$, *then the following asymptotic relation takes place, for* $j \in \mathbb{D}$:

$$\begin{aligned} E_{\varepsilon,\mathbb{D},k_{\bar{m}-1}j}/\check{u}_{\varepsilon,k_{\bar{m}-1}} &\to \bar{E}_{0,\mathbb{D},k_{\bar{m}-1}j} \\ &= E_{0,\mathbb{D},k_{\bar{m}-1}j} + \frac{{}_{\bar{k}_{\bar{m}}}\bar{w}_{0,k_{\bar{m}-1},k_{\bar{m}}}}{1 - {}_{\bar{k}_{\bar{m}}}\bar{w}_{0,k_{\bar{m}-1},k_{\bar{m}}}} E_{0,\mathbb{D},k_{\bar{m}}j} \text{ as } \varepsilon \to 0. \end{aligned} \tag{11.37}$$

(iii) *If the following is true:* **(e)** $k_{\bar{m}} \in {}_{\bar{k}_{\bar{m}-2}}\tilde{\mathbb{Y}}_{1,k_{\bar{m}-1}}$, ${}_{\bar{k}_{\bar{m}-2}}\tilde{p}_{0,k_{\bar{m}-1}k_{\bar{m}}} = 0$, *and* ${}_{\bar{k}_{\bar{m}}}\bar{w}_{0,k_{\bar{m}-1},k_{\bar{m}}} = 1$, *then the following asymptotic relation takes place, for* $j \in \mathbb{D}$ *such that* ${}_{\bar{k}_{\bar{m}-1}}\tilde{p}_{0,k_{\bar{m}}j} > 0$:

$$E_{\varepsilon,\mathbb{D},k_{\bar{m}-1}j}/\check{u}_{\varepsilon,k_{\bar{m}-1}} \to \infty \text{ as } \varepsilon \to 0. \tag{11.38}$$

Proof First, let us consider the case, where $k_{\bar{m}} \in {}_{\bar{k}_{\bar{m}-2}}\tilde{\mathbb{Y}}_{1,k_{\bar{m}-1}}$ and ${}_{\bar{k}_{\bar{m}-2}}\tilde{p}_{0,k_{\bar{m}-1}k_{\bar{m}}} > 0$, i.e., the following relation holds:

$$
{}_{\bar{k}_{\bar{m}-2}}\tilde{p}_{\varepsilon,k_{\bar{m}-1}k_{\bar{m}}} > 0, \text{ for } \varepsilon \in [0,1]. \tag{11.39}
$$

In this case, the asymptotic relation appearing in the condition $\mathbf{L}_{\mathbb{D},1}$ holds, with the limit ${}_{\bar{k}_{\bar{m}}}\bar{w}_{0,k_{\bar{m}-1},k_{\bar{m}}} \in (0,1]$. Indeed,

$$
\begin{aligned}
\frac{\bar{u}_{\varepsilon,k_{\bar{m}}}\ {}_{\bar{k}_{\bar{m}-2}}\tilde{p}_{\varepsilon,k_{\bar{m}-1}k_{\bar{m}}}}{\bar{u}_{\varepsilon,k_{\bar{m}-1}}} &= \Big(1 + \frac{{}_{\bar{k}_{\bar{m}-2}}\tilde{u}_{\varepsilon,k_{\bar{m}-1}}}{{}_{\bar{k}_{\bar{m}-1}}\tilde{u}_{\varepsilon,k_{\bar{m}}}\ {}_{\bar{k}_{\bar{m}-2}}\tilde{p}_{\varepsilon,k_{\bar{m}-1}k_{\bar{m}}}}\Big)^{-1} \\
&= \Big(1 + \frac{\big(1 - {}_{\bar{k}_{\bar{m}-1}}p_{\varepsilon,k_{\bar{m}}k_{\bar{m}}}\big)\, {}_{\bar{k}_{\bar{m}-2}}\tilde{u}_{\varepsilon,k_{\bar{m}-1}}}{{}_{\bar{k}_{\bar{m}-2}}\tilde{u}_{\varepsilon,k_{\bar{m}}}\ {}_{\bar{k}_{\bar{m}-2}}\tilde{p}_{\varepsilon,k_{\bar{m}-1}k_{\bar{m}}}}\Big)^{-1} \\
&\to \Big(1 + \frac{\big(1 - {}_{\bar{k}_{\bar{m}-1}}p_{0,k_{\bar{m}}k_{\bar{m}}}\big)\, {}_{\bar{k}_{\bar{m}-2}}w_{0,k_{\bar{m}-1},k_{\bar{m}}}}{{}_{\bar{k}_{\bar{m}-2}}\tilde{p}_{0,k_{\bar{m}-1}k_{\bar{m}}}}\Big)^{-1} \\
&= {}_{\bar{k}_{\bar{m}}}\bar{w}_{0,k_{\bar{m}-1}k_{\bar{m}}} \in (0,1] \text{ as } \varepsilon \to 0.
\end{aligned} \tag{11.40}
$$

In the case where the relation (11.39) holds, the normalisation function ${}_{\bar{k}_{\bar{m}}}\check{u}_{\varepsilon,k_{\bar{m}-1}} = {}_{\bar{k}_{\bar{m}-1}}\tilde{u}_{\varepsilon,k_{\bar{m}}}$ is used in the weak convergence relation given in Theorem 9.3. Obviously,

$$
\begin{aligned}
\frac{\bar{u}_{\varepsilon,k_{\bar{m}-1}}}{\check{u}_{\varepsilon,k_{\bar{m}-1}}} &= \frac{\bar{u}_{\varepsilon,k_{\bar{m}-1}}}{{}_{\bar{k}_{\bar{m}-1}}\tilde{u}_{\varepsilon,k_{\bar{m}}}} \\
&= \frac{\bar{u}_{\varepsilon,k_{\bar{m}-1}}\ {}_{\bar{k}_{\bar{m}-2}}\tilde{p}_{\varepsilon,k_{\bar{m}-1}k_{\bar{m}}}}{{}_{\bar{k}_{\bar{m}-1}}\tilde{u}_{\varepsilon,k_{\bar{m}}}\ {}_{\bar{k}_{\bar{m}-2}}\tilde{p}_{\varepsilon,k_{\bar{m}-1}k_{\bar{m}}}} \\
&\to \frac{{}_{\bar{k}_{\bar{m}-2}}\tilde{p}_{0,k_{\bar{m}-1}k_{\bar{m}}}}{{}_{\bar{k}_{\bar{m}}}\bar{w}_{0,k_{\bar{m}-1},k_{\bar{m}}}} \in (0,\infty) \text{ as } \varepsilon \to 0.
\end{aligned} \tag{11.41}
$$

Also, the relation (11.40) implies that,

$$
\frac{1 - {}_{\bar{k}_{\bar{m}}}\bar{w}_{0,k_{\bar{m}-1},k_{\bar{m}}}}{{}_{\bar{k}_{\bar{m}}}\bar{w}_{0,k_{\bar{m}-1},k_{\bar{m}}}} = \frac{\big(1 - {}_{\bar{k}_{\bar{m}-1}}p_{0,k_{\bar{m}}k_{\bar{m}}}\big)\, {}_{\bar{k}_{\bar{m}-2}}w_{0,k_{\bar{m}-1}k_{\bar{m}}}}{{}_{\bar{k}_{\bar{m}-2}}\tilde{p}_{0,k_{\bar{m}-1}k_{\bar{m}}}}. \tag{11.42}
$$

Finally, the relations (11.15), (11.18), (11.25), (11.41), and (11.42) imply that, in the case where the relation (11.39) holds, the following relation takes place, for $j \in \mathbb{D}$:

$$
\begin{aligned}
\frac{E_{\varepsilon,\mathbb{D},k_{\bar{m}-1}j}}{\check{u}_{\varepsilon,k_{\bar{m}-1}}} &= \frac{E_{\varepsilon,\mathbb{D},k_{\bar{m}-1}j}}{\bar{u}_{\varepsilon,k_{\bar{m}-1}}}\ \frac{\bar{u}_{\varepsilon,k_{\bar{m}-1}}}{\check{u}_{\varepsilon,k_{\bar{m}-1}}} \\
&\to \bar{E}_{0,\mathbb{D},k_{\bar{m}-1}j}\ \frac{{}_{\bar{k}_{\bar{m}-2}}\tilde{p}_{0,k_{\bar{m}-1}k_{\bar{m}}}}{{}_{\bar{k}_{\bar{m}}}\bar{w}_{0,k_{\bar{m}-1},k_{\bar{m}}}} \\
&= \big(1 - {}_{\bar{k}_{\bar{m}-1}}p_{0,k_{\bar{m}}k_{\bar{m}}}\big)\, {}_{\bar{k}_{\bar{m}-2}}w_{0,k_{\bar{m}-1}k_{\bar{m}}}\ {}_{\bar{k}_{\bar{m}-2}}\tilde{f}_{0,k_{\bar{m}-1}j}\ {}_{\bar{k}_{\bar{m}-2}}\tilde{p}_{0,k_{\bar{m}-1}j}
\end{aligned}
$$

$$
\begin{aligned}
&+ (1 - {}_{\bar{k}_{\bar{m}-1}}p_{0,k_{\bar{m}}k_{\bar{m}}})\, {}_{\bar{k}_{\bar{m}-2}}w_{0,k_{\bar{m}-1}k_{\bar{m}}}\, {}_{\bar{k}_{\bar{m}-2}}\tilde{f}_{0,k_{\bar{m}-1}k_{\bar{m}}}\, P_{0,\mathbb{D},k_{\bar{m}}j}\, {}_{\bar{k}_{\bar{m}-2}}\tilde{p}_{0,k_{\bar{m}-1}k_{\bar{m}}} \\
&+ E_{0,\mathbb{D},k_{\bar{m}}j}\, {}_{\bar{k}_{\bar{m}-2}}\tilde{p}_{0,k_{\bar{m}-1}k_{\bar{m}}} \\
&= {}_{\bar{k}_{\bar{m}-2}}\check{w}_{0,k_{\bar{m}-1},k_{\bar{m}}}\, {}_{\bar{k}_{\bar{m}-2}}\tilde{f}_{0,k_{\bar{m}-1}j}\, {}_{\bar{k}_{\bar{m}-2}}\tilde{p}_{0,k_{\bar{m}-1}j} \\
&+ {}_{\bar{k}_{\bar{m}-2}}\check{w}_{0,k_{\bar{m}-1}k_{\bar{m}}}\, {}_{\bar{k}_{\bar{m}-2}}\tilde{f}_{0,k_{\bar{m}-1}k_{\bar{m}}}\, P_{0,\mathbb{D},k_{\bar{m}}j}\, {}_{\bar{k}_{\bar{m}-2}}\tilde{p}_{0,k_{\bar{m}-1}k_{\bar{m}}} \\
&+ E_{0,\mathbb{D},k_{\bar{m}}j}\, {}_{\bar{k}_{\bar{m}-2}}\tilde{p}_{0,k_{\bar{m}-1}k_{\bar{m}}} \\
&= E_{0,\mathbb{D},k_{\bar{m}-1}j} \text{ as } \varepsilon \to 0.
\end{aligned} \tag{11.43}
$$

Second, let us assume that state $k_{\bar{m}} \notin {}_{\bar{k}_{\bar{m}-2}}\tilde{\mathbb{Y}}_{1,k_{\bar{m}-1}}$, that is,

$$
{}_{\bar{k}_{\bar{m}-2}}\tilde{p}_{\varepsilon,k_{\bar{m}-1}k_{\bar{m}}} = 0, \ \varepsilon \in [0, 1]. \tag{11.44}
$$

In this case, the normalisation function,

$$
\bar{u}_{\varepsilon,k_{\bar{m}-1}} = \check{u}_{\varepsilon,k_{\bar{m}-1}} = {}_{\bar{k}_{\bar{m}-2}}\tilde{u}_{\varepsilon,k_{\bar{m}-1}}. \tag{11.45}
$$

The asymptotic relation appearing in the condition $\mathbf{L}_{\mathbb{D},1}$ obviously holds, with the limit, ${}_{\bar{k}_{\bar{m}}}\bar{w}_{0,k_{\bar{m}-1},k_{\bar{m}}} = 0$.

The relation (11.25) takes, in this case, the following form, for $j \in \mathbb{D}$:

$$
\begin{aligned}
\frac{E_{\varepsilon,\mathbb{D},k_{\bar{m}-1}j}}{\check{u}_{\varepsilon,k_{\bar{m}-1}}} &= \frac{{}_{\bar{k}_{\bar{m}-2}}\tilde{f}_{\varepsilon,k_{\bar{m}-1}j}}{{}_{\bar{k}_{\bar{m}-2}}\tilde{u}_{\varepsilon,k_{\bar{m}-1}}}\, {}_{\bar{k}_{\bar{m}-2}}\tilde{p}_{\varepsilon,k_{\bar{m}-1}j} \\
&\to {}_{\bar{k}_{\bar{m}-2}}\tilde{f}_{0,k_{\bar{m}-1}j}\, {}_{\bar{k}_{\bar{m}-2}}\tilde{p}_{0,k_{\bar{m}-1}j} \\
&= \bar{E}_{0,\mathbb{D},k_{\bar{m}-1}j} = E_{0,\mathbb{D},k_{\bar{m}-1}j} < \infty \text{ as } \varepsilon \to 0.
\end{aligned} \tag{11.46}
$$

Third, let us assume that state $k_{\bar{m}} \in {}_{\bar{k}_{\bar{m}-2}}\tilde{\mathbb{Y}}_{1,k_{\bar{m}-1}}$ and ${}_{\bar{k}_{\bar{m}-2}}\tilde{p}_{0,k_{\bar{m}-1}k_{\bar{m}}} = 0$, that is,

$$
{}_{\bar{k}_{\bar{m}-2}}\tilde{p}_{\varepsilon,k_{\bar{m}-1}k_{\bar{m}}} > 0, \ \varepsilon \in (0, 1], \ \text{while } {}_{\bar{k}_{\bar{m}-2}}\tilde{p}_{0,k_{\bar{m}-1}k_{\bar{m}}} = 0. \tag{11.47}
$$

Note that, in this case,

$$
{}_{\bar{k}_{\bar{m}-2}}\tilde{p}_{\varepsilon,k_{\bar{m}-1}k_{\bar{m}}} \to 0 \text{ as } \varepsilon \to 0 \tag{11.48}
$$

and the normalisation function,

$$
\check{u}_{\varepsilon,k_{\bar{m}-1}} = {}_{\bar{k}_{\bar{m}-2}}\tilde{u}_{\varepsilon,k_{\bar{m}-1}}. \tag{11.49}
$$

In this case, the asymptotic relation appearing in the condition $\mathbf{L}_{\mathbb{D},1}$ holds, with the limit ${}_{\bar{k}_{\bar{m}}}\bar{w}_{0,k_{\bar{m}-1},k_{\bar{m}}}$, which can take any value in interval [0, 1]. Three cases should be considered, when limit ${}_{\bar{k}_{\bar{m}}}\bar{w}_{0,k_{\bar{m}-1},k_{\bar{m}}}$ equals 0, or takes value in interval (0, 1), or equals 1.

(1) If ${}_{\bar{k}_{\bar{m}}}\bar{w}_{0,k_{\bar{m}-1},k_{\bar{m}}} = 0$, then the asymptotic relation (11.25) takes the following form, for $j \in \mathbb{D}$:

$$\frac{E_{\varepsilon,\mathbb{D},k_{\bar{m}-1}j}}{\bar{u}_{\varepsilon,k_{\bar{m}-1}}} \to \bar{E}_{0,\mathbb{D},k_{\bar{m}-1}j}$$
$$= {}_{\bar{k}_{\bar{m}-2}}\tilde{f}_{0,k_{\bar{m}-1}j}\, {}_{\bar{k}_{\bar{m}-2}}\tilde{p}_{0,k_{\bar{m}-1}j} = E_{0,\mathbb{D},k_{\bar{m}-1}j} < \infty \text{ as } \varepsilon \to 0. \quad (11.50)$$

In this case,

$$\frac{\bar{u}_{\varepsilon,k_{\bar{m}-1}}}{\check{u}_{\varepsilon,k_{\bar{m}-1}}} = \frac{\bar{u}_{\varepsilon,k_{\bar{m}-1}}}{{}_{\bar{k}_{\bar{m}-2}}\tilde{u}_{\varepsilon,k_{\bar{m}-1}}} \to 1 \text{ as } \varepsilon \to 0. \quad (11.51)$$

The relations (11.50) and (11.51) imply that, for $j \in \mathbb{D}$,

$$\frac{E_{\varepsilon,\mathbb{D},k_{\bar{m}-1}j}}{\check{u}_{\varepsilon,k_{\bar{m}-1}}} \to {}_{\bar{k}_{\bar{m}-2}}\tilde{f}_{0,k_{\bar{m}-1}j}\, {}_{\bar{k}_{\bar{m}-2}}\tilde{p}_{0,k_{\bar{m}-1}j}$$
$$= \bar{E}_{0,\mathbb{D},k_{\bar{m}-1}j} = E_{0,\mathbb{D},k_{\bar{m}-1}j} < \infty \text{ as } \varepsilon \to 0. \quad (11.52)$$

(2) If ${}_{\bar{k}_{\bar{m}}}\bar{w}_{0,k_{\bar{m}-1},k_{\bar{m}}} \in (0,1)$, then the asymptotic relation (11.25) takes the following form, for $j \in \mathbb{D}$:

$$\frac{E_{\varepsilon,\mathbb{D},k_{\bar{m}-1}j}}{\bar{u}_{\varepsilon,k_{\bar{m}-1}}} \to \bar{E}_{0,\mathbb{D},k_{\bar{m}-1}j}$$
$$= (1 - {}_{\bar{k}_{\bar{m}}}\bar{w}_{0,k_{\bar{m}-1},k_{\bar{m}}})\, {}_{\bar{k}_{\bar{m}-2}}\tilde{f}_{0,k_{\bar{m}-1}j}\, {}_{\bar{k}_{\bar{m}-2}}\tilde{p}_{0,k_{\bar{m}-1}j}$$
$$+ {}_{\bar{k}_{\bar{m}}}\bar{w}_{0,k_{\bar{m}-1},k_{\bar{m}}}\, \bar{E}_{0,\mathbb{D},k_{\bar{m}}j} < \infty \text{ as } \varepsilon \to 0. \quad (11.53)$$

In this case,

$$\frac{\bar{u}_{\varepsilon,k_{\bar{m}-1}}}{\check{u}_{\varepsilon,k_{\bar{m}-1}}} = \frac{{}_{\bar{k}_{\bar{m}}}\bar{u}_{\varepsilon,k_{\bar{m}-1}}}{{}_{\bar{k}_{\bar{m}-2}}\tilde{u}_{\varepsilon,k_{\bar{m}-1}}} \to (1 - {}_{\bar{k}_{\bar{m}}}\bar{w}_{0,k_{\bar{m}-1},k_{\bar{m}}})^{-1} \text{ as } \varepsilon \to 0. \quad (11.54)$$

Therefore, the relation (11.53) implies that, for $j \in \mathbb{D}$,

$$\frac{E_{\varepsilon,\mathbb{D},k_{\bar{m}-1}j}}{\check{u}_{\varepsilon,k_{\bar{m}-1}}} \to {}_{\bar{k}_{\bar{m}-2}}\tilde{f}_{0,k_{\bar{m}-1}j}\, {}_{\bar{k}_{\bar{m}-2}}\tilde{p}_{0,k_{\bar{m}-1}j}$$
$$+ \frac{{}_{\bar{k}_{\bar{m}}}\bar{w}_{0,k_{\bar{m}-1},k_{\bar{m}}}}{1 - {}_{\bar{k}_{\bar{m}}}\bar{w}_{0,k_{\bar{m}-1},k_{\bar{m}}}} E_{0,\mathbb{D},k_{\bar{m}}j} \text{ as } \varepsilon \to 0. \quad (11.55)$$

According to the condition ${}_{\bar{k}_{\bar{m}-2}}\mathbf{K}_{\mathbb{D}}$, the expectation ${}_{\bar{k}_{\bar{m}-2}}\tilde{f}_{0,k_{\bar{m}-1}j} \in (0,\infty)$ and, thus, $E_{0,\mathbb{D},k_{\bar{m}}j} = {}_{\bar{k}_{\bar{m}-1}}\tilde{f}_{0,k_{\bar{m}}j}\, {}_{\bar{k}_{\bar{m}-1}}\tilde{p}_{0,k_{\bar{m}}j} \in (0,\infty)$ if and only if ${}_{\bar{k}_{\bar{m}-1}}\tilde{p}_{0,k_{\bar{m}}j} > 0$. In this case,

$${}_{\bar{k}_{\bar{m}-2}}\tilde{f}_{0,k_{\bar{m}-1}j}\, {}_{\bar{k}_{\bar{m}-2}}\tilde{p}_{0,k_{\bar{m}-1}j} + \frac{{}_{\bar{k}_{\bar{m}}}\bar{w}_{0,k_{\bar{m}-1},k_{\bar{m}}}}{1 - {}_{\bar{k}_{\bar{m}}}\bar{w}_{0,k_{\bar{m}-1},k_{\bar{m}}}} E_{0,\mathbb{D},k_{\bar{m}}j}$$
$$> {}_{\bar{k}_{\bar{m}-2}}\tilde{f}_{0,k_{\bar{m}-1}j}\, {}_{\bar{k}_{\bar{m}-2}}\tilde{p}_{0,k_{\bar{m}-1}j} = E_{0,\mathbb{D},k_{\bar{m}-1}j}. \quad (11.56)$$

(3) If ${}_{\bar{k}_{\bar{m}}}\bar{w}_{0,k_{\bar{m}-1},k_{\bar{m}}} = 1$, then the asymptotic relation (11.25) takes the following form, for $j \in \mathbb{D}$:

$$\frac{E_{\varepsilon,\mathbb{D},k_{\bar{m}-1}j}}{\bar{u}_{\varepsilon,k_{\bar{m}-1}}} \longrightarrow \bar{E}_{0,\mathbb{D},k_{\bar{m}-1}j}$$
$$= \bar{E}_{0,\mathbb{D},k_{\bar{m}}j} = E_{0,\mathbb{D},k_{\bar{m}}j} < \infty \text{ as } \varepsilon \to 0. \tag{11.57}$$

In this case,

$$\frac{\bar{u}_{\varepsilon,k_{\bar{m}-1}}}{\check{u}_{\varepsilon,k_{\bar{m}-1}}} = \frac{\bar{u}_{\varepsilon,k_{\bar{m}-1}}}{{}_{\bar{k}_{\bar{m}-2}}\tilde{u}_{\varepsilon,k_{\bar{m}-1}}} \longrightarrow \infty \text{ as } \varepsilon \to 0. \tag{11.58}$$

According to the condition ${}_{\bar{k}_{\bar{m}-2}}\mathbf{K}_{\mathbb{D}}$, the expectation ${}_{\bar{k}_{\bar{m}-2}}\tilde{f}_{0,k_{\bar{m}-1}j} \in (0,\infty)$ and, thus, $\bar{E}_{0,\mathbb{D},k_{\bar{m}}j} = E_{0,\mathbb{D},k_{\bar{m}}j} = {}_{\bar{k}_{\bar{m}-1}}\tilde{f}_{0,k_{\bar{m}}j}\, {}_{\bar{k}_{\bar{m}-1}}\tilde{p}_{0,k_{\bar{m}}j} \in (0,\infty)$ if and only if ${}_{\bar{k}_{\bar{m}-1}}\tilde{p}_{0,k_{\bar{m}}j} > 0$. In this case,

$$\frac{E_{\varepsilon,\mathbb{D},k_{\bar{m}-1}j}}{\check{u}_{\varepsilon,k_{\bar{m}-1}}} \longrightarrow \infty \text{ as } \varepsilon \to 0. \tag{11.59}$$

In the case, $E_{0,\mathbb{D},k_{\bar{m}-1}j} = {}_{\bar{k}_{\bar{m}-2}}\tilde{f}_{0,k_{\bar{m}-1}j}\, {}_{\bar{k}_{\bar{m}-2}}\tilde{p}_{0,k_{\bar{m}-1}j} < \infty$, while the expectations $E_{\varepsilon,\mathbb{D},k_{\bar{m}-1}j}$ normalised by the function $\check{u}_{\varepsilon,k_{\bar{m}-1}}$ (the normalisation function used in the weak convergence relation for distributions $G_{\varepsilon,\mathbb{D},k_{\bar{m}-1}j}(\cdot)$ given in Theorem 10.3, in the case where the probability ${}_{\bar{k}_{\bar{m}-2}}\tilde{p}_{0,k_{\bar{m}-1}k_{\bar{m}}} = 0$), converge to ∞ as $\varepsilon \to 0$. □

In conclusion, let us mention that, under the conditions of the statement **(iii)** of Theorem 11.4, the limit of normalised expectations $E_{\varepsilon,\mathbb{D},k_{\bar{m}-1}j}/\check{u}_{\varepsilon,k_{\bar{m}-1}}$ as $\varepsilon \to 0$ exists also for $j \in \mathbb{D}$ such that ${}_{\bar{k}_{\bar{m}-1}}\tilde{p}_{0,k_{\bar{m}}j} = 0$. Indeed, the relations (11.8) and (11.9) imply, in this case, that,

$$\begin{aligned}\frac{E_{\varepsilon,\mathbb{D},k_{\bar{m}-1}j}}{\check{u}_{\varepsilon,k_{\bar{m}-1}}} &= \frac{{}_{\bar{k}_{\bar{m}-2}}\tilde{f}_{\varepsilon,k_{\bar{m}-1}j}}{{}_{\bar{k}_{\bar{m}-2}}\tilde{u}_{\varepsilon,k_{\bar{m}-1}}}\, {}_{\bar{k}_{\bar{m}-2}}\tilde{p}_{\varepsilon,k_{\bar{m}-1}j} \\ &+ \frac{{}_{\bar{k}_{\bar{m}-2}}\tilde{f}_{\varepsilon,k_{\bar{m}-1}k_{\bar{m}}}}{{}_{\bar{k}_{\bar{m}-2}}\tilde{u}_{\varepsilon,k_{\bar{m}-1}}}\, {}_{\bar{k}_{\bar{m}-1}}\tilde{p}_{\varepsilon,k_{\bar{m}}j}\, {}_{\bar{k}_{\bar{m}-2}}\tilde{p}_{\varepsilon,k_{\bar{m}-1}k_{\bar{m}}} \\ &+ \frac{{}_{\bar{k}_{\bar{m}-1}}\tilde{f}_{\varepsilon,k_{\bar{m}}j}}{{}_{\bar{k}_{\bar{m}-1}}\tilde{u}_{\varepsilon,k_{\bar{m}}}}\, \frac{{}_{\bar{k}_{\bar{m}-1}}\tilde{u}_{\varepsilon,k_{\bar{m}}}}{{}_{\bar{k}_{\bar{m}-2}}\tilde{u}_{\varepsilon,k_{\bar{m}-1}}}\, {}_{\bar{k}_{\bar{m}-1}}\tilde{p}_{\varepsilon,k_{\bar{m}}j}\, {}_{\bar{k}_{\bar{m}-2}}\tilde{p}_{\varepsilon,k_{\bar{m}-1}k_{\bar{m}}}.\end{aligned} \tag{11.60}$$

Since the functions ${}_{\bar{k}_{\bar{m}-2}}\tilde{p}_{\cdot,k_{\bar{m}-1}j}$, ${}_{\bar{k}_{\bar{m}-2}}\tilde{p}_{\cdot,k_{\bar{m}-1}k_{\bar{m}}}$, ${}_{\bar{k}_{\bar{m}-1}}\tilde{u}_{\varepsilon,k_{\bar{m}}}$, and ${}_{\bar{k}_{\bar{m}-2}}\tilde{u}_{\varepsilon,k_{\bar{m}-1}}$ belong to the class of asymptotically comparable functions $\mathcal{H}$, there exist limits,

$$\lim_{\varepsilon\to 0} \frac{{}_{\bar{k}_{\bar{m}-1}}\tilde{u}_{\varepsilon,k_{\bar{m}}}}{{}_{\bar{k}_{\bar{m}-2}}\tilde{u}_{\varepsilon,k_{\bar{m}-1}}}\, {}_{\bar{k}_{\bar{m}-1}}\tilde{p}_{\varepsilon,k_{\bar{m}}j}\, {}_{\bar{k}_{\bar{m}-2}}\tilde{p}_{\varepsilon,k_{\bar{m}-1}k_{\bar{m}}} = {}_{\bar{k}_{\bar{m}}}\bar{w}_{0,k_{\bar{m}-1},\bar{m},j} \in [0,\infty]. \tag{11.61}$$

The above limits can be calculated using the operating formulas given in Lemma 8.2. In the case, where the family $\mathcal{H}$ is one of the families, $\mathcal{H}_1$, $\mathcal{H}_2$, or $\mathcal{H}_3$, the operating formulas given in Lemma 8.4, 8.6, or 8.8 should be used.

Since, ${}_{\bar{k}_{\bar{m}-1}}\tilde{e}_{\varepsilon,k_{\bar{m}}j}/\,{}_{\bar{k}_{\bar{m}-1}}\tilde{u}_{\varepsilon,k_{\bar{m}}} \to {}_{\bar{k}_{\bar{m}-1}}\tilde{f}_{0,k_{\bar{m}}j} \in (0,\infty)$ as $\varepsilon \to 0$, there also exist limits,

$$\lim_{\varepsilon\to 0} \frac{E_{\varepsilon,\mathbb{D},k_{\bar{m}-1}j}}{\check{u}_{\varepsilon,k_{\bar{m}-1}}} = {}_{\bar{k}_{\bar{m}-1}}\tilde{f}_{0,k_{\bar{m}}j}\, {}_{\bar{k}_{\bar{m}}}\bar{w}_{0,k_{\bar{m}-1},\bar{m},j} \in [0,\infty]. \tag{11.62}$$

A similar analysis regarding additional conditions, under which the expectations $E_{\varepsilon,\mathbb{D},k_nj}$ normalised by function $\check{u}_{\varepsilon,k_n}$ (the normalisation function used in the weak convergence relation for the distributions $G_{\varepsilon,\mathbb{D},k_nj}(\cdot\check{u}_{\varepsilon,k_n})$ given in Theorem 10.4) converge to the first moment of the corresponding limiting distribution $G_{0,\mathbb{D},k_nj}(\cdot)$, can be conducted for the general case with an arbitrary initial state in the domain $\bar{\mathbb{D}}$.

11.3 Asymptotics for Expectations of Return Times

In this section, we present conditions of convergence for normalised expectations of hitting and return times for the case, where an state $i \in \mathbb{D}$.

11.3.1 Asymptotic for Expectations of Hitting Times in the Case Where an Initial State Belongs to Domain $\mathbb{D}$

We assume that conditions $\mathbf{G}_{\mathbb{D}}$, $\mathbf{H}_{\mathbb{D}}$, $\mathbf{I}_{\mathbb{D}}$, $\mathbf{I}_{\mathbb{D},\mathcal{H}}$, $\mathbf{L}_{\mathbb{D}}$, $\mathbf{L}_{\mathbb{D},\mathcal{H}}$, $\mathbf{J}_{\mathbb{D}}$, $\mathbf{K}_{\mathbb{D}}$, ${}_{\bar{\mathbf{k}}_{\bar{m}-1}}\mathbf{M}_{\mathbb{D}}$ and $\mathbf{G}_{\mathbb{D}}$, $\mathbf{H}_{\mathbb{D}}$, $\mathbf{I}_{\mathbb{D}}$, $\mathbf{I}_{\mathbb{D},\mathcal{H}}$, $\mathbf{L}_{\mathbb{D}}$, $\mathbf{L}_{\mathbb{D},\mathcal{H}}$, $\mathbf{J}_{\mathbb{D}}$, $\mathbf{K}_{\mathbb{D}}$ are satisfied for the semi-Markov processes $\eta_\varepsilon(t)$.

The key role is played by the following relation, which takes place, for $r, j \in \mathbb{D}$ and $\varepsilon \in (0, 1]$:

$$
\begin{aligned}
E_{\varepsilon,\mathbb{D},rj} &= f_{\varepsilon,rj}p_{\varepsilon,rj} + \sum_{k_n \in \bar{\mathbb{D}}} (f_{\varepsilon,rk_n}P_{\varepsilon,\mathbb{D},k_nj} + E_{\varepsilon,\mathbb{D},k_nj})p_{\varepsilon,rk_n} \\
&= f_{\varepsilon,rj}p_{\varepsilon,rj} + \sum_{k_n \in \bar{\mathbb{D}}} f_{\varepsilon,rk_n}P_{\varepsilon,\mathbb{D},k_nj}p_{\varepsilon,rk_n} + \sum_{k_n \in \bar{\mathbb{D}}} E_{\varepsilon,\mathbb{D},k_nj}p_{\varepsilon,rk_n}.
\end{aligned} \tag{11.63}
$$

Note, first of all, that the above assumptions and relation (11.63) imply that $E_{\varepsilon,\mathbb{D},rj} < \infty$, for $r, j \in \mathbb{D}$ and $\varepsilon \in (0, 1]$.

Let us introduce the following normalising functions, for $r \in \mathbb{D}$:

$$
\bar{u}_{\varepsilon,r} = u_{\varepsilon,r} + \sum_{k_n \in \bar{\mathbb{D}}} \bar{u}_{\varepsilon,k_n}p_{\varepsilon,rk_n}. \tag{11.64}
$$

The conditions $\mathbf{G}$, $\mathbf{H}_{\mathbb{D}}$, $\mathbf{I}_{\mathcal{H}}$, $\mathbf{L}_{\mathcal{H}}$, and ${}_{\bar{\mathbf{k}}_{\bar{m}-1}}\mathbf{M}_{\mathbb{D}}$ imply that the following condition is satisfied:

$\mathbf{L}_{\mathbb{D},1}$: $\frac{\bar{u}_{\varepsilon,k_n}}{u_{\varepsilon,r}} \to {}_{\bar{k}_{\bar{m}}}\ddot{w}_{0,k_n,r} \in [0, \infty]$ as $\varepsilon \to 0$, for $k_n \in \bar{\mathbb{D}}, r \in \mathbb{D}$.

Since the functions $\bar{u}_{\cdot,k_n}$, $u_{\varepsilon,r}$, $p_{\cdot,rk_n}$, $k_n \in \bar{\mathbb{D}}, r \in \mathbb{D}$ belong to the class of asymptotically comparable functions $\mathcal{H}$, the limits appearing in the condition $\mathbf{L}_{\mathbb{D},1}$ can be computed with the use of the operating formulas given in Lemma 8.2. In the case, where the family $\mathcal{H}$ is one of the families $\mathcal{H}_1$, $\mathcal{H}_2$, or $\mathcal{H}_3$, the operating formulas given in Lemma 8.4, 8.6, or 8.8 should be used.

Note also that the normalisation functions $\bar{u}_{\cdot,r}, r \in \mathbb{D}$ belong to the family of asymptotically comparable functions $\mathcal{H}$.

The following theorem takes place.

Theorem 11.5 *Let conditions* $\mathbf{G}_{\mathbb{D}}$, $\mathbf{H}_{\mathbb{D}}$, $\mathbf{I}_{\mathbb{D}}$, $\mathbf{I}_{\mathbb{D},\mathcal{H}}$, $\mathbf{L}_{\mathbb{D}}$, $\mathbf{L}_{\mathbb{D},\mathcal{H}}$, $\mathbf{J}_{\mathbb{D}}$, $\mathbf{K}_{\mathbb{D}}$, ${}_{\bar{\mathbf{k}}_{\bar{m}-1}}\mathbf{M}_{\mathbb{D}}$ *and* $\mathbf{G}_{\mathbb{D}}$, $\mathbf{H}_{\mathbb{D}}$, $\mathbf{I}_{\mathbb{D}}$, $\mathbf{I}_{\mathbb{D},\mathcal{H}}$, $\mathbf{L}_{\mathbb{D}}$, $\mathbf{L}_{\mathbb{D},\mathcal{H}}$, $\mathbf{J}_{\mathbb{D}}$, $\mathbf{K}_{\mathbb{D}}$ *are satisfied for the semi-Markov processes* $\eta_\varepsilon(t)$. *Then*:

(i) *The following asymptotic relation takes place, for* $r, j \in \mathbb{D}$:

$$E_{\varepsilon,\mathbb{D},rj}/\bar{u}_{\varepsilon,r} \to \bar{E}_{0,\mathbb{D},rj} < \infty \text{ as } \varepsilon \to 0. \tag{11.65}$$

(ii) *The normalisation functions* $\bar{u}_{\cdot,r} \in (0,\infty), r \in \mathbb{D}$ *are given by relation (11.64) and belong to the family of asymptotically comparable functions* $\mathcal{H}$.

(iii) *The limits* $\bar{E}_{0,\mathbb{D},rj}$, $r, j \in \mathbb{D}$ *are given by the recurrent relation (11.68).*

(iv) $\bar{E}_{0,\mathbb{D},r} = \sum_{j\in\mathbb{D}} \bar{E}_{0,\mathbb{D},rj} \in (0,\infty)$, *for* $r \in \mathbb{D}$.

Proof The condition ${}_{\bar{\mathbf{k}}_{\bar{m}}}\mathbf{L}_{\mathbb{D}}$ implies that the following relations holds, for $r \in \mathbb{D}$:

$$\begin{aligned}\frac{u_{\varepsilon,r}}{\bar{u}_{\varepsilon,r}} &= 1 - \sum_{k_n\in\bar{\mathbb{D}}} \frac{\bar{u}_{\varepsilon,k_n} p_{\varepsilon,rk_n}}{\bar{u}_{\varepsilon,r}} \\ &\to 1 - \sum_{k_n\in\bar{\mathbb{D}}} {}_{\bar{k}_{\bar{m}}}\ddot{w}_{0,k_n,r} = {}_{\bar{k}_{\bar{m}}}\ddot{w}_{0,r,r} \in [0,1] \text{ as } \varepsilon \to 0.\end{aligned} \tag{11.66}$$

Obviously, the following relation holds, for $r \in \mathbb{D}$:

$${}_{\bar{k}_{\bar{m}}}\ddot{w}_{0,r,r} + \sum_{k_n\in\bar{\mathbb{D}}} {}_{\bar{k}_{\bar{m}}}\ddot{w}_{0,k_n,r} = 1. \tag{11.67}$$

Theorems 10.1–10.3, the condition $\mathbf{L}_{\mathbb{D},1}$, and the relations (11.63) and (11.66) imply that the following relation holds, for $r, j \in \mathbb{D}$,

$$\begin{aligned}E_{\varepsilon,\mathbb{D},rj}/\bar{u}_{\varepsilon,r} &= \frac{u_{\varepsilon,r}}{\bar{u}_{\varepsilon,r}}\frac{f_{\varepsilon,rj}}{u_{\varepsilon,r}} p_{\varepsilon,rj} + \sum_{k_n\in\bar{\mathbb{D}}} \frac{u_{\varepsilon,r}}{\bar{u}_{\varepsilon,r}}\frac{f_{\varepsilon,rk_n}}{u_{\varepsilon,r}} P_{\varepsilon,\mathbb{D},k_nj} p_{\varepsilon,rk_n} \\ &\quad + \sum_{k_n\in\bar{\mathbb{D}}} \frac{\bar{u}_{\varepsilon,k_n} p_{\varepsilon,rk_n}}{\bar{u}_{\varepsilon,r}} \frac{E_{\varepsilon,\mathbb{D},k_nj}}{\bar{u}_{\varepsilon,k_n}} \\ &\to {}_{\bar{k}_{\bar{m}}}\ddot{w}_{0,r,r}(f_{0,rj}p_{0,rj} + \sum_{k_n\in\bar{\mathbb{D}}} f_{0,rk_n} P_{0,\mathbb{D},k_nj} p_{0,rk_n}) \\ &\quad + \sum_{k_n\in\bar{\mathbb{D}}} {}_{\bar{k}_{\bar{m}}}\ddot{w}_{0,k_n,r}\bar{E}_{0,\mathbb{D},k_nj} \\ &= \bar{E}_{0,\mathbb{D},rj} \text{ as } \varepsilon \to 0.\end{aligned} \tag{11.68}$$

The relation (11.68) implies that the following relation holds, for $r \in \mathbb{D}$:

$$
\begin{aligned}
E_{\varepsilon,\mathbb{D},r}/\bar{u}_{\varepsilon,r} &= \sum_{j\in\mathbb{D}} E_{\varepsilon,\mathbb{D},rj}/\bar{u}_{\varepsilon,r} \\
&\to {}_{\bar{k}_{\bar{m}}}\ddot{w}_{0,r,r}(f_{0,rj}p_{0,rj} + \sum_{k_n\in\bar{\mathbb{D}}} f_{0,rk_n}p_{0,rk_n}) \\
&\quad + \sum_{k_n\in\bar{\mathbb{D}}} {}_{\bar{k}_{\bar{m}}}\ddot{w}_{0,k_n,r}\bar{E}_{0,\mathbb{D},k_n} \\
&= {}_{\bar{k}_{\bar{m}}}\ddot{w}_{0,r,r}f_{0,r} + \sum_{k_n\in\bar{\mathbb{D}}} {}_{\bar{k}_{\bar{m}}}\ddot{w}_{0,k_n,r}\bar{E}_{0,\mathbb{D},k_n} \\
&= \bar{E}_{0,\mathbb{D},r} \text{ as } \varepsilon \to 0,
\end{aligned} \tag{11.69}
$$

where

$$
f_{0,r} = \sum_{i\in\mathbb{X}} f_{0,ri}p_{0,rj}. \tag{11.70}
$$

The conditions of Theorem 11.5 and Theorems 11.1–11.3 imply that, for $r \in \mathbb{D}$,

$$
\bar{f} = f_{0,r} \wedge \min_{k_n\in\bar{\mathbb{D}}} \bar{E}_{0,\mathbb{D},k_n} > 0. \tag{11.71}
$$

Finally, the relations (11.69) and (11.71) imply that the following inequality holds, for $r \in \mathbb{D}$:

$$
\begin{aligned}
\bar{E}_{0,\mathbb{D},r} &= {}_{\bar{k}_{\bar{m}}}\ddot{w}_{0,r,r}f_{0,r} + \sum_{k_n\in\bar{\mathbb{D}}} {}_{\bar{k}_{\bar{m}}}\ddot{w}_{0,k_n,r}\bar{E}_{0,\mathbb{D},k_n} \\
&\geq \bar{f}\cdot({}_{\bar{k}_{\bar{m}}}\ddot{w}_{0,r,r} + \sum_{k_n\in\bar{\mathbb{D}}} {}_{\bar{k}_{\bar{m}}}\ddot{w}_{0,k_n,r}) = \bar{f} > 0.
\end{aligned} \tag{11.72}
$$

The proof is complete. □

11.3.2 Asymptotics of Expectations for Return Times to Domain $\mathbb{D}$

The return times $\dot{\tau}_{\varepsilon,\mathbb{D}}$ have been defined in Sect. 10.3.1, which also contains results concerning weak asymptotics for distributions of return times. Here, we present results concerning asymptotics for expectations of return times,

$$
\dot{E}_{\varepsilon,\mathbb{D},ij} = \mathsf{E}_i\dot{\tau}_{\varepsilon,\mathbb{D}},\ i,j \in \mathbb{D}. \tag{11.73}
$$

The relation (10.149) for the Laplace transforms of return times implies that the following relation holds for their expectations, for $\varepsilon \in (0,1]$ and $r, j \in \mathbb{D}$,

$$
\dot{E}_{\varepsilon,\mathbb{D},rj} = \sum_{i\in\bar{\mathbb{D}}}(E_{\varepsilon,\bar{\mathbb{D}},ri}P_{\varepsilon,\mathbb{D},ij} + E_{\varepsilon,\mathbb{D},ij}P_{\varepsilon,\bar{\mathbb{D}},ri}). \tag{11.74}
$$

Let us introduce the following normalising functions, for $r_l \in \mathbb{D}$:

$$\ddot{u}_{\varepsilon,r_l} = \bar{u}_{\varepsilon,r_l} + \sum_{k_n \in \bar{\mathbb{D}}} \bar{u}_{\varepsilon,k_n} P_{\varepsilon,\mathbb{D},r_l k_n}. \tag{11.75}$$

We assume that the conditions $\mathbf{G}_{\mathbb{D}}, \mathbf{H}_{\mathbb{D}}, \mathbf{I}_{\mathbb{D}}, \mathbf{I}_{\mathbb{D},\mathcal{H}}, \mathbf{L}_{\mathbb{D}}, \mathbf{L}_{\mathbb{D},\mathcal{H}}, \mathbf{J}_{\mathbb{D}}, \mathbf{K}_{\mathbb{D}}, {}_{\bar{\mathbf{k}}_{\bar{m}-1}}\mathbf{M}_{\mathbb{D}}$ and $\mathbf{G}_{\mathbb{D}}, \mathbf{H}_{\mathbb{D}}, \mathbf{I}_{\mathbb{D}}, \mathbf{I}_{\mathbb{D},\mathcal{H}}, \mathbf{L}_{\mathbb{D}}, \mathbf{L}_{\mathbb{D},\mathcal{H}}, \mathbf{J}_{\mathbb{D}}, \mathbf{K}_{\mathbb{D}}, {}_{\bar{\mathbf{r}}_{m-1}}\mathbf{M}_{\mathbb{D}}$ are satisfied for the semi-Markov processes $\eta_\varepsilon(t)$.

The conditions $\mathbf{G}_{\mathbb{D}}, \mathbf{H}_{\mathbb{D}}, \mathbf{I}_{\mathbb{D}}, \mathbf{I}_{\mathbb{D},\mathcal{H}}, \mathbf{L}_{\mathbb{D}}, \mathbf{L}_{\mathbb{D},\mathcal{H}}, {}_{\bar{\mathbf{k}}_{\bar{m}-1}}\mathbf{M}_{\mathbb{D}}$ and $\mathbf{G}_{\mathbb{D}}, \mathbf{H}_{\mathbb{D}}, \mathbf{I}_{\mathbb{D}}, \mathbf{I}_{\mathbb{D},\mathcal{H}}, \mathbf{L}_{\mathbb{D}}, \mathbf{L}_{\mathbb{D},\mathcal{H}}, {}_{\bar{\mathbf{r}}_{m-1}}\mathbf{M}_{\mathbb{D}}$ imply the fulfilment of following condition:

$\mathbf{L}_{\mathbb{D},2}$: $\frac{\bar{u}_{\varepsilon,k_n}}{\ddot{u}_{\varepsilon,r_l}} \to {}_{\langle \bar{k}_{\bar{m}}, \bar{r}_{\bar{m}} \rangle}\ddot{w}_{0,k_n,r_l} \in [0, \infty]$ as $\varepsilon \to 0$, for $r_l \in \mathbb{D}, k_n \in \bar{\mathbb{D}}$.

Since the functions $\bar{u}_{\cdot,k_n}, \ddot{u}_{\cdot,r_l}, k_n \in \bar{\mathbb{D}}, r_l \in \mathbb{D}$ belong to the class of asymptotically comparable functions $\mathcal{H}$, the limits appearing in the condition $\mathbf{L}_{\mathbb{D},2}$ can be calculated using the operating formulas given in Lemma 8.2. In the case, where the family $\mathcal{H}$ is one of the families, $\mathcal{H}_1$, $\mathcal{H}_2$, or $\mathcal{H}_3$, the operating formulas given in Lemma 8.4, 8.6, or 8.8 should be used.

Note also that the normalisation functions $\ddot{u}_{\cdot,r_l}, r_l \in \mathbb{D}$ belong to the family of asymptotically comparable functions $\mathcal{H}$.

The following theorem takes place.

Theorem 11.6 *Let the conditions* $\mathbf{G}_{\mathbb{D}}, \mathbf{H}_{\mathbb{D}}, \mathbf{I}_{\mathbb{D}}, \mathbf{I}_{\mathbb{D},\mathcal{H}}, \mathbf{L}_{\mathbb{D}}, \mathbf{L}_{\mathbb{D},\mathcal{H}}, \mathbf{J}_{\mathbb{D}}, \mathbf{K}_{\mathbb{D}}, {}_{\bar{\mathbf{k}}_{\bar{m}-1}}\mathbf{M}_{\mathbb{D}}$ *and* $\mathbf{G}_{\mathbb{D}}, \mathbf{H}_{\mathbb{D}}, \mathbf{I}_{\mathbb{D}}, \mathbf{I}_{\mathbb{D},\mathcal{H}}, \mathbf{L}_{\mathbb{D}}, \mathbf{L}_{\mathbb{D},\mathcal{H}}, \mathbf{J}_{\mathbb{D}}, \mathbf{K}_{\mathbb{D}}, {}_{\bar{\mathbf{r}}_{m-1}}\mathbf{M}_{\mathbb{D}}$ *be satisfied for the semi-Markov processes* $\eta_\varepsilon(t)$. *Then*:

(i) *The following asymptotic relation takes place, for* $r_l, j \in \mathbb{D}$,

$$\dot{E}_{\varepsilon,\mathbb{D},r_l j} / \ddot{u}_{\varepsilon,r_l} \to \dot{E}_{0,\mathbb{D},r_l j} < \infty \text{ as } \varepsilon \to 0. \tag{11.76}$$

(ii) *The normalisation functions* $\ddot{u}_{\cdot,r_l} \in (0, \infty), r_l \in \mathbb{D}$ *are given by relation (11.75) and belong to the family of asymptotically comparable functions* $\mathcal{H}$.

(iii) *The limits* $\dot{E}_{0,\mathbb{D},rj}, r, j \in \mathbb{D}$ *are given by the recurrent relation (11.79).*

(iv) $\dot{E}_{0,\mathbb{D},r_l} = \sum_{j \in \mathbb{D}} \dot{E}_{0,\mathbb{D},r_l j} \in (0, \infty)$, *for* $r_l \in \mathbb{D}$.

Proof The condition $\mathbf{L}_{\mathbb{D},3}$ implies that the following relation holds, for $r_l \in \mathbb{D}$:

$$\begin{aligned} \frac{\bar{u}_{\varepsilon,r_l}}{\ddot{u}_{\varepsilon,r_l}} &= 1 - \sum_{k_n \in \bar{\mathbb{D}}} \frac{\bar{u}_{\varepsilon,k_n} P_{\varepsilon,\mathbb{D},r_l k_n}}{\ddot{u}_{\varepsilon,r_l}} \\ &\to 1 - \sum_{k_n \in \bar{\mathbb{D}}} {}_{\langle \bar{k}_{\bar{m}}, \bar{r}_{\bar{m}} \rangle}\ddot{w}_{0,k_n,r_l} \\ &= {}_{\langle \bar{k}_{\bar{m}}, \bar{r}_{\bar{m}} \rangle}\ddot{w}_{0,r_l,r_l} \in [0, 1] \text{ as } \varepsilon \to 0. \end{aligned} \tag{11.77}$$

Obviously, the following relation holds, for $r_l \in \mathbb{D}$:

$${}_{\langle \bar{k}_{\bar{m}}, \bar{r}_{\bar{m}} \rangle}\ddot{w}_{0,r_l,r_l} + \sum_{k_n \in \bar{\mathbb{D}}} {}_{\langle \bar{k}_{\bar{m}}, \bar{r}_{\bar{m}} \rangle}\ddot{w}_{0,k_n,r_l} = 1. \tag{11.78}$$

Theorems 11.1–11.3, the condition $\mathbf{L}_{\mathbb{D},2}$, and the relations (10.149) and (11.77) imply that the following relation holds, for $r_l, j \in \mathbb{D}$:

$$
\begin{aligned}
\dot{E}_{\varepsilon,\mathbb{D},r_l j}/\ddot{u}_{\varepsilon,r_l} &= \sum_{k_n \in \bar{\mathbb{D}}} \frac{\bar{u}_{\varepsilon,r_l}}{\ddot{u}_{\varepsilon,r_l}} \frac{E_{\varepsilon,\bar{\mathbb{D}},r_l k_n}}{\bar{u}_{\varepsilon,r_l}} P_{\varepsilon,\mathbb{D},k_n j} \\
&\quad + \sum_{k_n \in \bar{\mathbb{D}}} \frac{\bar{u}_{\varepsilon,k_n} P_{\varepsilon,\bar{\mathbb{D}},r_l k_n}}{\ddot{u}_{\varepsilon,r_l}} \frac{E_{\varepsilon,\mathbb{D},k_n j}}{\bar{u}_{\varepsilon,k_n}} \\
&\longrightarrow {}_{\langle \bar{k}_{\bar{m}},\bar{r}_{\dot{m}} \rangle}\ddot{w}_{0,r_l,r_l} \sum_{k_n \in \bar{\mathbb{D}}} \bar{E}_{0,\bar{\mathbb{D}},r_l k_n} P_{0,\mathbb{D},k_n j} \\
&\quad + \sum_{k_n \in \bar{\mathbb{D}}} {}_{\langle \bar{k}_{\bar{m}},\bar{r}_{\dot{m}} \rangle}\ddot{w}_{0,k_n,r_l} \bar{E}_{0,\mathbb{D},k_n j} \\
&= \dot{E}_{0,\mathbb{D},r_l j} \text{ as } \varepsilon \to 0.
\end{aligned} \tag{11.79}
$$

The relation (11.79) implies that the following relation holds, for $r_l \in \mathbb{D}$:

$$
\begin{aligned}
\dot{E}_{\varepsilon,\mathbb{D},r_l}/\ddot{u}_{\varepsilon,r_l} &= \sum_{j \in \mathbb{D}} \dot{E}_{\varepsilon,\mathbb{D},r_l j}/\ddot{u}_{\varepsilon,r_l} \\
&\longrightarrow {}_{\langle \bar{k}_{\bar{m}},\bar{r}_{\dot{m}} \rangle}\ddot{w}_{0,r_l,r_l} \sum_{k_n \in \bar{\mathbb{D}}} \bar{E}_{0,\bar{\mathbb{D}},r_l k_n} \\
&\quad + \sum_{k_n \in \bar{\mathbb{D}}} {}_{\langle \bar{k}_{\bar{m}},\bar{r}_{\dot{m}} \rangle}\ddot{w}_{0,k_n,r_l} \bar{E}_{0,\mathbb{D},k_n} \\
&= {}_{\langle \bar{k}_{\bar{m}},\bar{r}_{\dot{m}} \rangle}\ddot{w}_{0,r_l,r_l} \bar{E}_{0,\bar{\mathbb{D}},r_l} + \sum_{k_n \in \bar{\mathbb{D}}} {}_{\bar{k}_{\bar{m}}}\ddot{w}_{0,k_n,r_l} \bar{E}_{0,\mathbb{D},k_n} \\
&= \dot{E}_{0,\mathbb{D},r_l} \text{ as } \varepsilon \to 0.
\end{aligned} \tag{11.80}
$$

The conditions of Theorem 11.6 and Theorems 11.1–11.3 imply that, for $r_l \in \mathbb{D}$,

$$
\dot{f} = \bar{E}_{0,\bar{\mathbb{D}},r_l} \wedge \min_{k_n \in \bar{\mathbb{D}}} \bar{E}_{0,\mathbb{D},k_n} > 0. \tag{11.81}
$$

Finally, the relations (11.69) and (11.71) imply that the following inequality holds, for $r \in \mathbb{D}$:

$$
\begin{aligned}
\bar{E}_{0,\mathbb{D},r} &= {}_{\langle \bar{k}_{\bar{m}},\bar{r}_{\dot{m}} \rangle}\ddot{w}_{0,r_l,r_l} \bar{E}_{0,\bar{\mathbb{D}},r_l} + \sum_{k_n \in \bar{\mathbb{D}}} {}_{\langle \bar{k}_{\bar{m}},\bar{r}_{\dot{m}} \rangle}\ddot{w}_{0,k_n,r_l} \bar{E}_{0,\mathbb{D},k_n} \\
&\geq \dot{f} \cdot \big({}_{\langle \bar{k}_{\bar{m}},\bar{r}_{\dot{m}} \rangle}\ddot{w}_{0,r,r} + \sum_{k_n \in \bar{\mathbb{D}}} {}_{\langle \bar{k}_{\bar{m}},\bar{r}_{\dot{m}} \rangle}\ddot{w}_{0,k_n,r}\big) = \dot{f} > 0.
\end{aligned} \tag{11.82}
$$

The proof is complete. □

Chapter 12
Generalisations and Examples of Limit Theorems for Hitting Times

In this chapter, we discuss some natural generalisations of the asymptotic results presented in Chaps. 8–11. We also describe the features of the asymptotic recurrent algorithms of phase space reduction for perturbed semi-Markov processes of the birth–death type. Finally, we give numerical examples illustrating the above algorithms and their applications to the asymptotic analysis of distributions and expectations of hitting times for singularly perturbed semi-Markov processes. This chapter includes three sections.

In Sect. 12.1, we discuss some generalisations of the perturbation conditions and asymptotic results for hitting times, in particular, the possibilities of obtaining analogous asymptotic results for vector- and real-valued hitting type functionals and obtaining some more general asymptotic results for perturbed semi-Markov processes with distributions of transition times belonging to domains of attraction for infinitely divisible laws.

In Sect. 12.2, the specific features of the asymptotic recurrent algorithms of phase space reductions for perturbed birth–death-type semi-Markov processes are commented. In this case, these algorithms preserve the structure of birth–death of reduced semi-Markov processes. This greatly simplifies the asymptotic analysis of hitting time.

In Sect. 12.3, we present some numerical examples illustrating particularities of the asymptotic recurrent algorithms of phase space reduction for distributions and expectations of hitting times for singularly perturbed semi-Markov processes.

12.1 Generalisations of Limit Theorems for Hitting Times

In this section, we give some additional comments and discuss some natural generalisations of the asymptotic results presented in Chaps. 8–11.

D. Silvestrov, *Perturbed Semi-Markov Type Processes I*,
https://doi.org/10.1007/978-3-030-92403-4_12

12.1.1 More General Perturbation Conditions

12.1.1.1. Regularity Conditions and Atoms at Zero for Limiting Distributions. Let us begin from the remark that the regularity condition $\mathbf{G}_{\mathbb{D}}$ **(b)** can be, in fact, omitted.

In this case, it is possible that the random variable $\nu_\varepsilon(t)$ can be improper random variables, i.e., take value ∞ with positive probabilities, and thus, the semi-Markov process $\eta_\varepsilon(t)$ is not well defined.

However, the condition **H** still guarantees the fulfilment of the relation (10.29), i.e., that the hitting times $\nu_{\varepsilon,\mathbb{D}}$ and $\tau_{\varepsilon,\mathbb{D}}$ are proper random variables. The random variable $\tau_{\varepsilon,\mathbb{D}}$ can be interpreted as the Markov reward accumulated on the trajectory of the Markov chain $\eta_{\varepsilon,n}$ until the first hitting of the Markov chain $\eta_{\varepsilon,n}$ in the domain $\mathbb{D}$.

The recurrent algorithms of phase space reduction presented in Chap. 9 and the asymptotic results for distributions and expectations of hitting times formulated in Chaps. 10 and 11 remain to be valid.

Another related generalisation is concerned the condition $\mathbf{J}_{\mathbb{D}}$ **(b)**, which requires that the corresponding limiting distributions for transition times are not concentrated at zero. This condition, together with other basic conditions, guarantees that the corresponding limiting distribution functions of hitting times ${}_{\bar{k}_n}G_{0,\mathbb{D},i}(\cdot)$ appearing in the theorems presented in Chap. 10 are also not concentrated at zero.

The recurrent algorithms of phase space reduction presented in Chap. 9 and the asymptotic results formulated in Chaps. 10 and 11 remain to be valid without the condition $\mathbf{J}_{\mathbb{D}}$ **(b)**. The only difference is that, in this case, there is no guarantee that the corresponding limiting distribution functions of hitting times ${}_{\bar{k}_n}G_{0,\mathbb{D},i}(\cdot)$ are not concentrated at zero.

An important example is connected with the functional, which can be defined for any domain $\mathbb{C} \subseteq \bar{\mathbb{D}}$,

$$\tau_{\varepsilon,\mathbb{C},\mathbb{D}} = \sum_{n=1}^{\nu_{\varepsilon,\mathbb{D}}} \kappa_{\varepsilon,n} \mathrm{I}(\eta_{\varepsilon,n-1} \in \mathbb{C}), \tag{12.1}$$

where

$$\nu_{\varepsilon,\mathbb{D}} = \min(n \geq 1 : \eta_{\varepsilon,n} \in \mathbb{D}). \tag{12.2}$$

The random variable $\tau_{\varepsilon,\mathbb{C},\mathbb{D}}$ can be interpreted as the accumulated reward in states from the domain $\mathbb{C}$ on the trajectory of the Markov chain $\eta_{\varepsilon,n}$ until the first hitting of this Markov chain in the domain $\mathbb{D}$.

In this case, one can consider the new Markov renewal process $(\eta_{\varepsilon,\mathbb{C},n}, \kappa_{\varepsilon,\mathbb{C},n}$, where $\eta_{\varepsilon,\mathbb{C},n} = \eta_{\varepsilon,n}$ and $\kappa_{\varepsilon,\mathbb{C},n} = \kappa_{\varepsilon,n}\mathrm{I}(\eta_{\varepsilon,n-1} \in \mathbb{C})$. Obviously, the basic conditions $\mathbf{H}_{\mathbb{D}}$, $\mathbf{I}_{\mathcal{H},\mathbb{D}}$, $\mathbf{L}_{\mathcal{H},\mathbb{D}}$ (and, thus, also $\mathbf{I}_{\mathbb{D}}$ and $\mathbf{L}_{\mathbb{D}}$), $\mathbf{J}_{\mathbb{D}}$, and $\mathbf{K}_{\mathbb{D}}$ remain to be satisfied for the new Markov renewal processes, with the only change that the distribution functions $F_{\varepsilon,ij}(t) = \mathrm{I}(t \geq 0), t \geq 0$ and expectations $f_{\varepsilon,ij} = 0$, for $i \in \bar{\mathbb{C}}, j \in \mathbb{X}$ and $\varepsilon \in (0, 1]$. One can choose the initial normalisation functions $u_{\varepsilon,i} \equiv 1$, for $i \in \bar{\mathbb{C}}$. In this case, the limiting distribution function $F_{0,ij}(t) = \mathrm{I}(t \geq 0), t \geq 0$ and the expectation $f_{0,ij} = 0$, for $i \in \bar{\mathbb{C}}, j \in \mathbb{X}$.

12.1.1.2 More General Perturbation Conditions. The question arises whether the use of the same normalisation function $u_{\varepsilon,i}$ in the asymptotic relations appearing in the condition **J** for the distribution functions of transition times $F_{\varepsilon,ij}(\cdot u_{\varepsilon,i})$ is a restrictive assumption.

In fact, the original model of the Markov renewal process $(\eta_{\varepsilon,n}, \kappa_{\varepsilon,n}), n = 0, 1, \ldots$, with the phase space $\mathbb{X}$ and the transition probabilities $Q_{\varepsilon,ij}(t) = p_{\varepsilon,ij}F_{\varepsilon,ij}(t)$, $t \geq 0, i, j \in \mathbb{X}$ can be replaced by its modification $(\hat{\eta}_{\varepsilon,n} = (\eta_{\varepsilon,n}, \eta_{\varepsilon,n+1}), \kappa_{\varepsilon,n}), n = 0, 1, \ldots$. The modified Markov renewal process $(\hat{\eta}_{\varepsilon,n}, \kappa_{\varepsilon,n})$ has the phase space $\hat{\mathbb{X}} \times [0, \infty)$, where $\hat{\mathbb{X}} = \{(i, j) : i \in \mathbb{X}, j \in \mathbb{Y}_{1,i}\}$, and the transition probabilities $\hat{Q}_{\varepsilon,(i,j)(k,r)}(t) = p_{\varepsilon,(i,j)(k,r)}F_{\varepsilon,(i,j)(k,r)}(t) = \mathrm{I}(k = j)p_{\varepsilon,jr}F_{\varepsilon,ij}(t)$, $t \geq 0$, $(i, j), (k, r) \in \hat{\mathbb{X}}$. In this case, the distribution function $F_{\varepsilon,(i,j)(k,r)}(t) = F_{\varepsilon,ij}(t)$ depends on the state (i, j) but does not depend on the state (k, r).

The hitting times $\nu_{\varepsilon,\mathbb{D}} = \min(n \geq 1 : \eta_{\varepsilon,n} \in \mathbb{D})$ and $\tau_{\varepsilon,\mathbb{D}}$ for the original Markov renewal process $(\eta_{\varepsilon,n}, \kappa_{\varepsilon,n})$ coincide with the hitting times, respectively, $\hat{\nu}_{\varepsilon,\hat{\mathbb{D}}} = \min(n \geq 1 : \hat{\eta}_{\varepsilon,n} \in \hat{\mathbb{D}})$ (here, $\hat{\mathbb{D}} = \{(i, j) : i \in \mathbb{D}, j \in \mathbb{Y}_{1,i}\}$) and $\hat{\tau}_{\varepsilon,\hat{\mathbb{D}}} = \sum_{n=1}^{\hat{\nu}_{\varepsilon,\hat{\mathbb{D}}}} \kappa_{\varepsilon,n}$ for the modified Markov renewal process $(\hat{\eta}_{\varepsilon,n}, \kappa_{\varepsilon,n})$.

The conditions $\mathbf{G}_{\mathbb{D}}$ **(a)**, $\mathbf{H}_{\mathbb{D}}$, and $\mathbf{I}_{\mathbb{D}}$, $\mathbf{I}_{\mathcal{H},\mathbb{D}}$, assumed to be satisfied for the original Markov renewal processes $(\eta_{\varepsilon,n}, \kappa_{\varepsilon,n})$, are also satisfied for the modified Markov renewal processes $(\hat{\eta}_{\varepsilon,n}, \kappa_{\varepsilon,n})$.

As far as the conditions $\mathbf{J}_{\mathbb{D}}$ and $\mathbf{K}_{\mathbb{D}}$ are concerned, they will take the following modified forms:

$\hat{\mathbf{J}}_{\mathbb{D}}$: **(a)** $F_{\varepsilon,ij}(\cdot\, u_{\varepsilon,ij}) \Rightarrow F_{0,ij}(\cdot)$ as $\varepsilon \to 0$, for $j \in \mathbb{Y}_{1,i}$, $i \in \mathbb{X}$, **(b)** $F_{0,ij}(\cdot)$, $j \in \mathbb{Y}_{1,i}, i \in \mathbb{X}$ are proper distribution functions such that $F_{0,ij}(0) < 1$, $j \in \mathbb{Y}_{1,i}, i \in \mathbb{X}$, **(c)** $u_{\varepsilon,ij} \in (0, \infty), \varepsilon \in (0, 1]$, for $j \in \mathbb{Y}_{1,i}$, $i \in \bar{\mathbb{D}}$.

and

$\hat{\mathbf{K}}_{\mathbb{D}}$: **(a)** $f_{\varepsilon,ij} = \int_0^\infty tF_{\varepsilon,ij}(dt) < \infty$, for $j \in \mathbb{Y}_{1,i}$, $i \in \mathbb{X}$, **(b)** $f_{\varepsilon,ij}/u_{\varepsilon,ij} \to f_{0,ij} = \int_0^\infty tF_{0,ij}(dt) < \infty$ as $\varepsilon \to 0$, for $j \in \mathbb{Y}_{1,i}, i \in \bar{\mathbb{D}}$.

Also, the conditions **L** and $\mathbf{L}_{\mathcal{H}}$ take the following modified forms:

$\hat{\mathbf{L}}_{\mathbb{D}}$: $u_{\varepsilon,ij} \to u_{0,ij} \in (0, \infty]$ as $\varepsilon \to 0$, for $j \in \mathbb{Y}_{1,i}, i \in \bar{\mathbb{D}}$

and

$\hat{\mathbf{L}}_{\mathcal{H},\mathbb{D}}$: The functions $u_{\cdot,ij}$, $j \in \mathbb{Y}_{1,i}, i \in \bar{\mathbb{D}}$ belong to the complete family of asymptotically comparable functions $\mathcal{H}$ appearing in the condition $\mathbf{I}_{\mathcal{H},\mathbb{D}}$.

The asymptotic recurrent algorithms of phase space reduction can be applied to the modified Markov renewal process $(\hat{\eta}_{\varepsilon,n}, \kappa_{\varepsilon,n})$. The corresponding asymptotic results will generalise asymptotic results presented in Chaps. 10 and 11 to the case, where conditions $\mathbf{J}_{\mathbb{D}}$, $\mathbf{K}_{\mathbb{D}}$, and $\hat{\mathbf{L}}_{\mathbb{D}}$, $\hat{\mathbf{L}}_{\mathcal{H},\mathbb{D}}$ take the above more general forms. Moreover, in this case, the corresponding asymptotic relations can be obtained for the distributions $\hat{G}_{\varepsilon,\mathbb{D},i(j,r)}(t) = \mathsf{P}_i\{\hat{\tau}_{\varepsilon,\hat{\mathbb{D}}} \leq t, \hat{\eta}_{\hat{\nu}_{\varepsilon,\hat{\mathbb{D}}}} = (j, r)\}$, which are more general than the distributions $G_{\varepsilon,\mathbb{D},ij}(t)$.

12.1.1.3 Perturbation Conditions Working for an Arbitrary Domain $\mathbb{D}$. It is easy to see that the conditions **G**, **H**, $\mathbf{I}_{\mathcal{H}}$, $\mathbf{L}_{\mathcal{H}}$ (and, thus, also **I** and **L**), **J**, and **K**

entail the fulfilment of the conditions $\mathbf{G}_{\mathbb{D}}$, $\mathbf{H}_{\mathbb{D}}$, $\mathbf{I}_{\mathcal{H},\mathbb{D}}$, $\mathbf{L}_{\mathcal{H},\mathbb{D}}$ (and, thus, also $\mathbf{I}_{\mathbb{D}}$ and $\mathbf{L}_{\mathbb{D}}$), $\mathbf{J}_{\mathbb{D}}$ and $\mathbf{K}_{\mathbb{D}}$ as well as the conditions $\mathbf{G}_{\mathbb{D}}$, $\mathbf{H}_{\mathbb{D}}$, $\mathbf{I}_{\mathcal{H},\mathbb{D}}$, $\mathbf{L}_{\mathcal{H},\mathbb{D}}$ (and, thus, also $\mathbf{I}_{\mathbb{D}}$ and $\mathbf{L}_{\mathbb{D}}$), $\mathbf{J}_{\mathbb{D}}$, and $\mathbf{K}_{\mathbb{D}}$ for any non-empty domain $\mathbb{D} \subset \mathbb{X}$.

12.1.2 More General Hitting Reward Functionals

12.1.2.1 Vector Hitting Rewards. In this case, we consider Markov renewal processes $(\eta_{\varepsilon,n}, \bar{\kappa}_{\varepsilon,n}) = (\eta_{\varepsilon,n}, (\kappa_{\varepsilon,1,n}, \ldots, \kappa_{\varepsilon,L,n})), n = 0, 1, \ldots$, with a phase space $\mathbb{X} \times \mathbb{R}_L^+$, where $\mathbb{X} = \{1, \ldots, M\}$ is a finite set and $\mathbb{R}_L^+ = [0, \infty) \times \cdots \times [0, \infty)$ is the L-product of the interval $[0, \infty)$.

The corresponding vector hitting reward functional is defined in the following way, for $\mathbb{D} \subseteq \mathbb{X}$,

$$\bar{\tau}_{\varepsilon,\mathbb{D}} = \sum_{n=1}^{\nu_{\varepsilon,\mathbb{D}}} \bar{\kappa}_{\varepsilon,n} = (\tau_{\varepsilon,1,\mathbb{D}}, \ldots, \tau_{\varepsilon,l,\mathbb{D}}), \tag{12.3}$$

where

$$\tau_{\varepsilon,l,\mathbb{D}} = \sum_{n=1}^{\nu_{\varepsilon,\mathbb{D}}} \kappa_{\varepsilon,l,n}, \ l = 1, \ldots, L. \tag{12.4}$$

In this case, one can reduce the problem to scalar case by using the method known as the Wold–Cramér device (see, for example, Silvestrov (2004) or Kallenberg (2021)).

Let us introduce the following scalar hitting reward functional, for $\bar{z} = (z_1, \ldots, z_l) \in \mathbb{R}_L^+$,

$$\tau_{\varepsilon,\bar{z},\mathbb{D}} = \sum_{n=1}^{\nu_{\varepsilon,\mathbb{D}}} \kappa_{\varepsilon,\bar{z},n}, \tag{12.5}$$

where

$$\kappa_{\varepsilon,\bar{z},n} = \sum_{l=1}^{L} z_l \kappa_{\varepsilon,l,n}, \ n \geq 1. \tag{12.6}$$

The L-dimensional Laplace transforms $\phi_{\varepsilon,ij}(\bar{z}) = \mathsf{E}\{e^{-\sum_{l=1}^{L} z_l \kappa_{\varepsilon,l,1}} / \eta_{\varepsilon,0} = i, \eta_{\varepsilon,1} = j\}$ and the scalar Laplace transforms $\phi_{\varepsilon,\bar{z},ij}(s) = \mathsf{E}\{e^{-s\kappa_{\varepsilon,\bar{z},1}} / \eta_{\varepsilon,0} = i, \eta_{\varepsilon,1} = j)$ are connected by the relation, $\phi_{\varepsilon,\bar{z},ij}(s) = \phi_{\varepsilon,ij}(s\bar{z})$, for $s \geq 0, \bar{z} \in \mathbb{R}_L^+$.

This property of Laplace transforms for transition times is invariant with respect to procedures of removing virtual transitions and one-step reduction of phase space.

This makes it possible to prove that the Laplace transforms of hitting times also have this property. This means that the L-dimensional Laplace transforms $\Psi_{\varepsilon,\mathbb{D},ij}(\bar{z}) = \mathsf{E}_i e^{-\sum_{l=1}^{L} z_l \tau_{\varepsilon,l,\mathbb{D}}} \mathrm{I}(\eta_{\varepsilon,\nu_{\varepsilon,\mathbb{D}}} = j)$ and the scalar Laplace transforms $\Psi_{\varepsilon,\bar{z},\mathbb{D},ij}(s) = \mathsf{E}_i e^{-s\tau_{\varepsilon,\bar{z},\mathbb{D}}} \mathrm{I}(\eta_{\varepsilon,\nu_{\varepsilon,\mathbb{D}}} = j)$ are also connected by the relation $\Psi_{\varepsilon,\bar{z},\mathbb{D},ij}(s) = \Psi_{\varepsilon,\mathbb{D},ij}(s\bar{z})$, for $s \geq 0, \bar{z} \in \mathbb{R}_L^+$.

The condition $\mathbf{J}^\circ$ **(a)** takes the form of convergence relations for the L-dimensional Laplace transforms, $\phi_{\varepsilon,ij}(\bar{z}/u_{\varepsilon,i})$, assumed to hold for $\bar{z} \in \mathbb{R}_L^+$. According to the Wold–Cramér device, this condition can be reformulated in the equivalent form of the corresponding convergence relations for the scalar Laplace transforms, $\phi_{\varepsilon,\bar{z},ij}(s/u_{\varepsilon,i})$, assumed to hold for $s \geq 0, \bar{z} \in \mathbb{R}_L^+$.

Similar remarks can be made with regard to expectations of transition and hitting times.

This makes it possible to apply results presented in Chaps. 8–10 and to get the corresponding convergence relations for the Laplace transforms $\Psi_{\varepsilon,\bar{z},\mathbb{D},ij}(s/\check{u}_{\varepsilon,i}), s \geq 0$, for $\bar{z} \in \mathbb{R}_L^+$, and then, using the Wold–Cramér device, rewrite them in the equivalent form of the corresponding convergence relations for the L-dimensional Laplace transforms $\Psi_{\varepsilon,\mathbb{D},ij}(\bar{z}/\check{u}_{\varepsilon,i}), \bar{z} \in \mathbb{R}_L^+$ and the corresponding weak convergence relations for the L-dimensional distributions of hitting times $G_{\varepsilon,\mathbb{D},ij}(\bar{t}\,\check{u}_{\varepsilon,i}), \bar{t} = (t_1, \ldots, t_L) \in \mathbb{R}_L^+$.

12.1.2.2 Real-Valued Hitting Rewards. Another natural generalisation is concerned the model with real-valued rewards. In this case, we consider Markov renewal processes $(\eta_{\varepsilon,n}, \kappa_{\varepsilon,n}), n = 0, 1, \ldots$ with a phase space $\mathbb{X} \times \mathbb{R}_1$, where $\mathbb{X} = \{1, \ldots, m\}$ is a finite set, and $\mathbb{R}_1 = (-\infty, \infty)$ is a real line.

The corresponding real-valued hitting reward functional can be defined in the standard way and represented as the difference of two non-negative hitting reward functionals, for $\mathbb{D} \subseteq \mathbb{X}$,

$$\tau_{\varepsilon,\mathbb{D}} = \sum_{n=1}^{\nu_{\varepsilon,\mathbb{D}}} \kappa_{\varepsilon,n} = \tau_{\varepsilon,\mathbb{D}}^+ - \tau_{\varepsilon,\mathbb{D}}^-, \tag{12.7}$$

where

$$\tau_{\varepsilon,\mathbb{D}}^{\pm} = \sum_{n=1}^{\nu_{\varepsilon,\mathbb{D}}} \kappa_{\varepsilon,n}^{\pm} \text{ and } \kappa_{\varepsilon,n}^{\pm} = \pm\kappa_{\varepsilon,n}\mathrm{I}(\pm\kappa_{\varepsilon,n} \geq 0),\ n \geq 1. \tag{12.8}$$

We can consider the vector hitting reward functional $\bar{\tau}_{\varepsilon,\mathbb{D}} = (\tau_{\varepsilon,\mathbb{D}}^+, \tau_{\varepsilon,\mathbb{D}}^-)$ and apply to this vector hitting reward functional the Wold–Cramér device, as described in Sect. 12.1.2.1, for obtaining weak convergence relations for the distributions $G_{\varepsilon,\mathbb{D},ij}(\bar{t}\,\check{u}_{\varepsilon,i}) = \mathsf{P}_i\{\check{u}_{\varepsilon,i}^{-1}\tau_{\varepsilon,\mathbb{D}}^+ \leq t_+, \check{u}_{\varepsilon,i}^{-1}\tau_{\varepsilon,\mathbb{D}}^- \leq t_-, \eta_{\varepsilon,\nu_{\varepsilon,\mathbb{D}}} = j\}, \bar{t} = (t_+, t_-) \in \mathbb{R}_2^+$. Such relations imply, in an obvious way, the corresponding weak convergence relations for the real-valued hitting reward functionals $\tau_{\varepsilon,\mathbb{D}} = \tau_{\varepsilon,\mathbb{D}}^+ - \tau_{\varepsilon,\mathbb{D}}^-$.

In order to escape some possible side effects, which can be caused by discontinuity of distribution functions of transition times in point 0, one can also use the more general splitting procedure for hitting times, which is based on random variables $\kappa_{\varepsilon,c,n}^{\pm} = \pm(\kappa_{\varepsilon,n} - c)\mathrm{I}(\pm\kappa_{\varepsilon,n} \geq \pm c), n \geq 1$, for some $c \geq 0$.

12.1.2.3 General Reward Functionals. Generalisations concerned vector hitting reward functionals (presented in Sect. 12.1.2.1) make it possible to obtain weak asymptotic relations for hitting frequencies. In this case, we consider the Markov renewal process $(\eta_{\varepsilon,n}, \bar{\kappa}_{\varepsilon,n}) = (\eta_{\varepsilon,n}, (\mathrm{I}(\eta_{\varepsilon,n-1} = k, \eta_{\varepsilon,n} = r), k, r \in \mathbb{X}), n = 0, 1, \ldots$ with the phase space $\mathbb{X} \times \mathbb{R}_L^+$, where $L = M^2$.

The corresponding vector hitting reward functional is defined in the following way,

$$\bar{\tau}_{\varepsilon,\mathbb{D}} = \sum_{n=1}^{\nu_{\varepsilon,\mathbb{D}}} \bar{\kappa}_{\varepsilon,n} = (\tau_{\varepsilon,kr,\mathbb{D}}, k, r \in \mathbb{X}), \tag{12.9}$$

where

$$\tau_{\varepsilon,kr,\mathbb{D}} = \sum_{n=1}^{\nu_{\varepsilon,\mathbb{D}}} \mathrm{I}(\eta_{\varepsilon,n-1} = k, \eta_{\varepsilon,n} = r). \tag{12.10}$$

In this case, the random variable $\tau_{\varepsilon,kr,\mathbb{D}}$ is the number of transitions of the form $k \to r$ made by the Markov chain $\eta_{\varepsilon,n}$ before hitting to domain $\mathbb{D}$.

Let us now consider the Markov renewal process $(\eta_{\varepsilon,n}, \gamma_{\varepsilon,n}), n = 0, 1, \ldots$, with a phase space $\mathbb{X} \times \mathbb{R}_1$ and define the hitting reward functional,

$$\tau_{\varepsilon,\mathbb{D}} = \sum_{n=1}^{\nu_{\varepsilon,\mathbb{D}}} \gamma_{\varepsilon,n}. \tag{12.11}$$

The following relation, based on conditional independence of random variables $\gamma_{\varepsilon,n}, n = 0, 1, \ldots$, with respect to the Markov chain $\eta_{\varepsilon,n}, n = 0, 1, \ldots$, represents the above functional in the form of a random sum,

$$\tau_{\varepsilon,\mathbb{D}} \stackrel{\mathsf{d}}{=} \sum_{k,r \in \mathbb{X}} \sum_{n=1}^{\tau_{\varepsilon,kr,\mathbb{D}}} \gamma_{\varepsilon,kr,n}, \tag{12.12}$$

where: (a) the random vector $\bar{\tau}_{\varepsilon,\mathbb{D}} = (\tau_{\varepsilon,kr,\mathbb{D}}, k, r \in \mathbb{X})$ and the random variables $\gamma_{\varepsilon,kr,n}, n \geq 1, k, r \in \mathbb{X}$ are mutually independent, (b) the random variable $\gamma_{\varepsilon,kr,n}$ has the distribution function $F_{\varepsilon,kr}(\cdot) = \mathsf{P}\{\gamma_{\varepsilon,kr,n} \leq t\} = \mathsf{P}\{\gamma_{\varepsilon,1} \leq \cdot / \eta_{\varepsilon,0} = k, \eta_{\varepsilon,1} = r\}$, for $n \geq 1$ and $k, r \in \mathbb{X}$.

The representation (12.12) let us formulate some natural conditions of weak convergence for the hitting reward functionals $\tau_{\varepsilon,\mathbb{D}}$ using the weak convergence theorems for randomly stopped stochastic processes given in Appendix A.

Let, for example: (c) $\eta_{\varepsilon,0} = i \in \mathbb{X}$, and the following asymptotic relation takes place $\bar{\tau}_{\varepsilon,\mathbb{D}}/\check{u}_{\varepsilon,i} \stackrel{\mathsf{d}}{\longrightarrow} \bar{\tau}_{0,i,\mathbb{D}}$ as $\varepsilon \to 0$, where $\bar{\tau}_{0,i,\mathbb{D}} = (\tau_{0,i,kr,\mathbb{D}}, k, r \in \mathbb{X})$ is a random vector with non-negative components, (d) $\check{u}_{\varepsilon,i} > 0$ is some normalisation function such that $\check{u}_{\varepsilon,i} \to \infty$ as $\varepsilon \to 0$, and (e) $\sum_{1 \leq n \leq t\check{u}_{\varepsilon,i}} \gamma_{\varepsilon,kr,n}/\hat{u}_{\varepsilon,i}, t \geq 0 \stackrel{\mathbf{d}}{\longrightarrow} \gamma_{0,kr}(t), t \geq 0$ as $\varepsilon \to 0$, for every $k, r \in \mathbb{X}$, where $\gamma_{0,i,kr}(t), t \geq 0$ is a Lévy process, (f) $\hat{u}_{\varepsilon,i} > 0$ is some normalisation function such that $\hat{u}_{\varepsilon,i} \to \infty$ as $\varepsilon \to 0$.

Then, relation (12.12) and Theorem A.4 imply that the following weak convergence relation takes place:

$$\tau_{\varepsilon,\mathbb{D}}/\hat{u}_{\varepsilon,i} \stackrel{\mathsf{d}}{\longrightarrow} \tau_{0,\mathbb{D}} = \sum_{k,r \in \mathbb{X}} \gamma_{0,i,kr}(\tau_{0,i,kr,\mathbb{D}}) \text{ as } \varepsilon \to 0, \tag{12.13}$$

where: (g) the random vector $\bar{\tau}_{0,i,\mathbb{D}}$ and the Lévy processes $\gamma_{0,i,kr}(\cdot), k, r \in \mathbb{X}$ are mutually independent.

The problem that should be commented on is caused by the fact that the normalisation functions $\check{u}_{\varepsilon,i}$, appearing in the above asymptotic relations (e), may differ for different $i \in \mathbb{X}$. This imposes some additional restrictions on distribution functions $F_{\varepsilon,kr}(\cdot)$, $k, r \in \mathbb{X}$.

The conditions **J** and **K** give a hint about one version of the model, in which the asymptotic relations of type (e) hold for any normalisation functions $0 < \check{u}_{\varepsilon,i} \to \infty$ and $\hat{u}_{\varepsilon,i} = \check{u}_{\varepsilon,i}$, with the corresponding limiting processes $\gamma_{0,i,kr}(t) = f_{0,kr}\, t, t \geq 0$, which are nonrandom linear functions. This is the case, where it is assumed that the distribution functions $\check{F}_{\varepsilon,kr}(\cdot) = F_{\varepsilon,kr}(\cdot \check{u}_{\varepsilon,i})$ converge weakly to some limiting distribution function $F_{0,kr}(\cdot)$ and the first moments $\check{f}_{\varepsilon,kr}$ of the distribution functions $\check{F}_{\varepsilon,kr}(\cdot)$ converge to the first moment $f_{0,kr} < \infty$ of the distribution function $F_{0,kr}(\cdot)$, as $\varepsilon \to 0$, for $k, r \in \mathbb{X}$.

Another example is where the random variables $\gamma_{\varepsilon,kr,n} = \gamma_{0,kr,n}$ do not depend on the perturbation parameter ε and their distribution functions $F_{\varepsilon,kr}(\cdot) = F_{0,kr}(\cdot)$ (which, in this case, also do not depend on ε) belong to the domain of attraction of some stable law. This means that the asymptotic relations of type (e) hold for any normalisation functions $0 < \check{u}_{\varepsilon,i} \to \infty$ and $\hat{u}_{\varepsilon,i} = \check{u}^{\alpha}_{\varepsilon,i} h(\check{u}_{\varepsilon,i})$ (here, $\alpha \in (0, 1]$ and $h(\cdot)$ is some slowly varying function), with some limiting stable Lévy processes $\gamma_{0,kr}(\cdot)$, $k, r \in \mathbb{X}$ (which, in this case, do not depend on i).

12.2 Hitting Times for Perturbed Birth–Death-Type Semi-Markov Processes

In this section, we describe procedures of removing virtual transition and phase space reduction for birth–death-type semi-Markov processes.

12.2.1 Perturbed Birth–Death-Type Semi-Markov Processes

12.2.1.1 Perturbation Conditions for Birth–Death-Type Semi-Markov Processes. Let $\mathbb{X} = \{0, \ldots, m\}$ be a finite set. Let also, $(\eta_{\varepsilon,n}, \kappa_{\varepsilon,n}), n = 0, 1, \ldots$ be, for every $\varepsilon \in (0, 1]$ a birth–death-type Markov renewal process, i.e., a homogeneous Markov chain, with a phase space $\mathbb{X} \times [0, \infty)$ and transition probabilities, for $(i, s), (j, t) \in \mathbb{X} \times [0, \infty)$,

$$Q_{\varepsilon,ij}(t) = \begin{cases} F_{\varepsilon,00}(t)p_{\varepsilon,00} & \text{for } i = 0, j = 0, \\ F_{\varepsilon,01}(t)p_{\varepsilon,01} & \text{for } i = 0, j = 1, \\ F_{\varepsilon,ii+1}(t)p_{\varepsilon,ii+1} & \text{for } 0 < i < m, j = i + 1, \\ F_{\varepsilon,ii}(t)p_{\varepsilon,ii} & \text{for } 0 < i < m, j = i, \\ F_{\varepsilon,ii-1}(t)p_{\varepsilon,ii-1} & \text{for } 0 < i < m, j = i - 1, \\ F_{\varepsilon,mm}(t)p_{\varepsilon,mm} & \text{for } i = m, j = m, \\ F_{\varepsilon,mm-1}(t)p_{\varepsilon,mm-1} & \text{for } i = m, j = m - 1, \\ 0 & \text{otherwise}, \end{cases} \tag{12.14}$$

where: (a) $F_{\varepsilon,ij}(\cdot)$ are distribution functions concentrated on $[0,\infty)$, for $0 \le i,j \le m$, $|j-i| \le 1$, (b) $p_{\varepsilon,ij} \ge 0$, for $0 \le i,j \le m$, $|j-i| \le 1$ and $p_{\varepsilon,00} + p_{\varepsilon,01} = 1$, $p_{\varepsilon,ii-1} + p_{\varepsilon,ii} + p_{\varepsilon,ii+1} = 1$, for $0 < i < m$, $p_{\varepsilon,mm-1} + p_{\varepsilon,mm} = 1$.

The first component of the above Markov renewal process, $\eta_{\varepsilon,n}, n = 0,1,\ldots$, is itself a homogeneous, so-called embedded, birth–death-type Markov chain, with the phase space $\mathbb{X}$ and the transition probabilities, for $i,j \in \mathbb{X}$,

$$Q_{\varepsilon,ij}(\infty) = \begin{cases} p_{\varepsilon,ij} & \text{for } 0 \le i,j \le m, |j-i| \le 1, \\ 0 & \text{otherwise.} \end{cases} \tag{12.15}$$

The above Markov renewal process is used to define the birth–death-type semi-Markov process,

$$\eta_\varepsilon(t) = \eta_{\varepsilon,\nu_\varepsilon(t)}, t \ge 0, \tag{12.16}$$

where $\zeta_{\varepsilon,n} = \kappa_{\varepsilon,1} + \cdots + \kappa_{\varepsilon,n}, n = 1,2,\ldots, \zeta_{\varepsilon,0} = 0$, are the corresponding moments of jumps, and $\nu_\varepsilon(t) = \max(n \ge 1 : \zeta_{\varepsilon,n} \le t)$ is the number of jumps in the interval $[0,t]$ for the above semi-Markov process.

The conditions **G**–**L** and $\mathbf{I}_{\mathcal{H}}$, $\mathbf{L}_{\mathcal{H}}$ take, in this case, forms presented below.

Let us recall the notation introduced in Sect. 9.1.

The distribution functions of sojourn times take the following form, for $i \in \mathbb{X}$ and $\varepsilon \in (0,1]$, and $t \ge 0$,

$$\begin{aligned} F_{\varepsilon,i}(t) &= \mathsf{P}_i\{\kappa_{\varepsilon,1} \le t\} \\ &= \begin{cases} F_{\varepsilon,00}(t)p_{\varepsilon,00} + F_{\varepsilon,01}(t)p_{\varepsilon,01} & \text{if } i = 0, \\ F_{\varepsilon,ii-1}(t)p_{\varepsilon,ii-1} + F_{\varepsilon,ii}(t)p_{\varepsilon,ii} + F_{\varepsilon,ii+1}(t)p_{\varepsilon,ii+1} & \text{if } 0 < i < m, \\ F_{\varepsilon,mm}(t)p_{\varepsilon,mm} + F_{\varepsilon,mn-1}(t)p_{\varepsilon,mm-1} & \text{if } i = m \end{cases} \end{aligned} \tag{12.17}$$

The conditions **G** and **H** take the following forms:

$\check{\mathbf{G}}$: **(a)** $p_{\varepsilon,ij} > 0$, $\varepsilon \in (0,1]$ or $p_{\varepsilon,ij} = 0$, $\varepsilon \in (0,1]$, for $0 \le i,j \le m, |j-i| \le 1$, **(b)** $F_{\varepsilon,i}(0) < 1$, $0 \le i \le m$, $\varepsilon \in (0,1]$

and

$\check{\mathbf{H}}$: $p_{1,ii+1} > 0$, for $i = 0,\ldots,m-1$ and $p_{1,ii-1} > 0$, for $i = 1,\ldots,m$.

Condition $\check{\mathbf{H}}$ implies that sets $\mathbb{Y}_{1,i} = \{i \in \mathbb{X} : p_{1,ij} > 0\}, i \in \mathbb{X}$ take the following forms:

$$\mathbb{Y}_{1,i} = \begin{cases} \{1\} & \text{for } i = 0, \text{ if } p_{1,00} = 0, \\ \{0,1\} & \text{for } i = 0, \text{ if } p_{1,00} > 0, \\ \{i-1,i+1\} & \text{for } 0 < i < m, \text{ if } p_{1,ii} = 0, \\ \{i-1,i,i+1\} & \text{for } 0 < i < m, \text{ if } p_{1,ii} > 0, \\ \{m-1\} & \text{for } i = m, \text{ if } p_{1,mm} = 0, \\ \{m-1,m\} & \text{for } i = m, \text{ if } p_{1,mm} > 0. \end{cases} \tag{12.18}$$

The perturbation conditions **I** and $\check{\mathbf{I}}_{\mathcal{H}}$ take the following forms:

$\check{\mathbf{I}}$: $p_{\varepsilon,ij} \to p_{0,ij}$ as $\varepsilon \to 0$, for $i,j \in \mathbb{X}$.

$\check{\mathbf{I}}_{\mathcal{H}}$: The functions $p_{\cdot,ij}$, $j \in \mathbb{Y}_{1,i}, i \in \mathbb{X}$ belong to a complete family of asymptotically comparable functions $\mathcal{H}$.

The perturbation condition **J** takes the following form:

$\check{\mathbf{J}}$: **(a)** $F_{\varepsilon,ij}(\cdot\, u_{\varepsilon,i}) \Rightarrow F_{0,ij}(\cdot)$ as $\varepsilon \to 0$, for $j \in \mathbb{Y}_{1,i}, i \in \mathbb{X}$, **(b)** $F_{0,ij}(\cdot)$ is a proper distribution function such that $F_{0,ij}(0) < 1$, for $j \in \mathbb{Y}_{1,i}, i \in \mathbb{X}$, **(c)** $u_{\varepsilon,i} \in (0, \infty), \varepsilon \in (0, 1]$, for $i \in \mathbb{X}$.

Let us recall the corresponding Laplace transforms, for $i, j \in \mathbb{X}$,

$$\phi_{\varepsilon,ij}(s) = \int_0^\infty e^{-st} F_{\varepsilon,ij}(dt),\ s \geq 0. \tag{12.19}$$

The condition $\check{\mathbf{J}}$ can be reformulated in the equivalent form in terms of the above Laplace transforms:

$\check{\mathbf{J}}^\circ$: **(a)** $\phi_{\varepsilon,ij}(su_{\varepsilon,i}) \to \phi_{0,ij}(s)$ as $\varepsilon \to 0$, for $s \geq 0$, $j \in \mathbb{Y}_{1,i}, i \in \mathbb{X}$, **(b)** $\phi_{0,ij}(s)$ is the Laplace transform of a proper distribution function $F_{0,ij}(\cdot)$ such that $F_{0,ij}(0) < 1$, for $j \in \mathbb{Y}_{1,i}, i \in \mathbb{X}$, **(c)** $u_{\varepsilon,i} \in (0, \infty), \varepsilon \in (0, 1]$, for $i \in \mathbb{X}$.

Let us also recall the expectations of transition times, for $i, j \in \mathbb{X}$ and $\varepsilon \in (0, 1]$,

$$f_{\varepsilon,ij} = \int_0^\infty t F_{\varepsilon,ij}(dt). \tag{12.20}$$

The perturbation condition **K** takes the following form:

$\check{\mathbf{K}}$: **(a)** $f_{\varepsilon,ij} < \infty$, $j \in \mathbb{Y}_{1,i}$, $i \in \mathbb{X}$, for every $\varepsilon \in (0, 1]$, **(b)** $f_{\varepsilon,ij}/u_{\varepsilon,i} \to f_{0,ij} = \int_0^\infty t F_{0,ij}(dt) < \infty$ as $\varepsilon \to 0$, for $j \in \mathbb{Y}_{1,i}, i \in \mathbb{X}$.

The conditions $\mathbf{I}_{\mathcal{H}}$, **L**, and $\mathbf{L}_{\mathcal{H}}$ take the following forms:

$\check{\mathbf{I}}_{\mathcal{H}}$: The functions $p_{\cdot,ij}$, $j \in \mathbb{Y}_{1,i}, i \in \mathbb{X}$ belong to a complete family of asymptotically comparable functions $\mathcal{H}$.

$\check{\mathbf{L}}$: $u_{\varepsilon,i} \to u_{0,i} \in (0, \infty]$ as $\varepsilon \to 0$, for $i \in \mathbb{X}$.

$\check{\mathbf{L}}_{\mathcal{H}}$: The functions $u_{\cdot,i}$, $i \in \mathbb{X}$ belong to the complete family of asymptotically comparable functions $\mathcal{H}$.

12.2.1.2 Examples of Perturbed Birth–Death-Type Semi-Markov Processes

Let us briefly give a number of examples of transition probabilities for stochastic systems of the birth–death type. For simplicity, we restrict ourselves to models of the Markov type, the transient characteristics of which are as follows:

$$Q_{\varepsilon,ij}(t) = \begin{cases} F_{\varepsilon,0,-}(t)p_{\varepsilon,0,-} & \text{for } j = 0, i = 0, \\ F_{\varepsilon,0,+}(t)p_{\varepsilon,0,+} & \text{for } j = 1, i = 0, \\ F_{\varepsilon,i,\pm}(t)p_{\varepsilon,i,\pm} & \text{for } j = i \pm 1, 1 < i < m, \\ F_{\varepsilon,m,-}(t)p_{\varepsilon,m,-} & \text{for } j = m - 1, i = m, \\ F_{\varepsilon,m,+}(t)p_{\varepsilon,m,+} & \text{for } j = m, i = m, \end{cases} \tag{12.21}$$

where, for $i = 1, \ldots, m$,

$$p_{\varepsilon,i,\pm} = \frac{\lambda_{\varepsilon,i,\pm}}{\lambda_{\varepsilon,i}}, \ F_{\varepsilon,i,\pm}(t) = 1 - e^{\lambda_{\varepsilon,i}t}, t \geq 0, \tag{12.22}$$

and

$$\lambda_{\varepsilon,i,\pm} \in (0, \infty), \ \lambda_{\varepsilon,i} = \lambda_{\varepsilon,i,+} + \lambda_{\varepsilon,i,-}. \tag{12.23}$$

First, let us mention some examples of perturbed bio-stochastic systems.

In perturbed population dynamics models, the following options for parameters are typical,

$$\lambda_{\varepsilon,i,\circ} = \begin{cases} \lambda_\varepsilon i + \nu_\varepsilon & \text{if } \circ = +, \\ \mu_\varepsilon i & \text{if } \circ = -, \end{cases} \tag{12.24}$$

where λ_ε and μ_ε are, respectively, birth and death intensities and ν_ε is an immigration rate.

In perturbed epidemic models, one of the possible options for parameters is as follows and has the following form:

$$\lambda_{\varepsilon,i,\circ} = \begin{cases} \lambda_\varepsilon i(N-i) + \nu_\varepsilon(N-i) & \text{if } \circ = +, \\ \mu_\varepsilon i & \text{if } \circ = -, \end{cases} \tag{12.25}$$

where λ_ε and μ_ε are, respectively, contamination and recovering intensities, and ν_ε is an external contamination intensity.

In perturbed population genetics models, parameters can, for example, take the following form:

$$\lambda_{\varepsilon,i,\circ} = \begin{cases} \lambda_\varepsilon i + \nu_\varepsilon(N-i) & \text{if } \circ = +, \\ \nu_\varepsilon i + \mu_\varepsilon(N-i) & \text{if } \circ = -, \end{cases} \tag{12.26}$$

where λ_ε and μ_ε are birth intensities of individual with genes of two different types, and ν_ε is a mutation intensity.

Let us also list some examples of perturbed queuing systems.

In the standard M/M(m/n) queuing systems, with the number of servers m and the number of repairing units $n \leq m$, the parameters take the following form:

$$\lambda_{\varepsilon,i,\circ} = \begin{cases} \mu_\varepsilon \min(i,n) & \text{if } \circ = +, \\ \lambda_\varepsilon i & \text{if } \circ = -, \end{cases} \tag{12.27}$$

where μ_ε and λ_ε are, respectively, the recovery and failure rates.

12.2.1.3 Hitting Times for Birth–Death-Type Semi-Markov Processes. The study of the hitting times $\tau_{\varepsilon,\mathbb{D}}$ for birth–death-type semi-Markov processes can be limited in either the cases, where the domain $\mathbb{D} = \{0\}$, $\mathbb{D} = \{m\}$, or $\mathbb{D} = \{0, m\}$. All other cases can easily be reduced to these three cases. Moreover, the second case is similar to the first and, therefore, does not require separate consideration.

12.2.2 Phase Space Reduction for Perturbed Birth–Death-Type Semi-Markov Processes

12.2.2.1 Removing of Virtual Transitions for Original Perturbed Birth–Death-Type Semi-Markov Processes. At this step, we transform the original semi-Markov process $\eta_\varepsilon(t)$ into a new semi-Markov process $\tilde{\eta}_\varepsilon(t)$ with virtual transitions removed, using the procedure described in Sect. 9.2.

In fact, the description of the corresponding algorithm and proofs are given in Sect. 9.2. We should only give the corresponding formulas for characteristics appearing in the conditions $\tilde{\mathbf{G}}$–$\tilde{\mathbf{L}}$ and $\tilde{\mathbf{I}}_{\mathcal{H}}$, $\tilde{\mathbf{L}}_{\mathcal{H}}$ for the semi-Markov process $\tilde{\eta}_\varepsilon(t)$, and the corresponding characteristics for the limiting semi-Markov process $\tilde{\eta}_0(t)$.

The transition probabilities $\tilde{p}_{\varepsilon,ij}, i, j \in \mathbb{X}$ for the embedded Markov chain $\tilde{\eta}_{\varepsilon,n}$ are given by the following relation, for $\varepsilon \in (0, 1]$,

$$\tilde{p}_{\varepsilon,ij} = \begin{cases} 1 & \text{for } j = 1, i = 0, \\ \frac{p_{\varepsilon,ii\pm1}}{\bar{p}_{\varepsilon,ii}} & \text{for } j = i \pm 1, 0 < i < m, \\ 1 & \text{for } j = m-1, i = m, \\ 0 & \text{otherwise.} \end{cases} \tag{12.28}$$

In this case, the sets $\tilde{\mathbb{Y}}_{1,i} = \{i \in \mathbb{X} : \tilde{p}_{1,ij} > 0\}, i \in \mathbb{X}$, take the following forms:

$$\tilde{\mathbb{Y}}_{1,i} = \begin{cases} \{1\} & \text{for } i = 0, \\ \{i-1, i+1\} & \text{for } 1 < i < m, \\ \{m-1\} & \text{for } i = m. \end{cases} \tag{12.29}$$

The Laplace transforms $\tilde{\phi}_{\varepsilon,ij}(s), j \in \tilde{\mathbb{Y}}_{1,i}, i \in \mathbb{X}$ for the distribution functions $\tilde{F}_{\varepsilon,ij}(t)$ take the following forms, for $\varepsilon \in (0, 1]$ and $s \ge 0$,

$$\tilde{\phi}_{\varepsilon,ij}(s) = \begin{cases} \frac{\phi_{\varepsilon,01}(s)\bar{p}_{\varepsilon,00}}{1-\phi_{\varepsilon,00}(s)p_{\varepsilon,00}} & \text{for } j = 1, i = 0, \\ \frac{\phi_{\varepsilon,ii\pm1}(s)\bar{p}_{\varepsilon,ii}}{1-\phi_{\varepsilon,ii}(s)p_{\varepsilon,ii}} & \text{for } j = i \pm 1, 1 < i < m, \\ \frac{\phi_{\varepsilon,mm-1}(s)\bar{p}_{\varepsilon,mm}}{1-\phi_{\varepsilon,mm}(s)p_{\varepsilon,mm}} & \text{for } j = m-1, i = m, \end{cases} \tag{12.30}$$

where $\bar{p}_{\varepsilon,ii} = 1 - p_{\varepsilon,ii}$, $i \in \mathbb{X}$.

The expectations $\tilde{f}_{\varepsilon,ij}, j \in \tilde{\mathbb{Y}}_{1,i}, i \in \mathbb{X}$ take the following forms, for $\varepsilon \in (0, 1]$,

$$\tilde{f}_{\varepsilon,ij} = \begin{cases} f_{\varepsilon,01} + \frac{p_{\varepsilon,00}f_{\varepsilon,00}}{\bar{p}_{\varepsilon,00}} & \text{for } j = 1, i = 0, \\ f_{\varepsilon,ii\pm1} + \frac{p_{\varepsilon,ii}f_{\varepsilon,ii}}{\bar{p}_{\varepsilon,ii}} & \text{for } j = i \pm 1, 0 < i < m, \\ f_{\varepsilon,mm-1} + \frac{p_{\varepsilon,mm}f_{\varepsilon,mm}}{\bar{p}_{\varepsilon,mm}} & \text{for } j = m-1, i = m. \end{cases} \tag{12.31}$$

The normalisation functions $\tilde{u}_{\varepsilon,i}$, $\varepsilon \in (0, 1]$ are given by the following relation:

$$\tilde{u}_{\varepsilon,i} = \bar{p}_{\varepsilon,ii}^{-1} u_{\varepsilon,i} \text{ for } i \in \mathbb{X}. \tag{12.32}$$

The transition probabilities $\tilde{p}_{0,ij}$ for the limiting embedded Markov chain $\tilde{\eta}_{0,n}$ are given by the following relation, for $i, j \in \mathbb{X}$,

$$\tilde{p}_{\varepsilon,ij} = \begin{cases} 1 & \text{for } j = 1, i = 0, \\ \tilde{p}_{0,ii\pm 1} & \text{for } j = i \pm 1, 0 < i < m, \\ 1 & \text{for } j = m - 1, i = m, \\ 0 & \text{otherwise.} \end{cases} \tag{12.33}$$

The limiting Laplace transforms $\tilde{\phi}_{0,ij}(s)$ for the distribution functions $\tilde{F}_{0,ij}(t)$ are given by the following relation, for $j \in \tilde{\mathbb{Y}}_{1,i}, i \in \mathbb{X}$ and $s \geq 0$,

$$\tilde{\phi}_{0,ij}(s) = \begin{cases} \frac{\phi_{\varepsilon,01}(s)\bar{p}_{\varepsilon,00}}{1-\phi_{\varepsilon,00}(s)p_{\varepsilon,00}} & \text{for } j = 1, i = 0, \text{ if } \bar{p}_{0,00} > 0, \\ \frac{1}{1+f_{0,00}s} & \text{for } j = 1, i = 0, \text{ if } \bar{p}_{0,00} = 0, \\ \frac{\phi_{0,ii\pm 1}(s)\bar{p}_{0,ii}}{1-\phi_{0,ii}(s)p_{0,ii}} & \text{for } j = i \pm 1, 0 < i < m, \text{ if } \bar{p}_{0,ii} > 0, \\ \frac{1}{1+f_{0,ii}s} & \text{for } j = i \pm 1, 0 < i < m, \text{ if } \bar{p}_{0,ii} = 0, \\ \frac{\phi_{0,mm-1}(s)\bar{p}_{0,mm}}{1-\phi_{0,mm}(s)p_{0,mm}} & \text{for } j = m - 1, i = m, \text{ if } \bar{p}_{0,mm} > 0, \\ \frac{1}{1+f_{0,mm}s} & \text{for } j = m - 1, i = m, \text{ if } \bar{p}_{0,mm} = 0. \end{cases} \tag{12.34}$$

The limiting expectations $\tilde{f}_{0,ij}$ are given by the following relation, for $j \in \tilde{\mathbb{Y}}_{1,i}, i \in \mathbb{X}$,

$$\tilde{f}_{0,ij} = \begin{cases} \bar{p}_{0,00}f_{0,01} + p_{0,00}f_{0,00} & \text{for } j = 1, i = 0, \\ \bar{p}_{0,ii}f_{0,ii\pm 1} + p_{0,ii}f_{0,ii} & \text{for } j = i \pm 1, 0 < i < m, \\ \bar{p}_{0,mm}f_{0,mm-1} + p_{0,mm}f_{0,mm} & \text{for } j = m - 1, i = m. \end{cases} \tag{12.35}$$

12.2.2.2 One-State Reduction of Phase Space for Perturbed Birth–Death-Type Semi-Markov Processes. At this step, we transform the semi-Markov process $\tilde{\eta}_\varepsilon(t)$ into a new semi-Markov process ${}_k\eta_\varepsilon(t)$ with the reduced phase space ${}_k\mathbb{X} = \mathbb{X} \setminus \{k\}$, using the procedure described in Sect. 9.3.

It should be noted that in the case where $0 \leq k \leq m$, the reduced phase space ${}_k\mathbb{X} = \{0, \ldots, k-1, k+1, \ldots, m\}$. By obvious re-numeration of states, $0 \to 0, \ldots, k-1 \to k-1, k+1 \to k, \ldots, m \to m-1$, the space ${}_k\mathbb{X}$ can be transformed into the standard form ${}_k\mathbb{X} = \{0, \ldots, m-1\}$.

Two cases should be considered.

The first case, where one of end states 0 or m is to be excluded from the phase space $\mathbb{X}$. These cases are similar, so we can restrict consideration, for example, to the latter case.

The second case, where some internal state $0 < k < m$ is to be excluded from the phase space $\mathbb{X}$.

In fact, the description of the corresponding algorithm and proofs are given in Sect. 9.3. We should only give the corresponding formulas for characteristics appearing in the conditions $\mathbf{{}_kG}$–$\mathbf{{}_kL}$ and $\mathbf{{}_kI}_{\mathcal{H}}$, $\mathbf{{}_kL}_{\mathcal{H}}$ for the semi-Markov processes

$\tilde{\eta}_\varepsilon(t)$ and the corresponding characteristics for the limiting semi-Markov process ${}_k\eta_0(t)$.

The transition probabilities ${}_mp_{\varepsilon,ij}, i, j \in {}_m\mathbb{X}$ for the embedded Markov chain ${}_m\eta_{\varepsilon,n}$ are given by the following relation, for $\varepsilon \in (0, 1]$,

$$
{}_mp_{\varepsilon,ij} = \begin{cases} 1 & \text{for } j = 1, i = 0, \\ \tilde{p}_{\varepsilon,ii\pm1} & \text{for } j = i \pm 1, 0 < i < m-1, \\ \tilde{p}_{\varepsilon,m-1m} & \text{for } j = m-1, i = m-1, \\ \tilde{p}_{\varepsilon,m-1m-2} & \text{for } j = m-2, i = m-1, \\ 0 & \text{otherwise.} \end{cases} \tag{12.36}
$$

The relation (12.36) shows that the embedded Markov chain ${}_m\eta_{\varepsilon,n}$ has the same birth–death type as the original embedded Markov chain $\eta_{\varepsilon,n}$. The Markov chain ${}_m\eta_{\varepsilon,n}$ has the reduced phase space ${}_m\mathbb{X} = \{0, \ldots, m-1\}$. Note also that this Markov chain admits virtual transitions of the form $m-1 \to m-1$.

The situation is slightly different in the case, where $0 < k < m$. The transition probabilities ${}_kp_{\varepsilon,ij}, i, j \in {}_k\mathbb{X}$ for the embedded Markov chain ${}_k\eta_{\varepsilon,n}$ are given by the following relation, for $\varepsilon \in (0, 1]$,

$$
{}_kp_{\varepsilon,ij} = \begin{cases} 1 & \text{for } j = 1, i = 0, \\ \tilde{p}_{\varepsilon,ii\pm1} & \text{for } j = i \pm 1, 0 < i < k-1, \\ \tilde{p}_{\varepsilon,k\pm1,k}\tilde{p}_{\varepsilon,kk\pm1} & \text{for } j = k \pm 1, i = k \pm 1, \\ \tilde{p}_{\varepsilon,k\pm1k\pm2} & \text{for } j = k \pm 2, i = k \pm 1, \\ \tilde{p}_{\varepsilon,k\pm1,k}\tilde{p}_{\varepsilon,kk\mp1} & \text{for } j = k \mp 1, i = k \pm 1, \\ \tilde{p}_{\varepsilon,ii\pm1} & \text{for } j = i \pm 1, k+1 < i < m, \\ 1 & \text{for } j = m-1, i = m, \\ 0 & \text{otherwise.} \end{cases} \tag{12.37}
$$

The relation (12.37) shows that the embedded Markov chain ${}_k\eta_{\varepsilon,n}$ has the same birth–death type as the initial embedded Markov chain $\eta_{\varepsilon,n}$. The Markov chain ${}_k\eta_{\varepsilon,n}$ has the reduced phase space ${}_m\mathbb{X} = \{0, \ldots, k-1, k+1, \ldots m\}$. Note also that this Markov chain admits virtual transitions of the forms $k+1 \to k+1$ and $k-1 \to k-1$.

The sets ${}_m\mathbb{Y}_{1,i} = \{i \in \mathbb{X} : {}_mp_{1,ij} > 0\}$, $i \in {}_m\mathbb{X}$ take the following forms,

$$
{}_m\mathbb{Y}_{1,i} = \begin{cases} \{1\} & \text{for } i = 0, \\ \{i-1, i+1\} & \text{for } 1 \le i < m-1, \\ \{m-2, m-1\} & \text{for } i = m-1. \end{cases} \tag{12.38}
$$

The sets ${}_k\mathbb{Y}_{1,i} = \{i \in \mathbb{X} : {}_kp_{1,ij} > 0\}$, $i \in {}_k\mathbb{X}$ take, the following forms, for $0 < k < m$,

$$
{}_k\mathbb{Y}_{1,i} = \begin{cases} \{1\} & \text{for } i = 0, \text{ if } 0 < k-1, \\ \{i-1, i+1\} & \text{for } 1 \le i < k-1, \\ \{k \pm 2, k \pm 1, k \mp 1\} & \text{for } i = k \pm 1, \\ \{i-1, i+1\} & \text{for } k+1 < i \le m-1, \\ \{m\} & \text{for } i = m, \text{ if } m > k-1. \end{cases} \tag{12.39}
$$

The Laplace transforms ${}_m\phi_{\varepsilon,ij}(s)$, $j \in {}_m\mathbb{Y}_{1,i}$, $i \in {}_m\mathbb{X}$ for the distribution functions ${}_mF_{\varepsilon,ij}(t)$ take the following forms, for $\varepsilon \in (0,1]$ and $s \ge 0$,

$$
{}_m\phi_{\varepsilon,ij}(s) = \begin{cases} \tilde{\phi}_{\varepsilon,10}(s) & \text{for } j = 1, i = 0, \\ \tilde{\phi}_{\varepsilon,ii\pm1}(s) & \text{for } j = i \pm 1, 1 < i < m-1, \\ \tilde{\phi}_{\varepsilon,m-1m-2}(s) & \text{for } j = m-2, i = m-1, \\ \tilde{\phi}_{\varepsilon,m-1m}(s)\tilde{\phi}_{\varepsilon,mm-1}(s) & \text{for } j = m-1, i = m-1. \end{cases} \tag{12.40}
$$

The Laplace transforms ${}_k\phi_{\varepsilon,ij}(s)$, $j \in {}_k\mathbb{Y}_{1,i}$, $i \in {}_k\mathbb{X}$ for the distribution functions ${}_kF_{\varepsilon,ij}(t)$ take the following forms, for $0 < k < m$, $\varepsilon \in (0,1]$, and $s \ge 0$,

$$
{}_k\phi_{\varepsilon,ij}(s) = \begin{cases} \tilde{\phi}_{\varepsilon,10}(s) & \text{for } j = 1, i = 0, \text{ if } 0 < k-1, \\ \tilde{\phi}_{\varepsilon,ii\pm1}(s) & \text{for } j = i \pm 1, 1 \le i < k-1, \\ \tilde{\phi}_{\varepsilon,k\pm1k\pm2}(s) & \text{for } j = k \pm 2, i = k \pm 1, \\ \tilde{\phi}_{\varepsilon,k\pm1k}(s)\tilde{\phi}_{\varepsilon,kk\pm1}(s) & \text{for } j = k \pm 1, i = k \pm 1, \\ \tilde{\phi}_{\varepsilon,k\pm1k}(s)\tilde{\phi}_{\varepsilon,kk\mp1}(s) & \text{for } j = k \mp 1, i = k \pm 1, \\ \tilde{\phi}_{\varepsilon,ii\pm1}(s) & \text{for } j = i \pm 1, \ k+1 < i \le m-1, \\ \tilde{\phi}_{\varepsilon,mm-1}(s) & \text{for } j = m-1, i = m, \text{ if } m > k-1. \end{cases} \tag{12.41}
$$

The expectations ${}_mf_{\varepsilon,ij}$, $j \in {}_k\mathbb{Y}_{1,i}$, $i \in \mathbb{X}$ take the following forms, for $\varepsilon \in (0,1]$,

$$
{}_mf_{\varepsilon,ij} = \begin{cases} \tilde{f}_{\varepsilon,10} & \text{for } j = 1, i = 0, \\ \tilde{f}_{\varepsilon,ii\pm1} & \text{for } j = i \pm 1, 1 < i < m-1, \\ \tilde{f}_{\varepsilon,m-1m-2} & \text{for } j = m-2, i = m-1, \\ \tilde{f}_{\varepsilon,m-1m} + \tilde{f}_{\varepsilon,mm-1} & \text{for } j = m-1, i = m-1. \end{cases} \tag{12.42}
$$

The expectations ${}_kf_{\varepsilon,ij}$, $j \in {}_k\mathbb{Y}_{1,i}$, $i \in \mathbb{X}$ take the following forms, for $0 < k < m$, $\varepsilon \in (0,1]$,

$$
{}_kf_{\varepsilon,ij} = \begin{cases} \tilde{f}_{\varepsilon,10} & \text{for } j = 1, i = 0, \text{ if } 0 < k-1, \\ \tilde{f}_{\varepsilon,ii\pm1} & \text{for } j = i \pm 1, 1 \le i < k-1, \\ \tilde{f}_{\varepsilon,k\pm1k\pm2} & \text{for } j = k \pm 2, i = k \pm 1, \\ \tilde{f}_{\varepsilon,k\pm1k} + \tilde{f}_{\varepsilon,kk\pm1} & \text{for } j = k \pm 1, i = k \pm 1, \\ \tilde{f}_{\varepsilon,k\pm1k} + \tilde{f}_{\varepsilon,kk\mp1} & \text{for } j = k \mp 1, i = k \pm 1, \\ \tilde{f}_{\varepsilon,ii\pm1} & \text{for } j = i \pm 1, \ k+1 < i \le m-1, \\ \tilde{f}_{\varepsilon,mm-1} & \text{for } j = m-1, i = m, \text{ if } m > k-1. \end{cases} \tag{12.43}
$$

The normalisation function ${}_k u_{\varepsilon,i}$, $\varepsilon \in (0, 1]$ is given by the following relation, for $i \in {}_k\mathbb{X}$,

$$ {}_k u_{\varepsilon,i} = \tilde{u}_{\varepsilon,i}. \tag{12.44} $$

The limiting transition probabilities ${}_m p_{0,ij}$ for the embedded Markov chain ${}_m\eta_{0,n}$ are given by the following relation:

$$ {}_m p_{0,ij} = \begin{cases} 1 & \text{for } j = 1, i = 0, \\ \tilde{p}_{0,ii\pm 1} & \text{for } j = i \pm 1, 1 < i < m-1, \\ \tilde{p}_{0,m-1m} & \text{for } j = m-1, i = m-1, \\ \tilde{p}_{0,m-1m-2} & \text{for } j = m-2, i = m-1, \\ 0 & \text{otherwise.} \end{cases} \tag{12.45} $$

The limiting transition probabilities ${}_k p_{0,ij}$ for the embedded Markov chain ${}_k\eta_{0,n}$ are given by the following relation, for $0 < k < m$,

$$ {}_k p_{0,ij} = \begin{cases} 1 & \text{for } j = 1, i = 0, \\ \tilde{p}_{0,ii\pm 1} & \text{for } j = i \pm 1, 0 < i < k-1, \\ \tilde{p}_{0,k\pm 1,k}\tilde{p}_{\varepsilon,kk\pm 1} & \text{for } j = k \pm 1, i = k \pm 1, \\ \tilde{p}_{0,k\pm 1k\pm 2} & \text{for } j = k \pm 2, i = k \pm 1, \\ \tilde{p}_{0,k\pm 1,k}\tilde{p}_{0,kk\mp 1} & \text{for } j = k \mp 1, i = k \pm 1, \\ \tilde{p}_{0,ii\pm 1} & \text{for } j = i \pm 1, k+1 < i < m, \\ 1 & \text{for } j = m-1, i = m, \\ 0 & \text{otherwise.} \end{cases} \tag{12.46} $$

As it was pointed out in Lemma 9.20, the conditions **G**, **H**, $\mathbf{I}_{\mathcal{H}}$, and $\mathbf{L}_{\mathcal{H}}$ imply that the condition $\hat{\mathbf{L}}$ is satisfied. This condition takes the following form:

$\check{\hat{\mathbf{L}}}$: $w_{\varepsilon,j,i} = \frac{\tilde{u}_{\varepsilon,j}}{\tilde{u}_{\varepsilon,i}} \longrightarrow w_{0,j,i} \in [0, \infty]$ as $\varepsilon \to 0$, for $i, j \in \mathbb{X}$.

Also, the absorbing rate condition $\mathbf{{}_kM}$ should be satisfied for the state k, in order the conditions $\mathbf{{}_kJ}_1$ would be satisfied for the semi-Markov processes ${}_k\eta_\varepsilon(t)$. By Remark 9.3, this condition can be replaced by the weaker local absorbing rate condition $\mathbf{{}_k\hat{M}}$, which takes, in this case, the following form:

$\mathbf{{}_k\check{\hat{M}}}$: $w_{0,k,i} \in [0, \infty)$, for $i \in \mathbb{X}$, $|i - k| \leq 1$.

In this case, the corresponding limiting Laplace transforms ${}_k\phi_{0,ij}(s)$, $j \in {}_k\mathbb{Y}_{1,i}$, $i \in \mathbb{X}$ for the distribution functions ${}_kF_{0,ij}(t)$ are given by the following relations, for the case $k = m$ and $s \geq 0$,

$$ {}_m\phi_{0,ij}(s) = \begin{cases} \tilde{\phi}_{0,10}(s) & \text{for } j = 1, i = 0, \\ \tilde{\phi}_{0,ii\pm 1}(s) & \text{for } j = i \pm 1, 1 < i < m-1, \\ \tilde{\phi}_{0,m-1m-2}(s) & \text{for } j = m-2, i = m-1, \\ \tilde{\phi}_{0,m-1m}(s)\tilde{\phi}_{0,mm-1}(w_{0,m,m-1}s) & \text{for } j = m-1, i = m-1 \end{cases} \tag{12.47} $$

and, for the case $0 < k < m$ and $s \geq 0$,

$$
{}_k\phi_{0,ij}(s) \\
= \begin{cases} \tilde{\phi}_{0,10}(s) & \text{for } j = 1, i = 0, \text{ if } 0 < k - 1, \\ \tilde{\phi}_{0,ii\pm1}(s) & \text{for } j = i \pm 1, 1 \leq i < k - 1, \\ \tilde{\phi}_{0,k\pm1k\pm2}(s) & \text{for } j = k \pm 2, i = k \pm 1, \\ \tilde{\phi}_{0,k\pm1k}(s)\tilde{\phi}_{\varepsilon,kk\pm1}(w_{0,k,\pm1}s) & \text{for } j = k \pm 1, i = k \pm 1, \\ \tilde{\phi}_{0,k\pm1k}(s)\tilde{\phi}_{\varepsilon,kk\mp1}(w_{0,k,\pm1}s) & \text{for } j = k \mp 1, i = k \pm 1, \\ \tilde{\phi}_{0,ii\pm1}(s) & \text{for } j = i \pm 1, k + 1 < i \leq m - 1, \\ \tilde{\phi}_{0,mm-1}(s) & \text{for } j = m - 1, i = m, \text{ if } m > k - 1. \end{cases} \tag{12.48}
$$

The limiting expectations ${}_kf_{0,ij}, j \in \tilde{\mathbb{Y}}_{1,i}, i \in \mathbb{X}$ are given by the following relations, for the case $k = m$,

$$
{}_mf_{0,ij} = \begin{cases} \tilde{f}_{0,10} & \text{for } j = 1, i = 0, \\ \tilde{f}_{0,ii\pm1} & \text{for } j = i \pm 1, 1 < i < m - 1, \\ \tilde{f}_{0,m-1m-2} & \text{for } j = m - 2, i = m - 1, \\ \tilde{f}_{0,m-1m} + \tilde{f}_{\varepsilon,mm-1}w_{0,m,m-1} & \text{for } j = m - 1, i = m - 1 \end{cases} \tag{12.49}
$$

and, for the case $0 < k < m$,

$$
{}_kf_{0,ij} = \begin{cases} \tilde{f}_{0,10} & \text{for } j = 1, i = 0, \text{ if } 0 < k - 1, \\ \tilde{f}_{0,ii\pm1} & \text{for } j = i \pm 1, 1 \leq i < k - 1, \\ \tilde{f}_{0,k\pm1k\pm2} & \text{for } j = k \pm 2, i = k \pm 1, \\ \tilde{f}_{0,k\pm1k} + \tilde{f}_{0,kk\pm1}w_{0,k,k\pm1} & \text{for } j = k \pm 1, i = k \pm 1, \\ \tilde{f}_{0,k\pm1k} + \tilde{f}_{0,kk\mp1}w_{0,k,k\mp1} & \text{for } j = k \mp 1, i = k \pm 1, \\ \tilde{f}_{0,ii\pm1} & \text{for } j = i \pm 1,\ k + 1 < i \leq m - 1, \\ \tilde{f}_{0,mm-1} & \text{for } j = m - 1, i = m, \text{ if } m > k - 1. \end{cases} \tag{12.50}
$$

12.2.2.3 The Asymptotic Recurrent Algorithm of Phase Space Reduction for Perturbed Birth–Death-Type Semi-Markov Processes. In Sects. 12.2.2.1 and 12.2.2.2, the first step of the asymptotic recurrent algorithm of phase space reduction for perturbed birth–death-type semi-Markov processes connected with the transformations of the original semi-Markov process was presented. These transformations can be represented by the symbolic diagram, $\eta_\varepsilon(t) \to \tilde{\eta}_\varepsilon(t) \to {}_k\eta_\varepsilon(t)$.

Importantly, both processes $\tilde{\eta}_\varepsilon(t)$ and ${}_k\eta_\varepsilon(t)$ are also birth–death-type semi-Markov processes.

The resulting process ${}_k\tilde{\eta}_\varepsilon(t)$ is a birth–death-type semi-Markov process with the reduced phase space $\mathbb{X}_k$.

The following steps of the asymptotic recurrent algorithm of phase space reduction are described in Sect. 9.4. In the case of birth–death-type semi-Markov processes these steps should be implemented using the relations given above in Sects. 12.2.2.1 and 12.2.2.2.

It should be noted that in the case of semi-Markov birth–death processes, the asymptotic algorithms for removing virtual transitions and reducing the phase space preserve the birth–death structure for reduced semi-Markov processes. This makes these algorithms computationally more efficient.

12.2.2.4 Limit Theorems for Hitting Times for Perturbed Birth–Death-Type Semi-Markov Processes. As mentioned in Sect. 12.2.1.3, hitting times $\tau_{\varepsilon,\mathbb{D}}$ are subject of studies for two cases, where domain $\mathbb{D} = \{0\}$ or $\mathbb{D} = \{0, m\}$.

The corresponding limit theorems based on the use of asymptotic recurrent algorithms of phase space reduction are presented in Chaps. 10 and 11.

12.3 Numerical Examples

In this section, we present numerical examples illustrating asymptotic recurrent algorithms for singularly perturbed semi-Markov processes and asymptotic results for hitting times, which are presented in Chaps. 10 and 11.

12.3.1 An Example of Singularly Perturbed Semi-Markov Processes with Two-State Domain $\bar{\mathbb{D}}$

12.3.1.1 Definition of the Model. Let us consider an example, which let us illustrate the asymptotic results presented in Chaps. 10 and 11.

In this example, the phase space $\mathbb{X} = \{1, 2, 3\}$, and the matrix $\mathbf{P}_\varepsilon = \|p_{\varepsilon,ij}\|$ of transition probabilities for the embedded Markov chain $\eta_{\varepsilon,n}$, has the following form, for $\varepsilon \in (0, 1]$,

$$\mathbf{P}_\varepsilon = \left\| \begin{array}{lll} 1 - \frac{1}{2}\varepsilon^\alpha - \frac{1}{2}\varepsilon^\beta & \frac{1}{2}\varepsilon^\alpha & \frac{1}{2}\varepsilon^\beta \\ \frac{1}{2}\varepsilon & 1 - \varepsilon & \frac{1}{2}\varepsilon \\ \frac{1}{3} & \frac{1}{3} & \frac{1}{3} \end{array} \right\|. \tag{12.51}$$

Here, parameters $\alpha, \beta \in [0, \infty)$.

Also, we assume that the distribution functions of transition times take the following forms, for $\varepsilon \in (0, 1]$,

$$F_{\varepsilon,ij}(\cdot) = \begin{cases} Con_{\varepsilon^{-\gamma}}(\cdot) & \text{for } i = 1,\ j = 1, 2, 3, \\ Con_1(\cdot) & \text{for } i = 2, 3,\ j = 1, 2, 3, \end{cases} \tag{12.52}$$

where the parameter $\gamma \in [0, \infty)$.

This means that the transition times from the state 1 take the value $\varepsilon^{-\gamma}$, while transition times from the states 2 and 3 take the value 1.

The semi-Markov process $\eta_\varepsilon(t)$ is constructed as described in Sect. 9.1.

Note that, in the case $\gamma = 0$, the semi-Markov process $\eta_\varepsilon(t)$ is a discrete time Markov chain embedded in continuous time.

12.3.1.2 Hitting Times. Let us consider the case, where domain $\mathbb{D} = \{3\}$ and, thus, domain $\bar{\mathbb{D}} = \{1, 2\}$, i.e., it is a two-state set. Since $\mathbb{D}$ is a one-state set, the indicator $\mathrm{I}(\eta_\varepsilon(\tau_{\varepsilon,\mathbb{D}}) = 3) = 1$, and, thus, for $i \in \bar{\mathbb{D}}$,

$$G_{\varepsilon,\mathbb{D},i3}(\cdot) = \mathsf{P}_i\{\tau_{\varepsilon,\mathbb{D}} \leq \cdot\}. \tag{12.53}$$

Let us describe the weak convergence asymptotics for the distribution functions of the hitting times $G_{\varepsilon,\mathbb{D},i3}(\cdot), i \in \bar{\mathbb{D}}$, that is, find the correct normalisation functions $\breve{u}_{\varepsilon,i}, i \in \bar{\mathbb{D}}$, the use of which leads to the limiting distributions $G_{0,\mathbb{D},i3}(\cdot), i \in \bar{\mathbb{D}}$ not concentrated in 0, and describe the corresponding asymptotics for the expectations of hitting times $E_{\varepsilon,\mathbb{D},i3}$.

It is useful to recall that, according to Lemma 10.1, the distribution functions $G_{\varepsilon,\mathbb{D},i3}(\cdot), i \in \bar{\mathbb{D}}$ are completely determined by the transition probabilities $Q_{\varepsilon,ij}(\cdot), i \in \bar{\mathbb{D}}, j \in \mathbb{X}$ for the semi-Markov process $\eta_\varepsilon(t)$ and, thus, the above distribution functions of hitting times coincide for semi-Markov processes with any form of transition probabilities $Q_{\varepsilon,3r}(\cdot), r \in \mathbb{X}$.

12.3.1.3 Perturbation Conditions. All basic perturbation conditions are satisfied for the semi-Markov processes $\eta_\varepsilon(t)$.

The condition **G** is satisfied. Moreover, the sets $\mathbb{Y}_{1,i} = \mathbb{X}$, for $i \in \mathbb{X}$.

The condition **H** is also satisfied.

The corresponding limiting matrix of transition probabilities $\mathbf{P}_0$ of the embedded Markov chain $\eta_{0,n}$ takes different forms for the following six cases:

$$\begin{aligned} &\mathbf{(1)}\ \alpha = \beta = 0,\ \mathbf{(2)}\ \alpha > \beta = 0,\ \mathbf{(3)}\ \beta > \alpha = 0, \\ &\mathbf{(4)}\ \alpha = \beta > 0,\ \mathbf{(5)}\ \alpha > \beta > 0,\ \mathbf{(6)}\ \beta > \alpha > 0. \end{aligned} \tag{12.54}$$

Let us denote the matrix $\mathbf{P}_0$ for these six cases, respectively, as $\mathbf{P}_0^{(h)}$, for $h = 1, \ldots, 6$. Obviously,

$$\mathbf{P}_0^{(1)} = \left\| \begin{matrix} 0 & \frac{1}{2} & \frac{1}{2} \\ 0 & 1 & 0 \\ \frac{1}{3} & \frac{1}{3} & \frac{1}{3} \end{matrix} \right\|,\ \mathbf{P}_0^{(2)} = \left\| \begin{matrix} \frac{1}{2} & 0 & \frac{1}{2} \\ 0 & 1 & 0 \\ \frac{1}{3} & \frac{1}{3} & \frac{1}{3} \end{matrix} \right\|,\ \mathbf{P}_0^{(3)} = \left\| \begin{matrix} \frac{1}{2} & \frac{1}{2} & 0 \\ 0 & 1 & 0 \\ \frac{1}{3} & \frac{1}{3} & \frac{1}{3} \end{matrix} \right\|, \tag{12.55}$$

and

$$\mathbf{P}_0^{(h)} = \left\| \begin{matrix} 1 & 0 & 0 \\ 0 & 1 & 0 \\ \frac{1}{3} & \frac{1}{3} & \frac{1}{3} \end{matrix} \right\|,\ h = 4, 5, 6. \tag{12.56}$$

In this model, the state 2 is an asymptotically absorbing state, while the state 1 is an asymptotically non-absorbing state, for the cases $h = 1, 2, 3$, or an asymptotically absorbing state, for the cases $h = 4, 5, 6$.

Obviously, the transition probabilities $p_{\varepsilon,ij}, i, j \in \mathbb{X}$ belong to the family of asymptotically comparable power-type functions $\mathcal{H}_1$ introduced in Sect. 8.2.1 and, thus, condition $\mathbf{I}_{\mathcal{H}_1}$ is satisfied.

In this case, it is natural to choose the initial normalisation functions,

$$u_{\varepsilon,1} = \varepsilon^{-\gamma},\ u_{\varepsilon,2} = u_{\varepsilon,3} = 1,\ \varepsilon \in (0, 1]. \tag{12.57}$$

Note, first of all that, in this case, the condition of asymptotic comparability **L** is satisfied for the normalisation functions $u_{\varepsilon,i}, i \in \mathbb{X}$, which belong to the complete family of asymptotically comparable functions $\mathcal{H}_1$ defined by the relation (8.14). Thus, the condition $\mathbf{L}_{\mathcal{H}_1}$ is satisfied.

The conditions **J** and **J°** are satisfied, with the corresponding limiting distribution functions $F_{0,ij}(t)$ and their Laplace transforms $\phi_{0,ij}(s)$, given by the following relations, for $i, j \in \mathbb{X}$,

$$F_{0,ij}(\cdot) = Con_1(\cdot), \tag{12.58}$$

and

$$\phi_{0,ij}(s) = e^{-s}, \ s \geq 0. \tag{12.59}$$

The condition **K** is also satisfied, and the corresponding limiting expectations take the following form, for $i, j \in \mathbb{X}$,

$$f_{0,ij} = 1. \tag{12.60}$$

12.3.2 A One-Step Asymptotic Reduction of Phase Space

The domain $\bar{\mathbb{D}}$ is a two-state set. Therefore, the asymptotic recurrent algorithm described in Chap. 9 includes only the chain of transformations $\eta_\varepsilon(t) \longrightarrow \tilde{\eta}_\varepsilon(t) \longrightarrow {}_{k_1}\eta_\varepsilon(t) \longrightarrow {}_{k_1}\tilde{\eta}_\varepsilon(t)$. As we will see, the state $k_1 = 1$ belongs to the set $\mathbb{W}_0$ of the least absorbing states and, thus, can be chosen for exclusion from the initial phase space $\mathbb{X}$.

12.3.2.1 Removing of Virtual Transitions for Semi-Markov Processes $\eta_\varepsilon(t)$. The procedure of asymptotic removing of virtual transitions described in Sect. 9.2 should be applied to the semi-Markov processes $\eta_\varepsilon(t)$.

The embedded Markov chain $\tilde{\eta}_{\varepsilon,n}$ has the transition probabilities $\tilde{p}_{\varepsilon,ij} = \mathrm{I}(j \neq i)\frac{p_{\varepsilon,ij}}{1-p_{\varepsilon,ii}}$, $i, j \in \mathbb{X}$ given by the relation (9.46). The matrix of transition probabilities for this Markov chain takes the following form, for $\varepsilon \in (0, 1]$,

$$\tilde{\mathbf{P}}_\varepsilon = \left\| \begin{matrix} 0 & \frac{\varepsilon^\alpha}{\varepsilon^\alpha + e^\beta} & \frac{\varepsilon^\beta}{\varepsilon^\alpha + e^\beta} \\ \frac{1}{2} & 0 & \frac{1}{2} \\ \frac{1}{2} & \frac{1}{2} & 0 \end{matrix} \right\|. \tag{12.61}$$

The corresponding limiting matrix of transition probabilities $\tilde{\mathbf{P}}_0$ for the Markov chain $\tilde{\eta}_{0,n}$ takes various forms for the cases **(1)–(6)**. We denote the matrix $\tilde{\mathbf{P}}_0$ for these cases, respectively, as $\tilde{\mathbf{P}}_0^{(h)}$, for $h = 1, \ldots, 6$. Obviously,

$$\tilde{\mathbf{P}}_0^{(1)} = \left\| \begin{matrix} 0 & \frac{1}{2} & \frac{1}{2} \\ \frac{1}{2} & 0 & \frac{1}{2} \\ \frac{1}{2} & \frac{1}{2} & 0 \end{matrix} \right\|, \ \tilde{\mathbf{P}}_0^{(2)} = \left\| \begin{matrix} 0 & 0 & 1 \\ \frac{1}{2} & 0 & \frac{1}{2} \\ \frac{1}{2} & \frac{1}{2} & 0 \end{matrix} \right\|, \ \tilde{\mathbf{P}}_0^{(3)} = \left\| \begin{matrix} 0 & 1 & 0 \\ \frac{1}{2} & 0 & \frac{1}{2} \\ \frac{1}{2} & \frac{1}{2} & 0 \end{matrix} \right\|, \tag{12.62}$$

and

$$\tilde{\mathbf{P}}_0^{(4)} = \begin{Vmatrix} 0 & \frac{1}{2} & \frac{1}{2} \\ \frac{1}{2} & 0 & \frac{1}{2} \\ \frac{1}{2} & \frac{1}{2} & 0 \end{Vmatrix}, \ \tilde{\mathbf{P}}_0^{(5)} = \begin{Vmatrix} 0 & 0 & 1 \\ \frac{1}{2} & 0 & \frac{1}{2} \\ \frac{1}{2} & \frac{1}{2} & 0 \end{Vmatrix}, \ \tilde{\mathbf{P}}_0^{(6)} = \begin{Vmatrix} 0 & 1 & 0 \\ \frac{1}{2} & 0 & \frac{1}{2} \\ \frac{1}{2} & \frac{1}{2} & 0 \end{Vmatrix}. \tag{12.63}$$

In this case, the normalisation functions $\tilde{u}_{\varepsilon,i} = (1 - p_{\varepsilon,ii})^{-1} u_{\varepsilon,i}, i \in \mathbb{X}$ take, according to the relation (9.60), the following forms:

$$\tilde{u}_{\varepsilon,1} = 2\varepsilon^{-\gamma}(\varepsilon^{\alpha} + \varepsilon^{\beta})^{-1}, \ \tilde{u}_{\varepsilon,2} = \varepsilon^{-1}, \ \tilde{u}_{\varepsilon,3} = 3/2, \ \varepsilon \in (0, 1]. \tag{12.64}$$

Let us define the parameter,

$$\delta = \min(\alpha, \beta) + \gamma. \tag{12.65}$$

If $\delta = 0$, then: $\tilde{u}_{\varepsilon,1} \equiv 1$, in the case **(1)**, while $\tilde{u}_{\varepsilon,1} \longrightarrow 2$ as $\varepsilon \longrightarrow 0$, in the cases **(2)** and **(3)**. If $\delta > 0$, then $\tilde{u}_{\varepsilon,1} \longrightarrow \infty$ as $\varepsilon \longrightarrow 0$, in the cases **(3)**–**(6)**.

Also, $\tilde{u}_{\varepsilon,2} \longrightarrow \infty$ as $\varepsilon \longrightarrow 0$, while $\tilde{u}_{\varepsilon,3} \equiv 3/2$, in the cases **(1)**–**(6)**.

The probability $p_{0,11} < 1$ in the cases **(1)**–**(3)**, i.e., if $\alpha \wedge \beta = 0$. While $p_{0,11} = 1$ in the cases **(4)**–**(6)**, i.e., if $\alpha \wedge \beta > 0$.

According to Theorem 9.1, the condition $\tilde{\mathbf{J}}^{\circ}$ is satisfied for the semi-Markov processes with removed virtual transitions $\tilde{\eta}(t)$.

The Laplace transform $\tilde{\phi}_{0,1j}(\cdot), j \in \mathbb{X}$ for the corresponding limiting distribution function $\tilde{F}_{0,1j}(\cdot)$ appearing in the condition $\tilde{\mathbf{J}}^{\circ}$ should be calculated using the relations (9.62) and (9.63), if $p_{0,11} < 1$, or the relation (9.73), if $p_{0,11} = 1$.

Thus, the relations (12.52), (12.55), and (12.59) imply that, for $j \in \mathbb{X}$ and $s \geq 0$,

$$\begin{aligned} \tilde{\phi}_{0,1j}(s) &= \frac{\phi_{0,1j}((1 - p_{0,11})s)(1 - p_{0,11})}{1 - p_{0,11}\phi_{0,1j}((1 - p_{0,11})s))} \\ &= \begin{cases} e^{-s} & \text{if } \alpha = \beta = 0, \\ \frac{1}{2}e^{-\frac{s}{2}}/(1 - \frac{1}{2}e^{-\frac{s}{2}}) & \text{if } \alpha > 0, \beta = 0 \text{ or } \alpha = 0, \beta > 0. \end{cases} \end{aligned} \tag{12.66}$$

Note that the distribution functions corresponding to the Laplace transforms appearing in the relation (12.66) are, respectively, $Con_1(\cdot)$, for the case $\alpha = \beta = 0$, or $Geo_{\frac{1}{2}}(2\cdot)$, for the cases $\alpha > 0, \beta = 0$ and $\alpha = 0, \beta > 0$.

If $\alpha \wedge \beta > 0$, the relations (12.52), (12.55), and (12.60) imply that, for $j \in \mathbb{X}$ and $s \geq 0$,

$$\tilde{\phi}_{0,1j}(s) = (1 + f_{0,11}s)^{-1} = (1 + s)^{-1}. \tag{12.67}$$

The probability $p_{0,22} = 1$, in the cases **(1)**–**(6)**.

Thus, the Laplace transform $\tilde{\phi}_{0,2j}(s), s \geq 0$ for the corresponding limiting distribution function $\tilde{F}_{0,2j}(\cdot), j = 1, 2, 3$ appearing in the condition $\tilde{\mathbf{J}}^{\circ}$ should be calculated using the relations (9.73) and (9.74).

In the case, the relations (12.52), (12.55), and (12.60) imply that, for $j \in \mathbb{X}$ and $s \geq 0$,

$$\tilde{\phi}_{0,2j}(s) = (1 + f_{0,22}s)^{-1} = (1 + s)^{-1}. \tag{12.68}$$

The distribution function corresponding to the Laplace transform appearing in the relations (12.67) and (12.68) is $Exp_1(\cdot)$.

According to Theorem 9.1, the condition $\tilde{\mathbf{K}}$ is satisfied for the semi-Markov processes $\tilde{\eta}(t)$.

Therefore, the limiting expectation $\tilde{f}_{0,ij}$ appearing in the condition $\tilde{\mathbf{K}}$ is the first moment of the distribution function $\tilde{F}_{0,ij}(\cdot)$ and, thus, for $i \in \bar{\mathbb{D}}$, $j \in \mathbb{X}$,

$$\tilde{f}_{0,ij} = 1. \tag{12.69}$$

As noted in Sect. 12.3.1.2, computing of the Laplace transforms $\tilde{\phi}_{0,3j}(\cdot)$, $j \in \mathbb{X}$ and the expectations $\tilde{f}_{0,3j}$, $j \in \mathbb{X}$ is not required.

12.3.2.2 Reduction of the Phase Space for Semi-Markov Processes $\tilde{\eta}_\varepsilon(t)$. The procedure of asymptotic one-step reduction of phase space, described in Sect. 9.3, should be applied to the semi-Markov processes $\tilde{\eta}_\varepsilon(t)$.

The relations (12.64) and (12.65) imply that,

$$\tilde{u}_{\varepsilon,1} = 2\varepsilon^{-\gamma}(\varepsilon^\alpha + \varepsilon^\beta)^{-1} \sim \frac{2}{1 + \mathrm{I}(\alpha = \beta)}\varepsilon^{-\delta} \text{ as } \varepsilon \to 0. \tag{12.70}$$

There are three cases to consider:

(i) $\delta \in [0, 1)$,

(ii) $\delta = 1$,

(iii) $\delta \in (1, \infty)$. (12.71)

Obviously,

$$w_{\varepsilon,1,2} = \frac{\tilde{u}_{\varepsilon,1}}{\tilde{u}_{\varepsilon,2}} \longrightarrow w_{0,1,2} \text{ as } \varepsilon \to 0, \tag{12.72}$$

where

$$w_{0,1,2} = \begin{cases} 0 & \text{if } \delta \in [0, 1), \\ \frac{2}{1+\mathrm{I}(\alpha=\beta)} & \text{if } \delta = 1, \\ \infty & \text{if } \delta \in (1, \infty). \end{cases} \tag{12.73}$$

Let us consider the case **(i)**, where $\delta \in [0, 1)$.

Note that

$$w_{\varepsilon,3,2} = \frac{\tilde{u}_{\varepsilon,3}}{\tilde{u}_{\varepsilon,2}} \longrightarrow 0 \text{ as } \varepsilon \to 0. \tag{12.74}$$

In this case, the conditions $\hat{\mathbf{L}}$ and ${}_1\mathbf{M}$ are satisfied, i.e., the states 1 and 3 are asymptotically less absorbing than the state 2.

That is why, the state 1 should be chosen for exclusion from the phase space $\mathbb{X}$, according to the procedure described in Sect. 9.3. In this case, the reduced phase space is ${}_1\mathbb{X} = \{2, 3\}$ and the domain ${}_1\bar{\mathbb{D}} = \{2\}$ is a one-state set.

The Markov chain ${}_1\eta_{\varepsilon,n}$ has the transition probabilities ${}_1p_{\varepsilon,ij} = \tilde{p}_{\varepsilon,ij} + \tilde{p}_{\varepsilon,i1}\tilde{p}_{\varepsilon,1j}$, $i, j = 2, 3$ given by the relation (9.87). The matrix of transition probabilities for this

Markov chain takes the following form, for $\varepsilon \in (0, 1]$,

$$ {}_1\mathbf{P}_\varepsilon = \left\| \begin{matrix} \frac{\varepsilon^\alpha}{2(\varepsilon^\alpha+\varepsilon^\beta)} & \frac{\varepsilon^\alpha+2\varepsilon^\beta}{2(\varepsilon^\alpha+\varepsilon^\beta)} \\ \frac{2\varepsilon^\alpha+\varepsilon^\beta}{2(\varepsilon^\alpha+\varepsilon^\beta)} & \frac{\varepsilon^\beta}{2(\varepsilon^\alpha+\varepsilon^\beta)} \end{matrix} \right\|. \quad (12.75) $$

The corresponding limiting matrix of transition probabilities ${}_1\mathbf{P}_0$ for the Markov chain ${}_1\eta_{0,n}$ takes three various forms in the cases: **(1)**, **(4)**, i.e., if $\alpha = \beta$; **(2)**, **(5)**, i.e., if $\alpha > \beta$; and **(3)**, **(6)**, i.e., if $\beta > \alpha$. We denote the matrix ${}_1\mathbf{P}_0$ for these cases, respectively, as ${}_1\mathbf{P}^{(h)}$, $h = 1, \ldots, 6$. Obviously,

$$ {}_1\mathbf{P}_0^{(1)} = \left\| \begin{matrix} \frac{1}{4} & \frac{3}{4} \\ \frac{3}{4} & \frac{1}{4} \end{matrix} \right\|, \ {}_1\mathbf{P}_0^{(2)} = \left\| \begin{matrix} 0 & 1 \\ \frac{1}{2} & \frac{1}{2} \end{matrix} \right\|, \ {}_1\mathbf{P}_0^{(3)} = \left\| \begin{matrix} \frac{1}{2} & \frac{1}{2} \\ 1 & 0 \end{matrix} \right\|. $$

$$ {}_1\mathbf{P}_0^{(4)} = \left\| \begin{matrix} \frac{1}{4} & \frac{3}{4} \\ \frac{3}{4} & \frac{1}{4} \end{matrix} \right\|, \ {}_1\mathbf{P}_0^{(5)} = \left\| \begin{matrix} 0 & 1 \\ \frac{1}{2} & \frac{1}{2} \end{matrix} \right\|, \ {}_1\mathbf{P}_0^{(6)} = \left\| \begin{matrix} \frac{1}{2} & \frac{1}{2} \\ 1 & 0 \end{matrix} \right\|. \quad (12.76) $$

In this case, the normalisation functions, ${}_1u_{\varepsilon,2}$ and ${}_1u_{\varepsilon,3}$, take, according to the relation (9.112), the following form:

$$ {}_1u_{\varepsilon,2} = \tilde{u}_{\varepsilon,2} = \varepsilon^{-1}, \ {}_1u_{\varepsilon,3} = \tilde{u}_{\varepsilon,3} = 3/2, \ \varepsilon \in (0, 1]. \quad (12.77) $$

According to Theorem 9.2, the condition ${}_1\mathbf{J}^\circ$ is satisfied.

By Lemma 9.19, the condition ${}_1\hat{\mathbf{I}}$ is satisfied and, for $j = 2, 3$,

$$ {}_1\hat{q}_{0,2j} = \lim_{\varepsilon\to 0} \frac{\tilde{p}_{\varepsilon,2j}}{{}_1p_{\varepsilon,2j}} = \begin{cases} 0 \text{ if } j = 2, \\ \frac{2}{3} \text{ if } j = 3, \ \alpha = \beta, \\ \frac{1}{2} \text{ if } j = 3, \ \alpha > \beta, \\ 1 \text{ if } j = 3, \ \beta > \alpha. \end{cases} \quad (12.78) $$

The Laplace transform ${}_1\phi_{0,2j}(s)$, $s \geq 0$ for the corresponding limiting distribution function ${}_1F_{0,2j}(\cdot)$ appearing in the condition ${}_1\mathbf{J}^\circ$ should be calculated using the relations (9.101) and (9.102), for $j = 2, 3$.

Using the relations (9.101), (9.102), (12.68), (12.72), (12.73), and (12.78) we get, taking into account that $w_{0,1,2} = 0$, that, for $j = 3$ and $s \geq 0$,

$$ \begin{aligned} {}_1\phi_{0,23}(s) &= \tilde{\phi}_{0,23}(s)\, {}_1q_{0,23} + \tilde{\phi}_{0,21}(s)\tilde{\phi}_{0,13}(w_{0,1,2}s)(1 - {}_1q_{0,23}) \\ &= (1 + s)^{-1}\, {}_1q_{0,23} + (1 + s)^{-1}(1 - {}_1q_{0,23}) = (1 + s)^{-1}, \end{aligned} \quad (12.79) $$

and, for $j = 2$ and $s \geq 0$,

$$ {}_1\phi_{0,22}(s) = \tilde{\phi}_{0,21}(s)\tilde{\phi}_{0,12}(w_{0,1,2}s) = (1 + s)^{-1}. \quad (12.80) $$

Thus, the distribution function ${}_1F_{0,2j}(\cdot) = Exp_1(\cdot)$, for $j = 2, 3$.

According to Theorem 9.2, the condition ${}_1\mathbf{K}$ is also satisfied.

Therefore, the limiting expectation ${}_1f_{0,2j}$ appearing in the condition ${}_1\mathbf{K}$ is the first moment of the distribution function ${}_1F_{0,2j}(\cdot)$ and, thus, for $j = 2, 3$,

$$ {}_1f_{0,2j} = 1. \tag{12.81} $$

According to the remarks made in Sect. 12.3.1.2, computing of the Laplace transforms ${}_1\phi_{0,3j}(\cdot), j \in \mathbb{X}$ and the expectations ${}_1f_{0,3j}, j \in \mathbb{X}$ is not required.

12.3.2.3 Removing of Virtual Transitions for the Reduced Semi-Markov Processes ${}_1\eta_\varepsilon(t)$. The procedure of asymptotic removing of virtual transitions described in Sect. 9.2 should be applied to the reduced semi-Markov process ${}_1\eta_\varepsilon(t)$.

The Markov chain ${}_1\tilde{\eta}_{\varepsilon,n}$ has the transition probabilities ${}_1p_{\varepsilon,ij} = \mathrm{I}(j \neq i)\frac{{}_1p_{\varepsilon,ij}}{1-{}_1p_{\varepsilon,ii}} = \mathrm{I}(j \neq i), i, j \in 2, 3$, and, thus, the matrix of its transition probabilities takes the following form, for $\varepsilon \in (0, 1]$,

$$ {}_1\tilde{\mathbf{P}}_\varepsilon = \begin{Vmatrix} 0 & 1 \\ 0 & 1 \end{Vmatrix}. \tag{12.82} $$

In this case, the normalisation function ${}_1\tilde{u}_{\varepsilon,2}$ takes, according to the relations (9.111), (12.51), and (12.75), the following form:

$$ \begin{aligned} {}_1\tilde{u}_{\varepsilon,2} &= (1 - {}_1p_{\varepsilon,22})^{-1}(1 - p_{\varepsilon,22})^{-1}u_{\varepsilon,2} \\ &= (1 - \frac{\varepsilon^\alpha}{2(\varepsilon^\alpha + e^\beta)})^{-1}\varepsilon^{-1} = \frac{2(\varepsilon^\alpha + \varepsilon^\beta)}{\varepsilon(\varepsilon^\alpha + 2\varepsilon^\beta)} \\ &\sim {}_1p_{0,23}^{-1}\,\varepsilon^{-1} \text{ as } \varepsilon \to 0, \end{aligned} \tag{12.83} $$

where the coefficient ${}_1p_{0,23}^{-1}$ takes, according to the relation (12.76), the following value,

$$ {}_1p_{0,23}^{-1} = \frac{4}{3} \cdot \mathrm{I}(\alpha = \beta) + 1 \cdot \mathrm{I}(\alpha > \beta) + 2 \cdot \mathrm{I}(\beta > \alpha). \tag{12.84} $$

According to Theorem 9.1, the condition ${}_1\tilde{\mathbf{J}}^\circ$ is satisfied for the semi-Markov processes ${}_1\tilde{\eta}(t)$ with removed virtual transitions.

The Laplace transform ${}_1\tilde{\phi}_{0,23}(s), s \geq 0$ for the corresponding limiting distribution function ${}_1\tilde{F}_{0,23}(\cdot)$ appearing in the condition ${}_1\tilde{\mathbf{J}}^\circ$ should be calculated using the relations (9.62) and (9.63), if ${}_1p_{0,22} < 1$, or the relation (9.73), if ${}_1p_{0,22} = 1$. These relations should be applied to the semi-Markov process ${}_1\eta_\varepsilon(t)$. The relation (12.76) shows that the first variant takes place.

Thus, using the relations (12.76), (12.79), and (12.80), we get, for $s \geq 0$,

$$ \begin{aligned} {}_1\tilde{\phi}_{0,23}(s) &= \frac{{}_1\phi_{0,23}((1 - {}_1p_{0,22})s)(1 - {}_1p_{0,22})}{1 - {}_1p_{0,22}\,{}_1\phi_{0,23}((1 - {}_1p_{0,22})s))} \\ &= \frac{1 - {}_1p_{0,22}}{1 + (1 - {}_1p_{0,22})s}\Big(1 - \frac{{}_1p_{0,22}}{1 + (1 - {}_1p_{0,22})s}\Big)^{-1} = \frac{1}{1+s}. \end{aligned} \tag{12.85} $$

According to Theorem 9.1, the condition ${}_1\tilde{\mathbf{K}}$ is also satisfied for the semi-Markov processes ${}_1\tilde{\eta}(t)$ with removed virtual transitions.

Also, according to the relations (12.76) and (12.81),

$$ {}_1\tilde{f}_{0,23} = (1 - {}_1p_{0,22})\,{}_1f_{0,23} + {}_1p_{0,22}\,{}_1f_{0,22} = 1. \tag{12.86} $$

12.3.3 Asymptotics for Distributions and Expectations of Hitting Times

Now we are ready to present asymptotic relations for the distribution functions $G_{\varepsilon,\mathbb{D},i3}(\cdot), i \in \bar{\mathbb{D}}$ and the expectations $E_{\varepsilon,\mathbb{D},i3}, i \in \bar{\mathbb{D}}$.

12.3.3.1 Weak Asymptotics for Distribution Functions of Hitting Times. First, let us describe the weak asymptotics for distribution functions $G_{\varepsilon,\mathbb{D},23}(\cdot)$.

According to Theorem 10.2, the normalisation function $\check{u}_{\varepsilon,2} = {}_1\tilde{u}_{\varepsilon,2}$, given by the relation (12.83), should be used in this case. Also, Theorem 10.2 and the relation (12.85) imply that the limiting Laplace transform $\Psi_{0,\mathbb{D},23}(s), s \geq 0$ takes the following form:

$$\Psi_{0,\mathbb{D},23}(s) = {}_1\tilde{\phi}_{0,23}(s) = \frac{1}{1+s}, \text{ for } s \geq 0. \tag{12.87}$$

Thus, Theorem 10.2 gives, in accordance with the relation (12.87), the following asymptotic relation:

$$G_{\varepsilon,\mathbb{D},23}(\cdot\, {}_1\tilde{u}_{\varepsilon,2}) \Rightarrow G_{0,\mathbb{D},23}(\cdot) = Exp_1(\cdot) \text{ as } \varepsilon \to 0. \tag{12.88}$$

Second, let us describe the weak asymptotics for distribution functions $G_{\varepsilon,\mathbb{D},13}(\cdot)$.

According to Theorem 10.3, the form of the corresponding weak asymptotics depends on the value of the probability $\tilde{p}_{0,12}$. It takes, by the relations (12.62) and (12.63), the following value:

$$\tilde{p}_{0,12} = \frac{1}{2} \cdot \mathrm{I}(\alpha = \beta) + 0 \cdot \mathrm{I}(\alpha > \beta) + 1 \cdot \mathrm{I}(\beta > \alpha). \tag{12.89}$$

In the cases $\alpha = \beta$ and $\beta > \alpha$, the probability $\tilde{p}_{0,12}$ takes positive values.

According to Theorem 10.3, the normalisation function $\check{u}_{\varepsilon,1} = {}_1\tilde{u}_{\varepsilon,2}$, given by relation (12.83), should be used in this case. Also, Theorem 10.3 and the relations (12.62), (12.63), (12.66), (12.67), (12.72), (12.73), (12.76), (12.87), and (12.89) imply that the limiting Laplace transform $\Psi_{0,\mathbb{D},13}(s), s \geq 0$ takes the following form, for $s \geq 0$,

$$\begin{aligned}\Psi_{0,\mathbb{D},13}(s) &= \tilde{\phi}_{0,13}((1 - {}_1p_{0,22})w_{0,1,2}s))\tilde{p}_{0,13} \\ &\quad + \Psi_{0,\mathbb{D},23}(s)\tilde{\phi}_{0,12}((1 - {}_1p_{0,22})w_{0,1,2}s)\tilde{p}_{0,12} \\ &= \tilde{p}_{0,13} + \frac{1}{1+s}\tilde{p}_{0,12} = \begin{cases} \frac{1}{2} + \frac{1}{2}\frac{1}{1+s} & \text{if } \alpha = \beta, \\ \frac{1}{1+s} & \text{if } \beta > \alpha. \end{cases}\end{aligned} \tag{12.90}$$

Thus, Theorem 10.3 gives, in accordance with the relation (12.90), the following asymptotic relation:

$$G_{\varepsilon,\mathbb{D},13}(\cdot\, {}_1\tilde{u}_{\varepsilon,2}) \Rightarrow G_{0,\mathbb{D},13}(\cdot) \text{ as } \varepsilon \to 0, \tag{12.91}$$

where

$$G_{0,\mathbb{D},13}(\cdot) = \begin{cases} \frac{1}{2}Con_0(\cdot) + \frac{1}{2}Exp_1(\cdot) & \text{if } \alpha = \beta, \\ Exp_1(\cdot) & \text{if } \beta > \alpha. \end{cases} \tag{12.92}$$

In the case $\alpha > \beta$, the probability $\tilde{p}_{0,12} = 0$.

According to Theorem 10.3, the normalisation function $\check{u}_{\varepsilon,1} = \tilde{u}_{\varepsilon,1}$, given by relation (12.70), should be used in this case. Also, Theorem 9.3 and the relations (12.62), (12.63), (12.66), (12.67), and (12.89) imply that, in this case, the limiting Laplace transform $\Psi_{0,\mathbb{D},13}(s)$ takes the following form, for $s \geq 0$,

$$\begin{aligned}\Psi_{0,\mathbb{D},13}(s) &= \tilde{\phi}_{0,13}(s)\tilde{p}_{0,13} \\ &= \begin{cases} \frac{1}{2}e^{-\frac{s}{2}}/(1 - \frac{1}{2}e^{-\frac{s}{2}}) & \text{if } \alpha > \beta = 0, \\ \frac{1}{1+s} & \text{if } \alpha > \beta > 0. \end{cases}\end{aligned} \tag{12.93}$$

Thus, Theorem 10.3 gives, in accordance with the relation (12.93), the following asymptotic relation:

$$G_{\varepsilon,\mathbb{D},13}(\cdot\, \tilde{u}_{\varepsilon,1}) \Rightarrow G_{0,\mathbb{D},13}(\cdot) \text{ as } \varepsilon \to 0, \tag{12.94}$$

where

$$G_{0,\mathbb{D},13}(\cdot) = \begin{cases} Geo_{\frac{1}{2}}(2\cdot) & \text{if } \alpha > \beta = 0, \\ Exp_1(\cdot) & \text{if } \alpha > \beta > 0. \end{cases} \tag{12.95}$$

12.3.3.2 Asymptotics for Expectations of Hitting Times. Remind that we consider the case **(i)**, where $\delta = \min(\alpha, \beta) + \gamma \in [0, 1)$.

In this case, the normalisation function ${}_1\tilde{u}_{\varepsilon,2}$, given by the relation (12.83) and used in the weak convergence relations (12.88), (12.91), and the normalisation function $\tilde{u}_{\varepsilon,1}$, given by the relation (12.70) and used in the weak convergence relation (12.94), are connected by the following relation:

$$\tilde{u}_{\varepsilon,1} = o({}_1\tilde{u}_{\varepsilon,2}) \text{ as } \varepsilon \to 0. \tag{12.96}$$

According to Theorem 11.1, the following asymptotic relation takes place:

$$E_{\varepsilon,\mathbb{D},23}/\, {}_1\tilde{u}_{\varepsilon,2} \to \bar{E}_{0,\mathbb{D},23} = E_{0,\mathbb{D},23} = 1 \text{ as } \varepsilon \to 0. \tag{12.97}$$

The situation is more complex in the case, where the initial state is 1.

In this case, the normalisation function used in Theorem 11.2 takes the following form:

$$\bar{u}_{\varepsilon,1} = \tilde{u}_{\varepsilon,1} + {}_1\tilde{u}_{\varepsilon,2}\,\tilde{p}_{\varepsilon,12} = \frac{2}{\varepsilon^{\gamma}(\varepsilon^{\alpha} + \varepsilon^{\beta})} + \frac{2(\varepsilon^{\alpha} + \varepsilon^{\beta})}{\varepsilon(\varepsilon^{\alpha} + 2\varepsilon^{\beta})}\frac{\varepsilon^{\alpha}}{\varepsilon^{\alpha} + e^{\beta}}. \tag{12.98}$$

The asymptotic comparability condition $\bar{\mathbf{L}}_{\mathbb{D},1}$ is satisfied and takes the following form:

$$\frac{{}_1\tilde{u}_{\varepsilon,2}\,\tilde{p}_{\varepsilon,12}}{\bar{u}_{\varepsilon,1}} = \Big(1 + \frac{\varepsilon(\varepsilon^{\alpha} + 2\varepsilon^{\beta})}{\varepsilon^{\alpha+\gamma}(\varepsilon^{\alpha} + \varepsilon^{\beta})}\Big)^{-1} \to \bar{u}_{\alpha,\beta,\gamma} \text{ as } \varepsilon \to 0, \tag{12.99}$$

where

$$\bar{u}_{\alpha,\beta,\gamma} = \begin{cases} 1 & \text{if } \alpha + \gamma < 1, \\ \frac{1}{3}\mathrm{I}(\alpha > \beta) + \frac{2}{5}\mathrm{I}(\alpha = \beta) + \frac{1}{2}\mathrm{I}(\alpha < \beta) & \text{if } \alpha + \gamma = 1, \\ 0 & \text{if } \alpha + \gamma > 1. \end{cases} \tag{12.100}$$

Recall that we are considering the case **(i)**, where the parameter $\delta = \min(\alpha, \beta) + \gamma \in [0, 1)$.

If $\alpha \leq \beta$, then $\delta = \alpha + \gamma < 1$, and, thus, $\bar{u}_{\alpha,\beta,\gamma} = 1$. In this case, the probability $\tilde{p}_{0,12} = \frac{1}{2}\mathrm{I}(\alpha = \beta) + 1\mathrm{I}(\alpha < \beta) > 0$ and $\bar{u}_{\varepsilon,1}/\,{}_1\tilde{u}_{\varepsilon,2} \to \tilde{p}_{0,12}$ as $\varepsilon \to 0$.

Therefore, according to Theorems 11.2 and 10.4, the following equivalent relations take place:

$$E_{\varepsilon,\mathbb{D},13}/\,\bar{u}_{\varepsilon,1} \longrightarrow \bar{E}_{0,\mathbb{D},13} = \bar{E}_{0,\mathbb{D},23} = 1 \text{ as } \varepsilon \to 0, \tag{12.101}$$

and

$$E_{\varepsilon,\mathbb{D},13}/\,{}_1\tilde{u}_{\varepsilon,2} \longrightarrow E_{0,\mathbb{D},13} \text{ as } \varepsilon \to 0 \tag{12.102}$$

where

$$E_{0,\mathbb{D},13} = \begin{cases} \frac{1}{2} & \text{if } \alpha = \beta, \\ 1 & \text{if } \alpha < \beta. \end{cases} \tag{12.103}$$

According to the relation (12.102), the expectations of hitting times $E_{\varepsilon,\mathbb{D},13}$ normalised by the function ${}_1\tilde{u}_{\varepsilon,2}$ (used in the corresponding weak convergence relations for distributions of hitting times, for the case where $\tilde{p}_{0,12} > 0$) converge to the first moment $E_{0,\mathbb{D},13}$ of the corresponding limiting distribution $G_{0,\mathbb{D},13}(\cdot)$.

If $\alpha > \beta$, then $\delta = \beta + \gamma < 1$, and thus, $\bar{u}_{\alpha,\beta,\gamma} = 1\mathrm{I}(\alpha + \gamma < 1) + \frac{1}{3}\mathrm{I}(\alpha + \gamma = 1) + 0\mathrm{I}(\alpha + \gamma > 1)$. In this case, the probability $\tilde{p}_{0,12} = 0$, and $\bar{u}_{\varepsilon,1}/\tilde{u}_{\varepsilon,1} \to (1 - \bar{u}_{\alpha,\beta,\gamma})^{-1}$ as $\varepsilon \to 0$.

Therefore, according to Theorems 11.2 and 11.4, the following relations take place:

$$\begin{aligned} E_{\varepsilon,\mathbb{D},13}/\bar{u}_{\varepsilon,1} \longrightarrow\ & \bar{E}_{0,\mathbb{D},13} \\ &= (1 - \bar{u}_{\alpha,\beta,\gamma})\tilde{f}_{0,13}\tilde{p}_{0,13} + \bar{u}_{\alpha,\beta,\gamma}\bar{E}_{0,\mathbb{D},23} \\ &= (1 - \bar{u}_{\alpha,\beta,\gamma}) + \bar{u}_{\alpha,\beta,\gamma} = 1 \text{ as } \varepsilon \to 0, \end{aligned} \tag{12.104}$$

and

$$\begin{aligned} E_{\varepsilon,\mathbb{D},13}/\tilde{u}_{\varepsilon,1} \longrightarrow\ & \bar{E}_{0,\mathbb{D},13}(1 - \bar{u}_{\alpha,\beta,\gamma})^{-1} \\ &= (1 - \bar{u}_{\alpha,\beta,\gamma})^{-1} \text{ as } \varepsilon \to 0. \end{aligned} \tag{12.105}$$

Three sub-cases should be considered.

(a) If $\beta + \gamma < 1 < \alpha + \gamma$, then $\bar{u}_{\alpha,\beta,\gamma} = 0$. In this case, $\bar{E}_{0,\mathbb{D},13} = \tilde{f}_{0,13}\tilde{p}_{0,13} = E_{0,\mathbb{D},13} = 1$ and $\bar{E}_{0,\mathbb{D},13}(1 - \bar{u}_{\alpha,\beta,\gamma})^{-1} = E_{0,\mathbb{D},13} = 1$.

Therefore, according to the relation (12.105), the expectations of hitting times $E_{\varepsilon,\mathbb{D},13}$ normalised by the function $\tilde{u}_{\varepsilon,1}$ (used in the corresponding weak convergence relation for distributions of hitting times, for the case where $\tilde{p}_{0,12} = 0$) converge to the first moment $E_{0,\mathbb{D},13}$ for the corresponding limiting distribution $G_{0,\mathbb{D},13}(\cdot)$.

(b) If $\beta+\gamma < \alpha+\gamma = 1$, then $\bar{u}_{\alpha,\beta,\gamma} = \frac{1}{3}$. In this case, $\bar{E}_{0,\mathbb{D},13} = \frac{2}{3}\tilde{e}_{0,13}\tilde{p}_{0,13} + \frac{1}{3} = \frac{2}{3}E_{0,\mathbb{D},13} + \frac{1}{3} = 1$, while $\bar{E}_{0,\mathbb{D},13}(1 - \bar{u}_{\alpha,\beta,\gamma})^{-1} = \tilde{f}_{0,13}\tilde{p}_{0,13} + \frac{1}{2} = E_{0,\mathbb{D},13} + \frac{1}{2} = \frac{3}{2}$.

Therefore, according to the relation (12.105), the expectations of hitting times $E_{\varepsilon,\mathbb{D},13}$ normalised by the function $\tilde{u}_{\varepsilon,1}$ (used in the corresponding weak convergence relation for distributions of hitting times, for the case where $\tilde{p}_{0,12} = 0$) converge to the constant, which differs from the first moment $E_{0,\mathbb{D},13}$ for the corresponding limiting distribution $G_{0,\mathbb{D},13}(\cdot)$.

(c) If $\beta + \gamma < \alpha + \gamma < 1$, then $\bar{u}_{\alpha,\beta,\gamma} = 1$. In this case, $\bar{E}_{0,\mathbb{D},13} = 1$, while $\bar{E}_{0,\mathbb{D},13}(1 - \bar{u}_{\alpha,\beta,\gamma})^{-1} = \infty$.

Therefore, according to the relation (12.105), the expectations of hitting times $E_{\varepsilon,\mathbb{D},13}$ normalised by the function $\tilde{u}_{\varepsilon,1}$ (used in the corresponding weak convergence relation for distributions of hitting times for the case where $\tilde{p}_{0,12} = 0$) converge to ∞.

Similar calculations can be performed in the cases **(ii)** $\delta = 1$ and **(iii)** $\delta \in (1, \infty)$.

In the case $\delta = 1$, the states 1 and 2 are equivalently asymptotically absorbing. Each state 1 or 2 can be used (in the procedure of one-step reduction of the phase space) in the above algorithm.

In the case of $\delta \in (1, \infty)$, the state 2 is asymptotically less absorbing than the state 1. Thus, the state 2 should be used (in the procedure of one-step reduction of phase space) in the above algorithm.

Appendix A
Limit Theorems for Randomly Stopped Stochastic Processes

In this appendix, we present some classical functional limit theorems for stochastic processes and limit theorems for randomly stopped stochastic processes, which are one of the main tools used to derive limit theorems for the first-rare-event times and the hitting times presented in the book.

A.1 Functional Limit Theorems for Càdlàg Stochastic Processes

We refer to the books by Gikhman and Skorokhod (1971), Whitt (2002), Jacod and Shiryaev (2002), Silvestrov (2004), Billingsley (2014), and and Kallenberg (2021), where readers can find a detailed exposition of the theory of functional limit theorems for stochastic processes.

A.1.1 Convergence of Càdlàg Stochastic Processes in Topology U

Let $\mathbb{D}_k$ be the space of càdlàg (possessing limits from the lefts and continuous from the right at any point) functions $\bar{x}(t) = (x_1(t), \ldots, x_k(t)), t \geq 0$, defined on the interval $[0, \infty)$ and taking values in the space $\mathbb{R}_k$.

Let $\bar{\xi}_\varepsilon(t) = (\xi_{\varepsilon,1}(t), \ldots, \xi_{\varepsilon,k}(t)), t \geq 0$, be, for each $\varepsilon \in [0, 1]$, a càdlàg stochastic process (its trajectories belong to the space $\mathbb{D}_k$) defined on a probability space $\langle \Omega_\varepsilon, \mathcal{F}_\varepsilon, \mathsf{P}_\varepsilon \rangle$.

Prokhorov (1956) deeply researched the natural U-topology in the space of continuous functions and gave necessary and sufficient conditions of U-convergence for continuous stochastic processes.

Let us formulate them in a slightly extended form, for càdlàg processes defined on a semi-infinite interval. This extension is known due to works by Skorokhod (1956), Stone (1963), and Lindwall (1973):

D. Silvestrov, *Perturbed Semi-Markov Type Processes I*,
https://doi.org/10.1007/978-3-030-92403-4

$\mathbf{Z}_1$: **(a)** $\bar{\xi}_\varepsilon(t), t \in U \xrightarrow{\mathsf{d}} \bar{\xi}_0(t), t \in U$ as $\varepsilon \to 0$, where U is some set dense in $[0, \infty)$ and containing point 0, and $\bar{\xi}_0(t), t \geq 0$, is a continuous process,

(b) $\lim_{c \to 0} \overline{\lim}_{\varepsilon \to 0} \mathsf{P}\{\Delta_{\mathsf{U}}(\bar{\xi}_\varepsilon(\cdot), c, T) \geq \delta\} = 0, \delta > 0$, for $T > 0$.

Here, $\Delta_{\mathsf{U}}(\bar{x}(\cdot), c, T)$ is the modulus of U-compactness defined by the following relation:

$$\Delta_{\mathsf{U}}(\bar{x}(\cdot), c, T) = \sup_{0 \vee (t''-c) \leq t' \leq t'' \leq (t'+c) \wedge T} |\bar{x}(t') - \bar{x}(t'')|. \tag{A.1}$$

The symbol of convergence in distribution used above means that the random vectors $(\bar{\xi}_\varepsilon(t_r), r = 1, \ldots, n) \xrightarrow{\mathsf{d}} (\bar{\xi}_0(t_r), r = 1, \ldots, n)$ as $\varepsilon \to 0$, for any $t_1, \cdots, t_n \in U, n \geq 1$.

Theorem A.1 *The condition* $\mathbf{Z}_1$ *is necessary and sufficient for the fulfilment of the following* U*-convergence relation,* $\bar{\xi}_\varepsilon(t), t \geq 0 \xrightarrow{\mathsf{U}} \bar{\xi}_0(t), t \geq 0$, as $\varepsilon \to 0$.

Remark A.1 It is useful noting that the condition $\mathbf{Z}_1$ **(b)** is implied by the condition $\mathbf{Z}_1$ **(a)** in the case, where all components of processes $\bar{\xi}_\varepsilon(t)$ are monotonic. In this case, the condition $\mathbf{Z}_1$ **(a)** is necessary and sufficient for the fulfilment of the U-convergence relation given in Theorem A.1.

Here we just note that the main meaning of the U-convergence of càdlàg stochastic processes is that it is equivalent to the following relation of convergence in distribution:

$$f(\bar{\xi}_\varepsilon(\cdot)) \xrightarrow{\mathsf{d}} f(\bar{\xi}_0(\cdot)) \text{ as } \varepsilon \to 0, \tag{A.2}$$

for any functional $f(\cdot)$ acting from the space $\mathbb{D}_k$ to $\mathbb{R}_1$ and U-continuous a.s. with respect to the measure generated by the limiting continuous process $\bar{\xi}_0(\cdot)$ on the Borel σ-algebra $\mathcal{B}_k$ of the space $\mathbb{D}_k$.

We again refer to the books pointed out in the preamble to this section for the detailed presentation of the theory.

A.1.2 Convergence of Càdlàg Stochastic Processes in Topology J

Skorokhod (1956) proposed a natural J-topology of convergence of càdlàg stochastic processes and gave necessary and sufficient condition for the J-convergence of càdlàgstochastic processes. This extension on the case of càdlàg stochastic processes defined on the interval $[0, \infty)$ is known due to works by Stone (1963) and Lindvall (1973):

$\mathbf{Z}_2$: **(a)** $\bar{\xi}_\varepsilon(t), t \in U \xrightarrow{\mathsf{d}} \bar{\xi}_0(t), t \in U$ as $\varepsilon \to 0$, where U is some set dense in $[0, \infty)$ and containing point 0,

(b) $\lim_{c \to 0} \overline{\lim}_{\varepsilon \to 0} \mathsf{P}\{\Delta_{\mathsf{J}}(\bar{\xi}_\varepsilon(\cdot), c, T) \geq \delta\} = 0, \delta > 0$, for $T > 0$.

Here, $\Delta_J(\bar{x}(\cdot), c, T)$ is the modulus of J-compactness defined by the following relation:

$$\Delta_J(\bar{x}(\cdot), c, T) = \sup_{0\vee(t-c)\leq t'\leq t\leq t''\leq(t+c)\wedge T} \min(|\bar{x}(t') - \bar{x}(t)|, |\bar{x}(t) - \bar{x}(t'')|). \quad \text{(A.3)}$$

Theorem A.2 *The condition* $\mathbf{Z}_2$ *is necessary and sufficient for the fulfilment of the following* J*-convergence relation,* $\bar{\xi}_\varepsilon(t), t \geq 0 \xrightarrow{\mathsf{J}} \bar{\xi}_0(t), t \geq 0,$ as $\varepsilon \to 0$.

Remark A.2 J-convergence and U-convergence of càdlàg processes are equivalent, in the case where the limiting process $\bar{\xi}_0(t), t \geq 0$, is continuous.

Here we just note that the main meaning of the J-convergence of càdlàg stochastic processes is that it is equivalent to the relation of convergence in distribution:

$$f(\bar{\xi}_\varepsilon(\cdot)) \xrightarrow{\mathsf{d}} f(\bar{\xi}_0(\cdot)) \text{ as } \varepsilon \to 0, \quad \text{(A.4)}$$

for any functional $f(\cdot)$ acting from the space $\mathbb{D}_k$ to $\mathbb{R}_1$ and J-continuous a.s. with respect to the measure generated by the limiting càdlàg process $\bar{\xi}_0(\cdot)$ on the Borel σ-algebra $\mathcal{B}_k$ of the space $\mathbb{D}_k$.

The conditions of a.s. J-continuity for various functionals such as the value of a càdlàg function and the value of jump at a point; the time moment and the value of a large jump; the maximal jump, the number and the sum of large jumps in a time interval; the maximum and the minimum of càdlàg process on a time interval; the exceeding time and the over-jump for a given level; the integral, and other functionals can be found, for example, in Silvestrov (2004).

We again refer to the books pointed out in the preamble to this section for the detailed presentation of the theory.

A.1.3 Convergence of Step-Sum Processes with Independent Increments in Topology J

Here, we would like to formulate one important result for step-sum processes with independent increments:

$$\bar{\xi}_\varepsilon(t) = \sum_{r\leq t v_\varepsilon} \bar{\xi}_{\varepsilon,r}, t \geq 0, \quad \text{(A.5)}$$

where: (a) $\bar{\xi}_{\varepsilon,r}, r = 1, 2, \ldots$, are i.i.d. random vectors and (b) $0 \leq v_\varepsilon \to \infty$ as $\varepsilon \to 0$.

Let us assume that the following condition holds:

$\mathbf{Z}_3$: $\bar{\xi}_\varepsilon(t), t \geq 0 \xrightarrow{\mathsf{d}} \bar{\xi}_0(t), t \geq 0,$ as $\varepsilon \to 0$.

Remark A.3 Since $\bar{\xi}_{\varepsilon,r} = (\xi_{\varepsilon,r,1}, \ldots, \xi_{\varepsilon,r,k}), r = 1, 2, \ldots$, are i.i.d. random vectors, the condition $\mathbf{Z}_3$ is implied by the following simpler relation of convergence in distribution, $\bar{\xi}_\varepsilon(u) \xrightarrow{\mathsf{d}} \bar{\xi}_0(u)$ as $\varepsilon \to 0$, assumed to hold for some $u > 0$.

As is well known, the limiting process $\bar{\xi}_0(t), t \geq 0$, is, in this case, a Lévy process (càdlàg homogeneous in time process with independent increments). The central criterion of convergence, which can be found, for example, in Loève (1977) and Skorokhod (1964, 1986) gives necessary and sufficient conditions for the fulfilment of the condition $\mathbf{Z}_3$.

The following theorem is also known, thanks to the works of Skorokhod (1957, 1964, 1986).

Theorem A.3 *Let the condition* $\mathbf{Z}_3$ *be satisfied. Then,* $\bar{\xi}_\varepsilon(t), t \geq 0 \stackrel{\mathsf{J}}{\longrightarrow} \bar{\xi}_0(t), t \geq 0$, as $\varepsilon \to 0$.

According to Theorem A.3, the convergence of step-sum processes $\bar{\xi}_\varepsilon(t), t \geq 0$, in distribution implies their J-convergence without any additional assumptions.

A.2 Limit Theorems for Randomly Stopped Stochastic Processes and Superpositions of Stochastic Processes

We refer to the book by Silvestrov (2004), where reader can find the detailed presentation of theory of limit theorems for randomly stopped stochastic processes.

A.2.1 Convergence in Distribution for Randomly Stopped Stochastic Processes

Let $\bar{\nu}_\varepsilon = (\nu_{\varepsilon,1}, \ldots, \nu_{\varepsilon,k})$ be, for each $\varepsilon \in [0, 1]$, a random vector with non-negative components, which is defined on the same probability space $\langle \Omega_\varepsilon, \mathcal{F}_\varepsilon, \mathsf{P}_\varepsilon \rangle$ as the càdlàg process $\bar{\xi}_\varepsilon(t), t \geq 0$.

The object of our interest are conditions of convergence in distribution for the random vectors:

$$\bar{\zeta}_\varepsilon = (\zeta_{\varepsilon,1}, \ldots, \zeta_{\varepsilon,k}) = (\xi_{\varepsilon,1}(\nu_{\varepsilon,1}), \ldots, \xi_{\varepsilon,k}(\nu_{\varepsilon,k})). \tag{A.6}$$

We extend the condition $\mathbf{Z}_2$ **(a)** and assume that the following condition is satisfied:

$\mathbf{Z}_4$: $(\bar{\nu}_\varepsilon, \bar{\xi}_\varepsilon(t)), t \in U \stackrel{\mathsf{d}}{\longrightarrow} (\bar{\nu}_0, \bar{\xi}_0(t)), t \in U$ as $\varepsilon \to 0$, where U is some set dense in $[0, \infty)$ and containing point 0.

Unfortunately, the conditions $\mathbf{Z}_2$ **(b)** and $\mathbf{Z}_4$ are not enough to ensure the convergence in distribution for the random vectors $\bar{\zeta}_\varepsilon$. Let us assume that the following continuity condition holds:

$\mathbf{Z}_5$: $\mathsf{P}\{\xi_{0,r}(\nu_{0,r}) = \xi_{0,r}(\nu_{0,r} - 0)\} = 1$, for $r = 1, \ldots, k$.

The condition $\mathbf{Z}_5$ means that the stopping moment $\nu_{0,r}$ is a point of continuity for the càdlàg process $\xi_{0,r}(t)$ with probability 1, for $r = 1, \ldots, k$.

It is worth noting that condition $\mathbf{Z}_5$ is automatically satisfied, if $\bar{\xi}_0(t), t \geq 0$, is a continuous process.

Also, the condition $\mathbf{Z}_5$ holds, if $\bar{\xi}_0(t), t \geq 0$, is a stochastically continuous càdlàg process independent of the random vector $\bar{\nu}_0$.

The following theorem is the simplified variant of results obtained in Silvestrov (1971c, 1972d, 1974, 2004).

Theorem A.4 *Let the conditions* $\mathbf{Z}_2$ **(b)**, $\mathbf{Z}_4$, *and* $\mathbf{Z}_5$ *be satisfied. Then,*

$$\bar{\zeta}_\varepsilon \stackrel{\mathsf{d}}{\longrightarrow} \bar{\zeta}_0 \text{ as } \varepsilon \to 0. \tag{A.7}$$

A.2.2 Convergence in Distribution for Superpositions of Càdlàg Processes

Let $\bar{\nu}_\varepsilon(t) = (\nu_{\varepsilon,1}(t), \ldots, \nu_{\varepsilon,k}(t)), t \geq 0$, be a vector càdlàg stochastic process with non-negative non-decreasing components, which is defined on the same probability space $\langle \Omega_\varepsilon, \mathcal{F}_\varepsilon, \mathsf{P}_\varepsilon \rangle$ as the càdlàg process $\bar{\xi}_\varepsilon(t), t \geq 0$, for every $\varepsilon \in [0, 1]$.

Consider also for $\varepsilon \in [0, 1]$ a càdlàg stochastic process, which is a vector superposition of stochastic processes $\bar{\xi}_\varepsilon(t), t \geq 0$, and $\bar{\nu}_\varepsilon(t), t \geq 0$:

$$\bar{\zeta}_\varepsilon(t) = (\zeta_{\varepsilon,1}(t), \ldots, \zeta_{\varepsilon,k}(t)) = (\xi_{\varepsilon,1}(\nu_{\varepsilon,1}(t)), \ldots, \xi_{\varepsilon,k}(\nu_{\varepsilon,k}(t))),\ t \geq 0. \tag{A.8}$$

The following condition is an analogue of the condition $\mathbf{Z}_4$ for stochastic processes:

$\mathbf{Z}_6$: $(\bar{\nu}_\varepsilon(t), \bar{\xi}_\varepsilon(s)), (t, s) \in U \times V \stackrel{\mathsf{d}}{\longrightarrow} (\bar{\nu}_0(t), \bar{\xi}_0(s)), U \times V$ as $\varepsilon \to 0$, where U and V are some sets dense in $[0, \infty)$ and containing point 0.

Also, the following condition is an analogue of the condition $\mathbf{Z}_5$ for stochastic processes:

$\mathbf{Z}_7$: $\mathsf{P}\{\xi_{0,r}(\nu_{0,r}(t)) = \xi_{0,r}(\nu_{0,r}(t) - 0)\} = 1$, for $r = 1, \ldots, k,\ t \in U$.

The following theorem is the simplified variant of results obtained in Silvestrov (1972d, 1974, 2004).

Theorem A.5 *Let the conditions* $\mathbf{Z}_2$ **(b)**, $\mathbf{Z}_6$, *and* $\mathbf{Z}_7$ *be satisfied. Then,*

$$\bar{\zeta}_\varepsilon(t), t \in U \stackrel{\mathsf{d}}{\longrightarrow} \bar{\zeta}_0(t), t \in U \text{ as } \varepsilon \to 0. \tag{A.9}$$

A.2.3 Convergence of Superpositions of Càdlàg Processes in Topology J

Let us also assume that the following condition holds:

$\mathbf{Z}_8$: $\bar{\nu}_0(t), t \geq 0$, is a continuous process.

The following theorem is the simplified variant of results obtained in Silvestrov (1972e, 1973, 1974, 2004).

Theorem A.6 *Let the conditions* $\mathbf{Z}_2$ **(b)**, $\mathbf{Z}_6$, $\mathbf{Z}_7$, *and* $\mathbf{Z}_8$ *be satisfied. Then,*

$$\bar{\zeta}_\varepsilon(t), t \geq 0 \xrightarrow{\mathsf{J}} \bar{\zeta}_0(t), t \geq 0 \text{ as } \varepsilon \to 0. \tag{A.10}$$

A.3 Supplementary Asymptotic Results

In this section we present some useful supplementary convergence results used in the book.

A.3.1 Slutsky Theorem and Related Results

Let $\bar{\xi}'_\varepsilon = (\xi'_{\varepsilon,1}, \ldots, \xi'_{\varepsilon,k})$ and $\bar{\xi}''_\varepsilon = (\xi''_{\varepsilon,1}, \ldots, \xi''_{\varepsilon,k})$ be, for every $\varepsilon \in [0, 1]$, random vectors defined on a probability space $\langle \Omega_\varepsilon, \mathcal{F}_\varepsilon, \mathsf{P}_\varepsilon \rangle$ and taking values in space $\mathbb{R}_k$.

Let us assume that the following condition holds:

$\mathbf{Z}_9$: **(a)** $\bar{\xi}'_\varepsilon \xrightarrow{\mathsf{d}} \bar{\xi}'_0$ as $\varepsilon \to 0$, **(b)** $\bar{\xi}''_\varepsilon \xrightarrow{\mathsf{d}} \bar{\xi}''_0$ as $\varepsilon \to 0$, where $\bar{\xi}''_0 \overset{a.s}{=} const \in \mathbb{R}_k$.

The following useful theorem takes place.

Theorem A.7 *Let condition* $\mathbf{Z}_9$ *hold. Then,*

$$(\bar{\xi}'_\varepsilon, \bar{\xi}''_\varepsilon) \xrightarrow{\mathsf{d}} (\bar{\xi}'_0, \bar{\xi}''_0) \text{ as } \varepsilon \to 0. \tag{A.11}$$

The relation (A.11) obviously implies that random variables $f(\bar{\xi}'_\varepsilon, \bar{\xi}''_\varepsilon) \xrightarrow{\mathsf{d}} f(\bar{\xi}'_0, \bar{\xi}''_0)$ as $\varepsilon \to 0$, for any continuous function $f(\bar{x}, \bar{y})$ acting from $\mathbb{R}_k \times \mathbb{R}_k$ to $\mathbb{R}_k$. For example, this relation holds for functions $f(\bar{x}, \bar{y}) = \bar{x} + \bar{y}$.

Let, for $\varepsilon \in [0, 1]$, the random vectors $\bar{\xi}^+_\varepsilon = \bar{\xi}'_\varepsilon + \bar{\xi}''_\varepsilon$.

The following theorem follows from Theorem A.7 and the above remark.

Theorem A.8 *Let the condition* $\mathbf{Z}_9$ *be satisfied. Then, the following relation holds:* $\bar{\xi}^+_\varepsilon \xrightarrow{\mathsf{d}} \bar{\xi}^+_0$ *as* $\varepsilon \to 0$. *In particular, if the random vector* $\bar{\xi}''_0 \overset{a.s}{=} \bar{0} = (0, \ldots, 0)$, *then* $\bar{\xi}^+_0 = \bar{\xi}'_0$.

A.3.2 Functional Analogues of Slutsky Theorem

Theorem A.7 has also an analogue for stochastic processes.

Let $\bar{\xi}'_\varepsilon(t) = (\xi'_{\varepsilon,1}(t), \ldots, \xi'_{\varepsilon,k}(t)), t \geq 0$, and $\bar{\xi}''_\varepsilon(t) = (\xi''_{\varepsilon,1}(t), \ldots, \xi''_{\varepsilon,k''}(t))$ be, for every $\varepsilon \in [0, 1]$, vector stochastic processes defined on a probability space $\langle \Omega_\varepsilon, \mathcal{F}_\varepsilon, \mathsf{P}_\varepsilon \rangle$, whose components are real-valued processes defined on the interval $[0, \infty)$.

The following condition is an analogue of the condition $\mathbf{Z}_9$ for stochastic processes.

$\mathbf{Z}_{10}$: **(a)** $\bar{\xi}'_\varepsilon(t), t \geq 0, \stackrel{\mathsf{d}}{\longrightarrow} \bar{\xi}'_0(t), t \geq 0$, as $\varepsilon \to 0$, **(b)** $\bar{\xi}''_\varepsilon(t), t \geq 0 \stackrel{\mathsf{d}}{\longrightarrow} \bar{\xi}''_0(t), t \geq 0$, as $\varepsilon \to 0$, where $\bar{\xi}''_0(t), t \geq 0$, is a.s. a non-random function.

The following theorem is an analogue of Theorem A.7.

Theorem A.9 *Let the condition* $\mathbf{Z}_{10}$ *be satisfied. Then,*

$$(\bar{\xi}'_\varepsilon(t), \bar{\xi}''_\varepsilon(t)), t \geq 0 \stackrel{\mathsf{d}}{\longrightarrow} (\bar{\xi}'_0(t), \bar{\xi}''_0(t)), t \geq 0 \text{ as } \varepsilon \to 0. \tag{A.12}$$

Let, for $\varepsilon \in [0, 1]$, the stochastic process $\bar{\xi}^+_\varepsilon(t) = \bar{\xi}'_\varepsilon(t) + \bar{\xi}''_\varepsilon(t), t \geq 0$.

The following useful theorem is a direct corollary of Theorem A.9.

Theorem A.10 *Let the condition* $\mathbf{Z}_{10}$ *be satisfied. Then, the following relations holds:* $\bar{\xi}^+_\varepsilon(t), t \geq 0 \stackrel{\mathsf{d}}{\longrightarrow} \bar{\xi}^+_0(t), t \geq 0$, *as* $\varepsilon \to 0$. *In particular, if* $\bar{\xi}''_0(t) \stackrel{a.s.}{=} \bar{0}, t \geq 0$, *then* $\bar{\xi}^+_0(t) = \bar{\xi}'_0(t), t \geq 0$.

Let $\bar{\xi}'_\varepsilon(t) = (\xi'_{\varepsilon,1}(t), \ldots, \xi'_{\varepsilon,k}(t)), t \geq 0$, and $\bar{\xi}''_\varepsilon(t) = (\xi''_{\varepsilon,1}(t), \ldots, \xi''_{\varepsilon,k''}(t))$ be, for every $\varepsilon \in [0, 1]$, càdlàg stochastic processes.

Let as also assume that the following condition holds:

$\mathbf{Z}_{11}$: **(a)** $\bar{\xi}'_\varepsilon(t), t \geq 0 \stackrel{\mathsf{J}}{\longrightarrow} \bar{\xi}'_0(t), t \geq 0$, as $\varepsilon \to 0$, **(b)** $\bar{\xi}''_\varepsilon(t), t \geq 0 \stackrel{\mathsf{U}}{\longrightarrow} \bar{\xi}''_0(t), t \geq 0$, as $\varepsilon \to 0$, where $\bar{\xi}''_0(t), t \geq 0$, is a non-random continuous function.

The following useful theorem is a functional analogue of Theorem A.10.

Theorem A.11 *Let the condition* $\mathbf{Z}_{11}$ *be satisfied. Then,* $\bar{\xi}^+_\varepsilon(t), t \geq 0 \stackrel{\mathsf{J}}{\longrightarrow} \bar{\xi}^+_0(t), t \geq 0$, *as* $\varepsilon \to 0$. *In particular, if* $\bar{\xi}''_0(t) = \bar{0}, t \geq 0$, *then* $\bar{\xi}^+_0(t) = \bar{\xi}'_0(t), t \geq 0$.

The proofs of Theorems A.7–A.11 as well as more general propositions can be found, for example, in Silvestrov (2004).

Appendix B
Methodological and Bibliographical Notes

This appendix contains methodological and bibliographic notes and comments on the new results presented in Volume 1. Some promising issues for future research in this area are also commented on.

B.1 Methodological Notes

B.1.1 A Survey of Works Related to Results Presented in the Book. In this subsection, we briefly comment on works related to the results presented in Volume 1.

Volume 1 is devoted to limit theorems for first-rare-event and hitting times for perturbed finite semi-Markov processes.

Random functionals similar to first-rare-event times and hitting times are also known by names such as first passage times, absorption times, in theoretical studies, as well as lifetimes, first failure times, extinction times, etc., in applications. Limit theorems for such functionals for perturbed Markov-type processes have been studied by many researchers.

The most deeply studied is the case of perturbed Markov chains and semi-Markov processes with finite phase spaces. The relevant works can be divided into two groups.

Regularly perturbed models with one class of communicative states for the corresponding limiting embedded Markov chains were studied in the works of Harris (1952), Belyaev (1963), Keilson (1966b, 1974, 1979, 1998), Darroch and Morris (1967), Korolyuk (1969), Silvestrov (1969b, 1970a, b , 1971a, b, 1972c, 1974, 1980a, 2016, 2019), Anisimov (1971a), Holst (1971, 1977, 1979), Koroljuk, Penev, and Turbin (1972, 1973), Masol and Silvestrov (1972), Zakusilo (1972a, b), Kovalenko (1973), Zubkov, and Mihaĭlov (1974), Masol (1974, 1976), Courtois (1975), Aldous (1982), Anisimov and Chernyak (1982), Korolyuk, D. (1983), Brown and Guang Ping (1984), Gut and Holst (1984), Aldous and Brown (1992, 1993), Asmussen (1994), Gyllenberg and Silvestrov (1994, 1998, 1999, 2008), Silvestrov and Drozdenko (2005, 2006a, b), Drozdenko (2007a, b ,2009), Benois, Landim,

D. Silvestrov, *Perturbed Semi-Markov Type Processes I*,
https://doi.org/10.1007/978-3-030-92403-4

and Mourragui (2013), Griffiths, Kang, Oliveira, and Patel (2014), Peres and Sousi (2015), Nardi, Zocca, and Borst (2016), and Mao Yong-Hua and Zhang Chi (2017).

Singularly perturbed models with several closed classes of communicative states for the corresponding limiting embedded Markov chains were studied in the works of Meshalkin (1958), Simon and Ando (1961), Hanen (1963d), Anisimov (1971b, 1988, 2008), Korolyuk and Turbin (1970, 1976a), Gusak and Korolyuk (1971), Kovalenko (1975, 1976), Gaĭtsgori and Pervozvanskiy(1975), Louchard and Latouche (1982), Coderch,Willsky, Sastry, and Castañon (1983), Delebecque, Quadrat, and Kokotović (1984), Cao and Stewart (1985), Rohlichek (1987), Rohlicek and Willsky (1988), Latouche (1991), Schweitzer (1991), Stewart (1994, 1998, 2001, 2003, and Yin and Zhang (2003, 2005, 2013).

Random functionals such as first-rare-event times and hitting times are natural generalisations of geometric type random sums. That is why, limit theorems for such random functionals are closely related to limit theorems for geometric type random sums and first-rare-event times for regenerative processes and related counting processes, which were the subject of research in works listed below in Sect. B.2.3. There exists also an extensive bibliography of works listed below in Sect. B.2.4, where limit theorems for hitting type functionals have been studied for perturbed Markov chains and semi-Markov type processes with countable and arbitrary phase spaces.

In Part I we present necessary and sufficient conditions for convergence in distribution for first-rare-event times and J-convergence for first-rare-event processes and the corresponding rare-event counting processes for regularly perturbed semi-Markov processes.

The main feature of most of the previous results is that they provide sufficient conditions of convergence for such functionals. As a rule, these conditions include assumptions that imply weak convergence for the distributions of sums of i.i.d. random variables distributed as the sojourn times for the semi-Markov processes (for each state) to some infinitely divisible laws plus some ergodicity conditions for the embedded Markov chains plus the condition of vanishing probabilities of a rare event occurring during one transition step for the semi-Markov processes.

The originating works containing such limit theorems for hitting times and similar stopping times for Markov chains and semi-Markov processes are Harris (1952), Meshalkin (1958), Gnedenko and Kovalenko (1964), Keilson (1966b), Solov'ev (1966), and and Korolyuk (1969). The most general sufficient conditions of this kind for convergence in distribution and convergence in topologies U and J for sum processes of real-valued random variables defined on regularly perturbed finite Markov chains stopped at hitting, renewal, and more general Markov-type moments can be found in Silvestrov (1970a, b, 1974).

The results presented in Part I relate to the model of regularly perturbed semi-Markov processes with a finite phase space. Instead of conditions based on "individual" distributions of sojourn times, we use more general and weaker conditions imposed on distributions of sojourn times averaged over the stationary distributions of the corresponding embedded Markov chains. Moreover, we show that these conditions are not only sufficient but also necessary conditions for convergence in distribution of first-rare-event times and convergence in the Skorokhod J-topology of rare-event processes and the corresponding rare-event counting processes.

These results are also illustrated by necessary and sufficient conditions of weak convergence for non-ruin distribution functions for perturbed risk processes in the models of diffusion and stable approximations as well as necessary and sufficient conditions of convergence in distribution of first-rare-event times for some perturbed M/M-type queuing systems.

The approach used in the book combines traditional methods based on cyclic representations of first-rare-event times in the form of geometric type random sums and general limit theorems for randomly stopped stochastic processes, developed in the works of Silvestrov (1971c, 1972c, d, e, 1973, 1974, 2004).

In the context of necessary and sufficient conditions of convergence in distribution for first-rare-event times, we would like also to mention the books Gnedenko and Korolev (1996) and Bening and Korolev (2002), where one can find some related results for geometric sums of random variables as well as the articles Korolyuk, D. and Silvestrov (1983, 1984) and Silvestrov and Velikiĭ (1988), where one can find some related results for similar functionals defined on Markov chains with arbitrary phase spaces.

In Part II we present new asymptotic recurrent algorithms of phase space reduction for regularly and singularly perturbed finite semi-Markov processes, efficient conditions for weak convergence of distributions and convergence of expectations for hitting times for regularly and singularly perturbed finite semi-Markov processes, and recurrent formulas for finding the corresponding normalisation functions, Laplace transforms for limiting distributions, and limits for expectations of hitting times.

Aggregation/disaggregation is one of the most effective and widely used approaches in the study of asymptotics for hitting times and similar functionals, especially for singularly perturbed models of Markov chains and semi-Markov processes. Selected works containing relevant asymptotic results for finite Markov chains and semi-Markov processes are Gambin, Krzyżanowski, and Pokarowski (2008), Marek, Mayer, and Pultarová (2009), Simon and Ando (1961), Anisimov (1971a, b, 1988, 2008), Courtois (1975), Gaĭtsgori and Pervozvanskiy (1975, 1983), Korolyuk and Turbin (1978), Voĭna (1979), Louchard and Latouche (1982), Coderch, Willsky, Sastry, and Castañon (1983), Chatelin and Miranker (1984), Delebecque, Quadrat, and Kokotović (1984), Cao and Stewart (1985), Haviv (1987, 1992, 1999), Rohlichek (1987), Rohlicek and Willsky (1988), Latouche (1991), Schweitzer (1984, 1991), Stewart (1994, 1998, 2001, 2003), Marek and Mayer (1998), Yin, Zhang, and Badowski (2000c, 2003), Yin and Zhang (2003, 2005, 2013), Korolyuk and Limnios (2005), Marek and Pultarová (2006), Gambin, Krzyżanowski, and Pokarowski (2008), and Marek, Mayer, and Pultarová (2009).

This approach is also widely used in limit theorems for hitting type functionals for perturbed Markov chains and semi-Markov type processes with countable and general phase spaces listed below in Sect. B.2.4, works containing results on rates of convergence and asymptotic expansions for distributions of hitting type functionals listed below in Sect. B.2.5, as well as in works containing asymptotic results for other functionals such as moments of hitting times, stationary and quasi-stationary distributions, eigenvalues, eigenvectors, Perron roots, coefficients of ergodicity, etc., which are listed below in Sect. B.2.6.

In Part II, we use the aggregation/disaggregation approach in a stochastic semi-Markov setting. Instead of algebraic reduction of the corresponding systems of linear equations, we use algorithms of phase space reduction for the corresponding semi-Markov processes based on the aggregation of their transition times. In such an algorithm, successive moments of hitting into new reduced phase spaces play the role of successive moments of jumps for new reduced semi-Markov processes. We also found natural perturbation conditions for the initial transition characteristics of the corresponding perturbed semi-Markov processes, which are inherited by the reduced semi-Markov processes.

The above asymptotic recurrent phase space reduction algorithm is based on the recurrent alternating application of two procedures: the removal of virtual transitions and one-step reduction of phase space for singularly perturbed semi-Markov processes. The first procedure provides a kind of asymptotic compression of the transition times and allows one to explicitly transform the initial normalisation functions and effectively compute the corresponding weak limits for distribution functions of new transition times for semi-Markov processes with removed virtual transitions. The second procedure allows one to reduce the phase space of the original semi-Markov processes and effectively compute the corresponding weak limits for distribution functions of transition times and the limiting transition probabilities of embedded Markov chains for the reduced semi-Markov processes. The above procedures preserve hitting times between states from the reduced phase spaces, that is, these times and, therefore, their distributions are the same for the original and the reduced semi-Markov processes. They allow one to effectively compute normalisation functions, weak limits for distributions of hitting times, and limits for their expectations for singularly perturbed semi-Markov processes as well as to track "switching" parameters determining the forms of these normalisation functions, weak limits for distributions of hitting times, and limits for their expectations.

Semi-Markov processes independently introduced by Lévy (1954), Smith (1955), and Takács (1954) naturally generalise discrete and continuous time Markov chains. These processes are a very flexible and efficient tool for analysing queuing, reliability and control systems, bio-stochastic systems, information networks, financial and insurance processes, and many other stochastic systems and processes.

The algorithms of phase space reduction presented in the book give one more example of effectiveness of the semi-Markov setting, which is an adequate and, moreover, a necessary element of the proposed method. Even in the case, where the initial process is a discrete or continuous time Markov chain, the procedure of phase space reduction leads to a semi-Markov process, since the times between successive hitting into the reduced phase space by the original process have distributions that can be neither geometric nor exponential.

Particular attention is paid to finding conditions under which the distribution functions of properly normalised hitting times for perturbed semi-Markov processes converge weakly to some distribution functions that are not concentrated at zero, and at the same time, expectations of hitting times normalised by the same functions converge to the first moments of the corresponding limiting distribution functions. The point is that this is not always true for singularly perturbed semi-Markov processes. At the same time, such simultaneous convergence plays an important role in applications to ergodic theorems for perturbed semi-Markov-type processes, in particular, to ergodic theorems for perturbed alternating regenerative processes modulating semi-Markov processes, which are the subject of research in Volume 2.

In this context, we would like to mention works of Hatori (1959, 1960), Leadbetter (1963), Silvestrov (1980a, b, 1996a, b, 2000a, 2007, 2010), Kohlas (1983), Abadov (1984), Gaver, Jacobs and Latouche (1984), Alsmeyer and Irle (1986), Silvestrov and Abadov (1984, 1991, 1993), Gyllenberg and Silvestrov (1998, 1999, 2008), Avrachenkov and Lasserre (1999), Avrachenkov and Haviv (2004), Hunter (2005), Gambin, Krzyżanowski, and Pokarowski (2008), Avrachenkov, Filar and Howlett (2013), Griffiths, Kang, Oliveira, and Patel (2014), Silvestrov,Manca, and Silvestrova (2014), Petersson (2016a, c), Silvestrov, D. and Silvestrov, S. (2016, 2017a, b, c, d), and Silvestrov and Manca (2017a, b) devoted to the study of moments of hitting times and related functionals, in particular stationary probabilities, for Markov-type processes including the corresponding asymptotic problems for perturbed processes.

Here, we have the opportunity to pay special attention to some of the latest works, where a deep asymptotic analysis for such functionals was carried out for regularly and singularly perturbed Markov chains and semi-Markov processes, based on asymptotically comparable power-type functions, in Gambin, Krzyżanowski, and Pokarowski (2008), Laurent series, in Avrachenkov, Filar and Howlett (2013), and Laurent asymptotic expansions without and with explicit upper bound for reminders, in Silvestrov, D. and Silvestrov, S. (2017c).

B.1.2 New Results Presented in the Book. In this section, we comment on the new results presented in Volume 1.

The main achievement of Part I is the necessary and sufficient conditions for convergence in distribution for first-rare-event times and processes and the corresponding rare-event counting processes. These conditions do not have a gap between their necessary and sufficient parts and give a kind of "final solution" for the limit theorems for the first-rare-event times and processes for regularly perturbed semi-Markov processes with finite phase spaces.

These results generalise and improve the results concerning necessary and sufficient conditions for convergence in distribution for first-rare-event times for semi-Markov processes obtained in papers by Silvestrov and Drozdenko (2005, 2006a, 2006b), and Drozdenko (2007a, 2007b, 2009).

First, weaker ergodic conditions for the embedded Markov chains are proposed. Instead of the traditional assumption about the convergence of transition probabilities to transition probabilities of some ergodic Markov chain, a new condition of asymptotically uniform ergodicity is used. Second, new proofs based on general limit theorems for stopping random processes, developed and presented in detail in Silvestrov (2004), are used instead of more traditional proofs based on cyclical representations of the first-rare-event times in the form of geometric-type random sums. This made it possible to formulate the results in a more general form of the corresponding functional limit theorems as well as to clarify the "random stopping" stricture of limiting random variables for the first-rare-event times.

The main limit theorems for the first-rare-event times and processes are presented in Chap. 2, which is an extended version of article Silvestrov (2016), and for rare-event counting processes in Chap. 3. Applications to perturbed risk processes given in Chap. 5 represent the corresponding results from Silvestrov and Drozdenko (2006b).

Generalisations of the limit theorems given in Chap. 2 to the vector first-rare-event times and rewards, the upper and low first-rare-event times, first-rare-event times for Markov renewal processes with transition period and extending phase spaces, as well as standard and directed hitting times given in Chap. 4, and the examples of applications to perturbed queuing systems given in Chaps. 6 and 7, are new results.

I would like to add comments related to the applications presented in Chaps. 5–7. The idea was to illustrate the theoretical results presented in the previous chapters by applying them to some classical models and to show that even for such models new results can be obtained. Namely, the necessary and sufficient conditions in their final form without gaps between their necessary and sufficient parts can be obtained in the corresponding limit theorems. Possible generalisations for more complex models are beyond the scope of this book.

In Part II, the aforementioned aggregation/disaggregation approach is used for constructing new asymptotic recurrent algorithms of phase space reduction and applying them to study of asymptotics for the hitting times for regularly and singularly perturbed finite semi-Markov processes.

The main achievement of Part II is the construction of new asymptotic recurrent algorithms of phase space reductions based on stochastic aggregation of transition times for perturbed semi-Markov processes. The key role here is played by new balanced forms of perturbation conditions imposed on transition characteristics of the original semi-Markov processes. These conditions are inherited by the reduced semi-Markov processes and are supplemented by effective recurrent formulas for finding transition characteristics of these reduced semi-Markov processes.

The proposed asymptotic recurrent algorithms are computationally efficient for models with regular and singular perturbations. They have a clear recurrent form well-prepared, for example, for efficient programming.

In this book, we continue the direction of research based on recurrent phase space reduction algorithms for semi-Markov processes, proposed in the article Silvestrov and Manca (2017), where a forward recurrent algorithm for computing the Laplace transforms of hitting times was presented, papers Silvestrov, D. and Silvestrov, S. (2016, 2017a, b, d) and book Silvestrov, D. and Silvestrov, S. (2017c), where the recurrent algorithms of phase space reduction combined with the techniques of Laurent asymptotic expansions are used for obtaining asymptotic expansions for moments of hitting times and stationary distributions for regularly and singularly perturbed semi-Markov processes.

In this book, the asymptotic recurrent algorithms for singularly perturbed semi-Markov processes are applied for solving a different problem, namely, obtaining weak asymptotics for distributions of hitting times. In the above works, the polynomial type perturbation models were researched. In this book, new type of perturbation conditions, based on the notion of complete families of asymptotically comparable functions, is proposed and thoroughly investigated. In particular, perturbation models can be based on general power functions and their natural combinations with exponential and logarithmic functions. Also, the forward recurrent algorithms of phase space reduction are supplemented by new backward recurrent algorithms, which allow one recurrently compute normalisation functions and Laplace transforms for limiting distributions of hitting times for arbitrary initial states. Finally, the analogous asymptotic results, based on more general perturbation conditions connected with the notion of complete families of asymptotically comparable functions, are obtained for expectations of hitting times. Part II is a significantly expanded version of the article Silvestrov (2019).

New asymptotic recurrent algorithms of phase space reduction for singularly perturbed semi-Markov processes and limit theorems for hitting times and their expectations presented in Part II play a key role in obtaining ergodic theorems for perturbed regenerative type processes modulated by regularly and singularly perturbed semi-Markov processes. Such ergodic theorems are the subject of research in Volume 2 of this monograph. The key role in such theorems is played by the asymptotics of the distributions and expectations of the hitting times (in particular, the return times) for the modulating semi-Markov processes. It is important that it was possible to obtain convergence conditions for the hitting times, which are based on the minimal convergence conditions for the transition characteristics of perturbed semi-Markov processes typical for ergodic theorems, and allows one to effectively compute normalisation functions, limiting distributions, expectations, and other characteristics determining the corresponding stationary distributions.

In conclusion it worth mentioning that continuous and discrete time Markov chains are particular cases of semi-Markov processes. That is why all limit theorems for perturbed semi-Markov processes presented in the book have obvious simplified analogues for regularly and singularly perturbed continuous and discrete time Markov chains.

B.1.3 New Problems. In this subsection, we formulate some new open problems related to the results presented in the book.

Part I presents necessary and sufficient conditions (without a gap between the necessary and sufficient parts) in limit theorems on convergence in distribution for the first-rare-event times and convergence in the Skorokhod J-topology for the rare-event processes and the counting processes generated by flows of rare events for the regularly perturbed finite semi-Markov processes.

It should be noted that obtaining necessary and sufficient conditions for convergence without a gap between their necessary and sufficient parts is usually a difficult task.

It would be natural to try to generalise these results to perturbed semi-Markov-type processes with countable and general phase spaces.

It would also be interesting to try to obtain necessary and sufficient conditions for convergence similar to those presented in Part I (based on the averaged characteristics of the corresponding semi-Markov processes) for the model of the first-rare-event reward functionals and processes with real-valued random summands for the case where the corresponding "external" limiting process is a general Lévy process with a Gaussian component.

A more ambitious open problem is related to finding similar necessary and sufficient convergence conditions in limit theorems for hitting times for singularly perturbed semi-Markov processes.

There is no doubt that the necessary and sufficient conditions for weak convergence of distribution functions without ruin, given in Chap. 5, can be generalised to more general models of risk processes, for example, risk processes with Markov-type modulation of input claim flows, risk processes with renewal type input claim flows, etc.

In Chaps. 6 and 7, some examples of necessary and sufficient conditions for the convergence in distribution for the first-rare-event times for perturbed M/M-type queuing systems are given. These results can certainly be generalised to perturbed M/G queuing systems and other models of perturbed queuing systems.

Part II presents phase space reduction algorithms and limit theorems for hitting times for singularly perturbed finite semi-Markov processes for the case when the time of reaching has finite expectations. In Chap. 12, we present some generalisation of the model in which the transition time distributions belong to the domains of attraction of stable laws.

Our hypothesis is that these results can be generalised to the case where the transition times belong to the regions of attraction of general infinitely divisible laws.

It would be natural to try to generalise the limit theorems for the hitting times for singularly perturbed semi-Markov processes, presented in Part II, to a more general form of the functional limit theorems.

Recurrent algorithms for obtaining asymptotic expansions with explicit estimates of the remainder terms for the hitting times for regularly and singularly perturbed semi-Markov processes, presented in the articles and the book of Silvestrov D. and Silvestrov S. (2017a, b, c, d), give hope that similar results with explicit estimates of the convergence rates can be obtained for distributions of the hitting times.

B.2 General Bibliographical Remarks

In this section, we briefly present a general bibliography of papers in the field of asymptotic problems for hitting times and related functionals, as well as for perturbed Markov, semi-Markov, and regenerative processes. These works can be divided into several groups.

B.2.1. The first works related to asymptotic problems for perturbed Markov and semi-Markov processes are Harris (1952), Meshalkin (1958), Simon, and Ando (1961), Belyaev (1963), Hanen (1963a, b, c, d), Kingman (1963), Kovalenko (1965, 1973, 1975), Keilson (1966a, b, 1974), Seneta (1967, 1968, 1973), Korolyuk (1969), Silvestrov (1969a, b, c, 1970a, b, 1971a, b, d, 1972a, b, c, 1974), Korolyuk and Turbin (1970), Anisimov (1971a, b, 1973a, b), Gusak and Korolyuk (1971), Turbin

(1971), Korolyuk, Penev, and Turbin (1972), Poliščuk and Turbin (1973), Korolyuk, Brodi, and Turbin (1974), Courtois (1975), and Gaĭtsgori and Pervozvanskiy (1975).

B.2.2. Limit theorems for distributions of hitting times and related problems for regularly and singularly perturbed finite Markov chains and semi-Markov-type processes were studied in Harris (1952), Meshalkin (1958), Simon and Ando (1961), Belyaev (1963), Hanen (1963d), Keilson (1966b, 1974, 1979, 1998), Darroch and Morris(1967), Korolyuk (1969), Silvestrov (1969b, 1970a, b, 1971a, b, d, 1974, 1980a, 2000a, 2016, 2019), Anisimov (1971a, b, 1988, 2008), Korolyuk and Turbin (1970, 1976a), Gusak and Korolyuk (1971), Holst (1971, 1977, 1979), Turbin (1971), Koroljuk, Penev, and Turbin (1972, 1973), Masol and Silvestrov (1972), Zakusilo (1972a, b), Kovalenko (1973, 1975, 1976), Zubkov, and Mihaĭlov (1974), Masol (1974, 1976), Courtois (1975), Gaĭtsgori and Pervozvanskiy (1975), Aldous (1982), Anisimov and Chernyak (1982), Louchard and Latouche (1982), Coderch, Willsky, Sastry, and Castañon (1983), Korolyuk, D. (1983), Brown and Guang Ping (1984), Delebecque, Quadrat, and Kokotović (1984), Gut and Holst (1984), Cao and Stewart (1985), Rohlichek (1987), Rohlicek and Willsky (1988), Latouche (1991), Schweitzer (1991), Aldous and Brown (1992, 1993), Asmussen (1994), Gyllenberg and Silvestrov (1994, 1998, 1999a, 2008), Stewart (1994, 1998, 2001, 2003), Silvestrov and Drozdenko (2005, 2006a, b), Drozdenko (2007a, b, 2009), Benois, Landim, and Mourragui (2013), Kartashov (2013a, 2015), Griffiths, Kang, Oliveira, and Patel (2014), Peres and Sousi (2015), Nardi, Zocca, and Borst (2016), Butler (2017), Mao Yong-Hua and Zhang Chi (2017), Petersson (2017), Clancy and Tjia (2018), Hössjer, Bechly, and Gauger (2018), Kumar and Gupta (2020), Lopes and Luczak (2020), and Cator and Don (2021).

B.2.3. Random functionals such as the first-rare-event times and the hitting times can be considered as natural generalisations of geometric-type random sums. That is why, limit theorems for such random functionals are closely connected with limit theorems for geometric-type random sums and the first-rare-event times for regenerative processes and related counting processes that were a subject of thorough studies in Rényi (1956), Belyaev (1963), Kovalenko (1965), Keilson (1966b, 1979), Gnedenko and Kovalenko (1964), Gnedenko and Frayer (1969), Silvestrov (1969a, b, 1972c, 1974, 1980a, 1995, 2000b, 2004, 2014a), Masol (1973, 1974), Solov'ev (1966, 1971), Jagers and Lindvall (1974), Kovalenko and Kuznetsov (1981, Korolyuk, D. (1982, 1983), Kalashnikov (1983, 1986, 1990, 1994b, 1997), Khusanbaev (1983, 1984), Kalashnikov and Vsekhsvyatskii (1985), Vsekhsvyatskii and Kalashnikov (1988), Hsu Guanghui and He Qiming (1990), Kruglov and Korolev (1990), Kartashov (1991), Gnedenko and Korolev (1996), Englund and Silvestrov (1997), Gyllenberg and Silvestrov (1999b, 2000a, b, 2008). Englund (2001), Bening and Korolev (2002), Silvestrov and Teugels (2004), Silvestrov and Drozdenko (2005, 2006a, b), Blanchet and Glynn (2007), Drozdenko (2007a, b, 2009), Glynn (2011), Ni (2011, 2014), Asmussen and Kortschak (2013), Asmussen and Foss (2014), Petersson (2014, 2016c), Korolev (2016), Clément and Landy (2017), Kushnir, O. and Kushnir, V. (2017), and Slepov (2021).

B.2.4. Similar asymptotic problems for Markov chains and processes of Markov type with a countable and general phase space were studied in Gaver (1964, 1976),

Keilson (1966a), Silvestrov (1969a, b, c, 1972a, b, c, 1974, 1979, 1980a, 1981), Gusak and Korolyuk (1971), Basu (1972), Anisimov (1973a, b, 1980, 1986, 1988, 2008), Solov'ev (1972), Gut (1974a, b, 1975, 1988), Mileshina (1975), Chow and Hsiung (1976), Korolyuk and Turbin (1976b, 1978), Kalashnikov (1978a, b, 1981, 1986, 1994a), Voĭna (1978, 1979), Kaplan (1979, 1980), Chow, Hsiung, and Yu (1980), Wang Zi Kun (1980), Kalashnikov and Anichkin (1981), Janson (1983), Bobrova (1983), Korolyuk, D. and Silvestrov (1983, 1984), Abadov (1984), Lalley (1984), Kartashov (1986, 1987, 1996, 2013b), Martin-Löf (1986), Silvestrov and Velikiĭ (1988), Aldous (1989), Korolyuk, D. (1990), Silvestrov and Abadov (1991, 1993), Korolyuk and Swishchuk (1995a, b), Sevast'yanov (1995, 1998, 2000, 2006), Motsa and Silvestrov (1996), Korolyuk (1997), Borovkov (1998a, b, 2015, 2016), Gasanenko (1999, 2001), Kesten and Maller (1999), Korolyuk, V.S. and Korolyuk, V.V. (1999), Asmussen (1998, 2000), Korolyuk and Limnios (2000, 2002, 2005, 2010), Alsmeyer and Hoefs (2001, 2004), Kupsa and Lacroix (2005), Jagers, Klebaner, and Sagitov (2007), Goldsheid (2008), Asmussen and Albrecher (2010), Aliyev, Rahimov, and Navidi (2011), Löpker and Stadje (2011), Glynn (2011), Bulinskaya (2012), Limnios (2012), Asmussen and Kortschak (2013), Serlet (2013, 2018), Fernandez, Manzo, Nardi, and Scoppola (2015), Vakhtel' and Denisov (2015), Vysotsky (2015), Afanasyev (2016, 2018), Alsmeyer and Marynych (2016), Bertoin and Kortchemski (2016), Fernandez, Manzo, Nardi, Scoppola, and Sohier (2016), Sun, Feuillet, and Robert (2016), Mao Yong-Hua and Zhang Chi (2017), Pène, Saussol, and Zweimüller (2017), Schulte-Geers and Stadje (2017), van Doorn (2017), Zakusilo and Matsak (2017), Denisov, Sakhanenko, and Wachtel (2018), D'Onofrio, Macci, and Pirozzi (2018), Ragimov, Farhadova, and Khalilov (2018), Aliev, Ragimov, Farhadova, and Khalilov (2019), Sloothaak,Wachtel, and Zwart (2018), Helali and Löwe (2019), Kuntz, Thomas, Stan, and Barahona (2019), Rahimov, Ibadova, and Farhadova (2019), Zweimüller (2019), Ascione, Pirozzi, and Toaldo (2020), Chatrabgoun, Daneshkhah, and Parham (2020), Labbé and Lacoin (2020), Grama, Lauvergnat, and Le Page (2020), and Buck (2021).

B.2.5. Convergence rates, asymptotic expansions in weak and large deviations limit theorems for distributions of hitting type functionals for Markov-type processes with finite, countable, and general phase spaces have been studied in Silvestrov (1969a, 1969b, 1970a, 1970b, 1971b, 1972b, 1995, 2000, 2007, 2010, 2014a), Korolyuk and Turbin (1970, 1976a, b, 1978), Gusak and Korolyuk (1971), Turbin (1971), Koroljuk, Penev, and Turbin (1972, 1973, 1981), Poliščuk and Turbin (1973, Courtois (1975), Kalashnikov (1981, 1983, 1986, 1990, 1994a, b, 1997), Kalashnikov and Anichkin (1981), Aldous (1982), Ahmad and Basu (1983), Bobrova (1983), Abadov (1984), Cao and Stewart (1985), Kalashnikov and Vsekhsvyatskii (1985), Kartashov (1986, 1987, 1991, 1996, 2013b), Vsekhsvyatskii and Kalashnikov (1988), Stewart and Sun (1990), Sudakova (1990), Silvestrov and Abadov (1991, 1993), Schweitzer (1991), Aldous and Brown (1992, 1993), Stewart (1994, 1998, 2001, 2003), Englund and Silvestrov (1997), Korolyuk (1997), Gyllenberg and Silvestrov (1998, 1999, 2000, 2008), Roberts, Rosenthal, and Schwartz (1998), Gasanenko (1999, 2001), Korolyuk, V.S. and Korolyuk, V.V. (1999), Bon and Kalashnikov (2001), Englund (2001), Aliyev and Abadov (2007), Ni (2011, 2014),

Ferré, Hervé, and Ledoux (2013), Griffiths, Kang, Oliveira, and Patel (2014), Petersson (2014, 2016c), Peres and Sousi (2015), Gast and Gaujal (2017), and Matsak (2021).

B.2.6. Asymptotic results including convergence, estimates for convergence rates, and asymptotic expansions for other related functionals such as moments of hitting times, stationary and quasi-stationary distributions, eigenvalues, eigenvectors, Perron roots, coefficients of ergodicity, etc., were studied in Hanen (1963a, b, c, d), Seneta (1967, 1968a, b, 1973, 1980, 1988a, b, 1991, 1993, 2006), Schweitzer (1968, 1984, 1986, 1987, 1991), Silvestrov (1969b, 2000, 2007, 2010), Stewart (1969, 1973, 1979, 1983, 1984a, b, 1990, 1991, 1993a, b, 1994, 1998, 2001, 2003), Turbin (1972), Golub and Seneta (1973, 1974), Gut (1974a, b, 1988), Courtois (1975, 1982), Gaĭtsgori and Pervozvanskiy (1975, 1983), Kovalenko (1975, 1976), Courtois and Louchard (1976), Korolyuk and Turbin (1976b, 1978), Allen, Anderssen and Seneta (1977), Kalashnikov (1978b, 1986, 1990, 1994a), Latouche and Louchard (1978), Meyer (1980, 1994, 2015), Tweedie (1980, 1998), Kalashnikov and Anichkin (1981), Delebecque (1983), Kohlas (1983), Chatelin and Miranker (1984), Courtois and Semal (1984a, b, c, 1991), Gaver, Jacobs and Latouche (1984), Haviv and Rothblum (1984), Haviv and Van der Heyden (1984), Koury, McAllister and Stewart (1984), McAllister, Stewart, G., and Stewart, W. (1984), Silvestrov and Abadov (1984, 1991, 1993), Cao and Stewart (1985), Funderlic and Meyer (1985), Kartashov (1985a, b, c, 1996), Vantilborgh (1985), Alsmeyer and Irle (1986), Hunter (1986, 1991a, b, 2005), Haviv (1986, 1987, 1988, 1992, 1999, 2006), Haviv and Ritov (1986, 1993), Burnley (1987), Gibson and Seneta (1987), Haviv, Ritov and Rothblum (1987); Haviv, Ritov, and Rothblum (1992), Rohlichek (1987), Meyer and Stewart (1988), Rohlicek and Willsky (1988), Sumita and Reiders (1988), Louchard and Latouche (1990), Stewart and Sun (1990), Stewart and Zhang (1991), Borovkov, Fayolle, and Korshunov (1992), Hassin and Haviv (1992), Meyn and Tweedie (1992, 1993a, b, 2009), Barlow (1993a, b), O'Cinneide (1993), Lasserre (1994), Pollett and Stewart (1994), Stewart, G., Stewart, W., and McAllister (1994), Eleĭko and Shurenkov (1996), Hoppensteadt, Salehi, and Skorokhod (1996), Cao (1998), Gyllenberg and Silvestrov (1998, 1999, 2000, 2008), Kovalenko, Birolini, Kuznetsov, and Shenbukher (1998), Marek and Mayer (1998), Yin and Zhang (1998, 2003, 2005, 2013), Roberts, Rosenthal, and Schwartz (1998), Avrachenkov (1999, 2000), Avrachenkov and Lasserre (1999), Yin, Zhang, and Badowski (2000a, b, c, 2003), Yin, G., Zhang, Yang, and Yin, K. (2001), Avrachenkov, Filar, and Haviv (2002), Craven (2003), Mitrophanov (2003, 2005a, b, 2006), Altman, Avrachenkov, and Núñez-Queija (2004), Avrachenkov and Haviv (2004), Zhang and Yin (2004), Mitrophanov, Lomsadze, and Borodovsky (2005), Guo (2006), Marek and Pultarová (2006), Sirl, Zhang and Pollett (2007), Gambin, Krzyżanowski, and Pokarowski (2008), Marek, Mayer, and Pultarová (2009), Barbour and Pollett (2010, 2012), Nonaka, Ono, Sadakane, and Yamashita (2010), Avrachenkov, Filar, and Haviv (2013), Petersson (2013, 2016a, b, c), Silvestrov, Manca, and Silvestrova (2014), Masuyama (2015), Betz and Le Roux (2016), Boxma, Mandjes, and Reed (2016), Hunter (2016, 2018), Silvestrov, D. and Silvestrov, S. (2016, 2017a, b, c, d), Silvestrov, Petersson, and Hössjer (2018), Abola, Biganda, Silvestrov, D., Silvestrov, S., Engström, Mango, and Kakuba (2019),

Silvestrov, D., Silvestrov, S., Abola, Biganda, Engström, Mango, and Kakuba(2019, 2020a,b), Hervé and Ledoux (2020), and and Issaadi (2020).

B.2.7. In conclusion, we would like to refer to books and review articles devoted to models of stochastic processes and asymptotic problems, which are the objects of study in the book.

First, we would like to point out several books and review articles that contain materials on semi-Markov-type processes, regenerative processes, and related modes of random processes. These are books Smith (1955), Cox (1962), Kemperman (1962), Feller (1968, 1971), Kingman (1972), Korolyuk, Brodi and Turbin (1974), Çinlar (1974, 1975), Gikhman and Skorokhod (1973), Cohen (1976), Silvestrov (1980a), Volker (1980), Revuz (1984), Kalashnikov (1994b), Korolyuk and Swishchuk (1995a, b), Kovalenko, Kuznetsov, and Shurenkov (1996), Kijima (1997), Rolski, Schmidli, Schmidt and Teugels (1999), Thorisson (2000), Limnios and Oprişan (2001), Janssen and Manca (2006, 2007), Barbu and Limnios (2008), Gyllenberg and Silvestrov (2008), Harlamov (2008), Serfozo (2009), Stewart, W. (2009), Collet, Martínez, and San Martín (2013), Ross (2014), Grabski (2015), and Kallenberg (2021).

We would like to mention books and review articles that include material on perturbed Markov, semi-Markov, and regenerative processes and related asymptotic problems. These are books Seneta (1973, 2006), Silvestrov (1974, 2004, 2014b, 2015), Korolyuk and Turbin (1976a, 1978), Courtois (1977), Kalashnikov (1978a, 1994b, 1997), Anisimov (1988, 2008), Gut (1988), Kruglov and Korolev (1990), Stewart and Sun (1990), Kovalenko(1994), Stewart (1994, 1998, 2001), Korolyuk and Swishchuk (1995a, b), Gnedenko and Korolev (1996), Kovalenko, Kuznetsov, and Shurenkov (1996), Kartashov (1996), Borovkov (1998c), Yin and Zhang (1998, 2005, 2013), Korolyuk, V.S. and Korolyuk, V.V. (1999), Bening and Korolev (2002), Bini, Latouche and Meini (2005), Korolyuk and Limnios (2005), Barbu and Limnios (2008), Gyllenberg and Silvestrov (2008), Meyn and Tweedie (2009), Avrachenkov, Filar and Howlett (2013), and Silvestrov, D. and Silvestrov, S. (2017c).

We would also like to mention some books and reviews in which results on perturbed Markov-type processes are applied to control theory, decision-making processes, the Internet, queuing theory, reliability theory, mathematical genetics, models of population dynamics and epidemics, insurance, and financial mathematics. These are books Gnedenko and Kovalenko (1964), Kovalenko (1975, 1994), Courtois (1977), Kalashnikov (1978a, 1994b, 1997), Iglehart and Shedler (1980), Silvestrov (1980a, 2014b, 2015), Woodroofe (1982), Anisimov, Zakusilo and Donchenko 1987), Shedler (1987, 1993), Abbad and Filar (1995), Kovalenko, Kuznetsov, and Shurenkov 1996), Kijima (1997), Kovalenko, Kuznetsov, and Pegg 1997), Asmussen (2000, 2003), Anisimov (2008), Barbu and Limnios (2008), Gyllenberg and Silvestrov (2008), Asmussen and Albrecher (2010), Avrachenkov, Filar and Howlett 2013), and Obzherin and Boyko (2015).

References

1. Abadov, Z.A. (1984). Asymptotical Expansions with Explicit Estimation of Constants for Exponential Moments of Sums of Random Variables Defined on a Markov Chain and their Applications to Limit Theorems for First Hitting Times. Candidate of Science dissertation, Kiev State University.
2. Abbad, M., Filar, J.A. (1995). Algorithms for singularly perturbed Markov control problems: A survey. In: Leondes, C.T. (Ed.) Techniques in Discrete-Time Stochastic Control Systems. Control and Dynamic Systems, **73**, Academic Press, New York, 257–289.
3. Abola, B., Biganda, P.S., Silvestrov, D., Silvestrov, S., Engstrröm, C., Mango, J.M., Kakuba, G. (2019). Nonlinearly perturbed Markov chains and information networks. In: Skiadas, C.H. (Ed.) Proceedings of the 18th Applied Stochastic Models and Data Analysis International Conference. Florence, Italy, 2019, ISAST: International Society for the Advancement of Science and Technology, 51–79.
4. Afanasyev, V.I. (2016). On the time of attaining a high level by a transient random walk in a random environment. Teor. Verojatnost. Primen., **61**, no. 2, 234–267 (English translation in Theory Probab. Appl., **61**, no. 2, 178–207).
5. Afanasyev, V.I. (2018). Two-boundary problem for a random walk in a random environment. Teor. Verojatnost. Primen., **63**, no. 3, 417–430 (English translation in Theory Probab. Appl., **63**, no. 3 339–350).
6. Ahmad, I.A., Basu, A.K. (1983). Equivalence of rates of convergence in the CLT between maximum partial sums and first passage times of random walks. Scand. Actuar. J., no. 2, 89–105.
7. Aldous, D.J. (1982). Markov chains with almost exponential hitting times. Stoch. Proces. Appl., **13**, 305–310.
8. Aldous, D. (1989). Hitting times for random walks on vertex-transitive graphs. Math. Proc. Cambridge Philos. Soc., **106**, no. 1, 179–191.
9. Aldous, D.J., Brown, M. (1992). Inequalities for rare events in time-reversible Markov chains. I. In: Shaked, M, Tong. Y.L (Eds.) Stochastic Inequalities. AMS-IMS-SIAM Joint Summer Research Conference held in Seattle, Washington, 1991. Institute of Mathematical Statistics Lecture Notes, Monograph Series, **22**, Hayward, CA, 1–16.
10. Aldous, D.J., Brown, M. (1993). Inequalities for rare events in time-reversible Markov chains. II. Stoch. Process. Appl., **44**, no. 1, 15–25.
11. Aliyev, S.A., Abadov, Z.A. (2007). A new approach for studying asymptotics of exponential moments of sums of random variables, connected in Markov chain. I. Trans. Natl. Acad. Sci. Azerb., Ser. Phys.-Tech. Math. Sci. **27**, no. 7, 17–32.
12. Aliev, S.A., Ragimov, F.H., Farhadova, A.D., Khalilov, V.S. (2019). Limit theorems for the family of the first passage time of the level by random walk described by a nonlinear function of the autoregressive process of order one. Uzbek Math. J., no. 1, 4–13.

D. Silvestrov, *Perturbed Semi-Markov Type Processes I*,
https://doi.org/10.1007/978-3-030-92403-4

13. Aliyev, S., Rahimov, F., Navidi, M. (2011). On asymptotic behavior of conditional probability of crossing the nonlinear boundary by a perturbed random walk. Theory Stoch. Process., **17**, no. 1, 5–11.
14. Allen, B., Anderssen, R.S., Seneta, E. (1977). Computation of stationary measures for infinite Markov chains. In: Neuts, M.F. (Ed.) Algorithmic Methods in Probability. Studies in the Management Sciences, **7**, North-Holland, Amsterdam, 13–23.
15. Alsmeyer, G., Hoefs, V. (2001). Markov renewal theory for stationary m-block factors. Markov Process. Related Fields, **7**, no. 2, 325–348.
16. Alsmeyer, G., Hoefs, V. (2004). Markov renewal theory for stationary $(m + 1)$-block factors: first passage time and overshoot. In: Semi-Markov Processes: Theory and Applications. Comm. Statist. Theory Methods, **33**, no. 3, 545–568.
17. Alsmeyer, G., Irle, A. (1986). Asymptotic expansions for the variance of stopping times in nonlinear renewal theory. Stoch. Process. Appl., **23**, 235–258.
18. Alsmeyer, G. Marynych, A. (2016). Renewal approximation for the absorption time of a decreasing Markov chain. J. Appl. Probab.. **53**, no. 3, 765–782.
19. Altman, E., Avrachenkov, K.E., Núñez-Queija, R. (2004). Perturbation analysis for denumerable Markov chains with application to queueing models. Adv. Appl. Probab., **36**, no. 3, 839–853.
20. Anisimov, V.V. (1971a). Limit theorems for sums of random variables on a Markov chain, connected with the exit from a set that forms a single class in the limit. Teor. Verojatnost. Mat. Stat., **4**, 3–17 (English translation in Theory Probab. Math. Statist., **4**, 1–13).
21. Anisimov, V.V. (1971b). Limit theorems for sums of random variables that are given on a subset of states of a Markov chain up to the moment of exit, in a series scheme. Teor. Verojatnost. Mat. Stat., **4**, 18–26 (English translation in Theory Probab. Math. Statist., **4**, 15–22).
22. Anisimov, V.V. (1973a). Limit theorems for sums of random variables that are given on a countable subset of the states of a Markov chain up to the first exit time. Teor. Verojatnost. Mat. Stat., **8**, 3–13 (English translation in Theory Probab. Math. Statist., **8**, 1–10).
23. Anisimov, V.V. (1973b). Asymptotical consolidation of states for random processes. Kibernetika, no. 3, 109–117.
24. Anisimov, V.V. (1980). Limit theorems for processes that allow the asymptotic enlargement of a component. Theor. Verojatnost. Mat. Stat., **22**, 3–15 (English translation in Theory Probab. Math. Statist., **22**, 1–13).
25. Anisimov, V.V. (1986). Approximation of Markov processes that can be asymptotically lumped. Teor. Verojatnost. Mat. Stat., **34**, 1–12 (English translation in Theory Probab. Math. Statist., **34**, 1–11).
26. Anisimov, V.V. (1988). Random Processes with Discrete Components. Vysshaya Shkola and Izdatel'stvo Kievskogo Universiteta, Kiev, 183 pp.
27. Anisimov, V.V. (2008). Switching Processes in Queueing Models. Applied Stochastic Methods Series. ISTE, London and Wiley, Hoboken, NJ, 345 pp.
28. Anisimov, V.V., Chernyak, A.V. (1982). Limit theorems for certain rare functionals on Markov chains and semi-Markov processes. Teor. Verojatnost. Mat. Stat., **26**, 3–8 (English translation in Theory Probab. Math. Statist., **26**, 1–6).
29. Anisimov, V.V., Zakusilo, O.K., Donchenko, V.S. (1987). Elements of Queueing and Asymptotical Analysis of Systems. Lybid', Kiev, 248 pp.
30. Ascione, G., Pirozzi, E., Toaldo, B. (2020). On the exit time from open sets of some semi-Markov processes. Ann. Appl. Probab., **30**, no. 3, 1130–1163.
31. Asmussen, S. (1994). Busy period analysis, rare events and transient behaviour in fluid flow models. J. Appl. Math. Stoch. Anal., **7**, no. 3, 269–299.
32. Asmussen, S. (1997). Phase-type distributions and related point processes: fitting and recent advances. In: Chakravarthy, S., Attahiru S. Alfa (Eds.) Matrix-analytic methods in stochastic models. Lecture Notes in Pure and Applied Mathematics, **183**, Dekker, New York, 137–149.
33. Asmussen, S. (1998). Subexponential asymptotics for stochastic processes: extremal behavior, stationary distributions and first passage probabilities. Ann. Appl. Probab., **8**, no. 2, 354–374.

34. Asmussen, S. (2000). Ruin probabilities. Advanced Series on Statistical Science & Applied Probability, **2**. World Scientific, River Edge, NJ, xii+385 pp.
35. Asmussen, S. (2003). Applied Probability and Queues. Second edition, Applications of Mathematics, **51**, Stochastic Modelling and Applied Probability. Springer, New York, xii+438 pp.
36. Asmussen, S., Albrecher, H. (2010). Ruin Probabilities. Second edition, Advanced Series on Statistical Science & Applied Probability, **14**, World Scientific, Hackensack, NJ, xviii+602 pp.
37. Asmussen, S., Foss, S. (2014). On exceedance times for some processes with dependent increments. J. Appl. Probab., **51**, no. 1, 136–151.
38. Asmussen, S. Kortschak, D. (2013). Second order corrections for the limits of normalized ruin times in the presence of heavy tails. Stoch. Syst., **3**, no. 2, 591–633.
39. Avrachenkov, K.E. (1999). Analytic Perturbation Theory and Its Applications. PhD Thesis, University of South Australia, 1999.
40. Avrachenkov, K.E. (2000). Singularly perturbed finite Markov chains with general ergodic structure. In: R. Boel, R., Stremersch, G. (Eds.) Discrete Event Systems. Analysis and Control. Kluwer International Series in Engineering and Computer Science, **569**, Kluwer, Boston, 429–432.
41. Avrachenkov, K.E., Filar, J., Haviv, M. (2002). Singular perturbations of Markov chains and decision processes. In: Feinberg, E.A., Shwartz, A. (Eds.) Handbook of Markov Decision Processes. Methods and Applications. International Series in Operations Research & Management Science, **40**, Kluwer, Boston, 113–150.
42. Avrachenkov, K.E., Filar, J.A., Howlett, P.G. (2013). Analytic Perturbation Theory and Its Applications. SIAM, Philadelphia, PA, xii+372 pp.
43. Avrachenkov, K.E., Haviv, M. (2004). The first Laurent series coefficients for singularly perturbed stochastic matrices. Linear Algebra Appl., **386**, 243–259.
44. Avrachenkov, K.E., Lasserre, J.B. (1999). The fundamental matrix of singularly perturbed Markov chains. Adv. Appl. Probab., **31**, no. 3, 679–697.
45. Barbour, A.D., Pollett, P.K. (2010). Total variation approximation for quasi-stationary distributions. J. Appl. Probab., **47**, no. 4, 934–946.
46. Barbour, A.D., Pollett, P.K. (2012). Total variation approximation for quasi-equilibrium distributions, II. Stoch. Process. Appl., **122**, no. 11, 3740–3756.
47. Barbu, V.S., Limnios, N. (2008). Semi-Markov Chains and Hidden Semi-Markov Models Toward Applications. Their Use in Reliability and DNA Analysis. Lecture Notes in Statistics, **191**, Springer, New York, xiv+224 pp.
48. Barlow, J. (1993a). Perturbation results for nearly uncoupled Markov chains with applications to iterative methods. Numer. Math., **65**, no. 1, 51–62.
49. Barlow, J.L. (1993b). Error bounds for the computation of null vectors with applications to Markov chains. SIAM J. Matrix Anal. Appl., **14**, no. 3, 798–812.
50. Basu, A.K. (1972). Invariance theorems for first passage time variables. Canad. Math. Bull., **15**, 171–176.
51. Belyaev, Yu.K. (1963). Limit theorems for flows subject to rarefaction. Teor. Verojatnost. Primen., **8**, 175–184 (English translation in Theory Probab. Appl., **8**, 165–173).
52. Bening, V.E., Korolev, V.Yu. (2002). Generalized Poisson Models and their Applications in Insurance and Finance. Modern Probability and Statistics, VSP, Utrecht, 432 pp.
53. Benois, O., Landim, C., Mourragui, M. (2013). Hitting times of rare events in Markov chains. J. Stat. Phys., **153**, no. 6, 967–990.
54. Bertoin, J., Kortchemski, I. (2016). Self-similar scaling limits of Markov chains on the positive integers. Ann. Appl. Probab., **26**, no. 4, 2556–2595.
55. Betz, V., Le Roux, S. (2016). Multi-scale metastable dynamics and the asymptotic stationary distribution of perturbed Markov chains. Stochastic Process. Appl., **126**, no. 11, 3499–3526.
56. Billingsley, P. (2014). Convergence of Probability Measures. Second edition. Wiley Series in Probability and Statistics, Wiley, New York, 296 pp. (First edition in 1968).
57. Bini, D.A., Latouche, G., Meini, B. (2005). Numerical Methods for Structured Markov Chains. Numerical Mathematics and Scientific Computation, Oxford Science Publications. Oxford University Press, New York, xii+327 pp.

58. Blanchet, J., Glynn, P. (2007). Uniform renewal theory with applications to expansions of random geometric sums. Adv. Appl. Probab., **39**, no. 4, 1070–1097.
59. Bobrova, A.F. (1983). Estimates of accuracy of an asymptotic consolidation of countable Markov chains. In Stability Problems for Stochastic Models. VNIISI, Moscow, 16–24.
60. Bon, J.L., Kalashnikov, V. (2001). Some estimates of geometric sums. Statist. Probab. Lett., **55**, no. 1, 89–97.
61. Borovkov, A.A. (1998a). On the asymptotics of distributions of first-passage times. I. Mat. Zametki, **75**, no. 1, 24–39 (English translation in Math. Notes, **75**, no. 1–2, 23–37).
62. Borovkov, A.A. (1998b). On the asymptotics of distributions of first-passage times. II. Mat. Zametki, **75**, no. 3, 350–359 (English translation in Math. Notes, **75**, no. 3–4, 322–330).
63. Borovkov, A.A. (1998c). Ergodicity and Stability of Stochastic Processes. Wiley Series in Probability and Statistics, **314**, Wiley, Chichester, xxiv+585 pp. (Translation from the 1994 Russian original).
64. Borovkov, A.A. (2015). Integral theorems for the first passage time of an arbitrary boundary by a compound renewal process. Sibirsk. Mat. Zh., **56**, no. 5, 961–981 (English translation in Sib. Math. J., **56**, no. 5, 765–782).
65. Borovkov, A.A. (2016). On the distribution of the first passage time of an arbitrary remote boundary by a random walk. Teor. Verojatnost. Primen., **61**, no. 2, 210–233 (English translation in Theory Probab. Appl., **61**, no. 2, 235–254).
66. Borovkov, A.A., Fayolle, G., Korshunov, D.A. (1992). Transient phenomena for Markov chains and applications. Adv. Appl. Probab., **24**, 322–342.
67. Boxma, O., Mandjes, M., Reed, J. (2016). On a class of reflected AR(1) processes. J. Appl. Probab., **53**, no. 3, 818–832.
68. Brown, M.G., Guang Ping (1984). On the waiting time for the first occurrence of a pattern. In: Reliability Theory and Models, Charlotte, N.C., 1983. Notes Rep. Comput. Sci. Appl. Math., **10**, Academic Press, Orlando, FL, 267–272.
69. Buck, M. (2021). Limit theorems for random walks with absorption. J. Theor. Probab., **34**, 241–263.
70. Bulinskaya, E.V. (2012). Hitting times in presence of taboo points for a random walk. Mat. Tr., **15**, no. 1, 3–26 (English translation in Siberian Adv. Math., **22** (2012), no. 4, 227–242.
71. Burnley, C. (1987). Perturbation of Markov chains. Math. Mag., **60**, no. 1, 21–30.
72. Butler, R.W. (2017). Asymptotic expansions and hazard rates for compound and first-passage distributions. Bernoulli, **23**, no. 4B, 3508–3536.
73. Cao, W.L., Stewart, W.J. (1985). Iterative aggregation/disaggregation techniques for nearly uncoupled Markov chains. J. Ass. Comp. Mach., **32**, 702–719.
74. Cao, X.R. (1998). The Maclaurin series for performance functions of Markov chains. Adv. Appl. Prob., **30**, 676–692.
75. Cator, E., Don, H. (2021). Explicit bounds for critical infection rates and expected extinction times of the contact process on finite random graphs. Bernoulli, **27**, no. 3, 1556–1582.
76. 2020) Chatelin, F., Miranker, W.L. (1984). Aggregation/disaggregation for eigenvalue problems. SIAM J. Numer. Anal., **21**, no. 3, 567–582.
77. Chatrabgoun, O., Daneshkhah, A., Parham, G. (2020). On the functional central limit theorem for first passage time of nonlinear semi-Markov reward processes. Comm. Statist. Theory, Meth., **49**, no. 19, 4737–4750.
78. Chow, Y.S., Hsiung, C.A. (1976). Limiting behaviour of $\max_{j\leq n} S_j/j^{-\alpha}$ and first passage time in random walk with positive drift. Bull. Inst. Math. Acad. Sinica, **4**, 35–44.
79. Chow, Y.S., Hsiung, C.A., Yu, K.F. (1980). Limit theorems for a positively drifting process and its related first passage times. Bull. Inst. Math. Acad. Sinica, **8**, 141–172.
80. Çinlar, E. (1974). Markov renewal theory: a survey. Manag. Sci., **21**, no. 7, 727–752.
81. Çinlar, E. (1975). Introduction to Stochastic Processes. Prentice-Hall, Englewood Cliffs, N.J., x+402 pp.
82. Clancy, D., Tjia, E. (2018). Approximating time to extinction for endemic infection models. Methodol. Comput. Appl. Probab., **20**, no. 4, 1043–1067.

83. Clément, D., Landy, R. (2017). High order expansions for renewal functions and applications to ruin theory. Ann. Appl. Probab., **27**, no. 4, 2342–2382.
84. Coderch, M., Willsky, A.S., Sastry, S.S., Castañon, D.A. (1983). Hierarchical aggregation of singularly perturbed finite state Markov processes. Stochastics, **8**, 259–289.
85. Cohen, J.W. (1976). On Regenerative Processes in Queueing Theory. Lecture Notes in Economics and Mathematical Systems, **121**, Springer, Berlin, ix+93 pp.
86. Collet, P., Martínez, S., San Martín, J. (2013). Quasi-Stationary Distributions. Markov Chains, Diffusions and Dynamical Systems. Probability and its Applications, Springer, Heidelberg, xvi+280 pp.
87. Courtois, P.J. (1975). Error analysis in nearly-completely decomposable stochastic systems. Econometrica, **43**, no. 4, 691–709.
88. Courtois, P.J. (1977). Decomposability: Queueing and Computer System Applications. ACM Monograph Series, Academic Press, New York, xiii+201 pp.
89. Courtois, P.J. (1982). Error minimization in decomposable stochastic models. In: Ralph L.D., Teunis J.O. (Eds). Applied Probability – Computer Science: the Interface, Vol. I. Progress in Computer Science, **2**, Birkhäuser, Boston, 189–210.
90. Courtois, P.J., Louchard, G. (1976). Approximation of eigen characteristics in nearly-completely decomposable stochastic systems. Stoch. Process. Appl., **4**, 283–296.
91. Courtois, P.J., Semal, P. (1984a). Error bounds for the analysis by decomposition of non-negative matrices. In: Iazeolla, G., Courtois, P.J., Hordijk, A. (Eds). Mathematical Computer Performance and Reliability. North-Holland, Amsterdam, 209–224.
92. Courtois, P.J., Semal, P. (1984b). Block decomposition and iteration in stochastic matrices. Philips J. Res., **39**, no. 4–5, 178–194.
93. Courtois, P.J., Semal, P. (1984c). Bounds for the positive eigenvectors of nonnegative matrices and for their approximations by decomposition. J. Assoc. Comput. Mach.. **31**, no. 4, 804–825.
94. Courtois, P.J., Semal, P. (1991). Bounds for transient characteristics of large or infinite Markov chains. In: Stewart, W.J. (Ed.) Numerical Solution of Markov Chains. Probability: Pure and Applied, **8**, Marcel Dekker, New York, 413–434.
95. Cox, D.R. (1962). Renewal Theory. Methuen, London and Wiley, New York, ix+142 pp.
96. Craven, B.D. (2003). Perturbed Markov processes. Stoch. Models, **19**, no. 2, 269–285.
97. Darroch, J.N., Morris, K.W. (1967). Some passage-time generating functions for discrete-time and continuous-time finite Markov chains. J. Appl. Probab., **4**, 496–507.
98. Delebecque, F. (1983). A reduction process for perturbed Markov chains. SIAM J. Appl. Math., **43**, 325–350.
99. Delebecque, F., Quadrat, J.P., Kokotović, P.V. (1984). A unified view of aggregation and coherency in networks and Markov chains. Inter. J. Control, **40**, 939–952.
100. Denisov, D., Sakhanenko, A., Wachtel, V. (2018). First-passage times over moving boundaries for asymptotically stable walks. Teor. Verojatnost. Primen. **63**, no. 4, 755–778 (Also in Theory Probab. Appl., **63**, no. 4, 613–633).
101. D'Onofrio, G., Macci, C., Pirozzi, E. (2018). Asymptotic results for first-passage times of some exponential processes. Methodol. Comput. Appl. Probab., **20**, no. 4, 1453–1476.
102. Drozdenko, M. (2007a). Weak convergence of first-rare-event times for semi-Markov processes. I. Theory Stoch. Process., **13**, no. 4, 29–63.
103. Drozdenko, M. (2007b). Weak Convergence of First-Rare-Event Times for Semi-Markov Processes. Doctoral dissertation **49**, Mälardalen University, Västerås.
104. Drozdenko, M. (2009). Weak convergence of first-rare-event times for semi-Markov processes. II. Theory Stoch. Process., **15**, no. 2, 99–118.
105. Eleĭko, Ya.I., Shurenkov, V.M. (1996) On the asymptotic representation of the Perron root of a matrix-valued stochastic evolution. Ukr. Mat. Zh., **48**, 35–43 (English translation in Ukr. Math. J., **48**, 38–47).
106. Englund, E. (2001). Nonlinearly Perturbed Renewal Equations with Applications. Doctoral dissertation, Umeå University.
107. Englund, E., Silvestrov, D.S. (1997). Mixed large deviation and ergodic theorems for regenerative processes with discrete time. In: Jagers, P., Kulldorff, G., Portenko, N., Silvestrov,

D. (Eds.) Proceedings of the Second Scandinavian–Ukrainian Conference in Mathematical Statistics, Vol. I, Umeå, 1997. Theory Stoch. Process., **3(19)**, no. 1–2, 164–176.

108. Feller, W. (1968). An Introduction to Probability Theory and Its Applications. Vol. I. Third edition, Wiley, New York, xviii+509 pp. (First edition in 1950).
109. Feller, W. (1971). An Introduction to Probability Theory and Its Applications, Vol. II. (Second edition, Wiley Series in Probability and Statistics, Wiley, New York, xxiv+669 pp. (First edition in 1966).
110. Fernandez, R., Manzo, F., Nardi, F.R., Scoppola, E. (2015). Asymptotically exponential hitting times and metastability: a pathwise approach without reversibility. Electron. J. Probab., **20**, Paper No. 122, 37 pp.
111. Fernandez, R., Manzo, F., Nardi, F.R., Scoppola, E., Sohier, J. (2016). Conditioned, quasi-stationary, restricted measures and escape from metastable states. Ann. Appl. Probab., **26**, no. 2, 760–793.
112. Ferré, D., Hervé, L., Ledoux, J. (2013). Regular perturbation of V-geometrically ergodic Markov chains. J. Appl. Probab., **50**, no. 1, 184–194.
113. Funderlic, R.E., Meyer, C.D.(1985). Sensitivity of the stationary distribution vector for an ergodic Markov chain. Linear Algebra Appl., **76**, 1–17.
114. Gaĭtsgori, V.G., Pervozvanskiy, A.A. (1975). Aggregation of states in a Markov chain with weak interaction. Kibernetika, no. 3, 91–98 (English translation in Cybernetics, **11**, no. 3, 441–450).
115. Gaĭtsgori, V.G., Pervozvanskiy, A. A. (1983). Decomposition and aggregation in problems with a small parameter. Izv. Akad. Nauk SSSR, Tekhn. Kibernet., no. 1, 33–46 (English translation in Engineer. Cybernetics, **2**, no. 1, 26–38).
116. Gambin, A., Krzyżanowski, P., Pokarowski, P. (2008). Aggregation algorithms for perturbed Markov chains with applications to networks modeling. SIAM J. Sci. Comput. **31**, no. 1, 45–73.
117. Gasanenko, V.A. (1999). Normed Poisson process in one-sided curvelinear boundary domain. In: Proceedings of the Third Ukrainian-Scandinavian Conference in Probability Theory and Mathematical Statistics, Kiev, 1999. Theory Stoch. Process., **5(21)**, no. 3–4, 64–71.
118. Gasanenko, V.A. (2001). On the asymptotics of the sojourn probability of a Poisson process between two divergent nonlinear boundaries. Ukr. Mat. Zh., **53**, 14–22 (English translation in Ukr. Math. J., **53**, 15–25).
119. Gast, N., Gaujal, B. (2017). Computing absorbing times via fluid approximations. Adv. Appl. Probab., **49**, no. 3, 768–790.
120. Gaver, D.P., Jr. (1964). An absorption probability problem. J. Math. Anal. Appl., **9**, 384–393.
121. Gaver, D.P. (1976). Random record models. J. Appl. Probab., **13**, 538–547.
122. Gaver, D.P., Jacobs, P.A., Latouche, G. (1984). Finite birth-and-death models in randomly changing environments. Adv. Appl. Probab., **16**, 715–731.
123. Gibson, D., Seneta, E. (1987). Augmented truncations of infinite stochastic matrices. J. Appl. Probab., **24**, no. 3, 600–608.
124. Gikhman, I.I., Skorokhod, A.V. (1971). Theory of Random Processes. 1. Probability Theory and Mathematical Statistics, Nauka, Moscow, 664 pp. (English edition: The Theory of Stochastic Processes. 1. Fundamental Principles of Mathematical Sciences, **210**, Springer, New York (1974) and Berlin (1980)).
125. Gikhman, I.I., Skorokhod, A.V. (1973). Theory of Random Processes. 2. Probability Theory and Mathematical Statistics, Nauka, Moscow, 639 pp. (English edition: The Theory of Stochastic Processes. II. Classics in Mathematics, Springer, Berlin (2004) and Fundamental Principles of Mathematical Sciences, **218**, Springer, Berlin (1975)).
126. Glynn, P. (2011). On exponential limit laws for hitting times of rare sets for Harris chains and processes. In: Glynn, P. Mikosch, T. and Rolski, T. (Eds.) New Frontiers in Applied Probability: a Festschrift for Søren Asmussen. J. Appl. Probab., Spec. Vol. **48A**, 319–326.
127. Gnedenko, B.V., Frayer, B. (1969). A few remarks on a result by I.N. Kovalenko. Litov. Mat. Sbornik, **9**, 463–470.

128. Gnedenko, B.V., Korolev, V.Yu. (1996). Random Summation. Limit Theorems and Applications. CRC Press, Boca Raton, FL, 288 pp.
129. Gnedenko, B.V., Kovalenko, I.N. (1964). Introduction to Queueing Theory. Nauka, Moscow (English edition: Introduction to Queuing Theory. Mathematical Modeling, **5**, Birkhäuser, Boston (1989), xi+315 pp.).
130. Goldsheid, I.Ya. (2008). Linear and sub-linear growth and the CLT for hitting times of a random walk in random environment on a strip. Probab. Theory Related Fields, **141**,, no. 3–4, 471–511.
131. Golub, G.H., Seneta, E. (1973). Computation of the stationary distribution of an infinite Markov matrix. Bull. Austral. Math. Soc., **8**, 333–341.
132. Golub, G. H., Seneta, E. (1974). Computation of the stationary distribution of an infinite stochastic matrix of special form. Bull. Austral. Math. Soc., **10**, 255–261.
133. Grabski, F. (2015). Semi-Markov Processes: Applications in System Reliability and Maintenance. Elsevier, Amsterdam, xiv+255 pp.
134. Grama, I., Lauvergnat, R., Le Page, É. (2020). Conditioned local limit theorems for random walks defined on finite Markov chains. Probab. Theory Related Fields, **176**, no. 1–2, 669–735.
135. Griffiths, S., Kang, R.J., Oliveira, R.I., Patel, V. (2014). Tight inequalities among set hitting times in Markov chains. Proc. Amer. Math. Soc., **142**, no. 9, 3285–3298.
136. Guo, D.Z. (2006). On the sensitivity of the solution of nearly uncoupled Markov chains. SIAM J. Matrix Anal. Appl., **14**, no. 4, 1112–1123.
137. Gusak, D.V., Korolyuk, V.S. (1971). Asymptotic behaviour of semi-Markov processes with a decomposable set of states. Teor. Verojatnost. Mat. Stat., **5**, 43–50 (English translation in Theory Probab. Math. Statist., **5**, 43–51).
138. Gut, A. (1974a). On the moments and limit distributions of some first passage times. Ann. Probab., **2**, 277–308.
139. Gut, A. (1974b). On convergence in r-mean of some first passage times and randomly indexed partial sums. Ann. Probab., **2**, 321–323.
140. Gut, A. (1975). Weak convergence and first passage times. J. Appl. Probab., **12**, 324–332.
141. Gut, A. (1988). Stopped Random Walks. Limit Theorems and Applications. Applied Probability. A Series of the Applied Probability Trust, **5**, Springer, New York, xiv+263 pp.
142. Gut, A., Holst, L. (1984). On the waiting time in a generalized roulette game. Statist. Probab. Lett., **2**, no. 4, 229–239.
143. Gyllenberg, M., Silvestrov, D.S. (1994). Quasi-stationary distributions of a stochastic metapopulation model. J. Math. Biol., **33**, 35–70.
144. Gyllenberg, M., Silvestrov, D.S. (1998). Quasi-stationary phenomena in semi-Markov models. In: Janssen, J., Limnios, N. (Eds.) Proceedings of the Second International Symposium on Semi-Markov Models: Theory and Applications, Compiègne, 1998, 87–93.
145. Gyllenberg, M., Silvestrov, D.S. (1999a). Quasi-stationary phenomena for semi-Markov processes. In: Janssen, J., Limnios, N. (Eds.) Semi-Markov Models and Applications. Kluwer, Dordrecht, 33–60.
146. Gyllenberg, M., Silvestrov, D.S. (1999b). Cramér-Lundberg and diffusion approximations for nonlinearly perturbed risk processes including numerical computation of ruin probabilities. In: Silvestrov, D., Yadrenko, M., Borisenko, O., Zinchenko, N. (Eds.) Proceedings of the Second International School on Actuarial and Financial Mathematics, Kiev, 1999. Theory Stoch. Process., **5(21)**, no. 1–2, 6–21.
147. Gyllenberg, M., Silvestrov, D.S. (2000a). Nonlinearly perturbed regenerative processes and pseudo-stationary phenomena for stochastic systems. Stoch. Process. Appl., **86**, 1–27.
148. Gyllenberg, M., Silvestrov, D.S. (2000b). Cramér–Lundberg approximation for nonlinearly perturbed risk processes. Insur. Math. Econom., **26**, 75–90.
149. Gyllenberg, M., Silvestrov, D.S. (2008). Quasi-Stationary Phenomena in Nonlinearly Perturbed Stochastic Systems. De Gruyter Expositions in Mathematics, **44**, Walter de Gruyter, Berlin, ix+579 pp.
150. Hanen, A. (1963a). Probème central limite dans le cas Markovien fini. La matrice limite n'a qu'une seule classe ergodique et pas d'état transitoire. C. R. Acad. Sci. Paris, **256**, 68–70.

151. Hanen, A. (1963b). Problème central limite dans le cas Markovien fini. II. La matrice limite a plusieurs classes ergodiques et pas détats transitoires. C. R. Acad. Sci. Paris, **256**, 362–364.
152. Hanen, A. (1963c). Problème central limite dans le cas Markovien fini. Cas général. C. R. Acad. Sci. Paris, **256**, 575–577.
153. Hanen, A. (1963d). Théorèmes limites pour une suite de chaînes de Markov. Ann. Inst. H. Poincaré, **18**, 197–301.
154. Harlamov, B. (2008). Continuous Semi-Markov Processes. Applied Stochastic Methods Series, ISTE, London and Wiley, Hoboken, NJ, 375 pp.
155. Harris, T.E. (1952). First passage and recurrence distributions. Trans. Amer. Math. Soc., **73**, 471–486.
156. Hassin, R., Haviv, M. (1992). Mean passage times and nearly uncoupled Markov chains. SIAM J. Disc. Math., **5**, 386–397.
157. Hatori, H. (1959). Some theorems in an extended renewal theory. I. Kōdai Math. Semin. Repts., **11**, no. 3, 139–146.
158. Hatori, H. (1960). Some theorems in an extended renewal theory. II. Kōdai Math. Semin. Repts., **12**, no. 1, 21–27.
159. Haviv, M. (1986). An approximation to the stationary distribution of a nearly completely decomposable Markov chain and its error analysis. SIAM J. Algebr. Discr. Meth., **7**, no. 4, 589–593.
160. Haviv, M. (1987). Aggregation/disaggregation methods for computing the stationary distribution of a Markov chain. SIAM J. Numer. Anal., **24**, no. 4, 952–966.
161. Haviv, M. (1988). Error bounds on an approximation to the dominant eigenvector of a nonnegative matrix. Linear and Multilinear Algebra, **23**, no. 2, 159–163.
162. Haviv, M. (1992). An aggregation/disaggregation algorithm for computing the stationary distribution of a large Markov chain. Comm. Statist. Stoch. Models, **8**, no. 3, 565–575.
163. Haviv, M. (1999). On censored Markov chains, best augmentations and aggregation/disaggregation procedures. Aggregation and disaggregation in operations research. Comput. Oper. Res., **26**, no. 10–11, 1125–1132.
164. Haviv, M. (2006). More on Rayleigh–Ritz refinement technique for nearly uncoupled stochastic matrices. SIAM J. Matrix Anal. Appl., **10**, no. 3, 287–293.
165. Haviv, M., Ritov, Y. (1986). An approximation to the stationary distribution of a nearly completely decomposable Markov chain and its error bound. SIAM J. Algebr. Discr. Methods, **7**, no. 4, 583–588.
166. Haviv, M., Ritov, Y. (1993). On series expansions and stochastic matrices. SIAM J. Matrix Anal. Appl., **14**, no. 3, 670–676.
167. Haviv, M., Ritov, Y., Rothblum, U.G. (1987). Iterative methods for approximating the subdominant modulus of an eigenvalue of a nonnegative matrix. Linear Algebra Appl., **87**, 61–75.
168. Haviv, M., Ritov, Y., Rothblum, U.G. (1992). Taylor expansions of eigenvalues of perturbed matrices with applications to spectral radii of nonnegative matrices. Linear Algebra Appl., **168**, 159–188.
169. Haviv, M., Rothblum, U.G. (1984). Bounds on distances between eigenvalues. Linear Algebra Appl., **63**, 101–118.
170. Haviv, M., Van der Heyden, L. (1984). Perturbation bounds for the stationary probabilities of a finite Markov chain. Adv. Appl. Probab., **16**, 804–818.
171. Helali, A., Löwe, M. (2019). Hitting times, commute times, and cover times for random walks on random hypergraphs. Statist. Probab. Lett., **154**, 108535, 6 pp.
172. Hervé, L., Ledoux, J. (2020). State-discretization of V-geometrically ergodic Markov chains and convergence to the stationary distribution. Methodol. Comput. Appl. Probab., **22**, 905–925.
173. Holst, L. (1971). Limit theorems for some occupancy and sequential occupancy problems. Ann. Math. Statist., **42**, 1671–1680.
174. Holst, L. (1977). Some asymptotic results for occupancy problems. Ann. Probab., **5**, 1028–1035.

175. Holst, L. (1979). A unified approach to limit theorems for urn models. J. Appl. Probab., **16**, 154–162.
176. Hoppensteadt, F., Salehi, H., Skorokhod, A. (1996). On the asymptotic behavior of Markov chains with small random perturbations of transition probabilities. In: Gupta, A.K. (Ed.) Multidimensional Statistical Analysis and Theory of Random Matrices: Proceedings of the Sixth Eugene Lukacs Symposium. Bowling Green, OH, 1996. VSP, Utrecht, 93–100.
177. Hössjer, O., Bechly, G., Gauger, A. (2018). Phase-Type Distribution Approximations of the Waiting Time Until Coordinated Mutations Get Fixed in a Population. In: Silvestrov, S., Rančić, M., Malyarenko, A. (Eds.) Stochastic Processes and Applications, Springer Proceedings in Mathematics & Statistics, **271**, Springer, Cham, Chapter 12, 245–314.
178. Hsu, Guanghui; He, Qiming (1990). The residual life for alternating renewal processes. Asia-Pac. J. Oper. Res., **7**, 76–81.
179. Hunter, J.J. (1986). Stationary distributions of perturbed Markov chains. Linear Algebra Appl., **82**, 201–214.
180. Hunter, J.J. (1991a). The computation of stationary distributions of Markov chains through perturbations. J. Appl. Math. Stoch. Anal., **4**, no. 1, 29–46.
181. Hunter, J.J. (1991b). A survey of generalized inverses and their use in applied probability. Math. Chronicle, **20**, 13–26.
182. Hunter, J.J. (2005). Stationary distributions and mean first passage times of perturbed Markov chains. Linear Algebra Appl., **410**, 217–243.
183. Hunter, J.J. (2016). The computation of key properties of Markov chains via perturbations. Linear Algebra Appl., **511**, 176–202.
184. Hunter, J.J. (2018). The computation of the mean first passage times for Markov chains. Linear Algebra Appl., **549**, 100–122.
185. Iglehart, D.L., Shedler, G.S. (1980). Regenerative Simulation of Response Times in Networks of Queues. Lecture Notes in Control and Information Sciences, **26**, Springer, Berlin, xii+204 pp.
186. Issaadi, B. (2020). Weak stability bounds for approximations of invariant measures with applications to queueing. Methodol. Comput. Appl. Probab., **22**, 371–400.
187. Jacod, J., Shiryaev, A.N. (2002). Limit Theorems for Stochastic Processes. Second edition. Fundamental Principles of Mathematical Sciences, **288**, Springer, Berlin, 684 pp.
188. Jagers, P., Lindvall, T. (1974). Thinning and rare events in point processes. *Z. Wahrsch. Verw. Gebiete*, **28**, 89–98.
189. Jagers, P, Klebaner, F.C., Sagitov, S. (2007). Markovian paths to extinction. Adv. in Appl. Probab., **39**, no. 2, 569–587.
190. Janson, S. (1983). Renewal theory for M-dependent variables. Ann. Probab. **11**, 558–568.
191. Janssen, J., Manca, R. (2006). Applied Semi-Markov Processes. Springer, New York, xii+309 pp.
192. Janssen, J., Manca, R. (2007). Semi-Markov Risk Models for Finance, Insurance and Reliability. Springer, New York, xvii+429 pp.
193. Jarner, S.F., Tweedie, R.L. (2003). Necessary conditions for geometric and polynomial ergodicity of random-walk-type Markov chains. Bernoulli, **9**, no. 4, 559–578.
194. Kalashnikov, V.V. (1978a). Qualitative Analysis of the Behaviour of Complex Systems by the Method of Test Functions. Series in Theory and Methods of Systems Analysis, Nauka, Moscow, 247 pp.
195. Kalashnikov, V.V. (1978b). Solution of the problem of approximating a denumerable Markov chain. Izv. Akad. Nauk SSSR Tekhn. Kibernet., no. 3, 92–95.
196. Kalashnikov, V.V. (1981). Methods of estimation of continuity in the problem of consolidation of states. In: Problems of Stability of Stochastic Models, VNIISI, Moscow, 48–53.
197. Kalashnikov, V.V. (1983). Estimation of the rate of convergence in Rénei's theorem. In: Stability Problems for Stochastic Models. VNIISI, Moscow, 48–57.
198. Kalashnikov, V.V. (1986). Approximation of some stochastic models. In: Semi-Markov Models, Plenum, New York, 319–335.

199. Kalashnikov, V.V. (1990). Regenerative queueing processes and their qualitative and quantitative analysis. Queueing Systems Theory Appl., **6**, no. 2, 113–136.
200. Kalashnikov, V.V. (1994a). Regeneration and general Markov chains. J. Appl. Math. Stoch. Anal., **7**, no. 3, 357–371.
201. Kalashnikov, V.V. (1994b). Topics on Regenerative Processes. CRC Press, Boca Raton, FL, 240 pp.
202. Kalashnikov, V.V. (1997). Geometric Sums: Bounds for Rare Events with Applications. Mathematics and its Applications, **413**, Kluwer, Dordrecht, ix+265 pp.
203. Kalashnikov, V.V., Anichkin, S.A. (1981). Continuity of random sequences and approximation of Markov chains. Adv. Appl. Probab., **13**, no. 2, 402–414.
204. Kalashnikov, V.V., Vsekhsvyatskii, S.Yu. (1985). Metric estimates of the first occurrence time in regenerative processes. In Stability Problems for Stochastic Models, Uzhgorod, 1984, Lecture Notes in Math., **1155**, Springer, Berlin, 102–130.
205. Kallenberg, O. (2021). Foundations of Modern Probability. Third Edition. Probability Theory and Stochastic Modelling, **99**, Springer, xii+946 pp.
206. Kaplan, E.I. (1979). Limit theorems for exit times of random sequences with mixing. Teor. Verojatnost. Mat. Stat., **21**, 53–59 (English translation in Theory Probab. Math. Statist., **21**, 59–65).
207. Kaplan, E.I. (1980). Limit Theorems for Sum of Switching Random Variables with an Arbitrary Phase Space of Switching Component. Candidate of Science dissertation, Kiev State University.
208. Kartashov, N.V. (1985a). Inequalities in stability and ergodicity theorems for Markov chains with a general phase space. I. Teor. Verojatnost. Primen., **30**, 230–240 (English translation in Theory Probab. Appl., **30**, 247–259).
209. Kartashov, N.V. (1985b). Inequalities in stability and ergodicity theorems for Markov chains with a general phase space. II. Teor. Verojatnost. Primen., **30**, 478–485 (English translation in Theory Probab. Appl., **30**, 507–515).
210. Kartashov, N.V. (1985c). Asymptotic expansions and inequalities in stability theorems for general Markov chains under relatively bounded perturbations. In: Stability Problems for Stochastic Models, Varna, 1985. VNIISI, Moscow, 75–85 (English translation in J. Soviet Math., **40**, no. 4, 509–518).
211. Kartashov, N.V. (1986). Inequalities in theorems of consolidation of Markov chains. Theor. Verojatnost. Mat. Stat., **34**, 62–73 (English translation in Theory Probab. Math. Statist., **34**, 67–80).
212. Kartashov, N.V. (1987). Estimates for the geometric asymptotics of Markov times on homogeneous chains. Teor. Verojatnost. Mat. Stat., **37**, 66–77 (English translation in Theory Probab. Math. Statist., **37**, 75–88).
213. Kartashov, N.V. (1991). Inequalities in Rénei's theorem. Teor. Ĭmovirn. Mat. Stat., **45**, 27–33 (English translation in Theory Probab. Math. Statist., **45**, 23–28).
214. Kartashov, M.V. (1996). Strong Stable Markov Chains. VSP, Utrecht and TBiMC, Kiev, 138 pp.
215. Kartashov, M.V. (2013a). The asymptotic behavior of rare Markov moments defined on time inhomogeneous Markov chains, Teor. Ĭmovirn. Mat. Stat., **88**, 83–94 (English translation in Theor. Probab. Math. Statist., **88**, 109–121).
216. Kartashov, M.V. (2013b). Quantitative and qualitative limits for exponential asymptotics of hitting times for birth-and-death chains in a scheme of series. Teor. Ĭmovirn. Mat. Stat., **89**, 40–50 (English translation in Theory Probab. Math. Statist, **89**, (2014), 45–56).
217. Kartashov, M.V. (2015). The asymptotic behavior of the distribution of Markov moments in time-inhomogeneous Markov chains and its application to a discrete Cramér-Lundberg model. Teor. Ĭmovirn. Mat. Stat., **92**, 41–60 (English translation in Theory Probab. Math. Statist, **92**, 37–58).
218. Keilson, J. (1966a). A technique for discussing the passage time distribution for stable systems. J. Royal Statist. Soc., Ser. B **28**, 477–486.

219. Keilson, J. (1966b). A limit theorem for passage times in ergodic regenerative processes. Ann. Math. Statist., **37**, 866–870.
220. Keilson, J. (1974). Sojourn times, exit times and jitter in multivariate Markov processes. Adv. Appl. Probab., **6**, 747–756.
221. Keilson, J. (1979). Markov Chain Models – Rarity and Exponentiality. Applied Mathematical Sciences, **28**, Springer, New York, xiii+184 pp.
222. Keilson, J. (1998). Covariance and relaxation time in finite Markov chains. J. Appl. Math. Stoch. Anal., **11**, 391–396.
223. Kemperman, J.H.B. (1962). The Passage Problem for a Stationary Markov Chain. Statistical Research Monographs, **1**, Univ. Chicago Press, 127 pp.
224. Kesten, H., Maller, R.A. (1999). Stability and other limit laws for exit times of random walks from a strip or a halfplane. Ann. Inst. H. Poincaré Probab. Statist., **35**, no. 6, 685–734.
225. Khusanbaev, Ya.M. (1983). Limit Theorems for Processes with Conditionally Independent Increments and Functionals of Additive Type for Regenerative Processes. Candidate of Science dissertation, Kiev State University.
226. Khusanbaev, Ya.M. (1984). Conditions for the convergence of random processes stopped at the time of underjump. Ukr. Mat. Zh., **36**, no. 1, 126–129 (English translation in Ukr. Math. J., **36**, no. 1, 116–119).
227. Kijima, M. (1997). Markov Processes for Stochastic Modelling. Stochastic Modeling Series. Chapman & Hall, London, x+341 pp.
228. Kingman, J.F. (1963). The exponential decay of Markovian transition probabilities. Proc. London Math. Soc., **13**, 337–358.
229. Kingman, J.F.C. (1972). Regenerative Phenomena. Wiley Series in Probability and Mathematical Statistics. Wiley, London, xii+190 pp.
230. Kohlas, J. (1983). Numerical computation of mean passage times and absorption probabilities in Markov and semi-Markov models, Zeitschrift Oper. Res., **30**, A197 – A207.
231. Korolev, V.Yu. (2016). Limit distributions for doubly stochastically rarefied renewal processes and their properties. Teor. Verojatnost. Primen., **61**, no. 4, 753–773 (English translation in Theory Probab. Appl., **61**, no. 4, 649–664).
232. Korolyuk, D.V. (1982). Limit distributions of additive functionals of regenerative processes up to the time of reaching asymptotically receding regions. Dokl. Akad. Nauk Ukr. SSR, Ser. A, no. 8, 16–18.
233. Korolyuk, D.V. (1983). Limit Theorems for Hitting Time Type Functionals Defined on Processes with Semi-Markov Switchings. Candidate of Science dissertation, Kiev State University.
234. Korolyuk, D.V. (1990). Asymptotics of the time of attaining a distant level by a random walk with delay. Teor. Verojatnost. Mat. Stat., **42**, 64–67 (English translation in Theory Probab. Math. Statist., **42**, 73–76).
235. Korolyuk, D.V., Silvestrov D.S. (1983). Time of reaching asymptotically receding domains for ergodic Markov chains. Teor. Verojatnost. Primen., **28**, 410–420 (English translation in Theory Probab. Appl., **28**, 432–442).
236. Korolyuk, D.V., Silvestrov D.S. (1984). Entry times into asymptotically receding regions for processes with semi-Markov switchings. Teor. Verojatnost. Primen., **29**, 539–544 (English translation in Theory Probab. Appl., **29**, 558–563).
237. Korolyuk, V.S. (1969). On asymptotical estimate for time of a semi-Markov process being in the set of states. Ukr. Mat. Zh., **21**, 842–845 (English translation in Ukr. Math. J., **21**, 705–710).
238. Korolyuk, V.S. (1997). Singularly perturbed stochastic systems. Ukr. Mat. Zh., **49**), 25–34 (English translation in Ukr. Math. J., **49**, 25–35).
239. Korolyuk, V.S., Brodi, S.M., Turbin, A.F. (1974). Semi-Markov processes and their application. Probability Theory. Mathematical Statistics. Theoretical Cybernetics, Vol. **11**, VINTI, Moscow, 1974, 47–97.
240. Korolyuk, V.S., Korolyuk, V.V. (1999). Stochastic Models of Systems. Mathematics and its Applications, **469**, Kluwer, Dordrecht, xii+185 pp.

241. Korolyuk, V.S., Limnios, N. (2000). Evolutionary systems in an asymptotic split phase space. In: Limnios, N., Nikulin, M. (Eds.) Recent Advances in Reliability Theory: Methodology, Practice and Inference. Birkhäuser, Boston, 145–161.
242. Korolyuk, V.S., Limnios, N. (2002). Markov additive processes in a phase merging scheme. In: Korolyuk, V., Prokhorov, Yu., Khokhlov, V., Klesov, O. (Eds.) Proceedings of the Conference Dedicated to the 90th Anniversary of Boris Vladimirovich Gnedenko, Kiev, 2002. Theory Stoch. Process., **8**, no. 3–4, 213–225.
243. Koroliuk, V.S., Limnios, N. (2005). Stochastic Systems in Merging Phase Space. World Scientific, Singapore, xv+331 pp.
244. Koroliuk, V.S., Limnios, N. (2010). Reliability of semi-Markov systems with asymptotic merging phase space. In: Rykov, V.V, Balakrishnan, N., Nikulin, M.S. (Eds.) Mathematical and Statistical Models and Methods in Reliability. Birkhäuser, Boston, 3–18.
245. Korolyuk, V.S., Penev, I.P., Turbin, A.F. (1972). The asymptotic behavior of the distribution of the absorption time of a Markov chain. Kibernetika, no. 2, 20–22.
246. Koroljuk, V.S., Penev, I.P., Turbin, A.F. (1973). Asymptotic expansion for the distribution of the absorption time of a Markov chain. Kibernetika, no. 4, 133–135.
247. Korolyuk, V.S., Penev, I.P., Turbin, A.F. (1981). Asymptotic expansion for the distribution of the absorption time of a weakly inhomogeneous Markov chain. In: Korolyuk, V.S. (Ed.) Analytic Methods of Investigation in Probability Theory. Akad. Nauk Ukr. SSR, Inst. Mat., Kiev, 97–105.
248. Korolyuk, V., Swishchuk, A. (1995a). Semi-Markov Random Evolutions. Mathematics and its Applications, **308**, Springer, Dordrecht, x+310 pp. (Revised English edition of Semi-Markov Random Evolutions. Naukova Dumka, Kiev, (1992). 254 pp.).
249. Korolyuk, V.S., Swishchuk, A.V. (1995b). Evolution of Systems in Random Media. CRC Press, Boca Raton, FL, 368 pp.
250. Korolyuk, V.S., Turbin, A.F. (1970). On the asymptotic behaviour of the occupation time of a semi-Markov process in a reducible subset of states. Teor. Verojatnost. Mat. Stat., **2**, 133–143 (English translation in Theory Probab. Math. Statist., **2**, 133–143).
251. Korolyuk, V.S., Turbin, A.F. (1976a). Semi-Markov Processes and its Applications. Naukova Dumka, Kiev, 184 pp.
252. Korolyuk, V.S., Turbin, A.F. (1976b). Asymptotic enlarging of semi-Markov processes with an arbitrary state space. In: Proceedings of the Third Japan-USSR Symposium on Probability Theory, Tashkent, 1975. Lecture Notes in Mathematics, **550**, Springer, Berlin, 297–315.
253. Korolyuk, V.S., Turbin, A.F. (1978). Mathematical Foundations of the State Lumping of Large Systems. Naukova Dumka, Kiev, 218 pp. (English edition: Mathematical Foundations of the State Lumping of Large Systems. Mathematics and its Applications, **264**, Kluwer, Dordrecht, 1993, x+278 pp.).
254. Koury, J.R., McAllister, D.F., Stewart, W.J. (1984). Iterative methods for computing stationary distributions of nearly completely decomposable Markov chains. SIAM J. Alg. Disc. Melh., **5**, 164–186.
255. Kovalenko, I.N. (1965). On the class of limit distributions for thinning flows of homogeneous events. Litov. Mat. Sbornik, **5**, 569–573.
256. Kovalenko, I.N. (1973). An algorithm of asymptotic analysis of a sojourn time of Markov chain in a set of states. Dokl. Acad. Nauk Ukr. SSR, Ser. A, no. 6, 422–426.
257. Kovalenko, I.N. (1975). Studies in the Reliability Analysis of Complex Systems. Naukova Dumka, Kiev, 210 pp.
258. Kovalenko, I.N. (1976). An analytic-statistical method for computing the characteristics of systems with high reliability. Kibernetika, no. 6, 82–92 (English translation in Cybernetics **12**, no. 6, 895–907).
259. Kovalenko, I.N. (1994). Rare events in queuing theory – a survey. Queuing Syst. Theory Appl., **16**, no. 1–2, 1–49.
260. Kovalenko, I.N., Birolini, A., Kuznetsov, N.Yu., Shenbukher, M. (1998). On the reduction of the phase space of a random process in analyzing the reliability of systems. Kibernet. Sistem. Anal., no. 4, 134–144 (English translation in Cybernet. Systems Anal., **34**, no. 4, 588–595).

261. Kovalenko, I.N., Kuznetsov, M.Ju. (1981). Renewal process and rare events limit theorems for essentially multidimensional queueing processes. Math. Operat. Statist., Ser. Statist., **12**, no. 2, 211–224.
262. Kovalenko, I.N., Kuznetsov, N.Yu., Pegg, P.A. (1997). Mathematical Theory of Reliability of Time Dependent Systems with Practical Applications. Wiley Series in Probability and Statistics, Wiley, New York, 316 pp.
263. Kovalenko, I.N., Kuznetsov, N.Yu., Shurenkov, V.M. (1996). Models of Random Processes. A Handbook for Mathematicians and Engineers. CRC Press, Boca Raton, FL, 446 pp. (A revised edition of the 1983 Russian original).
264. Kruglov, V.M., Korolev, V.Yu. (1990). Limit Theorems for Random Sums. Izdatel'stvo Moskovskogo Universiteta, Moscow, 269 pp.
265. Kumar, N., Gupta, U.C. (2020). A renewal generated geometric catastrophe model with discrete-time Markovian arrival process. Methodol. Comput. Appl. Probab.. **22**, 1293–1324.
266. Kuntz, J., Thomas, P., Stan, G., Barahona, M. (2019).The exit time finite state projection scheme: bounding exit distributions and occupation measures of continuous-time Markov chains. SIAM J. Sci. Comput., **41**, no. 2, A748–A769.
267. Kupsa, M., Lacroix, Y. (2005). Asymptotics for hitting times. Ann. Probab., **33**, no. 2, 610–619.
268. Kushnir, O.O., Kushnir, V.P. (2017). A study of characteristics of a highly reliable system with protection assuming that the renewal process is Poisson, Theor. Ĭmovirn. Mat. Stat. **96**, 125–130 (English translation in Theory Probab. Math. Statist. **96**, 127–132).
269. Labbé, C., Lacoin, H. (2020). Mixing time and cutoff for the weakly asymmetric simple exclusion process Ann. Appl. Probab., **30**, no. 4, 1847–1883.
270. Lalley, S.P. (1984). Limit theorems for first-passage times in linear and nonlinear renewal theory. Adv. Appl. Probab., **16**, 766–803.
271. Lasserre, J.B. (1994). A formula for singular perturbations of Markov chains. J. Appl. Probab., **31**, 829–833.
272. Latouche, G. (1991). First passage times in nearly decomposable Markov chains. In: Stewart, W.J. (Ed). Numerical Solution of Markov Chains. Probability: Pure and Applied, **8**, Marcel Dekker, New York, 401–411.
273. Latouche, G., Louchard, G. (1978). Return times in nearly decomposable stochastic processes. J. Appl. Probab., **15**, 251–267.
274. Leadbetter, M.R. (1963). On series expansion for the renewal moments. Biometrika, **50**, no. 1–2, 75–80.
275. Lévy, P. (1954/56). Processus semi-markoviens. In: Erven, P., Noordhoff, N.V. (Eds.) Proceedings of the International Congress of Mathematicians, Vol. III. Amsterdam, 1954. North-Holland, Amsterdam, 416–426.
276. Limnios, N. (2012). Reliability measures of semi-Markov systems with general state space. Methodol. Comput. Appl. Probab., **14**, no. 4, 895–917.
277. Limnios, N., Oprişan, G. (2001). Semi-Markov Processes and Reliability. Statistics for Industry and Technology, Birkhäuser, Boston, xii+222 pp.
278. Lindvall, T. (1973). Weak convergence of probability measures and random functions in the function space D[0, ∞). J. Appl. Probab., **10**, 109–121.
279. Loève, M. (1977). Probability Theory. I. Fourth edition. Graduate Texts in Mathematics, **45**, Springer, New York, xvii+425 pp. (First edition in 1955).
280. Lopes, F., Luczak, M. (2020). Extinction time for the weaker of two competing SIS epidemics. Ann. Appl. Probab., **30**, no. 6, 2880–2922.
281. Löpker, A., Stadje, W. (2011). Hitting times and the running maximum of Markovian growth-collapse processes. J. Appl. Probab., 48, no. 2, 295–312.
282. Louchard, G., Latouche, G. (1982). Random times in nearly-completely decomposable, transient Markov chains. Cahiers Centre Études Rech. Opér., **24**, no. 2–4, 321–352.
283. Louchard, G., Latouche, G. (1990). Geometric bounds on iterative approximations for nearly completely decomposable Markov chains. J. Appl. Probab., **27**, no. 3, 521–529.

284. Mao Yong-Hua, Zhang Chi (2017). Hitting time distributions for birth-death processes with bilateral absorbing boundaries. Probab. Engrg. Inform. Sci., **31**, no. 3, 345–356.
285. Marek, I., Mayer, P. (1998). Convergence analysis of an iterative aggregation/disaggregation method for computing stationary probability vectors of stochastic matrices. Numer. Linear Algebra Appl., **5**, no. 4, 253–274.
286. Marek, I., Mayer, P., Pultarová, I. (2009). Convergence issues in the theory and practice of iterative aggregation/disaggregation methods. Electron. Trans. Numer. Anal., **35**, 185–200.
287. Marek, I., Pultarová, I. (2006). A note on local and global convergence analysis of iterative aggregation-disaggregation methods. Linear Algebra Appl., **413**, no. 2–3, 327–341.
288. Martin-Löf, A. (1986). Entropy estimates for the first passage time of a random walk to a time dependent barrier. Scand. J. Statist., **13**, no. 3, 221–229.
289. Masol, V.I. (1973). On the asymptotic behaviour of the time of arrival at a high level by a regenerating stable process with independent increments. Teor. Verojatnost. Mat. Stat., **9**, 131–136 (English translation in Theory Probab. Math. Statist., **9**, 139–144).
290. Masol, V.I. (1974). Limit Distributions for Functionals of Additive Type of Regenerative Processes. Candidate of Science dissertation, Kiev State University.
291. Masol, V.I. (1976). Limit theorems for regenerating random processes that are controlled by a finite Markov chain. Dokl. Akad. Nauk Ukr. SSR, Ser. A, no. 3, 206–209.
292. Masol, V.I., Silvestrov, D.S. (1972). Record values of the occupation time of a semi-Markov process. Visnik Kiev. Univ., Ser. Mat. Meh., **14**, 81–89.
293. Masuyama, H. (2015). Error bounds for augmented truncations of discrete-time block-monotone Markov chains under geometric drift conditions. Adv. Appl. Probab., **47**, no. 1, 83–105.
294. McAllister, D.F., Stewart, G.W., Stewart, W.J. (1984). On a Rayleigh-Ritz refinement technique for nearly uncoupled stochastic matrices. Linear Algebra Appl., **60**, 1–25.
295. Meshalkin, L.D. (1958). Limit theorems for Markov chains with a finite number of states. Teor. Verojatnost. Primen., **3**, 361–385 (English translation in Theory Probab. Appl., **3**, 335–357).
296. Meyer, C.D., (1980). The condition of a finite Markov chain and perturbation bounds for the limiting probabilities. SIAM J. Algebr. Discr. Methods, **1**, no. 3, 273–283.
297. Meyer, C.D. (1994). Sensitivity of the stationary distribution of a Markov chain. SIAM J. Matrix Anal. Appl., **15**, no. 3, 715–728.
298. Meyer, C.D. (2015). Continuity of the Perron root. Linear Multilinear Algebra, **63**, no. 7, 1332–1336.
299. Meyer, C.D., Stewart, G.W. (1988). Derivatives and perturbations of eigenvectors. SIAM J. Numer. Anal., **25**, no. 3, 679–691.
300. Meyn, S.P., Tweedie, R.L. (1992). Stability of Markovian processes. I. Criteria for discrete-time chains. Adv. Appl. Probab., **24**, 542–574.
301. Meyn, S.P., Tweedie, R.L. (1993a). Stability of Markovian processes. II. Continuous-time processes and sampled chains. Adv. Appl. Probab., **25**, 487–517.
302. Meyn, S.P., Tweedie, R.L. (1993b). Stability of Markovian processes. III. Foster-Lyapunov criteria for continuous-time processes. Adv. Appl. Probab., **25**, 518–548.
303. Meyn, S.P., Tweedie, R.L. (2009). Markov Chains and Stochastic Stability. Cambridge University Press, xxviii+594 pp. (2nd edition of Markov Chains and Stochastic Stability. Communications and Control Engineering Series, Springer, London, 1993, xvi+ 548 pp.).
304. Mileshina, R.I. (1975). Limit Theorems for Controlled Random Walks. Candidate of Science dissertation, Kiev State University.
305. Mitrophanov, A.Yu. (2003). Stability and exponential convergence of continuous-time Markov chains. J. Appl. Probab., **40**, 970–979.
306. Mitrophanov, A.Yu. (2005a). Sensitivity and convergence of uniformly ergodic Markov chains. J. Appl. Probab., **42**, 1003–1014.
307. Mitrophanov, A.Yu. (2005b). Ergodicity coefficient and perturbation bounds for continuous-time Markov chains. Math. Ineq. Appl., **8**, no 1, 159–168.
308. Mitrophanov, A.Yu. (2006). Stability estimates for finite homogeneous continuous-time Markov chains. Theory Probab. Appl., **50**, no 2, 319–326.

309. Mitrophanov, A.Yu., Lomsadze, A., Borodovsky, M. (2005). Sensitivity of hidden Markov models. J. Appl. Probab., **42**, 632–642.
310. Motsa, A.I., Silvestrov, D.S. (1996). Asymptotics of extremal statistics and functionals of additive type for Markov chains. In: Klesov, O., Korolyuk, V., Kulldorff, G., Silvestrov, D. (Eds.) Proceedings of the First Ukrainian–Scandinavian Conference on Stochastic Dynamical Systems, Uzhgorod, 1995. Theory Stoch. Proces., **2(18)**, no. 1–2, 217–224.
311. Nardi, F.R., Zocca, A., Borst, S.C. (2016). Hitting time asymptotics for hard-core interactions on grids. J. Stat. Phys., **162**, no. 2, 522–576.
312. Ni, Y. (2011). Nonlinearly Perturbed Renewal Equations: Asymptotic Results and Applications. Doctoral dissertation, **106**, Mälardalen University, Västerås.
313. Ni, Y. (2014). Exponential asymptotical expansions for ruin probability in a classical risk process with non-polynomial perturbations. In: Silvestrov, D., Martin-Löf, A. (Eds.) Modern Problems in Insurance Mathematics. Chapter **6**, EAA series, Springer, Cham, 67–91.
314. Nonaka, Y., Ono, H., Sadakane, K., Yamashita, M. (2010). The hitting and cover times of Metropolis walks. Theoret. Comput. Sci., **411**, no. 16–18, 1889–1894.
315. Obzherin, Y.E., Boyko, E.G. (2015). Semi-Markov Models. Control of Restorable Systems with Latent Failures. Elsevier/Academic Press, Amsterdam, xiv+199 pp.
316. O'Cinneide, C.A. (1993). Entrywise perturbation theory and error analysis for Markov chains. Numer. Math., **65**, 109–120.
317. Pène, F., Saussol, B., Zweimüller, R. (2017). Return- and hitting-time limits for rare events of null-recurrent Markov maps. Ergodic Theory Dynam. Syst., **37**, no. 1, 244–276.
318. Peres, Y., Sousi, P. (2015). Mixing times are hitting times of large sets. J. Theoret. Probab., **28**, no. 2, 488–519.
319. Petersson, M. (2013). Quasi-stationary distributions for perturbed discrete time regenerative processes. Teor. Ĭmovirn. Mat. Stat., **89**, 140–155 (Also in Theor. Probab. Math. Statist., **89**, 153–168).
320. Petersson, M. (2014). Asymptotics of ruin probabilities for perturbed discrete time risk processes. In: Silvestrov, D., Martin-Löf, A. (Eds). Modern Problems in Insurance Mathematics. Chapter **7**, EAA series, Springer, Cham, 93–110.
321. Petersson, M. (2016a). Asymptotic expansions for moment functionals of perturbed discrete time semi-Markov processes. In: Silvestrov, S., Rančić, M. (Eds.) Engineering Mathematics II. Algebraic, Stochastic and Analysis Structures for Networks, Data Classification and Optimization, Chapter **8**, Springer Proceedings in Mathematics & Statistics **179**, Springer, Cham, 109—130.
322. Petersson, M. (2016b). Asymptotics for quasi-stationary distributions of perturbed discrete time semi-Markov processes. In: Silvestrov, S., Rančić, M. (Eds.) Engineering Mathematics II. Algebraic, Stochastic and Analysis Structures for Networks, Data Classification and Optimization, Chapter **8**, Springer Proceedings in Mathematics & Statistics **179**, Springer, Cham, 131—150.
323. Petersson, M. (2017). Quasi-stationary asymptotics for perturbed semi-Markov processes in discrete time. Methodol. Comput. Appl. Probab., **19**, 1047–1074.
324. Poliščuk, L.I., Turbin, A.F. (1973). Asymptotic expansions for certain characteristics of semi-Markov processes. Teor. Verojatnost. Mat. Stat., **8**, 122–127 (English translation in Theory Probab. Math. Statist., **8**, 121–126).
325. Pollett, P.K., Stewart, D.E. (1994). An efficient procedure for computing quasi-stationary distributions of Markov chains with sparse transition structure. Adv. Appl. Probab., **26**, 68–79.
326. Prokhorov, Yu.V. (1956). Convergence of random processes and limit theorems of probability theory. Teor. Verojatnost. Primen., **1**, 117–238 (English translation in Theory Probab. Appl., **1**, 157–214).
327. Ragimov, F.H., Farhadova, A.D., Khalilov, V.S. (2018). Limit theorem for the first passage time of the level by the random walk described by nonlinear function of the autoregression process of order one AR(1). Uzbek Math. J., no. 2, 127–135.

328. Rahimov, F H., Ibadova, I.A., Farhadova, A.D. (2019). Limit theorems for a family of the first passage times of a parabola by the sums of the squares autoregression process of order one (AR(1)). Uzbek Math. J. 2019, no. 2, 81–88.
329. Rényi, A. (1956). A characterization of Poisson processes. Magyar. Tud. Akad. Mat. Kutató Int. Kőzl., **1**, 519–527.
330. Revuz, D. (1984). Markov Chains. North-Holland Mathematical Library, **11**, North-Holland, Amsterdam, Elsevier, New York, 374 pp. (Revised version of 1st 1975 edition).
331. Roberts, G.O., Rosenthal, J.S., Schwartz, P.O. (1998). Convergence properties of perturbed Markov chains. J. Appl. Probab., **35**, 1–11.
332. Rohlichek, J.R., (1987). Aggregation and Time Scale Analysis of Perturbed Markov Systems. Ph.D. Thesis, Massachusetts Inst. Tech., Cambridge, MA.
333. Rohlicek, J.R., Willsky, A.S. (1988). Multiple time scale decomposition of discrete time Markov chains. Systems Control Lett., **11**, no. 4, 309–314.
334. Rolski, T., Schmidli, H., Schmidt, V., Teugels, J. (1999). Stochastic Processes for Insurance and Finance. Wiley Series in Probability and Statistics, Wiley, New York, xviii+654 pp.
335. Ross, S.M. (2014). Introduction to Probability Models. Eleventh edition. Elsevier/Academic Press, Amsterdam, xvi+767 pp.
336. Saas, D. (1972). Classes of limit distributions for sums of a random number of independent identically distributed random variables. Teor. Verojatnost. i Primenen., **17**, no. 3, 424–439 (English translation in Theory Probab. Appl., **17**, no. 3, 401–415).
337. Saas, D., Frayer, B. (1971). A certain problem of the theory of summing with a random index. Litov. Mat. Sb., **11**, 181–187.
338. Schulte-Geers, E., Stadje, W. (2017). Small drift limit theorems for random walks. J. Appl. Probab., **54**, no. 1, 199–212.
339. Schweitzer, P.J. (1968). Perturbation theory and finite Markov chains. J. Appl. Probab., **5**, 401–413.
340. Schweitzer, P.J. (1984). Aggregation methods for large Markov chains. In: Iazeolla, G., Courtois, P.J., Hordijk, A. (Eds.) Mathematical Computer Performance and Reliability. North-Holland, Amsterdam, 275–286.
341. Schweitzer, P.J. (1986). Posterior bounds on the equilibrium distribution of a finite Markov chain. Comm. Statist. Stoch. Models, **2**, no. 3, 323–338.
342. Schweitzer, P.J. (1987). Dual bounds on the equilibrium distribution of a finite Markov chain. J. Math. Anal. Appl., **126**, no. 2, 478–482.
343. Schweitzer, P.J. (1991). A survey of aggregation-disaggregation in large Markov chains. In: Stewart, W.J. (Ed). Numerical solution of Markov chains. Probability: Pure and Applied, **8**. Marcel Dekker, New York, 63–88.
344. Seneta, E. (1967). Finite approximations to infinite non-negative matrices. Proc. Cambridge Philos. Soc., **63**, 983–992.
345. Seneta, E. (1968a). Finite approximations to infinite non-negative matrices. II. Refinements and applications. Proc. Cambridge Philos. Soc., **64**, 465–470.
346. Seneta, E. (1968b). The principle of truncations in applied probability. Comment. Math. Univ. Carolinae, **9**, 237–242.
347. Seneta, E. (1973). Nonnegative Matrices. An Introduction to Theory and Applications. Wiley, New York, x+214 pp.
348. Seneta, E. (1980). Computing the stationary distribution for infinite Markov chains. Linear Algebra Appl. **34**, 259–267.
349. Seneta, E. (1988a). Sensitivity to perturbation of the stationary distribution: some refinements. Linear Algebra Appl., **108**, 121–126.
350. Seneta, E. (1988b). Perturbation of the stationary distribution measured by ergodicity coefficients. Adv. Appl. Probab., **20**, 228–230.
351. Seneta, E. (1991). Sensitivity analysis, ergodicity coefficients, and rank-one updates for finite Markov chains. In: Stewart, W.J. (Ed.) Numerical Solution of Markov Chains. Probability: Pure and Applied, **8**, Marcel Dekker, New York, 121–129.

352. Seneta, E. (1993). Sensitivity of finite Markov chains under perturbation. Statist. Probab. Lett., **17**, no. 2, 163–168.
353. Seneta, E. (2006). Nonnegative Matrices and Markov chains. Springer Series in Statistics, Springer, New York, 2006, xvi+287 pp. (A revised reprint of 2nd edition of Nonnegative Matrices and Markov Chains. Springer Series in Statistics. Springer, New York, 1981, xiii+279 pp.).
354. Serfozo, R. (2009). Basics of Applied Stochastic Processes. Probability and Its Applications. Springer, Berlin, xiv+443 pp.
355. Serlet, L. (2013). Hitting times for the perturbed reflecting random walk. Stoch. Process. Appl., **123**, no. 1, 110–130.
356. Serlet, L. (2018). Laurent Explicit laws for the records of the perturbed random walk on Z. Sém. Probab. XLIX, Lecture Notes in Math., **2215**, Springer, Cham, 495–519.
357. Sevast'yanov, B.A. (1995). Extinction probabilities of branching processes bounded from below. Teor. Verojatnost. Primen., **40**, 556–564 (English translation in Theory Probab. Appl., **40**, 495–502).
358. Sevast'yanov, B.A. (1998) Asymptotic behavior of extinction probabilities of stopped branching processes. Teor. Verojatnost. Primen., **43**, 390–397 (English translation in Theory Probab. Appl., **43**, 315–322).
359. Sevast'yanov, B.A. (2000). Sojourn times in a finite set of states of Markov branching processes and the probabilities of extinction of a modified Galton-Watson process. Diskret. Mat., **12**, no. 4, 39–45 (English translation in Discrete Math. Appl., **10**, no. 6, 535–541).
360. Sevast'yanov, B.A. (2006). Asymptotics of the survival probability of a bounded from below Markov critical process with continuous time and infinite variance. Diskret. Mat., **18**, no. 1, 3–8 (English translation in Discrete Math. Appl., **16**, no. 1, 1–5).
361. Shedler, G.S. (1987). Regeneration and Networks of Queues. Applied Probability. A Series of the Applied Probability Trust. Springer, New York, viii+224 pp.
362. Shedler, G.S. (1993). Regenerative Stochastic Simulation. Statistical Modeling and Decision Science. Academic Press, Boston, MA, x+400 pp.
363. Silvestrov, D.S. (1969a). Limit theorems for a non-recurrent walk connected with a Markov chain. Ukr. Mat. Ž., **21**, 790–804 (English translation in Ukr. Math. J., **21**, 657–669).
364. Silvestrov, D.S. (1969b). Limit Theorems for Semi-Markov Processes and their Applications to Random Walks. Candidate of Science dissertation, Kiev State University.
365. Silvestrov, D.S. (1969c). Asymptotic behaviour of the first passage time for sums of random variables controlled by a regular semi-Markov process. Dokl. Akad. Nauk SSSR, **189**, 949–951 (English translation in Soviet Math. Dokl., **10**, 1541–1544).
366. Silvestrov, D.S. (1970a). Limit theorems for semi-Markov processes and their applications.1. Teor. Verojatnost. Mat. Stat., **3**, 155–172 (English translation in Theory Probab. Math. Statist., **3**, 159–176).
367. Silvestrov, D.S. (1970b). Limit theorems for semi-Markov processes and their applications.2. Teor. Verojatnost. Mat. Stat., **3**, 173–194 (English translation in Theory Probab. Math. Statist., **3**, 177–198).
368. Silvestrov, D.S. (1971a). Limit theorems for semi-Markov summation schemes. 1. Teor. Verojatnost. Mat. Stat., **4**, 153–170 (English translation in Theory Probab. Math. Statist., **4**, 141–157).
369. Silvestrov, D. S. (1971b). Uniform estimates of the rate of convergence for sums of random variables defined on a finite homogeneous Markov chain with absorption. Theor. Verojatnost. Mat. Stat., **5**, 116–127 (English translation in Theory Probab. Math. Statist., **5**, 123–135).
370. Silvestrov, D.S. (1971c). Limit distributions for compositions of random functions. Dokl. Akad. Nauk SSSR, **199**, 1251–1252 (English translation in Soviet Math. Dokl., **12**, 1282–1285).
371. Silvestrov, D.S. (1971d). Limit theorems for semi-Markov processes. Dokl. Acad. Nauk USSR. Ser. A, no. 11, 987–989.

372. Silvestrov, D.S. (1972a). Limit distributions for sums of random variables that are defined on a countable Markov chain with absorption. Dokl. Acad. Nauk Ukr. SSR, Ser. A, no. 4, 337–340.
373. Silvestrov, D.S. (1972b). Estimation of the rate of convergence for sums of random variables that are defined on a countable Markov chain with absorption. Dokl. Acad. Nauk Ukr. SSR, Ser. A, no. 5, 436–438.
374. Silvestrov, D.S. (1972c). Limit Theorems for Composite Random Functions. Doctor of Science dissertation, Kiev State University.
375. Silvestrov, D.S. (1972d). Remarks on the limit of composite random function. Teor. Verojatnost. Primen., **17**, no. 4, 707–715 (English translation in Theory Probab. Appl., **17**, no. 4, 669–677).
376. Silvestrov, D.S. (1972e). On convergence of complex random function in the J topology. Dokl. Akad. Nauk SSSR, **202**, 539–540 (English translation in Soviet Math. Dokl., **13**, 152–154).
377. Silvestrov, D.S. (1973). Conditions for convergence of the superposition of random processes in the J topology. Teor. Verojatnost. Primen., **18**, 605–607 (English translation in Theory Probab. Appl., **18**, 579–582).
378. Silvestrov, D.S. (1974). Limit Theorems for Composite Random Functions. Vysshaya Shkola and Izdatel'stvo Kievskogo Universiteta, Kiev, 318 pp.
379. Silvestrov, D.S. (1979). A remark on limit distributions for times of attainment for asymptotically recurrent Markov chains. Theor. Sluch. Proces., **7**, 106–109.
380. Silvestrov, D.S. (1980a). Semi-Markov Processes with a Discrete State Space. Library for the Engineer in Reliability, Sovetskoe Radio, Moscow, 272 pp.
381. Silvestrov, D.S. (1980b). Mean hitting times for semi-Markov processes, and queueing networks. Elektron. Infor. Kybern., **16**, 399–415.
382. Silvestrov, D.S. (1981). Theorems of large deviations type for entry times of a sequence with mixing. Teor. Verojatnost. Mat. Stat., **24**, 129–135 (English translation in Theory Probab. Math. Statist., **24**, 145–151).
383. Silvestrov, D.S. (1995). Exponential asymptotic for perturbed renewal equations. Teor. Ĭmovirn. Mat. Stat., **52**, 143–153 (English translation in Theory Probab. Math. Statist., **52**, 153–162).
384. Silvestrov, D.S. (1996a). Recurrence relations for generalised hitting times for semi-Markov processes. Ann. Appl. Probab., **6**, 617–649.
385. Silvestrov, D.S. (1996b). Hitting times for semi-Markov dynamical systems. Theory Stoch. Process. **2(18)**, no. 1–2, 260–268.
386. Silvestrov, D.S. (2000a). Nonlinearly perturbed Markov chains and large deviations for lifetime functionals. In: Limnios, N., Nikulin, M. (Eds.) Recent Advances in Reliability Theory: Methodology, Practice and Inference. Birkhäuser, Boston, 135–144.
387. Silvestrov, D.S. (2000b). Limit theorems for generalised exceeding times, renewal and risk type processes. Theory Stoch. Process. **6(22)**, no. 3–4, 177–238.
388. Silvestrov D.S. (2004). Limit Theorems for Randomly Stopped Stochastic Processes. Probability and Its Applications, Springer, London, xvi+398 pp.
389. Silvestrov, D.S. (2007). Asymptotic expansions for quasi-stationary distributions of nonlinearly perturbed semi-Markov processes. Theory Stoch. Process., **13**, no. 1–2, 267–271.
390. Silvestrov D.S. (2010). Nonlinearly perturbed stochastic processes and systems. In: Rykov, V., Balakrishnan, N., Nikulin, M. (Eds). Mathematical and Statistical Models and Methods in Reliability, Chapter **2**, Birkhäuser, 19–38.
391. Silvestrov D.S. (2014a). Improved asymptotics for ruin probabilities. In: Silvestrov, D., Martin-Löf, A. (Eds). Modern Problems in Insurance Mathematics, Chapter **5**, EAA series, Springer, Cham, 93–110.
392. Silvestrov D.S. (2014b). American-Type Options. Stochastic Approximation Methods, Volume 1. De Gruyter Studies in Mathematics, **56**, Walter de Gruyter, Berlin, x+509 pp.
393. Silvestrov D.S. (2015). American-Type Options. Stochastic Approximation Methods, Volume 2. De Gruyter Studies in Mathematics, **57**, Walter de Gruyter, Berlin, xi+558 pp.

394. Silvestrov, D. (2016). Necessary and sufficient conditions for convergence of first-rare-event times for perturbed semi-Markov processes. Theor. Ĭmovirn. Mat. Stat. **95**, 119–137 (Also in Theory Probab. Math. Statist. **95**, 135–151).
395. Silvestrov, D. (2019). Algorithms of phase space reduction and asymptotics of hitting times for perturbed semi-Markov processes. Research Report 2019:12, Department of Mathematics, Stockholm University, 153 pp.
396. Silvestrov, D.S., Abadov, Z.A. (1984). Asymptotic behaviour for exponential moments of sums of random variables defined on exponentially ergodic Markov chains. Dokl. Acad. Nauk Ukr. SSR, Ser. A, no. 4, 23–25.
397. Silvestrov, D.S., Abadov, Z.A. (1991). Uniform representations of exponential moments of sums of random variables defined on a Markov chain, and of distributions of passage times. 1. Teor. Verojatnost. Mat. Stat., **45**, 108–127 (English translation in Theory Probab. Math. Statist., **45**, 105–120).
398. Silvestrov, D.S., Abadov, Z.A. (1993). Uniform representations of exponential moments of sums of random variables defined on a Markov chain, and of distributions of passage times. 2. Teor. Verojatnost. Mat. Stat., **48**, 175–183 (English translation in Theory Probab. Math. Statist., **48**, 125–130).
399. Silvestrov, D.S., Drozdenko, M.O. (2005). Necessary and sufficient conditions for the weak convergence of the first-rare-event times for semi-Markov processes. Dopov. Nac. Akad. Nauk Ukr., Mat. Prirodozn. Tekh. Nauki, no. 11, 25–28.
400. Silvestrov, D.S., Drozdenko, M.O. (2006a). Necessary and sufficient conditions for weak convergence of first-rare-event times for semi-Markov processes. I. Theory Stoch. Process., **12(28)**, no. 3–4, 151–186.
401. Silvestrov, D.S., Drozdenko, M.O. (2006b). Necessary and sufficient conditions for weak convergence of first-rare-event times for semi-Markov processes. II. Theory Stoch. Process., **12(28)**, no. 3–4, 187–202.
402. Silvestrov, D., Manca, R. (2017a). Recurrent Algorithms for mixed power-exponential moments of hitting times for semi-Markov processes. Proceedings of the 17th Applied Stochastic Models and Data Analysis International Conference, London, 2017. ISAST: International Society for the Advancement of Science and Technology, 905–923.
403. Silvestrov, D., Manca, R. (2017b). Reward algorithms for semi-Markov processes. Methodol. Comput. Appl. Probab., **19**, no. 4, 1191–1209.
404. Silvestrov, D., Manca, R., Silvestrova, E. (2014). Computational algorithms for moments of accumulated Markov and semi-Markov rewards. Comm. Statist. Theory, Meth., **43**, no. 7, 1453–1469.
405. Silvestrov, D., Petersson, M., Hössjer, O. (2018). Nonlinearly perturbed birth-death-type models. In: Silvestrov, S., Rančić, M., Malyarenko, A. (Eds.) Stochastic Processes and Applications, Chapter **11**, Springer Proceedings in Mathematics & Statistics, **271**, Springer, Cham, 189–244.
406. Silvestrov, D., Silvestrov, S. (2016). Asymptotic expansions for stationary distributions of perturbed semi-Markov processes. In: Silvestrov, S., Rančić, M. (Eds.) Engineering Mathematics II. Algebraic, Stochastic and Analysis Structures for Networks, Data Classification and Optimization, Chapter **10**, Springer Proceedings in Mathematics & Statistics, **179**, Springer, Cham, 151–222.
407. Silvestrov, D., Silvestrov, S. (2017a). Asymptotic expansions for stationary distributions of nonlinearly perturbed semi-Markov processes.1. Methodol. Comput. Appl. Probab., https://doi.org/10.1007/s11009-017-9605-0, 20 pp., **21**, no. 3 (2019), 945–964.
408. Silvestrov, D., Silvestrov, S. (2017b). Asymptotic expansions for stationary distributions of nonlinearly perturbed semi-Markov processes.2. Methodol. Comput. Appl. Probab., https://doi.org//10.1007/s11009-017-9607-yh, 20 pp., **21**, no. 3 (2019), 965–984.
409. Silvestrov, D., Silvestrov, S. (2017c). Nonlinearly Perturbed Semi-Markov Processes. Springer Briefs in Probability and Mathematical Statistics, Springer, Cham, xiv+143 pp.

410. Silvestrov, D., Silvestrov, S. (2017d). Asymptotic expansions for power-exponential moments of hitting times for nonlinearly perturbed semi-Markov processes. Theor. Ĭmovirn. Mat. Stat. **97**, 171–187 (Also in Theory Probab. Math. Statist. **97**, 183–200).
411. Silvestrov, D., Silvestrov, S., Abola, B., Biganda, P.S., Engstrröm, C., Mango, J.M., Kakuba, G. (2020a). Perturbation analysis for stationary distributions of Markov chains with damping component. In: Silvestrov, S., Rančić, M., Malyarenko, A. (Eds.) Algebraic Structures and Applications, Chapter **38**. Springer Proceedings in Mathematics & Statistics, **317**, Springer, Cham, 903–934.
412. Silvestrov, D., Silvestrov, S., Abola, B., Biganda, P.S., Engstrröm, C., Mango, J.M., Kakuba, G. (2020b). Perturbed Markov chains with damping component. Methodol. Comput, Appl, Probab., https://doi.org/10.1007/s11009-020-09815-9, 31 pp., **23**, no. 1 (2021), 369–397.
413. Silvestrov, D.S., Stenflo, Ö. (1998). Ergodic theorems for iterated function systems controlled by regenerative sequences. J. Theor. Probab., **11**, 589–608.
414. Silvestrov, D.S., Teugels, J.L. (2004). Limit theorems for mixed max-sum processes with renewal stopping. Ann. Appl. Probab., **14**, 1838–1868.
415. Silvestrov, D.S., Velikiĭ, Yu.A. (1988). Necessary and sufficient conditions for the convergence of attainment times. In: Zolotarev, V.M., Kalashnikov, V.V. (Eds.) Stability Problems for Stochastic Models. Trudy Seminara, VNIISI, Moscow, 129–137 (English translation in J. Soviet. Math., **57**, (1991), 3317–3324).
416. Simon, H.A., Ando, A. (1961). Aggregation of variables in dynamic systems. Econometrica, **29**, 111–138.
417. Sirl, D., Zhang, H., Pollett, P. (2007). Computable bounds for the decay parameter of a birth-death process. J. Appl. Probab., **44**, no. 2, 476–491.
418. Skorokhod, A.V. (1956). Limit theorems for stochastic processes. Teor. Verojatnost. Primen., **1**, 289–319 (English translation in Theory Probab. Appl., **1**, 261–290).
419. Skorokhod, A.V. (1957). Limit theorems for processes with independent increments. Teor. Verojatnost. Primen., **2**, 145–177 (English translation in Theory Probab. Appl., **2**, 138–171)
420. Skorokhod, A.V. (1964). Random Processes with Independent Increments. Probability Theory and Mathematical Statistics, Nauka, Moscow, 278 pp. (English edition: Nat. Lending Library for Sci. and Tech., Boston Spa, 1971).
421. Skorokhod, A.V. (1986). Random Processes with Independent Increments. Second edition, Probability Theory and Mathematical Statistics, Nauka, Moscow, 320 pp. (English edition: Mathematics and its Applications, **47**, Kluwer, Dordrecht, 1991, xii+279 pp.).
422. Slepov, N.A. (2021). The convergence rate of random geometric sum distribution to the Laplace law. Teor. Verojatnost. Primen., **66**, no. 1, 149–174.
423. Slоothaak, F., Wachtel, V., Zwart, B. (2018). First-passage time asymptotics over moving boundaries for random walk bridges. J. Appl. Probab., **55**, no. 2, 627–651.
424. Smith, W.L. (1955). Regenerative Stochastic Processes. Proceedings of the Royal Society, Ser. A: Mathematical, Physical and Engineering Sciences, **232**, 6–31.
425. Solov'ev, A.D. (1966). A combinatorial identity and its application to the problem on the first occurrence of a rare event. Teor. Verojatnost. Primen., **11**, 313–320 (English translation in Theory Probab. Appl., **11**, 276–282).
426. Solov'ev, A.D. (1971). Asymptotical behaviour of the first occurrence time of a rare event in a regenerative process. Izv. Akad. Nauk SSSR, Tekhn. Kibernet., no. 6, 79–89 (English translation in Engrg. Cybern., **9**, no. 6, 1038–1048).
427. Solov'ev, A.D. (1972). Asymptotic distribution of the moment of first crossing of a high level by a birth and death process. In: Proceedings of the Sixth Berkeley Symposium on Mathematical Statistics and Probability, Berkeley, CA, 1970/1971, Vol. III: Probability theory. Univ. California Press, Berkeley, CA, 71–86.
428. Stewart, G.W. (1969). On the continuity of the generalized inverse. SIAM J. Appl. Math., **17**, 33–45.
429. Stewart, G.W. (1973). Error and perturbation bounds for subspaces associated with certain eigenvalue problems. SIAM Rev., **15**, 727–764.

430. Stewart, G.W. (1979). Perturbation bounds for the definite generalized eigenvalue problem. Linear Algebra Appl., **23**, 69–85.
431. Stewart, G.W. (1983). Computable error bounds for aggregated Markov chains. J. Assoc. Comput. Mach., **30**, no. 2, 271–285.
432. Stewart, G.W. (1984a). On the structure of nearly uncoupled Markov chains. In: Iazeolla, G, Courtois, P.J., Hordijk, A. (Eds). Mathematical Computer Performance and Reliability. North-Holland, Amsterdam, 287–302.
433. Stewart, G.W. (1984b). A second order perturbation expansion for small singular values. Linear Algebra Appl., **56**, 231–235.
434. Stewart, G.W. (1990). Stochastic perturbation theory. SIAM Rev., **32**, no. 4, 579–610.
435. Stewart, G.W. (1991). On the sensitivity of nearly uncoupled Markov chains. In: Stewart, W.J. (Ed.) Numerical Solution of Markov Chains. Probability: Pure and Applied, **8**, Marcel Dekker, New York, 105–119.
436. Stewart, G.W. (1993a). Gaussian elimination, perturbation theory, and Markov chains. In: Meyer, C.D., Plemmons, R.J. (Eds.) Linear Algebra, Markov Chains, and Queueing Models. IMA Volumes in Mathematics and its Applications, **48**, Springer, New York, 59–69.
437. Stewart, G.W. (1993b). On the perturbation of Markov chains with nearly transient states. Numer. Math., **65**, no. 1, 135–141.
438. Stewart, W.J. (1994). Introduction to the Numerical Solution of Markov Chains. Princeton University Press, Princeton, NJ, xx+539 pp.
439. Stewart, G.W. (1998). Matrix Algorithms. Vol. I. Basic Decompositions. SIAM, Philadelphia, PA, xx+458 pp.
440. Stewart, G.W. (2001). Matrix Algorithms. Vol. II. Eigensystems. SIAM, Philadelphia, PA, xx+469 pp.
441. Stewart, G.W. (2003). On the powers of a matrix with perturbations. Numer. Math., **96**, no. 2, 363–376.
442. Stewart, G.W., Stewart, W.J., McAllister, D.F. (1994). A two-stage iteration for solving nearly completely decomposable Markov chains. In: Golub, G., Greenbaum, A., Luskin, M. (Eds.) Recent Advances in Iterative Methods. IMA Volumes in Mathematics and its Applications, **60**, Springer, New York, 201–216.
443. Stewart, G.W., Sun, J.G. (1990). Matrix Perturbation Theory. Computer Science and Scientific Computing. Academic Press, Boston, xvi+365 pp.
444. Stewart, G.W., Zhang, G. (1991). On a direct method for the solution of nearly uncoupled Markov chains. Numer. Math., **59**, no. 1, 1–11.
445. Stewart, W.J. (2009). Probability, Markov Chains, Queues, and Simulation: The Mathematical Basis of Performance Modeling. Princeton University Press, Princeton, NJ, xviii+758 pp.
446. Stone, C. (1963). Weak convergence of stochastic processes defined on semi-infinite time intervals. Proc. Amer. Math. Soc., **14**, 694–696.
447. Sudakova, E.V. (1990). Exponential asymptotics of sums of geometrical number of independent random variables. Theor. Ĭmovirn. Mat. Stat., **42**, 161–166 (English translation in Theory Probab. Math. Statist., **42**, 135–139).
448. Sumita, U., Reiders, M. (1988). A new algorithm for computing the ergodic probability vector for large Markov chains: Replacement process approach. Probab. Eng. Infor. Sci., **4**, 89–116.
449. Sun, W., Feuillet, M., Robert, P. (2016). Analysis of large unreliable stochastic networks. Ann. Appl. Probab., **26**, no. 5, 2959–3000.
450. Takács, L. (1954). Some investigations concerning recurrent stochastic processes of a certain type. Magyar. Tud. Akad. Mat. Kutató Int. Kőzl., **3**, 114–128.
451. Thorisson, H. (2000). Coupling, Stationarity and Regeneration. Probability and its Applications, Springer, New York, xiv+517 pp.
452. Turbin, A.F. (1971). On asymptotic behavior of time of a semi-Markov process being in a reducible set of states. Linear case. Teor. Verojatnost. Mat. Stat., **4**, 179–194 (English translation in Theory Probab. Math. Statist., **4**, 167–182).
453. Turbin, A.F. (1972). An application of the theory of perturbations of linear operators to the solution of certain problems that are connected with Markov chains and semi-Markov

processes. Teor. Verojatnost. Mat. Stat., **6**, 118–128 (English translation in Theory Probab. Math. Statist., **6**, 119–130).

454. Tweedie, R.L. (1980). Perturbations of countable Markov chains and processes. Ann. Inst. Statist. Math., **32**, no. 2, 283–290.
455. Tweedie, R.L. (1998). Truncation approximations of invariant measures for Markov chains. J. Appl. Probab., **35**, 517–536.
456. Vakhtel', V.I., Denisov, D.È. (2015). Exact asymptotics for the instant of crossing a curved boundary by an asymptotically stable random walk. Teor. Verojatnost. Primen., **60**, no. 3, 459–481 (English translation in Theory Probab. Appl., **60**, no. 3, 481–500).
457. van Doorn, E.A. (2017). An orthogonal-polynomial approach to first-hitting times of birth-death processes. J. Theoret. Probab., **30**, no. 2, 594–607.
458. Vantilborgh, H. (1985). Aggregation with an error of O(ϵ^2). J. Assoc. Comput. Mach., **32**, no. 1, 162–190.
459. Voĭna, A.A. (1978). Limit Theorems for Shemes of Summation on Processes with Arbitrary Sets of States and their Applications. Candidate of Science dissertation, Kiev State University.
460. Voĭna, A.A. (1979). Sojourn time in a fixed subset and the consolidation of states of random processes with an arbitrary phase space. Theor. Verojatnost. Mat. Stat., **20**, 30–39 (English translation in Theory Probab. Math. Statist., **20**, 35–44).
461. Volker, N. (1980). Semi-Markov Processes. Scientific Paperbacks, Mathematics/Physics Series, **260**. Akademie-Verlag, Berlin, 162 pp.
462. Vsekhsvyatskii, S.Yu., Kalashnikov, V.V. (1988). Estimates in the Rényi theorem in terms of renewal theory. Teor. Verojatnost. Primen., **33**, no. 2, 369–373 (English translation in Theory Probab. Appl., **33**, no. 2, 347–352).
463. Vysotsky, V. (2015). Limit theorems for random walks that avoid bounded sets, with applications to the largest gap problem. Stoch. Process. Appl., **125**, no. 5, 1886–1910.
464. Wang Zi Kun (1980). Sojourn times and first passage times for birth and death processes. Sci. Sinica, **23**, no. 3, 269–279.
465. Whitt, W. (1980). Some useful functions for functional limit theorems, Math. Oper. Res., **5**, 67–85.
466. Whitt, W. (2002). Stochastic-Processes Limits. An Introduction to Stochastic-Process Limits and their Applications to Queues. Springer Series in Operations Research, Springer, New York, 602 pp.
467. Woodroofe, M. (1982). Nonlinear Renewal Theory in Sequential Analysis. CBMS-NSF Regional Conference Series in Applied Mathematics, **39**, Soc. Industr. Appl. Math., Philadelphia, v+113 pp.
468. Yin, G.G., Zhang, Q. (1998). Continuous-time Markov Chains and Applications. A Singular Perturbation Approach. Applications of Mathematics, **37**, Springer, New York, ivx+349 pp.
469. Yin, G., Zhang, Q. (2003). Discrete-time singularly perturbed Markov chains. In: Yao, D.D., Zhang, H., Zhou, X.Y. (Eds). Stochastic Modeling and Optimization. Springer, New York, 1–42.
470. Yin, G.G., Zhang, Q. (2005). Discrete-Time Markov Chains. Two-Time-Scale Methods and Applications. Stochastic Modelling and Applied Probability, Springer, New York, xix+348 pp.
471. Yin, G.G., Zhang, Q. (2013). Continuous-Time Markov Chains and Applications. A Two-Time-Scale Approach. Stochastic Modelling and Applied Probability, **37**. Springer, New York, xxii+427 pp. (2nd revised edition of Continuous-Time Markov Chains and Applications. A Singular Perturbation Approach. Applications of Mathematics, **37**, Springer, New York, 1998. xvi+349 pp.).
472. Yin, G., Zhang, Q., Badowski, G. (2000a). Asymptotic properties of a singularly perturbed Markov chain with inclusion of transient states. Ann. Appl. Probab., **10**, 549–572.
473. Yin, G., Zhang, Q., Badowski, G. (2000b). Occupation measures of singularly perturbed Markov chains with absorbing states. Acta Math. Sin., **16**, no. 1, 161–180.
474. Yin, G., Zhang, Q., Badowski, G. (2000c). Singularly perturbed Markov chains: convergence and aggregation. J. Multivar. Anal., **72**, no. 2, 208–229.

475. Yin, G., Zhang, Q., Badowski, G. (2003). Discrete-time singularly perturbed Markov chains: aggregation, occupation measures, and switching diffusion limit. Adv. Appl. Probab., **35**, no. 2, 449–476.
476. Yin, G., Zhang, Q., Yang, H., Yin, K. (2001). Discrete-time dynamic systems arising from singularly perturbed Markov chains. In: Proceedings of the Third World Congress of Nonlinear Analysts, Part 7, Catania, 2000. Nonlinear Anal., **47**, no. 7, 4763–4774.
477. Zakusilo, O.K. (1972a). Thinning semi-Markov processes. Teor. Verojatnost. Mat. Stat., **6**, 54–59 (English translation in Theory Probab. Math. Statist., **6**, 53–58).
478. Zakusilo, O.K. (1972b). Necessary conditions for convergence of semi-Markov processes that thin. Teor. Verojatnost. Mat. Stat., **7**, 65–69 (English translation in Theory Probab. Math. Statist., **7**, 63–66).
479. Zakusilo, O.K., Matsak, Ī.K. (2017). On extreme values of some regenerative processes. Theor. Ĭmovirn. Mat. Stat. **97**, 58–71 (English translation in Theory Probab. Math. Statist. **97**, 57–71).
480. Zakusilo, O.K., Matsak, Ī.K. (2021). Estimates for the rate of convergence in the limit theorem for extreme values of regenerative processes. Ukraïn. Mat. Zh., **72**, no. 8, 1064–1081 (English translation in Ukrainian Math. J., **72**, no. 8, 1230–1250).
481. Zhang, Q., Yin, G. (2004). Exponential bounds for discrete-time singularly perturbed Markov chains. J. Math. Anal. Appl., **293**, no. 2, 645–662.
482. Zubkov, A.M., Mihaĭlov, V.G. (1974). Limit distributions of random variables that are connected with long duplications in a sequence of independent trials. Teor. Verojatnost. Primen., **19**, 173–181 (English translation in Theory Probab. Appl., **19**, 172–179).
483. Zweimüller, R. (2019). Hitting-time limits for some exceptional rare events of ergodic maps. Stochastic Process. Appl., **129**, no. 5, 1556–1567.

Index

D. Silvestrov, *Perturbed Semi-Markov Type Processes I*,
https://doi.org/10.1007/978-3-030-92403-4

The manufacturer's authorised representative in the EU is Springer Nature Customer Service Centre GmbH, Europaplatz 3, 69115 Heidelberg, Germany. If you have any concerns regarding our products, please contact ProductSafety@springernature.com

Printed and bound by CPI Group (UK) Ltd, Croydon, CR0 4YY
30/07/2026
02182056-0007